HANDBOOK OF NEURAL ENGINEERING

HANDBOOK OF NEURAL ENGINEERING

Edited by

METIN AKAY

IEEE Engineering in Medicine and Biology Society, *Sponsor*

IEEE PRESS

WILEY-INTERSCIENCE
A JOHN WILEY & SONS, INC., PUBLICATION

For general information on our other products and services or for technical support, please contact our Customer Care Department within the United States at (800) 762-2974, outside the United States at (317) 572-3993 or fax (317) 572-4002.

Wiley also publishes its books in a variety of electronic formats. Some content that appears in print may not be available in electronic formats. For more information about Wiley products, visit our web site at www.wiley.com.

Library of Congress Cataloging-in-Publication Data is available.

ISBN 0-978-0-470-05669-1

Printed in the United States of America

10   9   8   7   6   5   4   3   2   1

CONTENTS

PREFACE

Neuroscience has become more quantitative and information-driven science since emerging implantable and wearable sensors from macro to nano and computational tools facilitate collection and analysis of vast amounts of neural data. Complexity analysis of neural systems provides physiological knowledge for the organization, management, and mining of neural data by using advanced computational tools since the neurological data are inherently complex and nonuniform and collected at multiple temporal and spatial scales. The investigations of complex neural systems and processes require an extensive colloboration among biologists, mathematicians, computer scientists, and engineers to improve our understanding of complex neurological process from system to gene.

Neural engineering is a new discipline which coalesces engineering including electronic and photonic technologies, computer science, physics, chemistry, mathematics with cellular, molecular, cognitive and behavioral neuroscience to understand the organizational principles and underlying mechanisms of the biology of neural systems and to study the behavior dynamics and complexity of neural systems in nature. Neural engineering deals with many aspects of basic and clinical problems associated with neural dysfunction including the representation of sensory and motor information, electrical stimulation of the neuromuscular system to control muscle activation and movement, the analysis and visualization of complex neural systems at multiscale from the single cell to system levels to understand the underlying mechanisms, development of novel electronic and photonic devices and techniques for experimental probing, the neural simulation studies, the design and development of human–machine interface systems and artificial vision sensors and neural prosthesis to restore and enhance the impaired sensory and motor systems and functions.

To highlight this emerging discipline, we devoted this edited book to neural engineering related to research. This handbook highlights recent advances in wearable and implantable neural sensors/probes and computational neural science and engineering. It incorporates fundamentals of neuroscience, engineering, mathematical, and information sciences. As a primer, educational material, technical reference, research and development resource, this book in terms of its intellectual substance and rigor is peer-reviewed. The contributors have been invited from diverse disciplinary groups representing academia, industry, private, and government organizations. The make-up of participants represents the geographical distribution of neural engineering and neuroscience activity around the world.

I am very confident that it will become the unique Neural Engineering resource that contributes to the organization of the neural engineering and science knowledge domain and facilitates its growth and development in content and in participation.

I am very grateful for all the contributors and their strong support for the initiative. I thank Mrs. Jeanne Audino of the IEEE Press and Lisa Van Horn of Wiley for their strong support, help, and hard work during the entire time of editing this handbook. Working in concert with them and the contributors really helped me with the content development and to manage the peer-review process. I am grateful to them.

Finally, many thanks to my wife, Dr. Yasemin M. Akay, and our son, Altug R. Akay, for their support, encouragement, and patience. They have been my driving source.

METIN AKAY

Scottsdale, Arizona
November 2006

CONTRIBUTORS

N. Acikgoz   Department of Biomedical Engineering, University of Miami, Coral Gables, Florida

Ashish Ahuja   Department of Electrical Engineering, University of Southern California, Los Angeles, California

Kenzo Akazawa   Department of Architecture and Computer Technology, University of Granada, Granada, Spain

Carlos Alberola-López   Department of Teoría de la Señal y Communicaciones e Ingeniería Telemática, Universidad de Valladolid, Valladolid, Spain

David Anderson   ECE Department, Michigan State University, East Lansing, Michigan

Juan Ignazio Arribas   Department of Teoría de la Señal y Communicaciones e Ingeníeria Telemática, Universidad de Valladolid, Valladolid, Spain

L. Astolfi   IRCCS Fondazione Santa Lucia, Rome, Italy

G. Asuni   Advanced Robotics and Technology and Laboratory, Scuola Superiore Sant' Anna, Pisa, Italy

Bruno Azzerboni   Universitá Mediterranea di Reggio Calabria, Reggio Calabria, Italy

F. Babiloni   IRCCS Fondazione Santa Lucia, Rome, Italy

Kyunigim Baek   Department of Information and Computer Sciences, University of Hawaii at Manoa, Honolulu, Hawaii

T. Bajd   Faculty of Electrical Engineering, University of Ljubljana, Ljubljana, Slovenia

Thomas C. Baker   Department of Bioengineering, University of Illinois, Chicago, Illinois

G. Baselli   Polytechnic University, Milan, Italy

H. E. Bedell   Department of Electrical and Computer Engineering, University of Houston, Houston, Texas

Theodore W. Berger   Department of Biomedical Engineering, University of Southern California, Los Angeles, California

Laura Bonzano   Neuroengineering and Bio-nano Technologies (NBT) Group, Department of Biophysical and Electronic Engineering—DIBE, University of Genoa, Genoa, Italy

Nashaat N. Boutros   Department of Psychiatry, Wayne State University, Detroit, Michigan

B. G. Breitmeyer   Department of Electrical and Computer Engineering University of Houston, Houston, Texas

Philip Byrnes-Preston   Graduate School of Biomedical Engineering, University of New South Wales, Sydney, Australia

M. C. Carrozza   Advanced Robotics Technology and Systems Laboratory, Scuola Superiore Sant' Anna, Pisa, Italy

S. Cerutti   Polytechnic University, Milan, Italy

Spencer C. Chen Graduate School of Biomedical Engineering, University of New South Wales, Sydney, Australia

Michela Chiappalone Neuroengineering and Bio-nano Technologies (NBT) Group, Department of Biophysical and Electronic Engineering—DIBE, University of Genoa, Genoa, Italy

Jesús Cid-Sueiro Department of Teoría de la Señal y Communicaciones e Ingeníería Telemática, Universidad de Valladolid, Valladolid, Spain

I. Cikajlo Institute for Rehabilitation, Republic of Slovenia, Ljubljana, Slovenia

F. Cincotti IRCCS Fondazione Santa Lucia, Rome, Italy

Spiros H. Courellis Department of Biomedical Engineering, University of Southern California, Los Angeles, California

Andrea D' Avella Department of Neuromotor Physiology, Fondazione Santa Lucia, Rome, Italy

P. Dario Center for Research in Microengineering, Scuola Superiore Sant' Anna, Pisa, Italy

Wolfgang Delb Key Numerics, Scientific Computing & Computational Intelligence, Saarbruecken, Germany

R. E. Delgado Department of Biomedical Engineering, University of Miami, Coral Gables, Florida

Silvia Delsanto Department of Automation and Information, Politecnico di Torino, Torino, Italy

A. Demosthenous Institute of Biomedical Engineering, Imperial College of London, United Kingdom

Socrates Dokos Graduate School of Biomedical Engineering, University of New South Wales, Sydney, Australia

N. Donaldson Institute of Biomedical Engineering, Imperial College of London, Institute of Biochemical Engineering, United Kingdom

P. D. Einziger Department of Electrical Engineering, Israel Institute of Technology, Haifa, Israel

Gopal Erinjippurath Department of Biomedical Engineering, University of Southern California, Los Angeles, California

F. Fazio Institute of Molecular Bioimaging and Physiology (IBFM) CNR, Milan, Italy

Eduardo Fernández Department of Architecture and Computer Technology, University of Granada, Granada, Spain

Guglielmo Foffani Drexel University, School of Biomedical Engineering, Science and Health Systems, Philadelphia, Pennsylvania

H. Fotowat Department of Electrical and Computer Engineering, University of Houston, Houston, Texas

Gonzalo A. García Department of Architecture and Computer Technology, University of Granada, Granada, Spain

Adam D. Gerson Department of Biomedical Engineering, Columbia University, New York, New York

Ghassan Gholmieh Department of Biomedical Engineering, University of Southern California, Los Angeles, California

Maysam Ghovanloo Center for Wireless Integrated Microsystems, University of Michigan, Ann Arbor, Michigan

A. Gonzáles Macros Universidad Politecnica de Madrid, Madrid, Spain

John J. Granacki Department of Biomedical Engineering, University of Southern California, Los Angeles California

Rylie Green Graduate School of Biomedical Engineering, University of New South Wales, Sydney, Australia

E. Guglielmelli Advanced Robotics Technology and System Laboratory, Scuola Superiore Sant' Anna, Pisa, Italy

Luke E. Hallum Graduate School of Biomedical Engineering, University of New South Wales, Sydney, Australia

Anant Hedge Center for Neuro-Engineering and Cognitive Science, University of Houston, Houston, Texas

C. Heneghan Department of Electrical, Electronic, and Mechanical Engineering, University College, Dublin, Ireland

John R. Hetling Department of Bioengineering, University of Illinois, Chicago, Illinois

Min Chi Hsaio Department of Biomedical Engineering, University of Southern California, Los Angeles, California

Khan M. Iftekharuddin Intelligent Systems and Image Processing Laboratory, Institute of Intelligent Systems, Department of Electrical and Computer Engineering, University of Memphis, Memphis, Tennessee

Maurizio Ipsale Universitá Mediterranea di Reggio Calabria, Reggio Calabria, Italy

Ben H. Jansen Center for Neuro-Engineering and Cognitive Science, University of Houston, Houston, Texas

Erik Weber Jensen Danmeter A/S Research Group, Odense, Denmark

Fabio LaForesta Universitá Mediterranea di Reggio Calabria, Reggio Calabria, Italy

Jeff LaCoss Information Sciences Institute, University of Southern California, Los Angeles, California

Fabrizio Lamberti Department of Automation and Information, Politecnico di Torino, Torino, Italy

C. Laschi Advanced Robotics Technology and Systems Laboratory, Scuola Superiore Sant' Anna, Pisa, Italy

Torsten Lehmann School of Electrical Engineering and Telecommunications, University of New South Wales, Sydney, Australia

Klaus Lehnertz Department of Epileptology, Neurophysics Group, University of Bonn, Bonn, Germany

Steven C. Leiser Department of Neurobiology and Anatomy, Drexel University College of Medicine, Philadelphia, Pennsylvania

F. Leoni Advanced Robotics Technology and Systems Laboratory, Scuola Superiore Sant' Anna, Pisa, Italy

Nicholas A. Lesica Division of Engineering and Applied Sciences, Havard University, Cambridge, Massachusetts

Hector Litvan Hospital Santa Creu i Santa Pau, Department of Cardiac Anesthesia, Barcelona, Spain

L. M. Livshitz Department of Biomedical Engineering, Israel Institute of Technology, Haifa, Israel

Nigel H. Lovell Graduate School of Biomedical Engineering, University of New South Wales, Sydney, Australia

Henrick Hautop Lund   University of Southern Denmark, Odense, Denmark

E. Marani   Biomedical Signals and Systems Department, Faculty of Electrical Engineering, Mathematics and Computer Science/Institute for Biomedical Technology, University of Twente, Enschede, The Netherlands

Vasilis Z. Marmarelis   Department of Biomedical Engineering, University of Southern California, Los Angeles, California

Antonio Martínez   Department of Architecture and Computer Technology, University of Granada, Granada, Spain

Pablo Martinez   Danmeter A/S Research Group, Odense, Denmark

Sergio Martinoia   Neuroengineering and Bio-nano Technologies (NBT) Group, Department of Biophysical and Electronic Engineering—DIBE, University of Genoa, Genoa, Italy

J. A. Martín-Pereda   Universidad Politecnica de Madrid, Madrid, Spain

Z. Matjačić   Institute for Rehabilitation, Republic of Slovenia, Ljubljana, Slovenia

D. Mattia   IRCCS Fondazione Santa Lucia, Rome, Italy

M. Mattiocco   IRCCS Fondazione Santa Lucia, Rome, Italy

G. McDarby   Department of Electrical, Electronic, and Mechanical Engineering, University College, Dublin, Ireland

A. Menciassi   Center for Research in Microengineering, Scuola Superiore Sant' Anna, Pisa, Italy

S. Micera   Advanced Robotics Technology and Systems Laboratory, Scuola Superiore Sant' Anna, Pisa, Italy

J. Mizrahi   Department of Biomedical Engineering, Israel Institute of Technology, Haifa, Israel

Francesco Carlo Morabito   Universitá Mediterranea di Reggio Calabria, Reggio Calabria, Italy

Christian A. Morillas   Department of Architecture and Computer Technology, University of Granada, Granada, Spain

Lia Morra   Department of Automation and Information, Politecnico di Torino, Torino, Italy

Karen A. Moxon   School of Biomedical Engineering, Science and Health Systems, Drexel University, Philadelphia, Pennsylvania

Rolf Müller   The Maersk McKinney Moller Institute for Production Technology, University of Southern Denmark, Odense, Denmark

Andrew J. Myrick   Department of Bioengineering, University of Illinois, Chicago, Illinois

Khalil Najafi   Center for Wireless Integrated Microsystems, University of Michigan, Ann Arbor, Michigan

Patrick Nasiatka   Department of Electrical Engineering, University of Southern California, Los Angeles, California

Jacob Nielsen   University of Southern Denmark, Odense, Denmark

P. Nolan   Department of Electrical, Electronic, and Mechanical Engineering, University College, Dublin, Ireland

H. Öğmen   Department of Electrical and Computer Engineering University of Houston, Houston, Texas

Ryuhei Okuno   Department of Architecture and Computer Technology, University of Granada, Granada, Spain

Karim Oweiss   ECE Department, Michigan State University, East Lansing, Michigan

O. Ozdamar   Department of Biomedical Engineering, University of Miami, Coral Gables, Florida

Carlos Parra   Intelligent Systems and Image Processing Laboratory, Institute of Intelligent Systems, Department of Electrical and Computer Engineering, University of Memphis, Memphis, Tennessee

Lucas C. Parra   New York Center for Biomedical Engineering, City College of New York, New York, New York

Francisco J. Pelayo   Department of Architecture and Computer Technology, University of Granada, Granada, Spain

D. Perani   Scientific Institute H San Raffaele, Milan, Italy

Herbert Peremans   University of Antwerp, Antwerpen, Belgium

Peter K. Plinkert   Department of Otorhinolaryngology, Saarland University Hospital, Homburg/Saar, Germany

Laura Poole-Warren   Graduate School of Biomedical Engineering, University of New South Wales, Sydney, Australia

Dejan B. Popovic   Center for Sensory Motor Interaction, Department of Health Science and Technology, Aalborg University, Aalborg, Denmark

Philip Preston   Graduate School of Biomedical Engineering, University of New South Wales, Sydney, Australia

Jose C. Principe   Computational NeuroEngineering Laboratory, University of Florida, Gainesville, Florida

M. S. Rahal   Institute of Biomedical Engineering, Imperial College of London, United Kingdom

G. Rizzo   Institute of Molecular Bioimaging and Physiology (IBFM) CNR, Milan, Italy

Bernardo Rodriguez   Danmeter A/S Research Group, Odense, Denmark

B. H. Roelofsen   Biomedical Signals and Systems Department, Faculty of Electrical Engineering, Mathematics and Computer Science/Institute for Biomedical Technology, University of Twente, Enschede, The Netherlands

Samuel Romero   Department of Architecture and Computer Technology, University of Granada, Granada, Spain

T. G. Ruardij   Biomedical Signals and Systems Department, Faculty of Electrical Engineering, Mathematics and Computer Science/Institute for Biomedical Technology, University of Twente, Enschede, The Netherlands

Jacob Ruben   Center for Neuro-Engineering and Cognitive Science, University of Houston, Houston, Texas

W. L. C. Rutten   Biomedical Signals and Systems Department, Faculty of Electrical Engineering, Mathematics and Computer Science/Institute for Biomedical Technology, University of Twente, Enschede, The Netherlands

Paul Sajda   Department of Biomedical Engineering, Columbia University, New York, New York

Justin C. Sanchez   Computational NeuroEnginerring Laboratory, University of Florida, Gainesville, Florida

P. Scifo   Scientific Institute H San Raffaele, Milan, Italy

R. Shouldice   Department of Electrical, Electronic, and Mechanical Engineering, University College, Dublin, Ireland

Thomas Sinkjær   Center for Sensory Motor Interaction, Department of Health Science and Technology, Aalborg University, Aalborg, Denmark

Dong Song   Department of Biomedical Engineering, University of Southern California, Los Angeles, California

Vijay Srinivasan   Department of Electrical Engineering, University of Southern California, Los Angeles, California

Garrett B. Stanley   Division of Engineering and Applied Sciences, Harvard University, Cambridge, Massachusetts

A. Starita   Department of Informatics, Scuola Superiore Sant'Anna, Pisa, Italy

Daniel J. Strauss   Key Numerics, Scientific Computing & Computational Intelligence, Saarbruecken, Germany

Michele M.R.F. Struys   Department of Anesthesia, Ghent University Hospital, Ghent, Belgium

Gregg J. Suaning   School of Engineering, University of Newcastle, Newcastle, Australia

Armand R. Tanguay, Jr.   Department of Biomedical Engineering, University of Southern California, Los Angeles, California

M. Tettamanti   Scientific Institute of San Raffaele, Milan, Italy

Matthew Tresch   Biomedical Engineering and Physical Medicine and Rehabilitation, Northwestern University, Evanston, Illinois

I. F. Triantis   Institute of Biomedical Engineering, Imperial College of London, United Kingdom

Banu Tutunculer   Department of Neurobiology and Anatomy, Drexel University College of Medicine, Philadelphia, Pennsylvania

Alessandro Vato   Neuroengineering and Bio-nano Technologies (NBT) Group, Department of Biophysical and Electrical Engineering—DIBE, University of Genoa, Genoa, Italy

Hugo Vereecke   Department of Anesthesia, Ghent University Hospital, Ghent, Belgium

P. Vitali   Scientific Institute H San Raffaele, Milan, Italy

S. Ward   Department of Electrical, Electronic, and Mechanical Engineering, University College, Dublin, Ireland

Jack Wills   Department of Electrical Engineering, University of Southern California, Los Angeles, California

E. Yavuz   Department of Biomedical Engineering, University of Miami, Coral Gables, Florida

PART I

NEURAL SIGNAL AND IMAGE PROCESSING AND MODELING

THIS PART focuses on the analysis and modeling of neural activity and activities related to electroencephalography (EEG) using the nonlinear and nonstationary analysis methods, including the chaos, fractal, and time-frequency and time-scale analysis methods. It focuses on measuring functional, physiological, and metabolic activities in the human brain using current and emerging medical imaging technologies, including functional magnetic resonance imaging (fMRI), MRI, single photon emission computed tomography (SPECT), and positron emission somography (PET).

CHAPTER 1

OPTIMAL SIGNAL PROCESSING FOR BRAIN–MACHINE INTERFACES

Justin C. Sanchez and Jose C. Principe

1.1 INTRODUCTION

Several landmark experimental paradigms have shown the feasibility of using neuroprosthetic devices to restore motor function and control in individuals who are "locked in" or have lost the ability to control movement of their limbs [1–19]. In these experiments, researchers seek to both rehabilitate and augment the performance of neural–motor systems using brain–machine interfaces (BMIs) that transfer the intent of the individual (as collected from the cortex) into control commands for prosthetic limbs and/or computers. Brain–machine interface research has been strongly motivated by the need to help the more than 2 million individuals in the United States suffering from a wide variety of neurological disorders that include spinal cord injury and diseases of the peripheral nervous system [20]. While the symptoms and causes of these disabilities are diverse, one characteristic is common in many of these neurological conditions: Normal functioning of the brain remains intact. If the brain is spared from injury and control signals can be extracted, the BMI problem becomes one of finding *optimal* signal processing techniques to efficiently and accurately convert these signals into operative control commands.

A variety of noninvasive and invasive techniques have been used to collect control signals from the cortex. Some of the earliest brain–computer interfaces (BCIs) utilized electrical potentials collected from the scalp through the use of electroencephalography (EEG) [21]. This approach to interfacing individuals with machines has the appeal of a low threshold of clinical use since no surgical procedures are required to install the sensors detecting the control signals. Additionally, EEG hardware technology is at a stage where it is relatively inexpensive, portable, and easy to use. In terms of ease of access to the neurophysiology, EEG is an attractive choice; however, in terms of signal processing, this approach has been limited to basic communication capabilities with a computer screen and suffers from a low bandwidth (a few bits per second). The fundamental difficulty of using EEG for BCI is due to the "spatiotemporal filtering" of neuronal activity resulting from the different conductivities of the scalp, skull, and dura, which limits the signal-to-noise ratio of the time series and blurs the localization of the neural population firings [22]. Long training sessions are often required for BCI users to learn to modulate their neuronal activity, and users can respond only after minutes of concentration. Another approach to noninvasively collecting control signals is through the use

Handbook of Neural Engineering. Edited by Metin Akay

of near-infrared (NIR) spectroscopy, which uses the scattering of light to detect blood oxygenation [23]. Like EEG, NIR spectroscopy is also a noninvasive technology with relatively course spatial resolution, however the temporal resolution is on the order of tens of milliseconds. To overcome the difficulties of recording control signals through the skull, researchers have moved to using a more invasive technique of measuring EEG on the surface of the brain with the electrocorticogram (ECoG). By placing dense arrays directly upon the motor cortex, this approach has the appeal of increasing the spatial resolution of EEG, and studies are now underway to assess the utility of the collected signals [24]. Recently, invasive techniques that utilize multiple arrays of microelectrodes that are chronically implanted into the cortical tissue have shown the most promise for restoring motor function to disabled individuals [10]. It has been shown that the firing rates of single cells collected from multiple cortices contain precise information about the motor intent of the individual. A variety of experimental paradigms have demonstrated that awake, behaving primates can learn to control external devices with high accuracy using optimal signal processing algorithms to interpret the modulation of neuronal activity as collected from the microelectrode arrays. A recent special issue of the *IEEE Transactions on Biomedical Engineering* (June 2004) provides a very good overview of the state of the art.

A conceptual drawing of a BMI is depicted in Figure 1.1, where neural activity from hundreds of cells is recorded (step 1), conditioned (step 2), and translated (step 3) directly into hand position (HP), hand velocity (HV), and hand gripping force (GF) of a prosthetic arm or cursor control for a computer. The focus of this chapter is centered on step 3 of the diagram where we will use optimal signal processing techniques to find the functional relationship between neuronal activity and behavior. From an optimal signal processing point of view, BMI modeling in step 3 is a challenging task because of several factors: the intrinsic partial access to the motor cortex information due to the spatial subsampling of the neural activity, the unknown aspects of neural coding, the huge dimensionality of the problem, the noisy nature of the signal pickup, and the need for real-time signal processing algorithms. The problem is further complicated by the need for good generalization in nonstationary environments, which is dependent upon model topologies, fitting criteria, and training algorithms. Finally, we must contend with reconstruction accuracy, which is linked to our choice of linear-versus-nonlinear and feedforward-versus-feedback models.

Since the basic biological and engineering challenges associated with optimal signal processing for BMI experiments requires a highly interdisciplinary knowledgebase involving neuroscience, electrical and computer engineering, and biomechanics, the BMI modeling problem will be addressed in several steps. First, an overview of the pioneering modeling approaches will give the reader depth into what has been accomplished in this area of research. Second, we will familiarize the reader with characteristics of the

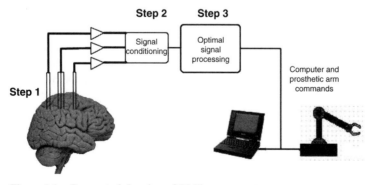

Figure 1.1   Conceptual drawing of BMI components.

neural recordings that the signal processing methods utilize. Third, we cover in detail the current approaches to modeling in BMIs. Finally, real implementations of optimal models for BMIs will be presented and their performance compared.

1.2 HISTORICAL OVERVIEW OF BMI APPROACHES/MODELS

The foundations of BMI research were probed in the early 1980s by E. Schmidt and E. Fetz, who were interested in finding out if it was possible to use neural recordings from the motor cortex of a primate to control an external device [1, 25]. In this pioneering work, Schmidt measured how well primates could be conditioned to modulate the firing patterns of *single* cortical cells using a series of eight target lamps each symbolizing a cellular *firing rate* that the primate was required to produce. The study did confirm that a primate was able to modulate neural firing to match the target rates and additionally estimated the information transfer rate in the neural recordings to be half that of using the intact motor system as the output. With this result, Schmidt proposed that engineered interfaces could be designed to use modulations of neural firing rates as control signals.

Shortly after Schmidt [1] published his results, Georgopoulos et al. [2] presented a theory for neural *population* coding of hand kinematics as well as a method for reconstructing hand trajectories called the population vector algorithm (PVA) [3]. Using center-out reaching tasks, Georgopoulos proposed that each cell in the *motor cortex* has a "preferred hand direction" for which it fires maximally and the distribution of cellular firing over a range of movement directions could be characterized by a simple cosine function [2]. In this theory, arm movements were shown to be constructed by a population "voting" process among the cells; each cell makes a vectoral contribution to the overall movement in its preferred direction with magnitude proportional to the cell's average firing rate [26].

Schmidt's proof of concept and Georgopoulos et al.'s BMI application to reaching tasks spawned a variety of studies implementing "out-of-the-box" signal processing modeling approaches. One of the most notable studies by Chapin et al. [6] showed that a recurrent neural network could be used to translate the neural activity of 21–46 neurons of rats trained to obtain water by pressing a lever with their paw. The usefulness of this BMI was demonstrated when the animals routinely stopped physically moving their limbs to obtain the water reward. Also in the neural network class, Lin et al. [27] used a self-organizing map (SOM) that clustered neurons with similar firing patters which then indicated movement directions for a spiral drawing task. Borrowing from control theory, Kalaska et al. [28] proposed the use for forward- and inverse-control architectures for reaching movements. Also during this period other researchers presented interpretations of population coding which included a probability-based population coding from Sanger [29] and muscle-based cellular tuning from Mussa-Ivaldi [30].

Almost 20 years after Schmidt's [1] and Georgopoulos et al.'s initial experiments, Wessberg and colleagues [10] presented the next major advancement in BMIs [2, 3] by demonstrating a real (nonportable) neuroprosthetic device in which the neuronal activity of a primate was used to control a robotic arm [10]. This research group hypothesized that the information needed for the BMI is distributed across several cortices and therefore neuronal activity was collected from 100 cells in *multiple* cortical areas (premotor, primary motor, and posterior parietal) while the primate performed a three-dimensional (3D) feeding (reaching) task. Linear and nonlinear signal processing techniques including a frequency-domain Wiener filter (WF) and a time delay neural network (TDNN) were used to estimate hand position. Trajectory estimates were then transferred via the Internet to a local robot and a robot located at another university.

In parallel with the work of Nicolelis [20], Serruya and colleagues [14] presented a contrasting view of BMIs by showing that a 2D computer cursor control task could be achieved using only a few neurons (7–30) located only in the primary motor cortex of a primate. The WF signal processing methodology was again implemented here; however this paradigm was *closed loop* since the primate received instant visual feedback from the cursor position output from the WF. The novelty of this experiment results from the primate's opportunity to incorporate the signal processing model into its motor processing.

The final BMI approach we briefly review is from Andersen's research group, which showed that the endpoint of hand reaching can be estimated using a Bayesian probabilistic method [31]. Neural recordings were taken from the parietal reach region (PRR) since they are believed to encode the planning and target of hand motor tasks. Using this hypothesis this research group devised a paradigm in which a primate was cued to move its hand to rectangular grid target locations presented on a computer screen. The neural-to-motor translation involves computing the likelihood of neural activity given a particular target. While this technique has been shown to accurately predict the endpoint of hand reaching, it differs for the aforementioned techniques by not accounting for the hand trajectory.

1.3 CHARACTERISTICS OF NEURAL RECORDINGS

One of the most important steps in implementing optimal signal processing technique for *any* application is data analysis. Here the reader should take note that optimality in the signal processing technique is predicated on the matching between the statistics of the data match and the a priori assumptions inherent in any signal processing technique [32]. In the case of BMIs, the statistical properties of the neural recordings and the analysis of neural ensemble data are not fully understood. Hence, this lack of information means that the neural–motor translation is not guaranteed to be the best possible, even if optimal signal processing is utilized (because the criterion for optimality may not match the data properties). Despite this reality, through the development of new neuronal data analysis techniques we can improve the match between neural recordings and BMI design [33, 34]. For this reason, it is important for the reader to be familiar with the characteristics of neural recordings that would be encountered.

The process of extracting signals from the motor, premotor, and parietal cortices of a behaving animal involves the implantation of subdural microwire electrode arrays into the brain tissue (usually layer V) [10]. At this point, the reader should be aware that current BMI studies involve the sampling of a minuscule fraction of motor cortex activity (tens to hundreds of cortical cells recorded from motor-related areas that are estimated to contain 100 million neurons) [35]. Each microwire measures the potentials (*action potentials*) resulting from ionic current exchanges across the membranes of neurons locally surrounding the electrode. Typical cellular potentials as shown in Figure 1.2*a* have magnitudes ranging from hundreds of microvolts to tens of millivolts and time durations of tens to a couple of milliseconds [34]. Since action potentials are so short in duration, it is common to treat them as point processes where the continuous voltage waveform is converted into a series of *time stamps* indicating the instance in time when the spike occurred. Using the time stamps, a series of pulses (*spikes*—zeros or ones) can be used to visualize the activity of each neuron; this time series shown in Figure 1.2*b* is referred to as a *spike train*. The spike trains of neural ensembles are sparse, nonstationary, and discontinuous. While the statistical properties of neural recordings can vary depending on the sample area, animal, and behavior paradigm, in general spike trains are assumed to have a Poisson distribution [33]. To reduce the sparsity in neuronal recordings, a method

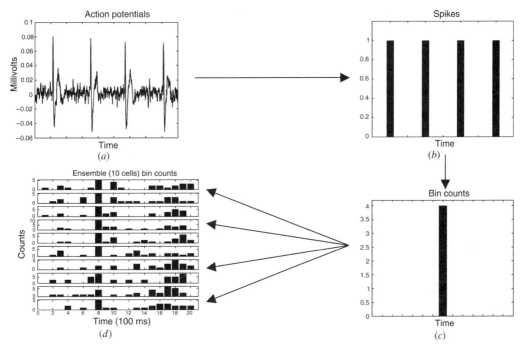

Figure 1.2 Spike-binning process: (*a*) cellular potentials; (*b*) a spike train; (*c*) bin count for single cell; (*d*) ensemble of bin counts.

of binning is used to count the number of spikes in 100-ms nonoverlapping windows as shown in Figure 1.2*c*. This method greatly reduces the number of zeros in the digitized time series and also provides a time-to-amplitude conversion of the firing events. Even with the binning procedure, the data remain extremely sparse. In order for the reader to assess the degree of sparsity and nonstationarity in BMI data, we present in Table 1.1 observations from a 25-min BMI experiment recorded at the Nicolelis's laboratory at Duke University. From the table we can see that the percentage of zeros can be as high as 80% indicating that the data are extremely sparse. Next we compute the firing rate for each cell in nonoverlapping 1-min windows and compute the average across all cells. The ensemble of cells used in this analysis primarily contains low firing rates given by the small ensemble average. Additionally we can see the time variability of the 1-min ensemble average given by the associated standard deviation.

In Figure 1.3, the average firing rate of the ensemble (computed in nonoverlapping 60-s windows) is tracked for a 38-min session. From minute to minute, the mean value in the firing rate can change drastically depending on the movement being performed. Ideally we would like our optimal signal processing techniques to capture the changes observed in Figure 1.3. However, the reader should be aware that any of the out-of-the-box signal processing techniques such as WFs and artificial neural networks assume stationary statistics

TABLE 1.1 Neuronal Activity for 25-min Recording Session

	Percentage of zeros	Average firing rate (spikes/cell/min)
3D reaching task (104 cells)	86	0.25 ± 0.03
2D cursor control task (185 cells)	60	0.69 ± 0.02

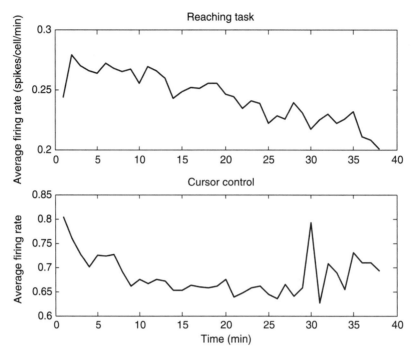

Figure 1.3   Time-varying statistics of neuronal recordings for two behaviors.

over time, which means that the derived neural-to-motor mapping will not be optimal unless the window size is very well controlled. More importantly, any performance evaluations and model interpretations drawn by the experimenter can be biased by the mismatch between data and model type.

1.4   MODELING PROBLEM

The models implemented in BMIs must learn to interpret neuronal activity and accurately translate it into commands for a robot that mimic the intended motor behavior. By analyzing recordings of neural activity collected simultaneously with behavior, the aim is to find a *functional* relationship between neural activity and the kinematic variables of position, velocity, acceleration, and force. An important question here is how to choose the class of functions and model topologies that best match the data while being sufficiently powerful to create a mapping from neuronal activity to a variety of behaviors. As a guide, prior knowledge about the nervous system can be used to help develop this relationship. Since the experimental paradigm involves modeling two related multidimensional time variables (neural firing and behavior), we are directed to a general class of input–output (I/O) models. Within this class there are several candidate models available, and based on the amount of neurophysiological information that is utilized about the system, an appropriate modeling approach can be chosen. Three types of I/O models based on the amount of prior knowledge exist in the literature [36]:

- "White Box" The model is perfectly known (except for the parameters) and based on physical insight and observations.
- "Gray Box" Some physical insight is available but the model is not totally specified and other parameters need to be determined from the data.

- "Black Box" Little or no physical insight is available or used to choose the model, so the chosen model is picked on other grounds (e.g., robustness, easy implementation).

The choice of white, gray, or black box is dependent upon our ability to access and measure signals at various levels of the motor system as well as the computational cost of implementing the model in our current computing hardware.

The first modeling approach, white box, would require the highest level of physiological detail. Starting with behavior and tracing back, the system comprises muscles, peripheral nerves, the spinal cord, and ultimately the brain. This is a daunting task of system modeling due to the complexity, interconnectivity, and dimensionality of the involved neural structures. Model implementation would require the parameterization of a complete motor system [37] that includes the cortex, cerebellum, basal ganglia, thalamus, corticospinal tracts, and motor units. Since all of the details of each component/subcomponent of the described motor system remain unknown and are the subject of study for many neurophysiological research groups around the world, it is not presently feasible to implement white-box BMIs. Even if it was possible to parameterize the system to some high level of detail, the task of implementing the system in our state-of-the-art computers and digital signal processors (DSPs) would be an extremely demanding task.

The gray-box model requires a reduced level of physical insight. In the gray-box approach, one could take a particularly important feature of the motor nervous system, incorporate this knowledge into the model, and then use data to determine the rest of the unknown parameters. Two examples of gray-box models can be found in the BMI literature. One of the most common examples is Georgopoulos's PVA [3]. Using observations that cortical neuronal firing rates were dependent on the direction of arm movement, a model was formulated to incorporate the weighted sum of the neuronal firing rates. The weights of the model are then determined from the neural and behavioral recordings. A second example is given by Todorov, who extended the PVA by observed multiple correlations of M1 firing with movement position, velocity, acceleration, force exerted on an object, visual target position, movement preparation, and joint configuration [7, 8]. With these observations, Todorov [42] proposed a minimal, linear model that relates the delayed firings in M1 to the sum of many mechanistic variables (position, velocity, acceleration, and force of the hand). Todorov's model is intrinsically a *generative model* [16, 43]. Using knowledge about the relationship between arm kinematics and neural activity, the *states* (preferably the feature space of Todorov) of linear or nonlinear *dynamical* systems can be assigned. This methodology is supported by a well-known training procedure developed by Kalman for the linear case [44] and has been recently extended to the nonlinear case under the graphical models or Bayesian network frameworks. Since the formulation of generative models is recursive in nature, it is believed that the model is well suited for learning about motor systems because the states are all intrinsically related in time.

The last I/O model presented is the black-box model for BMIs. In this case, it is assumed that no physical insight is available for the model. The foundations of this type of time-series modeling were laid by Norbert Wiener for applications of gun control during World War II [45]. While military gun control applications may not seem to have a natural connection to BMIs, Wiener provided the tools for building models that correlate *unspecified* time-series inputs (in our case neuronal firing rates) and outputs (hand/arm movements). While this WF is topographically similar to the PVA, it is interesting to note that it was developed more than 30 years before Georgopoulos was developing his linear model relating neuronal activity to arm movement direction.

The three I/O modeling abstractions have gained large support from the scientific community and are also a well-established methodology in control theory for system identification [32]. Here we will concentrate on the last two types, which have been applied by engineers for many years to a wide variety of applications and have proven that the methods produce viable phenomenological descriptions when properly applied [46, 47]. One of the advantages of the techniques is that they quickly find, with relatively simple algorithms, optimal mappings (in the sense of minimum error power) between different time series using a nonparametric approach (i.e., without requiring a specific model for the time-series generation). These advantages have to be counterweighted by the abstract (nonstructural) level of the modeling and the many difficulties of the method, such as determining what a reasonable fit, a model order, and a topology are to appropriately represent the relationships among the input and desired response time series.

1.4.1 Gray Box

1.4.1.1 Population Vector Algorithm

The first model discussed is the PVA, which assumes that a cell's firing rate is a function of the velocity vector associated with the movement performed by the individual. The PVA model is given by

$$s_n(\mathbf{V}) = b_0^n + b_x^n v_x + b_y^n v_y + b_z^n v_z = \mathbf{B} \cdot \mathbf{V} = |\mathbf{B}||\mathbf{V}| \cos \theta \tag{1.1}$$

where the firing rate s for neuron n is a weighted ($b_{x,y,z}^n$) sum of the vectoral components ($v_{x,y,z}$) of the unit velocity vector $\mathbf{V}$ of the hand plus the mean firing rate b_0^n. The relationship in (1.1) is the inner product between the velocity vector of the movement and the weight vector for each neuron. The inner product (i.e., spiking rate) of this relationship becomes maximum when the weight vector $\mathbf{B}$ is collinear with the velocity vector $\mathbf{V}$. At this point, the weight vector $\mathbf{B}$ can be thought of as the cell's preferred direction for firing since it indicates the direction for which the neuron's activity will be maximum. The weights b^n can be determined by multiple regression techniques [3]. Each neuron makes a vectoral contribution w in the direction of $\mathbf{P}_i$ with magnitude given in (1.2). The resulting population vector or movement is given by (1.3), where the reconstructed movement at time t is simply the sum of each neuron's preferred direction weighted by the firing rate:

$$w_n(\mathbf{V}, t) = s_n(\mathbf{V}) - b_0^n \tag{1.2}$$

$$\mathbf{P}(\mathbf{V}, t) = \sum_{n=1}^{N} w_n(\mathbf{V}, t) \frac{\mathbf{B}_n}{\|\mathbf{B}_n\|} \tag{1.3}$$

It should be noted that the PVA approach includes several assumptions whose appropriateness in the context of neural physiology and motor control will be considered here. First, each cell is considered independently in its contribution to the kinematic trajectory. The formulation does not consider feedback of the neuronal firing patterns—a feature found in real interconnected neural architectures. Second, neuronal firing counts are linearly combined to reproduce the trajectory. At this point, it remains unknown how the neural activation of nonlinear functions will be necessary for complex movement trajectories.

1.4.1.2 Todorov's Mechanistic Model

An extension to the PVA has been proposed by Todorov [42], who considered multiple correlations of M1 firing with movement velocity and acceleration, position, force exerted on an object, visual target position, movement preparation, and joint configuration [2, 4, 10, 12–15, 38–41]. Todorov advanced the alternative hypothesis that the neural correlates with kinematic variables are epiphenomena of muscle activation stimulated by neural activation. Using studies showing that M1 contains multiple, overlapping representations of arm muscles and forms dense corticospinal projections to the spinal cord and is involved with the triggering of motor programs and modulation of spinal reflexes [35], Todorov [42] proposed a minimal, linear model that relates the delayed firings in M1 to the *sum* of mechanistic variables (position, velocity, acceleration, and force of the hand). Todorov's model takes the form

$$\mathbf{Us}(t - \Delta) = \mathbf{F}^{-1}\mathbf{GF}(t) + m\mathbf{HA}(t) + b\mathbf{HV}(t) + k\mathbf{HP}(t) \tag{1.4}$$

where the neural population vector $\mathbf{U}$ is scaled by the neural activity $\mathbf{s}(t)$ and is related to the scaled kinematic properties of gripping force $\mathbf{GF}(t)$, hand acceleration $\mathbf{HA}(t)$, velocity $\mathbf{HV}(t)$, and position $\mathbf{HP}(t)$.[1] From the BMI experimental setup, spatial samplings (in the hundred of neurons) of the input $\mathbf{s}(t)$ and the hand position, velocity, and acceleartion are collected synchronously; therefore the problem is one of finding the appropriate constants using a system identification framework [32]. Todorov's model in (1.4) assumes a first-order force production model and a local linear approximation to multijoint kinematics that may be too restrictive for BMIs. The mechanistic model for neural control of motor activity given in (1.4) involves a dynamical system where the output variables, position, velocity, acceleration, and force, of the motor system are driven by an high-dimensional input signal that is comprised of delayed ensemble neural activity [42]. In this interpretation of (1.4), the neural activity can be viewed as the *cause* of the changes in the mechanical variables and the system will be performing the decoding. In an alternative interpretation of Eq. (1.4), one can regard the neural activity as a distributed *representation* of the mechanical activity, and the system will be performing generative modeling. Next, a more general state space model implementation, the Kalman filter, will be presented. This filter corresponds to the *representation* interpretation of Todorov's model for neural control. Todorov's model clearly builds upon the PVA and can explain multiple neuronal and kinematic correlations; however, it is still a linear model of a potentially nonlinear system.

1.4.1.3 Kalman Filter A variety of Bayesian encoding/decoding approaches have been implemented in BMI applications [12, 16, 48]. In this framework, model designs have been developed based upon various assumptions that include the Kalman filter (linear, Gaussian model), extended Kalman filter (EKF, nonlinear, Gaussian model), and particle filter (PF, linear/nonlinear, Poisson model). Our discussion will begin with the most basic of the Bayesian approaches: the Kalman filter. This approach assumes a linear relationship between hand motion states and neural firing rates as well as Gaussian noise in the observed firing activity. The Kalman formulation attempts to estimate the state $\mathbf{x}(t)$ of a linear dynamical system as shown in Figure 1.4. For BMI applications, we define the states as the hand position, velocity, and acceleration, which are governed by a linear dynamical equation, as shown in

$$\mathbf{x}(t) = [\mathbf{HP}(t) \quad \mathbf{HV}(t) \quad \mathbf{HA}(t)]^{\mathrm{T}} \tag{1.5}$$

where $\mathbf{HP}$, $\mathbf{HV}$, and $\mathbf{HA}$ are the hand position, velocity, and acceleration vectors,[2] respectively. The Kalman formulation consists of a generative model for the data specified by the linear dynamic equation for the state in

$$\mathbf{x}(t + 1) = \mathbf{Ax}(t) + \mathbf{u}(t) \tag{1.6}$$

---

[1]The mechanistic model reduces to the PVA if the force, acceleration, and position terms are removed and neurons are independently considered.

[2]The state vector is of dimension $9 + N$; each kinematic variable contains an x, y, and z component plus the dimensionality of the neural ensemble.

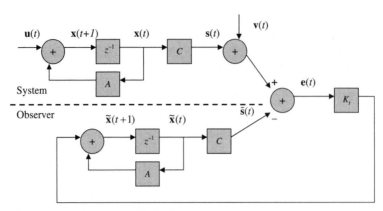

Figure 1.4 Kalman filter block diagram.

where $\mathbf{u}(t)$ is assumed to be a zero-mean Gaussian noise term with covariance $\mathbf{U}$. The output mapping (from state to spike trains) for this BMI linear system is simply

$$\mathbf{s}(t) = \mathbf{C}\mathbf{x}(t) + \mathbf{v}(t) \tag{1.7}$$

where $\mathbf{v}(t)$ is the zero-mean Gaussian measurement noise with covariance $\mathbf{V}$ and $\mathbf{s}$ is a vector consisting of the neuron firing patterns binned in nonoverlapping windows. In this specific formulation, the output-mapping matrix $\mathbf{C}$ has dimensions $N \times 9$. Alternatively, we could have also included the spike counts of N neurons in the state vector as $\mathbf{f}_1, \ldots, \mathbf{f}_N$. This specific formulation would exploit the fact that the future hand position is a function of not only the current hand position, velocity, and acceleration but also the current cortical firing patterns. However, this advantage comes at the cost of large training set requirements, since this extended model would contain many more parameters to be optimized. To train the topology given in Figure 1.4, L training samples of $\mathbf{x}(t)$ and $\mathbf{s}(t)$ are utilized, and the model parameters $\mathbf{A}$ and $\mathbf{U}$ as given in (1.6) are determined using least squares. The optimization problem to be solved is given by

$$\mathbf{A} = \arg \min_{\mathbf{A}} \sum_{t=1}^{L-1} \|\mathbf{x}(t+1) - \mathbf{A}\mathbf{x}(t)\|^2 \tag{1.8}$$

The solution to this optimization problem is found to be

$$\mathbf{A} = \mathbf{X}_1 \mathbf{X}_0^T (\mathbf{X}_1 \mathbf{X}_1^T)^{-1} \tag{1.9}$$

where the matrices are defined as $\mathbf{X}_0 = [\mathbf{x}_1 \quad \cdots \quad \mathbf{x}_{L-1}]$, $\mathbf{X}_1 = [\mathbf{x}_2 \quad \cdots \quad \mathbf{x}_L]$. The estimate of the covariance matrix $\mathbf{U}$ can then be obtained using

$$\mathbf{U} = \frac{(\mathbf{X}_1 - \mathbf{A}\mathbf{X}_0)(\mathbf{X}_1 - \mathbf{A}\mathbf{X}_0)^T}{L-1} \tag{1.10}$$

Once the system parameters are determined using least squares on the training data, the model obtained $(\mathbf{A}, \mathbf{C}, \mathbf{U})$ can be used in the Kalman filter to generate estimates of hand positions from neuronal firing measurements. Essentially, the model proposed here assumes a linear dynamical relationship between current and future trajectory states. Since the Kalman filter formulation requires a reference output from the model, the spike counts are assigned to the output, as they are the only available signals.

The Kalman filter is an adaptive state estimator (observer) where the observer gains are optimized to minimize the state estimation error variance. In real-time operation, the Kalman gain matrix K (1.12) is updated using the projection of the error covariance in (1.11) and the error covariance update in (1.14). During model testing, the Kalman gain correction is a powerful method for decreasing estimation error. The state in (1.13) is updated by adjusting the current state value by the error multiplied with the Kalman gain:

$$\mathbf{P}^-(t+1) = \mathbf{AP}(t)\mathbf{A}^\mathrm{T} + \mathbf{U} \tag{1.11}$$

$$\mathbf{K}(t+1) = \mathbf{P}^-(t+1)\mathbf{C}^\mathrm{T}(\mathbf{CP}^-(t+1)\mathbf{C}^\mathrm{T})^{-1} \tag{1.12}$$

$$\tilde{\mathbf{x}}(t+1) = \mathbf{A}\tilde{\mathbf{x}}(t) + \mathbf{K}(t+1)(\mathbf{S}(t+1) - \mathbf{CA}\tilde{\mathbf{x}}(t)) \tag{1.13}$$

$$\mathbf{P}(t+1) = (\mathbf{I} - \mathbf{K}(t+1)\mathbf{C})\mathbf{P}^-(t+1) \tag{1.14}$$

While the Kalman filter equations provide a closed-form decoding procedure for linear Gaussian models, we have to consider the fact that the relationship between neuronal activity and behavior may be nonlinear. Moreover, measured neuronal firing often follows Poisson distributions. The consequences for such a mismatch between the model and the real system will be expressed as additional errors in the final position estimates. To cope with this problem, we need to go beyond the linear Gaussian model assumption. In principle, for an arbitrary nonlinear dynamical system with arbitrary known noise distributions, the internal states (**HP**, **HV**, and **HA**) can be estimated from the measured outputs (neuronal activity). Algorithms designed to this end include the EKF, the unscented Kalman filter (UKF), the PF, and their variants. All of these algorithms basically try the complicated recursive Bayesian state estimation problem using various simplifications and statistical techniques. In the literature, BMI researchers have already implemented the PF approach [16].

In this most general framework, the state and output equations can include nonlinear functions $f_1(\cdot)$ and $f_2(\cdot)$ as given in

$$\mathbf{x}(t+1) = f_1(\mathbf{x}(t)) + \mathbf{u}(t) \qquad \mathbf{s}(t) = f_2(\mathbf{x}(t)) + \mathbf{v}(t) \tag{1.15}$$

Experimental observations have shown that the measured spike trains typically follow a Poisson distribution, which is given in

$$p(\mathbf{s}(t) \mid \mathbf{x}(t)) = \prod_{n=1}^{N} \frac{e^{-\lambda_i(\mathbf{x}(t))}[\lambda_i(\mathbf{x}(t))]^{y_i(t)}}{[s_i(t)]!} \tag{1.16}$$

The tuning function of the ith neuron is denoted by λ_i and $s_i(t)$ is the firing count of the ith neuron at time instant t. In order to decode neuronal activity into hand kinematics for the nonlinear observation equations and Poission spiking models, the recursive Bayesian estimator called the PF can be used. In this approach, we seek to recursively update the posterior probability of the state vector given the neuronal measurements as shown in (1.17), where $\mathbf{S}_t$ is the entire history of neuronal firing up to time t:

$$p(\mathbf{x}(t)|\mathbf{S}_t) = \mu p(\mathbf{s}(t)|x(t))p(\mathbf{x}(t)|\mathbf{S}_{t-1}) \tag{1.17}$$

After some manipulations, the equivalent expression in (1.18) can be obtained [49]:

$$p(\mathbf{x}(t)|\mathbf{S}_t) = \mu p(\mathbf{s}(t)|\mathbf{x}(t)) \int p(\mathbf{x}(t)|\mathbf{x}(t-1))p(\mathbf{x}(t-1)|\mathbf{S}_{t-1}) \, d\mathbf{x}(t-1) \tag{1.18}$$

Notice that (1.18) is essentially the recursion of the conditional state distribution that we seek from $p(\mathbf{x}(t-1)|\mathbf{S}_{t-1})$ to $p(\mathbf{x}(t)|\mathbf{S}_t)$. Analytical evaluation of the integral on the right-hand side is, in general, impossible. Therefore, the following simplifying assumption is made: The conditional state distribution can be approximated by a delta train, which

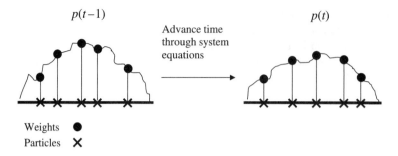

Figure 1.5   Estimation of continuous distributions by sampling (particles).

samples the continuous distribution at appropriately selected values called particles. Each particle also has a weight associated with it that represents the probability of that particle, as shown in Figure 1.5. The elegance of the PF approach lies in the simplification of the integral in (1.18) to a summation. These simplifications lead to the following update equations for the particles, their weights, and the state estimate:

$$x_i(t+1) = f_1(x_i(t)) \tag{1.19}$$

$$\tilde{w}_i(t+1) = P(s(t)|x_i(t))w_i(t) \tag{1.20}$$

$$w_i(t+1) = \frac{\tilde{w}_i(t+1)}{\sum_i \tilde{w}_i(t+1)} \tag{1.21}$$

$$\tilde{\mathbf{x}}(t+1) = \sum_{i=1}^{N} w_i(t+1)x_i(t+1) \tag{1.22}$$

We have shown that, using measured position, velocity, and acceleration as states and neuronal firing counts as model outputs within this recursive, probabilistic framework, this approach may seem to be the best state-of-the-art method available to understand the encoding and decoding between neural activity and hand kinematics. Unfortunely, for BMIs this particular formulation is faced with problems of parameter estimation. The generative model is required to find the mapping from the low-dimensional kinematic parameter state space to the high-dimensional output space of neuronal firing patterns (100+ dimensions). Estimating model parameters from the collapsed space to the high-dimensional neural state can be difficult and yield multiple solutions. Moreover, in the BMI literature the use of the PF has only produced marginal improvements over the standard Kalman formulation, which may not justify the extra computational complexity [16]. For this modeling approach, our use of physiological knowledge and choice of modeling framework actually complicates the mapping process. As an alternative, one could disregard any knowledge about the system being modeled and use a strictly data-driven methodology to build the model.

1.4.2   Black Box

1.4.2.1   Wiener Filters   The first black-box model we will discuss assumes that there exists a linear mapping between the desired hand kinematics and neuronal firing counts. In this model, the delayed versions of the firing counts, $s(t-l)$, are the bases that construct the output signal. Figure 1.6 shows the topology of the multiple input–multiple output (MIMO) WF where the output y_j is a weighted linear combination of the l most recent

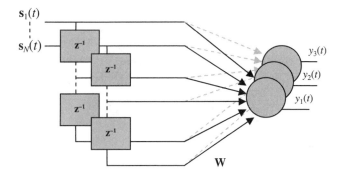

Figure 1.6   FIR filter topology. Each neuronal input $\mathbf{s}_N$ contains a tap delay line with l taps.

values[3] of neuronal inputs $\mathbf{s}$ given in (1.23) [32]. Here y_j can be defined to be any of the single coordinate directions of the kinematic variables **HP**, **HV**, **HA**, or **GF**. The model parameters are updated using the optimal linear least-squares (LS) solution that matches the Wiener solution. The Wiener solution is given by (1.24), where $\mathbf{R}$ and $\mathbf{P}_j$ are the auto-correlation and cross-correlation functions, respectively, and $\mathbf{d}_j$ is hand trajectory, velocity, or gripping force:[4]

$$y_j(t) = \mathbf{W}_j \mathbf{s}(t) \tag{1.23}$$

$$\mathbf{W}_j = \mathbf{R}^{-1} \mathbf{P}_j = E(\mathbf{s}^{\mathrm{T}} \mathbf{s})^{-1} E(\mathbf{s}^{\mathrm{T}} \mathbf{d}_j) \tag{1.24}$$

The autocorrelation matrix $\mathbf{R}$ and the cross-correlation matrix $\mathbf{P}$ can be estimated directly from the data using either the autocorrelation or the covariance method [32]. Experimentally we verified that the size of the data block should contain at least 10 min of recordings for better performance. In this MIMO problem, the autocorrelation matrix is not Toeplitz, even when the autocorrelation method is employed. One of the real dangers of computing the WF solution to BMIs is that $\mathbf{R}$ may not be full rank [50]. Instead of using the Moore–Penrose inverse, we utilize a regularized solution substituting $\mathbf{R}^{-1}$ by $(\mathbf{R} + \lambda \mathbf{I})^{-1}$, where λ is the regularization constant estimated from a cross-validation set. Effectively this solution corresponds to ridge regression [51]. The computational complexity of the WF is high for the number of input channels used in our experiments. For 100 neural channels using 10 tap delays and three outputs, the total number of weights is 3000. This means that one must invert a 1000×1000 matrix every N samples, where N is the size of the training data block.

As is well known in adaptive filter theory [32], search procedures can be used to find the optimal solution using gradient descent or Newton-type search algorithms. The most widely used algorithm in this setting is the least mean-square (LMS) algorithm, which utilizes stochastic gradient descent [32]. For real data we recommend the normalized LMS algorithm instead,

$$\mathbf{W}_j(t+1) = \mathbf{W}_j(t) + \frac{\eta}{\|s(t)\|} \mathbf{e}_j(t)\mathbf{s}(t) \tag{1.25}$$

where η is the step size or learning rate, $e(t) = d_x(t) - y_x(t)$ is the error, and $\|\mathbf{s}(t)\|$ is the power of the input signal contained in the taps of the filter. Using the normalized LMS

---

[3]In our studies we have observed neuronal activity correlated with behavior for up to 10 lags.

[4]Each neuronal input and desired trajectory for the WF was preprocessed to have a mean value of zero.

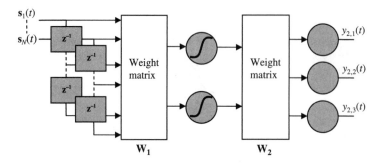

Figure 1.7   Time delay neural network topology.

algorithm (NLMS), the model parameters are updated incrementally at every new sample and so the computation is greatly reduced. The step size must be experimentally determined from the data. One may think that the issues of nonstationarity are largely resolved since the filter is always being updated, tracking the changing statistics. However, during testing the desired response is not available so the weights of the filter have to be frozen after training. Therefore, NLMS is still subject to the same problems as the Wiener solution, although it may provide slightly better results when properly trained (when the data are not stationary, the LMS and the Wiener solution do not necessarily coincide [32]).

Linear filters trained with mean-square error (MSE) provide the best linear estimate of the mapping between neural firing patterns and hand position. Even though the solution is guaranteed to converge to the global optimum, the model assumes that the relationship between neural activity and hand position is linear, which may not be the case. Furthermore, for large input spaces, including memory in the input introduces many extra degrees of freedom to the model, hindering generalization capabilities.

1.4.2.2   *Time Delay Neural Network*   Spatiotemporal nonlinear mappings of neuronal firing patterns to hand position can be constructed using a TDNN [52]. The TDNN architecture consists of a tap delay line memory structure at the input in which past neuronal firing patterns in time can be stored, followed by a one-hidden-layer perceptron with a linear output as shown in Figure 1.7. The output of the first hidden layer of the network can be described with the relation $y_1(t) = f(W_1 s(t))$, where $f(\cdot)$ is the hyperbolic tangent nonlinearity [$\tanh(\beta x)$].[5] The input vector s includes l most recent spike counts from N input neurons. In this model the nonlinear weighted and delayed versions of the firing counts $s(t - l)$ construct the output of the *hidden layer*. The number of delays in the topology should be set so that there is significant coupling between the input and desired signal. The number of hidden processing elements (PEs) is determined through experimentation. The output layer of the network produces the hand trajectory $y_2(t)$ using a linear combination of the hidden states and is given by $y_2(t) = W_2 y_1(t)$. The weights (W_1, W_2) of this network can be trained using static backpropagation[6] with the MSE as the learning criterion. This is the great advantage of this artificial neural network.

This topology is more powerful than the linear finite impulse response (FIR) filter because it is effectively a nonlinear combination of FIR filters (as many as the number of hidden PEs). Each of the hidden PE outputs can be thought of as a basis function of the output space (nonlinearly adapted from the input) utilized to project the

---

[5]The logistic function is another common nonlinearity used in neural networks.

[6]Backpropagation is a simple application of the chain rule, which propagates the gradients through the topology.

high-dimensional data. While the nonlinear nature of the TDNN may seem as an attractive choice for BMIs, putting memory at the input of this topology presents difficulties in training and model generalization. Adding memory to the high-dimensional neural input introduces many free parameters to train. For example, if a neural ensemble contains 100 neurons with 10 delays of memory and the TDNN topology contains five hidden PEs, 5000 free parameters are introduced in the input layer alone. Large data sets and slow learning rates are required to avoid overfitting. Untrained weights can also add variance to the testing performance, thus decreasing accuracy.

1.4.2.3 Nonlinear Mixture of Competitive Local Linear Models

1.4.2.3 *Nonlinear Mixture of Competitive Local Linear Models*   The next model topology that we will discuss is in general similar to the TDNN; however, the training procedure undertaken here is significantly different. This modeling method uses the divide-and-conquer approach. Our reasoning is that a complex nonlinear modeling task can be elucidated by dividing it into simpler linear modeling tasks and combining them properly [53]. Previously, this approach was successfully applied to nonstationary signal segmentation, assuming that a nonstationary signal is a combination of piecewise stationary signals [54]. Hypothesizing that the neural activity will demonstrate varying characteristics for different localities in the space of the hand trajectories, we expect the multiple-model approach, in which each linear model specializes in a local region, to provide a better overall I/O mapping. However, here the problem is different since the goal is not to segment a signal but to segment the joint input/desired signal space. The overall system archtitecture is depicted in Figure 1.8.

The local linear models can be conceived as a committee of experts each specializing in one segment of the hand trajectory space. The multilayer perceptron (MLP) is introduced to the system in order to nonlinearly combine the predictions generated by all the linear models. Experiments demonstrated that this nonlinear mixture of competitive local linear models is potentially more powerful than a gated mixture of linear experts, where only the winning expert's opinion or their weighted sum is considered. In addition,

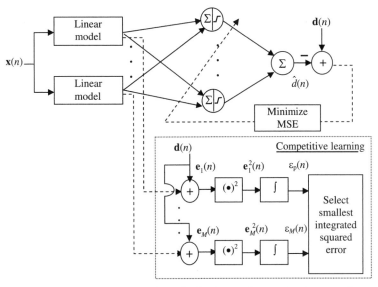

Figure 1.8   This topology consists of selecting a winner using integrated squared errors from each linear model. Outputs from M trained linear models are then fed to a MLP.

for BMI applications, it has the advantage that it does not require a selection scheme in testing, which can only be accurately done using the desired output. For example, in a prosthetic arm application, the desired hand position is not available in practice.

The topology allows a two-stage training procedure that can be performed independently: first competitive learning for the local linear models and then error back-propagation learning for the MLP. In off-line training, this can be done sequentially, where first the local linear models are optimized and then their outputs are used as training inputs for the following MLP. It is important to note that in this scheme both the local linear models and the MLP are trained to approximate the same desired response, which is the hand trajectory of the primate.

The training of the multiple linear models is accomplished by competitively (hard or soft) updating their weights in accordance with previous approaches using the normalized least-mean-square (NLMS) algorithm [32]. The winning model is determined by comparing the (leaky) integrated squared errors of all competing models and selecting the model that exhibits the least integrated error for the corresponding input [54]. In competitive training, the leaky integrated squared error for the ith model is given by

$$\varepsilon_i(t) = (1 - \mu)\varepsilon_i(t - 1) + \mu e_i^2(t) \qquad i = 1, \ldots, M \tag{1.26}$$

where M is the number of models and μ is the time constant of the leaky integrator. If hard competition is employed, then only the weight vector of the winning model is updated. Specifically, the hard-competition update rule for the weight vector of the winning model is

$$\mathbf{w}_{\text{winner}}(t + 1) = \mathbf{w}_{\text{winner}}(t) + \frac{\eta e_{\text{winner}}(t)\mathbf{s}(t)}{\gamma + \|\mathbf{s}(t)\|^2} \tag{1.27}$$

where $\mathbf{w}_{\text{winner}}$ is the weight vector, $\mathbf{s}(t)$ is the current input, $e_{\text{winner}}(t)$ is the instantaneous error of the winning model, η is the learning rate, and γ is the small positive constant. If soft competition is used, a Gaussian weighting function centered at the winning model is applied to all competing models. Every model is then updated proportional to the weight assigned to that model by this Gaussian weighting function:

$$\mathbf{w}_i(t + 1) = \mathbf{w}_i(t) + \frac{\eta(t)\Lambda_{i,j}(t)e_t(t)\mathbf{x}(t)}{\gamma + \|\mathbf{x}(t)\|^2} \qquad i = 1, \ldots, M \tag{1.28}$$

where $\mathbf{w}_i$ is the weight vector of the ith model, the jth model is the winner, and $\Lambda_{i,j}(t)$ is the weighting function

$$\Lambda_{i,j}(t) = \exp\left(-\frac{d_{i,j}^2}{2\sigma^2(t)}\right) \tag{1.29}$$

where $d_{i,j}$ is the Euclidean distance between index i and j which is equal to $|j - i|$, $\eta(t)$ is the annealed learning rate, and $\sigma^2(t)$ is the kernel width, which decreases exponentially as t increases. The learning rate also decreases exponentially with time.

Soft competition preserves the topology of the input space by updating the models neighboring the winner; thus it is expected to result in smoother transitions between models specializing in topologically neighboring regions (of the state space). However, in the experimental results, it was shown that the hard-competition rule comparisons on data sets utilized in BMI experiments did not show any significant difference in generalization performance (possibly due to the nature of the data set used in these experiments).

The competitive training of the first layer of linear models to match the hand trajectory using the neural activity creates a set of basis signals from which the following single hidden-layer MLP can generate accurate hand position predictions. However, this performance is at the price of an increase in the number of free model parameters which results from each of the local linear models (100 neurons × 10 tap delays × 3 coordinates × 10 models = 30,000 parameters in the input). Additional experimentation is necessary to evaluate the long-term performance (generalization) of such a model in BMI applications.

1.4.2.4 Recurrent Multilayer Perceptron The final black-box BMI model discussed is potentially the most powerful because it not only contains a nonlinearity but also includes dynamics through the use of feedback. The recurrent multilayer perceptron (RMLP) architecture (Fig. 1.9) consists of an input layer with N neuronal input channels, a fully connected hidden layer of PEs (in this case tanh), and an output layer of linear PEs. Each hidden layer PE is connected to every other hidden PE using a unit time delay. In the input layer equation (1.30), the state produced at the output of the first hidden layer is a nonlinear function of a weighted combination (including a bias) of the current input and the previous state. The feedback of the state allows for continuous representations on multiple time scales and effectively implements a short-term memory mechanism. Here, $f(\cdot)$ is a sigmoid nonlinearity (in this case *tanh*), and the weight matrices $\mathbf{W}_1$, $\mathbf{W}_2$, and $\mathbf{W}_f$ as well as the bias vectors $\mathbf{b}_1$ and $\mathbf{b}_2$ are again trained using synchronized neural activity and hand position data. Each hidden PE output can be thought of as a nonlinear adaptive basis of the output space utilized to project the high-dimensional data. These projections are then linearly combined to form the outputs of the RMLP that will predict the desired hand movements. One of the disadvantages of the RMLP when compared with the Kalman filter is that there is no known closed-form solution to estimate the matrices $\mathbf{W}_f$, $\mathbf{W}_1$, and $\mathbf{W}_2$ in the model; therefore, gradient descent learning is used. The RMLP can be trained with backpropagation through time (BPTT) or real-time recurrent learning (RTRL) [55]:

$$\mathbf{y}_1(t) = f(\mathbf{W}_1\mathbf{s}(t) + \mathbf{W}_f\mathbf{y}_1(t-1) + \mathbf{b}_1) \tag{1.30}$$

$$\mathbf{y}_2(t) = \mathbf{W}_2\mathbf{y}_1(t) + \mathbf{b}_2 \tag{1.31}$$

If the RMLP approach for the BMI is contrasted with Todorov's model, one can see that the RMLP accepts the neuronal activity (also as binned spike counts) as input and generates a prediction of the hand position using first-order internal dynamics. Although the model output $\mathbf{y}_2$ can consist of only the hand position, the RMLP must learn to build an efficient internal dynamical representation of the other mechanical variables (velocity,

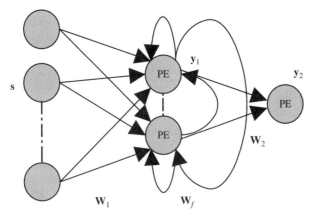

Figure 1.9 Fully connected, state recurrent neural network.

acceleration, and force) through the use of feedback. In fact, in this model, the hidden state vector ($\mathbf{y}_1$) can be regarded as the RMLP representation of these mechanical variables driven by the neural activity in the input ($\mathbf{s}$). Hence, the dynamical nature of Todorov's model is implemented through the nonlinear feedback in the RMLP. The output layer is responsible for extracting the position information from the representation in $\mathbf{y}_1$ using a linear combination. An interesting analogy exists between the output layer weight matrix $\mathbf{W}_2$ in the RMLP and the matrix $\mathbf{U}$ in Todorov's model. This analogy stems from the fact that each column of $\mathbf{U}$ represents a direction in the space spanning the mixture of mechanical variables to which the corresponding individual neuron is cosine tuned, which is a natural consequence of the inner product. Similarly, each column of $\mathbf{W}_2$ represents a direction in the space of hand position to which a nonlinear mixture of neuronal activity is tuned. In general, the combination of Todorov's theory with the use of nonlinearity and dynamics gives the RMLP a powerful approximating capability.

1.4.3   Generalization/Regularization/Weight Decay/ Cross Validation

The primary goal in BMI experiments is to produce the best estimates of HP, HV, and GF from neuronal activity that has not been used to train the model. This testing performance describes the generalization ability of the models. To achieve good generalization for a given problem, the two first considerations to be addressed are the choice of model topology and training algorithm. These choices are especially important in the design of BMIs because performance is dependent upon how well the model deals with the large dimensionality of the input as well as how the model generalizes in nonstationary environments. The generalization of the model can be explained in terms of the bias-variance dilemma of machine learning [56], which is related to the number of free parameters of a model. The MIMO structure of BMIs built for the data presented here can have as few as several hundred to as many as several thousand free parameters. On one extreme if the model does not contain enough parameters, there are too few degrees of freedom to fit the function to be estimated, which results in bias errors. On the other extreme, models with too many degrees of freedom tend to overfit the function to be estimated. In terms of BMIs, models tend to err on the latter because of the large dimensionality of the input. The BMI model overfitting is especially a problem in topologies where memory is implemented at the input layer. With each new delay element, the number of free parameters will scale with the number of input neurons as in the FIR filter and TDNN.

To handle the bias-variance dilemma, one could use the traditional Akaike or BIC criteria; however, the MIMO structure of BMIs excludes these approaches [57]. As a second option, during model training regularization techniques could be implemented that attempt to reduce the value of unimportant weights to zero and effectively prune the size of the model topology [58].

In BMI experiments we are not only faced with regularization issues but also we must consider ill-conditioned model solutions that result from the use of finite data sets. For example, computation of the optimal solution for the linear WF involves inverting a poorly conditioned input correlation matrix that results from sparse neural firing data that are highly variable. One method of dealing with this problem is to use the pseudoinverse. However, since we are interested in both conditioning and regularization, we chose to use ridge regression (RR) [51] where an identity matrix is multiplied by a white-noise variance and is added to the correlation matrix. The criterion function of RR is given by

$$J(\mathbf{w}) = E[\|\mathbf{e}\|^2] + \delta\|\mathbf{w}\|^2 \tag{1.32}$$

where **w** are the weights, **e** is the model error, and the additional term $\delta\|\mathbf{w}\|^2$ smooths the cost function. The choice of the amount of regularization (δ) plays an important role in the generalization performance and for larger deltas performance can suffer because SNR is sacrificed for smaller condition numbers. It has been proposed by Larsen et al. that δ can be optimized by minimizing the generalization error with respect to δ [47]. For other model topologies such as the TDNN, RMLP, and the LMS update for the FIR, weight decay (WD) regularization is an on-line method of RR to minimize the criterion function in (1.32) using the stochastic gradient, updating the weights by

$$\mathbf{w}(n+1) = \mathbf{w}(n) + \eta\nabla(J) - \delta\mathbf{w}(n) \tag{1.33}$$

Both RR and WD can be viewed as the implementations of a Bayesian approach to complexity control in supervised learning using a zero-mean Gaussian prior [59].

A second method that can be used to maximize the generalization of a BMI model is called cross validation. Developments in learning theory have shown that during model training there is a point of maximum generalization after which model performance on unseen data will begin to deteriorate [60]. After this point the model is said to be *overtrained*. To circumvent this problem, a cross-validation set can be used to indicate an early stopping point in the training procedure. To implement this method, the training data are divided into a training set and a cross-validation set. Periodically during model training, the cross-validation set is used to test the performance of the model. When the error in the validation set begins to increase, the training should be stopped.

1.5 EXAMPLES

Four examples of how optimal signal processing can be used on real behavioral and neural recordings for the development of BMIs will now be given. The examples are focused on comparing the performance of linear, generative, nonlinear, feedforward, and dynamical models for the hand-reaching motor task. The four models include the FIR Wiener filter, TDNN (the nonlinear extension to the WF), Kalman filter, and RMLP. Since each of these models employs very different principles and has different mapping power, it is expected that they will perform differently and the extent to which they differ will be quantitatively compared. With these topologies, it will be shown how BMI performance is affected by the number of free model parameters, computational complexity, and nonlinear dynamics. In the following examples, the firing times of single neurons were recorded by researchers at Duke University while a primate performed a 3D reaching task that involved a right-handed reach to food and subsequent placing of the food in the mouth, as shown in Figure 1.10 [10]. Neuronal firings, binned (added) in nonoverlapping windows of 100 ms, were directly used as inputs to the models. The primate's hand position (HP), used as the desired signal, was also recorded (with a time-shared clock) and digitized with a 200-Hz sampling rate. Each model was trained using 20,010 consecutive time bins (2001 s) of data.

One of the most difficult aspects of modeling for BMIs is the dimensionality of the neuronal input (in this case 104 cells). Because of this large dimensionality, even the simplest models contain topologies with thousands of free parameters. Moreover, the BMI model is often trying to approximate relatively simple trajectories resembling sine waves which practically can be approximated with only two free parameters. Immediately, we are faced with avoiding overfitting the data. Large dimensionality also has an impact on the computational complexity of the model, which can require thousands more multiplications, divisions, and function evaluations. This is especially a problem if we wish to

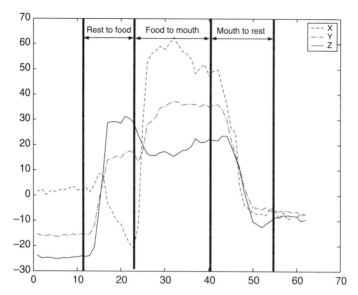

Figure 1.10   Reaching movement trajectory.

implement the model in low-power portable DSPs. Here we will assess each of the four BMI models in terms of their number of free parameters and computational complexity.

Model overfitting is often described in terms of prediction risk (PR), which is the expected performance of a topology when predicting new trajectories not encountered during training [61]. Several estimates of the PR for linear models have been proposed in the literature [62–65]. A simple way to develop a formulation for the prediction risk is to assume the quadratic form in (1.34), where e is the training error for a model with Φ parameters and N training samples. In this quadratic formulation, we can consider an optimal number of parameters, Φ^{Opt}, that minimizes the PR. We wish to estimate how the PR will vary with Φ, which can be given by a simple Taylor series expansion of (1.34) around Φ^{Opt} as performed in [65]. Manipulation of the Taylor expansion will yield the general form in (1.35). Other formulations for the PR include the generalized cross validation (GCV) and Akaike's final prediction error (FPE) given in (1.36) and (1.37). The important characteristic of (1.35)–(1.37) is that they all involve the interplay of the number of model parameters to the number of training samples. In general, the PR increases as the number of model parameters increases:

$$PR = E[e^2(\Phi_N)] \tag{1.34}$$

$$PR \approx e^2\left(1 + \frac{\Phi}{N}\right) \tag{1.35}$$

$$GCV = \frac{e^2}{(1 - \Phi/N)^2} \tag{1.36}$$

$$FPE = e^2\left(\frac{1 + \Phi/N}{1 - \Phi/N}\right) \tag{1.37}$$

The formulations for the PR presented here have been extended to nonlinear models [66]. While the estimation of the PR for linear models is rather straightforward, in the case nonlinear models the formulation is complicated by two factors. First, the nonlinear formulation involves computing the effective number of parameters (a number that differs from the true number of parameter in the model), which is nontrivial to estimate since it depends on the amount of model bias, model nonlinearity, and amount of regularization used in training [66]. Second, the formulation involves computing the noise covariance matrix of the desired signal, another parameter that is nontrivial to compute, especially in the context of BMI hand trajectories.

For the reaching task data set, all of the models utilize 104 neuronal inputs, as shown in Table 1.2. The first encounter with an explosion in the number of free parameters occurs for both the WF and TDNN since they contain a 10-tap delay line at the input. Immediately the number of inputs is multiplied by 10. The TDNN topology has the greatest number of free parameters, 5215, of the feedforward topologies because the neuronal tap delay memory structure is also multiplied by the 5 hidden processing elements following the input. The WF, which does not contain any hidden processing elements, contains 3120 free parameters. In the case of the Kalman filter, which is the largest topology, the number of parameters explodes due to the size of the $\mathbf{A}$ and $\mathbf{C}$ matrices since they both contain the square of the dimensionality of the 104 neuronal inputs. Finally, the RMLP topology is the most frugal since it moves its memory structure to the hidden layer through the use of feedback, yielding a total of 560 free parameters.

To quantify how the number of free parameters affects model training time, a Pentium 4 class computer with 512 MB DDR RAM, the software package NeuroSolutions for the neural networks [67], and MATLAB for computing the Kalman and Wiener solution were used to train the models. The training times of all four topologies are given in Table 1.2. For the WF, the computation of the inverse of a 1040×1040 autocorrelation matrix took 47 s in MATLAB, which is optimized for matrix computations. For the neural networks, the complete set of data is presented to the learning algorithm in several iterations called epochs. In NeuroSolutions, whose programming is based on C, 20,010 samples were presented 130 and 1000 times in 22 min, 15 s and 6 min, 35 s for the TDNN and RMLP, respectively [67]. The TDNN was trained with backpropagation and the RMLP was trained with BPTT [55] with a trajectory of 30 samples and learning rates of 0.01, 0.01, and 0.001 for the input, feedback, and output layers, respectively. Momemtum learning was also implemented with a rate of 0.7. One hundred Monte

TABLE 1.2   Model Parameters

	WF	TDNN	Kalman filter	RMLP
Training time	47 s	22 min, 15 s	2 min, 43 s	6 min, 35 s
Number of epochs	1	130	1	1000
Cross validation	N/A	1000 pts.	N/A	1000 pts.
Number of inputs	104	104	104	104
Number of tap delays	10	10	N/A	N/A
Number of hidden PEs	N/A	5	113 (states)	5
Number of outputs	3	3	9	3
Number of adapted weights	3120	5215	12073	560
Regularization	0.1 (RR)	1×10^{-5} (WD)	N/A	1×10^{-5} (WD)
Learning rates	N/A	1×10^{-4} (input)	N/A	1×10^{-2} (input)
		1×10^{-5} (output)		1×10^{-2} (feedback)
				1×10^{-3} (output)

Carlo simulations with different initial conditions were conducted of neuronal data to improve the chances of obtaining the global optimum. Of all the Monte Carlo simulations, the network with the smallest error achieved a MSE of 0.0203 ± 0.0009. A small training standard deviation indicates the network repeatedly achieved the same level of performance. Neural network training was stopped using the method of cross validation (batch size of 1000 pts.), to maximize the generalization of the network [60]. The Kalman proved to be the slowest to train since the update of the Kalman gain requires several matrix multiplies and divisions. In these simulations, the number of epochs chosen was based upon performance in a 1000-sample cross-validation set, which will be discussed in the next section. To maximize generalization during training, ridge regression, weight decay, and slow learning rates were also implemented.

The number of free parameters is also related to the computational complexity of each model given in Table 1.3. The number of multiplies, adds, and function evaluations describe how demanding the topology is for producing an output. The computational complexity especially becomes critical when implementing the model in a low-power portable DSP, which is the intended outcome for BMI applications. In Table 1.3 define N_0, t, d, and N_1 to be the number of inputs, tap delays, outputs, and hidden PEs, respectively. In this case, only the number of multiplications and function evaluations are presented since the number of additions is essentially identical to the number of multiplications. Again it can be seen that demanding models contain memory in the neural input layer. With the addition of each neuronal input the computational complexity of the WF increases by 10 and the TDNN by 50. The Kalman filter is the most computationally complex $[O((N_0 + 9)^3)]$ since both the state transition and output matrix contain dimensionality of the neuronal input. For the neural networks, the number of function evaluations is not as demanding since they contain only five for both the TDNN and RMLP. Comparing the neural network training times also exemplifies the computational complexity of each topology; the TDNN (the most computationally complex) requires the most training time and allows only a hundred presentations of the training data. As a rule of thumb, to overcome these difficulties, BMI architectures should avoid the use of memory structures at the input.

In testing, all model parameters were fixed and 3000 consecutive bins (300 s) of novel neuronal data were fed into the models to predict new hand trajectories. Figure 1.11 shows the output of the three topologies in the test set with 3D hand position for one reaching movement. While only a single movement is presented for simplicity, it can be shown that during the short period of observation (5 min) there is no noticeable degradation of the model fitting across time. From the plots it can be seen that qualitatively all three topologies do a fair job at capturing the reach to the food and the initial reach to the mouth. However, both the WF and TDNN cannot maintain the peak values of HP at the mouth position. Additionally the WF and TDNN have smooth transitions between the food and mouth while the RMLP sharply changes its position in this region. The traditional way to quantitatively report results in BMI is through the correlation coefficient (CC) between

TABLE 1.3   Model Computational Complexity

	Multiplications	Function evaluations
WF	$N_0 \times t \times d$	N/A
TDNN	$N_0 \times t \times N_1 \times d$	N_1
Kalman filter	$O((N_0 + 9)^3)$	N/A
RMLP	$N_0 \times N_1 + N_1 \times d + N_1 \times N_1$	N_1

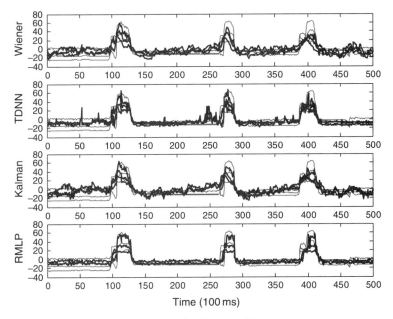

Figure 1.11   Testing performance for three reaching movements.

the actual and estimated hand position trajectories as shown in Table 1.4 which was computed for the entire trajectory. The WF and TDNN perform similarly on average in their CC values while the RMLP has significantly greater performance. In general, the overall experimental performance can also depend on the movement trajectory studied, animal, daily recording session, and variability in the neurons probed.

The reaching task which consists of a reach to food and subsequent reach to the mouth is embedded in periods where the animal's hand is at rest as shown by the flat trajectories to the left and right of the movement. Since we are interested in how the models perform in each mode of the movement we present CC for movement and rest periods. The performance metrics are also computed using a sliding window of 40 samples (4 s) so that an estimate of the standard deviation could be quantified. The window length of 40 was selected because each movement spans about 4 s.

The reaching task testing metrics are presented in Table 1.5. It can be seen in the table that the CC can give a misleading perspective of performance since all the models produce approximately the same values. Nevertheless, the Kolmogorov–Smirnov (K–S) for a p value of 0.05 is used to compare the correlation coefficients with the simplest model, the WF. The TDNN, Kalman, and RMLP all produced CC values that were significantly different than the FIR filter and the CC values itself can be used to gauge if

TABLE 1.4   Model Performance

	Correlation coefficient		
	X	Y	Z
WF	0.52	0.60	0.64
TDNN	0.46	0.56	0.58
Kalman filter	0.56	0.64	0.65
RMLP	0.67	0.76	0.83

TABLE 1.5   **Comparison of Reaching vs. Resting Movements**

	WF	TDNN	Kalman filter	RMLP
Correlation coefficient (movement)	0.83 ± 0.09	0.08 ± 0.17	0.83 ± 0.11	0.84 ± 0.15
CC K–S test (movement)	0	1	1	1
Correlation coefficient (rest)	0.10 ± 0.29	0.04 ± 0.25	0.03 ± 0.26	0.06 ± 0.25
CC K–S test (rest)	0	1	1	1

the difference is significantly better. All four models have poor resting CC values which can be attributed to the output variability in the trajectory (i.e. there is not a strong linear relationship between the output and desired trajectories).

1.6   PERFORMANCE DISCUSSION

With the performance results reported in these experiments we can now discuss practical considerations when building BMIs. By far the easiest model to implement is the WF. With its quick computation time and straightforward linear algebra mathematics it is clearly an attractive choice for BMIs. We can also explain its function in terms of simple weighted sums of delayed versions of the ensemble neuronal firing (i.e., it is correlating neuronal activity with HP). However, from the trajectories in Figure 1.11, subplot 1, the output is noisy and does not accurately capture the details of the movement. These errors may be attributed, first, to the solution obtained from inverting a poorly conditioned autocorrelation matrix and, second, to the number of free parameters in the model topology. While we may think that by adding nonlinearity to the WF topology as in the TDNN we can obtain a more powerful tool, we found that the large increase in the number of free parameters overshadowed the increase in performance. We have also found that training the TDNN for this problem is slow and tedious and subject to getting trapped in local minima. The next model studied, the Kalman filter, was the most computationally complex to train and contained the largest number of free parameters (see comparison in Tables 1.2 and 1.3), which resulted in noisy trajectories similar to the linear model. Training this model involved the difficult mapping from a lower dimensional kinematic state space to the neuronal output space as well as initial estimates of the noise covariances, which are unknown to the operator. In contrast, many of these training and performance issues can be overcome in the RMLP. With the choice of moving the memory structure to the hidden layer, we immediately gain a reduction in the number of free parameters. This change is not without a cost since the BPTT training algorithm is more difficult to implement than, for example, the WF. Nevertheless, using a combination of dynamics and nonlinearity in the hidden layer also allowed the model to accurately capture the quick transitions in the movement as well as maintain the peak hand positions at the mouth. Capturing these positions resulted in larger values in the correlation coefficient. While the RMLP was able to outperform the other two topologies, it is not free from error; the output is still extremely noisy for applications of real BMIs (imagine trying to grasp a glass of water). Additionally the negative sloping trajectories for the reach to the food were not accurately captured. The search for the right modeling tools and techniques to overcome the errors presented here is the subject of future research for optimal signal processing for BMIs.

REFERENCES

1. SCHMIDT, E. M. (1980). "Single neuron recording from motor cortex as a possible source of signals for control of external devices." *Ann. Biomed. Eng.* 8:339–349.

2. GEORGOPOULOS, A., J. KALASKA, ET AL. (1982). "On the relations between the direction of two-dimensional arm movements and cell discharge in primate motor cortex." *J. Neurosci.* 2:1527–1537.

3. GEORGOPOULOS, A. P., A. B. SCHWARTZ, ET AL. (1986). "Neuronal population coding of movement direction." *Science* 233(4771):1416–1419.

4. GEORGOPOULOS, A. P., J. T. LURITO, ET AL. (1989). "Mental rotation of the neuronal population vector." *Science* 243(4888):234–236.

5. ANDERSEN, R. A., L. H. SNYDER, ET AL. (1997). "Multimodal representation of space in the posterior parietal cortex and its use in planning movements." *Annu. Rev. Neurosci.* 20:303–330.

6. CHAPIN, J. K., K. A. MOXON, ET AL. (1999). "Real-time control of a robot arm using simultaneously recorded neurons in the motor cortex." *Nature Neurosci.* 2(7):664–670.

7. MORAN, D. W. AND A. B. SCHWARTZ (1999). "Motor cortical activity during drawing movements: Population representation during spiral tracing." *J. Neurophysiol.* 82(5):2693–2704.

8. MORAN, D. W. AND A. B. SCHWARTZ (1999). "Motor cortical representation of speed and direction during reaching." *J. Neurophysiol.* 82(5):2676–2692.

9. SHENOY, K. V., D. MEEKER, ET AL. (2003). "Neural prosthetic control signals from plan activity." *NeuroReport* 14:591–597.

10. WESSBERG, C. R., J. D. STAMBAUGH, ET AL. (2000). "Real-time prediction of hand trajectory by ensembler or cortical neurons in primates." *Nature* 408:361–365.

11. SCHWARTZ, A. B., D. M. TAYLOR, ET AL. (2001). "Extraction algorithms for cortical control of arm prosthetics." *Curr. Opin. Neurobiol.* 11(6):701–708.

12. SANCHEZ, J. C., D. ERDOGMUS, ET AL. (2002). A comparison between nonlinear mappings and linear state estimation to model the relation from motor cortical neuronal firing to hand movements. SAB Workshop on Motor Control in Humans and Robots: on the Interplay of Real Brains and Artificial Devices, University of Edinburgh, Scotland.

13. SANCHEZ, J. C., S. P. KIM, ET AL. (2002). Input-output mapping performance of linear and nonlinear models for estimating hand trajectories from cortical neuronal firing patterns. International Work on Neural Networks for Signal Processing, Martigny, Switzerland, IEEE.

14. SERRUYA, M. D., N. G. HATSOPOULOS, ET AL. (2002). "Brain-machine interface: Instant neural control of a movement signal." *Nature* 416:141–142.

15. TAYLOR, D. M., S. I. H. TILLERY, AND A. B. SCHWARTZ (2002). "Direct cortical control of 3D neuroprosthetic devices." *Science* 296:1829–1832.

16. GAO, Y., M. J. BLACK, ET AL. (2003). A quantitative comparison of linear and non-linear models of motor cortical activity for the encoding and decoding of arm motions. The 1st International IEEE EMBS Conference on Neural Engineering, Capri, Italy, IEEE.

17. KIM, S. P., J. C. SANCHEZ, ET AL. (2003). Modeling the relation from motor cortical neuronal firing to hand movements using competitive linear filters and a MLP. International Joint Conference on Neural Networks, Portland, OR, IEEE.

18. KIM, S. P., J. C. SANCHEZ, ET AL. (2003). "Divide-and-conquer approach for brain machine interfaces: Nonlinear mixture of competitive linear models." *Neural Networks* 16(5/6):865–871.

19. SANCHEZ, J. C., D. ERDOGMUS, ET AL. (2003). Learning the contributions of the motor, premotor AND posterior parietal cortices for hand trajectory reconstruction in a brain machine interface. IEEE EMBS Neural Engineering Conference, Capri, Italy.

20. NICOLELIS, M. A. L. (2003). "Brain-machine interfaces to restore motor function and probe neural circuits." *Nature Rev. Neurosci.* 4:417–422.

21. WOLPAW, J. R., N. BIRBAUMER, ET AL. (2002). "Brain-computer interfaces for communication and control." *Clin. Neurophys.* 113:767–791.

22. NUNEZ, P. L. (1981). *Electric Fields of the Brain: The Neurophysics of EEG.* New York, Oxford University Press.

23. JASDZEWSKI, G., G. STRANGMAN, ET AL. (2003). "Differences in the hemodynamic response to event-related motor and visual paradigms as measured by near-infrared spectroscopy." *NeuroImage* 20:479–488.

24. ROHDE, M. M., S. L. BEMENT, ET AL. (2002). "Quality estimation of subdurally recorded event-related potentials based on signal-to-noise ratio." *IEEE Trans. Biomed. Eng.* 49(1):31–40.

25. FETZ, E. E. AND D. V. FINOCCHIO (1975). "Correlations between activity of motor cortex cells and arm muscles during operantly conditioned response patterns." *Exp. Brain Res.* 23(3):217–240.

26. GEORGOPOULOS, A. P., R. E. KETTNER, ET AL. (1988). "Primate motor cortex and free arm movements to visual targets in three-dimensional space. II. Coding of the direction of movement by a neuronal population." *J. Neurosci. Official J. Soc. Neurosci.* 8(8): 2928–2937.

27. LIN, S., J. SI, ET AL. (1997). "Self-organization of firing activities in monkey's motor cortex: trajectory computation from spike signals." *Neural Computation* 9:607–621.

28. KALASKA, J. F., S. H. SCOTT, ET AL. (1997). Cortical control of reaching Movements. *Curr. Opin. Neurobiol.* 7:849–859.

29. SANGER, T. D. (1996). "Probability density estimation for the interpretation of neural population codes." *J. Neurophysiol.* 76(4):2790–2793.

30. MUSSA-IVALDI, F. A. (1988). "Do neurons in the motor cortex encode movement directions? An alternative hypothesis." *Neurosci. Lett.* 91:106–111.

31. SHENOY, K. V., D. MEEKER, ET AL. (2003). Neural prosthetic control signals from plan activity. *NeuroReport* 14:591–597.

32. HAYKIN, S. (1996). *Adaptive Filter Theory*. Upper Saddle River, NJ, Prentice-Hall International.

33. RIEKE, F. (1996). *Spikes: Exploring the Neural Code*. Cambridge, MA, MIT Press.

34. NICOLELIS, M. A. L. (1999). *Methods for Neural Ensemble Recordings*. Boca Raton, FL, CRC Press.

35. LEONARD, C. T. (1998). *The Neuroscience of Human Movement*. St. Louis: Mosby, 1998.

36. LJUNG, L. (2001). Black-box models from input-output measurements. IEEE Instrumentation and Measurement Technology Conference, Budapest, Hungary, IEEE.

37. KANDEL, E. R., J. H. SCHWARTZ, ET AL., Eds. (2000). *Principles of Neural Science*. New York, McGraw-Hill.

38. THATCH, W. T. (1978). "Correlation of neural discharge with pattern and force of muscular activity, joint position, and direction of intended next movement in motor cortex and cerebellum." *J. Neurophys.* 41:654–676.

39. FLAMENT, D. AND J. HORE (1988). "Relations of motor cortex neural discharge to kinematics of passive and active elbow movements in the monkey." *J. Neurophysiol.* 60(4):1268–1284.

40. KALASKA, J. F., D. A. D. COHEN, ET AL. (1989). "A comparison of movement direction-related versus load direction-related activity in primate motor cortex, using a two-dimensional reaching task." *J. Neurosci.* 9(6):2080–2102.

41. SCOTT, S. H. AND J. F. KALASKA (1995). "Changes in motor cortex activity during reaching movements with similar hand paths but different arm postures." *J. Neurophysiol.* 73(6):2563–2567.

42. TODOROV, E. (2000). "Direct cortical control of muscle activation in voluntary arm movements: a model." *Nature Neurosci.* 3:391–398.

43. WU, W., M. J. BLACK, ET AL. (2002). "Inferring hand motion from multi-cell recordings in motor cortex using a Kalman filter." In *SAB Workshop on Motor Control in Humans and Robots: On the Interplay of Real Brains and Artificial Devices*. University of Edinburgh, Scotland, 66–73.

44. KALMAN, R. E. (1960). "A new approach to linear filtering and prediction problems." *Trans. ASME J. Basic Eng.* 82(Series D):35–45.

45. WIENER, N. (1949). *Extrapolation, Interpolation, and Smoothing of Stationary Time Series with Engineering Applications*. Cambridge, MA, MIT Press.

46. HAYKIN, S. (1994). *Neural Networks: A Comprehensive Foundation*. New York and Toronto, Macmillan and Maxwell Macmillan Canada.

47. ORR, G. AND K.-R. MÜLLER (1998). *Neural Networks: Tricks of the Trade*. Berlin and New York, Springer.

48. BROCKWELL, A. E., A. L. ROJAS, ET AL. (2003). "Recursive Bayesian decoding of motor cortical signals by particle filtering." *J. Neurophysiol.* 91:1899–1907.

49. MASKELL, S. AND N. GORDON (2001). "A tutorial on particle filters for on-line nonlinear/non-Gaussian Bayesian tracking." *Target Tracking: Algorithms and Applications* 2:1–15.

50. SANCHEZ, J. C., J. M. CARMENA, ET AL. (2003). "Ascertaining the importance of neurons to develop better brain machine interfaces." *IEEE Trans. Biomed. Eng.* 61(6):943–953.

51. HOERL, A. E. AND R. W. KENNARD (1970). "Ridge regression: Biased estimation for nonorthogonal problems." *Technometrics* 12(3):55–67.

52. SANCHEZ, J. C. (2004). From cortical neural spike trains to behavior: Modeling and Analysis. Department of Biomedical Engineering, University of Florida, Gainesville.

53. FARMER, J. D. AND J. J. SIDOROWICH (1987). "Predicting chaotic time series." *Phys. Rev. Lett.* 50:845–848.

54. FANCOURT, C. AND J. C. PRINCIPE (1996). *Temporal Self-Organization Through Competitive Prediction*. International Conference on Acoustics, Speech, and Signal Processing, Atlanta.

55. PRÍNCIPE, J. C., N. R. EULIANO, ET AL. (2000). *Neural and Adaptive Systems: Fundamentals Through Simulations*. New York, Wiley.

56. GEMAN, S., E. BIENENSTOCK, ET AL. (1992). "Neural networks and the bias/variance dilemma." *Neural Computation* 4:1–58.

57. AKAIKE, H. (1974). "A new look at the statistical model identification." *IEEE Trans. Auto. Control* 19:716–723.

58. WAHBA, G. (1990). *Spline Models for Observational Data*. Montpelier, Capital City Press.

59. NEAL, R. (1996). *Bayesian Learning for Neural Networks*. Cambridge, Cambridge University Press.

60. VAPNIK, V. (1999). *The Nature of Statistical Learning Theory*. New York, Springer Verlag.

61. MOODY, J. (1994). Prediction risk and architecture selection for neural networks. In V. CHERKASSKY, J. H. FRIEDMAN, AND H. WECHSLER, Eds., *From Statistics to Neural Networks: Theory and Pattern Recognition Applications*. New York, Springer-Verlag.

62. AKAIKE, H. (1970). "Statistical predictor identification." *Ann. Inst. Statist. Math.* 22:203–217.

63. CRAVEN, P. AND G. WAHBA (1979). "Smoothing noisy data with spline functions: Estimating the correct degree of smoothing by the method of generalized cross-validation." *Numer. Math.* 31:377–403.

64. GOLUB, G., H. HEATH, ET AL. (1979). "Generalized cross validation as a method for choosing a good ridge parameter." *Technometrics* 21:215–224.

65. SODERSTROM, T. AND P. STOICA (1989). *System Identification*. New York, Prentice-Hall.

66. MOODY, J. (1992). The effective number of parameters: An analysis of generalization and regularization in nonlinear learning systems. Advances in Neural Information Processing Systems, San Meteo, CA.

67. LEFEBVRE, W. C., J. C. PRINCIPE, ET AL. (1994). *NeuroSolutions*. Gainesville, NeuroDimension.

CHAPTER 2

MODULATION OF ELECTROPHYSIOLOGICAL ACTIVITY IN NEURAL NETWORKS: TOWARD A BIOARTIFICIAL LIVING SYSTEM

Laura Bonzano, Alessandro Vato, Michela Chiappalone, and Sergio Martinoia

2.1 INTRODUCTION

The unique property of the brain to learn and remember is what makes the nervous system different from any other. Its functional plasticity enables a large range of possible responses to stimuli, allowing the nervous system to integrate high-resolution, multimodal sensory information and to control extremely precise motor actions. These same properties can be studied in ex vivo cultured networks of neurons, where, at a simplified level of organization, the collective and functional electrophysiological properties emerge and can be experimentally characterized, so contributing to a better understanding of how the brain processes information [1, 2]. Consequently, cultured networks can be considered as simplified neurobiological systems leading to the theoretical analysis of neurodynamics.

2.2 MATERIALS AND METHODS

Nowadays, networks of dissociated neurons can be cultured and kept in healthy conditions for a long time (from weeks up to months) as experimental preparations [3]. In such in vitro neurobiological systems, the neuronal physiology and the efficacy of synaptic connections between neurons can be quantitatively characterized, exploiting activity-dependent network modifications and investigating the time-dependent interactions among the nervous cells that might be used in the brain to represent information [4, 5]. In addition, such networks offer new assay and sensing systems that lie between biochemistry and whole-animal experiments and provide rapid and quantitative information on neurophysiological responses to chemicals and toxins [6–9] or specific electrical stimulating waveforms [10–12].

Handbook of Neural Engineering. Edited by Metin Akay
Copyright © 2007 The Institute of Electrical and Electronics Engineers, Inc.

In order to study neuronal dynamics at the network level, it is necessary to record electrical activity at multiple sites simultaneously for long periods. Among the promising electrophysiological experimental approaches, planar microelectrode arrays (MEAs) appear the best choice for such measurements. After many pioneering studies started in the 1980s [6, 13], MEAs have become a powerful tool in the framework of in vitro electrophysiology, allowing multisite, long-term and noninvasive extracellular recordings from tens of sites and simultaneous electrical stimulation from different regions of the array [3, 8, 14].

By using multisite recording of extracellular signals with MEAs, we monitored the patterns of neural activity arising from cultured cortical neurons of rat embryo in different experimental conditions, studying the network moving in a variety of distinct behavioral "states."

2.2.1   General Procedure for Culturing Cortical Neurons

2.2.1.1   Preculture Preparation   Microelectrode arrays are sterilized and pretreated with adhesion factors, such as poly-D/L-lysine and laminin, in order to improve the neuron–electrode coupling.

2.2.1.2   Dissection   Cortical neurons are obtained from E-18/19 rat embryos, removing the whole uterus of a pregnant Wistar rat and placing it in a Petri dish with phosphate-buffered saline (PBS) or Hank's balanced salt solution (HBSS). Pups are then removed from their own embryonic sac and decapitated. We isolate the cortical hemisphere, remove the meninges and olfactory bulbs, cut off the hippocampus and basal ganglia, and collect all cortical hemispheres in a Petri dish of 35 mm diameter with PBS on ice.

Afterward the cortices must be chopped into small pieces, adding trypsin solution, and the tissue pieces are transferred, with a sylanized Pasteur pipette, to a 15-mL centrifuge tube containing 2–3 mL of trypsin solution for each cortex. The tube is placed into the water bath and incubated for 25–30 min at 37°C.

After removing the tube from the bath, it is necessary to take away as much trypsin as possible with a pipette; the proteolytic digestion is blocked adding to the tissue 5 mL of Dulbecco's modified Eagles's medium (DMEM)–F12 medium (or neurobasal medium) containing 10% serum. This operation is repeated twice, drawing off the supernatant and suspending the tissue in 3 mL of the same medium, tritrating 5–10 times with a flame-narrowed pipette and continuing until no more clumps of tissue are visible. Cell suspension is centrifuged for 5 min at 1000 rpm and the cells are suspended in 10 mL of neurobasal medium supplemented with 2% B-27 and 1% glutamax-I.

The number of cells is counted by using a hemocytometer, and the cortical cell suspension is diluted to a final concentration of 800,000 cells/mL; 100–150 μL of the cell suspension solution is placed in each of the prepared microarray culture wells obtaining a final concentration of 450,000 cells/mL (see Fig. 2.1). The total number of cell plated on the MEA in a 6.5-mm × 6.5-mm region is about 1500 cells/mm^2.

2.2.1.3   Cell Maintenance   The MEAs are placed in a humidified incubator having an atmosphere of 5% CO_2 at 37°C; part of the media is replaced with a fresh one each week.

At first, neurons and glial cells, living in coculture, improve the development of synaptic connections of neuronal network, then glial cells continue to proliferate, so four days after the cell plating, Ara-C, which acts as an inhibitor of replication and overgrowth of nonneural cells, is added.

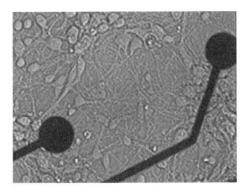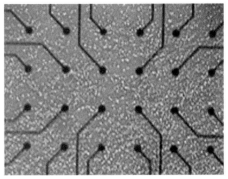

Figure 2.1 Cultures of cortical neurons extracted from rat embryos coupled to MEAs after 15 DIV (days in vitro).

During experimental sessions, measurements were carried out in physiological "basal" medium, made up of the following components: NaCl 150 mM, $CaCl_2$ 1.3 mM, $MgCl_2$ 0.7 mM, KCl 2.8 mM, glucose 10 mM, and N-(2-hydroxyethyl) piperazine-N'-2-ethanesulfonic acid (HEPES) buffer 10 mM.

2.2.2 Experimental Setup

The experimental setup currently adopted in our laboratories is schematically represented in Figure 2.2; it is based on the MEA60 system (Multichannel Systems—MCS Reutlingen, Germany, http://www.multichannelsystems.com) and consists of the following:

- The microelectrode array (MEA1060), a planar array of 60 TiN electrodes (30 μm diameter, 200 μm spaced) on glass substrate
- A mounting support with integrated 60-channel pre- and filter amplifier with a gain of 1200×
- A personal computer equipped with the MC_Card, a personal computer interface (PCI) analog-to-digital (A/D)/digital-to-analog (D/A) board with a maximum of 128 recording channels, 12 bits of resolution, and a maximum sampling frequency of 50 kHz/channel (we use 10 kHz/channel)

Figure 2.2 Experimental setup currently used in our laboratories, based on MEA60 system (Multichannel Systems).

- The software MCRack, used for real-time signal displaying and multichannel acquisition
- Custom-made software neural signal manager (NSM) and MATLAB routines for data analysis [15]
- The stimulus generator (STG) 1008 stimulus generator, capable of stimulating the network through up to eight channels
- The software MCStimulus, used to drive the stimulus generator
- An inverted microscope (connected to a TV camera)
- An antivibration table and a Faraday cage

2.3 RESULTS

Networks of neurons extracted from the developing central nervous system (CNS) are spontaneously active and show typical electrophysiological activity patterns ranging from apparently random spiking to more organized and densely packed spike activity called "bursts." In order to represent these behaviors from a quantitative point of view, we described the activity at both the burst and at spike levels, analyzing the electrophysiological pattern in different experimental conditions employing custom-developed algorithms of burst analysis and standard statistical procedures for spike analysis [15].

2.3.1 Network Activity Modulated by Chemical Agents

The aim of chemical stimulation experiments is to study how drugs affect the network behavior on the basis of some simple parameters [e.g., spike rate, burst rate, interspike interval (ISI), interburst interval (IBI), burst duration] and, therefore, to study how the network can be driven to a more excited or inhibited state. One of the main objectives is to control the excitability level of the network, thus increasing or decreasing the sensitivity of neurons to external electrical stimuli. According to the experimental setup currently in use, chemicals are not delivered in a localized way, to specific and identified subpopulations of neurons, but they are added to the culture bath, so that the entire network experiences the same concentration of the added chemical agents.

Focusing, in the present context, on glutamate, the most important excitatory neurotransmitter in the CNS, manipulations of the network were achieved through agonists and antagonists of glutamatergic ionotropic N-methyl-D-aspartate and non-NMDA receptors [9, 16, 17].

2.3.1.1 Experimental Protocols We investigated the effect of APV (D-2-amino-5-phosphonopentanoic acid) and CNQX (6-cyano-7-nitroquinoxaline-2, 3-dione), competitive antagonists respectively of NMDA and non-NMDA channels, on the electrical activity in cultured networks.

At first, the two selected substances (i.e., APV and CNQX) were applied to our cultures separately: Different MEAs were used for the two chemicals, delivered at increasing concentrations for studying a simple dose–response profile. By refining these procedures, we should have been able to define the optimal concentration that allows us to keep the neuronal culture in a certain reproducible state (e.g., with a defined mean firing rate).

In the following, the adopted experimental procedures are reported; they include separated sessions for the concentrations of different drugs and some intermediate "basal" (i.e., where no drugs were applied) phases, which show the reversibility of the process; that is, after the washout, we obtained a mean global behavior of the network close to the one observed at the beginning of the experiment:

1. *For CNQX*:
   - The MEA is positioned in the experimental setup, the medium is changed (i.e., from neurobasal to physiological—see Section 2.2.1.3), and the culture is allowed to stabilize for 20 min.
   - Basal condition: 20 min recording.
   - CNQX1 (10 μM): 20 min recording.
   - Basal 2: 5 min recording.
   - CNQX2 (50 μM): 20 min recording.
   - Basal 3: 5 min recording.
   - CNQX3 (100 μM): 20 min recording.
   - Basal 4: 5 min recording.

2. *The same experimental procedure used for APV with concentrations of 25, 50, and 100 μM*: It turned out that the application of APV has a remarkable effect on the spontaneous firing activity, as could be expected and as is illustrated by looking at the raw data (see Fig. 2.3a, where the most active sites before the application of APV are highlighted) and at the ISI histograms (see Fig. 2.3b). Figure 2.3 shows a comparison between the control (i.e., physiological) condition and the 100 μM APV condition; as can be seen, the spontaneous activity in the network is almost abolished.

In terms of the global behavior of the network, the effects of drugs can be better observed by their dose–response curves, representing the relationship between the mean firing rate (i.e., number of spikes per second) computed on the 60 channels and drug concentration. The experimental findings are displayed in Figure 2.4 for CNQX and APV application: The expected results of a decreasing activity in response to the increasing concentration of the drug were obtained.

In particular, Figure 2.4a shows the results of two experiments performed on 22-DIV-old cultures treated with CNQX; they both follow the same trend.

In Figure 2.4b, the results of two experiments with APV are shown; in this case the two examined cultures have the age of 15 DIV, and even if the initial spontaneous activity is different, it reaches almost coinciding values for high concentrations of the drug (50–100 μM).

These results are also confirmed by analyzing the bursting activity: The mean number of bursts occurring in 5 min is calculated over the most active channels for each experimental phase and reported in Figures 2.4c,d, respectively, for CNQX and APV addition. The values decrease while the inhibitory drug concentration increases as a further consequence of the reduced mean firing rate.

The two drugs act on different receptors (CNQX on non-NMDA channels, APV on NMDA ones) and they affect the network through different pathways. However, in both cases, CNQX and APV treatment reduces the global activity but the network is not totally silent. Considering that in the network the two types of receptors coexist, to have a silent network the application of both receptor blockers is needed.

Thus, considering the obtained dose–response curves, we performed some experiments applying the two substances at a concentration of 50 μM for APV and 10 μM

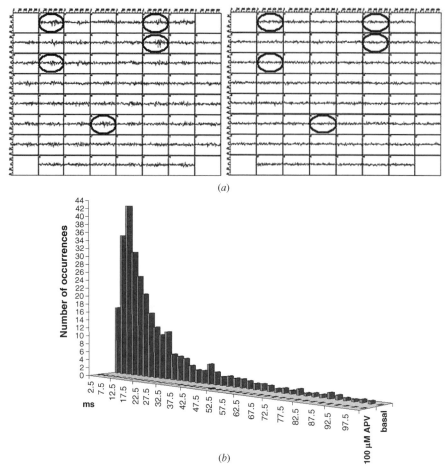

Figure 2.3 (*a*) Electrophysiological activity at 21 DIV recorded from 60 electrodes and monitored in real time, in basal medium (left) and after addition of 100 μM APV (right): In each window is shown 1 s of the recorded signals in a voltage range of ±40 μV. Note that some of the most active electrodes have been circled to highlight the different behaviors in the two experimental conditions. (*b*) The same results can also be appreciated looking at the ISI histogram representing the basal medium vs. 100 μM APV condition.

for CNQX to investigate the combined effect of the two blockers on the network activity.

The utilized experimental protocol for CNQX and APV is as follows:

- The MEA is positioned in the experimental setup, the medium is changed (i.e., from neurobasal to physiological—see Section 2.2.1.3), and the culture is allowed to stabilize for 20 min.
- Basal condition: 20 min recording.
- MIX1 (10 μM CNQX, 50 μM APV): 20 min recording.
- Basal 2: 20 min recording.
- MIX2 (10 μM CNQX, 50 μM APV): 20 min recording.
- Basal 3: 20 min recording.

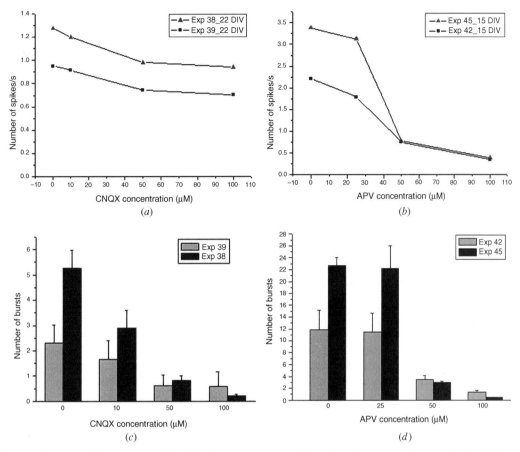

Figure 2.4   Mean firing rate calculated for different concentrations of (*a*) CNQX and (*b*) APV and (*c, d*) number of bursts calculated for the same experiments.

The obtained results are summarized in Figure 2.5. The mean firing rate values corresponding to the APV–CNQX treatment are now very close to zero for all 60 electrodes, while after a single washout the network shows a significant activity.

2.3.2   Network Activity Modulated by Electrical Stimulation

It is well known that in vitro neuronal network activity can be modulated by electrical stimulation [10, 11]. It has also been shown that activity-dependent modifications in the network reflect changes in the synaptic efficacy and that this fact is widely recognized as a cellular basis of learning, memory, and developmental plasticity [11, 18–20]. As a preliminary step toward the elucidations of simple and basic properties of neural adaptability (i.e., plasticity), we started to characterize the electrophysiological activity of developing neuronal networks in response to applied electrical stimuli.

The objective is to find out pathways activated by external stimuli, exploring half of the microelectrodes as stimulating channels, and to select some of them as possible input for inducing synaptic weight changes in the network.

Our experimental protocols for electrical stimulation have been adapted from the literature [10, 11, 18]. Stimuli consist of trains of bipolar pulses at low frequency

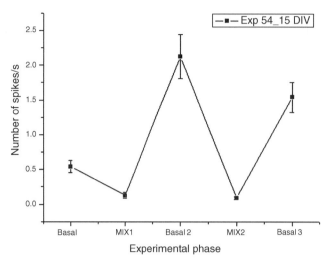

Figure 2.5   Mean firing rate calculated for different phases: basal 1, basal 2, basal 3 (physiological conditions) and MIX1, MIX2 (10 μM CNQX–50 μM APV treatment).

(about 0.2 Hz, ±1.0 V, 250 μs); in the experiments presented here, 70 stimuli were delivered in about 7 min through 20/30 stimulating sites over a total of 60 electrodes in order to explore the whole network and to discover the existence of possible neural pathways embedded in the networks, where the signals propagate from not directly connected nervous cells. Preliminary results show remarkable differences in the electrophysiological activity of the network in different stimulation conditions that encourage us to think that it might be possible to induce controlled plastic changes in the network and to "force" the network to move toward particular behavioral states. This is an important experimental evidence attesting the feasibility of imposing a kind of "adaptive learning" in a biological nonlinear system.

To quantify the response with respect to the delivered stimulus of a specific recording site, we used the well-known representation called poststimulus time histogram (PSTH) [21] and we noticed different shapes for different stimulation sites. The first consideration is that it is possible to modulate the electrical responses of the cell by only changing the stimulation site (see Fig. 2.6) and that there are also specific neural pathways in the network.

By looking at the probability of firing after the stimulation, it is possible to identify two principal shapes in the PSTH (see Figs. 2.6a,b): the "early response," where first spikes are evoked immediately after the stimulus (they overlap to the last part of the stimulus artifact), and the "delayed response," where spike activity is evoked 50–100 ms after the stimulus and which consists of a bursting activity lasting about 300 ms. The distinction between these two types of evoked response is also evident looking at the raw signals recorded during a stimulation phase through different electrodes (see Figs. 2.6c,d).

Very recently, it has been shown that, looking inside this delayed response, it is possible to induce selective learning, activating path-specific plastic changes [18].

Several parameters were extracted to better characterize spontaneous network behavior and to analyze changes induced by electrical stimulation. In particular we focused our attention on the analysis of the mean firing rate, the IBI, and the duration of single bursts.

In each experimental condition, corresponding to a different stimulation site, the mean firing rates were calculated for two groups of recording electrodes: the least active and the most active ones during the basal condition (Figs. 2.7a,b), in order to

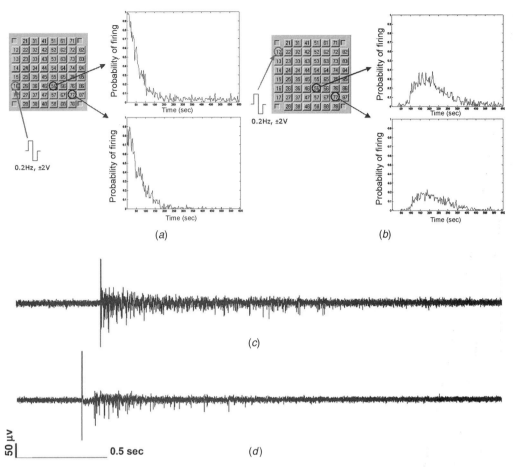

Figure 2.6 Poststimulus time histograms for two recording sites (56 and 77): (*a*) an "early" response is clearly visible; (*b*) by changing the stimulating site, a "delayed" response is obtained. The two types of responses can be seen in the electrophysiological activity recorded from one of the two recording electrodes presented above: During electrical stimulation from two different stimulating sites the burst following the stimulus has a different pattern in terms of delay from the stimulus that is almost absent in (*c*), while it is about 50–100 ms in (*d*).

underline the different evoked behaviors with respect to the spontaneous activity. As far as the least active sites are concerned, an increase in firing activity can be noticed for all the stimulation conditions. We obtained the opposite effect on the other group of electrodes, for which the mean firing rate decreased as a consequence of the stimulation: This result could be explained as a rupture of the high-frequency spontaneous activity and, at the same time, as an adaptation to the stimulation pattern.

Besides, looking at the IBI histogram (Figs. 2.7c,d), we can see how externalstimuli can tune the bursting spontaneous activity and lock it around the stimulus frequency (5–6 s).

Further experiments were performed by slightly changing the experimental protocol (i.e., amplitude and duration of the biphasic stimulating signal, voltage-vs.-current stimulation) and similar results were obtained (data not shown). Thus, in accordance with other works [11, 22], it has been demonstrated that the electrophysiological activity is stimulus dependent, showing a rich repertoire of evoked responses with a specific dependence on the stimulating site.

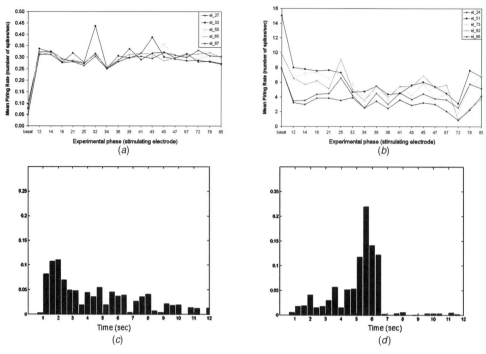

Figure 2.7   Mean firing rate in different experimental conditions (basal medium and stimulation through different electrodes) for (*a*) the least active sites and (*b*) the most active ones. Mean IBI in (*c*) spontaneous conditions and (*d*) during electrical stimulation.

2.4   CONCLUSIONS

In the present work we showed that simple protocols of chemical and electrical stimulation can induce, in neuronal networks, different patterns of activity that can be quantified by looking at the firing rate on each recording channel and at the overall spiking activity. Moreover the analysis of the burst activity, in terms of burst duration, IBI, and temporal distribution of evoked bursts, can meaningfully identify different states in the network dynamics.

Summarizing the obtained results, based on preliminary performed experimental sessions, we can state that the network response is modulated by chemical agents that are known to act as antagonist of glutamatergic synapses, thus demonstrating that the network dynamics is dominated, as far as the excitatory connections are concerned, by NMDA and non-NMDA receptors. Yet, the electrical stimulation is a spatially dependent phenomenon, since different stimulating sites evoke different responses (distinct "patterns" or "states") on the same recording electrodes. Therefore, there are different functional pathways in the network responsible for the signal activation and propagation that can be identified by ad hoc algorithms (such as the presented PSTH). Further analysis devoted to the identification of specific pathways of connections inside the networks could be performed also by looking at the latency of the responses evoked by the stimulating site on different recording electrodes.

From the experimental point of view, future steps will be devoted to identify a subset of electrodes capable of tuning the transmission properties of the network and to implement specific stimulation protocols aimed at modifying the synaptic connections and at inducing long term-phenomena such as longterm potentiation (LTP) and longterm depression (LTD).

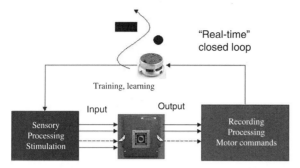

Figure 2.8   Scheme for real-time closed-loop system between dissociated culture of cortical neurons and small mobile robot.

Taking advantage of the intrinsic plasticity of neural networks, our final goal is to obtain desired "arbitrary" responses through a process of learning driven by external electrical and chemical stimulation.

2.5   FUTURE TRENDS

In this work we addressed the problem of monitoring and modulating the collective behavior (i.e., the electrophysiological activity) of large population of neurons cultured in vitro and coupled to MEAs. Many mechanisms underlying the emerging changes in the dynamics of such experimental preparations still need to be deeply investigated, but we think that the widespread use of MEAs or, generally, multichannel microdevices will be of great relevance for studying the spatiotemporal dynamics of living neurons on a long-term basis. Moreover, as recently reported by Potter and co-workers [23, 24], we are convinced that the possibility of studying ex vivo neuronal preparations coupled to artificial systems (e.g., to a mobile robot), so providing to the neuronal system a kind of "body," can be considered a new experimental paradigm with many implications in the field of neural engineering. As sketched in Figure 2.8, we can think of having a neuronal network bidirectionally connected to a mobile robot. The robot is moving in a controlled environment (i.e., a designed playground) and it gives the neuronal system input stimuli as a function of its actual behavior (e.g., input coming from the infrared position sensors). From the other side, the neuronal network responds to the input stimuli and provides the robot (by means of an appropriate coding of the recorded electrophysiological activity) with the motor commands for moving in the developed playground. This new proposed experimental paradigm, where, in a sense, a neuronal system is "embodied" and "situated," could allow the study of the adaptive properties (i.e., plastic changes) of a neurobiological system in a more realistic way by trying to train a neuronal network to support and control a desired task that has to be accomplished by the robot.

ACKNOWLEDGMENTS

The authors wish to thank Brunella Tedesco for her help in the cell culture preparation and maintenance, Antonio Novellino for his support in the development of software tools and data analysis, and Marco Bove for his useful indications and suggestions for designing the experimental protocols.

This work has been partially supported by the Neurobit project (EU Contract No. IST-2001-33564): "A Bioartificial Brain with an Artificial Body: Training a Cultured Neural Tissue to Support the Purposive Behavior of an Artificial Body."

REFERENCES

1. J. STREIT, A. TSCHERTER, M. O. HEUSCHKEL, AND P. RENAUD, "The generation of rhythmic activity in dissociated cultures of rat spinal cord," *European Journal of Neuroscience*, vol. 14, pp. 191–202, 2001.

2. A. TSCHERTER, M. O. HEUSCHKEL, P. RENAUD, AND J. STREIT, "Spatiotemporal characterization of rhythmic activity in spinal cord slice cultures," *European Journal of Neuroscience*, vol. 14, pp. 179–190, 2001.

3. S. M. POTTER AND T. B. DEMARSE, "A new approach to neural cell culture for long-term studies," *Journal of Neuroscience Methods*, vol. 110, pp. 17–24, 2001.

4. M. A. L. NICOLELIS, E. E. FANSELOW, AND A. A. GHAZANFAR, "Hebb's dream: the resurgence of cell assemblies," *Neuron*, vol. 19, pp. 219–221, 1997.

5. G. BI AND M. POO, "Distributed synaptic modification in neural networks induced by patterned stimulation," *Nature*, vol. 401, pp. 792–796, 1999.

6. G. W. GROSS, B. K. RHOADES, AND R. J. JORDAN, "Neuronal networks for biochemical sensing," *Sensors and Actuators*, vol. 6, pp. 1–8, 1992.

7. G. W. GROSS, "Internal dynamics of randomized mammalian networks in culture," in *Enabling Technologies for Cultured Neural Networks*, D. A. STENGER AND T. M. MCKENNA eds., Academic Press, New York, 1994, Chapter 13, 277–317.

8. M. CHIAPPALONE, A. VATO, M. T. TEDESCO, M. MARCOLI, F. A. DAVIDE, AND S. MARTINOIA, "Networks of neurons coupled to microelectrode arrays: A neuronal sensory system for pharmacological Applications," *Biosensors & Bioelectronics*, vol. 18, pp. 627–634, 2003.

9. M. CANEPARI, M. BOVE, E. MAEDA, M. CAPPELLO, AND A. KAWANA, "Experimental analysis of neuronal dynamics in cultured cortical networks and transitions between different patterns of activity," *Biological Cybernetics*, vol. 77, pp. 153–162, 1997.

10. Y. JIMBO, H. P. C. ROBINSON, AND A. KAWANA, "Strengthening of synchronized activity by tetanic stimulation in cortical cultures: Application of planar electrode arrays," *IEEE Transactions on Biomedical Engineering*, vol. 45, pp. 1297–1304, 1998.

11. Y. JIMBO, Y. TATENO, AND H. P. C. ROBINSON, "Simultaneous induction of pathway-specific potentiation and depression in networks of cortical neurons," *Biophysical Journal*, vol. 76, pp. 670–678, 1999.

12. A. NOVELLINO, M. CHIAPPALONE, A. VATO, M. BOVE, M. T. TEDESCO, AND S. MARTINOIA, "Behaviors from an electrically stimulated spinal cord neuronal network cultured on microelectrode arrays," *Neurocomputing*, vol. 52–54C, pp. 661–669, 2003.

13. G. W. GROSS AND J. M. KOWALSKI, "Experimental and theoretical analyses of random networks dynamics," in *Neural Networks, Concepts, Application and Implementation*, vol. 4, P. Antognetti and V. Milutinovic, eds., Prentice-Hall, Englewood Cliffs, NJ, 1991, pp. 47–110.

14. M. BOVE, M. GRATTAROLA, AND G. VERRESCHI, "In vitro 2D networks of neurons characterized by processing the signals recorded with a planar microtransducer Array," *IEEE Transactions on Biomedical Engineering*, vol. 44, pp. 964–977, 1997.

15. D. H. PERKEL, G. L. GERSTEIN, AND G. P. MOORE, "Neuronal spike train and stochastic point processes I. The single spike train," *Biophysical Journal*, vol. 7, pp. 391–418, 1967.

16. G. W. GROSS, H. M. E. AZZAZY, M. C. WU, AND B. K. RHODES, "The use of neuronal networks on multielectrode arrays as biosensors," *Biosensors & Bioelectronics*, vol. 10, pp. 553–567, 1995.

17. T. HONORÉ, S. N. DAVIES, J. DREJER, E. J. FLETCHER, P. JACOBSEN, D. LODGE, AND F. E. NIELSEN, "Quinoxalinediones: Potent competitive non-NMDA glutamate receptor antagonists," *Science*, vol. 241, pp. 701–703, 1988.

18. G. SHAHAF AND S. MAROM, "Learning in networks of cortical neurons," *Journal of Neuroscience*, vol. 21, pp. 8782–8788, 2001.

19. S. MAROM AND G. SHAHAF, "Development, learning and memory in large random networks of cortical neurons: Lessons beyond anatomy," *Quarterly Reviews of Biophysics*, vol. 35, pp. 63–87, 2002.

20. L. C. KATZ AND C. J. SHATZ, "Synaptic activity and the construction of cortical circuits," *Science*, vol. 274, pp. 1133–1138, 1996.

21. F. RIEKE, D. WARLAND, R. D. R. VAN STEVENINCK, AND W. BIALEK, *Spikes: Exploring the Neural Code*, MIT Press, Cambridge, MA, 1997.

22. H. P. C. ROBINSON, M. KAWAHARA, Y. JIMBO, K. TORIMITSU, Y. KURODA, AND A. KAWANA, "Periodic synchronized bursting in intracellular calcium transients elicited by low magnesium in cultured cortical neurons," *Journal of Neurophysiology*, vol. 70, pp. 1606–1616, 1993.

23. S. M. POTTER, "Distributed processing in cultured neural networks," *Progress in Brain Research*, vol. 130, pp. 49–62, 2001.

24. T. B. DEMARSE, D. A. WAGENAAR, A. W. BLAU, AND S. M. POTTER, "The neurally controlled animal: Biological brains acting with simulated bodies," *Autonomous Robots*, vol. 11, pp. 305–310, 2001.

CHAPTER *3*

ESTIMATION OF POSTERIOR PROBABILITIES WITH NEURAL NETWORKS: APPLICATION TO MICROCALCIFICATION DETECTION IN BREAST CANCER DIAGNOSIS

Juan Ignacio Arribas, Jesús Cid-Sueiro, and Carlos Alberola-López

3.1 INTRODUCTION

Neural networks (NNs) are customarily used as classifiers aimed at minimizing classification error rates. However, it is known that the NN architectures that compute soft decisions can be used to estimate posterior class probabilities; sometimes, it could be useful to implement general decision rules other than the maximum a posteriori (MAP) decision criterion. In addition, probabilities provide a confidence measure of the classifier decisions, a fact that is essential in applications in which a high risk is involved.

This chapter is devoted to the general problem of estimating posterior class probabilities using NNs. Two components of the estimation problem are discussed: Model selection, on one side, and parameter learning, on the other. The analysis assumes an NN model called the generalized softmax perceptron (GSP), although most of the discussion can be easily extended to other schemes, such as the hierarchical mixture of experts (HME) [1], which has inspired part of our work, or even the well-known multilayer perceptron.

The use of posterior probability estimates is applied in this chapter to a medical decision support system; the testbed used is the detection of microcalcifications (MCCs) in mammograms, which is a key step in breast cancer early diagnosis.

The chapter is organized as follows: Section 3.2 discusses the estimation of posterior class probabilities with NNs, with emphasis in a medical application; Section 3.3 discusses learning and model selection algorithms for the GSP networks; Section 3.4 proposes a system for MCC detection based on the GSP; Section 3.5 shows some simulation results on detection performance using a mammogram database; and Section 3.6 provides some conclusions and future trends.

Handbook of Neural Engineering. Edited by Metin Akay
Copyright © 2007 The Institute of Electrical and Electronics Engineers, Inc.

3.2   NEURAL NETWORKS AND ESTIMATION OF A POSTERIORI PROBABILITIES IN MEDICAL APPLICATIONS

3.2.1   Some Background Notes

A posteriori (or posterior) probabilities, or generally speaking, posterior density functions, constitute a well-known and widely applied discipline in estimation and detection theory [2, 3] and, more widely speaking, in pattern recognition. The concept underlying posteriors is to update one's prior knowledge about the state of nature (with the term *nature* adapted to the specific problem one is dealing with) according to the observation of reality that one may have.

Specifically, recall the problem of a binary decision, that is, the need of choosing either hypothesis H_0 or H_1 having as information both our prior knowledge about the probabilities of each hypothesis being the right one [say $P(H_i)$, $i = 0, 1$] and how the observation $\mathbf{x}$ probabilistically behaves according to each of the two hypotheses [say $f(\mathbf{x}|H_i)$, $i = 0, 1$]. In this case, it is well known that the optimum decision procedure is to compare the likelihood ratio, that is, $f(\mathbf{x}|H_1)/f(\mathbf{x}|H_0)$ with a threshold. The value of the threshold is, from a Bayesian perspective, a function of the two prior probabilities and a set of costs (set at designer needs) associated to the two decisions. Specifically

$$\frac{f(\mathbf{x}|H_1)}{f(\mathbf{x}|H_0)} \underset{H_0}{\overset{H_1}{\gtrless}} \frac{P(H_0)(C_{10} - C_{00})}{P(H_1)(C_{01} - C_{11})} \tag{3.1}$$

with C_{ij} the cost of choosing H_i when H_j is correct.

A straightforward manipulation of this equation leads to

$$\frac{P(H_1|\mathbf{x})}{P(H_0|\mathbf{x})} \underset{H_0}{\overset{H_1}{\gtrless}} \frac{C_{10} - C_{00}}{C_{01} - C_{11}} \tag{3.2}$$

with $P(H_i|\mathbf{x})$ the a posteriori probability that hypothesis H_i is true, $i = 0, 1$. For the case $C_{ij} = 1 - \delta_{ij}$, with δ_{ij} the Kronecker delta function, the equation above leads to the general MAP rule, which can be rewritten, now extended to the M-ary detection problem, as

$$H^* = \arg \max_{H_i} P(H_i|\mathbf{x}) \qquad i = 1, \dots, M \tag{3.3}$$

It is therefore clear that posterior probabilities are sufficient statistics for MAP decision problems.

3.2.2   Posterior Probabilities as Aid to Medical Diagnosis

Physicians are used to working with concepts of sensitivity and specificity of a diagnosis procedure [4, 5]. Considering H_0 the absence of a pathology and H_1 the presence of a pathology, the *sensitivity*[1] of the test is defined as $P(H_1|H_1)$ and the *specificity* of the test is $P(H_0|H_0)$. These quantities indicate the probabilities of a test doing fine. However, a test may draw false positives (FPs), with probability $P(H_1|H_0)$, and false negatives (FNs), with probability $P(H_0|H_1)$.

---

[1]In this case, probability $P(H_i|H_j)$ should be read as the probability of deciding H_i when H_j is correct.

Clearly, the ideal test is that for which $P(H_i|H_j) = \delta_{ij}$. However, for real situations, this is not possible. Sensitivity and specificity are not independent but are related by means of a function known as receiver operating characteristics (ROCs). The physician should therefore choose at which point of this curve to work, that is, how to trade FPs (also known as 1-specificity) and FNs (also known as the sensitivity) according to the problem at hand.

Two tests, generally speaking, will perform differently. The way to tell whether one of them outperforms the other is to compare their ROCs. If one ROC is always above the other, that is, if setting the specificity then the sensitivity of one test is always above that of the other, then the former is better. Taking this as the optimality criterion, the key question is the following: Is there a way to design an optimum test? The answer is also well known [3, 6] and is called the Neyman–Pearson (NP) lemma, and a testing procedure that is built upon this principle has been called the *ideal observer* elsewhere [7].

The NP lemma states that the optimum test is given by comparing the likelihood ratio mentioned before with a threshold. Note that this is operationally similar to the procedure described in Eq. (3.1). However, what changes now is the value of the threshold. This is set by assuring that the specificity will take on a particular value, set at the physician's will.

The reader may then wonder whether a testing procedure built upon a ratio of posterior probabilities, as the one expressed in Eq. (3.2), is optimum in the sense stated in this section. The answer is yes: The test described in Eq. (3.1) only differs in the point in the optimum ROC on which it is operating; while a test built upon the NP lemma sets where to work in the ROC, the test built upon posteriors does not fix where to work. Its motivation is different: It changes prior knowledge into posterior knowledge by means of the observations and decides upon the costs given to each decision and each state of nature. This assures that the overall risk (in a Bayesian sense) is minimum [2].

The question is therefore: Why should a physician use a procedure built upon posteriors? Here are the answers:

- The test indicated in (3.2) also implements and ideal observer [7]. So, no loss of optimality is committed.

- As we will point out in the chapter, it is simple to build NN architectures that estimate posteriors out of training data. For nonparametric models, the estimation of both posteriors and likelihood functions is essentially equivalent. However, for parametric models, the procedure to estimate posteriors reflects the objectives of the estimation, which is not the case for the estimation of likelihood functions. Therefore, it seems more natural to use this former approach.

- Working with posteriors, the designer is not only provided with a hard decision about the presence or absence of a pathology but is also provided with a measure of confidence about the decision itself. This is an important piece of additional information that is also encoded in a *natural* range for human reasoning (i.e., within the interval [0, 1]). This reminds us of well-known fuzzy logic procedures often used in medical environments [8].

- As previously stated, posteriors indicate how our prior knowledge changes according to the observations. The prior knowledge, in particular $P(H_1)$, is in fact the prevalence [5] of an illness or pathology. Assume an NN is trained with data from a region with some degree of prevalence but is then used with patients from some other region where the prevalence of the pathology is known to be different. In this case, accepting that symptoms (i.e., observations) are conditionally equally distributed (CED) regardless of the region where the patient is from, it is simple to adapt the posteriors from one region to the other without retraining the NN.

Specifically, denote by H_i^j the hypothesis $i(i = 0, 1)$ in region $j(j = 1, 2)$. Then

$$P(H_i^1|\mathbf{x}) = \frac{f(\mathbf{x}|H_i^1)P(H_i^1)}{f^1(\mathbf{x})}$$

$$P(H_i^2|\mathbf{x}) = \frac{f(\mathbf{x}|H_i^2)P(H_i^2)}{f^2(\mathbf{x})}$$

(3.4)

Accepting that symptoms are CED, then $f(\mathbf{x}|H_i^j) = f(\mathbf{x}|H_i)$. Therefore

$$P(H_0^2|\mathbf{x}) = \frac{f^1(\mathbf{x})}{f^2(\mathbf{x})}\frac{P(H_0^1|\mathbf{x})}{P(H_0^1)}P(H_0^2) = \lambda\frac{P(H_0^1|\mathbf{x})}{P(H_0^1)}P(H_0^2)$$

$$P(H_1^2|\mathbf{x}) = \lambda\frac{P(H_1^1|\mathbf{x})}{P(H_1^1)}P(H_1^2)$$

(3.5)

where $\lambda = f^1(\mathbf{x})/f^2(\mathbf{x})$. Since

$$P(H_0^2|\mathbf{x}) + P(H_1^2|\mathbf{x}) = 1$$

(3.6)

then

$$\lambda\left[\frac{P(H_0^1|\mathbf{x})}{P(H_0^1)}P(H_0^2) + \frac{P(H_1^1|\mathbf{x})}{P(H_1^1)}P(H_1^2)\right] = 1$$

(3.7)

which leads to

$$P(H_0^2|\mathbf{x}) = \frac{[P(H_0^1|\mathbf{x})/P(H_0^1)]P(H_0^2)}{[P(H_0^1|\mathbf{x})/P(H_0^1)]P(H_0^2) + [P(H_1^1|\mathbf{x})/P(H_1^1)]P(H_1^2)}$$

(3.8)

which gives the procedure to update posteriors using as inputs the posteriors of the originally trained NN and the information about the prevalences in the two regions, information that is in the public domain.

The reader should notice that we have never stated that decisions based on posteriors are preferable to those based on likelihood functions. Actually, the term *natural* has been used twice in the reasons to back up the use of posteriors for decision making. This *naturalness* should be balanced by the physician to make the final decision. In any case, the procedure based on posteriors can be very easily forced to work, if needed, as the procedure based on likelihoods: Simply tune the threshold on the right-hand side of Eq. (3.2) during the training phase so that some value of specificity is guaranteed.

3.2.3   Probability Estimation with NN

Consider a sample set $S = \{(\mathbf{x}^k, d^k), k = 1, \ldots, K\}$, where $\mathbf{x}^k \in \Re^\Re$ is an observation vector and $\mathbf{d}^k \in U_L = \{\mathbf{u}_0, \ldots, \mathbf{u}_{L-1}\}$ is an element in the set of possible target classes. Class i label $\mathbf{u}_i \in \Re^L$ has all components null but the one at the ith, row, which is unity (i.e., classes are mutually exclusive).

In order to estimate posterior probabilities in an L-class problem, consider the structure shown in Figure 3.1, where the soft classifier is an NN computing a nonlinear mapping $\mathbf{g}_w : \chi \to \mathcal{P}$, with parameters $\mathbf{w}$, χ being the input feature space and $\mathcal{P} = \{\mathbf{y} \in [0,1]^L | 0 \leq y_i \leq 1, \sum_{i=1}^L y_i = 1\}$ a probability space, such that the soft decision satisfies the

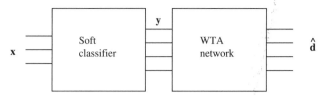

Figure 3.1   Soft-decision multiple output and WTA network.

probability constraints

$$0 \le y_j \le 1 \qquad \sum_{j=1}^{L} y_j = 1 \tag{3.9}$$

Assuming that soft decisions are posterior probability estimates, the hard decision $\widehat{d}$ should be computed according to the decision criterion previously established. For instance, under the MAP criterion, the hard decision becomes a *winner-takes-all* (WTA) function.

Training pursues the calculation of the appropriate weights to estimate probabilities. It is carried out by minimizing some average, based on samples in $\mathcal{S}$, of a cost function $\mathcal{C}(\mathbf{y}, \mathbf{d})$.

The first attempts to estimate a posteriori probabilities [9–12] were based on the application of the square error (SE) or Euclidean distance, that is,

$$C_{\text{SE}}(\mathbf{y}, \mathbf{d}) = \sum_{i=1}^{L} (d_i - y_i)^2 = \|\mathbf{y} - \mathbf{d}\|^2 \tag{3.10}$$

It is not difficult to show that, over all nonlinear mappings $\mathbf{y} = g(\mathbf{x})$, the one minimizing $E\{C_{\text{SE}}(\mathbf{y}, \mathbf{d})\}$ is given by $g(\mathbf{x}) = E\{\mathbf{d}|\mathbf{x}\}$. Since $E\{d_i|\mathbf{x}\} = P(H_i|\mathbf{x})$, it is clear that minimizing any empirical estimate of $E\{C_{\text{SE}}\}$ based on samples in $\mathcal{S}$, estimates of posterior probabilities are obtained.

Also, the cross entropy (CE), given by

$$C_{\text{CE}}(\mathbf{y}, \mathbf{d}) = -\sum_{i=1}^{L} d_i \log y_i \tag{3.11}$$

is known to provide class posterior probability estimates [13–15]. In fact, there is an infinite number of cost functions satisfying this property: Extending previous results from other authors, Miller et al. [9] derived the necessary and sufficient conditions for a cost function to minimize to a probability, providing a closed-form expression for these functions in binary problems. The analysis has been extended to the multiclass case in [16–18], where it is shown that any cost function that provides posterior probability estimates—generically called strict sense Bayesian (SSB)—can be written in the form

$$C(\mathbf{y}, \mathbf{d}) = h(\mathbf{y}) + (\mathbf{d} - \mathbf{y})^{\text{T}} \nabla_{\mathbf{y}} h(\mathbf{y}) \tag{3.12}$$

where $h(\mathbf{y})$ is any strictly convex function in $\mathcal{P}$ which can be interpreted as an entropy measure. As a matter of fact, this expression is also a sufficient condition: Any cost function expressed in this way provides probability estimates. In particular, if h is the Shannon entropy, $h(\mathbf{y}) = -\sum_i y_i \log y_i$, then Eq. (3.12) becomes the CE, so it can be easily proved by the reader that both CE and SE cost functions are SSB.

In order to choose an SSB cost function to estimate posterior probabilities, one may wonder what is the best option for a given problem. Although the answer is not clear, because there is not yet any published comparative analysis among general SSB cost functions, there is some (theoretical and empirical) evidence that learning algorithms based on CE tend to show better performance than those based on SE [10, 19, 20]. For this reason, the CE has been used in our simulations shown later.

3.3 POSTERIOR PROBABILITY ESTIMATION BASED ON GSP NETWORKS

This section discusses the problem of estimating posterior probability maps in multiple-hypotheses classification with the particular type of architecture that has been used in the experiments: the GSP. First, we present a functional description of the network. Second, we discuss training algorithms for parameter estimation and, lastly we discuss the model selection problem in GSP networks.

3.3.1 Neural Architecture

Consider the neural architecture that, for input feature vector $\mathbf{x}$, produces output y_i, given by

$$y_i = \sum_{j=1}^{M_i} y_{ij} \tag{3.13}$$

where y_{ij} are the outputs of a *softmax* nonlinear activation function given by

$$y_{ij} = \frac{\exp(o_{ij})}{\sum_{k=1}^{L} \sum_{m=1}^{M_k} \exp(o_{km})} \qquad i = 1, 2, \ldots, L \qquad j = 1, 2, \ldots, M_i \tag{3.14}$$

and $o_{ij} = \mathbf{w}_{ij}^{\mathrm{T}}\mathbf{x} + b_{ij}$, where $\mathbf{w}_{ij}$ and b_{ij} are the weight vectors and biases, respectively, L is the number of classes, and M_i is the number of softmax outputs aggregated to compute y_i.[2]

The network structure is illustrated in Figure 3.2. In [16, 17, 21], this network is named the *generalized softmax perceptron* (GSP). It is easy to understand that Eqs. (3.13) and (3.14) ensure that outputs y_i satisfy probability constraints given in (3.9).

Since we are interested in the application of MCC detection, we have only considered binary networks ($L = 2$) in the simulations. In spite of this, we discuss here the general multiclass case.

The GSP is a universal posterior probability network, in the sense that, for values of M_i sufficiently large, it can approximate any probability map with arbitrary precision. This fact can be easily proved by noting that the GSP can compute exact posterior probabilities when data come from a mixture model based on Gaussian kernels with the same variance matrix. Since a Gaussian mixture can approximate any density function, the GSP can approximate any posterior probability map.

In order to estimate class probabilities from samples in training set S, an SSB cost function and a search method must be selected. In particular, the stochastic gradient

---

[2]Note that, when the number of inputs is equal to 2, the softmax is equivalent to a pair of sigmoidal activation functions.

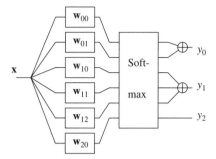

Figure 3.2   The GSP neural architecture.

learning rules to minimize the CE, Eq. (3.11), are given by

$$\mathbf{w}_{ij}^{k+1} = \mathbf{w}_{ij}^{k} - \rho^{k} \frac{y_{ij}^{k}}{y_{i}^{k}} (d_{i}^{k} - y_{i}^{k}) \mathbf{x}^{k} \tag{3.15}$$

$$b_{ij}^{k+1} = b_{ij}^{k} - \rho^{k} \frac{y_{ij}^{k}}{y_{i}^{k}} (d_{i}^{k} - y_{i}^{k}) \qquad 1 \le i \le L \qquad 1 \le j \le M_{i} \tag{3.16}$$

where $\mathbf{w}_{ij}^{k}$ and b_{ij}^{k} are the weight vectors and biases, respectively, y_{ij}^{k} are the softmax outputs for input $\mathbf{x}^{k}$, d_{i}^{k} are the labels for sample $\mathbf{x}^{k}$, and ρ^{k} is the learning step at iteration k.

In the following section we derive these rules from a different perspective, based on maximum-likelihood (ML) estimation, which provides an interpretation for the softmax outputs and some insight into the behavior of the GSP network and which suggests some ways to design other estimation algorithms.

3.3.2   Probability Model Based on GSP

Since the GSP outputs satisfy the probability constraints, we can say that any GSP implements a multinomial probability model of the form

$$P(\mathbf{d}|\mathbf{W}, \mathbf{x}) = \prod_{i=1}^{L} (y_{i})^{d_{i}} \tag{3.17}$$

where matrix $\mathbf{W}$ encompasses all GSP parameters (weight vector and biases) and y_{i} is given by Eqs. (3.13) and (3.14). According to this, we arrive at $y_{i} = P(\mathbf{u}_{i}|\mathbf{W}, \mathbf{x})$. This probabilistic interpretation of the network outputs is useful for training: The model parameters that best fit the data in $\mathcal{S}$ can be estimated by ML as

$$\widehat{\mathbf{W}} = \arg \max_{\mathbf{W}} l(\mathcal{S}, \mathbf{W}) \tag{3.18}$$

where

$$l(\mathcal{S}, \mathbf{W}) = \sum_{k=1}^{K} \log P(\mathbf{d}^{k}|\mathbf{W}, \mathbf{x}^{k}) \tag{3.19}$$

Following an analysis similar to that of Jordan and Jacobs [1] for HME, this maximization can be done iteratively by means of the expectation maximization (EM) algorithm.

To do so, let us assume that each class is partitioned into several *subclasses*. Let M_{i} be the number of subclasses inside class i. The subclass for a given sample $\mathbf{x}$ will be represented by a subclass label vector $\mathbf{z}$ with components $z_{ij} \in \{0, 1\}$, $i = 1, \ldots, L$, $j = 1, \ldots, M_i$, such that one and only one of them is equal to 1 and

$$d_i = \sum_{j=1}^{M_i} z_{ij} \qquad 1 \le i \le L \tag{3.20}$$

Also, let us assume that the joint probability model of $\mathbf{d}$ and $\mathbf{z}$ is given by

$$P(\mathbf{d}, \mathbf{z} | \mathbf{W}, \mathbf{x}) = I_{\mathbf{d}, \mathbf{z}} \prod_{i=1}^{L} \prod_{j=1}^{M_i} (y_{ij})^{z_{ij}} \tag{3.21}$$

where $I_{\mathbf{d,z}}$ is an indicator function equal to 1 if $\mathbf{d}$ and $\mathbf{z}$ satisfy Eq. (3.20) and equal to 0 otherwise.

Note that the joint model equation (3.21) is consistent with that in Eq. (3.17), in the sense that the latter results from the former by marginalization,

$$\sum_{\mathbf{z}} P(\mathbf{d}, \mathbf{z} | \mathbf{W}, \mathbf{x}) = P(\mathbf{d} | \mathbf{W}, \mathbf{x}) \tag{3.22}$$

Since there is no information about subclasses in $\mathcal{S}$, learning can be supervised at the class level but must be unsupervised at the subclass level. The EM algorithm based on hidden variables $\mathbf{z}^k$ to maximize log-likelihood $l(\mathcal{S}, \mathbf{W})$ in (3.19) provides us with a way to proceed. Consider the *complete* data set $\mathcal{S}_c = \{(\mathbf{x}^k, \mathbf{d}^k, \mathbf{z}^k), k = 1, \ldots, K\}$, which extends $\mathcal{S}$ by including subclass labels. According to Eq. (3.21), the *complete data likelihood* is

$$\ell_c(\mathcal{S}_c, \mathbf{W}) = \sum_{k=1}^{K} \log P(\mathbf{d}^k, \mathbf{z}^k | \mathbf{W}, \mathbf{x}^k) = \sum_{k=1}^{K} \sum_{i=1}^{L} \sum_{j=1}^{M_i} z_{ij}^k \log y_{ij}^k \tag{3.23}$$

The EM algorithm for ML estimation of the GSP posterior model proceeds, at iteration t, in two steps:

1. **E-step**: Compute $Q(\mathbf{W}, \mathbf{W}^t) = E\{l_c(\mathcal{S}_c, \mathbf{W}) | \mathcal{S}, \mathbf{W}^t\}$.
2. **M-step**: Compute $\mathbf{W}^{t+1} = \arg\max_{\mathbf{W}} Q(\mathbf{W}, \mathbf{W}^t)$.

In order to apply the E-step to l_c, note that, given $\mathcal{S}$ and assuming that true parameters are $\mathbf{W}^t$, the only unknown components in ℓ_c are the hidden variables. Therefore,

$$Q(\mathbf{W}, \mathbf{W}^t) = \sum_{k=1}^{K} \sum_{i=1}^{L} \sum_{j=1}^{M_i} E\{z_{ij}^k | \mathcal{S}, \mathbf{W}^t\} \log y_{ij}^k \tag{3.24}$$

Let us use compact notation y_{ij}^{kt} to refer to the softmax output (relative to class i and subclass j) for input $\mathbf{x}^k$ at iteration t (and the same for z_{ij}^{kt} and y_i^{kt}). Then

$$E\{z_{ij}^k | \mathcal{S}, \mathbf{W}^t\} = E\{z_{ij}^k | \mathbf{x}^k, \mathbf{d}^k, \mathbf{W}^t\} = d_i^k \frac{y_{ij}^{kt}}{y_i^{kt}} \tag{3.25}$$

Substituting (3.25) in (3.24), we get

$$Q(\mathbf{W}, \mathbf{W}^t) = \sum_{k=1}^{K} \sum_{i=1}^{L} d_i^k \sum_{j=1}^{M_i} \frac{y_{ij}^{kt}}{y_i^{kt}} \log y_{ij}^k \tag{3.26}$$

Therefore, the M-step reduces to

$$\mathbf{W}^{t+1} = \arg\max_{\mathbf{W}} \sum_{k=1}^{K} \sum_{i=1}^{L} d_i^k \sum_{j=1}^{M_i} \frac{y_{ij}^{kt}}{y_i^{kt}} \log y_{ij}^k \tag{3.27}$$

Note that, during the M-step, only y_{ij}^k depends on $\mathbf{W}$. Maximization can be done in different ways. For the HME, the iteratively reweighted least squares algorithm [1] or the Newton–Raphson algorithm [22] has been explored. A more simple solution (that is also suggested in [1]) consists of replacing the M-step by a single iteration of a gradient search rule,

$$\mathbf{W}^{t+1} = \mathbf{W}^{t} - \rho^t \sum_{k=1}^{K} \sum_{i=1}^{L} d_i^k \sum_{j=1}^{M_i} \frac{1}{y_i^{kt}} \nabla_{\mathbf{W}} y_{ij}^k \bigg|_{\mathbf{W}=\mathbf{W}_t} \tag{3.28}$$

which leads to

$$\mathbf{w}_{mn}^{t+1} = \mathbf{w}_{mn}^{t} - \rho^t \sum_{k=1}^{K} \sum_{i=1}^{L} d_i^k \sum_{j=1}^{M_i} \frac{y_{ij}^{kt}}{y_i^{kt}} (\delta_{m-i}\delta_{n-j} - y_{mn}^{kt})\mathbf{x}^k$$

$$= \mathbf{w}_{mn}^{t} - \rho^t \sum_{k=1}^{K} \frac{y_{mn}^{kt}}{y_m^{kt}} (d_m^k - y_m^{kt})\mathbf{x}^k \tag{3.29}$$

$$b_{mn}^{t+1} = b_{mn}^{t} - \rho^t \sum_{k=1}^{K} \frac{y_{mn}^{kt}}{y_m^{kt}} (d_m^k - y_m^{kt}) \tag{3.30}$$

where ρ^t is the learning step. A further simplification can be done by replacing the gradient search rule by an incremental version. In doing so, rules (3.15) and (3.16) result.

3.3.3 A Posteriori Probability Model Selection Algorithm

While the number of classes is fixed and assumed known, the number of subclasses inside each class is unknown and must be estimated from samples during training: It is a well-known fact that a high number of subclasses may lead to data overfitting, a situation in which the cost averaged over the training set is small but the cost averaged over a test set with new samples is high.

The problem of determining the optimal network size, also known as the *model selection* problem, is as well known as, in general, difficult to solve; see [16, 17, 23]. The selected architecture must find a balance between the approximation power of large networks and the usually higher generalization capabilities of small networks.

A review of model selection algorithms is beyond the scope of this chapter. The algorithm we propose to determine the GSP configuration, which has been called the *a posteriori probability model selection* (PPMS) algorithm [16, 17], belongs to the family of growing and pruning algorithms [24]: Starting from a pre-defined architecture, subclasses are added to or removed from the network during learning according to needs. The PPMS algorithm determines the number of subclasses by seeking a balance between generalization capability and learning toward minimal output errors.

Although PPMS could be easily extended to other networks, like the HME [1], the formulation presented here assumes a GSP architecture. The PPMS algorithm combines pruning, splitting, and merging operations in a similar way to other algorithms proposed in the literature, mainly for estimating Gaussian mixtures (see [25] or [26] for instance). In particular, PPMS can be related to the model selection algorithm proposed in [25] where an approach to automatically growing and pruning HME [1, 27] is proposed, obtaining better generalization performance than traditional static and balanced hierarchies. We observe several similarities between their procedure and the algorithm proposed here because they also compute posterior probabilities in such a way that a path is pruned if the instantaneous node probability of activation falls below a certain threshold.

The fundamental idea behind PPMS is the following: According to the GSP structure and its underlying probability model, the posterior probability of each class is a sum of the subclass probabilities. The importance of a subclass in the sum can be measured by means of its prior probability,

$$P_{ij}^t = P\{z_{ij} = 1 | \mathbf{W}^t\} = \int P\{z_{ij} = 1 | \mathbf{W}^t, \mathbf{x}\} p(\mathbf{x}) \, d\mathbf{x} = \int y_{ij} p(\mathbf{x}) \, d\mathbf{x} \qquad (3.31)$$

which can be approximated using samples in $\mathcal{S}$ as[3]

$$\widehat{P}_{ij}^t = \frac{1}{K} \sum_{k=1}^{K} y_{ij}^{kt} \qquad (3.32)$$

A small value of $\widehat{P}_{ij}^t$ is an indicator that only a few samples are being *captured* by subclass j in class i, so its elimination should not affect the whole network performance in a significant way, at least in average terms. On the contrary, a high subclass prior probability may indicate that the subclass captures too many input samples. The PPMS algorithm explores this hypothesis by dividing the subclass into two halves in order to represent this data distribution more accurately. Finally, it has also been observed that, under some circumstances, certain weights in different subclasses of the same class tend to follow a very similar time evolution to other weights within the same class. This fact suggests a new action: merging similar subclasses into a unique subclass.

The PPMS algorithm implements these ideas via three actions, called *prune*, *split*, and *merge*, as follows:

1. *Prune*   Remove a subclass by eliminating its weight vector if its a priori probability estimate is below a certain pruning threshold μ_{prune}. That is,

$$\widehat{P}_{ij}^t < \mu_{\text{prune}} \qquad (3.33)$$

2. *Split*   Add a newsubclass by splitting in two a subclass whose prior probability estimate is greater than split threshold μ_{split}. That is,

$$\widehat{P}_{ij}^t < \mu_{\text{split}} \qquad (3.34)$$

Splitting subclass (i, j) is done by removing weight vector $\mathbf{w}_{ij}^t$ and constructing a pair of new weight vectors $\mathbf{w}_{ij}^{t+1}$ and $\mathbf{w}_{ij'}^{t+1}$ such that, at least initially, the posterior

---

[3]Since $\mathbf{W}^t$ is computed based on samples in $\mathcal{S}$, prior probability estimates based on the same sample set are biased. Therefore, priors should be estimated by averaging from a validation set different from $\mathcal{S}$. In spite of this, in order to reduce the data demand, our simulations are based on only one sample set.

probability map is approximately the same:

$$\mathbf{w}_{ij}^{t+1} = \mathbf{w}_{ij}^{t} + \Delta$$

$$\mathbf{w}_{ij'}^{t+1} = \mathbf{w}_{ij}^{t} - \Delta$$

$$b_{ij}^{t+1} = b_{ij}^{t+1} - \log 2 \tag{3.35}$$

$$b_{ij'}^{t+1} = b_{ij}^{t+1} - \log 2$$

It is not difficult to show that, for $\Delta = 0$, probability maps $P(\mathbf{d}|\mathbf{x}, \mathbf{W}^t)$ and $P(\mathbf{d}|\mathbf{x}, \mathbf{W}^{t+1})$ are identical. In order that new weight vectors can evolve differently during time, using a small nonzero value of Δ is advisable. The log 2 constant has a halving effect.

3. *Merge*   Mix or fuse two subclasses into a single one. That is, the respective weight vectors of both subclasses are fused into a single one if they are close enough. Subclasses j and j' in class i are merged if $\mathcal{D}(\mathbf{w}_{ij}, \mathbf{w}_{ij'}) < \mu_{\mathrm{merge}} \forall i$ and $\forall j, j'$, where $\mathcal{D}$ represents a distance measure and μ_{merge} is the merging threshold. In our simulations, $\mathcal{D}$ was taken to be the Euclidean distance.

After merging subclasses j and j', a new weight vector $\mathbf{w}_{ij}^{t+1}$ and bias b_{ij}^{t+1} are constructed according to

$$\mathbf{w}_{ij}^{t+1} = \tfrac{1}{2}(\mathbf{w}_{ij}^{t} + \mathbf{w}_{ij'}^{t})$$

$$b_{ij}^{t+1} = \log[\exp(b_{ij}^{t}) + \exp(b_{ij'}^{t})] \tag{3.36}$$

3.3.4   Implementation of PPMS

An iteration of PPMS can be implemented after each M-step during learning or after several iterations of the EM algorithm.

In our simulations, we have explored the implementation of PPMS after each M-step, where the M-step is reduced to a single iteration of rules (3.15) and (3.16). Also, to reduce computational demands, prior probabilities are not estimated using Eq. (3.32) but are updated iteratively as $\widehat{P}_{ij}^{t+1} = (1 - \alpha^t)\widehat{P}_{ij}^{t+1} + \alpha^t y_{ij}^t$, where $0 \leq \alpha^t \leq 1$. Note that, if prior probabilities are initially nonzero and sum to 1, the updating rule preserves the satisfaction of these probability constraints.

After each application of PPMS, the prior estimates must also be updated. This is done as follows:

1. *Pruning*   When a subclass is removed, the a priori probability estimates of the remaining classes do not sum to 1. This inconsistency can be resolved by redistributing the prior estimate of the removed subclass proportionally among the rest of the subclasses. In particular, given that condition (3.33) is true, the new prior probability estimates $\widehat{P}_{ij'}^{t+1}$ are computed from $\widehat{P}_{ij'}^{t}$, after pruning subclass j of class i as follows:

$$\widehat{P}_{ij'}^{t+1} = \widehat{P}_{ij'}^{t} \frac{\sum_{n=1}^{M_i} \widehat{P}_{in}^{t}}{\sum_{n=1, n \neq j}^{M_i} \widehat{P}_{in}^{t}} \tag{3.37}$$

$$1 \leq i \leq L \qquad 1 \leq j' \leq M_i \qquad j' \neq j$$

2. *Splitting*   The a priori probability of the split subclass is also divided into two equal parts that will go to each of the descendants. That is, if subclass j in class i is split into

subclasses j and j', prior estimates are assigned to new subclasses as

$$\widehat{P}_{ij}^{t+1} = \tfrac{1}{2}\widehat{P}_{ij}^{t} \qquad \widehat{P}_{ij'}^{t+1} = \tfrac{1}{2}\widehat{P}_{ij'}^{t} \tag{3.38}$$

3. *Merging*   Probability vectors are modified in accordance with probabilistic laws; see (3.9). Formally speaking, if subclasses j and j' in class i are merged into subclass j,

$$\widehat{P}_{ij}^{t+1} = \widehat{P}_{ij}^{t} + \widehat{P}_{ij'}^{t} \tag{3.39}$$

3.4   MICROCALCIFICATION DETECTION SYSTEM

3.4.1   Background Notes in Microcalcification Automatic Detection

Breast cancer is a major public health problem. It is known that about 40–75 new cases per 100,000 women are diagnosed each year in Spain, [28]. Similar quantities have been observed in the United States; see [4].

Mammography is the current clinical procedure followed to the early detection of cancer but ultrasound imaging of the breast is gaining importance nowadays. Radiographic findings in breast cancer can be divided into two categories: MCCs and masses [4]. About the former, debris and necrotic cells tend to calcify, so they can be detected by radiography. It has been shown that between 30 and 50% of breast carcinomas detected radiographically showed clustered MCCs, while 80% revealed MCCs after microscopic examination. About the latter, most cancers that present as masses are invasive cancers (i.e., tumors that have extended beyond the duct lumen) and are usually harder to detect due to the similarity of mass lesions with the surrounding normal *parenchymal* tissue.

Efforts in automatically detecting either of the two major symptoms of breast cancer started in 1967 [29] and are numerous since then. Attention has been paid to image enhancement, feature definition and extraction, and classifier design. The reader may want to consult [4] for an interesting introduction to these topics and follow the numerous references included there. What we would like to point out is that, to the best of our knowledge, the idea of using posteriors to build a classifier and to provide the physician with a confidence in the decision based on posterior probabilities has not been explored so far. This is our goal in what follows.

3.4.2   System Description

In Figure 3.3, we show a general scheme of the system we propose in order to detect MCCs; the system will be window oriented, and, for each window, the a posteriori probability that a MCC is present will be computed; to that end, we make use of the GSP and the PPMS algorithm, introduced respectively in Sections 3.3.1 and 3.3.3. The NN will be fed with features calculated from the pixel intensities within each window. A brief description of these features is given in the forthcoming section.

The tessellation of the image causes MCC detections to obey a window-oriented pattern; in addition, since decisions are made independently in each window, no spatial coherence is forced so, due to FNs, clustered MCCs may show up as a number of separated positive decisions. In order to obtain realistic results about the MCC shapes we have defined a MCC segmentation procedure; this is a regularization procedure consisting of two stages: First, a minimum distance is defined between detected MCCs so as to consider

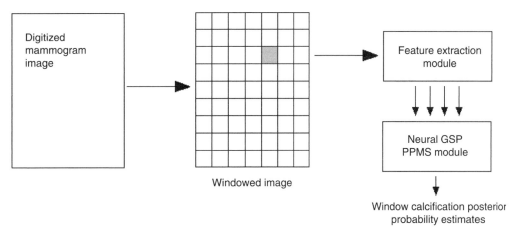

Figure 3.3   The MCC system block diagram: an overview.

that detected MCCs within this distance belong to the same cluster. Then an approximation to the *convex hull* of this cluster is found by drawing the minimum rectangle (with sides parallel to the image borders) that encloses all the detected MCCs within the cluster. The final stage is a pixel-level decision: A decision for each pixel within the rectangle just drawn (and not only within the detected MCCs) to belong or not to the cluster is made. This is done by testing whether the pixel intensity falls inside the range of the intensities of the detected MCCs within the cluster.

3.4.3   Definition of Input Features

In [30], the authors define, on both the spatial and the frequency domains, a set of four features as inputs to an NN for MCC detection. The two spatial features are the *pixel intensity variance* and the *pixel energy variance*. Both are calculated as sample variances within each processing window, and the *pixel energy* is defined as the square of the pixel intensity. The two frequency features are the *block activity* and the *spectral entropy*. The former is the summation of the absolute values of the coefficients of the discrete cosine transform (DCT) of the pixels within the block (i.e., the window), with the direct-current coefficient excluded. The latter is the classical definition of the entropy, calculated in this case on the normalized DCT coefficients, where the normalization constant is the *block activity*.

This set of features constitute the four-dimensional input vector that we will use for our neural GSP architecture. Needless to say, these features compute search for MCC boundaries; MCC interiors, provided that they fit entirely within an inspection window, are likely to be missed. However, the regularization procedure will eliminate these losses.

3.5   RESULTS

The original mammogram database images were in the *Lumiscan75* scanner format. Images had an optical spatial resolution of 150 dpi or, equivalently, about 3500 pixels/ cm^2, which is a low resolution for today's systems and close to the lower limit of the resolution needed to detect most MCCs to some authors' judgment. We internally used all the 12 bits depth during computations. The database consisted of images of the whole breast area and others of more detailed areas specified by expert radiologists.

The window size was set to 8×8 square pixels, obtaining the best MCC detection and segmentation results with that value after some trial and error testing. We defined both a training set and a test set. All the mammogram images shown here belong to the test set. We made a double-windowed scan of the original mammograms with half-window interleave in the detection process in order to increase the detection capacity of the system.

During simulations, the number of subclasses per class is randomly initialized as well as the weight matrix **W** of the GSP NN; the learning step ρ^t was decreased to progressively freeze learning as the algorithm evolves. The PPMS parameters were empirically determined based on trial and error (no critical dependencies were observed in these values, though), and the initial values were $\mu^0_{\text{prune}} = 0.025$, $\mu^0_{\text{split}} = 0.250$, and $\mu^0_{\text{merge}} = 0.1$. After applying PPMS we obtained a GSP network complexity with one subclass for the class MCC-present and seven subclasses for the class MCC-absent.

In Figure 3.4 a number of regions of interest (ROIs) have been drawn by an expert; the result of the algorithm is shown on the right, in which detected MCCs have been marked in white on the original mammogram background. Figure 3.5 shows details of the ROI in the bottom-right corner of Figure 3.4; Figure 3.5a shows the original mammogram, Figure 3.5b shows the detected MCCs in white, and Figure 3.5c shows the posterior probabilities of a single pass indicated in each inspection window. As is clear from the image, if an inspection window fits within a MCC, the posterior probability is fairly small. However, this effect is removed by subsequent regularization.

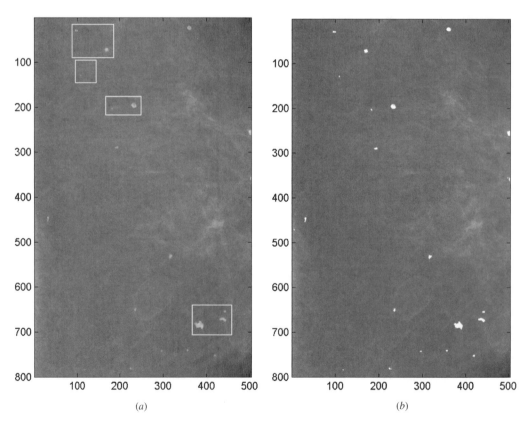

(a)   (b)

Figure 3.4   (a) Original mammogram with four ROIs specified by the expert radiologist; (b) segmented MCCs on the original image (necrosis *grasa* case).

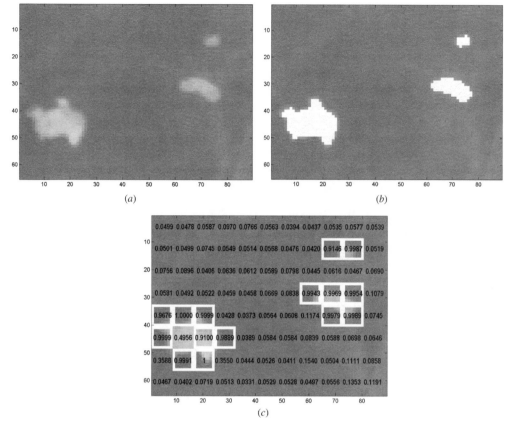

Figure 3.5 Details of ROI on bottom-left corner of Figure 3.4: (*a*) original portion of mammogram; (*b*) segmented output (MCC in white on the original mammogram); (*c*) posterior probabilities of a single pass of the algorithm on each inspection window (necrosis *grasa* case).

Figure 3.6 shows the results of the algorithm on a second example. In this case a mammogram with a carcinoma is shown (original image in Fig. 3.6*a*); the results of the first pass of the algorithm are shown in Figures 3.6*b,c*. Figure 3.6*b* shows the detected MCCs superimposed on the original image and Figure 3.6*c* shows the posterior probabilities on each inspection window. Figures 3.6*d,e* show the result of the second pass of the algorithm with the same ordering as in the first pass. In this example, the sparsity of the MCCs causes the MCCs to not fit entirely within each inspection window, so in those inspection windows where MCCs are present, large variance values are observed, leading to high posterior probabilities. Table 3.1 summarizes the performance of the proposed MCC detection system by indicating both the FP and FN rates in our test set[4] and both including and excluding the borders of the mammogram images.[5] Just as a rough comparison we have taken the results reported in [30] as a reference. The comparison, however, is not totally conclusive since the test images used in this chapter and in [30] do not coincide. We have used this paper as benchmark since the features used in our

---

[4]When computing FP and FN in Table 3.1, we have considered the total number of windows, which is not always a common way to compute those rates.

[5]We have not used any procedure to first locate the breast within the mammogram. So by "excluding borders" we mean manually delineating the breast and running the algorithm within. Therefore, excluding borders means excluding sources of errors such as the presence of letters within the field of inspection, the breast boundary, etc.

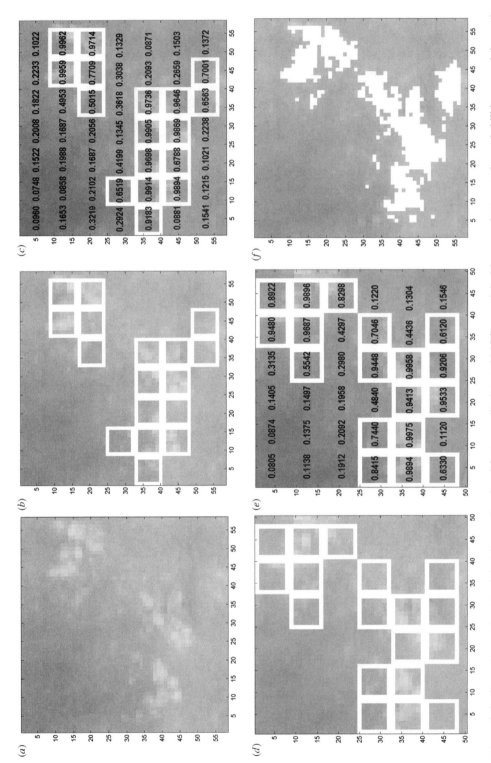

Figure 3.6 Second example: (a) original image. Results of first pass: (b) detected MCCs on original image and (c) posterior probabilities on each inspection window. Results of second pass: (d) detected MCCs on original image and (e) posterior probabilities on each inspection window; (f) final segmentation (carcinoma case).

TABLE 3.1.1   Total Error and Error Rates in % (FP and FN) for the Full Mammograms and Mammograms with Borders Excluded Borders of Mammograms in the Test Set Over a Population of 21,648 and 19,384, Respectively, Windowed 8 × 8 Pixel Images

Case	Total error windows	% Total	FP	%FP	FN	% FN
Test, full	413	1.91	104	0.48	309	1.43
Test, no borders	246	1.27	29	0.15	217	1.12

work have been taken from it. Specifically, denoting by TN the true negatives and by TP the true positives, we have, with border excluded, $TN = 19,384 - 29 = 19,355$ windows and $TP = 19,384 - 217 = 19,167$ windows or, equivalently, $TN = 99.85\%$ and $TP = 98.88\%$. Consequently for our test set we have obtained $TP = 98.88\%$ and $FP = 0.1\%$. The best results reported in [30] $TP = 90.1\%$, $FP = 0.71\%$. The result, although not conclusive, suggests that the system proposed here may constitute an approach worth taking.

3.6   FUTURE TRENDS

We have presented a neural methodology to make decisions by means of posterior probabilities learned from data. In addition, a complexity selection algorithm has been proposed, and the whole system has been applied to detect and segment MCCs in mammograms. We show results on images from a hospital database, and we obtain posterior probabilities for the presence of a MCC with the use of the PPMS algorithm. Our understanding is that probability estimates are a valuable piece of information; in particular, they give an idea of the certainty in the decisions made by the computer, which can be weighed by a physician to make the final diagnosis. As of today we are not aware of similar approaches applied to this problem. In addition, the generation of ROC curves are being carried out in order to compare with others, although the reader should take into account that the images from the database considered here are private, so other standard public-domain mammogram databases should be considered. Needless to say, image enhancement may dramatically improve the performance of any decision-making procedure. This constitutes part of our current effort, since various enhancement algorithms are being tested, together with a validation of the system described here against standard public mammogram databases and the application of the presented algorithm and NN methodology in other medical problems of interest [31].

ACKNOWLEDGMENTS

The authors wish to first express their sincere gratitude to Martiniano Mateos-Marcos for the help received while preparing part of the figures included in the manuscript and in programming some simulations. They also thank the Radiology Department personnel at the Asturias General Hospital in Oviedo for the help received while evaluating the performance of the system and to the radiologists at Puerta de Hierro Hospital in Madrid for providing the mammogram database used in training the NN. Juan I. Arribas wants to thank Dr. Luis del Pozo for reviewing the manuscript and Yunfeng Wu and Dr. Gonzalo García for their help. This work was supported by grant numbers FP6-507609 SIMILAR Network of Excellence from the European Union and TIC01-3808-C02-02, TIC02-03713, and TEC04-06647-C03-01 from Comisión Interministerial de Ciencia y Tecnología, Spain.

REFERENCES

1. M. I. JORDAN AND R. A. JACOBS. Hierarchical mixtures of experts and the EM algorithm. *Neural Computation*, 6(2):181–214, 1994.

2. H. L. VAN TREES. *Detection, Estimation and Modulation Theory*. Wiley, New York, 1968.

3. S. M. KAY. *Fundamentals of Statistical Signal Processing. Detection Theory*. Prentice-Hall, Englewood Cliffs, NJ, 1998.

4. M. L. GIGER, Z. HUO, M. A. KUPINSKI, AND C. J. VYBORNY. Computer-aided diagnois in mammography. In M. SONKA AND J. M. FITZPATRICK, Eds., *Handbook of Medical Imaging*, Vol. II. The International Society for Optical Engineering, Bellingham, WA, 2000.

5. B. ROSNER. *Fundamentals of Biostatistics*. Duxbury Thomson Learning, Pacific Grove, CA, 2000.

6. S. M. KAY. *Fundamentals of Statistical Signal Processing. Estimation Theory*. Prentice-Hall, Englewood Cliffs, NJ, 1993.

7. M. A. KUPINSKI, D. C. EDWARDS, M. L. GIGER, AND C. E. METZ. Ideal observer approximation using Bayesian classification neural networks. *IEEE Trans. Med. Imag.*, 20(9):886–899, September 2001.

8. P. S. SZCZEPANIAK, P. J. LISBOA, AND J. KACPRZYK. *Fuzzy Systems in Medicine*. Physica-Verlag, Heidelberg, 2000.

9. J. MILLER, R. GODMAN, AND P. SMYTH. On loss functions which minimize to conditional expected values and posterior probabilities. *IEEE Trans. Neural Networks*, 4(39):1907–1908, July 1993.

10. B. S. WITTNER AND J. S. DENKER. Strategies for teaching layered neural networks classification tasks. In W. V. OZ AND M. YANNAKAKIS, Eds., *Neural Information Processing Systems*, vol. 1. CA, 1988, pp. 850–859.

11. D. W. RUCK, S. K. ROGERS, M. KABRISKY, M. E. OXLEY, AND B. W. SUTER. The multilayer perceptron as an approximation to a Bayes optimal discriminant function. *IEEE Trans. Neural Networks*, 1(4):296–298, 1990.

12. E. A. WAN. Neural Network classification: A Bayesian interpretation. *IEEE Trans. Neural Networks*, 1(4):303–305, December 1990.

13. R. G. GALLAGER. *Information Theory and Reliable Communication*. Wiley, New York, 1968.

14. E. B. BAUM AND F. WILCZEK. Supervised learning of probability distributions by neural networks. In D. Z. ANDERSON, Ed., *Neural Information Processing Systems*. American Institute of Physics, New York, 1988, pp. 52–61.

15. A. EL-JAROUDI AND J. MAKOUL. A new error criterion for posterior probability estimation with neural nets. *International Joint Conference on Neural Nets*, San Diego, 1990.

16. J. I. ARRIBAS AND J. CID-SUEIRO. A model selection algorithm for a posteriori probability estimation with neural networks. *IEEE Trans. Neural Networks*, 16(4):799–809, July 2005.

17. J. I. ARRIBAS. Neural networks for a posteriori probability estimation: structures and algorithms. Ph.D.

dissertation, Electrical Engineering Department, Universidad de Valladolid, Spain, 2001.

18. J. CID-Sueiro, J. I. ARRIBAS, S. URBÁN-MUÑOZ, AND A. R. FIGUEIRAS-VIDAL. Cost functions to estimate a posteriori probability in multiclass problems. *IEEE Trans. Neural Networks*, 10(3):645–656, May 1999.

19. S. I. AMARI. Backpropagation and stochastic gradient descent method. *Neuro computing*, 5:185–196, 1993.

20. B. A. TELFER AND H. H. SZU. Energy functions for minimizing misclassification error with minimum-complexity networks. *Neural Networks*, 7(5):809–818, 1994.

21. J. I. ARRIBAS, J. CID-Sueiro, T. ADALI, AND A. R. FIGUEIRAS-VIDAL. Neural architectures for parametric estimation of a posteriori probabilities by constrained conditional density functions. In Y. H. HU, J. LARSEN, E. WILSON, AND S. DOUGLAS, Eds., *Neural Networks for Signal Processing*. IEEE Signal Processing Society, Madison, WI, pp. 263–272.

22. K. CHEN, L. XU, AND H. CHI. Improved learning algorithm for mixture of experts in multiclass classification. *Neural Networks*, 12:1229–1252, June 1999.

23. V. N. VAPNIK. An overview of statistical learning theory. *IEEE Trans. Neural Networks*, 10(5):988–999, September 1999.

24. R. REED. Pruning algorithms—a survey. *IEEE Trans. Neural Networks*, 4(5):740–747, September 1993.

25. J. FRITSCH, M. FINKE, AND A. WAIBEL. Adaptively growing hierarchical mixture of experts. In *Advances in Neural Information Processing Systems*, vol. 6. Morgan Kaufmann Publishers, CA, 1994.

26. J. L. ALBA, L. DOCIO, D. DOCAMPO, AND O. W. MARQUEZ. Growing Gaussian mixtures network for classification applications. *Signal Proc.*, 76(1):43–60, 1999.

27. L. XU, M. I. JORDAN, AND G. E. HINTON. An alternative model for mixtures of experts. In T. LEEN, G. TESAURO, AND D. TOURETZKY, Eds., *Advances in Neural Information Processing Systems*, vol. 7. MIT Press, Cambridge, MA, 1995, pp. 633–640.

28. Dirección General de Salud Pública. *Population Screening of Breast Cancer in Spain* (in Spanish). Ministerio de Sanidad y Consumo, Madrid, Spain, 1998.

29. F. WINSBERG, M. ELKIN, J. MACY, V. BORDAZ, AND W. WEYMOUTH. Detection of radiographic abnormalities in mammograms by means of optical scanning and computer analysis. *Radiology*, 89:211–215, 1967.

30. B. ZHENG, W. QIAN, AND L. CLARKE. Digital mammography: Mixed feature neural network with spectral entropy decision for detection of microcalcifications. *IEEE Trans. Med. Imag.*, 15(5):589–597, October 1996.

31. A. TRISTÁN AND J. I. ARRIBAS. A radius and ulna skeletal age assessment system. In V. CALHOUN and T. ADALI, Eds, *Machine Learning for Signal Processing*, IEEE Signal Processing Society, Mystic, CN, 2005, pp. 221–226.

CHAPTER 4

IDENTIFICATION OF CENTRAL AUDITORY PROCESSING DISORDERS BY BINAURALLY EVOKED BRAINSTEM RESPONSES

Daniel J. Strauss, Wolfgang Delb, and Peter K. Plinkert

4.1 INTRODUCTION

Binaural interaction in auditory brainstem responses (ABRs) was demonstrated in animals by Jewett [1], who compared the binaurally evoked response with the sum of the monaurally evoked responses on both ears and noted a decreased amplitude of the binaural waveform. Later, binaural interaction in ABRs was also shown to be present in humans by Dobie and Berlin [2], Hosford et al. [3], and Dobie and Norton [4]. These authors introduced the computation of the binaural interaction component (BIC) as the arithmetical difference between the sum of the monaurally evoked ABRs and the binaurally evoked ABR. According to Furst et al. [5] and Brantberg et al. [6, 7], the so-called β-wave is the most consistent part of the BIC waveform. Furst et al. [5] showed that the β-wave is related to directional hearing.

Possible applications of the β-wave analysis are the diagnosis of the central auditory processing disorder (CAPD) and the objective examination of directional hearing in bilateral cochlear implant users. In the latter mentioned application, it might be possible to predict the clinical outcome preoperatively by using BIC measurements. So far, mainly applications of BIC measurements in view of the CAPD have been reported in the literature [8, 9].

The American Speech–Language and Hearing Association Task Force on Central Auditory Processing Consensus Development defined the CAPD as a deficit in one or more of the following central auditory processes: sound localization/lateralization, auditory discrimination, auditory pattern recognition, temporal aspects of auditory processing, and performance deficits when the auditory signal is embedded in competing acoustic signals or when the auditory signal is degraded [10]. Deficits in binaural processing are a part of this definition and also in [11–13] the importance of binaural testing in patients with CAPDs is highlighted. Thus with the relation of the β-wave to directional hearing, it may serve as an objective criterion in the CAPD diagnosis.

Handbook of Neural Engineering. Edited by Metin Akay
Copyright © 2007 The Institute of Electrical and Electronics Engineers, Inc.

However, the identification of this wave still remains a challenge due to a poor signal quality which is affected by noise and a lack of objective detection criteria. As a consequence of this, several detection criteria have been developed. Furst et al. [5] searched for the largest positive peak in the time period between 1 ms before and 1 ms after the wave in the monaural brainstem response. Brantberg et al. [6, 7] considered a β-wave as being present if a reproducible peak could be observed during the downslope of the wave V in the binaurally evoked brainstem response. Stollmann et al. [14] tried an objective detection of BIC peaks using a template method and did not show any advantages compared to a method in which the detection of the BIC peaks is based on the signal-to-noise ratio. It seems obvious that in many cases the peaks identified as β-waves would not be identical comparing the cited methods of detection. Furthermore, the signal features of the binaurally evoked responses as well as of the sum of the monaurally evoked brainstem potentials that result in the computation of a typical β-wave do not seem to be well defined. Brantberg et al. [6, 7] reported on latency differences between the wave V of the monaural sum and the binaurally evoked potentials that account for the formation of the β-wave while others mainly found amplitude differences. A closer look at the signals of many subjects shows that in reality amplitude differences as well as latency differences can be the reason for the formation of a β-wave. Sometimes, even differences in the downslope of wave V are accounted for its generation. It seems hard to believe that all these signal modifications are caused by the same physiological process. Rather, one would suspect different physiological processes or, in the case of amplitudes, even accidental signal fluctuations underlying the formation of the BIC waveform.

All this along with the fact that expert knowledge is commonly used for the identification of the β-wave [5–7, 9] leads to a loss of objectivity and certainty in the evaluation of ABRs in the diagnosis of the CAPD. Therefore, we investigate a different approach to the examination of directional hearing and the objective detection of the CAPD using an evaluation of the binaurally evoked brainstem response directly. In particular, we present two fully objective detection methods which are based on adapted time-scale feature extractions in these potentials. We show that such a direct approach is truly objective, reduces measurement cost significantly, and provides at least comparable results as the β-wave identification for the discrimination of patients being at risk for CAPD and patients not being at risk for CAPD.

4.2 METHODS

4.2.1 Data

Auditory evoked potentials were recorded using a commercially available device (ZLE—Systemtechnik, Munich, Germany) in a sound-proof chamber (filter 0.1–5 kHz, sampling frequency 20 kHz, amplification factor 150,000). In each measurement, 4000 clicks of alternating polarity were presented binaurally or monaurally at an intensity of 65 dB hearing level (HL) with an interstimulus interval of 60 ms. The response to each first click was written in memory 1 whereas the response to every second click was written in memory 2. In the remainder of this chapter, we consider the averaged version of the responses in memory 1 and memory 2 if the separation is not explicitly stated. The potentials were obtained using electrodes placed at the neck, the vertex, and the upper forehead, respectively. After averaging, data were processed using a personal computer system. In the binaural measurements interaural time delay (ITD) (stimulus on the left side being delayed) varied between 0.0 and 1.0 ms (0.0, 0.4, 0.6, 0.8, and 1.0 ms).

4.2.2 Expert Analysis

The BIC is computed by subtracting the binaurally evoked response from the sum of the monaurally evoked responses (see Fig. 4.1). Before the summation of the monaural responses, the left monaural response is shifted in time according to the ITD; see Brantberg et al. [7]. As usual, the analysis of the BIC in the time domain is based on a visual analysis using expert knowledge. We use one of the detection criteria given in Brantberg et al. [7]. Accordingly, a β-wave is considered to be present if a peak is observed during the downslope of wave V. We explicitly impose the reproducibility of this peak, that is, a high concurrence of its representation in memory 1 and memory 2.

4.2.3 MLDB: Feature Extraction Approach

Feature extraction by some linear or nonlinear transform of the data with subsequent feature selection is an attractive tool for reducing the dimensionality of the problem to tackle with the *curse of dimensionality* [15]. Recently, time–frequency analysis methods using wavelets with no special focus on the data have been suggested for feature extraction (e.g., see [16, 17]). Data-dependent schemes are also known which mainly rely on an adjustment of the decomposition tree in wavelet packet decompositions; see [18] for a recent review and comparison of several approaches.

Among these methods, the *local discriminant basis* (LDB) algorithm of Saito and Coifman [19] is a well-accepted scheme which utilizes a tree adjustment and relies on the *best-basis* paradigm known in signal compression [20]. This algorithm selects a basis from a dictionary that illuminates the dissimilarities among classes and has very recently shown its state-of-the-art performance for real-world pattern recognition tasks

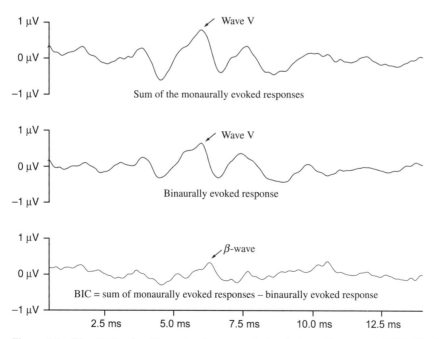

Figure 4.1   The ABRs after binaural and monaural stimulation with computed BIC. The β-wave arises during the downslope of *wave V* of the responses.

in [18]. The objective of the LDB algorithm is to adjust the decomposition tree which corresponds to a fixed two-channel filter bank building block.

The theory of signal-adapted filter banks has been developed in signal compression in recent years (e.g., see [21–23]). Up to now, the underlying ideas mainly stick on this restricted area although they may have merit in other application fields such as pattern recognition. In recent papers, we have shown that an adaptation technique from signal compression is an effective tool for real-world pattern recognition tasks when using appropriate class separability criteria, that is, discrimination criteria instead of compression conditions (e.g., see [24, 25]).

Recently, we have constructed shape-adapted LDBs, which we called morphological LDBs (MLDBs), by using signal-adapted filter banks [26]. Compared to LDBs, our MLDBs utilize additionally to the tree adjustment an adaptation of the shape of the analyzing wavelet packets, that is, an adjustment of the two-channel filter bank building block. In this way, much more discriminant information can be extracted among signal classes. We used these MLDBs in Strauss et al. [27] to illuminate discriminant information between the binaural waveforms of probands with normal binaural processing and children with deficits in the binaural processing. Here we apply these MLDBs to capture discriminant information between the sum of the monaural waveforms and the binaural waveform of adults being not at risk for pathological binaural interaction. Our overall objective here is the identification of features which correspond to binaural interaction in the binaurally evoked waveforms which are not included in the sum of the monaural waveforms.

4.2.3.1 MLDB Algorithm

The time-scale energy map of a set of M waveforms $\mathbf{x}_i \in \Omega_{0,0} \subset \mathbb{R}^d (i = 1, \ldots, M)$ is defined by

$$\Upsilon(j, k, m) := \sum_{i=1}^{M} y_{i,j,k}^2[m] \|\mathbf{x}_{i,l}\|_2^{-2} \qquad (4.1)$$

for $j = 0, \ldots, J, k = 0, \ldots, 2^j - 1$, and $m \in \mathcal{T}_j$, where $y_{i,j,k}^2[\cdot]$ denote the wavelet packet expansion coefficients and $\mathcal{T}_j$ an appropriate index set; see Appendix A.

In our further discussions, we use overlined symbols to distinguish the quantities corresponding to the sum of the monaural waveforms from the binaural waveform, that is, we use $\bar{\mathbf{x}}$, $\bar{y}_{j,k}$, and $\overline{\Upsilon}(j, k, m)$ for denoting the waveform, the expansion coefficients in Eq. (4.17), and the time-scale energy map in Eq. (4.1), respectively, for the sum of the monaural waveforms.

We introduce the set $B_{j,k} = \{\mathbf{q}_{j,k}^m : m \in \mathcal{T}_j\}$ which contains all translations of the atom corresponding to $\Omega_{j,k}$. Let $A_{j,k}$ represent the LDB (see below) restricted to the span of $B_{j,k}$ and let $\Delta_{j,k}$ be a working array. We define $\mathcal{D}_2(\mathbf{a}, \mathbf{b}) = \|\mathbf{a} - \mathbf{b}\|_2^2$ and set $A_{J,k} = B_{J,k}$ and $\Delta_{J,k} = \mathcal{D}_2(\Upsilon(j,k,m), \overline{\Upsilon}(j,k,m))$ for $k = 0, \ldots, 2^j - 1$. Then the best subsets $A_{j,k}$ for $j = J - 1, \ldots, 0, k = 0, \ldots, 2^j - 1$ are determined by the following rule:

Set $\Delta_{j,k} = \sum_{m \in \mathcal{T}_j} \mathcal{D}_2(\Upsilon(j, k, m), \overline{\Upsilon}(j, k, m))$

If $\Delta_{j,k} \geq \Delta_{j+1,2k} + \Delta_{j+1,2k+1}$

Then $A_{j,k} = B_{j,k}$

Else $A_{j,k} = A_{j+1,2k} \cup A_{j+1,2k+1}$ and $\Delta_{j,k} = \Delta_{j+1,2k} + \Delta_{j+1,2k+1}$

By this selection rule, $\Delta_{0,0}$ becomes the largest possible discriminant value. The morphology of the atoms $\mathbf{q}_{j,k}^m$ is defined via Eqs. (4.14) and (4.15), respectively, and thus by the underlying two-channel paraunitary filter bank. The MLDB algorithm utilizes the lattice parameterization of such filter banks for a morphological adaptation of the atoms. For this, the polyphase matrix of the analysis bank

$$\mathbf{H}_{\text{pol}}(z) := \begin{pmatrix} H_{00}(z) & H_{01}(z) \\ H_{10}(z) & H_{11}(z) \end{pmatrix}$$

with entries from the polyphase decomposition

$$H_i(z) = H_{i0}(z^2) + z^{-1} H_{i1}(z^2) \qquad i = 0, 1$$

is decomposed into

$$\mathbf{H}_{\text{pol}}(z) = \left(\prod_{k=0}^{K-1} \begin{pmatrix} \cos \vartheta_k & \sin \vartheta_k \\ -\sin \vartheta_k & \cos \vartheta_k \end{pmatrix} \begin{pmatrix} 1 & 0 \\ 0 & z^{-1} \end{pmatrix} \right) \begin{pmatrix} \cos \vartheta_K & \sin \vartheta_K \\ -\sin \vartheta_K & \cos \vartheta_K \end{pmatrix} \tag{4.2}$$

where $\vartheta_K \in [0, 2\pi)$ and $\vartheta_k \in [0, \pi)$ $(k = 0, \ldots, K-1)$ for FIR filters of order $2K + 1$. Let ϑ_K be the residue of $\pi/4 - \sum_{k=0}^{K-1} \vartheta_k$ modulo 2π in $[0, 2\pi)$. Then the space

$$\mathcal{P}^K := \{ \boldsymbol{\vartheta} = (\vartheta_0, \ldots, \vartheta_{K-1}) \colon \vartheta_k \in [0, \pi) \}$$

can serve to parameterize all two-channel far infrared (FIR) paraunitary filter banks with at least one vanishing moment of the high-pass filter, that is, a zero mean.

Now we have to solve the optimization problem

$$\hat{\boldsymbol{\vartheta}} = \arg \max_{\boldsymbol{\vartheta} \in \mathcal{P}^K} \Delta_{0, 0}(\boldsymbol{\vartheta}) \tag{4.3}$$

by a genetic algorithm. The wavelet packet basis associated with $\hat{\boldsymbol{\vartheta}}$ in (4.3) is called the MLDB. We use a 40-bit encoding for each angle in $[0, \pi]$ where we set $K = 2$ in (4.2). An initial population of 100 is generated randomly. The probabilities for crossover and mutation are set to $p_c = 0.95$ and $p_m = 0.005$, respectively. The MLDB algorithm is summarized in Figure 4.2.

4.2.3.2 *Study Group I* The study group (study group I) considered for this feature extraction study consists of 14 adults with normal hearing (threshold <10 dB HL between 500 and 6000 Hz), without any history of peripheral and central auditory disorders and without intellectual deficit. The individuals exhibited normal directional hearing as judged from a localization task in a setting where seven sound sources were located in a half circle around the patient. Normal speech detection in noise was verified by means of the binaural intelligibility difference using a commercially available test (BIRD test, Starkey laboratories, Germany). In all subjects, wave V latencies in the monaurally evoked brainstem responses of both ears did not show differences of more than 0.2 ms. All probands showed a reproducible β-wave in the BIC at an ITD of 0.4 ms by the expert analysis described in Section 4.2.2.

4.2.4 Hybrid Wavelet: Machine Learning Approach

The MLDB feature extraction approach which we have presented in Section 4.2.3 allows for the extraction of discriminant features in the time-scale domain between the sum of the monaural responses and the binaural response. The extracted features are clearly defined signal properties, localized in time and frequency. They may be correlated with other known representatives of binaural interaction in ABRs such as the β-wave. From a clinical point of view, this is the major advantage of such a "white-box" feature extraction approach.

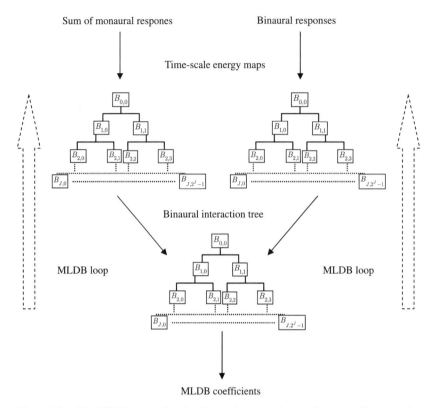

Figure 4.2 The MLDB algorithm for discrimination of sum of monaurally evoked response from binaural waveform. The arrows denote the MLDB loop of the filter bank optimization.

The MLDB algorithm can easily be applied if signals to be analyzed exhibit a relatively homogeneous morphology and a good reproducibility of the particular features in time, such as the centered ABR waveforms of adults in Section 4.2.3. However, recent studies have shown [28] that ABRs of children (mean age 8 years) exhibit a much higher heterogeneity in the signal morphology as the ABRs of adults do (mean age 21 years). Therefore, we present a more robust scheme which takes this fact into account for the discrimination of children being at risk for CAPD and children being not at risk for CAPD which was first proposed by Strauss et al. [29] and thoroughly studied in [30]. Moreover, the presented scheme is shift invariant such that a centering of ABRs prior the analysis is not necessary.

In particular, we present a hybrid wavelet–machine learning scheme to the detection of the CAPD. This scheme consists of adapted wavelet decompositions and support vector machine (SVM) classifiers. The adapted wavelet decomposition serves again for the feature extraction, but in contrast to the MLDB algorithm, shift-invariant subband features are used. This feature extraction procedure is optimized to provide a large margin of the subsequent SVM classifier such that our whole hybrid scheme is embedded in statistical learning theory by the large margin theorem. It is worth emphasizing that we do not restrict our interest to binaural interaction here, which reflects only a part of the definition of the CAPD. Thus this "black-box" detection approach also covers the heterogeneous definition of the CAPD which might be a further advantage.

4.2.5 Basics of SVMs

The SVM is a novel type of learning machine and very promising for pattern recognition [31]. Basically, the SVM relies on the well-known optimal hyperplane classification, that is, the separation of two classes of points by a hyperplane such that the distance of distinct points from the hyperplane, the so-called *margin*, is maximal. The SVMs utilize this linear separation method in very high dimensional feature spaces induced by reproducing kernels to obtain a nonlinear separation of original patterns. In contrast to many other learning schemes, for example, feedforward backpropagation neural networks, training a SVM yields a global solution as it is based on a quadratic programming (QP) problem (see Appendix B). Moreover, the complexity of a SVM is automatically adapted to the data. In general, there are only a few parameters to adjust; see [31, 32] for detailed discussions.

Let $\mathcal{X}$ be a compact subset of $\mathbb{R}^d$ containing the data to be classified. We suppose that there exists an underlying unknown function t, the so-called *target function*, which maps $\mathcal{X}$ to the binary set $\{-1, 1\}$. Given a training set

$$\mathcal{A} := \{(\mathbf{x}_i, y_i) \in \mathcal{X} \times \{-1, 1\}: \quad i = 1, \ldots, M\} \tag{4.4}$$

of M associations, we are interested in the construction of a real-valued function f defined on $\mathcal{X}$ such that $\text{sgn}(f)$ is a "good approximation" of t. If f classifies the training data correctly, then we have that $\text{sgn}(f(\mathbf{x}_i)) = t(\mathbf{x}_i) = y_i$ for all $i = 1, \ldots, M$ [$\text{sgn}(f(\mathbf{x})) := 1$, if $f(\mathbf{x}) \geq 0$, and -1 otherwise]. We will search for f in some reproducing kernel Hilbert spaces (RKHSs) $\mathcal{H}_K$ (see Appendix B) and a regularization problem in RKHSs arises. For a given training set (4.4) we intend to construct a function $f \in \mathcal{H}_K$ which minimizes

$$\lambda \sum_{i=1}^{M} [1 - y_i f(\mathbf{x}_i)]_+ + \tfrac{1}{2} \|f\|_{\mathcal{H}_K}^2 \tag{4.5}$$

where $(\tau)_+$ equals τ if $\tau \geq 0$ and zero otherwise. This unconstrained optimization problem can be rewritten as a constrained optimization problem in the SVM feature space $\mathcal{F}_K \subset \ell^2$ using the feature map $\boldsymbol{\Phi}: \mathcal{X} \to \mathcal{F}_K$ (see Appendix B) of the form: Find $\mathbf{w} \in \mathcal{F}_K$ and $u_i (i = 1, \ldots, M)$ to minimize

$$\lambda \left(\sum_{i=1}^{M} u_i \right) + \tfrac{1}{2} \|\mathbf{w}\|_{\mathcal{F}_K}^2 \tag{4.6}$$

subject to

$$\begin{aligned} y_i \langle \mathbf{w}, \boldsymbol{\Phi}(\mathbf{x}_i) \rangle_{\mathcal{F}_K} &\geq 1 - u_i & i = 1, \ldots, M \\ u_i &\geq 0 & i = 1, \ldots, M \end{aligned} \tag{4.7}$$

In general, the feature space $\mathcal{F}_K$ is infinitely dimensional. For the sake of simplicity and an easier illustration, we assume for a moment that $\mathcal{F}_K \subset \mathbb{R}^n$. Then the function $\tilde{f}_{\mathbf{w}}(\mathbf{v}) := \langle \mathbf{w}, \mathbf{v} \rangle_{\mathcal{F}_K}$ defines a hyperplane $H_{\mathbf{w}} := \{\mathbf{v} \in \mathcal{F}_K : \tilde{f}_{\mathbf{w}}(\mathbf{v}) = 0\}$ in $\mathbb{R}^n$ through the origin and an arbitrary point $\mathbf{v}_i \in \mathcal{F}_K$ has the distance $|\langle \mathbf{w}, \mathbf{v}_i \rangle_{\mathcal{F}_K}| / \|\mathbf{w}\|_{\mathcal{F}_K}$ from $H_{\mathbf{w}}$. Note that $\tilde{f}_{\mathbf{w}}(\boldsymbol{\Phi}(\mathbf{x})) = f_{\mathbf{w}}(x)$. Thus, the constraints $y_i \langle \mathbf{w}, \boldsymbol{\Phi}(\mathbf{x}_i) \rangle_{\mathcal{F}_K} / \|\mathbf{w}\|_{\mathcal{F}_K} \geq 1 / \|\mathbf{w}\|_{\mathcal{F}_K} - u_i / \|\mathbf{w}\|_{\mathcal{F}_K}$ $(i = 1, \ldots, M)$ in (4.7) require that every $\boldsymbol{\Phi}(\mathbf{x}_i)$ must at least have the distance $1 / \|\mathbf{w}\|_{\mathcal{F}_K} - u_i / \|\mathbf{w}\|_{\mathcal{F}_K}$ from $H_{\mathbf{w}}$.

If there exists $\mathbf{w} \in \mathcal{F}_K$ so that (4.7) can be fulfilled with $u_i = 0$ $(i = 1, \ldots, M)$, then we say that our training set is linearly separable in $\mathcal{F}_K$. In this case, the optimization problem (4.6) can be further simplified to: Find $\mathbf{w} \in \mathcal{F}_K$ to minimize

$$\frac{1}{2}\|\mathbf{w}\|^2_{\mathcal{F}_K} \tag{4.8}$$

subject to

$$y_i\langle\mathbf{w}, \boldsymbol{\Phi}(\mathbf{x}_i)\rangle_{\mathcal{F}_K} \geq 1 \qquad i = 1,\ldots,M$$

Given $\mathcal{H}_K$ and $\mathcal{A}$, the optimization problem above has a unique solution $f_{\mathbf{w}^*}$. In our hyperplane context $H_{\mathbf{w}^*}$ is exactly the hyperplane which has maximal distance γ from the training data, where

$$\gamma := \frac{1}{\|\mathbf{w}^*\|_{\mathcal{F}_K}} = \frac{1}{\|f_{\mathbf{w}^*}\|_{\mathcal{H}_K}} = \max_{\mathbf{w}\in\mathcal{F}_K} \min_{i=1,\ldots,M} \left\{ \frac{|\langle\mathbf{w}, \boldsymbol{\Phi}(\mathbf{x}_i)\rangle_{\mathcal{F}_K}|}{\|\mathbf{w}\|_{\mathcal{F}_K}} \right\} \tag{4.9}$$

The value γ is called the *margin* of $f_{\mathbf{w}^*}$ with respect to the training set $\mathcal{A}$. See Figure 4.3 for an illustration of the mapping and the separation procedure.

4.2.5.1 *Adaptation in Feature Spaces*   Now we introduce our adaptation strategy for feature spaces that is based on wavelets and filter banks that was originally proposed in [25] and extended in [30] for the inclusion of morphological features.

The original wavelet-support vector classifier as proposed in [25] relies on *multilevel concentrations* $\xi(\cdot) = \| \cdot \|^p_{\ell^p} (1 \leq p < \infty)$ of coefficient vectors of adapted wavelet or frame decompositions as feature vectors, that is, scale features.

When using frame decompositions by nonsubsampled filter as suggested by Strauss and Steidl [25], the decompositions becomes invariant to shifts of the signal. This is an important fact here as we are interested in a shift-invariant decomposition which needs no centering of the signals; see [33] for detailed discussions on the shift variance of orthogonal decompositions. The implementation of frame decomposition is closely

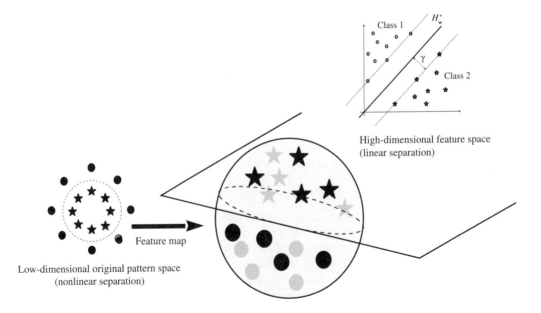

Figure 4.3   The points are mapped by the feature map from the original space to the feature space where the linear separation by the hyperplane takes plan with margin γ.

related to the orthogonal decompositions described in Appendix A. The only thing we have to change is to drop the multirate sampling operations and slightly modify the filter coefficients for generating the basis functions (see [25, 34] for details). As a consequence of this, the lattice parameterization described in Section 4.2.3 can still be applied when the filters are modified appropriately [25].

These feature vectors incorporate the information about local instabilities in time as a priori information. For the classification of ABRs, we also include the morphological information of the waveforms as a feature as the discriminant information which separates the normal group from the CAPD group may also be reflected in the transient evolution of ABRs.

Since we are interested in a shift-invariant classification scheme, we may only evaluate the morphology of ABRs as a whole and not the exact latency of transient features. One possible way to realize this is by the use of entropy which is already used to evaluate the subbands of wavelet and wavelet packet decompositions for the purpose of signal compression [20]. When using an appropriate entropy in connection with the tight frame decomposition described above, it is invariant to shifts of the ABRs. We define the entropy of a sequence $\mathbf{x} \in \ell^2$ by

$$E(\mathbf{x}) = -\sum_{n \in \mathbb{Z}} \frac{|x[n]|^2}{\|\mathbf{x}\|_{\ell^2}^2} \ln \frac{|x[n]|^2}{\|\mathbf{x}\|_{\ell^2}^2} \tag{4.10}$$

Let $\mathbf{d}_j^{\vartheta} = \tilde{\mathbf{y}}_{j,1}^{\vartheta} (J = 1, \ldots, J)$, where $\tilde{\mathbf{y}}_{j,1}$ are the coefficients of a parameterized octave-band frame decomposition, that is, the coefficients of a subtree of the full binary tree (see Appendix A) without sampling operations and parameterized by the lattice angle vector ϑ. For a fixed ABR waveform $\mathbf{x}$ we define the function

$$\boldsymbol{\zeta}_\mathbf{x}(\vartheta) = (\zeta_1(\vartheta), \ldots, \zeta_{2J}(\vartheta))$$
$$= \left(\|\mathbf{d}_1^{\vartheta}\|_{\ell^1}, \ldots, \|\mathbf{d}_J^{\vartheta}\|_{\ell^1}, E(\mathbf{d}_1^{\vartheta}), \ldots, E(\mathbf{d}_J^{\vartheta}) \right)$$

set $\boldsymbol{\zeta}_i(\vartheta) := \boldsymbol{\zeta}_{\mathbf{x}_i}(\vartheta)(i = 1, \ldots, M)$, and normalize $\boldsymbol{\zeta}_i$ in ℓ^1.

The number J is the decomposition depth, that is, the number of octave bands. We restrict ourselves here to an octave-band decomposition as in [29] but, of course, any binary tree can be used here. The first J elements of this feature vector carry multilevel concentration of the subbands in ℓ^1, that is, a scale information. The second J elements carry the morphological information reflected in the entropy as defined in (4.10). Note that $\boldsymbol{\zeta}_i(\vartheta)$ is totally invariant against shifts of the ABRs. The shift invariance does not deteriorate the analysis as the latency of wave V does not provide discriminant information between the waveform groups (see [9, 35] and Section 4.3.2).

Now we intend to find ϑ so that

$$\mathcal{A}(\vartheta) = \{\boldsymbol{\zeta}_i(\vartheta), y_i) \in \mathcal{X} \subset \mathbb{R}^{2J} \times \{\pm 1\} : i = 01, \ldots, M\}$$

is a "good" training set for a SVM. Note that we restrict our interest to the hard-margin SVM here, that is, $\lambda = \infty$ in (4.5).

As we have described earlier, it is the fundamental concept of SVMs that we expect a good generalization performance if they have a large margin defined by (4.9). Therefore, our strategy is now to obtain feature vectors that are mapped to far-apart points in the SVM feature space $\mathcal{F}_K$ for the distinct classes and result in a large margin of the SVM in this space. Consequently, we try to find $\hat{\vartheta}$ such that

$$\hat{\vartheta} = \arg \max_{\vartheta \in \mathcal{P}} \left\{ \min_{i \in M_+, j \in M_-} \| \Phi(\zeta_i(\vartheta)) - \Phi(\zeta_j(\vartheta)) \|_{\mathcal{F}_K}^2 \right\}$$

where the $\Phi(\cdot)$ denotes the SVM feature map, that is, the map from the original space $\mathcal{X}$ to the feature space $\mathcal{F}_K$. For kernels arising from radial basis functions, this problem can be transformed to an optimization problem in the original space [25] such that

$$\hat{\vartheta} = \arg \max_{\vartheta \in \mathcal{P}} \left\{ \min_{i \in M_+, j \in M_-} \| \zeta_i(\vartheta) - \zeta_j(\vartheta) \|_2 \right\} \tag{4.11}$$

To simplify the solution, the patterns belonging to different classes can be averaged such that the distance between the centers is maximized; for details see [30]. Thus we have transformed the problem from the feature space induced by the reproducing kernel to a solvable problem in the original space where the octave-band features live. Equivalent to Section 4.2.3, this optimization problem is again solved by a genetic algorithm.

4.2.5.2 Study Group II   The study group (study group II) considered for this machine learning study consisted of 60 children who were examined for a possible CAPD (aged between 6 and 12 years). All the children showed normal peripheral hearing (pure-tone threshold <15 dB between 500 and 6000 Hz) and normal monaural speech discrimination for monosyllables (monosyllable test of the German speech intelligibility test Freiburger Sprachtest >80% at 60 dB HL). Patients with diagnosed attention-deficit hyperactivity disorder and low intellectual performance were excluded from the study.

These patients were divided into two groups according to the subjective testing procedure described in [9]. By this separation, the normal group, that is, not at risk for CAPD, consisted of 29 patients with a mean age of 8.8 years (standard deviation 1.5 years). All subjects in this group showed at least average intellectual performance as judged from their reading and writing skills and their performance in school. The CAPD group, that is, at risk for CAPD, consisted of 20 patients with a mean age of 8.9 years (standard deviation 1.5 years). There was no statistically significant age difference in comparison to the normal group.

4.3 RESULTS

4.3.1 Study Group I: Feature Extraction Study

4.3.1.1 Time-Domain Analysis   Typical BIC waveforms could be identified in all of our 14 individuals for an ITD of 0.4 ms. As further calculations using the MLDB algorithm are done with a test set of 10 subjects, the analysis in the time domain is also done exclusively in these individuals. The number of subjects with clearly visible β-peaks changed with the interaural delay of the binaurally applied stimuli. While β-waves could be identified in 8 out of 10 cases using an ITD of 0.0 ms, a β-wave was seen in every subject at an ITD of 0.4 ms. With higher ITDs the percentage of clearly visible β-waves gradually decreased as shown in Figure 4.4a. The β-latencies increased as ITDs increased by approximately ITD$/2$ in the range of 0–0.8 ms. With higher ITDs the latency shift decreased (Fig. 4.4b).

4.3.1.2 MLDB Feature Extraction   For determining the MLDB by (4.3), we used the binaural waveforms and the sums of the monaural waveforms for an ITD of 0.0 ms of 10

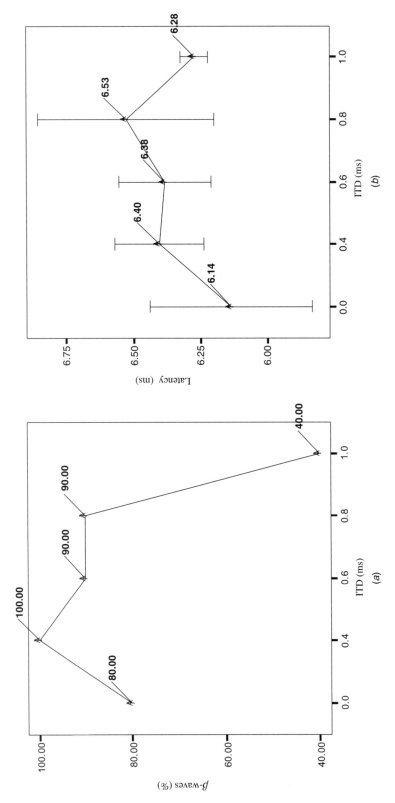

Figure 4.4 (a) Percentage of visible β-waves in BIC at different ITDs. (b) Latencies of BIC β-wave at increasing ITDs.

individuals (i.e., $M = 10$) and a decomposition depth of $J_{\max}$. To cope with shift variance which may deteriorate an automated analysis by wavelet packets, we centered the binaural waveforms by wave V of the binaural responses before the analysis.

The most discriminant MLDB feature which we call coefficient 1 is exclusively considered for the further investigation. It corresponds to a particular cell in the time–frequency domain and is specified by a triplet (j, k, m) that we denote by (j', k', m') in the following. For this time-scale cell, the mean of the induced energy by the binaural waveform and the sum of the monaural waveforms differs significantly.

Furst et al. [5] demonstrated that the β-wave of the BIC provides information on the binaural interaction and that the latencies increase with increasing ITD. With ITDs higher than 1.2 ms, the β-wave of the BIC is rarely present [36]. Thus to show that the time-scale feature of the binaural response represented by coefficient 1 reflects binaural interaction similar to the β-wave of the BIC, we observed its behavior for increasing ITDs. Figure 4.5 shows the difference $d' := |y_{j',k'}[m']| - |\bar{y}_{j',k'}[m']|$ for the individual subject. It is noticeable that this difference decreases significantly for ITDs larger than 0.6 ms. However, the difference was positive in 6 out of 10 cases with an ITD of 0.8 ms. Note that d' is always positive for ITD < 0.8 ms. In other words, the magnitude of coefficient 1 for the binaural waveforms is larger than that of the sums of the monaural waveforms in all subjects for ITD < 0.8 ms. Thus the condition $|y_{j',k'}[m']| - |\bar{y}_{j',k'}[m']| > 0$ can be applied for detecting the binaural interaction for ITD < 0.8 ms in a machine analysis without any utilization of expert knowledge and interference.

As we have identified a dissimilarity of the energy distribution in the time-scale domain among the binaural waveform and the sum of the monaural waveforms, we can use this feature for the detection of binaural hearing from binaural waveforms only without any comparison to the sum of the monaural waveforms. It seems obvious that

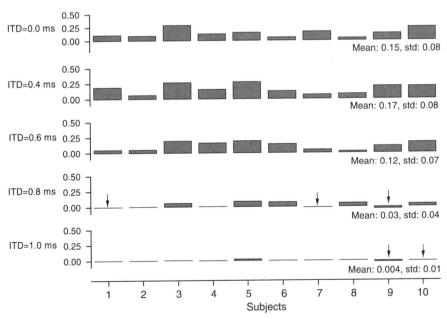

Figure 4.5 Differences d' of most discriminant MLDB coefficient for binaural waveform and sum of monaural waveforms of individual subject. Negative differences ($d' < 0$) are marked by the down arrow.

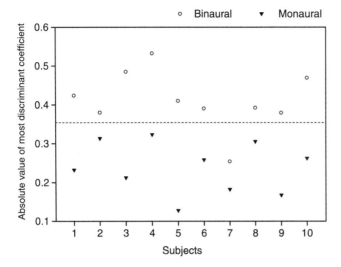

Figure 4.6   Absolute value of most discriminant MLDB feature for (○) binaural and (▼) sum of monaural waveforms.

the information on binaural interaction is included in the binaural response itself as the sum of the monaurally evoked responses does not contain any information on binaural interaction. In Figure 4.6 we have shown a simple threshold classification of the binaural and the sum of the monaural waveforms for ITD $= 0.4$ ms, that is, the ITD where the highest dissimilarity of these waveforms appears; see Figure 4.5. It is noticeable that the binaural waveforms and the sums of the monaural waveforms are separated without overlap with the exception of one subject.

4.3.2   Study Group II: Machine Learning Study

4.3.2.1   Time-Domain Analysis   The analysis of the β-wave, as described in Section 4.2.2, allows for a discrimination of the normal from the CAPD group with 71% specificity and 65% sensitivity in average. Additional to this conventional analysis technique, we examined other easy-to-capture time-domain features involving the binaural waveform only. The results of this analysis are given in Table 4.1. None of these parameters showed a difference which is statistically significant between the normal and the CAPD group.

4.3.2.2   Hybrid Classification Scheme   For the described frame decompositions we use a maximal decomposition depth of 7 and discard the first two levels as they contain noise or very detailed features. In this way, we have an input dimension of 10 for the

TABLE 4.1   Time-domain Parameters of Wave V of Binaural ABR for Normal and CAPD Groups

Parameter	Amplitude (μV)	Latency (ms)	Slope (μV/ms)
Normal	1.32 ± 0.3	5.90 ± 0.25	0.86 ± 0.62
CAPD	1.58 ± 0.4	5.85 ± 0.18	0.79 ± 0.70

Note: Given as mean $\pm$ standard deviation.

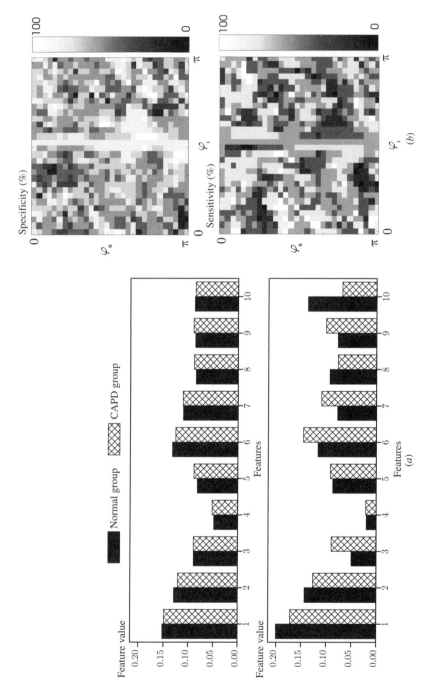

Figure 4.7 (*a*) Feature vectors of normal and CAPD groups for nonadapted angles (top) and for adapted angles (bottom). (*b*) Specificity and sensitivity in lattice parameter space.

SVM. For the adaptation, we parameterize filters of order 5, which leads to two free angles $\vartheta = (\vartheta_0, \vartheta_1)(\vartheta_0, \vartheta_1 \in [0, \pi])$.

Let us start with some experiments and primarily examinations. In Figure 4.7*a* we have shown the normalized feature vectors of the normal and the CAPD group, respectively, for a nonadapted angle pair which performs worse (Fig. 4.7*a*) and in the adapted case (Fig. 4.7*b*). The first five features represent the multilevel concentration whereas the remaining five features represent the morphological information in terms of entropy. It is noticeable that the feature vectors of the normal and the CAPD groups, respectively, show a high degree of dissimilarity in the adapted case. Of course, here we expect a much better performance of the classifier than in the nonadapted case.

Next we examine the feature extraction above in combination with a classifier. The specificity and sensitivity of our hybrid approach in the lattice parameter space given by $\{\varphi_0, \varphi_1\}$ are shown in Figure 4.7*b* for a fixed test set. As clearly noticeable, many angle pairs lead to a good performance of the SVM classifier (Gaussian kernel with standard deviation 1), in the way that they separate the normal from the CAPD group.

In practice, we have to determine the free parameters a priori from the training set. For the hybrid approach, we have an objective function motivated by statistical learning theory to determine the optimal angles a priori by (4.11). When approaching (4.11) as described in [30], we achieve an average specificity of 75% and a sensitivity of 80% (Gaussian kernel with standard deviation 1). Even though there are angles in the parameter space which perform better (see Fig. 4.7), we achieve a good performance by choosing the angles by (4.11), that is, the large-margin criterion for the adaptation.

In view of the results expected from the subjective tests, the hybrid approach is at least comparable in its performance to the β-wave detection. However, it has the major advantage that it requires the binaurally evoked brainstem potential only and is fully machine based and thus truly objective.

4.4 DISCUSSION

As the identification of a defined signal feature describing the dissimilarity between the summed monaural brainstem evoked potentials and the binaurally evoked brainstem responses was one aim of the present study, it has to be proved whether the analyzed data are likely to contain information on binaural interaction and are in accordance to the data regarding the BIC reported in the literature. The analysis of the BIC in the time domain was restricted to the β-peak as it has been shown to be the most consistent part of the BIC. The latencies of the β-wave at an ITD of 0.0 ms were in the same range as reported by Brantberg et al. [6], who used the same detection criteria that were used here. Also a gradual increase of β-latencies as reported by others [5, 7, 36] was observed.

The increase of β-latencies with increasing ITD was in the range of ITD/2 as compared to the latencies obtained with an ITD of 0.0 ms. This finding is in accordance with the assumption of a delay line coincidence detector as a basic mechanism for sound localization as described by Jeffress [37]. His model served as a basis for a number of succeeding models explaining binaural interaction [38–41]. In this model an array of neurons is innervated by collaterals of second-order neurons which bring excitatory input from each ear. The fibers from the two sides run in the opposite direction along the array of neurons. Each neuron of the array fires only if the contralateral and ipsilateral excitations coincide in time. This temporal coincidence is only possible if a certain population of neurons is located in such a way that the combination of time delays connected with a given localization of a sound source leads to an excitation at approximately the same time.

The model would predict that the β-wave of the binaural difference waveform is delayed according to the ITD applied. The latency delay of the β-wave expected from a given ITD would be exactly as high as ITD/2 as compared to an ITD of 0 ms. The data of Jones and Van der Poel [36] as well as Brantberg et al. [6] and our own data support this model as an increase of β-latencies on the order of ITD/2 was found. However, even though there is some morphological evidence ([42] and the review in [43]), it still remains unclear whether there is a true delay line coincidence detection mechanism implemented in the medial superior olive of humans [44] as the majority of evidence for this model is provided by using data of studies on the owl.

One of the major problems with the assumption of a pure coincidence detection mechanism is the fact that the Jeffress [37] model does not include inhibitory input that can explain the formation of the BIC. On the other hand, there is considerable morphological evidence [44–46] that there are inhibitory neurons involved in the processing of ITDs within the medial superior olive. Furthermore, the calculation of the BIC is based on inhibitory rather than excitatory mechanisms taking place during binaural signal processing. However, as there is also considerable electrophysiological evidence for the presence of some sort of a delay line mechanism, it seems likely that this type of signal processing is at least a part of the detection mechanism for interaural time delays taking place in the brainstem [40, 41, 47].

A major feature of the β-wave in the BIC connected with directional hearing is the increase in latency with ITD. However, for ITDs exceeding 0.8 ms the increase in latency was less pronounced. A closer look at data reported in the literature also shows that there is a decrease in latency shift of the β-peak at high (>0.8-ms) ITDs [7]. As the temporo-temporal diameter of a human head rarely exceeds 22 cm, the physiological range of ITDs is from 0 ms to approximately 0.7 ms, which explains the mentioned decrease in latency shift.

One aim of the present study was to identify signal features that differentiate the summed monaural signals from the binaurally evoked brainstem potentials. Using this signal feature, it should be possible to judge whether a binaural interaction is present or not by just analyzing the binaurally evoked potentials without calculating the difference between the summed monaural potentials and the binaurally evoked potentials. We tried to discover such features in the time-scale domain.

A major issue when employing time–frequency decompositions for signal discrimination is the choice of suitable features. To realize an automated and systematic detection of such features, we applied the recently developed MLDB algorithm [26], which has shown to be superior to other algorithms used before for waveform recognition. By means of this algorithm, we extracted the most discriminant time-scale features that exhibit a dissimilarity between the sum of the monaural waveforms and the binaural response, which we called coefficient 1. As the magnitude of coefficient 1 differs in binaurally and monaurally evoked potentials, the calculation of the difference in magnitude of coefficient 1 similar to the calculation of the BIC can serve as a measure of binaural interaction. To prove this, the resulting difference d' should behave in the same way as described for the BIC.

In our settings this means that this difference should be positive up to an ITD of at least 0.6 ms. As shown in Figure 4.5, the difference d' is positive in every case up to an ITD of 0.6 ms and in most cases when an ITD of 0.8 ms is used. This result principally replicates the results obtained from the analysis of the β-wave in the time domain [28, 35]; see also Section 4.3.1.1. Thus the MLDB feature extraction approach which we have presented in Section 4.2.3 allows for the extraction of discriminant features in the time-scale domain between the sum of the monaural responses and the binaural response which is correlated with binaural interaction and might be used for the objective detection of the CAPD. From a clinical point of view, this is the major advantage of such a "white-box" feature extraction approach. However, the relation of the extracted MLDB features to directional hearing has to be proved in further studies using patients with defects in the brainstem that result in impaired directional hearing. The MLDB algorithm can easily be applied if signals to be analyzed exhibit a relatively homogeneous morphology and a good reproducibility of the particular features in time such as centered ABR waveforms of adults in Section 4.2.3. However, recent studies have shown [28] that ABRs of children (mean age 8 years) exhibit a much higher heterogeneity in the signal morphology than the ABRs of adults do (mean age 21 years). Therefore, we presented a more robust scheme which takes this fact into account for the discrimination of children being at risk for CAPD and children being not at risk for CAPD, namely a hybrid wavelet–machine learning approach to the detection of the CAPD.

This scheme consisted of adapted wavelet decompositions and SVM classifiers. The adapted wavelet decomposition again serves for the feature extraction, but in contrast to the MLDB algorithm, shift-invariant subband features are used. This feature extraction procedure was optimized to provide a large margin of the subsequent SVM classifier such that our whole hybrid scheme was embedded in statistical learning theory by the large-margin theorem. It is worth emphasizing that we do not restrict our interest to binaural interaction here, which reflects only a part of the definition of the CAPD. Thus this "black-box" detection approach also covered the heterogeneous definition of the CAPD, which might be a further benefit.

The sensitivity and specificity of the hybrid approach were at least comparable to the conventional β-wave detection in view of the results expected from subjective tests, at least when using the group definition given in [9]. However, it has the major advantage that it reduces the measurement cost by two-thirds and is truly objective, in the way that it needs no expert interference for the evaluation.

A disadvantage in contrast to our white-box MLDB feature extraction approach is that the features involved are hard to correlate with physiological processes. However, it is more reliable due to the use of this more abstract but robust signal features.

4.5 CONCLUSION

We have summarized recent work that we have done to identify the CAPD by using the binaurally evoked ABR directly. In particular, we presented an automated feature extraction scheme in the time-scale domain and hybrid machine learning approach. We showed that the direct use of the binaurally evoked response is truly objective, reduces measurement cost significantly, and provides at least comparable results as the β-wave identification for the discrimination of patients being at risk for CAPD and patients not being at risk for CAPD. We conclude that the identification of the CAPD by binaurally evoked brainstem responses is efficient and superior to the β-wave detection due to reasons of implementation.

APPENDIX A: WAVELET PACKET DECOMPOSITIONS

Let $H_0(z) := \sum_{k \in \mathbb{Z}} h_0[k] z^{-k}$ be the z-transform of the analysis low-pass filter and $H_1(z) := \sum_{k \in \mathbb{Z}} h_1[k] z^{-k}$ the z-transform of the analysis high-pass filter of a two-channel filter bank with real-valued filter coefficients. Throughout this chapter, we used a capital letter to denote a function in the z-transform domain and the corresponding small letter to denote its time-domain version. We assume that the high-pass filter has a zero mean, that is, we have that $H_1(1) = 0$. A two-channel filter bank with analysis filters $H_0(z)$ and $H_1(z)$ is called *paraunitary* (sometimes also referred as *orthogonal*) if they satisfy

$$H_0(z^{-1})H_0(z) + H_1(z^{-1})H_1(z) = 2 \tag{4.12}$$

$$H_0(z^{-1})H_0(-z) + H_1(z^{-1})H_1(-z) = 0 \tag{4.13}$$

The corresponding synthesis filters are given by

$$G_0(z) = H_0(z^{-1}) \qquad G_1(z) = H_1(z^{-1})$$

For the implementation of a wavelet packet decomposition, we arrange such two-channel paraunitary filter banks in a binary tree of decomposition depth J. Binary trees can be expressed by their equivalent parallel structure. Let us define $Q_{0,0}(z) := 1$ and $Q_{0,1}(z) := 1$. Then the synthesis filters of the equivalent parallel structure are given by the recursion

$$Q_{j+1,2k}(z) = G_0(z^{2^j})Q_{j,k}(z) \tag{4.14}$$

$$Q_{j+1,2k+1}(z) = G_1(z^{2^j})Q_{j,k}(z) \tag{4.15}$$

for $j = 0,\ldots,J$, $k = 0,1,\ldots,2^j - 1$. Let $\mathbf{q}_{j,k}^m = (q_{j,k}[n - 2^j m])_{n \in \mathbb{Z}}$ denote the translation of the impulse responses of these filters by $2^j m$ samples and let ℓ^2 denote the Hilbert space of all square summable sequences. Then the space $\Omega_{0,0} := \ell^2$ is decomposed in mutually orthogonal subspaces such that

$$\Omega_{j,k} = \Omega_{j+1,2k} \oplus \Omega_{j+1,2k+1} \tag{4.16}$$

where $\Omega_{j,k} = \overline{\mathrm{span}\{\mathbf{q}_{j,k}^m : m \in \mathbb{Z}\}}$ $(j = 1,\ldots,J, k = 0,1,\ldots,2^j - 1)$. We can define the wavelet packet projection operator

$$P_{j,k} : \Omega_{0,0} \longrightarrow \Omega_{j,k} \qquad j = 1,\ldots,J \qquad k = 0,\ldots,2^j - 1$$

with

$$P_{j,k}\mathbf{x} = \sum_{m \in \mathbb{Z}} y_{j,k}[m]\mathbf{q}_{j,k}^m$$

where the expansion coefficients are given by

$$y_{j,k}[m] = \langle \mathbf{x}, \mathbf{q}_{j,k}^m \rangle_{\ell^2} \tag{4.17}$$

For applying this concept to finite-length signals $\Omega_{0,0} \subset \mathbb{R}^d$, we employ the wraparound technique; see [48]. We will exclusively deal with signals with dimension d a power of 2. With respect to the downsamling operation, we a define a maximal decomposition depth by $J_{\max} = \log_2 d$. For a fixed level j and maximal decomposition depth $J_{\max}$, we define the set of indices

$$\mathcal{T}_j = \{0, 1, \ldots, 2^{J_{\max}-j} - 1\}$$

APPENDIX B

Here we present some definitions used in the SVM context. First we define exactly the feature space of a SVM and then we turn to the solution of the optimization problem associated with the learning a SVM classifier.

B1   Feature Spaces of SVMs

Let $K : \mathcal{X} \times \mathcal{X} \longrightarrow \mathbb{R}$ ($\mathcal{X}$ is a compact subset of $\mathbb{R}^d$) be a positive-definite symmetric function in $L^2(\mathcal{X} \times \mathcal{X})$. For a given K, there exists a *reproducing kernel Hilbert space*

$$\mathcal{H}_K = \overline{\text{span}\{K(\tilde{\mathbf{x}}, \cdot) : \tilde{\mathbf{x}} \in \mathcal{X}\}}$$

of real-valued functions on $\mathcal{X}$ with inner product determined by

$$\langle K(\tilde{\mathbf{x}}, \mathbf{x}), K(\tilde{\mathbf{x}}, \mathbf{x}) \rangle_{\mathcal{H}_K} = K(\tilde{\mathbf{x}}, \tilde{\mathbf{x}})$$

which has the reproducing kernel K, that is, $\langle f(\cdot), K(\tilde{\mathbf{x}}, \cdot) \rangle_{\mathcal{H}_K} = f(\tilde{\mathbf{x}})$, $f \in \mathcal{H}_K$. By *Mercer's theorem*, the reproducing kernel K can be expanded in a uniformly convergent series on $\mathcal{X} \times \mathcal{X}$,

$$K(\mathbf{x}, \mathbf{y}) = \sum_{j=1}^{\infty} \eta_j \varphi_j(\mathbf{x}) \varphi_j(\mathbf{y}) \tag{4.18}$$

where $\eta_j \geq 0$ are the eigenvalues of the integral operator $T_K : L^2(\mathcal{X}) \to L^2(\mathcal{X})$ with

$$T_K f(\mathbf{y}) = \int_{\mathcal{X}} K(\mathbf{x}, \mathbf{y}) f(\mathbf{x}) \, d\mathbf{x}$$

and where $\{\varphi_j\}_{j \in \mathbb{N}}$ are the corresponding $L^2(\mathcal{X})$-orthonormalized eigenfunctions. We restrict our interest to functions K that arise from a radial basis function (RBF). In other words, we assume that there exists a real-valued function k on $\mathbb{R}$ such that

$$K(\mathbf{x}, \mathbf{y}) = k(\|\mathbf{x} - \mathbf{y}\|_2) \tag{4.19}$$

where $\| \cdot \|_2$ denotes the Euclidean norm on $\mathbb{R}^d$.

We introduce a so-called *feature map* $\boldsymbol{\Phi} : \mathcal{X} \to \ell^2$ by

$$\boldsymbol{\Phi}(\cdot) = \left(\sqrt{\eta_j} \varphi_j(\cdot) \right)_{j \in \mathbb{N}}$$

Let ℓ^2 denote the Hilbert space of real-valued quadratic summable sequences $a = (a_i)_{i \in \mathbb{N}}$ with inner product $\langle a, b \rangle_{\ell^2} = \sum_{i \in \mathbb{N}} a_i b_i$. By (4.18), we have that $\boldsymbol{\Phi}(\mathbf{x}), \mathbf{x} \in \mathcal{X}$, is an element in ℓ^2 with

$$\|\boldsymbol{\Phi}(\mathbf{x})\|_{\ell^2}^2 = \sum_{j=1}^{\infty} \eta_j \varphi_j^2(\mathbf{x}) = K(\mathbf{x}, \mathbf{x}) = k(0)$$

We define the *feature space* $\mathcal{F}_K \subset \ell^2$ by the ℓ^2-closure of all finite linear combinations of elements $\boldsymbol{\Phi}(\mathbf{x})$ ($\mathbf{x} \in \mathcal{X}$),

$$\mathcal{F}_K = \overline{\text{span}\{\boldsymbol{\Phi}(\mathbf{x}) : \mathbf{x} \in \mathcal{X}\}}$$

Then $\mathcal{F}_K$ is a Hilbert space with $\| \cdot \|_{\mathcal{F}_K} = \| \cdot \|_{\ell^2}$. The feature space $\mathcal{F}_K$ and the reproducing kernel Hilbert space $\mathcal{H}_K$ are isometrically isomorphic with isometry $\iota : \mathcal{F}_K \to \mathcal{H}_K$ defined by $\iota(\mathbf{w}) = f_{\mathbf{w}}(\mathbf{x}) = \langle \mathbf{w}, \boldsymbol{\Phi}(\mathbf{x}) \rangle_{\ell^2} = \sum_{j=1}^{\infty} w_j \sqrt{\eta_j} \varphi_j(\mathbf{x})$.

B2 Solving the SVM Optimization Problem

By the *representer theorem* [32, 49], the minimizer of (4.6) has the form

$$f(\mathbf{x}) = \sum_{j=1}^{M} c_j K(\mathbf{x}, \mathbf{x}_j) \tag{4.20}$$

Setting $\mathbf{f} := (f(\mathbf{x}_1), \dots, f(\mathbf{x}_M))^{\mathrm{T}}$, $\mathbf{K} := (K(\mathbf{x}_i, \mathbf{x}_j))_{i,j=1}^{M}$, and $\mathbf{c} := (c_1, \dots, c_M)^{\mathrm{T}}$, we obtain that

$$\mathbf{f} = \mathbf{K}\mathbf{c}$$

Note that $\mathbf{K}$ is positive definite. Further, let $\mathbf{Y} := \mathrm{diag}(y_1, \dots, y_M)$ and $\mathbf{u} := (u_1, \dots, u_M)^{\mathrm{T}}$. By $\mathbf{0}$ and $\mathbf{e}$ we denote the vectors with M entries 0 and 1, respectively. Then the optimization problem (4.6) can be rewritten as

$$\min_{\mathbf{u},\mathbf{c}} \lambda \mathbf{e}^{\mathrm{T}}\mathbf{u} + \tfrac{1}{2}\mathbf{c}^{\mathrm{T}}\mathbf{K}\mathbf{c} \tag{4.21}$$

subject to

$$\mathbf{u} \geq \mathbf{e} - \mathbf{Y}\mathbf{K}\mathbf{c} \qquad \mathbf{u} \geq 0.$$

The dual problem with Lagrange multipliers $\boldsymbol{\alpha} = (\alpha_1, \dots, \alpha_M)^{\mathrm{T}}$ and $\boldsymbol{\beta} = (\beta_1, \dots, \beta_M)^{\mathrm{T}}$ reads

$$\max_{\mathbf{c},\mathbf{u},\alpha,\beta} L(\mathbf{c}, \mathbf{u}, \boldsymbol{\alpha}, \boldsymbol{\beta})$$

where

$$L(\mathbf{c}, \mathbf{u}, \boldsymbol{\alpha}, \boldsymbol{\beta}) := \lambda \mathbf{e}^{\mathrm{T}}\mathbf{u} + \tfrac{1}{2}\mathbf{c}^{\mathrm{T}}\mathbf{K}\mathbf{c} - \boldsymbol{\beta}^{\mathrm{T}}\mathbf{u} + \boldsymbol{\alpha}^{\mathrm{T}}\mathbf{e} - \boldsymbol{\alpha}^{\mathrm{T}}\mathbf{Y}\mathbf{K}\mathbf{c} - \boldsymbol{\alpha}^{\mathrm{T}}\mathbf{u}$$

subject to

$$\frac{\partial L}{\partial \mathbf{c}} = 0 \qquad \frac{\partial L}{\partial \mathbf{u}} = 0 \qquad \boldsymbol{\alpha} \geq 0 \qquad \boldsymbol{\beta} \geq 0$$

Now $0 = \partial L/\partial \mathbf{c} = \mathbf{K}\mathbf{c} - \mathbf{K}\mathbf{Y}\boldsymbol{\alpha}$ yields

$$\mathbf{c} = \mathbf{Y}\boldsymbol{\alpha} \tag{4.22}$$

Further we have by $\partial L/\partial \mathbf{u} = 0$ that $\boldsymbol{\beta} = \lambda \mathbf{e} - \boldsymbol{\alpha}$. Thus, our optimization problem becomes

$$\max_{\alpha}\left(-\tfrac{1}{2}\boldsymbol{\alpha}^{\mathrm{T}}\mathbf{Y}\mathbf{K}\mathbf{Y}\boldsymbol{\alpha} + \mathbf{e}^{\mathrm{T}}\boldsymbol{\alpha}\right) \tag{4.23}$$

subject to

$$0 \leq \boldsymbol{\alpha} \leq \lambda \mathbf{e}$$

This QP problem is usually solved in the SVM literature. The *support vectors* (SVs) are those training patterns $\mathbf{x}_i$ for which α_i does not vanish. Let I denote the index set of the support vectors $I := \{i \in \{1, \dots, M\} : \alpha_i \neq 0\}$; then by (4.20) and (4.22), the function f has the sparse representation

$$f(\mathbf{x}) = \sum_{i \in I} c_i K(\mathbf{x}_i, \mathbf{x}) = \sum_{i \in I} y_i \alpha_i K(\mathbf{x}_i, \mathbf{x})$$

which depends only on the SVs. With respect to the margin we obtain by (4.9) and that

$$\gamma = (\|f\|_{\mathcal{H}_K})^{-1} = (\mathbf{c}^T \mathbf{K} \mathbf{c})^{-1/2} = \left(\sum_{i \in I} y_i \alpha_i f(\mathbf{x}_i) \right)^{-1/2}$$

Due to the Kuhn–Tucker conditions [50] the solution f of the QP problem (4.21) has to fulfill

$$\alpha_i (1 - y_i f(\mathbf{x}_i) - u_i) = 0 \qquad i = 1, \ldots, M$$

In case of hard-margin classification with $u_i = 0$ this implies that $y_i f(\mathbf{x}_i) = 1, i \in I$, so that we obtain the following simple expression for the margin:

$$\gamma = \left(\sum_{i \in I} \alpha_i \right)^{-1/2} \tag{4.24}$$

REFERENCES

1. D. L. JEWETT. Volume conducted potentials in response to auditory stimuli as detected by averaging in the cat. *Electroenceph. Clin. Neurophysiol.*, 28:609–618, 1970.

2. R. A. DOBIE AND C. I. BERLIN. Binaural interaction in brainstem evoked response. *Arch. Otolaryngol.*, 105:391–398, 1979.

3. H. L. HOSFORD, B. C. FULLERTON, AND R. A. LEVINE. Binaural interaction in human and cat brainstem auditory responses. *Acoust. Soc. Am.*, 65(Suppl. 1):86, 1979.

4. R. A. DOBIE AND S. J. NORTON. Binaural interaction in human auditory evoked potentials. *Electroencephal. Clin. Neurophysiol.*, 49:303–313, 1980.

5. M. FURST, R. A. LEVINE, AND P. M. McGAFFIGAN. Click lateralization is related to the β-component of the dichotic brainstem auditory evoked potentials of human subjects. *J. Acoust. Soc. Am.*, 78:1644–1651, 1985.

6. K. BRANTBERG, H. HANSSON, P. A. FRANSSON, AND U. ROSENHALL. The binaural interaction component in human ABR is stable within the 0 to 1 ms range of interaural time difference. *Audiol. Neurootol.*, 4:8–94, 1997.

7. K. BRANTBERG, P. A. FRANSSON, H. HANSSON, AND U. ROSENHALL. Measures of the binaural interaction component in human auditory brainstem response using objective detection criteria. *Scand. Audiol.*, 28:15–26, 1999.

8. V. K. GOPAL AND K. PIEREL. Binaural interaction component in children at risk for central auditory processing disorders. *Scand. Audiol.*, 28:77–84, 1999.

9. W. DELB, D. J. STRAUSS, G. HOHENBERG, AND K. P. PLINKERT. The binaural interaction component in children with central auditory processing disorders. *Int. J. Audiol.*, 42:401–412, 2003.

10. American Speech–Language–Hearing Association. Central auditory processing: Current status of research and implications for clinical practice. Task force on central auditory processing disorders consensus development. *Am. J. Audiol.*, 5:41–54, 1996.

11. J. JERGER, K. JOHNSON, S. JERGER, N. COKER, F. PIROZZOLO, AND L. GRAY. Central auditory processing disorder: A case study. *J. Am. Acad. Audiol.*, 2:36–54, 1991.

12. J. JERGER, R. CHMIEL, R. TONINI, E. MURPHY, AND M. KENT. Twin study of central auditory processing disorder. *J. Am. Acad. Audiol.*, 10:521–528, 1999.

13. American Academy of Audiology. Consensus Conference on the Diagnosis of Auditory Processing Disorders in School–Aged Children, Dallas, TX, April 2000.

14. M. H. STOLLMANN, A. F. SNIK, G. C. HOMBERGEN, AND R. NIEWENHUYS. Detection of binaural interaction component in auditory brainstem responses. *Br. J. Audiol.*, 30:227–232, 1996.

15. R. E. BELLMAN. *Adaptive Control Process*. Princeton University Press, Princeton, NJ, 1961.

16. L. J. TREJO AND M. J. SHENSA. Feature extraction of event–related potentials using wavelets: An application to human performance monitoring. *Brain and Language*, 66:89–107, 1999.

17. K. ENGLEHART, B. HUDGINS, P. A. PARKER, AND M. STEVENSON. Classification of the myoelec-tric signal using time–frequency based representations. *Med. Eng. Phys.*, 21:431–438, 1999.

18. G. RUTLEDGE AND G. McLEAN. Comparison of several wavelet packet feature extraction algorithms. *IEEE Trans. Pattern Recognition Machine Intell.*, submitted for publication.

19. N. SAITO AND R. R. COIFMAN. Local discriminant bases. In A. F. LAINE AND M. A. UNSER, Ed., *Wavelet Applications in Signal and Image Processing*, vol. II, A. F. LAINE and M. A. UNSER, eds., Proceedings of SPIE, San Diego, CA, 27–29, July 1994, vol. 2303.

20. R. R. COIFMAN AND M. V. WICKERHAUSER. Entropy based algorithms for best basis selection. *IEEE. Trans. Inform. Theory*, 32:712–718, 1992.

21. P. H. DELSARTE, B. MACQ, AND D. T. M. SLOCK. Signal adapted multiresolution transforms for image coding. *IEEE Trans. Inform. Theory*, 38:897–903, 1992.

22. P. MOULIN AND K. MIHÇAK. Theory and design of signal-adapted FIR paraunitary filter banks. *IEEE Trans. Signal Process.*, 46:920–929, 1998.

23. P. VAIDYANATHAN AND S. AKKARAKARAN. A review of the theory and applications of principal component filter banks. *J. Appl. Computat. Harmonic Anal.*, 10:254–289, 2001.

24. D. J. STRAUSS, J. JUNG, A. RIEDER, AND Y. MANOLI. Classification of endocardial electrograms using adapted wavelet packets and neural networks. *Ann. Biomed. Eng.*, 29:483–492, 2001.

25. D. J. STRAUSS AND G. STEIDL. Hybrid wavelet-support vector classification of waveforms. *J. Computat. Appl. Math.*, 148:375–400, 2002.

26. D. J. STRAUSS, G. STEIDL, AND W. DELB. Feature extraction by shape-adapted local discriminant bases. *Signal Process.*, 83:359–376, 2003.

27. D. J. STRAUSS, W. DELB, AND P. K. PLINKERT. A time-scale representation of binaural interaction components in auditory brainstem responses. *Comp. Biol. Med.* 34:461–477, 2004.

28. D. J. HECKER, W. DELB, F. CORONA, AND D. J. STRAUSS. Possible macroscopic indicators of neural maturation in subcortical auditory pathways in school-age children. In *Proceedings of the 28th Annual International of the IEEE Engineering in Medicine and Biology Society*, September 2006, New York, NY, 1173–1179, 2006.

29. D. J. STRAUSS, W. DELB, AND P. K. PLINKERT. Identification of central auditory processing disorders by scale and entropy features of binaural auditory brainstem potentials. In *Proceedings of the First International IEEE EMBS Conference on Neural Engineering*, Capri Island, Italy, IEEE, 2003, pp. 410–413.

30. D. J. STRAUSS, W. DELB, AND P. K. PLINKERT. Objective detection of the central auditory processing disorder: A new machine learning approach. *IEEE Trans. Biomed. Eng.*, 51:1147–1155, 2004.

31. V. VAPNIK. *The Nature of Statistical Learning Theory*. Springer, New York, 1995.

32. G. WAHBA. Support vector machines, reproducing kernel Hilbert spaces and the randomized GACV. In B. SCHÄOLKOPF, C. BURGES, AND A. J. SMOLA, Eds., *Advances in Kernel Methods—Support Vector Learning*. MIT Press, Cambridge, MA, 1999, pp. 293–306.

33. E. P. SIMONCELLI, W. T. FREEMAN, E. H. ADELSON, AND D. J. HEGGER. Shiftable multiscale transforms. *IEEE Trans. Inform. Theory*, 38:587–608, 1992.

34. Z. CVETKOVIĆ AND M. VETTERLI. Oversampled filter banks. *IEEE Trans. Signal Process.*, 46:1245–1255, 1998.

35. W. DELB, D. J. STRAUSS, AND K. P. PLINKERT. A time–frequency feature extraction scheme for the automated detection of binaural interaction in auditory brainstem responses. *Int. J. Audiol.*, 43:69–78, 2004.

36. S. J. JONES AND J. C. VAN DER POEL. Binaural interaction in the brain stem auditory evoked potential: Evidence for a delay line coincidence detection mechanism. *Electroenceph. Clin. Neurophysiol.*, 77:214–224, 1990.

37. L. A. JEFFRESS. A place theory of sound localization. *J. Comp. Physiol. Psychol.*, 41:35–39, 1948.

38. S. YOUNG AND E. W. RUBEL. Frequency specific projections of individual neurons in chick auditory brainstem nuclei. *J. Neurosci.*, 7:1373–1378, 1983.

39. P. UNGAN, S. YAGCIOGLU, AND B. ÖZMEN. Interaural delay-dependent changes in the binaural difference potential in cat auditory brainstem response: Implications about the origin of the binaural interaction component. *Hear. Res.*, 106:66–82, 1997.

40. J. BREEBAART. Binaural processing model based on contralateral inhibition. I. Model structure. *J. Acoust. Soc. Am.*, 110:1074–1088, 2001.

41. V. AHARONSON AND M. FURST. A model for sound lateralization. *J. Acoust. Soc. Am.*, 109:2840–2851, 2001.

42. T. C. T. YIN AND J. C. CHAN. Interaural time sensitivity in the medial superior olive of the cat. *J. Neurophysiol.*, 645:465–488, 1990.

43. J. K. MOORE. Organization of the superior olivary complex. *Micros. Res. Tech.*, 51:403–412, 2000.

44. B. GROTHE. The evolution of temporal processing in the medial superior olive, an auditory brainstem structure. *Prog. Neurobiol.*, 61:581–610, 2000.

45. A. BRAND, O. BEHREND, T. MARQUARDT, D. MCALPINE, AND B. GROTHE. Precise inhibition is essential for microsecond interaural time difference coding. *Nature*, 417:543–547, 2002.

46. D. MCALPINE AND B. GROTHE. Sound localization and delay lines—Do mammals fit the model? *Trends Neurosci.*, 26:347–350, 2003.

47. H. CAI, L. H. CARNEY, AND H. S. COLBURN. A model for binaural response properties of inferior colliculus neurons. II. A model with interaural time difference-sensitive excitatory and inhibitory inputs and an adaptation mechanism. *J. Acoust. Soc. Am.*, 103:494–506, 1998.

48. G. STRANG AND T. NGUYEN. *Wavelets and Filter Banks*. Wellesley–Cambridge Press, Wellesley, MA, 1996.

CHAPTER 5

FUNCTIONAL CHARACTERIZATION OF ADAPTIVE VISUAL ENCODING

Nicholas A. Lesica and Garrett B. Stanley

5.1 INTRODUCTION

Our visual system receives, encodes, and transmits information about the outside world to areas of our brain that process the information and govern our behavior. Despite decades of research, the means by which these tasks are performed by the underlying neuronal circuitry remain a mystery. This lack of understanding is due in part to the overwhelming complexity of the neuronal circuitry as well as to the elusive nature of the encoding that results from adaptive mechanisms which constantly alter the function of the circuitry based on current external conditions. The ubiquitous nature of adaptive mechanisms throughout the nervous system leads one to infer that adaptive encoding may indeed be a guiding principle of information transmission in the brain. Investigations of visual encoding, therefore, must directly address these adaptive mechanisms to develop an understanding of their basic neurobiological function as well as the implications of their function on the design of engineered interfaces that seek to enhance or replace neuronal function lost to trauma or disease [1, 2].

The functional characterization of encoding in the visual system was first outlined in the pioneering work of Hartline, Barlow, and Kuffler [3–5], who described the relationship between the intensity of the visual stimulus projected onto the photoreceptors of the retina and the firing rate of downstream neurons within the retina. A systematic approach to functional characterization was subsequently provided by Marmarelis and Marmarelis [6], involving the determination of a series of filters (linear or nonlinear) that describes the responses of visual neurons to a white-noise stimulus designed to efficiently probe the system in question. This approach has been used extensively to characterize the basic function of neurons in a variety of visual and nonvisual sensory areas [7–9]. However, several assumptions are generally involved in this approach, namely that the stimulus is stationary (drawn from a fixed statistical distribution) and the encoding properties of the neuron are time invariant. While these assumptions may be valid under artificial laboratory conditions, studies of visual responses under natural conditions have revealed encoding strategies by which they are directly violated.

In a natural setting, the statistical distribution of the stimulus is constantly changing. For example, the mean intensity of light incident upon the retina can vary over many orders of magnitude as a result of changes in illumination or eye movements across the

Handbook of Neural Engineering. Edited by Metin Akay

visual scene. However, over any short interval, the distribution of light intensities will only occupy a small subset of this range. Because of this nonstationarity, the visual system must employ adaptive encoding strategies to optimize the processing of any one subset of stimuli without sacrificing the ability to process another. For example, retinal neurons maintain a relatively small operating range which is shifted and scaled to maximize differential sensitivity over the current statistical distribution of the stimulus [10], enabling maximal flow of visual information to downstream neurons [11, 12].

Initial reports of adaptive encoding focused primarily on the modulation of gain, demonstrating changes in the sensitivity of the response to the intensity of the visual stimulus. This adaptation was observed in response to changes in both the mean and variance of the stimulus on both fast (millisecond) and slow (second) time scales [10, 13–15]. In addition to changes in gain, recent investigations have also revealed further effects of adaptation, such as changes in spatial and temporal filtering properties [16, 17] and modulation of the baseline membrane potential [18, 19], which are also thought to play important roles in visual processing under natural conditions. Models of visual encoding that are based on responses to stationary stimuli are insufficient to characterize visual function in the natural environment, as they do not reflect the function of these adaptive mechanisms and reflect only the average behavior of the system over the interval of investigation. In this chapter, a new framework for the functional characterization of adaptive visual encoding under nonstationary stimulus conditions via adaptive estimation is developed. Within this framework, the function of multiple adaptive encoding mechanisms can be isolated and uniquely characterized.

The remainder of the chapter is structured as follows: In Section 5.2, a simple model of adaptive visual encoding consisting of a cascade of a time-varying receptive field (RF) and a rectifying static nonlinearity is developed. In Section 5.3, a recursive least-squares (RLS) approach to adaptively estimating the parameters of the time-varying RF from stimulus/response data is presented. In Section 5.4, the shortcomings of RLS are identified and an extended recursive least-squares (ERLS) approach with an adaptive learning rate is developed to provide improved tracking of a rapidly changing RF. In Section 5.5, the encoding model is expanded to include an adaptive offset before the static nonlinearity that serves to modulate the operating point of the neuron with respect to the rectification threshold, and the effects of changes in this operating point on the estimation of the parameters of the RF are investigated. In Section 5.6, the results are summarized and directions for future research are suggested.

5.2 MODEL OF VISUAL ENCODING

The framework for the analysis of adaptive encoding developed in this chapter is based on the properties of neurons in the early visual pathway. However, the general nature of the framework ensures that the concepts apply to later stages of processing in the visual pathway as well as to other sensory systems. The mapping from stimulus light intensity to firing rate response in a visual neuron can be represented by a cascade of a linear filter and a rectifying static nonlinearity. Cascade encoding models have been shown to provide accurate predictions of the responses of visual neurons under dynamic stimulation [20–22]. A schematic diagram of a cascade encoding model is shown in Figure 5.1. The components of the encoding model are intended to correspond to underlying neural mechanisms. However, because the model is functional in nature and designed to characterize firing rate responses to visual stimuli rather than modulations in membrane potential due to synaptic input currents, the correspondence between model parameters and intracellular quantities is indirect.

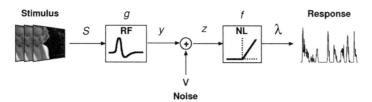

Figure 5.1 Simple cascade model of visual encoding. The spatiotemporal visual stimulus s is passed through a time-varying linear filter g (the spatiotemporal RF) to yield the intermediate signal y. This signal is then combined with additive, independent, Gaussian noise v to yield the generating function z and passed through a rectifying static nonlinearity f to produce the nonnegative firing rate λ. (Adapted, with permission, from [33]. Copyright © 2005 by IEEE.)

The input to the cascade encoding model is the spatiotemporal signal $s[p,n]$. For computer-driven visual stimuli discretized in space–time, p represents the grid index of a pixel on the screen and n is the time sample. (Note that pixel refers not to the atomic display units of the monitor but, for instance, to squares in a white-noise checkerboard.) No assumptions are made about the statistics of the stimulus, as natural signals are often nonstationary and correlated. To produce the intermediate signal y, which reflects the stimulus-related modulations in the membrane potential of the neuron, the stimulus is passed through the linear filter $g_n[p,m]$ (convolution in time, integration in space) representing P (total pixels in stimulus) separate temporal filters each with M parameters. This filter is known as the spatiotemporal RF and captures the spatial and temporal integration of the stimulus that occurs within the visual pathway. The subscript n denotes the time-varying nature of the RF. If s and g_n are organized appropriately, then this discrete-time operation can be written as a dot product $y[n] = s_n^{\mathrm{T}} g_n$, where s_n and g_n are the column vectors:

$$s_n = [s[P, n-M+1], s[P-1, n-M+1], \ldots, s[1, n-M+1],$$
$$s[P, n-M+2], \ldots, s[1,n]]^{\mathrm{T}}$$

$$g_n = [g_n[P,M], g_n[P-1,M], \ldots, g_n[1,M], g_n[P, M-1], \ldots, g_n[1,1]]^{\mathrm{T}}$$

and T denotes matrix transpose.

Before being passed through the static nonlinearity, the filtered stimulus y is combined with additive noise v to yield z, which is known as the *generating function*. The noise v represents spontaneous fluctuations in the membrane potential of the neuron, which are reflected in the variability of firing rate responses to repeated presentations of the same stimulus. These fluctuations have been shown to be uncorrelated over time and independent of the membrane potential of the neuron, with a distribution that is approximately Gaussian [23]. Thus, the noise v is assumed to be independent of the stimulus, nonstationary, and Gaussian with zero mean $\mathcal{N}(0, \sigma_v^2[n])$.

The generating function z is passed through a static nonlinearity $f(\cdot)$ to yield the nonnegative firing rate λ. This static nonlinearity captures the rectifying properties of the transformation from the membrane potential of the neuron to its observed firing rate. The adaptive estimation framework presented below is developed for a general static nonlinearity $f(\cdot)$ and the specific form of the function should be chosen based on the properties of the system under investigation. A common model for the static nonlinearity present in visual neurons is linear half-wave rectification [24]:

$$f(z) = \begin{cases} z & z \geq 0 \\ 0 & z < 0 \end{cases} \tag{5.1}$$

The linear rectifying static nonlinearity implies that the neuron is silent when the membrane potential is below a threshold (in this case zero) and that modulations in the membrane potential above that threshold are reflected as proportional modulations in firing rate.

5.3 RECURSIVE LEAST-SQUARES ESTIMATION

Given the model structure described in the previous section, there are a variety of ways in which observations of stimulus–response data can be used to identify the parameters of the RF that provide the best functional characterization of a given neuron. The traditional approach is to estimate a single time-invariant RF using linear least-squares estimation (also known as reverse correlation). However, a time-invariant RF captures only the average behavior of the neuron and is insufficient to describe the adaptive encoding of nonstationary stimuli. Thus, to accurately characterize adaptive visual encoding in a natural setting where the statistics of the stimulus are constantly changing, an adaptive estimation approach must be employed.

The basic premise of adaptive estimation is that the estimate of the model parameters at time $n + 1$ is computed by combining the previous estimate from time n with an update based on the observation of the response at time $n + 1$ and prior knowledge of the evolutionary dynamics of the parameters. One particular form of adaptive estimation that has been used successfully to characterize the encoding properties of visual neurons is RLS [25, 26]. In fact, RLS was designed to provide an on-line estimate of time-invariant model parameters, but its design also allows for the tracking of time-varying parameters [27]. The RLS algorithm is based on the minimization of the prediction error, which is defined as the difference between the observed firing rate in an interval and the expected firing rate in the interval given the current stimulus and the estimated model parameters. A derivation of the RLS approach to estimating the parameters of the RF based on the model shown in Figure 5.1 is given in the Appendix.

We have previously detailed an RLS technique for estimating the RFs of visual neurons [25]. Assuming the model structure shown in Figure 5.1, the RLS algorithm for the estimation of the RF parameters can be written as follows:

$$e[n] = \lambda[n] - f(s_n^T \hat{g}_{n|n-1}) \qquad \text{Prediction error}$$

$$G_n = \frac{\gamma^{-1} K_{n|n-1} s_n}{\gamma^{-1} s_n^T K_{n|n-1} s_n + 1} \qquad \text{Update gain}$$

$$\hat{g}_{n+1|n} = \hat{g}_{n|n-1} + G_n e[n] \qquad \text{Update parameter estimates}$$

$$K_{n+1|n} = \gamma^{-1} K_{n|n-1} - \gamma^{-1} G_n s_n^T K_{n|n-1} \qquad \text{Update inverse of stimulus autocovariance}$$

where $0 \ll \gamma \le 1$ serves to downweight past information and is therefore often called the *forgetting factor*. At each time step, the gain G is calculated based on the estimate of the inverse of the stimulus autocovariance matrix, denoted as K, and combined with the prediction error e to update the RF parameter estimate $\hat{g}$. Note, however, that the recursive framework avoids the explicit inversion of the stimulus autocovariance matrix. The subscript $n|n-1$ denotes an estimate at time n given all observations up to and including time $n - 1$.

The prediction error e is the difference between the observed and predicted firing rates. Given the stimulus and the current estimate of the RF, the expected firing rate is

$$
\begin{aligned}
E\{\lambda[n]|s_n, \hat{g}_{n|n-1}\} &= \int_\lambda \lambda[n] p(\lambda[n]|s_n, \hat{g}_{n|n-1}) \, d\lambda[n] \\
&= \int_v f(s_n^{\mathrm{T}} \hat{g}_{n|n-1} + v[n]) p(v[n]) \, dv[n]
\end{aligned}
\tag{5.2}
$$

where $p(\lambda[n]|s_n, \hat{g}_{n|n-1})$ is the probability density function of the predicted response conditioned on the current stimulus and estimated model parameters. For small v relative to $s_n^{\mathrm{T}} \hat{g}_{n|n-1}$, the expectation can be approximated as $E\{\lambda[n]|s_n, \hat{g}_{n|n-1}\} \approx f(s_n^{\mathrm{T}} \hat{g}_{n|n-1})$ through a series expansion about $s_n^{\mathrm{T}} \hat{g}_{n|n-1}$. This approximation is valid when the signal-to-noise ratio (SNR) is large, as is typically the case in visual neurons under dynamic stimulation. In the event that this approximation is not valid, the integral expression for the expected firing rate can be evaluated at each time step.

5.4 EXTENDED RECURSIVE LEAST-SQUARES ESTIMATION

The dynamics of the system shown in Figure 5.1 can be represented by the following state-space model:

$$
g_{n+1} = F_n g_n + q_n
\tag{5.3}
$$

$$
\lambda[n] = f(s_n^{\mathrm{T}} g_n + v[n])
\tag{5.4}
$$

where F_n (known as the *state evolution matrix*) and q_n (known as the *state evolution noise*) specify the deterministic and stochastic components of the evolutionary dynamics of the RF parameters, respectively. Investigation of the model underlying RLS reveals that the technique is designed to estimate time-invariant RF parameters ($F_n = I$ and $q_n = 0$) based on the assumption that the variance of the noise in the observed response decreases exponentially over time ($\sigma_v^2[n] \propto \gamma^n$) [28]. This assumption causes the current observations to be weighted more heavily in the computation of the parameter estimates than those in the past and provides RLS with the ability to track a slowly varying RF. However, as the parameters change more quickly, the ability of RLS to track them decreases. The tracking behavior of the RLS algorithm can be greatly improved by assuming a model in which the RF parameters are time varying and their evolution is treated as a more general stochastic process.

The optimal algorithm for tracking the time-varying parameters of a state-space model, in terms of minimizing the mean-squared error (MSE) between the predicted and observed responses, is the Kalman filter [29]. However, implementation of the Kalman filter requires exact knowledge of the quantities $F_n, \Sigma_q[n]$ (the covariance of the state evolution noise), and $\sigma_v^2[n]$, which are generally unknown during the experimental investigation of neural systems, and noise properties that cannot be guaranteed for the dynamics in question. Fortunately, some of the tracking ability of the Kalman filter can be transferred to the RLS framework by replacing the deterministic model which underlies RLS estimation with a stochastic one that approximates that which underlies the Kalman filter. The result, known as the ERLS, was developed in [28] based on the correspondence between RLS and the Kalman filter presented in [30]. Here, a particular form of ERLS is developed that is designed to track adaptation of visual encoding properties in response to changes in the statistical properties of the stimulus.

The model underlying the Kalman filter assumes that the RF parameters evolve according to the general model $g_{n+1} = F_n g_n + q_n$, where q_n is a vector of nonstationary Gaussian white-noise $\mathcal{N}(0, \Sigma_q[n])$. To simplify the incorporation of this model into the RLS framework, assume that the parameter evolution is completely stochastic ($F_n = I$) and that the parameters evolve independently and at equal rates ($\Sigma_q[n] = \sigma_q^2[n]I$). In this stochastic model, the evolution of the parameter estimates is constrained only by the variance $\sigma_q^2[n]$, and this parameter can be used to control the tracking behavior of the algorithm based on knowledge of the underlying system. When the prediction error is likely to be the result of changing encoding properties (during adaptation to changes in the statistics of the stimulus), a large value of $\sigma_q^2[n]$ is used to allow the estimate to track these changes. Conversely, if the parameters are not likely to be changing, a small value of $\sigma_q^2[n]$ is used to avoid tracking the noise in the observed response. Thus, $\sigma_q^2[n]$ functions as an adaptive learning rate.

The value of $\sigma_q^2[n]$ is adjusted based on knowledge of how features of the stimulus affect the parameters. For example, adaptation in the visual system generally occurs in the interval directly following a change in a feature of the stimulus (mean, variance, etc.). With this knowledge, $\sigma_q^2[n]$ is increased following a stimulus transition, allowing the parameter estimate to change quickly. Similarly, if the statistics of the stimulus have been stationary for some time and the underlying parameters are not likely to be adapting, $\sigma_q^2[n]$ is decreased. The dynamics of the adaptive learning rate should be based on knowledge of the adaptive properties of the system under investigation, reflecting the expected rate of change (per estimation time step) of the parameters under the given stimulus conditions. For a situation where the relevant stimulus feature or the adaptive properties are not known a priori, $\sigma_q^2[n]$ should be set to a relatively small constant value throughout the trial. This gives the estimate some degree of adaptability (although very fast changes in parameter values will likely be missed), while keeping the steady-state noise level in a reasonable range. The initial estimate provides some information about the adaptive behavior of the system, and the estimation can be performed again with a more appropriate choice of $\sigma_q^2[n]$.

The ERLS algorithm for the model in Figure 5.1 is as follows:

$$e[n] = \lambda[n] - f(s_n^T \hat{g}_{n|n-1}) \qquad \text{Prediction error}$$

$$G_n = \frac{K_{n|n-1} s_n}{s_n^T K_{n|n-1} s_n + 1} \qquad \text{Update gain}$$

$$\hat{g}_{n+1|n} = \hat{g}_{n|n-1} + G_n e[n] \qquad \text{Update parameter estimates}$$

$$K_{n+1|n} = K_{n|n-1} - G_n s_n^T K_{n|n-1} + \sigma_q^2[n]I \qquad \text{Update inverse of stimulus}$$
$$\text{autocovariance}$$

Again, the estimate is generated by solving the above equations sequentially at each time step. To initialize the algorithm, the initial conditions $\hat{g}_{0|-1} = 0$ and $K_{0|-1} = \delta \times I$ are used. The regularization parameter δ affects the convergence properties and steady-state error of the ERLS estimate by placing a smoothness constraint on the parameter estimates, removing some of the error introduced by highly correlated natural stimuli [31]. For estimation from responses to uncorrelated white-noise stimuli in the examples below, δ was set to 10^{-4}. For estimation from responses to correlated naturalistic stimuli, δ was set to 10^{-2}.

5.4.1 Examples

In the following simulations, examples of adaptive encoding in retinal ganglion cells are used to demonstrate the ability of ERLS to track changes in RF parameters during nonstationary stimulation. Ganglion cells are the output neurons of the retina and provide the only pathway for the transmission of visual information from the retina to the brain. Responses to nonstationary stimuli were simulated using the cascade encoding model shown in Figure 5.1, which has been shown to provide accurate predictions of ganglion cell responses [19]. In the first example, the tracking performance of ERLS is compared to that of standard RLS during a contrast switching experiment. In the second example, ERLS is used to track adaptive RF modulations from responses to a naturalistic stimulus in which the contrast is constantly varying. Note that the examples presented here utilize realistic simulations of retinal ganglion cells which provide carefully controlled scenarios for the investigation of the adaptive estimation techniques. For examples of the application of these techniques to experimentally recorded responses in the retina, thalamus, and cortex, see our previously published findings [25, 32–34].

5.4.1.1 *Comparison of RLS and ERLS*   A biphasic temporal RF typical of retinal ganglion cells with a time course of 300 ms was used in the cascade encoding model shown in Figure 5.1 to simulate responses to a contrast switching, spatially uniform Gaussian white-noise stimulus. Note that this can be directly interpreted as the impulse response of the system, mapping the visual stimulus intensity to the modulations in neuronal firing rate. A new luminance value for the stimulus was chosen every 10 ms and the root-mean-square (RMS, ratio of standard deviation to mean) contrast was switched from 0.05 to 0.30 every 10 ms. Contrast gain control has been shown to modulate the RF gain of visual neurons in response to changes in stimulus contrast, with a time course that is approximately equal to the integration time of the neuron [13, 35]. To simulate the adaptive changes that have been observed experimentally, the gain of the RF (magnitude of peak value) was increased by a factor of 2 following a decrease in contrast and decreased by a factor of 2 following an increase in contrast. The variance of the noise v was adjusted to produce responses with an SNR of 5. This value is consistent with those measured in the experimental responses of retinal ganglion cells [33, 34]. ERLS and standard RLS were used to track changes in the RF parameters from the simulated responses at a temporal resolution of 30 ms. The results are shown in Figure 5.2.

Figure 5.2*a* shows the gain of the actual RF (gray) along with the gain of the RLS RF estimate (black). The RLS estimate was generated with forgetting factor $\gamma = 0.96$ (which corresponds to a memory time constant of approximately 7 s). This value was optimal in the sense that it yielded the lowest MSE in the RF estimate (10.4% of the variance of the actual RF) over the entire trial for all $0 \ll \gamma \leq 1$. Figure 5.2*b* shows the gain of the actual RF, along with the gain of the ERLS RF estimate, computed with a fixed learning rate $\sigma_q^2[n] = 10^{-5}$. This value of $\sigma_q^2[n]$ was also chosen to minimize the MSE in the RF estimate over the entire trial. The MSE in the ERLS estimate with fixed learning rate (7.6%) is lower than that of the optimal RLS estimate, illustrating the enhanced tracking ability that results from incorporating the stochastic model of parameter evolution. The tracking performance of the ERLS estimate can be further improved by using an adaptive learning rate to exploit the relationship between the stimulus and the adaptive nature of the system. Because contrast gain control only modulates the gain of the RF in the short interval following a change in contrast, the learning rate $\sigma_q^2[n]$ is set to a large value in those intervals to allow the estimate to adapt quickly and to a small value otherwise, so that noise in the observed response is not attributed to changes in the encoding properties of the neuron.

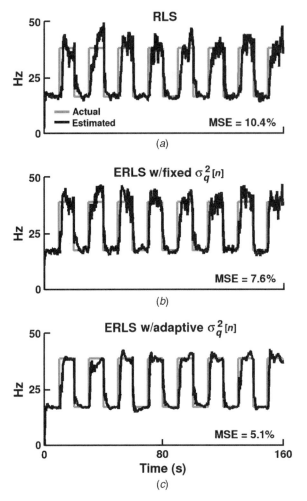

Figure 5.2   Comparison of RLS and ERLS. (*a*) Gain of actual RF (gray) and gain of RLS estimate (black) for 160-s segment of contrast switching white-noise stimulus. The value of the forgetting factor γ used to compute the RLS estimate was 0.96. The MSE in the RF estimate over the entire trial was 10.4% of the variance of the actual RF. (*b*) Gains of actual RF and ERLS estimate computed with $\sigma_q^2[n] = 10^{-5}$ for entire trial. The MSE in the RF estimate over the entire trial was 7.6%. (*c*) Gains of actual RF and ERLS estimate computed with $\sigma_q^2[n] = 10^{-4}$ for 1 s following each contrast transition and 10^{-6} at all other times. The MSE in the RF estimate over the entire trial was 5.1%. (Reproduced, with permission, from [33]. Copyright © 2005 by IEEE.)

Accordingly, $\sigma_q^2[n]$ was set to 10^{-4} during the transient intervals (1 s following each contrast transition) and 10^{-6} during steady-state intervals (all other times). The gain of the resulting RF estimate is shown in Figure 5.2*c*. The adaptive learning rate allows the estimate to closely track the fast changes in gain while maintaining a low steady-state error between transitions. The MSE in the ERLS estimate with adaptive learning rate (5.1%) is half of that in the standard RLS estimate. The values of $\sigma_q^2[n]$ used to generate the ERLS estimate with an adaptive learning rate were chosen based on the adaptive dynamics of the simulated neuron but were not optimized. Similar results were obtained with a range of values for $\sigma_q^2[n]$ during the transient and steady-state intervals (not shown), indicating the robust improvement in tracking provided by the adaptive learning rate.

5.4.1.2 *Tracking RF Changes during Natural Stimulation*

In a natural setting, tracking RF changes is complicated by the lack of clear transitions in the relevant stimulus features (as opposed to the contrast switching example above). The following contrast gain control simulation demonstrates the ability of ERLS to track RF modulation from responses to a stimulus which is continuously nonstationary. The stimulus was the temporal intensity of one pixel of a gray-scale natural-scene movie recorded in the forest with a home video camera, updated every 30 ms, as shown in Figure 5.3*a*.

For more details regarding the natural-scene movies, see [36]. For this example, a stimulus was chosen in which the mean intensity was relatively constant over time while the contrast was constantly changing. The response of a retinal ganglion cell was simulated as above. The gain of the temporal RF was varied inversely with the contrast of the stimulus and, thus, varied continuously throughout the trial. At each time step, the contrast was defined as the RMS contrast of the previous 300-ms segment of the stimulus, in accordance with the time course of contrast gain control (Fig. 5.3*b*), and the gain of the

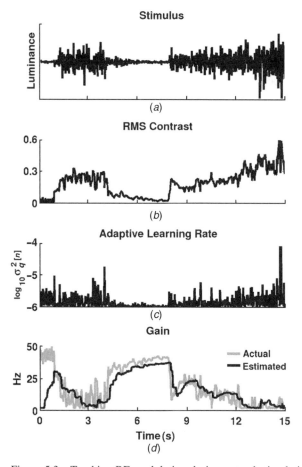

Figure 5.3   Tracking RF modulation during natural stimulation with ERLS. (*a*) The stimulus was spatially uniform and the luminance was modulated according to the intensity of a typical pixel in a natural-scene movie, updated every 30 ms. (*b*) The RMS contrast of natural stimulus throughout trial. (*c*) Value of learning rate $\sigma_q^2[n]$ throughout trial. (*d*) Gain of actual RF during simulation (gray) and gain of ERLS RF estimate (black). (Reproduced, with permission, from [33]. Copyright © 2005 by IEEE.)

RF was set to twice the inverse of the contrast. Because the contrast of the stimulus was constantly varying and transitions were not well defined, the value of the adaptive learning rate $\sigma_q^2[n]$ was proportional to the derivative of the stimulus contrast, as shown in Figure 5.3c. At each time step, $\sigma_q^2[n]$ was defined as 10^{-4} times the absolute value of the first-order difference in the contrast of the stimulus. Figure 5.3d shows the results of the estimation. The gain of the ERLS RF estimate (black) closely tracks that of the actual RF (gray). Aside from the error associated with the initial conditions and some of the very fast transients, the ERLS RF estimate captures most of the gain changes in the actual RF.

5.5 IDENTIFICATION OF MULTIPLE ADAPTIVE MECHANISMS

As demonstrated in the previous section, ERLS provides accurate tracking of RF changes during nonstationary stimulation. In addition to the RF, recent studies of adaptive function in the visual system have demonstrated a second locus of adaptation. Intracellular recordings of retinal, thalamic, and cortical responses to nonstationary stimuli have revealed modulation of the baseline membrane potential on both fast (millisecond) and slow (second) time scales in responses to changes in features of the stimulus [18, 19, 37]. These changes have an important functional role, as the baseline membrane potential determines the size of the stimulus that is necessary to evoke a spike response, thereby setting the *operating point* of the neuron with respect to the spike threshold. For example, the same stimulus and RF can result in a high firing rate if the membrane is depolarized (and the potential is already close to the spike threshold), or no spikes at all if the membrane is hyperpolarized.

To reflect these changes in baseline membrane potential, the cascade encoding model must be expanded. In the cascade model shown in Figure 5.4, an offset θ is added to the filtered stimulus y before the static nonlinearity. This offset shifts the operating point of the model with respect to the rectification threshold, capturing the effects of changes in the baseline membrane potential. It is important to note that the offset captures only those changes in the membrane potential that are not accounted for by the filtering of the visual stimulus in the RF. For example, although a decrease in the mean of the stimulus would result in a direct decrease in the mean of the membrane potential, this change would be reflected in the filtered stimulus y, not in the offset. However, if this decrease in the

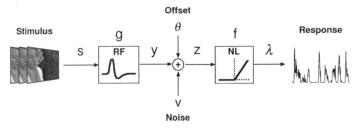

Figure 5.4 Expanded model of visual encoding. The spatiotemporal visual stimulus s is passed through a time-varying linear filter g (the spatiotemporal RF) to yield the intermediate signal y. This signal is then combined with additive, independent, Gaussian noise v and and time-varying offset θ to yield the generating function z and passed through a rectifying static nonlinearity f to produce the nonnegative firing rate λ. (Adapted, with permission, from [33]. Copyright © 2005 by IEEE.)

mean of the stimulus also causes a change in the baseline membrane potential via some indirect adaptive mechanism, that change would be reflected in the offset.

To track adaptive changes in both the RF and the offset, the ERLS technique described in the previous section must be expanded. If the offset is not included in the estimation process, its interactions with the static nonlinearity can influence the estimation of the RF. Because the RF and offset may be changing simultaneously, it is imperative that the estimation technique be able to identify changes in model parameters uniquely. If the model structure underlying the parameter estimation process is misspecified by neglecting the offset (or assuming it to be zero), changes in the baseline membrane potential, or even in the statistics of the input, can be reflected as changes in the gain of the RF.

Consider a reduced encoding model defined by the mapping from the stimulus s to the generating function z (with zero offset) via a time-invariant RF g. Based on observations of s and z, the linear least-squares RF estimate $\hat{g}_1$ that minimizes the MSE between the predicted generating function $\hat{z}$ and the actual generating function z is given by $\hat{g}_1 = \Phi_{ss}^{-1}\phi_{sz}$, where Φ_{ss} is the Toeplitz matrix of the stimulus autocovariance at different time lags and ϕ_{sz} is the cross covariance between the stimulus and response [6]. In the absence of noise, the estimate $\hat{g}_1$ will equal the actual RF g, and in the presence of noise, $\hat{g}_1$ will converge to g as more data are observed. However, when the observed response is not the generating function z but, for example, the rectified firing rate λ, there is a mismatch between the model assumed in linear least-squares estimation and the actual system.

The mapping from s to λ consists of a cascade of two elements, the linear RF g and the static nonlinearity f. Because the generating function z undergoes additional processing in the static nonlinearity, the linear least-squares RF estimate from observations of s and λ, which is $\hat{g}_2 = \Phi_{ss}^{-1}\phi_{s\lambda}$, does not necessarily equal $\hat{g}_1$, the RF estimate from observations of s and z. In fact, according to the result of a theorem by Bussgang [38], $\hat{g}_2$ is a scaled version of $\hat{g}_1$. Bussgang's theorem states that the cross covariance between the input to a static nonlinearity and the output of a static nonlinearity, in this case $\phi_{z\lambda}$, is proportional to the autocovariance of the input to the static nonlinearity, in this case ϕ_{zz}. Thus, the linear least-squares estimate of the mapping from z to λ is a constant $C = \phi_{z\lambda}/\phi_{zz}$ and the best linear estimate of the two-element cascade mapping s to λ is $\hat{g}_2 = C\Phi_{ss}^{-1}\phi_{sz} = C\hat{g}_1$.

Consider $\hat{g}_1$, the RF estimated using linear least-squares from observations of the generating function z, and $\hat{g}_2$, the RF estimated using linear least-squares from observations of the rectified response λ. As we have previously described [25], the scaling constant C relating $\hat{g}_2$ to $\hat{g}_1$ is a function of the fraction of the distribution of z that is rectified. Assuming that the distribution of y (before offset) is symmetric with zero mean, then the fraction of the generating function z (after offset) that will be rectified is a function of the ratio of the offset to the standard deviation of z, θ/σ_z. For zero offset and a linear half-wave rectifying static nonlinearity, half of the generating function is rectified and the scaling constant C is equal to 0.5. Because the predicted response in the RLS and ERLS algorithms defined above is also rectified, the scaling in the RF estimate with zero offset is accounted for and the RF estimate matches the actual RF. However, because there is no offset in the model underlying the RLS and ERLS algorithms as defined above, the RF estimate is vulnerable to effects that result from nonzero offsets. For positive offsets, the scaling constant C approaches 1 as less of the signal is rectified, and for negative offsets, the scaling constant C approaches zero as more of the signal is rectified. Thus, estimation of the RF using the RLS or ERLS algorithms as defined above for nonzero offsets will yield a scaled version of the actual RF. It should be noted that, although the scaled version of the RF that results from estimating the RF without considering a nonzero offset minimizes the MSE between the actual response and the predicted response of the encoding model without an offset, the result is not functionally equivalent to the model containing the actual RF and offset.

When using the RLS or ERLS technique as defined above to estimate the RF of the neuron from observations of the rectified firing rate λ, changes in the ratio of the mean of z to its standard deviation will result in a change in the fraction of z that is rectified and apparent changes in the gain of the RF. When the offset is zero and the stimulus is zero mean, these effects are avoided as the ratio of the mean of z to its standard deviation is always zero, even if the standard deviation of the stimulus or the gain of the RF is changing. However, when the offset or the mean of the stimulus is nonzero, changes in the gain of the RF estimate can be induced by an actual change in the gain of the RF or by changes in the offset θ or the mean or standard deviation of the stimulus. This result has important implications for the analysis of adaptive encoding, as the stimulus is nonstationary, and changes in both the gain and offset of the neuron have been reported in experimental observations.

The confounding effects on the estimation of the RF caused by changes in the operating point of the neuron can be avoided by including the offset in the estimation process. The generating function (without noise) $z[n] = s_n^T g_n + \theta[n]$ can be written as the dot product $z[n] = \mathbf{s}_n^T \mathbf{g}_n = [s_n\ 1]^T[g_n\ \theta[n]]$. Because the parameter vector $\mathbf{g}_n$ is a linear function of the augmented stimulus vector $\mathbf{s}_n$, the RF and offset can be estimated simultaneously within the ERLS framework described above simply by substituting the augmented stimulus and parameter vectors $\mathbf{s}_n$ and $\mathbf{g}_n$ for the original stimulus and parameter vectors s_n and g_n.

5.5.1   Examples

In the following examples, simulated responses are used to understand the effects of changes in the operating point of the neuron on the estimation of the RF parameters. Both steady-state and adaptive responses are used to estimate RFs with and without simultaneous estimation of the offset. In both examples, estimation of the RFs without simultaneous estimation of the offset produces misleading results, while accurate estimates of the RF and offset are obtained when the expanded ERLS technique is used to estimate both simultaneously.

5.5.1.1   Effects of Operating Point on RF Estimation

The response of a retinal ganglion cell (RGC) to a single trial of spatially uniform, zero-mean, stationary white noise was simulated as described in Section 5.4. The simulated responses were used to estimate the parameters of the RF with and without simultaneous estimation of the offset. During the estimation process, the distribution of the generating function z is inferred, based on observations of the response and the structure of the underlying encoding model, and the gain of the RF estimate is based on the spread of this inferred distribution. For a given stimulus, a narrow distribution of z corresponds to an RF with a small gain, while a wide distribution of z corresponds to an RF with a large gain. Comparing the actual and inferred distributions of the generating function under various conditions can provide some insight into the effects of the offset on the estimation of the RF parameters. In Figure 5.5, the distributions of the actual generating function z of the model neuron (with $\theta/\sigma_z = -0.5$) are compared to the distributions of the inferred generating function $\hat{z}$, generated from RFs estimated with and without simultaneous estimation of the offset.

Figure 5.5a shows the distribution of the actual generating function z (gray) and the inferred generating function $\hat{z}$ (black) when the offset is not included in the estimation process. The fraction of the actual generating function that is present in the observed response, after offset and rectification, is shaded. Because the offset is neglected during

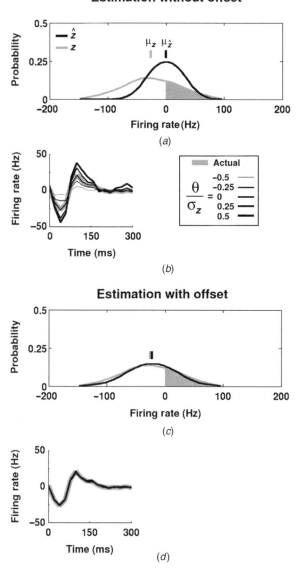

Figure 5.5   Effects of operating point on estimation of RF. The response of a RGC to spatially uniform, zero-mean, stationary white noise was simulated and the responses were used to estimate the RF of the simulated neuron, with and without simultaneous estimation of the offset.
(a) Probability distributions of actual generating function of simulated neuron z (gray) and inferred generating function $\hat{z}$ (black) generated by encoding model with RF estimated without simultaneous estimation of offset. The fraction of the actual generating function that is present in the observed response after offset and rectification is shaded. The mean of each distribution is indicated by the vertical bars above each distribution. (b) RF estimates (black, thickness corresponds to offset value, see legend) when RF is estimated without simultaneous estimation of offset for variety of offset values. The actual RF is also shown (gray). (c) Distributions of actual generating function and inferred generating function generated by encoding model with RF estimated while simultaneously estimating offset. (d) RF estimates when RF and offset are estimated simultaneously for same range of offset values presented in (b). (Adapted, with permission, from [34]. Copyright © 2006 by the Taylor and Francis Group.)

the estimation process, the mean of the inferred generating function $\mu_{\hat{z}}$ is constrained to be zero (for a zero-mean stimulus). Thus, the distribution of $\hat{z}$ is centered around zero, while the distribution of z is centered around the negative offset. Because the offset is not included in the estimation process and the generating function is assumed to be zero mean, the RF estimate is scaled as if exactly half of the actual generating function was rectified. When the actual offset is less than zero, as in this example, this results in an RF estimate with a gain that is smaller than that of the actual RF, while, when the actual offset is greater than zero, this results in an RF estimate with a gain that is larger than that of the actual RF. This is evidenced by the RF estimates shown in Figure 5.5*b*.

The RFs estimated without simultaneous estimation of the offset are shown for a variety of offset values (ratio of offset to standard deviation of generating function, θ/σ_z, between -0.5 and 0.5). The effects described above are visible in the scaling of the RF estimates (black, thickness corresponds to offset value) relative to the actual RF (gray). For zero offset, the gain of the RF estimate matches that of the actual RF. For nonzero offsets the effects of the interaction between the offset and the static nonlinearity are visible as a scaling of the RF estimate. When the RF and offset are estimated simultaneously, the distribution of the inferred generating function $\hat{z}$ matches that of the actual generating function z (Fig. 5.5*c*), and the RF estimates are accurate across the entire range of offset values (Fig. 5.5*d*).

5.5.1.2 Simultaneous Tracking of Changes in RF and Offset

The interaction between the offset and the static nonlinearity described in the previous example can have a significant impact on the estimation of the parameters of the encoding model during adaptive function, potentially masking, confounding, or creating the illusion of adaptive function. In this example, the response of a RGC to spatially uniform white noise was simulated as above. However, in this simulation, the contrast of the stimulus was increased midway through the 60-s trial. Baccus and Meister [19] found that such changes in stimulus contrast were followed by fast changes in gain and temporal dynamics in RGCs (over the time course of approximately 100 ms) as well as changes in baseline membrane potential with opposing fast and slow (over the time course of approximately 10 s) dynamics. To model these changes, simulations were conducted in which the RGC responded to the contrast switch with corresponding changes in gain (defined as the peak amplitude of the RF) and/or offset. Again, ERLS is used to estimate the parameters of the RF and offset from stimulus–response data. As described above, if the RF and offset are not estimated simultaneously, then changes in the offset or in the statistics of the stimulus can be reflected as changes in the gain of the RF estimate.

In the first example, both gain and offset remained fixed while the stimulus was increased from low to high contrast. The results of estimating the RF of the simulated neuron with and without simultaneous estimation of the offset are shown in Figure 5.6*a*. While the gain of the RF estimated with simultaneous estimation of the offset (solid black) is similar to the actual gain (dashed black) and remains relatively constant throughout the trial, the gain of the RF estimated without simultaneous estimation of the offset (gray) decreases after the contrast switch. The increase in the standard deviation of the stimulus results in a decrease in the ratio θ/σ_z from 0.5 to 0.25, which affects the scaling of the RF estimate when the offset is not estimated simultaneously. Because the response of the neuron has been rectified and has a nonzero offset that is neglected during the estimation process, changes in the standard deviation of the stimulus are reflected as changes in the gain of the RF estimate. Although the encoding properties of the neuron are completely stationary, the RF estimated without simultaneous estimation of the offset appears to adapt due to the interaction between the offset and the static

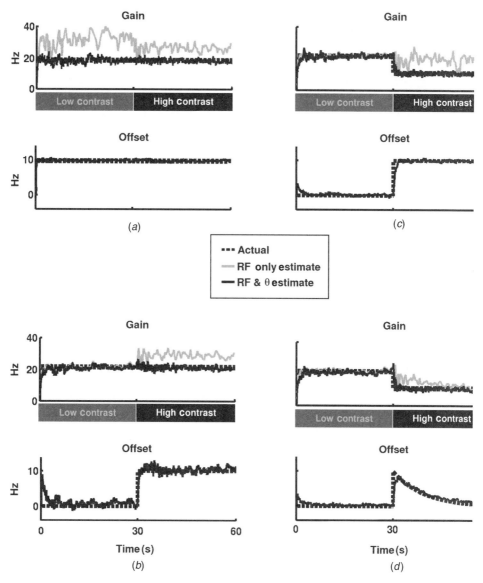

Figure 5.6   Simultaneous tracking of changes in RF and offset. The response of a RGC to spatially uniform, contrast switching white noise was simulated and the responses were used to estimate the RF and offset of the simulated neuron. The contrast switch in the stimulus was accompanied by corresponding changes in the gain and offset of the simulated neuron. (*a*) RF and offset estimates for simulated neuron with gain and offset held fixed throughout trial. The RF estimates with (black) and without (gray) simultaneous estimation of the offset are shown in the top plot, along with the offset estimate in the bottom plot. In both plots, the actual value of the quantity to be estimated is also shown (dashed black). The contrast of the stimulus is indicated under the time axis of the top plot. Similar plots are shown for examples in which the simulated neuron responded to the contrast switch with a fast change in offset (*b*), fast changes in gain and offset (*c*), as well as an additional slow change in offset (*d*). (Reproduced, with permission, from [34]. Copyright © 2006 by the Taylor and Francis Group.)

nonlinearity. Note also the increased variability in the RF estimated without simultaneous estimation of the offset.

In the second example, the gain remained fixed after the switch from low to high contrast, while the offset θ was increased from 0 to 10 over the 300 ms following the switch (θ/σ_z increased from 0 to 0.25). The results of estimating the RF of the simulated neuron with and without simultaneous estimation of the offset are shown in Figure 5.6*b*. While the gain of the RF estimated with simultaneous estimation of the offset (solid black) is similar to the actual gain (dashed black) throughout the trial, the gain of the RF estimated without simultaneous estimation of the offset (gray) increases after the contrast switch. Because the response of the neuron has been rectified and the offset is neglected during the estimation process, changes in the offset are mistaken for changes in gain. In fact, in this case, this effect is mitigated somewhat by the effects of the change in the standard deviation of the stimulus contrast described above.

In the third example, the fast increase in offset following the switch from low to high contrast was accompanied by a fast decrease in gain. This results in an increase of the ratio θ/σ_z from 0 to 0.5. The results of estimating the RF of the simulated neuron with and without simultaneous estimation of the offset are shown in Figure 5.6*c*. While the gain of the RF estimated with simultaneous estimation of the offset (solid black) tracks the decrease in the actual gain (dashed black), the gain of the RF estimated without simultaneous estimation of the offset remains relatively constant following the contrast switch. In this case, the actual decrease in gain is countered by the apparent increase in gain that results from neglecting the offset during the estimation process. Thus, the interaction between the offset and the static nonlinearity causes the adaptive changes to be completely masked.

Finally, in the fourth example, a slow decrease in offset is added to the fast changes in gain and offset in the previous simulation, to more closely approximate experimental observations. In this example, the offset decreases exponentially toward its original value of zero in the 30 s following the change in contrast and corresponding fast increases in gain and offset, reflecting the adaptive behavior observed in actual RGCs. The ratio θ/σ_z increases from 0 to 0.5 immediately following the switch and gradually returns to zero. The results of estimating the RF of the simulated neuron with and without simultaneous estimation of the offset are shown in Figure 5.6*d*. While the gain of the RF estimated with simultaneous estimation of the offset (solid black) tracks the fast decrease in the actual gain (dashed black), the gain of the RF estimated without simultaneous estimation of the offset decreases slowly after the contrast switch. In this case, the interaction between the offset and the static nonlinearity results in the fast adaptive changes being completely masked and the slow change in offset being reflected as a slow change in gain.

5.6 CONCLUSIONS

In this chapter, a new framework for the functional characterization of adaptive sensory systems under nonstationary stimulus conditions via adaptive estimation was developed. A simple nonlinear model of adaptive visual encoding with a time-varying RF was introduced and an RLS approach to estimating the parameters of the RF from stimulus–response data was presented. The RLS approach has several drawbacks that limit its ability to track fast RF changes, namely the dynamics of its underlying state-space model and its fixed learning rate. The ERLS approach was introduced to provide improved tracking of fast RF changes by including a stochastic model of parameter evolution based on that which underlies the Kalman filter and replacing the fixed learning rate with an adaptive one that is dependent upon changes in the features of the stimulus that elicit adaptive behavior. Based on experimental observations of underlying subthreshold membrane properties during adaptation, the encoding model was extended to include an adaptive offset to capture changes in the operating point of the neuron with respect to the rectification threshold and the ERLS approach was modified to allow simultaneous tracking of adaptive changes in the RF and the offset. By formulating the problem in this manner, the spatiotemporal integration properties of the RF can be decoupled from the intrinsic membrane properties of the neuron captured by the offset and the static nonlinearity, and each adaptive component of the model can be uniquely identified. It was demonstrated that, if not properly accounted for, changes in the operating point of the neuron can have a significant impact on the estimation of the parameters of the encoding model during adaptive function, potentially masking, confounding, or creating the illusion of adaptive function.

The encoding models and parameter estimation techniques presented here provide a framework for the investigation of adaptation under complex stimulus conditions, so that comprehensive models of visual function in the natural environment can be developed. By observing adaptive visual function while systematically varying stimulus features, the features that evoke adaptation and their effects on the encoding properties of the pathway can be precisely characterized. Although the basic properties of adaptation have been studied extensively at the level of the retina, thalamus, and primary visual cortex, the interactions between different forms of adaptation have not been adequately described. Finally, in addition to enhancing our understanding of the basic neurobiological function provided by the early sensory pathways, adaptive mechanisms must be understood and characterized for the design of engineering applications that seek to enhance or replace neuronal function lost due to trauma or disease. Rudimentary gain control mechanisms have been successfully implemented in cochlear prosthetics, and the development of similar mechanisms is imperative for the success of analogous devices for the visual pathway.

APPENDIX: DERIVATION OF THE RLS ALGORITHM

This Appendix provides a derivation of the RLS algorithm for estimation of the RF g from observations of y (see Fig. 5.1). The derivation is based on that found in [27]. The goal of RLS is to recursively generate $\hat{g}$, the RF estimate that minimizes the weighted MSE between the observed response y and the predicted response of the model $\hat{y} = s_n^T \hat{g}_n$. At time n, the cost function J can be written as

$$J[n] = \sum_{i=M}^{n} \gamma^{n-i} |e[i]|^2 \tag{5.5}$$

where

$$e[i] \triangleq y[i] - s_i^T \hat{g}_i \tag{5.6}$$

and γ is a positive constant between zero and 1, often referred to as the *forgetting factor*, that determines the weight of each observation. The vectors s_i and $\hat{g}_i$ are defined as

$$s_i \triangleq [s[P, i-M+1], s[P-1, i-M+1], \dots, s[1, i-M+1],$$
$$s[P, i-M+2], \dots, s[1, i]]^T$$

$$\hat{g}_i \triangleq [\hat{g}_i[P, M], \hat{g}_i[P-1, M], \dots, \hat{g}_i[1, M], \hat{g}_i[P, M-1], \dots, \hat{g}_i[1, 1]]^T$$

where T denotes matrix transpose and P and M are the number of spatial and temporal elements in the RF, respectively. The value $\gamma = 1$ causes all observations to be weighted equally, while $\gamma < 1$ causes the current observation to be weighted the most heavily and past observations to be successively downweighted.

The estimate of the RF that minimizes the cost function in Eq. (5.5) at time n is $\hat{g}_n = \Phi_{s_n s_n}^{-1} \phi_{s_n y[n]}$, where $\Phi_{s_n s_n}$ is the Toeplitz matrix of the stimulus autocovariance at different lags and $\phi_{s_n y[n]}$ is the vector of the cross covariance between the stimulus and response at different lags:

$$\Phi_{s_n s_n} = \sum_{i=M}^{n} \gamma^{n-i} s_i s_i^{\mathrm{T}} \qquad \phi_{s_n y[n]} = \sum_{i=M}^{n} \gamma^{n-i} s_i y[i]$$

The terms corresponding to the current observation can be isolated to obtain recursive expressions for $\Phi_{s_n s_n}$ and $\phi_{s_n y[n]}$:

$$\Phi_{s_n s_n} = \gamma \left[\sum_{i=M}^{n-1} \gamma^{n-1-i} s_i s_i^{\mathrm{T}} \right] + s_n s_n^{\mathrm{T}} = \gamma \Phi_{s_{n-1} s_{n-1}} + s_n s_n^{\mathrm{T}}$$

$$\phi_{s_n y[n]} = \gamma \left[\sum_{i=M}^{n-1} \gamma^{n-1-i} s_i y[i] \right] + s_n y[n] = \gamma \phi_{s_{n-1} y[n-1]} + s_n y[n]$$

(5.7)

These expressions can be used to form a recursive estimate of the RF. This requires a recursive expression for $\Phi_{s_n s_n}^{-1}$, which can be obtained using the matrix inversion lemma. The matrix inversion lemma states that if $A = B^{-1} + CD^{-1}C^{\mathrm{T}}$, then $A^{-1} = B - BC(D + C^{\mathrm{T}} BC)^{-1} C^{\mathrm{T}} B$. Applying the matrix inversion lemma to the recursive expression for $\Phi_{s_n s_n}$ in Eq. (5.7) yields

$$\Phi_{s_n s_n}^{-1} = \gamma^{-1} \Phi_{s_{n-1} s_{n-1}}^{-1} - \frac{\gamma^{-2} \Phi_{s_{n-1} s_{n-1}}^{-1} s_n s_n^{\mathrm{T}} \Phi_{s_{n-1} s_{n-1}}^{-1}}{\gamma^{-1} s_n^{\mathrm{T}} \Phi_{s_{n-1} s_{n-1}}^{-1} s_n + 1}$$

(5.8)

Let $K_n \triangleq \Phi_{s_n s_n}^{-1}$ and

$$G_n \triangleq \frac{\gamma^{-1} K_{n-1} s_n}{\gamma^{-1} s_n^{\mathrm{T}} K_{n-1} s_n + 1}$$

(5.9)

Equation (5.8) can be simplified to obtain

$$K_n = \gamma^{-1} K_{n-1} - \gamma^{-1} G_n s_n^{\mathrm{T}} K_{n-1}$$

(5.10)

Equation (5.9) can be rearranged to obtain a simplified expression for G_n:

$$G_n = \gamma^{-1} K_{n-1} s_n - \gamma^{-1} G_n s_n^{\mathrm{T}} K_{n-1} s_n = K_n s_n$$

Now, a recursive expression for the RF estimate can be written as

$$\begin{aligned}
\hat{g}_n &= K_n \phi_{s_n y[n]} = \gamma K_n \phi_{s_{n-1} y[n-1]} + K_n s_n y[n] \\
&= K_{n-1} \phi_{s_{n-1} y[n-1]} - G_n s_n^{\mathrm{T}} K_{n-1} \phi_{s_{n-1} y[n-1]} + K_n s_n y[n] \\
&= \hat{g}_{n-1} - G_n s_n^{\mathrm{T}} \hat{g}_{n-1} + K_n s_n y[n] = \hat{g}_{n-1} + G_n [y[n] - s_n^{\mathrm{T}} \hat{g}_{n-1}] \\
&= \hat{g}_{n-1} + G_n e[n]
\end{aligned}$$

(5.11)

Thus, the estimate of the RF can be computed recursively at each time step by solving Eq. (5.6) and (5.9)–(5.11) in the following sequence:

$$e[n] = y[n] - s_n^T \hat{g}_{n-1} \qquad \text{Prediction error}$$

$$G_n = \frac{\gamma^{-1} K_{n-1} s_n}{\gamma^{-1} s_n^T K_{n-1} s_n + 1} \qquad \text{Update gain}$$

$$\hat{g}_n = \hat{g}_{n-1} + G_n e[n] \qquad \text{Update parameter estimates}$$

$$K_n = \gamma^{-1} K_{n-1} - \gamma^{-1} G_n s_n^T K_{n-1} \quad \text{Update inverse of stimulus autocovariance}$$

If, instead of observations of y, the estimation is being performed with observations of the firing rate λ, then the predicted response in Eq. (5.6), $s_n^T \hat{g}_{n-1}$, is replaced by $f(s_n^T \hat{g}_{n-1})$ to avoid the scaling effects of the static nonlinearity as described in the text.

REFERENCES

1. R. A. NORMANN, E. M. MAYNARD, P. J. ROUSCHE, AND D. J. WARREN. A neural interface for a cortical vision prosthesis. *Vision Res.*, 39:2577–2587, 1999.

2. M. S. HUMAYUN, E. DE JUAN, G. DAGNELIE, R. J. GREENBERG, R. H. PROPST, AND D. H. PHILLIPS. Visual perception elicited by electrical stimulation of retina in blind humans. *Arch. Opthalmol.*, 114(1):40–46, 1996.

3. H. K. HARTLINE. The response of single optic nerve fibres of the vertebrate eye to illumination of the retina. *Am. J. Physiol.*, 121:400–415, 1938.

4. H. B. BARLOW. Summation and inhibition in the frog's retina. *J. Physiol.*, 119:69–88, 1953.

5. S. W. KUFFLER. Discharge patterns and functional organisation of the mammalian retina. *J. Neurphysiol.*, 16:37–68, 1953.

6. P. Z. MARMARELIS AND V. Z. MARMARELIS. *Analysis of Physiological Systems.* Plenum, New York, 1978.

7. H. M. SAKAI, K. NAKA, AND M. I. KORENBERG. White-noise analysis in visual neuroscience. *Visual Neurosci.*, 1:287–296, 1988.

8. J. J. DICARLO, K. O. JOHNSON, AND S. S. HSIAO. Structure of receptive fields in area 3b of primary somatosensory cortex in the alert monkey. *J. Neurosci.*, 18:2626–2645, 1998.

9. R. L. JENISON, J. H. W. SCHNUPP, R. A. REALE, AND J. F. BRUGGE. Auditory space-time receptive field dynamics revealed by spherical white-noise analysis. *J. Neurosci.*, 21:4408–4415, 2001.

10. R. SHAPLEY AND C. ENROTH-CUGELL. Visual adaptation and retinal gain controls. *Prog. Ret. Res.*, 3:263–346, 1984.

11. N. BRENNER, W. BIALEK, AND R. DE RUYTER VAN STEVENINCK. Adaptive rescaling maximizes information transmission. *Neuron*, 26:695–702, 2000.

12. A. L. FAIRHALL, G. D. LEWEN, W. BIALEK, AND R. R. DE RUYTER VAN STEVENICK. Efficiency and ambiguity in an adaptive neural code. *Nature*, 412:787–790, 2001.

13. R. M. SHAPLEY AND J. D. VICTOR. The effect of contrast on the transfer properties of cat retinal ganglion cells. *J. Physiol.*, 285:275–298, 1978.

14. D. G. ALBRECHT, S. B. FARRAR, AND D. B. HAMILTON. Spatial contrast adaptation characteristics of neurons recorded in the cat's visual cortex. *J. Physiol.*, 347:713–739, 1984.

15. S. M. SMIRNAKIS, M. J. BERRY, D. K. WARLAND, W. BIALEK, AND M. MEISTER. Adaptation of retinal processing to image contrast and spatial scale. *Nature*, 386:69–73, 1997.

16. M. P. SCENIAK, D. L. RINGACH, M. J. HAWKEN, AND R. SHAPLEY. Contrast's effect on spatial summation by macaque v1 neurons. *Nature Neurosci.*, 2:733–739, 1999.

17. J. B. TROY, D. L. BOHNSACK, AND L. C. DILLER. Spatial properties of the cat x-cell receptive field as a function of mean light level. *Visual Neurosci.*, 16:1089–1104, 1999.

18. M. CARANDINI AND D. FERSTER. A tonic hyperpolarization underlying contrast adaptation in cat visual cortex. *Science*, 276:949–952, 1997.

19. S. A. BACCUS AND M. MEISTER. Fast and slow contrast adaptation in retinal circuitry. *Neuron*, 36:909–919, 2002.

20. M. J. BERRY AND M. MEISTER. Refractoriness and neural precision. *J. Neurosci.*, 18:2200–2211, 1988.

21. J. KEAT, P. REINAGEL, R. C. REID, AND M. MEISTER. Predicting every spike: A model for the responses of visual neurons. *Neuron*, 30:830–817, 2001.

22. N. A. LESICA AND G. B. STANLEY. Encoding of natural scene movies by tonic and burst spikes in the lateral geniculate nucleus. *J. Neurosci.*, 24:10731–10740, 2004.

23. I. LAMPL, I. REICHOVA, AND D. FERSTER. Synchronous membrane potential fuctuations in neurons of the cat visual cortex. *Neuron*, 22:361–374, 1999.

24. P. DAYAN AND L. F. ABBOTT. *Theoretical Neuroscience.* MIT Press, Cambridge, MA, 2001.

25. G. B. STANLEY. Adaptive spatiotemporal receptive field estimation in the visual pathway. *Neural Computation*, 14:2925–2946, 2002.

26. D. L. RINGACH, M. J. HAWKEN, AND R. SHAPLEY. Receptive field structure of neurons in monkey visual cortex revealed by stimulation with natural image sequences. *J. Vision*, 2:12–24, 2002.

27. S. HAYKIN. *Adaptive Filter Theory*, 4th ed. Prentice-Hall, Upper Saddle River, NJ, 2002.

28. S. HAYKIN, A. H. SEYED, J. R. ZEIDLER, P. YEE, AND P. C. WEI. Adaptive tracking of linear time-variant systems by extended RLS algorithms. *IEEE Trans. Signal Process.*, 45:1118–1128, 1997.

29. R. E. KALMAN. A new approach to linear filtering and prediction problems. *Trans. ASME J. Basic Eng.*, 82:35–45, 1960.

30. A. H. SAYED AND T. KAILATH. A state-space approach to adaptive RLS filtering. *IEEE Signal Process. Mag.*, 11:18–60, 1994.

31. B. WILLMORE AND D. SMYTH. Methods for first-order kernel estimation: Simple-cell receptive fields from responses to natural scenes. *Network: Comput. Neural Syst.*, 14:553–577, 2003.

32. N. A. LESICA, A. S. BOLOORI, AND G. B. STANLEY. Adaptive encoding in the visual pathway. *Network: Comput. Neural Syst.*, 14:119–135, 2003.

33. N. A. LESICA AND G. B. STANLEY. Tracking receptive field modulation during natural stimulation. *IEEE Trans. Neural Syst. Rehab. Eng.*, 13:194–200, 2005.

34. N. A. LESICA AND G. B. STANLEY. Decoupling functional mechanisms of adaptive encoding. *Network: Comput. Neural Syst.*, 17:43–60, 2006.

35. J. D. VICTOR. The dynamics of the cat retinal x cell centre. *J. Physiol.*, 386:219–246, 1987.

36. G. B. STANLEY, F. F. LI, AND Y. DAN. Reconstruction of natural scenes from ensemble responses in the lateral geniculate nucleus. *J. Neurosci.*, 19(18):8036–8042, 1999.

37. M. V. SANCHEZ-VIVES, L. G. NOWAK, AND D. A. MCCORMICK. Membrane mechanisms underlying contrast adaptation in cat area 17 in vivo. *J. Neurosci.*, 20:4267–4285, 2000.

38. J. J. BUSSGANG. Crosscorrelation functions of amplitude distorted gaussian signals. *MIT Res. Lab. Elec. Tech. Rep.*, 216:1–14, 1952.

CHAPTER 6

DECONVOLUTION OF OVERLAPPING AUDITORY BRAINSTEM RESPONSES OBTAINED AT HIGH STIMULUS RATES

O. Ozdamar, R. E. Delgado, E. Yavuz, and N. Acikgoz

6.1 INTRODUCTION

Auditory-evoked potentials (AEPs) are generally recorded at low stimulation rates to prevent the overlap of the responses. The neurons in the auditory pathway, however, have long been known to respond to stimuli at very high rates with little adaptation. Rates up to 1000 Hz (corresponding to the refractory period of neurons) can be coded by different populations of auditory neurons. Such high rate responses, however, cannot be recorded with conventional evoked potential techniques due to the overlap of responses, which may cover a long period after stimulus onset. The middle-latency response (MLR) typically lasts up to about 50–80 ms, which limits the maximum recordable rates to 12.5–20 Hz [1]. The auditory brainstem response (ABR) is an earlier response with a duration of about 12–15 ms and is generally recorded at rates below 67–83 Hz.

Eysholdt and Schreiner [2] first offered a solution to this problem in 1982 using special properties of the maximum-length sequences (MLSs) to deconvolve the overlapping ABR responses obtained at rates higher than 100 Hz. Later, other special sequences such as Legendre series were introduced to expand the choice of sequences [3–5]. One major limitation of these special techniques has always been the limited set of available stimulus sequences, which never exceeded a handful in number. In addition, the instantaneous rates of these sequences cover most of the frequencies with a heavy representation in high frequencies (see Fig. 6.1, top left, for an MLS example). A new frequency-domain method for deconvolution was recently introduced by Jewett and his colleagues [6], but this technique was also limited to a small selection of specially designed sequences which are generated on a trial-by-error basis.

Recently, we have developed a new time-domain method called continuous-loop averaging deconvolution (CLAD) for high-stimulation-rate recording [7–9] which enables the deconvolution of overlapping responses to a very large set of stimulus sequences as long as such a deconvolution is mathematically possible and signal-to-noise ratio (SNR) conditions are favorable. The CLAD method generalizes the deconvolution

Handbook of Neural Engineering. Edited by Metin Akay

process in the time domain to most arbitrary sequences. In this study, we present CLAD ABR recordings made with stimulation rates up to 800 Hz. An analysis of the responses and comparison with previous studies are made.

6.2 METHODOLOGY

6.2.1 CLAD Method

The CLAD method requires data to be acquired using a continuous acquisition loop buffer $\mathbf{v}[t]$ containing the arithmetic sum of all the individual responses starting at their respective triggering positions. It assumes that the individual responses $\mathbf{a}[t]$ to each stimuli are independent of each other and that the measured complex response is the arithmetic sum of the individual overlapping responses. Given these constraints, the proposed technique can deconvolve the resulting complex response acquired to provide the actual response to each stimulus (or triggering event). Data in response to a desired stimulus presentation sequence are continuously acquired throughout the sequence until the sequence starts again. The time duration of the acquisition loop determines the duration of the deconvolved response to the triggering event.

The convoluted measured response vector $\mathbf{v}[t]$ is related to the desired deconvoluted response vector $\mathbf{a}[t]$ with the following simple matrix equation:

$$\mathbf{v}[t] = \mathbf{M}\mathbf{a}[t] \tag{6.1}$$

where $\mathbf{M}$ is a multidiagonal matrix with a sequence of 1's and 0's related to the stimulus sequence. Deconvoluted response to individual stimuli can be obtained simply by solving for $\mathbf{a}[t]$ as follows:

$$\mathbf{a}[t] = \mathbf{M}^{-1}\mathbf{v}[t] \tag{6.2}$$

Not all stimulation sequences will result in equations that generate unique solutions. The existence of $\mathbf{M}^{-1}$ determines if the deconvoluted response can be obtained. As expected, the traditionally used uniform-rate stimulation (isochronous sequence pattern) does not produce an invertable matrix. Unlike other methods, however, CLAD allows the generation of sequences that are nearly isochronous by varying only the position of the final stimulus. One very important property of the inverse matrix is that it retains its multidiagonal property. All the coefficients repeat diagonally across the entire matrix. This allows the computer to store only the first row of the matrix $[\mathbf{M}^{-1}]_r$ as a row vector $\mathbf{m}^{-1}[t]$ for use during deconvolution resulting in a considerable data storage savings [8].

6.2.2 Stimuli

The stimuli used consisted of rarefaction clicks (100 μs duration) presented at 11.1 Hz for conventional averaging or seven CLAD sequences, shown in Table 6.1. These sequences were designed to cover a range of mean stimulation rates from 58.6 to 800.8 Hz. The mean rate for each sequence is calculated by averaging the instantaneous rates whose probability distributions are displayed in Figure 6.1. As shown, CLAD sequences can be distributed according to any arbitrary pattern. In the lower rate sequences used in this study (58.6–156.3 Hz), instantaneous rates were concentrated near their mean rates, thus providing near-isochronous stimulation patterns. The instantaneous rates for the higher frequency sequence (195.3–800.8 Hz) instantaneous rates were distributed more broadly. They

TABLE 6.1   Statistical Descriptors for the CLAD Sequences used in this Study

Mean rate (Hz)	Stimuli per cycle N	Median rate (Hz)	Min rate (Hz)	Max rate (Hz)	Stimulus sequence (base 256)
58.6	3	56.18	55.56	64.94	1, 90, 180
97.7	5	100.00	87.72	102.04	1, 50, 100, 150, 200
156.3	8	151.52	151.52	192.31	1, 33, 66, 99, 132, 165, 198, 231
195.3	10	192.31	106.38	714.29	1, 25, 51, 98, 125, 152, 175, 202, 225, 250
293.0	15	312.50	192.31	454.55	1, 17, 33, 53, 67, 85, 9, 119, 139, 150, 171, 187, 200, 220, 231
507.8	26	454.55 (EST)	312.50	1666.67	1, 9, 12, 19, 32, 41, 52, 59, 70, 78, 91, 100, 111, 119, 131, 140, 153, 159, 171, 182, 190, 201, 209, 222, 230, 241
800.8	41	1000.0	292.12	1666.67	1, 6, 11, 19, 22, 31, 36, 40, 47, 52, 59, 67, 71, 80, 83, 91, 94, 100, 109, 113, 120, 125, 133, 137, 141, 150, 155, 163, 168, 171, 181, 185, 191, 196, 200, 210, 213, 221, 228, 233, 240

were, however, more centrally distributed as compared to the corresponding MLS sequences.

6.2.3   Subjects and Recordings

Auditory brainstem response recordings from six young-adult subjects (four males, two females, ages 21–30 years) with normal hearing were made using the Smart-EP (Intelligent Hearing Systems, Miami) evoked potential system. All recordings were obtained with the stimulus presented to the right ear using insert earphones (ER3A) while the subjects were lying down on a bed in a sound-attenuated room. Conventional recording parameters (gain 100,000; filters 10–1500 Hz, 6 dB/octave) and electrode placements (positive: upper forehead; negative: right mastoid; ground: left mastoid) were used.

Each subject was first tested using the standard ABR averaging technique at a stimulation rate of 11.1 Hz. The subjects were then tested using the seven CLAD sequences with clicks at 60 dB HL. The order of sequence presentation was randomized. For two additional sequences (97.7 and 195.3 Hz) eight click levels (from 0 to 70 dB HL in 10-dB steps) were used. Similar recordings were obtained using MLS and Legendre sequences (LGSs) as well. At least two recordings were acquired for each stimulation sequence and click level.

The CLAD buffer contained 2048 data points collected at 25 μs sampling time. This sampling time provided a recording period of 51.2 ms. Each averaged recording consisted of two buffers implemented in a split-sweep technique containing 512 stimulus sequence. For deconvolution purposes, each buffer was downsampled to 256 points and the response deconvolution calculated in real time as described by Delgado and Ozdamar [8]. Both convolved and deconvolved recordings were stored for later analysis.

The ABR waves III and V were labeled in each recording and the amplitudes were measured using the falling slope. Wave I was not measured since it was not present in all recordings. The amplitudes, peak latencies, and interpeak intervals were plotted for analysis.

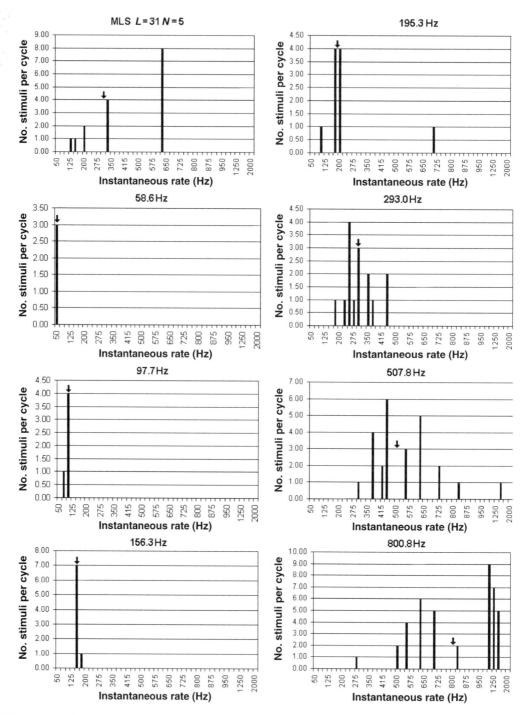

Figure 6.1   Distribution of MLS and CLAD instantaneous rates used in this study. Top left: rate distribution of MLS sequence $L = 31$. The remaining diagrams depict the instantaneous rates for each CLAD sequence. The mean rates for all CLAD sequences are marked with arrows.

6.3 RESULTS

6.3.1 Stimulation Rate Experiments

For the rate experiment, recordings to 60-dB HL rarefaction clicks at seven stimulation rates (mean rates 58.6, 97.7, 156.3, 195.3, 293.0, 507.8, and 800.7 Hz) were acquired from five ears. The ABR recordings obtained at all seven rates produced identifiable wave V components at all high and moderate click levels. Examples of convolved and deconvolved recordings using 60-dB HL clicks at different rates are shown in Figure 6.2. As can be observed, waves I, III, and V are readily discernible in both convolved and deconvolved recordings obtained at rates less than 100 Hz. For higher rates, ABR components start overlapping with each other, producing complex tracings. Deconvolved responses obtained with CLAD shows wave V component clearly even at 507.8 and 800.7 Hz as observed. As expected, at such high rates wave V is largely diminished in amplitude and prolonged in latency. Earlier components I and III are less discernible at such rates.

Wave III and V latencies, amplitudes, and latency intervals for all five ears for all recordings are plotted in Figure 6.3. The top-left portion of the diagram shows the latencies of peaks III and V and below the interpeak III–V interval. As expected, the latencies become prolonged with increasing stimulation rates. The interpeak III–V latency also increases with increasing stimulation rate. The change in amplitude with

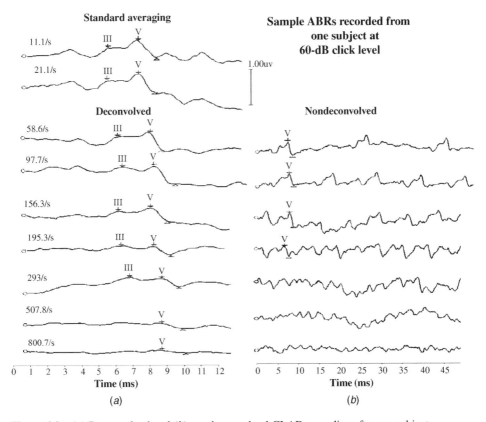

Figure 6.2 (*a*) Deconvolved and (*b*) nondeconvolved CLAD recordings for one subject acquired at stimulation sound level of 60 dB HL using seven different rates. Waves III and V are labeled accordingly. Two top-left recordings were acquired at rates of 11.1 and 21.1 Hz using standard averaging.

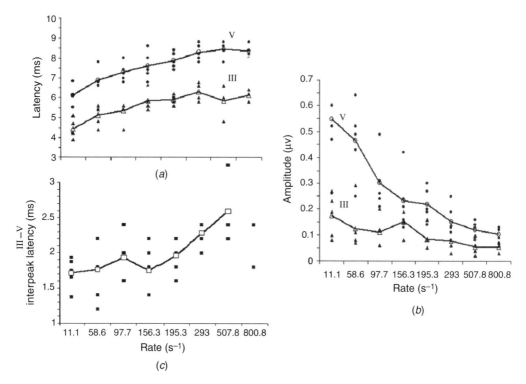

Figure 6.3 (*a*) Latencies and (*b*) amplitudes of waves III and V and (*c*) III–V interpeak latencies as function of stimulation rate. Data points correspond to individual subjects. The mean values obtained from five ears are shown with open symbols and connected from one rate to another. The first rate value (11.1 Hz) was obtained through conventional averaging while the others were obtained using CLAD.

respect to stimulation rate for both waves III and V is shown on the right side. A large reduction in amplitude can be observed for wave V as the stimulation rate is increased.

6.3.2 Click-Level Experiment

For click-level experiments, ABR data from five ears were recorded from 0 to 70 dB HL in 10-dB steps at rates of 97.7 and 195.3 Hz. Decreasing click levels also produced amplitude attenuation and latency prolongation in both rates. An example set of recordings obtained at 195.3 Hz to increasing click levels is shown in Figure 6.4. For this set of recordings, wave V was observed down to 10 dB HL. Wave V was discernible down to 20 dB HL in all five ears and at 10 dB HL in three ears.

Figure 6.5 shows the average latency and amplitude characteristics with respect to click level for two stimulation rates (97.7 and 195.3 Hz). An increase in click level shows a decrease in peak latency and an increase in peak amplitude. The mean III–V interpeak interval shows a small increase (about 0.2 ms) with increasing stimulation level.

6.4 DISCUSSION

The data acquired in this study demonstrate that CLAD can be used for real-time acqui-sition of clinical-quality ABR recordings at high stimulation rates up to 800.8 Hz. The experiments also show that this technique can be utilized to obtain ABR responses at

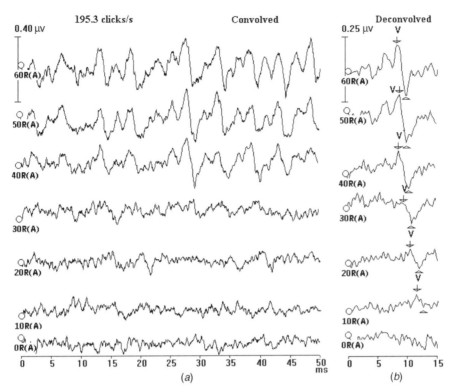

Figure 6.4 Representative ABR recordings from one subject obtained at decreasing click levels. (*a*) Nondeconvolved ABR recordings obtained at rate of 195.3 Hz. (*b*) Corresponding deconvolved ABR recordings obtained with CLAD.

rates close to the absolute refractory period (about 1 kHz) of auditory brainstem fibers. As expected, the ABR recordings displayed the typical pattern of increases in peak latency and decreases in peak amplitude with both increasing stimulation rate and decreasing stimulation intensity. All major ABR components were clearly discernible. Comparison of the CLAD recordings with those acquired using MLS and LGS sequences showed that the CLAD recordings were very similar with respect to overall morphology, further validating this technique, and compared favorably with the previously reported latency and amplitude measurements made using standard averaging and the MLS technique [10–13]. The CLAD data reported in this study further expand the stimulation rates up to 800 Hz, not reported in previous studies. Due to a wide selection of stimulus sequences, CLAD provides a suitable methodology to study adaptation effects at very high stimulus rates. Recently we have formulated the CLAD technique in the frequency domain which provided a clearer insight on the SNR characteristics of the deconvolution process [14].

This new technique also provides additional insight as to ABR peak components and generators. The trend in latencies, interpeak intervals, and amplitudes with respect to stimulation rate was as expected. Wave III amplitudes did not show a very significant change due to the manner in which they were measured (falling phase of the peak). As rate increases, wave III and V amplitudes tend to change differently. These changes could be due to the differential interactions in the fast and slow components of ABR as a function of stimulation rate. Increasing the stimulation intensity and the averaging count is expected to sufficiently increase the SNR level of the recording to also study wave I characteristics in the future.

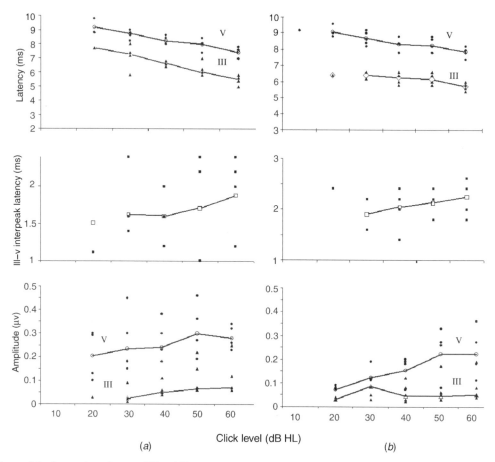

Figure 6.5   Latencies of waves III and V (top) and III–V interpeak intervals (middle) and amplitudes of waves III and V (bottom) as function of stimulation intensity for data acquired at presentation rates of (*a*) 97.7 Hz and (*b*) 195.3 Hz from one representative ear.

Unlike previously developed techniques, the ability to generate CLAD sequences that are almost isochronous provides a method that can be more easily compared to standard recordings acquired with traditional averaging methods at lower stimulation rates. The CLAD method provides a wider range of stimulation sequences and better control of the interstimulus intervals with less stimulus jitter. The ability to generate almost any arbitrary stimulation sequence using CLAD greatly expands the clinical potential of ABRs for early clinical evaluation of neurological conditions such as multiple sclerosis (MS) and other myelination disorders. This study is only a demonstration of the possible applications of this technique. The ability to generate stimulation sequences that are nearly isochronous allows researchers to study the auditory pathway in a more precise manner than previously possible with MLS sequences. This could also have more direct influence on late evoked potentials in which the stimulus sequence is commonly jittered. By contrast, MLS cannot provide the necessary small constant jitter that is required to collect late evoked potentials but rather provides abrupt jumps in instantaneous rates. Since near isochronous rates can be achieved using CLAD, different adaptation effects of the broadly distributed MLS rates can be avoided.

This study demonstrates that CLAD can be used for real-time acquisition of overlapping ABRs obtained at high stimulation rates. This procedure can also be implemented for any evoked potential application or technique requiring synchronized time averaging. This technique may enable us to study the early effects of demyelinating diseases on evoked responses and differentiation of adaptation along the auditory pathway which may be useful as a diagnostic and monitoring tool.

REFERENCES

1. OZDAMAR, O. AND KRAUS, N., Auditory middle latency responses in humans, *Audiology*, 1983, 22:34–49.
2. EYSHOLDT, U. AND SCHREINER, C., Maximum length sequences—A fast method for measuring brainstem evoked responses, *Audiology*, 1982, 21:242–250.
3. BURKARD, R., SHI, Y. AND HECOX, K. E., A comparison of maximum length sequences and Legendre sequences for the derivation of brainstem auditory evoked responses at rapid rates of stimulation, *J. Acoust. Soc. Am.*, 1990, 87:1656–1664.
4. SHI, Y. AND HECOX, K. E., Nonlinear system identification by *m*-pulse sequences: Application to brainstem auditory evoked responses, *IEEE Trans. Biomed. Eng.*, 1991, 38:834–845.
5. BELL, S. L., ALLEN, R. AND LUTMAN, M. E., Optimizing the acquisition time of the middle latency response using maximum length sequences and chirps, *J. Acoust. Soc. Am.*, 2002, 112: 2065–2073.
6. JEWETT, D. L., LARSON-PRIOR, L. S., AND BAIRD, W., A novel techniques for analysis of temporally-overlapped neural responses, Evoked Response Audiometry XVII Biennial Symposium IERASG, 2001 (A), p. 31.
7. DELGADO, R. E. AND OZDAMAR, O., New methodology for acquisition of high stimulation rate evoked responses: Continuous loop averaging deconvolution (CLAD), Conf. Assoc. Res. Otolaryngol. (ARO), 2003 (A).
8. DELGADO, R. E. AND OZDAMAR, O., Deconvolution of evoked responses obtained at high stimulus rates, *J. Acoust. Soc. Am.*, 2004, 115:2065–2073.
9. OZDAMAR, O., DELGADO, R. E., YAVUZ, E., THOMBRE, K. V., AND ACIKGOZ, N., Proc. First Int. IEEE EMBS Conf. on Neural Engineering, Capri, 2003.
10. DON, M., ALLEN, A. R., AND STARR, A., Effect of click rate on the latency of auditory brainstem responses in humans, *Ann. Otol.*, 1977, 86:186–195.
11. PALUDETTI, G., MAURIZI, M., AND OTTAVIANI, F., Effects of stimulus repetition rate on the auditory brainstem responses (ABR), *Am. J. Otol.*, 1983, 4:226–234.
12. LASKY, R. E., Rate and adaptation effects on the auditory evoked brainstem response in human newborns and adults, *Hear. Res.*, 1997, 111:165–176.
13. BURKARD, R. F. AND SIMS, D., The human auditory brainstem response to high click rates: Aging effects, *Am. J. Audiol.*, 2001, 10:53–61.
14. OZDAMAR, O. AND BOHORQUEZ, J., Signal-to-noise ratio and frequency analysis of continuous loop averaging deconvolution (CLAD) of overlapping evoked potentials, *J. Acoust. Soc. Am.*, 2006, 119: 429–438.

CHAPTER 7

AUTONOMIC CARDIAC MODULATION AT SINOATRIAL AND ATRIOVENTRICULAR NODES: OBSERVATIONS AND MODELS

S. Ward, R. Shouldice, C. Heneghan,
P. Nolan, and G. McDarby

7.1 INTRODUCTION

The complete sequence of successive atrial and ventricular electromechanical events produced by the conduction of electrical impulses in the heart is known as the cardiac cycle. Two main contributors to timing variations within the cardiac cycle are the spontaneous firing rates of the sinoatrial (SA) pacemaker cells and the conduction time through the atrioventricular (AV) node, both of which are primarily regulated by the autonomic nervous system (ANS). The influence of the ANS on the SA node causes alterations in the spontaneous firing rate of the primary pacemaker and will therefore have a direct influence on the overall heart rate as assessed by interbeat (PP or RR) interval measurements obtained from the surface lead electrocardiogram (ECG). The study of this interbeat variation is well established and is known as heart rate variability (HRV) analysis, which has been traditionally used as a marker of cardiovascular dysfunction [1]. The ANS also influences the conduction time through the AV node which may be assessed through intrabeat (PR) interval measurements (under the assumption that the propagation time from the SA node to the AV node is relatively fixed). Research into the area of ANS effects on AV conduction time (AVCT) assessed through PR interval variation has intensified recently due to both increased interest in noninvasive assessment of ANS activity at the AV node and the development of more accurate and reliable PR interval estimators [2–8]. The ANS affects the AV and SA nodes in a similar manner, though neural interactions with intrinsic AV nodal properties can lead to highly complex AVCT dynamics. The refractory period is one such property, which will slow down conduction if the AV cells have recently conducted and hence tries to produce long PR intervals in response to short PP cycles. The purpose of this chapter is to noninvasively examine the relation between ANS activity at the SA node and AV node in normal subjects in vivo and to provide a physiologically plausible model of how ANS activity modulates both overall cardiac cycle and AVCT. Our experimental observations are based on

Handbook of Neural Engineering. Edited by Metin Akay

the autonomic response generated by a simple supine-to-standing transition. We show that an integrate and fire model can be used to account for variation in the observed PP and PR intervals. Moreover, we wish to address a significant unresolved issue in the field, namely, is autonomic activity at the AV node strongly coupled with autonomic activity at the SA node? In this context "strong coupling" refers to the scenario where changes in parasympathetic and sympathetic innervations at the SA node are faithfully repeated at the AV node (i.e., if parasympathetic activity at the SA node goes up by 50%, then parasympathetic activity at the AV node goes up by 50%). Conversely, we can postulate physiology in which autonomic activity at the AV node is unrelated to that at the SA node (e.g., it will be perfectly normal for parasympathetic outflow to increase at the SA node and decrease at the AV node simultaneously). This is illustrated schematically in Figure 7.1, where we

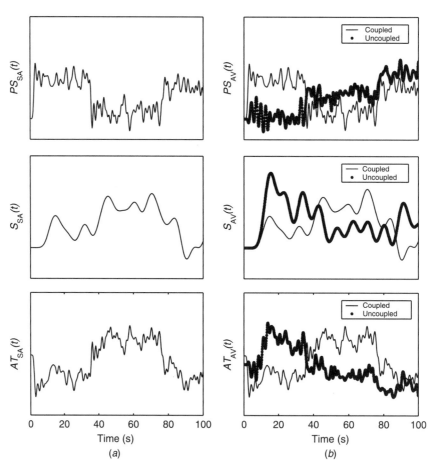

Figure 7.1 Schematic representation of concept of "coupled" and "uncoupled" innervation of the SA and AV nodes. In all plots, x axis units represent time in seconds and the y axis is in arbitrary units, corresponding to "neural modulation strength." (a) Time-varying neural activity at SA node over period of 100 s. Here $PS_{SA}(t)$ is the time-varying component of the parasympathetic system at the SA node, $S_{SA}(t)$ is the time-varying component of the sympathetic system at the SA node, and $AT_{SA}(t)$ is the overall autonomic stimulation [expressed as $AT_{SA}(t)S_{SA}(t) - PS_{SA}(t)$]. ($b$) corresponding time-varying signals for autonomic modulation at AV node for two cases. For the coupled case, the individual components and their sum are exactly proportional to modulations at the SA node. For the uncoupled curves, there is no relation (other than chance) between the signal at the SA node and the signal at the AV node.

model parasympathetic and sympathetic innervations as time-varying signals and form an overall autonomic signal as the sum of the sympathetic outflow and the negative of the parasympathetic outflow. This figure shows that in the perfectly coupled case the overall autonomic effects at the SA and AV nodes is highly correlated; in the uncoupled case, these two signals can be totally uncorrelated. In the literature coupled behavior is also referred to as "dependent" and uncoupled as "independent." Pure-coupled or pure-uncoupled behavior is unlikely to be seen in nature, but they are useful concepts to broadly describe the potential behavior of autonomic neural activity. Indeed, our initial experimental and modeling results indicate that in fact both types of behavior (or a mixture) are observed and can be modeled by an integrate-and-fire model.

7.2   PHYSIOLOGICAL BACKROUND

Both efferent and afferent innervations interact with intracardiac innervations to maintain adequate cardiac output [9]. The two main divisions of the ANS are the parasympathetic and sympathetic branches, both of which innervate the heart. The parasympathetic nervous system affects the heart through cholinergic neurotransmitters while the sympathetic system alters the cardiac cycle through adrenergic neurotransmitters. They influence the heart in a time-dependent manner with parasympathetic effects having a much shorter latency period than sympathetic effects. The activity of both of these divisions is modulated by external factors such as respiration and blood pressure. These modulatory effects have periodic properties reflected in their spectra of interval variations where high-frequency (HF) components (0.15–0.5 Hz) are ascribed solely to the modulation of the parasympathetic system and low-frequency (LF) components (0.05–0.15 Hz) are attributed to both parasympathetic and sympathetic modulatory effects [10, 11]. Shortened cycle lengths or reduced conduction times are normally associated with a diminution of parasympathetic and/or an increase in sympathetic activity. Increased cycle lengths or conduction times are normally associated with opposing changes in neural activity. The rate at which the heart beats in the absence of neurohumoral influences is referred to as the intrinsic heart rate. However, in the normal healthy individual, the parasympathetic influence, also referred to as vagal influence, predominates at the SA node, significantly reducing the heart rate from its intrinsic rate to a slower rate. The exact effects of neural influence on the AV node in relation to AVCT remain somewhat unclear [12–15]. The human AV node, like the SA node, possesses a high innervation density, but unlike the SA node, which has an extremely dense cholinergic (parasympathetic-mediated) innervation compared to other neural subpopulations, the compact region of the AV node has been observed to possess similar densities of cholinergic and adrenergic (sympathetic-mediated) neural subpopulations [9, 16]. In addition the behavior of preceding cycle and conduction rates for many prior beats will affect the timings within the current beat. These intrinsic nodal effects can be summarized as refractoriness, facilitation, and fatigue. Refractory effects at the AV node provide a significant intrinsic source of variation and determine the recovery time of the node (assessed using RP intervals). The AV node has a relative refractory period after activation during which its conduction velocity is reduced. Encroachment upon the relative refractory period of the AV node has the paradoxical effect of lengthening AVCT even as cycle length shortens. Vagal stimulation of the AV node slows nodal recovery while sympathetic stimulation accelerates recovery and hence increases AV conduction velocity. More subtle effects relating to stimulation history have also been seen in the AV node. Facilitation is the process that allows a shorter than expected AV nodal conduction time to occur after a long AV conduction period that was preceded by a short refractory period. Fatigue is the time-dependent prolongation in AV nodal conduction time during rapid, repetitive excitation. Although facilitation and fatigue may not contribute as much to AV conduction variation as refractory effects, nonlinear interaction between these three intrinsic properties may result in increasingly complex AV nodal conduction dynamics [17, 18].

As a result of this complex physiological system and the experimental difficulties in accurately acquiring AVCT, relatively little work has been carried out on characterizing and modeling AVCT. Moreover, to our knowledge there is no definitive answer as to whether overall ANS activity at the SA and AV nodes is tightly coupled under normal physiological circumstances or whether a high degree of uncoupled innervation takes place at these two nodes. Indeed, the published literature in this area is somewhat in disagreement. Leffler et al. concluded from their results that variation in cycle length and AVCT is derived from a common origin [5]. In their work, they accounted for rate recovery effects in the AV node in order to isolate the autonomic induced variability in RR and PR intervals and hence concluded that in general the autonomic influence on the SA and AV nodes is tightly coupled. In contrast, Kowallik and Meesmann concluded from their studies, involving spontaneous changes in heart rate during sleep, that independent autonomic innervation of the SA and AV nodes occur, even accounting for rate recovery effects [7]. During sleep, they showed that major body movements were accompanied by sudden decreases in PP interval. However, PR interval change was not consistent—shortening, no change, and lengthening were observed. Furthermore, tonic changes in the PR interval occurred over 15-min periods during which the range of PP intervals was constant. Recovery-adjusted PR intervals and cycle lengths (RR intervals) were negatively correlated for some periods, suggesting some form of independent autonomic modulation of the SA and AV nodes. Finally, Forester et al. proposed a midway position in which there is both coupled and decoupled activity; they postulated that standing reduces the vagal outflow to the SA node and possibly increases the overall sympathetic tone, while the vagal input to the AV node remains relatively unchanged [19].

7.3 EXPERIMENTAL METHODS AND MEASURES

One possible way of unraveling the parasympathetic and sympathetic effects present in interbeat and intrabeat intervals is to analyze the interval variations as a result of postural transitions [5, 6, 19]. For example, a supine-to-standing transition will produce an autonomic response whose effect at the SA node is well understood. Therefore, we designed a simple experimental procedure which included this postural transition and sufficient time in both supine and standing positions to assess statistical variations in a "steady-state" condition. The protocol also included sections of deep paced respiration, since this is known to increase parasympathetic modulation, but the observations for paced respiration are not discussed or modeled in this chapter.

7.3.1 Protocol

Data were collected from 20 normal healthy male subjects in sinus rhythm with a mean age of 25 [standard deviation (SD) 3.28 years, range 22–34 years]. Subjects had no history of cardiac or respiratory illness and were not taking medications with autonomic effects. The protocol also stipulated no alcohol or caffeine for 8 h prior to the study, no food 2 h prior, a normal night's sleep prior to the study, and no unusual stressors. Informed consent was obtained from all subjects prior to data collection and the laboratory was closed to others for the duration of the experimental protocol with unobtrusive background

TABLE 7.1   Experimental Protocol: Interventions and Onset Times

Time (min)	Intervention
0	Supine acclimatization period
15	Supine paced deep breathing, 6 vital capacity breaths per minute
17	10 min supine rest
27	Standing rapidly and remaining as motionless as possible
37	Standing paced deep breathing, 6 vital capacity breaths per minute
39	Normal breathing
42	End

music played. The data consist of standard bipolar leads I, II, and III and Lewis lead ECG signals for 42 min duration. The Lewis lead (modified lead II) electrode configuration is considered in this work due to the fact that it emphasizes atrial activity and therefore simplifies the task of automated P-wave detection. Signals are amplified and bandpass filtered in hardware at acquisition time (Grass P155, Astro-Med, Slough, UK), with a passband between 0.01 and 100 Hz. They are digitized at a sampling rate of 1000 Hz using a CED μ1401 interface and Spike 2 software (Cambridge Electronic Design, Cambridge, UK). A 50-Hz digital notch filter is applied to attenuate power line interference. Lung tidal volume is monitored simultaneously by uncalibrated AC-coupled respiratory inductance plethysmography (Respitrace, Ambulatory Monitoring, Ardsley, NY). Signals are recorded in both supine and standing positions, with two sections of deep breathing. Table 7.1 outlines the onset times of the various experimental interventions.

7.3.2   Results

The PP (P-wave onset to following P-wave onset), PR (P-wave onset to following QRS onset), and RP (QRS onset to following P-wave onset) intervals were extracted using a wavelet-based technique [20]. In [20], we showed that the accuracy of automated PP and PR interval estimation was comparable to that of a human expert. To capture the "average" changes provoked by the supine-to-standing transition, Figure 7.2a shows the median PP and PR interval for all 20 subjects on a beat-by-beat basis in a 300-beat window centered around the standing event. Figure 7.2b shows the mean PP and PR intervals over a 40-beat window. To illustrate that the group median and mean responses reflect individual subjects, Figure 7.2c shows the same response for subject number RM170802P1, also over a 40-beat window. In general, immediately after the event, a pronounced shortening of the PP interval is evident; in contrast the PR interval lengthens. To illustrate the complexities of explaining variations in AVCT, consider that this lengthening could be ascribed to either (a) refractory period effects (i.e., AV nodal recovery period increased by previous short RP interval) or (b) a paradoxical transient increase of parasympathetic activation of the AV node. After about 10 beats have elapsed, the PR interval duration returns to near preevent baseline value; PP continues to shorten. However, the temporal decoupling of the PP and PR intervals at the transition is indicative of a difference in the underlying time constant of autonomic influence of the SA and AV nodes and may suggest that the relative sympathovagal innervation of each node is different. The effect is visible in the group mean, trimmed (at 10%) mean, and median of the interval values on a beat-by-beat basis. From beat numbers 20–35 in Figure 7.2a the

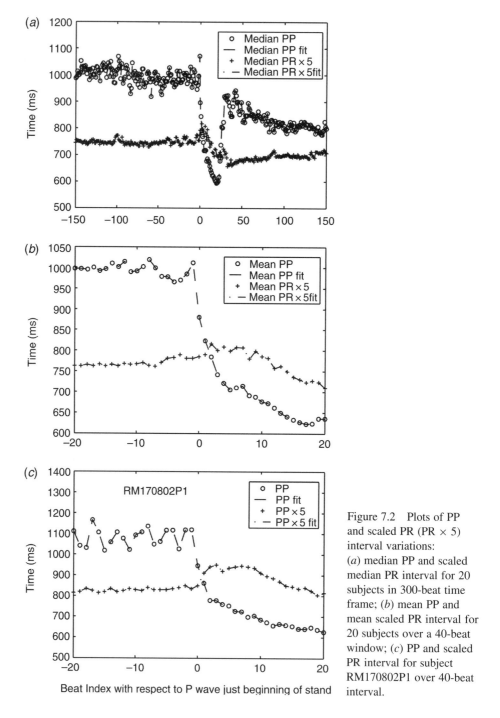

Figure 7.2 Plots of PP and scaled PR (PR × 5) interval variations: (a) median PP and scaled median PR interval for 20 subjects in 300-beat time frame; (b) mean PP and mean scaled PR interval for 20 subjects over a 40-beat window; (c) PP and scaled PR interval for subject RM170802P1 over 40-beat interval.

PP value is observed to return to a pattern similar to that seen before the transient event, albeit at a lower (i.e., higher rate) baseline value. The overall PR interval trend is similar to that of PP, with the notable exception of the early 0–5 beat number index lengthening. An interpretation of these results is that the transition from supine to standing causes (a) a reduction in parasympathetic activation of the SA node, (b) no significant change in the parasympathetic activation of the AV node, and (c) an increase in sympathetic activation of both the SA and AV nodes. This is consistent with the findings of [19]. In our later modeling, we will show that such a scenario leads to model results similar to those experimentally observed.

As a second experimental observation, we will consider values of PP, PR, and RP in the steady-state supine and standing positions. If PR interval variability was purely controlled by AV autonomic activity directly coupled to SA autonomic activity, then plots of PP versus PR (or RP vs. PR) should be approximately linear. Conversely, if the only effect on AVCT is refractoriness, then PR should be a hyperbolic function of RP. In practice we can expect some mixture of these two behaviors. Figures 7.3a,b,c show plots of recovery period, $R_{n-1}P_n$, versus intrabeat interval, P_nR_n, on a steady-state posture basis for three subjects (YK020702P1, CH280602P1, and DM130702P1). These plots are representative of the types of patterns seen. Figure 7.3a shows a subject who can vary his PR and RP intervals over quite a large range. No significant refractory influence is seen, even for quite short RP intervals (i.e., there is no increase in PR for small RP). The standing and supine positions lead to quite distinct "operating" points for PR. An interpretation of these clusters is that the overall autonomic influence at the SA and AV nodes is tightly coupled (though it cannot answer the question whether the individual parasympathetic and sympathetic components are also tightly coupled). Figure 7.3b shows a subject who achieves a similar range of RP and PR intervals, but in this case, there are not two distinct clusters of values. There is no clear-cut refractory effect, though in general we see no very short PR intervals for the shortest RP intervals. Since the RP and PR intervals are not clustered, an interpretation is that the overall autonomic tone at the SA and AV nodes are not tightly coupled. Note that the PR interval changes are not an artifact induced by P-wave morphology change as there is no instantaneous change in PR interval after the transient event, for example, due to electrode movement. The separation of the clusters in Figure 7.3c again shows a subject who has distinct operating points in the supine and standing positions. However, in this case refractory effects seem to dominate ANS effects with reduced RP intervals resulting in an increase in PR intervals for both postural positions. To give some idea of population distribution, 10 of the 20 subjects display clear clusters as in Figure 7.3a; 5 of the subjects have totally overlapping RP versus PR plots (as in Fig. 7.3b), while the remaining 5 cannot be easily classified. Our conclusion from consideration of these plots is that coupled and uncoupled innervation of the SA and AV nodes occurs, and the degree of influence of refractoriness is highly subject dependent.

As an additional experimental observation, we show typical interval-based PP and PR spectra produced for the steady-state data. These spectra differ widely in character for different subjects. For instance, Figures 7.4a,b show a LF increase and HF decrease when standing in PP but an overall decrease in PR power. The peak in the PP spectra at 0.1 Hz can probably be attributed to baroreceptor reflexes when standing; this peak is not clearly defined in the PR spectrum. These spectra therefore would be consistent with decoupled autonomic innervation of the SA and AV nodes, since it appears that there is autonomic activity at the SA node which is absent at the AV node. These spectra are consistent with those in [6] but not with those of [5]. However, other subjects show spectra in which there are common changes in both the PP and PR spectra which would be in agreement with [5] (see Figs. 1.4c,d. Our conclusion from consideration of the spectra is that both coupled and uncoupled innervation of the SA and AV nodes occurs.

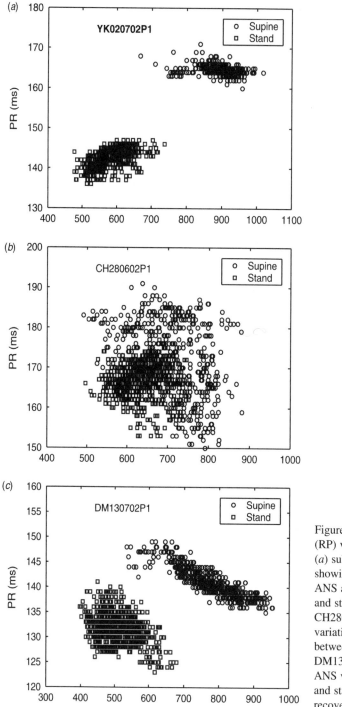

Figure 7.3   Recovery time (RP) versus PR interval: (*a*) subject YK020702P1, showing large variation in ANS activity between supine and standing; (*b*) subject CH280602P1, showing little variation in ANS activity between postures; (*c*) subject DM130702P1, showing large ANS variation between supine and standing but with dominant recovery effects over ANS effects.

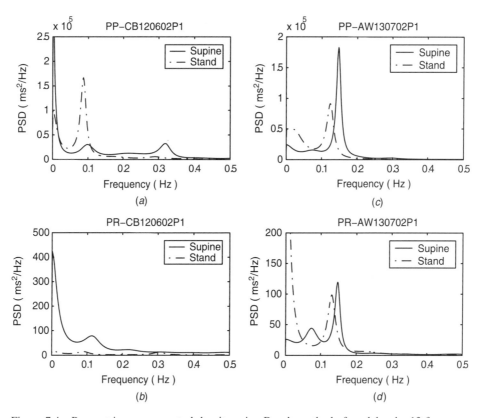

Figure 7.4 Parametric power spectral density using Burg's method of model order 13 for supine and standing positions: (*a*) PP interval of subject CB120602P1; (*b*) PR interval of subject CB120602P1; (*c*) PP interval of subject AW130702P1; (*d*) PR interval of subject AW130702P1.

7.4 MODELS

7.4.1 Integrate-and-Fire-Based Model

Given the complexity of observed changes in AVCT, we decided to develop a simple physiologically plausible model to explore the range of behavior possible. An ultimate motivation for this modeling is to be able to account for experimental observations seen in both normal conditions and under various experimentally induced autonomic blockades. In this chapter, we will restrict ourselves to trying to account for the noninvasive transient and steady-state in vivo observations induced by postural change from supine to standing. Our modeling approach is based on the well-known integral pulse frequency modulation (IPFM) model, also known as integrate-to-threshold (ITT) models. These are physiologically plausible models for the transformation of a continuous input signal, representing ANS activity at the SA node, into a series of cardiac events which represent firings of the SA node pacemaker cells [21–25]. The model simulates the firing of pacemaker action potentials at the SA node, therefore representing the P onset on the surface ECG. Traditionally the IPFM model has been used in cardiac modeling to provide realistic models of HRV with RR interval variability used as a surrogate for PP variability, as QRS peaks are easier to detect. The upper section of Figure 7.5 shows a conventional IPFM model for the firing of the SA node. The input signal consists of a positive steady-state value (in our simulations we use the value 1) and a time-varying component $m_{SA}(t)$,

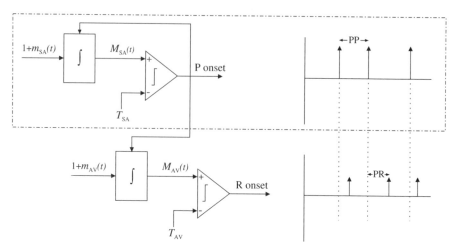

Figure 7.5   Extended integrate-and-fire model. Top panel illustrates conventional IPFM model.

which can be interpreted as the overall autonomic outflow at the SA node, comprised of a mixture of sympathetic–parasympathetic influences. The input signal is integrated until its integrated value $M_{SA}(t)$ exceeds the reference threshold value T_{SA}, at which point an action potential is initiated and the integrator is reset to zero. Here, $M_{SA}(t)$ represents the transmembrane potential of the cell and T_{SA} is the threshold level of the cell. Sympathetic activity is prevalent when $m_{SA}(t)$ is positive, which increases the rate of rise of $M_{SA}(t)$ and therefore decreases the interfiring times of the node. A preponderance of parasympathetic influences is represented by a negative $m_{SA}(t)$, which decreases the rate of rise of $M_{SA}(t)$ and prolongs SA node firing times. The IPFM model has been widely used to investigate the dynamics of the ANS at the SA node. In order to investigate the relationship of ANS activity at both the SA and AV nodes, we apply the conventional IPFM model to generate spike trains representing the SA node firings (P-wave onsets) and then extend the model with a second integrate-and-fire stage to generate a second spike train representing ventricular contraction and QRS onsets, as shown in the lower section of Figure 7.5. The AV integrator does not commence until the SA node has fired and is effectively insensitive to the input signal prior to the SA node firing and subsequent to the end of the PR interval. When the SA node has fired, the input signal is integrated, and when the threshold is reached, an event is generated to mark the end of the PR interval. In line with regression models detailed in [5, 7] the PR interval representing AV conduction time is restricted to being a function of ANS activity at the AV node and the refractory period of the conduction tissue and does not include more subtle effects such as facilitation and fatigue. The refractory period is assessed using the previous RP interval which is measured as the time elapsed from the previous QRS complex to the current P-wave. Therefore, if the previous RP interval is short, then this RP term will not contribute as significantly to the integrator input and will slow down the time until the next firing. Therefore the overall modulating signal $m_{AV}(t)$ to the second integrator represents both the influence of the ANS and the intrinsic refractory effects of the node. The autonomic inputs to the model integrators are mathematically expressed as

$$m_{SA}(t) = -k_0 VPS_{SA}(t) + k_1 VS_{SA}(t) - APS_{SA} + AS_{SA}$$
$$m_{AV}(t) = -k_2 VPS_{AV}(t) + k_3 VS_{AV}(t) - APS_{AV} + AS_{AV} + k_4 RP_{n-1}$$

where

$VPS_{SA}(t)$ = time-varying parasympathetic influence at SA node

$VS_{SA}(t)$ = time-varying sympathetic influence at SA node

APS_{SA} = tonic parasympathetic level at SA node

AS_{SA} = tonic sympathetic level at SA node

$VPS_{AV}(t)$ = time-varying parasympathetic influence at AV node

$VS_{AV}(t)$ = time-varying sympathetic influence at AV node

APS_{AV} = tonic parasympathetic level at AV node

AS_{AV} = tonic sympathetic level at AV node

RP_{n-1} = previous QRS complex to P-wave interval

k_0 = weight of time-varying parasympathetic influence at SA node

k_1 = weight of time-varying sympathetic influence at SA node

k_2 = weight of time-varying parasympathetic influence at AV node

k_3 = weight of time-varying sympathetic influence at AV node

k_4 = weight of recovery effect

The overall parasympathetic and sympathetic effects at each node are therefore defined as

$$PS_{SA}(t) = k_0 VPS_{SA}(t) + APS_{SA}$$
$$S_{SA}(t) = k_1 VS_{SA}(t) + AS_{SA}$$
$$PS_{AV}(t) = k_2 VPS_{AV}(t) + APS_{AV}$$
$$S_{AV}(t) = k_3 VS_{AV}(t) + AS_{AV}$$

Parameter reduction techniques could be applied to this model, but for the sake of ease of physiological interpretation, we have presented an overparameterized model.

7.4.2 Simulation

Dependent and independent neural innervations are simulated to test the hypothesis that (a) ANS activity at the SA and AV nodes comes from a common origin and (b) ANS activity at the two nodes is independent. The paradoxical PP and PR interval behavior during the transition from supine to standing, which is evident in Figure 7.2c for subject RM170802P1, is examined along with data from subject PS190602P1 showing minimal autonomic variation of AV conduction between the steady-state supine and standing positions. For the transient analysis, 20 beats before and after the beginning of the transition from supine to standing are generated. For the steady-state analysis 8 min of data are generated for both the supine and standing positions.

Zero-mean unit-variance Gaussian random sequences are used to generate the signals. In general the parasympathetic signals are obtained by filtering a sequence using a unity-gain low-pass filter with a cutoff frequency at around 0.5 Hz. Respiratory effects observed in the spectra of the experimental data are also added using a band-pass-filtered Gaussian sequence at the required cutoff frequencies. Sympathetic signals are generated by filtering random sequences using low-pass filters and a cutoff frequency below 0.1 Hz to reflect the slower time dynamics of sympathetic innervation. For the steady-state simulation of 8 min duration the low-pass filters have $1/f$ (spectra inversely proportional to their frequency f) frequency characteristics. The filter cutoff frequencies and model parameters are adjusted to obtain similar interval time series and spectra to those obtained using experimental data. To examine scenario (a), we define coupled innervation as follows:

$$PS_{AV}(t) = \alpha PS_{SA}(t) \qquad S_{AV}(t) = \alpha S_{SA}(t)$$

The autonomic inputs at the AV and SA nodes are therefore comparable except for some gain factor α. This means we can remove k_2, k_3, APS_{AV}, and AS_{AV} from the model equation above and add the gain parameter α so the input into the AV node looks as follows:

$$m_{AV}(t) = \alpha[(-k_0 VPS_{SA}(t) - APS_{SA}) + (k_1 VS_{SA}(t) + AS_{SA})] + k_4 RP_{n-1}$$

To simulate scenario (b), the autonomic signals at the SA node are generated independently from the autonomic signals at the AV node. We define uncoupled innervation as

$$E[PS_{SA}(t)PS_{AV}(t)] = E[PS_{SA}(t)]E[PS_{AV}(t)]$$
$$E[S_{SA}(t)S_{AV}(t)] = E[S_{SA}(t)]E[S_{AV}(t)]$$

TABLE 7.2 Transient Analysis: $T_{SA} = 0.893$, $T_{AV} = 0.225$

Var	Dependent		Independent	
	Sup	Std	Sup	Std
k_0	0.20	0.10	0.20	0.05
k_1	0.20	0.10	0.10	0.05
k_2	—	—	0.10	0.10
k_3	—	—	0.10	0.10
k_4	0.40	0.40	0.05	0.05
APS_{SA}	0.40	0.10	0.38	0.10
AS_{SA}	0.20	0.50	0.20	0.50
APS_{AV}	—	—	0.15	0.30
AS_{AV}	—	—	0.50	0.60
α	0.10	0.10	—	—

Note: Var, model input variables; Sup, supine position; Std, standing position.

TABLE 7.3 Steady-State Analysis: $T_{SA} = 0.975$, $T_{AV} = 0.178$

Var	Dependent		Independent	
	Sup	Std	Sup	Std
k_0	0.65	0.62	0.70	0.67
k_1	0.40	0.40	0.35	0.35
k_2	—	—	0.29	0.29
k_3	—	—	0.10	0.14
k_4	0.02	0.02	0.02	0.02
APS_{SA}	0.43	0.25	0.43	0.25
AS_{SA}	0.32	0.32	0.32	0.32
APS_{AV}	—	—	0.20	0.20
AS_{AV}	—	—	0.15	0.18
α	0.35	0.35	—	—

Note: Var, model input variables; Sup, supine position; Std, standing position.

The parameter values for the transient analysis simulation are shown in Table 7.2 and values for the steady-state analysis are shown in Table 7.3.

7.4.3 Results

Figure 7.6a is the same plot as shown in Figure 7.2c and represents the PP and scaled PR interval variations of subject RM170802P1 for 20 beats preceding and subsequent to the transition from supine to standing. As described earlier, the PP and PR intervals show a contrasting variation during the postural transition followed by the PR interval returning to its pretransition value, varying in a similar manner to the PP interval. Figures 7.6b,c show the results from the dependent and independent innervation simulations. These simulation results seem to support the suggestion that the lengthening of the PR interval in response to the pronounced shortening of the PP interval may be explained by either refractory effects for the dependent case or the paradoxical increase in parasympathetic activation of the AV node for the independent case. For dependent innervation, we simulate a subject with dominant refractory effects which can result in a paradoxical lengthening of the PR interval in response to a significant shortening of the PP interval. However, if the refractory effect was indeed dominant for the subject, then we may expect the experimental data to result in a similar plot to Figure 7.6b, where the PR interval does not return to similar pretransition values and would continue to increase with the continued shortening of the PP intervals. This difference in the model results and experimental data may be due to the fact that for the model the refractory period is not a function of parasympathetic or sympathetic stimulation, which seems to be the case physiologically. For the independent innervation model, the contrasting PP and PR interval variations immediately during the transition followed by the return of the PR interval to pretransition levels and variations are due solely to independently different tonic and modulatory innervations at the AV node with refractory effects playing a less significant role.

Figure 7.7a shows the plot of recovery period, previous RP interval, versus the PR interval taken from a subject in steady-state supine and standing positions. It depicts the case for a subject who exhibits very little autonomic variation between supine and standing positions; this is evident from the overlapping data clusters. The subject in the supine position shows little reduction in PR interval for reduced RP intervals, which suggests some sort of balance between autonomic and refractory effects, although there is a large variation in the PR interval range for any given RP value. The steady-state standing position, however, shows increased autonomic activity over refractory effects as shorter RP intervals result in more significant decreases in PR intervals. The results shown in Figure 7.7b indicate that the closely coupled approach will result in an approximately linear relationship between the RP and PR intervals and similar PP and PR interval spectra, which is evident from Figures 7.8f,h. Although the experimental data did show an increased linear RP-versus-PR relationship for the standing position, the dependent innervation PP and PR spectra do not agree with those of the experimental data (Figs 7.8b,d), which do not show similar spectra for the PP and PR interval variations. Also, in this closely coupled simulation case, where autonomic effects are dominant, parallel changes in PP and PR intervals are evident in both the supine and standing positions as shown in Figures 7.8e,g. In contrast, the ECG interval time series in Figures 7.8a,c show no definitive dependency of PR on PP for the supine position, which may indicate either independent autonomic activity at the AV node, especially in relation to the parasympathetic outflow, or a more complex relationship between autonomic inputs, cycle length, and nodal refractory effects. Independent innervation can result in similar RP-versus-PR distributions and spectra, as shown in Figures 7.7c and 7.8j,l, but the time series in this case show no significant parallel or inverse changes for either position. The correlation coefficients for

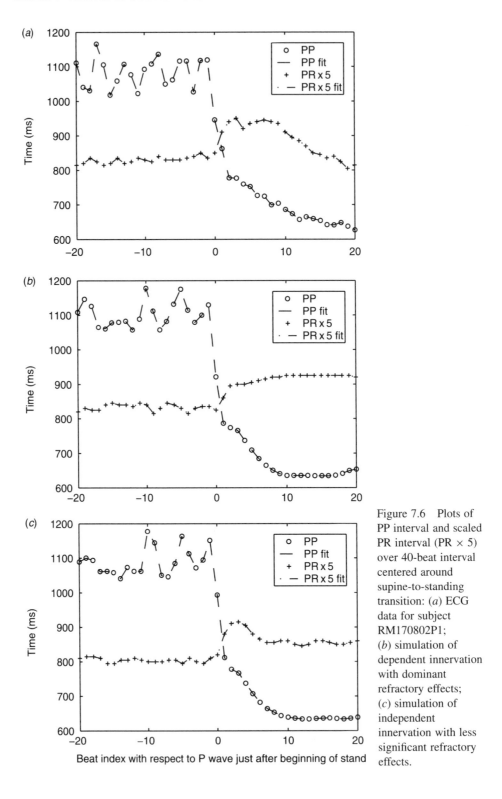

Figure 7.6 Plots of PP interval and scaled PR interval (PR × 5) over 40-beat interval centered around supine-to-standing transition: (a) ECG data for subject RM170802P1; (b) simulation of dependent innervation with dominant refractory effects; (c) simulation of independent innervation with less significant refractory effects.

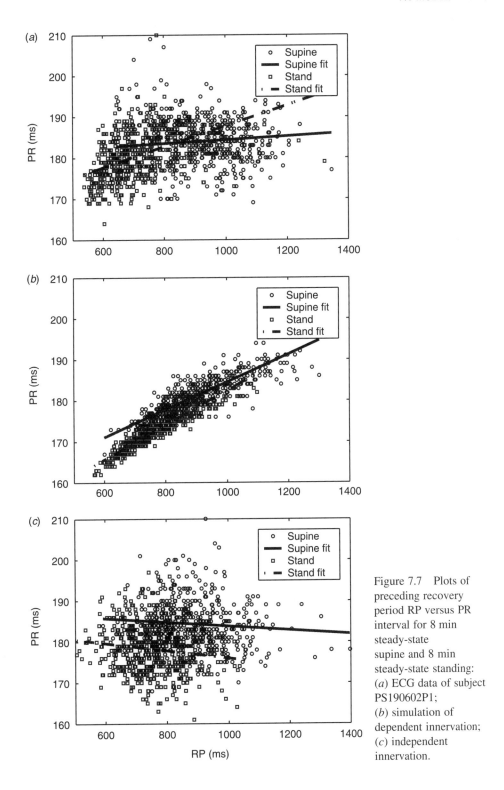

Figure 7.7   Plots of preceding recovery period RP versus PR interval for 8 min steady-state supine and 8 min steady-state standing: (*a*) ECG data of subject PS190602P1; (*b*) simulation of dependent innervation; (*c*) independent innervation.

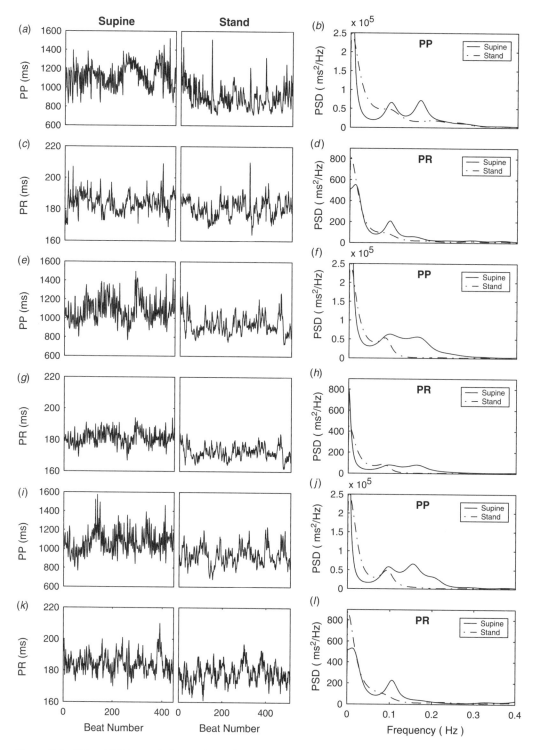

Figure 7.8 PP and PR interval time series and spectra for 8 min steady-state supine and standing for ECG data of subject PS190602P1 in (a), (b), (c), and (d); simulation results using dependent innervation in (e), (f), (g), and (h); and independent innervation results in (i), (j), (k), and (l).

TABLE 7.4   Correlation Coefficients for PP and PR Intervals for Subject PS190602P1

	Supine	Standing
ECG: subject data	0.138	0.547
Simulation		
Dependent innervation	0.932	0.968
Independent innervation	−0.040	−0.091

the PP and PR intervals for subject PS190602P1 and for the dependent and independent innervation simulations are shown in Table 7.4. The correlation results for this subject suggest there may be a combination of dependent and independent effects taking place. Similar to the spectral results outlined earlier the correlation coefficients vary widely for each subject. Although some subjects show either a strong inverse or parallel dependency between intervals, this relationship is not uniform across subjects with others showing weak interval correlations in either or both postural positions.

7.5   DISCUSSION

The argument as to whether independent modulation effects occur at the AV node or whether neural influence at both nodes comes from a common origin is still unresolved. The fact that the SA node is primarily innervated by the right vagus and the AV node by the left vagus suggests that independent modulation is a plausible hypothesis. However, the complicated interaction between neural innervation, heart rate, and intrinsic nodal properties makes assessment of this behavior in a noninvasive manner quite difficult. Leffler et al. [5] concluded that cycle length and conduction time may be related in either a parallel or an inverse manner. A parallel relationship was presumed to indicate predominant ANS over recovery effects whereas an inverse relationship was associated with predominant refractory over ANS effects. In Leffler's study all standing and 5 of 11 supine subjects showed a significant parallel relation and only 4 of 11 supine subjects showed a significant inverse relation. These results are supported by Donnerstein et al. [4], who suggest a more significant inverse relation occurs with increased vagal activity. However, their results do differ somewhat from both the results obtained in our study and those in [7] which show some cases of high interval dependence but also show cases which suggest little dependence of PR on PP in either a parallel or inverse manner. One possible explanation is the complex intrinsic nodal dynamics at the AV node. As already mentioned, autonomic stimulation will alter intrinsic nodal effects, which may result in the dependent and independent effects seen in previous studies and account for the intersubject variations. However, as suggested in [7] these independent effects may not be apparent under circumstances in which vagal tone is reduced. Another explanation is that independent autonomic outflow occurs at the AV node which becomes more evident with increased vagal outflow or during a transient change in vagal activity, as shown in Figure 7.2. Further studies of PP and PR interval variation in situations of reduced vagal activity (e.g., exercise) and increased vagal activity (e.g., sleep) would be beneficial.

When assessing ANS behavior at the AV node noninvasively the limitations inherent in PR interval measurements should be considered. Physiologically the measurement is only an estimation of AVCT and assumes relatively constant atrial, His-Purkinje, and ventricular activation times. It has been suggested that these activation times may increase due to premature beats and may introduce an important source of error in assessing AV nodal function [17]. Quantification and assessment of these errors would also be beneficial.

7.6 FUTURE TRENDS

Assessment of ANS behavior in humans using pharmacological intervention has provided some useful insights [2, 5]. Further detailed studies of pharmacological intervention effects on the SA and AV nodes and on intrinsic functional nodal properties could be of interest. Also, further immunohistochemical and histochemical studies may provide more detailed information about the variation and distribution of neural subpopulations in the human heart. The importance of the intrinsic nervous system in the heart is becoming evident. Further investigations into the anatomy and electrophysiological properties of intracardiac neurons and their contribution to cardiac innervation would also be beneficial. Models describing conduction through the AV node and heart rate simulation at the SA node have been well established with improvements being made continually. The integrate-and-fire model outlined in this chapter provides a physiologically plausible model for examining possible neural and nodal functional interactions at both the SA and AV node and may provide useful insights which could be beneficial to pacemaker design where it is not clear what the optimal AV conduction delay should be set at. To further elucidate neural behavior at both nodes and further develop the model, it would be beneficial to model PR interval variations during pharmacological interventions, exercise, and sleep. It is envisaged that future gains made in the understanding of the anatomy and electrophysiological properties of the heart will be accompanied by more accurate and representative models. A greater understanding of the physiology of the heart, development and assessment of noninvasive experimental methods, and improved mathematical modeling of the heart are needed. Such development could result in improved assessment and diagnosis of cardiac dysfunction and pathology and provide a greater understanding of both physiological and emotional responses.

REFERENCES

1. Task Force of the European Society of Cardiology and the North American Society of Pacing and Electrophysiology, "Heart rate variability—Standards of measurement, physiological interpretation and clinical use," *European Heart Journal*, vol. 17, pp. 354–381, 1996.
2. S. G. CARRUTHERS, B. McCALL, B. A. CORDELL, AND R. WU, "Relationships between heart rate and PR interval during physiological and pharmacological interventions," *British Journal of Clinical Pharmacology*, vol. 23, pp. 259–265, 1987.
3. J. M. RAWLES, G. R. PAI, AND S. R. REID, "A method of quantifying sinus arrhythmia: Parallel effect of respiration on P-P and P-R intervals," *Clinical Science*, vol. 76, pp. 103–108, 1989.
4. R. L. DONNERSTEIN, W. A. SCOTT, AND T. R. LLOYD, "Spontaneous beat-to-beat variation of PR interval in normal children," *American Journal of Cardiology*, vol. 66, pp. 753–754, 1990.
5. C. T. LEFFLER, J. P. SAUL, AND R. J. COHEN, "Rate-related and autonomic effects on atrioventricular conduction assessed through beat-to-beat PR interval and cycle length variability," *Journal of Cardiovascular Electrophysiology*, vol. 5, pp. 2–15, 1994.
6. G. NOLLO, M. D. GRECO, F. RAVELLI, AND M. DISERTORI, "Evidence of low- and high-frequency oscillations in human AV interval variability: Evaluation and spectral analysis," *American Journal of Physiology*, vol. 267, no. 4, pp. 1410–1418, 1994.
7. P. KOWALLIK AND M. MEESMANN, "Independent autonomic modulation of the human sinus and AV nodes: Evidence from beat-to-beat measurements of PR and PP intervals during sleep," *Journal of Cardiovascular Electrophysiology*, vol. 6, pp. 993–1003, 1995.
8. R. SHOULDICE, C. HENEGHAN, AND P. NOLAN, "Methods of quantifying respiratory modulation in human PR electrocardiographic intervals," *Proceedings*

of the 24th Annual International Conference on Engineering in Medicine and Biology, vol. 2, pp. 1622–1623, 2002.

9. G. J. T. HORST, *The Nervous System and the Heart*, 1st ed., New Jersey, Humana Press, 2000.

10. S. AKSELROD, D. GORDON, F. A. UBEL, D. C. SHANNON, A. C. BARGER, AND R. J. COHEN, "Power spectrum analysis of heart rate fluctuation: A quantitative probe of beat-to-beat cardiovascular control," *Science*, vol. 213, pp. 220–222, 1981.

11. A. MALLIANI, M. PAGANI, F. LOMBARDI, AND S. CERUTTI, "Cardiovascular neural regulation explored in the frequency domain," *Circulation*, vol. 84, pp. 482–492, 1991.

12. P. MARTIN, "The influence of the parasympathetic nervous system on atrioventricular conduction," *Circulation Research*, vol. 41, pp. 593–599, 1977.

13. D. W. WALLICK, P. J. MARTIN, Y. MASUDA, AND M. N. LEVY, "Effects of autonomic activity and changes in heart rate on atrioventricular conduction," *American Journal of Physiology*, vol. 243, pp. H523–527, 1982.

14. F. URTHALER, B. H. NEELY, G. R. HAGEMAN, AND L. R. SMITH, "Differential sympathetic-parasympathetic interactions in sinus node and AV junction," *American Journal of Physiology*, vol. 250, pp. H43–H51, 1986.

15. Y. FURUKAWA, M. TAKEI, M. NARITA, Y. KARASAWA, A. TADA, H. ZENDA, AND S. CHIBA, "Different sympathetic-parasympathetic interactions on sinus node and atrioventricular conduction in dog hearts," *European Journal of Pharmacology*, vol. 334, pp. 191–200, 1997.

16. S. J. CRICK, J. WHARTON, M. N. SHEPPARD, D. ROYSTON, M. H. YACOUB, R. H. ANDERSON, AND J. M. POLAK, "Innervation of the human cardiac conduction system: Aquantitative immunohistochemical and histochemical study," *Circulation*, vol. 89, pp. 1697–1708, 1994.

17. J. BILLETTE AND S. NATTEL, "Dynamic behavior of the atrioventricular node: A functional model of interaction between recovery facilitation and fatigue," *Journal of Cardiovascular Electrophysiology*, vol. 5, pp. 90–102, 1994.

18. D. J. CHRISITINI, K. M. STEIN, S. M. MARKOWITZ, S. MITTAL, D. J. SLOTWINER, S. IWAI, AND B. B. LERMAN, "Complex AV nodal dynamics during ventricular-triggered atrial pacing in humans," *American Journal of Physiology (Heart Circulation and Physiology)*, vol. 281, pp. H865–H872, 2001.

19. J. FORESTER, H. BO, J. W. SLEIGH, AND J. D. HENDERSON, "Variability of R-R, P-wave-to-R-wave, and R wave-to-T wave intervals," *American Journal of Physiology*, vol. 273, no. 6, pp. 2857–2860, 1997.

20. R. SHOULDICE, C. HENEGHAN, P. NOLAN, P. G. NOLAN, AND W. MCNICHOLAS, "Modulating effect of respiration on atrioventricular conduction time assessed using PR interval variation," *Medical and Biological Engineering and Computing*, vol. 40, pp. 609–617, 2002.

21. E. J. BAYLY, "Spectral analysis of pulse frequency modulation in the nervous systems," *IEEE Transactions on Biomedical Engineering*, vol. 15, pp. 257–265, 1968.

22. R. W. DE BOER, J. M. KAREMAKER, and J. STRACKEE, "Description of heart rate variability data in accordance with a physiological model for the genesis of heartbeats," *Psychophysiology*, vol. 22, no. 2, pp. 147–155, 1985.

23. R. G. TURCOTT AND M. C. TEICH, "Fractal character of the electrocardiogram: Distinguishing heart-failure and normal patients," *Annals of Biomedical Engineering*, vol. 24, pp. 269–293, 1996.

24. G. B. STANLEY, K. POOLLA, AND R. A. SIEGAL, "Threshold modeling of autonomic control of heart rate variability," *IEEE Transactions on Biomedical Engineering*, vol. 47, pp. 1147–1153, 2000.

25. J. MATEO AND P. LAGUNA, "Improved heart rate variability signal analysis from the beat occurrence times according to the IPFM model," *IEEE Transactions on Biomedical Engineering*, vol. 47, pp. 985–996, 2000.

CHAPTER 8

NEURAL NETWORKS AND TIME–FREQUENCY ANALYSIS OF SURFACE ELECTROMYOGRAPHIC SIGNALS FOR MUSCLE CEREBRAL CONTROL

Bruno Azzerboni, Maurizio Ipsale,
Fabio La Foresta, and Francesco Carlo Morabito

8.1 INTRODUCTION

Control of motor units (MUs) is one of the most complex tasks of the brain. The involved signal, which starts from some neurons of the brain and arrives at the muscle, is managed by a very complicated control system. Usually, the path of this signal is crossed through the body, that is, a signal starting from the left brain lobe controls muscles located on the right side of the body, whereas a signal starting from the right brain lobe controls muscles located on the left. Because of this process, if a stroke appears in one brain lobe, the pathological patient cannot control the opposite side of the body [1]. However, this behavior does not work in some type of muscle, such as the postural ones, since a stroke in one lobe side does not imply the inhibition of these muscles and the pathological patient still assumes a right posture [1]. This experimental observation suggests that a common drive could start at both brain lobes and affect muscles located in both body sides. The main concept proposed here, aimed at validating the last sentence, exploits the correlation techniques to show that they can be used to investigate dependencies between neurophysiological signals and to determine signal pathways in the central nervous system.

In recent years, there has been a resurgence in the use of spectral methods, such as Fourier transform to process biomedical signals [1–3]; a spectral-based approach can have advantages over the time-domain approach since it extends the information that can be extracted from experimental readings by permitting the study of interactions between two simultaneously recorded signals. A useful parameter for characterizing the linear interaction in the frequency domain is the coherence function [3], whose estimate, typically carried out on digital experimental data, is able to provide a bounded measure of the linear dependency between two processes. The coherence function shows two advantages with respect to time-domain measures of association: (a) it is a bounded measure, constrained within the range (0,1), where zero is attained in the case of signal

Handbook of Neural Engineering. Edited by Metin Akay

independence while the maximum value is achieved in the case of a perfect linear relationship; (b) it is a measure that does not depend on the units of measurement [3].

Thus, the coherence spectral analysis carried out on two myoelectric signals simultaneously recorded [without electroencephalographic (EEG) acquisition] can be performed to investigate the existence of a control common drive of muscular activity [1]. The main disadvantage of this approach is that the presence of some artifacts in the myoelectric signal recordings can strongly corrupt coherence analysis; to overcome this practical problem, some advanced techniques, such independent-component analysis (ICA) [4], wavelet transform [5], and a combination of them, are here proposed.

In this work, we show that the preprocessing artifact removal step and the coherence analysis reveal the presence of the underlying common drive only in the axial muscles. In particular, in Section 8.2, we will describe the preliminary step of data acquisition by means of a surface electromyography (sEMG) [6] analysis; in Section 8.3 we will discuss the method used in order to perform coherence analysis. In Sections 8.4 and 8.5, we will explain in detail the artifact removal procedure based on neural networks and time-frequency analysis. Finally, in Section 8.6, we will report and comment on the achieved results that seem in full agreement with some physiological studies, which assume the existence of this common drive only for axial muscles, in order to help the postural task. As previously mentioned, the distal muscles do not need this kind of control; accordingly, the coherence analysis shows a low value for all frequencies when applied on two first dorsal interosseous muscles.

8.2 DATA ACQUISITION: SURFACE ELECTROMYOGRAPHY

Electrical muscle activity monitoring can be employed to study neuroscience problems, such as motor control by the brain. From the clinical point of view, classical needle electromyography is commonly used to make a diagnosis in muscles and in peripheral nerves. Since the nervous system controls simultaneously a set of muscles, a detailed study must consider multichannel recordings of muscle activity. For these reasons, sEMG is more adequate to monitor the muscle activity. Unfortunately, it suffers by various drawbacks, one being the cross-talk phenomenon that is a part of the EMG signal acquired on a muscle which is generated by another one. A solution to this problem is ICA, a technique proposed in recent years to overcome the latter problem [7].

Here, seven active electrodes performing a sEMG were attached to the body of a healthy cooperating human subject. The aim was to investigate the common drive (a signal starting from the brain) by processing only the reading, that is, the myoelectric signal, recorded by some surface electrodes attached to the muscles. In our application, we attached four electrodes to the pectoral muscles and three electrodes to intrinsic hand muscles, such as the first dorsal interosseous (FDI). Finally, we attached two electrodes to the right muscle and the other electrode to the left one. In Figure 8.1 the electrode mapping and the sEMG recordings of the pectoral muscles are shown, while in Figure 8.2 we show the electrode mapping and the surface sEMG recordings of the FDI muscles.

Electrode mapping is designed to properly carry out the task of the coherence analysis. In the FDI muscles we attached three electrodes (two to the right muscle and the other to the left one) in order to perform the calculation of two kinds of coherence: the ipsilateral one (intrinsic coherence between one muscle and itself) and the bilateral one (coherence between two opposite muscles). For the pectoral muscles we must use at least four electrodes because the coherence analysis must be preceded by the artifact removal

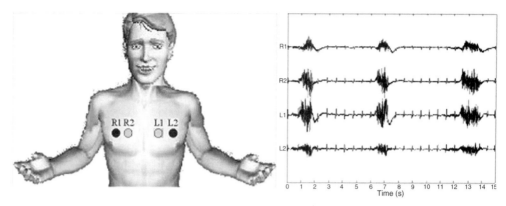

Figure 8.1   Pectoral muscles: electrode mapping and example of sEMG recordings.

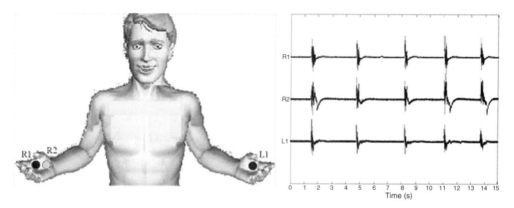

Figure 8.2   First dorsal interosseous: electrode mapping and example of sEMG recordings.

processing. Actually, the signals read by the electrodes on the pectoral muscles are highly corrupted by the cardiac artifact. Since this kind of artifact has a repetition frequency centerd on the same spectral range where we investigate for the presence of the common drive, the coherence analysis could be highly corrupted by the artifact. In particular, it could be possible to find coherence in this spectral range caused by the artifact presence, rather than by the common drive that we are investigating [8].

By using the proposed pattern of electrodes, we can perform different kinds of coherence measures in order to validate the existence of the common drive, as we would like to infer. Various cycles of simultaneous muscle contraction (both side muscles are contracted synchronously) are recorded by the electrode mapping.

During the registration session a 50-Hz notch filter and a low-pass filter (cutoff frequency 500 Hz) were applied; throughout the experiments the sampling frequency $f_s = 1$ kHz.

8.3   METHODS EMPLOYED: SPECTRAL AND COHERENCE ANALYSIS

In this section we present the technique employed to perform the spectral and coherence analysis. Figure 8.3 shows the multistep procedure used to estimate the existence of the

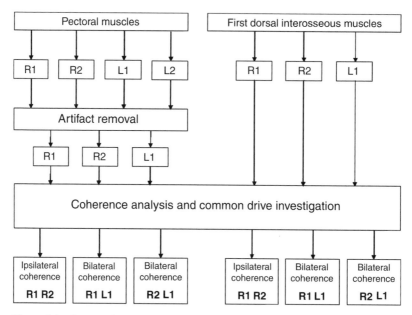

Figure 8.3 Proposed method block scheme; it executes the procedure employed to investigate the existence of a common drive.

common drive. The first step of signal processing, carried out only on the pectoral muscle recordings, is artifact removal, since the recorded signals are corrupted by cardiac activity. This extra signal can invalidate the conclusions that we can carry out from the results of the proposed approach. We will describe the artifact removal process in the next section. Thus, the filtered signals, which contain only meaningful information about muscle activity, are processed by coherence analysis in order to investigate the presence of a common drive.

Six different experiments were performed, three for each muscle type. Ipsilateral or bilateral coherence in both muscle types were investigated. Two electrodes are attached on the right side (R1 and R2) and only one electrode on the left side (L1): The fourth recording in the pectoral muscles was rejected by the artifact removal stage. We conjecture that the common drive exists only in postural muscles, such as pectoral ones, whereas its presence could not be found in counterlateral muscles, such as the FDI. Obviously, in FDI coherence, the three calculated functions (R1–R2, R1–L1, R2–L1) need to be almost similar to each other, whereas this similarity could not be found in the three calculated coherence functions for the pectoral muscles.

The coherence analysis was demonstrated to be very meaningful in studying the presence of the common drive. The coherence algorithm is a spectral analysis that allows us to determine how much two different signals are correlated in spectral information. The magnitude-squared coherence between two signals $x(n)$ and $y(n)$ is the following:

$$C_{xy}(\omega) = \frac{|S_{xy}(\omega)|^2}{S_{xx}(\omega)S_{yy}(\omega)} \tag{8.1}$$

where $S_{xx}(\omega)$ and $S_{yy}(\omega)$ are respectively the power spectra of $x(n)$ and $y(n)$, and $S_{xy}(\omega)$ is the cross-spectral density function. The coherence measures the correlation between $x(n)$ and $y(n)$ at the angular frequency ω. Particular attention must be paid when computing this function. First, the signals that we study (the myoelectric signals) cannot be considered stationary and then a simple spectral estimation cannot be applied without incurring failing results. Actually, the nonstationarity can create a cross-interference phenomenon that can artificially raise the coherence value. This high value is not meaningful, because it is not related to the brain activity. In order to minimize these effects, we estimate the spectra (cross-correlation spectrum and self-correlation spectrum or power spectrum density) by using the Welch method, which consists in windowing the signal before the fast Fourier transform (FFT) calculus and calculating a mean spectrum over all these (overlapped) windows [9].

8.4 ARTIFACT REMOVAL: NEURAL NETWORKS AND TIME–FREQUENCY ANALYSIS

A key topic in biomedical signal processing is artifact removal. The artifact is a superimposed signal that can hide some useful information in a measured signal. It is usual to classify the artifacts according to their features in both the time and frequency domains. Some artifacts are well localized in frequency, whereas their influence is spread over all the time axis of the original signal. Others typologies of artifacts, instead, are confined to a small temporal area, while their spectral components cover almost all the frequency spectrum of the original signal. The easiest to filter are those artifacts that are well localized in the frequency domain and their spectral components do not overlap with the spectral content of the original signal.

Neglecting the last condition, in which artifact removal is performed by a simple finite impulse response (FIR) or infinite impulse response (IIR) digital filter (stop band), implemented by classical filtering techniques, our attention was on the other categories of artifacts, in which it is not possible to remove the artifact influence by a simple filter. This can mainly occur for two reasons: (i) the artifact and the signal are overlapped in the frequency domain; (ii) the artifact could distort the entire signal frequency representation. In these cases the use of a stop-band filter could remove some meaningful components of the original signal or it would not be able to remove the artifact.

In this section we present three methods of artifact removal: using (a) a wavelet filter, (b) a neural ICA filter, and (c) the mixed wavelet–ICA filter; the last method is the main contribution of this work. We describe in detail the methodologies, and in the last section numerous simulations are proposed to substantiate the work.

8.4.1 Time–Frequency Analysis: Wavelet Filter

Wavelet analysis is a special time–frequency representation [10] that was introduced in order to overcome the limitations of both time and frequency resolution of the classical Fourier techniques, unavoidable also in their proposed evolutions, such as the short-time Fourier transform (STFT). The classical Fourier transform (FT) allows us to represent a signal in two different domains: the time and spectral domains. However, while in the former all the frequency information is hidden by the time content, in the latter the temporal information is lost in order to extract all the frequency content. This results from the choice of the basis Fourier functions, the exponential functions, which are nonzero in the entire time axis. In the STFT, the original signal is windowed in a time interval before the decomposition over the same basis functions. This method allows a time–frequency representation. However, a new problem arises using this method: How wide should the window be selected? If we select a very narrow window, we can optimize the time resolution but we lose frequency resolution and vice versa when we choose a very large window, coming back to the classical FT.

A possible solution to this problem is yielded by the wavelet transform (WT), which allows us to obtain better time resolution for the highest frequency components and poorer time resolution (with better frequency resolution) for the lowest frequency components. Most of the signals that we will meet are basically in agreement with this simple analysis: Their lowest frequencies last for long time intervals, while their highest frequencies last for short time intervals. Actually, most biomedical signals are in agreement with this practical rule. The wavelet transform is a multiresolution analysis (MRA). Here, a scaling function $\varphi(t)$ is used to create a series of approximations of a signal, each differing by a factor 2 (or by another fixed factor) from its nearest-neighbor approximations. Additional functions $\psi(t)$, called *wavelets*, are then used to encode the difference between adjacent approximations.

In its discrete version, the WT is implemented by a bank of bandpass filters each having a frequency band and a central frequency half that of the previous one [10, 11]. First the original signal $s(t)$ is passed through two filters, a low-pass one and a high-pass one. From the low-pass filter an approximation signal is extracted, $A(t)$, whereas from the high-pass signal a detail signal, $D(t)$, is taken out. In the standard tree of decomposition only the approximation signal is passed again through the second stage of filters, and so on until the last level of the decomposition. For each level the frequency band of the signal and the sampling frequency are halved. The wavelet series expansion of a signal $s(t) \in L^2(\mathbf{R})$, where $L^2(\mathbf{R})$ denotes the set of all measurable square-integrable functions, can be expressed as

$$s(t) = \sum_{k} c_{j_0 k}\varphi_{j_0 k}(t) + \sum_{j=j_0}^{\infty}\sum_{k} d_{jk}\psi_{jk}(t) \tag{8.2}$$

where j_0 is an arbitrary starting scale,

$$d_{jk} = \int x(t)\psi_{jk}^*(t)\,dt \tag{8.3}$$

are called the detail or wavelet coefficients, and

$$\psi_{jk}(t) = \frac{1}{\sqrt{2^j}}\psi\left(\frac{t - k2^j}{2^j}\right) \tag{8.4}$$

are the wavelet functions. The approximation or scaling coefficients are

$$c_{jk} = \int x(t)\varphi_{jk}^*(t)\,dt \tag{8.5}$$

where

$$\varphi_{jk}(t) = \frac{1}{\sqrt{2^j}}\varphi\left(\frac{t - k2^j}{2^j}\right) \tag{8.6}$$

are the scaling functions. The details and the approximations are defined as

$$D_j(t) = \sum_{k} d_{jk} \cdot \psi_{jk}(t)$$

$$A_j(t) = \sum_{k} c_{jk} \cdot \varphi_{jk}(t) \tag{8.7}$$

and the final reconstruction of the original signal can be computed by the details and the approximations and it can be described by the following equation for fixed N:

$$s(t) = A_N(t) + D_1(t) + D_2(t) + \cdots + D_N(t) \tag{8.8}$$

The wavelet analysis can be used to perform artifact removal [5]. Its practical application is based on the spectral separation between the original signal and the artifact: A good removal is possible only if the artifact spectral content is well localized (compactly supported).

In the case of multichannel recordings artifact removal must be performed separately, channel by channel, by applying the same algorithm for each channel recording. The process of artifact removal by using the WT can be resumed as follows:

1. Wavelet decomposition of single channel recording. From an original corrupted signal s, we obtain the approximation and the detail signals, $s(t) = A_N(t) + D_1(t) + D_2(t) + \cdots + D_N(t)$.

2. Identification of the detail signals that represent the artifact, for instance $D_i(t)$, $D_j(t)$, $D_k(t)$, ...

3. Thresholding of the detail signals that represent the artifact, yielding $\check{D}_i(t)$, $\check{D}_j(t)$, $\check{D}_k(t)$, ...

4. Wavelet reconstruction of the cleaned data: $\check{s}(t) = A_N(t) + \check{D}_1(t) + \check{D}_2(t) + \cdots + \check{D}_N(t)$, where $\check{D}_i(t) = D_i(t)$ if the detail $D_i(t)$ does not contain any artifact contributions.

8.4.2 Independent Component Analysis: Neural ICA Filter

The ICA [12] is a method for solving the *blind source separation* (BSS) problem: to recover N independent source signals, $\mathbf{s} = \{s_1(t), s_2(t), \ldots, s_N(t)\}$ from M linear mixtures, $\mathbf{x} = \{x_1(t), x_2(t), \ldots, x_M(t)\}$, modeled as the result of multiplying the matrix of source activity waveforms by an unknown matrix $\mathbf{A}$:

$$\mathbf{x} = \mathbf{As} \tag{8.9}$$

The basic blind signal processing techniques do not use any training data and do not assume any a priori knowledge about the parameters of the mixing systems (i.e., absence of knowledge about the matrix $\mathbf{A}$). The ICA resolves the BSS problem under the hypothesis that the sources are statistically independent of each other. In the last 10 years many algorithms have been proposed to perform the ICA. We use here the simple Bell–Sejnowski Infomax algorithm, developed in 1995, which can be described by the block scheme depicted in Figure 8.4. In this algorithm, maximizing the joint entropy $H(\mathbf{y})$, of the output of a neural processor is equivalent to minimizing the mutual information among the output components, $\mathbf{y}_i = \varphi(\mathbf{u}_i)$, where $\varphi(\mathbf{u}_i)$ is an invertible bounded nonlinearity [12–14]. In particular the algorithm estimates a matrix $\mathbf{W}$ such that

$$\mathbf{s} = \mathbf{Wx} \tag{8.10}$$

Recently, Lee et al. [15] extended the ability of the Infomax algorithm to perform BSS on linear mixtures of sources having either sub- or super-Gaussian distributions. The Infomax principle applied to the source separation problem consists of maximizing the information transfer through a system of the general type

$$I(\mathbf{y}, \mathbf{u}) = H(\mathbf{y}) - H(\mathbf{y} \mid \mathbf{u}) \tag{8.11}$$

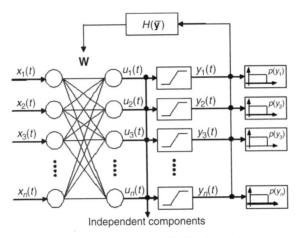

Figure 8.4 Infomax algorithm: block scheme. It uses an unsupervised learning procedure under the constraint of entropy maximization of the nonlinear outputs.

where $H(\mathbf{y})$ is the entropy of the output nonlinearities and $H(\mathbf{y} \mid \mathbf{u})$ is the residual entropy that did not come from the input and has the lowest possible value.

It can be viewed that the maximization of the information transfer is equivalent to the maximization of the entropy of the sigmoid outputs, which is the reason the Infomax algorithm is also interpreted as a variant of the well-known maximum-entropy method (MEM). The output nonlinearity is defined as

$$\varphi(\mathbf{u}) = -\frac{1}{p(\mathbf{u})}\frac{\partial p(\mathbf{u})}{\partial \mathbf{u}} \tag{8.12}$$

where $p(\mathbf{u})$ is the estimated probability density function of the independent sources. The learning rule is

$$\Delta \mathbf{W} \propto [\mathbf{I} - \varphi(\mathbf{u})\,\mathbf{u}^{\mathrm{T}}]\,\mathbf{W} \tag{8.13}$$

where the nonlinearity is often selected as the hyperbolic tangent (only for super-Gaussian sources):

$$\varphi(\mathbf{u}) = \tanh(\mathbf{u}) \tag{8.14}$$

The algorithm ends revealing the independent sources $\mathbf{u}$ and the unmixing matrix $\mathbf{W}$. To reconstruct the mixed signals, it is necessary to compute the matrix $\mathbf{A}$ (that is the pseudoinverse of $\mathbf{W}$): $\mathbf{x} = \mathbf{A}\mathbf{u}$. The application of ICA on the artifact removal has been successfully proposed in biomedical signal analysis [7, 16–19], particularly on brain signals [4]. This technique allowed, for instance, separation of the unwanted contribution of electrooculography (EOG) and electrocardiography (ECG) in raw electroencephalographic (EEG) signal and in sEMG recordings [17, 18].

Unlike wavelet analysis, the ICA approach can be applied only to multichannel recordings. Consider some M-channel recordings $\mathbf{x}_i$, $i = 1, \ldots, M$. Artifact removal by ICA is done as follows:

1. The ICA over all channels of multichannel recordings: We obtain N independent components $\mathbf{u}_1, \ldots, \mathbf{u}_N$ and a mixing matrix $\mathbf{A}$.

2. Identification of the artifact sources, for instance, $\mathbf{u}_i, \mathbf{u}_j, \ldots, \mathbf{u}_k$.

3. Elimination of the artifact sources. It is sufficient to set to zero the columns i, j, k of the matrix $\mathbf{A}$, obtaining the new mixing matrix $\hat{\mathbf{A}}$.

4. Artifact component removal and data reconstruction, $\mathbf{x}_{\text{rec}} = \hat{\mathbf{A}} \cdot \mathbf{u}$.

The ICA can often be simplified by means of the principal-component analysis (PCA) preprocessing [20] in such a way that the computational burden can be reduced, thus reducing the computational time. The PCA is a well-known method of extracting from a signal mixing some uncorrelated but not independent sources, since two signals can be uncorrelated without being independent; conversely, if two signals are independent, they are also uncorrelated. The PCA can be used to reduce the data dimensionality and/or to perform whitening to obtain some unit variance components.

8.4.3 Proposed Mixed Wavelet–ICA Filter

The proposed algorithm, described by the block scheme shown in Figure 8.5, encompasses the properties of the wavelet filter with those of the neural ICA filter. In a wavelet–ICA filter a preprocessing step based on a discrete wavelet transform (DWT) is applied. We perform the wavelet decomposition at a fixed level for each channel. Details that concern the spectral range where the artifact is localized are selected. Then, the ICA block minimizes a measure of statistical independence of the new data set by determining the minima of a suitable objective function. This new algorithm works as follows:

1. Wavelet decomposition of every channel of multichannel recordings.

2. Selection of the details that contain some artifact component:

  a. PCA and/or whitening to lighten the computational charge.

  b. ICA by means of the above introduced Infomax algorithm.

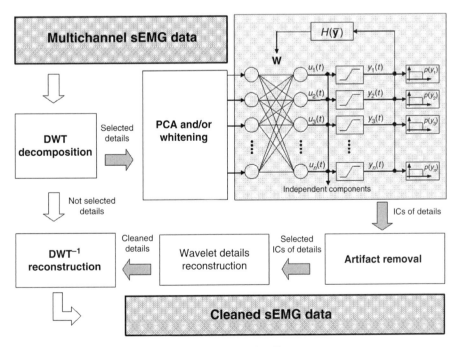

Figure 8.5 Block scheme of proposed wavelet–ICA filter.

    c. Artifact removal by ICA as described in the previous section.

    d. ICA reconstruction to obtain cleaned details.

3. Inverse discrete wavelet analysis (wavelet reconstruction) using the cleaned details revealed in step 2d and the unselected details obtained in step 1. The output of this last step is the cleaned signal mapping.

8.5   SIMULATIONS: FILTER PERFORMANCE

The simulations of the artifact removal by means of wavelet, ICA, and wavelet–ICA filters are shown in this section. We test the performances of the above-mentioned algorithms by means of some simulations designed on specially synthesized signals. The artifacts are also synthesized in order to test the quality of the different approaches. A 120-s sEMG signal was generated. Each signal was first subtracted by its mean value; thus each processed signal is at zero mean. Then, for each one of the above-described approaches, a signal is mixed with a different artifact signal and finally the outputs of the three procedures are investigated.

In order to give a quantitative measure of the goodness of the algorithm, we use the covariance matrix (i.e., for zero-mean signals, the correlation matrix). The entries of this matrix (given in percentage), which represent a comparison between the original signal (when the artifact is not yet added) and the reconstructed signal after artifact removal, are the cross-correlation coefficients. A similar performance parameter is calculated computing the covariance matrix between the spectrum of the original signal and that of the reconstructed sEMG after artifact removal. To avoid any possible misunderstanding, we call the first parameter the time cross-correlation coefficient and the second the spectral cross-correlation coefficient.

8.5.1   Wavelet Filter Performance

We synthesizes a sEMG signal corrupted by an artifact signal. The artifact is a periodic high-frequency burst. This kind of artifact can be viewed as a stimulus signal that replicates itself with an established time period. Moreover, its frequency content is well localized and it is almost entirely separated by the frequency content of the original signal. The mean frequency of the artifact spectral content is around 30% higher than the maximum frequency present in the original signal. This is a kind of signal quite identifiable by the WT algorithm. In Figure 8.6 we show the original synthesized sEMG signal, the synthesized artifact signal, and the corrupted sEMG signal (original signal mixed with the artifact).

In Figure 8.7 the complete wavelet decomposition (performed by the DWT algorithm) is shown. This figure confirms the high ability of this filter to separate the sEMG signal by the artifact, which is almost all included in the first detail (which includes the highest frequencies). The kind of wavelet function used in this application and the respective scaling function are called Daubechies 4; these two functions are shown in Figures 8.7*b,c*.

Figure 8.8 shows the wavelet artifact removal application by comparing the original signal with the reconstructed one after artifact removal. The time cross-correlation coefficient calculated by means of the covariance matrix is very high (96.4%), showing that this algorithm works very well with this kind of artifact. Figure 8.9 shows the coherence function (as it was defined in Section 8.4) for these two signals (the original one and the filtered one); this figure also shows the high performance of the algorithm, revealing a value that is unity almost everywhere.

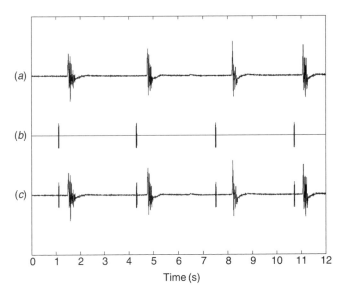

Figure 8.6 (*a*) Original synthesized sEMG signal. (*b*) Synthesized artifact. (*c*) Corrupted sEMG signal obtained by mixing (*a*) and (*b*).

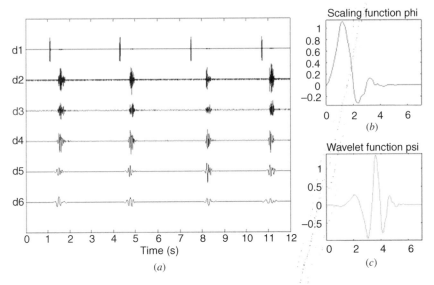

Figure 8.7 (*a*) Wavelet decomposition (details) of synthesized corrupted sEMG signal. The first detail (associated by the higher frequency content) contains almost all the artifact contributions. (*b*) Scaling Daubechies 4 function. (*c*) Wavelet Daubechies 4 function.

Finally, Table 8.1 resumes the performances of this approach to artifact removal, showing also the spectral cross-correlation coefficient (95.9%), which reveals again the goodness of the algorithm.

8.5.2 Neural ICA Filter Performance

In this application the original sEMG recording is made of three channel recordings. The artifact signal is a similar ECG signal. This kind of artifact is more difficult to deal with.

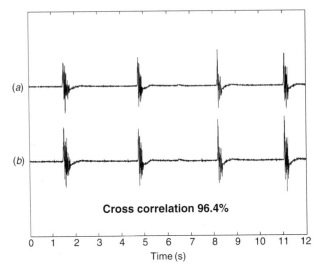

Figure 8.8 (*a*) Original synthesized sEMG signal and (*b*) reconstructed signal after artifact removal. The cross-correlation between the two signals is very high (96.4%).

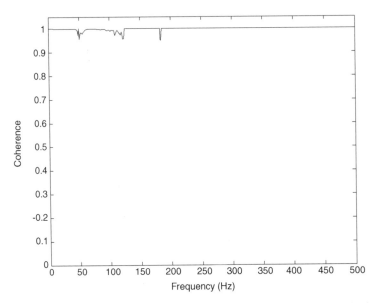

Figure 8.9 Coherence function (defined in Section 8.3) between signals shown in Figure 8.8. This value is 1 almost everywhere, showing the high performance of the wavelet approach for this kind of artifact removal.

TABLE 8.1 Artifact Removal: Correlation

Channel of sEMG	Wavelet filter	
	T_{corr}	F_{corr}
CH1	0.964	0.959

Note: T_{corr} is the time cross-correlation between original channel and reconstructed channel. F_{corr} is the spectral cross-correlation between original channel and reconstructed channel.

First, its spectral content is not compactly supported, but it is nonzero in almost all the frequency axis. Moreover, its time shape is different for each recording channel. This is the most suitable kind of signal identifiable by an ICA approach. Figure 8.10 shows the synthesis of a corrupted sEMG. In Figure 8.10a, three sEMG synthesized signals are shown (1, 2, 3) together with a similar ECG synthesized artifact (4). In Figure 8.10b each sEMG signal is mixed with the artifact, thus generating the corrupted sEMG signal.

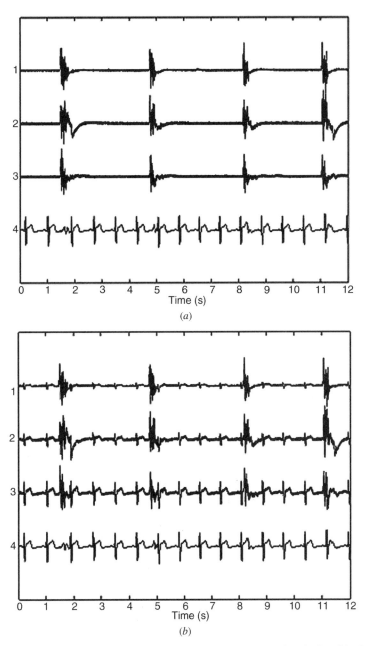

Figure 8.10 (a) Original synthesized sEMG signal. The fourth signal is the synthesized artifact, a similar ECG signal. (b) Corrupted sEMG signal obtained by mixing recordings 1, 2, and 3 of (a) with artifact 4 of (a).

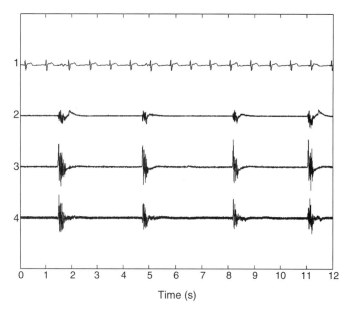

Figure 8.11   Independent components: results of ICA application to signals shown in Figure 8.10*b*. The algorithm was able to separate the artifact from the corrupted signals.

Figure 8.11 shows the independent components extracted by means of the neural ICA algorithm. We can see that this approach is able to separate the artifact from the recordings. In this figure the artifact is at the first channel, while in the original synthesized signals it was at the fourth channel. This is caused by one of the two ambiguities of ICA: We cannot determine the order of the independent components (the other ambiguity is represented by the inability to determine the variances and then the energies of the independent components).

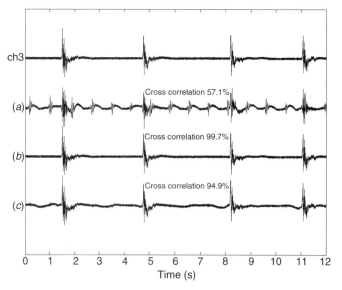

Figure 8.12   Performance of different approaches to artifact removal showing the third recording and the reconstructed signals after artifact removal: (*a*) wavelet filter; (*b*) neural ICA filter; (*c*) wavelet–ICA filter.

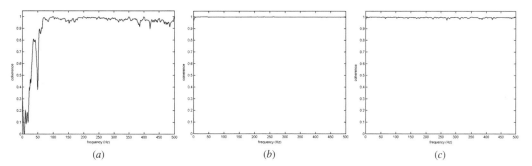

Figure 8.13   Coherence between original signal and reconstructed one after artifact removal: (*a*) wavelet filter; (*b*) neural ICA filter; (*c*) wavelet–ICA filter.

To test the performance of the implemented filters, we selected the least favorable recording channel (in this case, channel 3), and we computed the cross-correlation coefficient after the application of the three different approaches presented in this chapter.

Figure 8.12 resumes this performance test. The first raw signal represents the original recording of the third channel. The second raw signal is the reconstructed signal after the artifact removal by the wavelet filter described Section 8.4.1: The performance is not good (the cross-correlation coefficient is 57.1%), according to our previous choice of the artifact signal. The third raw signal is the reconstructed signal after artifact removal by the neural ICA filter described in Section 8.4.2: The performance is very good (the cross-correlation coefficient is 99.7%), revealing the high quality of this approach.

Finally, the fourth raw signal is the reconstructed signal after artifact removal by means of the wavelet–ICA filter described in Section 8.4.3: Its performance is good (the cross-correlation coefficient is 94.9%), even if it is not as good as the neural ICA approach. It is important to observe that the ICA filter works very well because in the corrupted signals we include also the artifact signal and thus separation by the algorithm becomes very easy.

Figure 8.13 shows the coherence functions for the three approaches in the described application. Here we can confirm the high quality of the two last approaches (unitary coherence almost everywhere) and the low performance of the wavelet approach. Table 8.2 presents the results shown in the figures, showing the quality parameters (time and spectral cross-correlation coefficients) also in the other two recordings.

8.5.3   Wavelet–ICA Filter Performance

In this application, the same original sEMG recordings are corrupted by a new kind of artifact that encompasses the characteristics of the artifacts shown respectively in

TABLE 8.2   Artifact Removal: Correlation

Channel of sEMG	Wavelet filter		Neural ICA filter		Wavelet–ICA filter	
	T_{corr}	F_{corr}	T_{corr}	F_{corr}	T_{corr}	F_{corr}
CH1	0.833	0.797	0.999	0.999	0.997	0.995
CH2	0.877	0.801	0.999	0.999	0.997	0.993
CH3	0.571	0.556	0.997	0.995	0.949	0.991

Note: T_{corr} is time cross-correlation between original channel and reconstructed channel. F_{corr} is spectral cross-correlation between original channel and reconstructed channel.

Figures 8.6 and 8.10*a*. In effect, this artifact is composed by a similar ECG signal mixed with some muscle activity bursts. For these corrupted signals the new approach has revealed the best ability to separate the original signal by the artifact.

Figure 8.14 shows the synthesis of a corrupted sEMG. In Figure 8.14*a* three sEMG synthesized signals are shown (1, 2, 3), together with a synthesized artifact (4). In

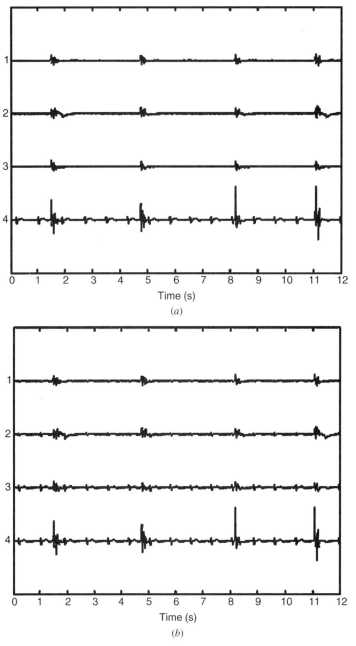

(*a*)

(*b*)

Figure 8.14 (*a*) Original synthesized sEMG signals. The fourth signal is the synthesized artifact. (*b*) Corrupted sEMG signals obtained by mixing recordings 1, 2, and 3 of (*a*) with artifact 4 of (*a*).

Figure 8.14*b* each sEMG signal is mixed with the artifact, generating the corrupted sEMG signal.

Figure 8.15*a* shows the synthesised artifact compared with its identification by the wavelet–ICA filter, while in Figure 8.15*b* all the removed artifacts for each channel of Figure 8.14*b* are shown. To test the performance of the implemented filters, we selected

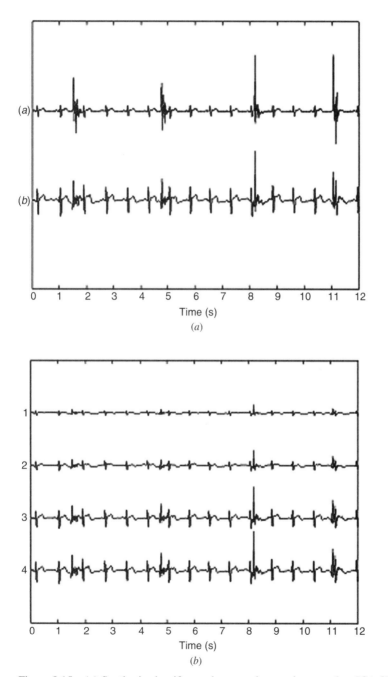

Figure 8.15 (*a*) Synthesized artifact and removed one using wavelet–ICA filter. (*b*) Removed artifacts for each recording channel.

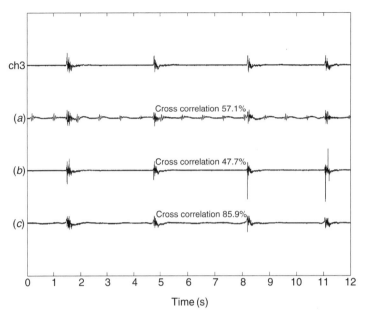

Figure 8.16   Performance of different approach to artifact removal. Shown are the third recording and the reconstructed signals after artifact removal: (*a*) by wavelet filter; (*b*) by neural ICA filter; (*c*) by wavelet–ICA filter.

the least favorable recording channel (in this case, channel 3), and we computed the cross-correlation coefficient after the application of the three different approaches presented in this chapter.

Figure 8.16 resumes this performance test. The first raw signal represents the original recording of the third channel.

The second raw signal is the reconstructed signal after artifact removal by the wavelet filter described in Section 8.4.1: The performance is not so good (the cross-correlation coefficient is 57.1%), but we already knew that this kind of artifact was not suitable for a wavelet approach.

The third raw signal is the reconstructed signal after artifact removal by the neural ICA filter described in Section 8.4.2: The performance is very poor (the cross-correlation coefficient is 47.7%), revealing the poor reliability of this approach to remove such

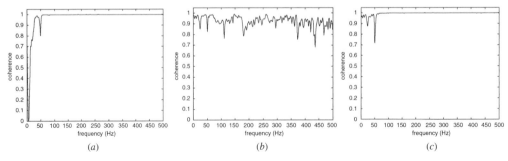

Figure 8.17   Coherence between original signal and reconstructed one after artifact removal: (*a*) by wavelet filter; (*b*) by neural ICA filter; (*c*) by wavelet–ICA filter.

TABLE 8.3 Artifact Removal: Correlation

Channel of sEMG	Wavelet filter		Neural ICA filter		Wavelet–ICA filter	
	T_{corr}	F_{corr}	T_{corr}	F_{corr}	T_{corr}	F_{corr}
CH1	0.833	0.785	0.974	0.951	0.985	0.979
CH2	0.877	0.813	0.972	0.951	0.985	0.979
CH3	0.571	0.553	0.477	0.455	0.859	0.851

Note: T_{corr} is time cross-correlation between original channel and reconstructed channel. F_{corr} is spectral cross-correlation between original channel and reconstructed channel.

artifacts. Finally, the fourth raw signal is the reconstructed signal after artifact removal by the wavelet–ICA filter described in Section 8.4.3: Its performance is very good (the cross-correlation coefficient is 85.9%), showing that this is the best approach to remove such artifacts.

Figure 8.17 shows the coherence functions for the three described approaches in the same application. Here, we can confirm both the high quality of the last approach (unitary coherence almost everywhere) and the bad performance of the first two approaches. Table 8.3 presents the results shown in the figures, showing the quality parameters (time and spectral cross-correlation coefficients) also in the others two recordings.

8.6 APPLICATION: COMMON DRIVE DETECTION

In this section, we present the coherence analysis applied on the contractions cycles acquired as described in Section 8.2 and the related implications in the medical task that we described in the first section: the common drive detection. We first perform artifact removal by means of the wavelet–ICA filter on the sEMG recordings related to the pectoral muscles activity.

8.6.1 Wavelet–ICA Filter Preprocessing: Artifact Removal

Figure 8.18 shows the recordings of the sEMG signals acquired on the pectoral muscles. It is now evident that an ECG artifact strongly affects the acquisition of the true sEMG signal. This artifact is present in all recording channels. Artifact removal is a required pre-processing analysis in order to make meaningful the results of the medical task. The features of this artifact are comparable with the peculiarities of the synthesized artifact shown in Figure 8.15a. Thus, based on the considerations carried out in the previous section, we perform the artifact removal using the wavelet–ICA filter, computing the filtered sEMG signals shown in Figure 8.19. To test the performance of this approach for the artifact removal in these real sEMG signals, we compare the ECG reduction performed by the three different approaches presented in the previous section. The results are shown in Figure 8.20 for the most corrupted electrode, and they are presented in Table 8.4. Yet, the best approach is the wavelet–ICA filter, revealing a 98% ECG reduction, with respect to the 90% for the neural ICA filter and the 73% of the wavelet filter (for the third channel).

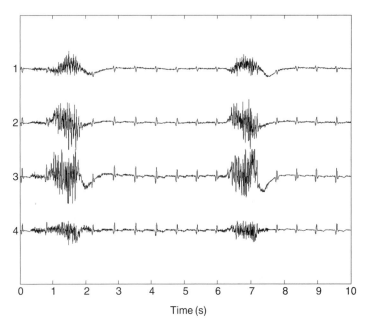

Figure 8.18   Surface EMG recordings of pectoral muscles. The location of the electrodes is shown in Figure 8.1.

8.6.2   Common Drive Detection: Coherence Analysis

In Figure 8.21 (respectively, Fig. 8.21*a* for the pectoral muscles and Fig. 8.21*b* for the FDI muscles) we show the coherence between two electrodes attached to the same

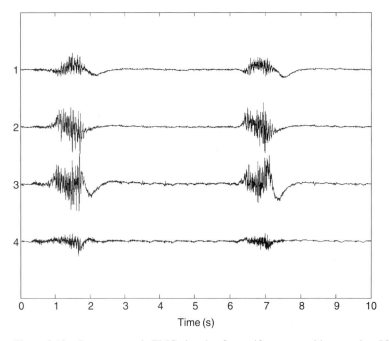

Figure 8.19   Reconstructed sEMG signals after artifact removal by wavelet–ICA filter.

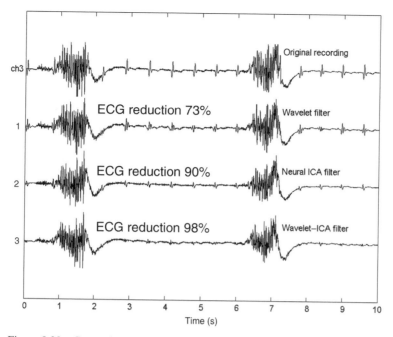

Figure 8.20   Comparison between ECG reduction in three approaches of artifact removal presented in this chapter.

TABLE 8.4   Artifact Removal: ECG Reduction

Channel of sEMG	ECG reduction (%)		
	Wavelet filter	Neural ICA filter	Wavelet–ICA filter
CH1	80	93	90
CH2	78	90	99
CH3	73	90	98
CH4	78	85	98

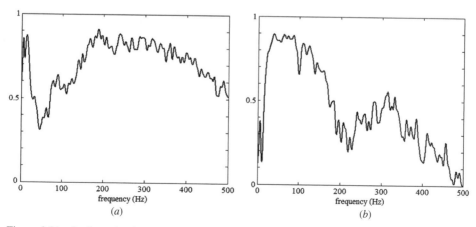

Figure 8.21   Ipsilateral coherence: (*a*) pectoral muscles; (*b*) FDI muscles.

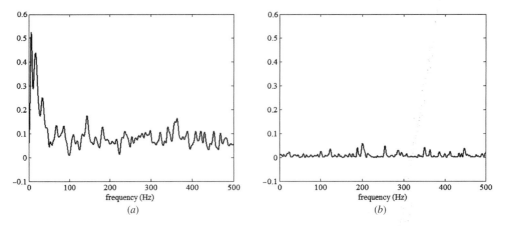

Figure 8.22 Bilateral coherence: (*a*) pectoral muscles; (*b*) FDI muscles.

muscle (ipsilateral coherence). As conjectured, we found that there is a high coherence value for almost all frequencies of the spectrum. This condition is true for both muscle types.

In Figure 8.22 the coherence between electrodes attached to different sides (bilateral coherence) is shown. The analysis was performed for both muscle types. In Figure 8.22*a*, we investigate the coherence between pectoral muscles. There is a meaningful high coherence value for low frequencies (<4 Hz). Figure 8.22*b* shows the same analysis performed for the signals recorded by FDI muscles. We can observe that there is not a high coherence value for the whole frequency axis.

The comparison between calculated functions for the two different kinds of muscles is the result we were searching for: The high coherence value for low frequencies in

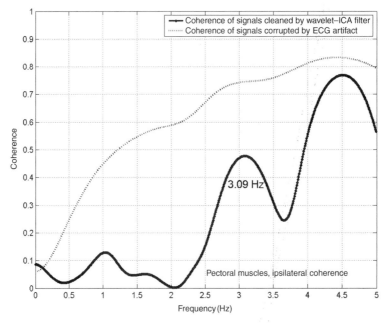

Figure 8.23 Pectoral muscles: ipsilateral coherence. Continuous line: after artifact removal preprocessing. Dashed line: without artifact removal preprocessing.

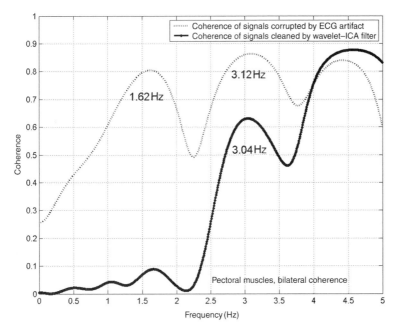

Figure 8.24   Pectoral muscles: bilateral coherence. Continuous line: after artifact removal preprocessing. Dashed line: without artifact removal preprocessing.

pectoral muscles is exactly the common drive that starts from both brain lobes. The absence of this coherence peak in the FDI muscles is in agreement with physiological studies. A low-frequency zoom of coherence analysis is shown in Figures 8.23, 8.24 (continuous line), and 8.25.

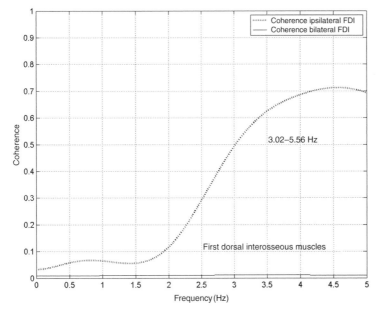

Figure 8.25   FDI muscles: coherence analysis. Continuous line: bilateral coherence. Dashed line: ipsilateral coherence.

TABLE 8.5   Coherence Peak (<4 Hz)

Muscle	Ipsilateral		Bilateral	
	Coherence	Frequency (Hz)	Coherence	Frequency (Hz)
Pectoral	~0.5	3.09	>0.5	3.04
First dorsal interosseous	>0.5	3.02–5.56	~0.0	0–5

We have shown the need for artifact removal preprocessing in Figures 8.23 and 8.24. If we analyze the coherence function between two electrodes attached to both pectoral muscles, we find a high coherence value for almost all the low frequencies. This result is not meaningful, as we can demonstrate by analyzing Figure 8.24, where the spectrum of the contraction for low frequencies is shown. We can observe that there are two dominant peaks, the first at 1.62 Hz and the second at 3.12 Hz. It seems obvious to identify the first peak with the cardiac activity. The high value of the coherence is due to the presence of these two activities in both muscles. But this is not what we are searching for. Our aim was to investigate the presence of a common drive starting from the brain.

Table 8.5 presents the most important results achieved in the present work, that is, the presence of a common drive, starting from the brain, only for the pectoral muscles (and not for the FDI ones), revealed by a high value of the coherence function located at the frequency of 3.09 Hz.

8.7   CONCLUSIONS

This chapter presented a method of detecting common drives for axial muscle control. The coherence analysis has been shown to be fully adaptable in order to achieve these results. A novel procedure for processing sEMG data has been proposed that encompasses ICA, wavelet transform, and coherence analysis. It allows us to extract the information needed for the task. The wavelet–ICA filter was used to properly clean the raw signals from the artifacts that unavoidably corrupt the real signals.

The comparison between coherence analysis in axial muscles and in distal ones reveals the presence of the searched common drive only in the first category of muscles, as supposed by previous clinical studies.

REFERENCES

1. J. F. MARSDEN, S. F. FARMER, D. M. HALLIDAY, J. R. ROSENBERG, AND P. BROWN, "The unilateral and bilateral control of motor unit pairs in the first dorsal interosseous and paraspinal muscles in man," *Journal of Physiology*, vol. 521.2, pp. 553–564, 1999.
2. J. D. BRONZINO, *The Biomedical Engineering Handbook*, vols. I and II, 2nd ed., CRC Press and IEEE Press, Boca Raton, FL, 2000.
3. A. M. AMJAD, D. M. HALLIDAY, J. R. ROSENBERG, AND B. A. CONWAY, "An extended difference of coherence test for comparing and combining several independent coherence estimates: Theory and application to the study of motor units and physiological tremor,"

Journal of Neuroscience Methods, vol. 73, pp. 69–79, 1997.
4. T. JUNG, S. MAKEIG, M. WESTERFIELD, J. TOWNSEND, E. COURCHESNE, AND T. SEJNOWSKI, "Removal of eye activity artifacts from visual event-related potentials in normal and clinical subjects," *Clinical Neurophysiology*, vol. 111, pp. 1745–1758, 2000.
5. H. LIANG AND Z. LIN, "Stimulus artifact cancellation in the serosal recordings of gastric myoelectric activity using wavelet transform," *IEEE Transactions on Biomedical Engineering*, vol. 49, pp. 681–688, 2002.
6. J. S. KARLSSON, B. GERDLE, AND M. AKAY, "Analyzing surface myoelectric signals recorded during

isokinetic contractions," *IEEE Engineering in Medicine and Biology Magazine*, vol. 20, pp. 97–105, 2001.

7. B. AZZERBONI, G. FINOCCHIO, M. IPSALE, F. LA FORESTA, M. J. MCKEOWN, AND F. C. MORABITO, "Spatio-temporal analysis of surface electromyography signals by independent component and time-scale analysis," in *Proceedings of the Second Joint EMBS/BMES Conference*, Houston, TX, 2002, pp. 112–113.

8. B. AZZERBONI, M. IPSALE, F. LA FORESTA, AND F. C. MORABITO, "Common drive detection for axial muscles cerebral control and coherence analysis of surface electromyography by neural networks," in *Neural Engineering, 2003, Proceedings of the First International IEEE EMBS Conference*, 2003, pp. 352–355.

9. S. M. KAY, *Modern Spectral Estimation: Theory and Application*, Prentice-Hall, Englewood Cliffs, NJ, 1988.

10. M. AKAY, *Time Frequency and Wavelets in Biomedical Signal Processing*, IEEE Press, New York, 1997.

11. R. C. GONZALEZ AND R. E. WOODS, *Digital Image Processing*, Prentice-Hall, Upper Saddle River, NJ, 2002.

12. A. HYVARINEN, J. KARHUNEN, E. OJA, *Independent Component Analysis*, Wiley, New York, 2001.

13. A. J. BELL AND T. J. SEJNOWSKI, "An information-maximisation approach to blind separation and blind deconvolution," *Neural Computation*, vol. 7, pp. 1129–1159, 1995.

14. I. KOPRIVA AND H. SZU, "Space-time variant blind source separation with additive noise," Lecture Notes in Computer Science, vol. 3195, pp. 240–247, 2004.

15. T.-W. LEE, M. GIROLAMI, AND T. SEJNOWSKI, "Independent component analysis using an extended infomax algorithm for mixed sub-Gaussian and super-Gaussian sources," *Neural Computation*, vol. 11, no. 2, pp. 606–633, 1999.

16. B. AZZERBONI, G. FINOCCHIO, M. IPSALE, F. LA FORESTA, AND F. C. MORABITO, "A new approach to detection of muscle activation by independent component analysis and wavelet transform," *Lecture Notes in Computer Science*, vol. 2486, pp. 109–116, 2002.

17. B. AZZERBONI AND F. LA FORESTA, "Clinical applications of myoelectric signal processing by neural network and spectral analysis," in *Proceedings of The 4th Annual IEEE EMBS Special Topic Conference on Information Technology Applications in Biomedicine (ITAB 2003)*, 2003, pp. 265–268.

18. B. AZZERBONI, M. CARPENTIERI, F. LA FORESTA, AND F. C. MORABITO, "Neural–ICA and wavelet transform for artifacts removal in surface EMG," *Proceedings of The 2004 International Joint Conference on Neural Networks (IJCNN 2004)*, 2004, pp. 3223–3228.

19. F. LA FORESTA, N. MAMMONE, AND F. C. MORABITO, "Independent component and wavelet analysis for fECG extraction: The ST waveform evaluation," in *Proceedings of The 2005 IEEE International Conference on Computational Intelligence for Measurement System and Applications (CIMSA 2005)*, 2005, pp. 86–90.

20. B. AZZERBONI, M. IPSALE, F. LA FORESTA, AND F. C. MORABITO, "PCA and ICA for the extraction of EEG dominant components in cerebral death assessment," in *Proceedings of The 2005 International Joint Conference on Neural Networks (IJCNN 2005)*, 2005, pp. 2532–2537.

CHAPTER 9

MULTIRESOLUTION FRACTAL ANALYSIS OF MEDICAL IMAGES

Khan M. Iftekharuddin and Carlos Parra

9.1 INTRODUCTION

The fractal concept developed by Mandelbrot [1, 2], who coined the term *fractal* from the Latin *fractus*, provides a useful tool to explain a variety of naturally occurring phenomena. A fractal is an irregular geometric object with an infinite nesting of structure at all scales. Fractal objects can be found everywhere in nature, such as coastlines, fern trees, snowflakes, clouds, mountains, and bacteria. Some of the most important properties of fractals are self-similarity, chaos, and noninteger fractal dimension (FD). Fractals are self-similar, which means that structures are repeated at different scales of size. The FD offers a quantitative measure of self-similarity and scaling of texture [3]. The texture represents important characteristic of an image such as ruggedness and underlying structural self-similarity patterns. For medical image processing, texture may be fundamental to differentiate tissues as well as to detect and analyze pathologies or lesions [4]. The factors contributing to image texture include the random nature of the object itself (tissues and body structures), noise, imaging modality, imaging limitations, and a variety of measurement parameters. Thus, texture analysis offers a promising approach to robust differentiation of healthy tissues and tumor.

The fundamental fractal theory that models texture [1, 2] also provides the framework for the modeling of different kinds of fractal motion. One of the most useful applications of this theory is the generation of fractional Brownian motion (fBm), which is appropriate for the visual simulation of fractal structures and processes frequently found in nature [5] as well as in biomedical imaging. In addition to fractal theory, stochastic signal processing may play a vital role for the spectrum computation and parameter estimation of nonstationary processes. Extensive research [6–8] related to time–frequency description of fractal processes has been pursued successfully.

The fBm analysis of natural texture scenes is immensely enhanced with the development of both wavelet analysis and Mallat's multiresolution analysis (MRA) [9, 10]. The MRA offers a common signal processing framework wherein the statistical and spectral properties of the fBms may be exploited to estimate FD and hence the texture content of an image. Heneghan et al. [11] propose an important formulation of a 2D model that exploits both the spectral properties and correlation function of an fBm and estimates FD using the statistical properties of the continuous wavelet transform (CWT) of an fBm. Wornell [12] describes a detailed demonstration on how $1/f$ processes may be optimally represented in orthonormal wavelet bases, which may be useful for discrete wavelet

Handbook of Neural Engineering. Edited by Metin Akay
Copyright © 2007 The Institute of Electrical and Electronics Engineers, Inc.

formulations of the fBm. Considering the nonstationary charactereristics, alternative techniques have been proposed to estimate the spectral density of both 1D [13] and 2D [14–16] models of the fBm processes. Although the spectral estimation of fBm using Wigner–Ville distribution (WVD) suggests better performance over that of the traditional fast Fourier transform (FFT) estimation, the high computational cost of WVD may be prohibitive.

The fundamental goals of this work are the formal development of theoretical fBm models, formulation of corresponding algorithms, and evaluation of algorithms for robust identification of brain tumors in tomographic images. For the verification of our algorithmic framework, we generate and use synthetic fractal as well real computerized tomography (CT) images. The formal development of the 2D models for the statistical and spectral estimation of FD of these images is pursued and compared.

9.2 BACKGROUND REVIEW

The fBms are a part of the set of $1/f$ processes, corresponding to a generalization of the ordinary Brownian motion $B_H(s)$. They are nonstationary zero-mean Gaussian random functions, defined as [1, 2]

$$B_H(t) - B_H(s) = \frac{1}{\Gamma(H + 0.5)} \left\{ \int_{-\infty}^{0} \left[(t - s)^{H-0.5} - s^{H-0.5} \right] dB(s) \right. $$
$$\left. + \int_{0}^{\infty} (t - s)^{H-0.5} dB(s) \right\} \tag{9.1}$$

where the Hurst coefficient H, restricted to $0 < H < 1$, is the parameter that characterizes fBm, t and s correspond to different observation times of the process B_H, Γ is Euler's gamma function, and $B_H(0) = 0$. Despite the nonstationary characteristic of fBm processes, their increments are *stationary* and *self-similar* [8]. This observation suggests that the time–frequency signal decomposition techniques such as MRA or WVD [6] are well suited for the fBm signal analysis. Thus, the MRA and WVD may be exploited to obtain a framework for the efficient spectral computation of H and hence the FD of the corresponding fractal process [5] as follows:

$$FD = E_u + 1 - H \tag{9.2}$$

where E_u is the Euclidean dimension that contains the fBm (i.e., the position of each point of the process is described with the vector $\vec{x} = (x_1, \ldots, x_E)$

In general, fBms are nonstationary processes, given that the associated correlation function is not exclusively a function of the difference of observation times but is defined as [6]

$$r_{B_H}(t, s) = E[B_H(t)B_H(s)] = \frac{V_H}{2} \left(|t|^{2H} + |s|^{2H} - |t - s|^{2H} \right) \qquad 0 < H < 1 \tag{9.3}$$

with

$$V_H = \Gamma(1 - 2H) \frac{\cos(\pi H)}{\pi H} \tag{9.4}$$

which is a function only of the Hurst coefficient H. The nonstationary property suggests that spectral estimation of fBm may not be obtained by exploiting the standard spectral density and, hence, alternative techniques are needed to estimate the power content of these signals. However, the *increments* of the *fBm processes* are stationary, as can be seen in their variance function, which depends only on the time interval between the observations [7],

$$E\left[|B_H(t) - B_H(s)|^2\right] = V_H(t - s)^{2H} \tag{9.5}$$

with V_H defined as in (9.4). The *fBm increments* are also self-similar, as shown in the following equality of probability distributions at any scale α [7]:

$$B_H(t + \alpha\tau) - B_H(t) = \alpha^H B_H(t) \tag{9.6}$$

The previous properties can be extended to any dimension. For the 2D case, let $B(\vec{u})$ be an fBm, where $\vec{u}$ corresponds to the position (u_x, u_y) of a point in a 2D process satisfying the following conditions [11]:

(i) The process is nonstationary given that its correlation is not simply a function of $|\vec{u} - \vec{v}|$,

$$r_{B_H}(\vec{u},\ \vec{v}) = E\left[B_H(\vec{u})B_H(\vec{v})\right]\ = \frac{V_H}{2}\left(|\vec{u}|^{2H} + |\vec{v}|^{2H} - |\vec{u} - \vec{v}|^{2H}\right) \tag{9.7}$$

(ii) The increments of the process $\Delta B(\vec{u}) = B(\vec{u} + \Delta\vec{u}) - B(\vec{u})$ form a stationary, zero-mean Gaussian process.

(iii) The variance of the increments $\Delta B(\vec{u})$ depends only on the distance $\Delta u = \sqrt{\Delta u_x^2 + \Delta u_y^2}$ such that

$$E\left[|\Delta B_H(\vec{u})|^2\right] \propto \Delta u^H \tag{9.8}$$

In the Fourier domain, it is known that some of the most frequently observed structures in fractal geometry, generally known as $1/f$ processes, show a power spectrum following the power law relationship,

$$S(\omega) \propto \frac{k}{|\omega|^\gamma} \tag{9.9}$$

where ω corresponds to the spatial frequency and $\gamma = 2H + 1$. This type of spectrum is associated to statistical properties that are reflected in a scaling behavior (self-similarity), in which the process is statistically invariant to dilations or contractions, as described in the equation

$$S(\omega) = |a|^\gamma S_X(a\omega) \tag{9.10}$$

where a is a constant.

9.3 PROPOSED ANALYTICAL MODELS

Our proposed models are broadly grouped into two types such as the statistical variance method and the power spectrum method. For the power spectrum method, we implement both FFT and WVD approaches.

9.3.1 Statistical Variance Method

In this section, we define statistical properties of the fBm and consider its implications for signal analysis. We present the description of 1D and 2D models as well as an algorithm for the estimation of the H coefficients.

9.3.1.1 One-Dimensional Case

The essential characteristic of nonstationarity in fBm is manifested in its low-frequency components [6]. When the fBm is decomposed using a MRA, the low- and high-frequency components may be separated into a nonstationary approximation and a stationary detail part, given the low-pass and bandpass character of the respective analyzing wavelets. For a specific approximation resolution 2^J, the multiresolution representation of an fBm process is given by [9]

$$B_H(t) = 2^{-J/2} \sum_n a_J[n]\phi(2^{-J}t - n) \; + \sum_j 2^{-j/2} \sum_n nd_j[n]\psi(2^{-j}t - n) \qquad (9.11)$$

with $j = -J, \ldots, \infty$ and $n = -\infty, \ldots, \infty$ and the basic wavelet satisfying the admissibility condition [7]

$$\int_{-\infty}^{\infty} \psi(t)\, dt = 0 \qquad (9.12)$$

The orthonormal wavelet decomposition (details) of the fBm at resolution j is [7]

$$d_j[n] = 2^{-j/2} \int_{-\infty}^{\infty} B_H(t)\psi(2^{-j}t - n)\, dt \qquad (9.13)$$

From (9.3) and (9.13), the variance of the detail wavelet coefficients is related to the specific analyzing wavelet and the H coefficient of the fBm processes as [7]

$$E\big[|d_j[n]|^2\big] = \frac{V_H}{2} V_\psi(H)(2^j)^{2H+1} \qquad (9.14)$$

where $V_\psi(H)$ is a scale-independent constant that depends only on the inner product of the selected mother wavelet $\psi(t)$ and the value of H [6] such that

$$V_\psi(H) = -\int_{-\infty}^{\infty} \psi(t)\psi(s)\, dt\, |\tau|^{2H} d\tau \qquad (9.15)$$

with $\tau = t - s$. By applying the logarithm on both sides of Eq. (9.14), the resulting equation is linear [18] and given as

$$\log_2 E\big[|d_j[n]|^2\big] = (2H + 1)j + C_1 \qquad (9.16)$$

with the constant C_1 defined as

$$C_1 = \log_2 \frac{V_H}{2} V_\psi(H) \qquad (9.17)$$

The Hurst coefficient H (and the dimension) of a fBm process can be calculated from the slope of this variance and plotted as a function of the resolution in a log-log plot.

9.3.1.2 Two-Dimensional Case

The wavelet filter used to obtain the equation for the high-frequency components at a specific resolution j is described in [9]. From

(9.13), it may be shown that the 2D extension of the detail coefficients, at resolution j and $(n, m) \in Z^2$, is given as

$$D_{2^j}^3[\vec{\eta}] = \left(2^{-j} \int_{-\infty}^{\infty} B_H(\vec{u}) \cdot \Psi_{2^j}^3(\vec{u} - 2^{-j}\vec{\eta}) \, d\vec{u}\right)_{(\eta \in Z^2)} \tag{9.18}$$

where η corresponds to the position $[n, m]$ and $\Psi_{2^j}^3$ satisfies the admissibility condition of (9.12). The variance of the detail coefficients in Eq. (9.19) is obtained similar to the continuous-wavelet-based approach [11]. In 2D, the expression for the variance is

$$E\left[|D_{2^j}^3[\vec{\eta}]|^2\right] = 2^{-2j} \int_{\vec{u}} \int_{\vec{v}} \left(\Psi_{2^j}^{3*}(\vec{u} - 2^{-j}\vec{\eta}) \cdot \Psi_{2^j}^3(\vec{v} - 2^{-j}\vec{\eta}) E[B(\vec{u})B(\vec{v})]\right) d\vec{u} d\vec{v} \tag{9.19}$$

Considering the definition of the covariance function of a process in (9.7), the previous equation can be expanded as,

$$E\left[|D_{2^j}^3[\vec{\eta}]|^2\right] = \frac{V_H}{2} 2^{-2j}$$

$$\times \int_{\vec{u}} \Psi_{2^j}^3(\vec{v} - 2^{-j}\vec{\eta}) \, d\vec{v} \cdot \int_{\vec{v}} \Psi_{2^j}^{3*}(\vec{u} - 2^{-j}\vec{\eta}) \cdot |\vec{u}|^{2H} d\vec{u} \, d\vec{v}$$

$$+ \int_{\vec{u}} \Psi_{2^j}^{3*}(\vec{u} - 2^{-j}\vec{\eta}) \, d\vec{u} \cdot \int_{\vec{v}} \Psi_{2^j}^3(\vec{v} - 2^{-j}\vec{\eta}) |\vec{v}|^{2H} d\vec{u} \, d\vec{v}$$

$$+ \int_{\vec{u}} \int_{\vec{v}} \Psi_{2^j}^{3*}(\vec{u} - 2^{-j}\vec{\eta}) \cdot \Psi_{2^j}^3(\vec{v} - 2^{-j}\vec{\eta}) |\vec{u} - \vec{v}|^{2H} d\vec{u} \, d\vec{v} \tag{9.20}$$

Given that $\Psi_{2^j}^3$ satisfies the admissibility condition (9.12), it is possible to show that

$$\int_{\vec{v}} \Psi_{2^j}^3(\vec{v} - 2^{-j}\vec{\eta}) \, d\vec{v} = \int_{\vec{u}} \Psi_{2^j}^{3*}(\vec{u} - 2^{-j}\vec{\eta}) \, d\vec{u} = 0 \tag{9.21}$$

Thus, (9.19) can be written as

$$E\left[|D_{2^j}^3[\vec{\eta}]|^2\right] = \frac{V_H}{2} 2^{-2j} \cdot \left(\Psi_{2^j}^{3*}(\vec{u} - 2^{-j}\vec{\eta}) \cdot \int_{\vec{u}} \int_{\vec{v}} \Psi_{2^j}^3(\vec{v} - 2^{-j}\vec{\eta}) |\vec{u} - \vec{v}|^{2H} \, d\vec{u} \, d\vec{v}\right) \tag{9.22}$$

By substituting $\vec{p} = \vec{u} - \vec{v}$ and $\vec{q} = \vec{v} - 2^{-j}\vec{\eta}$, we may rewrite Eq. (9.19) as [11]

$$V_{\Psi_{2^j}^3} = \int_{\vec{p}} \int_{\vec{q}} \Psi_{2^j}^{3*}(\vec{p} + \vec{q}) \cdot \Psi_{2^j}^3(\vec{q}) |\vec{p}|^{2H} \, d\vec{p} \, d\vec{q} \tag{9.23}$$

Then, the integral in q is the wavelet transform of the wavelet itself, at a resolution j, similar to the scale-independent constant derived in (9.15). The variance of the 2D detail signal $D_{2^j}^3[n, m]$ can be considered as a power law of the scale 2^j and can be used to calculate H in a similar way to (9.16) as

$$\log_2 E\left[|D_{2^j}^3[n, m]|^2\right] = (2H + 2)j + C_2 \tag{9.24}$$

with

$$C_2 = \log_2 \frac{V_H}{2} V_{\Psi_{2^j}^3}(H) \tag{9.25}$$

The FD value of the 2D fBm process can be extracted from the slope of the variance in (9.25) plotted as a function of the resolution j in a log-log plot.

9.3.2 Power Spectrum Method

A complimentary approach to compute FD of an fBm uses the wavelet representation of its power spectrum. In [6], Flandrin shows that the spectrum of an fBm follows the power law of fractional order shown in (9.9) using either a time–frequency description or a time-scale description. Similar to the development of the statistical method, we describe a general 1D case and its formal extension to 2D.

9.3.2.1 One-Dimensional Case

If the frequency signal $S(\omega)$ is filtered with a wavelet filter $\psi(u)$, the resulting spectrum at the specific resolution is [9]

$$S_{2^j}(\omega) = S(\omega)\left|\bar{\psi}\left(2^{-j}\omega\right)\right|^2 \tag{9.26}$$

where

$$\bar{\psi}(\omega) = e^{-i\omega}\overline{\hat{h}(\omega + \pi)} \tag{9.27}$$

where $H(\omega)$ corresponds to the discrete-time fourier transform of the corresponding scaling function $\phi(x)$ and defined in terms of its coefficients $h(n)$:

$$\hat{h}(\omega) = \sum_{n=-\infty}^{\infty} h(n)e^{-in\omega} \tag{9.28}$$

Using the sampling for the discrete detail description of a function f in [9],

$$D_{2^j} = \left(\left(f(u) * \psi_{2^j}(-u)\right)\left(2^{-j}n\right)\right) \tag{9.29}$$

which contains the coefficients of the high-frequency details of the function. The spectrum of the discrete detail signal can be written as [7]

$$S_{2^j}^d(\omega) = 2^j \sum_{k=-\infty}^{\infty} S_{2^j}(\omega + 2^j 2k\pi) \tag{9.30}$$

The energy of the detail function at a specific resolution j is defined as [9]

$$\sigma_{2^j}^2 = \frac{2^{-j}}{2\pi} \int_{-2^j\pi}^{2^j\pi} S_{2^j}^d(\omega)\, d\omega \tag{9.31}$$

This equation describes the support of the wavelet in the frequency domain [10] for a specific resolution j. Finally, it can be shown that the solution of the integral leads to an expression that relates the energy content in two consecutive resolution filtering operations [9],

$$\sigma_{2^j}^2 = 2^{2H}\sigma_{2^{j+1}}^2 \tag{9.32}$$

From this expression, the Hurst coefficient H can be derived as

$$H = \frac{1}{2}\log_2\left(\frac{\sigma_{2^j}^2}{\sigma_{2^{j+1}}^2}\right) \tag{9.33}$$

9.3.2.2 Two-Dimensional Case

We follow the same steps of the previous section to extend the analysis to the 2D case. For a 2D fBm, the power spectral density is defined as [11]

$$S(\vec{\omega}) = S(\omega_x, \omega_y) \propto \frac{K}{(\omega_x^2 + \omega_y^2)^{(2H+2)/2}} \qquad (9.34)$$

where K is a constant and $\vec{\omega}$ is a vector describing frequency, whose components are ω_x and ω_y, which in turn correspond to frequency in channels x and y, respectively. For this part of the study, $S(\vec{\omega})$ is computed as the power spectral density of a stationary signal, which may not apply strictly to our 2D fBm, which is a nonstationary process. However, this approach leads to good results. The 2D spectrum is given as

$$S(\omega_x, \omega_y) = |\text{FFT(image)}|^2 \qquad (9.35)$$

If the frequency-domain signal is filtered with a wavelet filter, the resulting spectrum at the specific resolution is [9]

$$S_{2^j}(\vec{\omega}) = S(\vec{\omega}) \left| \bar{\Psi}_{2^j}^3 \left(2^{-j}(\omega_x, \omega_y) \right) \right|^2 \qquad (9.36)$$

where

$$\left| \Psi(\omega_x, \omega_y) \right|^2 = \left| \psi(\omega_x) \right|^2 \left| \psi(\omega_y) \right|^2 \qquad (9.37)$$

The discrete version of this spectrum is given as [9]

$$S_{2^j}^d(\omega_x, \omega_y) = 2^{2j} \sum_{l=-\infty}^{\infty} \sum_{k=-\infty}^{\infty} S_{2^j}\left(\omega_x + 2^j 2k\pi, \omega_y + 2^j 2l\pi \right) \qquad (9.38)$$

The energy of the detail function at a specific resolution j may be computed by integration of the support vector $\bar{\Psi}_j^3(\omega)$ of a chosen wavelet filter [10, 17], as shown in Figure 9.1.

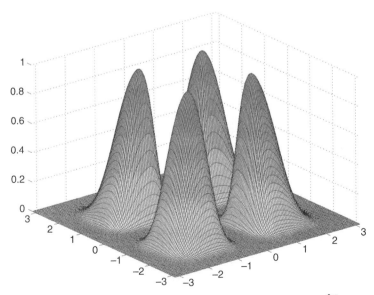

Figure 9.1   Fourier transform of the high frequency 2D wavelet, $\hat{\Psi}^3(\omega_x, \omega_y)$ for Daubechies 6.

This is described by the equation

$$\sigma_{2^j}^2 = \frac{2^{-2j}}{4\pi^2} \int_{-2^j\pi}^{2^j\pi} \int_{-2^j\pi}^{2^j\pi} S_{2^j}^d(\omega_x, \omega_y)\,d\omega_x\,d\omega_y \tag{9.39}$$

After an appropriate change of variable in the previous equation, we may obtain [9]

$$\sigma_{2^j}^2 = 2^{2H}\sigma_{2^{j+1}}^2 \tag{9.40}$$

Thus, in the case of 2D signals, the ratio of the energy corresponding to the detail signals at successive resolutions provides a solution for the computation of H similar to the 1D derivation presented in (9.33),

$$H = \frac{1}{2}\log_2\left(\frac{\sigma_{2^j}^2}{\sigma_{2^{j+1}}^2}\right) \tag{9.41}$$

9.3.3 Wigner–Ville Distribution Method

Previous work related to the modeling and estimation of fBm parameters shows that the WVD is a valid method for the analysis of $1/f$ processes. In the particular case of a 1D process, f, its WVD is given by [13, 18]

$$W_f(t, \omega) = \int_{-\infty}^{\infty} f\left(t + \frac{t_0}{2}\right) f^*\left(t - \frac{t_0}{2}\right) \exp\left(-j\omega t_0\right) d\omega_0 \tag{9.42}$$

where t corresponds to time or space and ω to frequency. The corresponding equation in the frequency domain is [13]

$$W_F(t, \omega) = \int_{-\infty}^{\infty} F\left(\omega + \frac{\omega_0}{2}\right) F^*\left(\omega - \frac{\omega_0}{2}\right) \exp\left(-j\omega_0 t\right) dt_0 \tag{9.43}$$

where $F(\omega)$ corresponds to the Fourier transform of $f(t)$ and $W_f(t, \omega) = W_F(t, \omega)$. There is a complete description of the properties of this time–frequency distribution in [14]. The extension to 2D is described with the equation [15]

$$W_f(x, y, u, v) = \int_{R^2} f\left(x + \frac{\alpha}{2}, y + \frac{\beta}{2}\right) f^*\left(x - \frac{\alpha}{2}, y - \frac{\beta}{2}\right)$$
$$\times \exp\left(-2\pi j(u\alpha + v\beta)\right) d\alpha\,d\beta \tag{9.44}$$

where x and y are time (or space) variables; u and v are their respective frequency components and α and β are the corresponding dummy integration variables that describe the shifting in each direction. For an image of size $N_X \times N_Y$, the discrete time–frequency version of the 2D WVD is given as [16]

$$W(n_1, n_2, m_1, m_2) = \frac{1}{4N_XN_Y} \sum_{l_1=0}^{N_X-1} \sum_{l_2=0}^{N_Y-1} f(l_1, l_2)f^*(n_1 - l_1, n_2 - l_2)$$
$$\times \exp\left(-\pi j\left(\frac{m_1}{N_X}(2l_1 - n_1) + \frac{m_2}{N_Y}(2l_2 - n_2)\right)\right) \tag{9.45}$$

where n_1 and n_2 are time (or space) discrete variables, m_1 and m_2 are their respective frequency components, and l_1 and l_2 are the variables that describe the shifting, in each direction, of the conjugate function. Let $\bar{f}$ be the original image sampled in both the time (space) and frequency domains. The WVD for this signal is defined as [16]

$$W_{\tilde{f}}(x, y, u, v) = \frac{1}{XY} \sum_{n_1, n_2, m_1, m_2} W(n_1, n_2, m_1, m_2) \delta\left(x - n_1 \frac{X}{2}\right) \delta\left(y - n_2 \frac{Y}{2}\right)$$

$$\times \delta\left(u - \frac{m_1}{2 N_x X}\right) \delta\left(v - \frac{m_2}{2 N_y Y}\right) \tag{9.46}$$

For a given fBm process, the local power spectrum is computed directly from its WVD as [13]

$$P_{B_H}(t, \omega) = |W_{B_H}(t, \omega)| \tag{9.47}$$

To compute FD in the image, we propose similar steps as described in Eqs. (9.18)–(9.25) and (9.34)–(9.39) for the 2D case.

9.4 RESULTS

We test our proposed models for computing H, and hence FD, using both synthetic and real images. For the synthetic-image case, we generate fractal images using two different algorithms, while for the real image, we use a CT image.

9.4.1 Synthetic Image

We generate synthetic fBm input images with a known FD values. We implement two procedures to synthesize 2D fBm images [5]. The first image generation method explores the *midpoint displacement* algorithm while the second method uses the *spectral* properties of the fBm. The statistics derived from each of the fBm generation algorithms suggest that the power spectrum method offers FD values closer to the theoretical values that are used to generate each fBm image.

9.4.1.1 Variance Method We use synthetic images obtained using two fBm synthesis algorithms as described above. For each experiment using our proposed FD estimation models, we specify the synthetic fBm images as the input. The average of these measurements is computed for each of the 20 uniformly distributed values of H, with $0 < H < 1$. The results are summarized in Figures 9.2a,b respectively. Figure 9.3

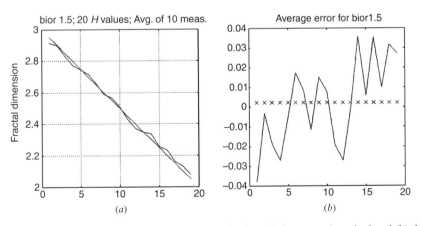

Figure 9.2 FD results for images synthesized using (a) the spectral method and (b) the difference of actual and theoretical estimation values.

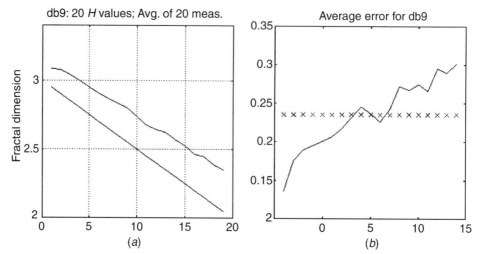

Figure 9.3 FD results for images synthesized using (a) the midpoint approach and (b) the spectral method.

shows an average difference of 0.24 between the theoretical and the estimated FD. However, the estimated values show a linear trend wherein an appropriate shift and slope corrections obtained from the statistical characterization of each method may be exploited to obtain the actual FD values.

9.4.1.2 Spectral Method Our proposed spectral method offers far more consistent FD estimation results when compared to the variance method. Further, the estimated FD values are also similar to the theoretical values and more coherent in all the detail scales. Figures 9.4a,b present the worst and best cases for this approach, respectively. The lines, from lower to upper, describe the ratio of energy content [see Eq. (9.44)] between resolutions 1–2, 2–3, 3–4, and 4–5. The analysis with the Daubechies 2 wavelet offers relatively poor results. However, the rest of the Daubechies family and the other families of wavelets show consistent results similar to the ones shown in Figure 9.4b.

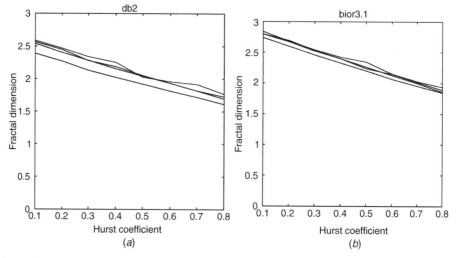

Figure 9.4 Spectral estimation of FD using a (a) Daubechies 2 and (b) biorthogonal wavelets.

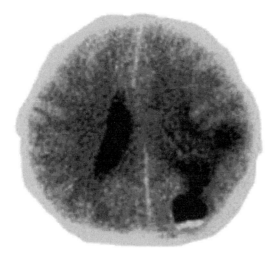

Figure 9.5 CT image used to estimate FD.

9.4.2 Real Medical Image

We apply our three estimation models to a brain CT image in Figure 9.5. For this study, 16 and 32 pixel grids are used to analyze the CT image.

9.4.2.1 Variance Method
We apply our variance algorithm to the input CT image to identify tumor tissue. The FD results for 32- and 16-pixel grids are shown in Figures 9.6*a,b*, respectively. In the 32-pixel-grid case, positions (3,4) and (4,3) correspond to high values of FD. This trend is observed with almost all the wavelets in the family. A similar observation is noticed for the other tumor locations in position (6,3) or (7,5). When a finer grid is considered, it is possible to observe more tissue details corresponding to FD estimation.

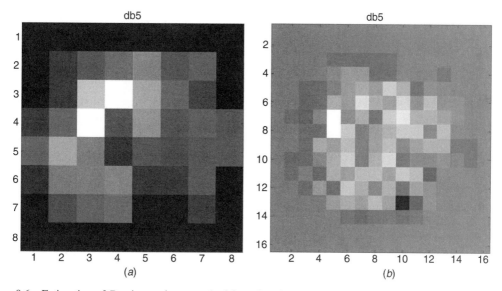

Figure 9.6 Estimation of D using variance method for using the wavelet Daubechies 5 for a grid of (*a*) 32 pixels; and (*b*) 16 pixels.

Although the localization of certain areas of the image associated to tumor tissue is consistent, the results for this method depend, to an important extent, on the analyzing wavelet. Another drawback of this method is seen in the relationship between observations of different grain. There is not a direct correspondence between different grid sizes and tumor locations, as can be seen in Figures 9.6a,b. The variance method is efficient in terms of time of execution since wavelet decomposition and the computation of the variance of the detail coefficients are relatively faster.

9.4.2.2 *Spectral Method*

We also apply our FFT-based spectral method to the CT image as shown in Figure 9.5. These results are shown in Figures 9.7a,b, respectively. In this experiment, the position (4,3) of Figure 9.7a shows the highest FD estimate in the CT image. This can be clearly associated to positions (7,5) and (8,5) in Figure 9.7b. Similar results are observed in position (7,5) for the 32-pixel grid and the corresponding low values in (13,10) and (13,11) for the 16-pixel grid of Figures 9.8a,b, respectively. Contrary to the variance method, the spectral method shows a relative independence from the selected analyzing wavelet wherein almost the same features are identified with any wavelet that is used to perform the filtering. Further, spectral method results are independent of the grid size. It is also observed that the computation times are considerably higher for the spectral FFT algorithm compared to that of the variance method.

9.4.2.3 *WVD Method*

Our last experiment shows the FD results using the spectral WVD method. The results clearly show that for the 32-pixel grid in Figure 9.8a, the pixel location (4,3) corresponds to a high FD value. For the 16-pixel-grid analysis in Figure 9.8b, all three algorithms correctly identify the low H location such as (13,10). The FD measurement values obtained using all three algorithms, specifically for the 32-pixel-grid position (3,4), are summarized in Table 9.1. The graphical results for the 32-pixel grid are very similar to the ones obtained in the previous methods. The FD

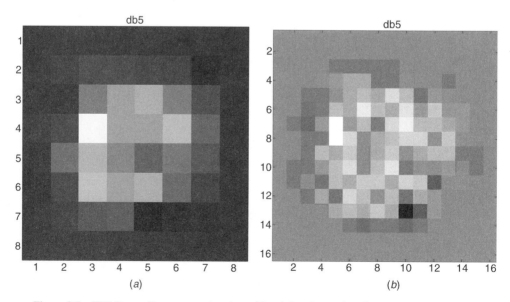

Figure 9.7 FFT Power Spectrum estimation of local D values using the wavelet Daubechies 5 for a grid of (*a*) 32 pixels; and (*b*) 16 pixels.

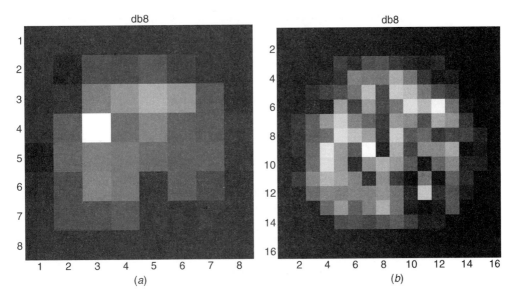

Figure 9.8   WVD spectral estimation of FD values using the Daubechies 5 wavelet for a grid of (a) 32 pixels; (b) 16 pixels.

TABLE 9.1   Comparison of FD Estimates at Position (4,3) Using the
Three Analysis Models

	General	Daubechies	Biorthogonal	Coiflet
Variance	3.08	3.09	3.13	3.03
FFT spectrum	2.52	2.53	2.49	2.54
WVD spectrum	2.50	2.57	2.42	2.55

values obtained from the variance method analysis are, on the average, 0.4 higher than the theoretical values. On the other hand, the estimated fractal dimension from the FFT method tends to be below the nominal FD. The WVD results are, in general, closer to the theoretical results. Despite this relative difference, all three models are robust and may be used to obtain FD values from fBm images. However, appropriate correction factors (shift and slope corrections) to estimated FD values may render the analysis of the images more accurate. This can be exploited in the differentiation of tissues and hence to the location of abnormal processes identifiable in a tomographic image.

9.5   CONCLUSION

We propose three formal mathematical models to compute the Hurst coefficient, and hence the FD, of synthetic fBm and real CT images, respectively. Our first model is based on the ratio of statistical variance of the diagonal detail wavelet coefficients at different scales of the multiresolution decomposition of the analyzed image. The other two models involve the ratio of energy content at subsequent resolutions, derived from the wavelet-filtered spectra of the image. The two frequency-domain processing models are obtained via Blackman–Tukey spectral estimation and WVD methods, respectively. Experimental results suggest that the lesions in the CT image may be effectively localized using high or low values of the FD. Further, the spectral fBm analysis of tomographic images provides a more consistent framework for the multiresolution analysis when compared to the variance method. However, the spectral analysis of fBm is more computationally involved than that of the variance for an image.

9.6 FUTURE TRENDS

The algorithm for tumor location identification has been an area of intense research. The multiresolution-fractal-based approach offers a promising direction in this effort. Our work in this chapter represents preliminary results in multiresolution-fractal analysis of medical images. We are interested in developing classification algorithms based on the three techniques proposed in this chapter. We also plan to further validate our results using additional synthetic as well as real medical images with different brain pathologies. We expect to develop a more generalized fBm model known as multifractional Brownian motion (mBm) wherein the Hurst coefficient H is modeled as a function of time or position for more detailed tissue recognition. Finally, we plan to optimize the structure of the algorithms to obtain the minimum processing times such that these procedures may be used in a clinical setting.

ACKNOWLEDGMENT

The authors wish to thank the Whitaker Foundation for partially supporting this work through a Biomedical Engineering Research Grant (RG-01-0125).

REFERENCES

1. MANDELBORT, B. B., AND VAN NESS, J. W., Fractional Brownian motions, fractional noises and applications. *SIAM Review*, 10(4), 422–437, Oct. 1968.
2. MANDELBORT, B. B., *The Fractal Geometry of Nature*. San Francisco: W. H. Freeman, 1982.
3. DAVIES, S., AND HALL, P., Fractal analysis of surface roughness by using spatial data, *Journal of The Royal Statistical Society Series, B Statistical Methodology*, 61(1), 3–29, 1999.
4. ZOOK, J., AND IFTEKHARUDDIN, K. M., Statistical analysis of fractal-based techniques in brain tumor detection. *Magnetic Resonance Imaging*, 23(5), 671–678, 2005.
5. PEITGEN, H.-O., AND SAUPE, D., In M. F. Barnsley et al., Eds. *The Science of Fractal Images*. New York: Springer-Verlag, 1987, p. 64.
6. FLANDRIN, P., On the spectrum of fractional Brownian motions. *IEEE Transactions on Information Theory*, 35(1), 197–199, Jan. 1989.
7. FLANDRIN, P., Wavelet analysis and synthesis of fractional Brownian motion. *IEEE Transactions on Information Theory*, 38(2, Part 2), 910–917, Mar. 1992.
8. TEWFIK, A. H., AND KIM, M., Correlation structure of the discrete wavelet coefficients of fractional Brownian motion. *IEEE Transactions on Information Theory*, 38(2, Part 2), 904–909, Mar. 1992.
9. MALLAT, S. G., A theory for multiresolution signal decomposition: The wavelet representation. *IEEE Transactions on Pattern Analysis and Machine Intelligence*, 11(7), 674–693, July 1989.
10. MALLAT, S. G., *A Wavelet Tour of Signal Processing*. San Diego: Academic Press, 1998.
11. HENEGHAN, C., LOWEN, S. B., AND TEICH, M. C., Two-dimensional fractional Brownian motion: wavelet analysis and synthesis. In *Proceedings of the IEEE Southwest Symposium on Image Analysis and Interpretation*, 1996, pp. 213–217.
12. WORNELL, G. W., Wavelet-based representations for the $1/f$ family of fractal processes. In *Proceedings of the IEEE*, 81(10), 1428–1450, Oct. 1993.
13. WEN, C.-Y. AND ACHARYA, R., Self-similar texture characterization using Wigner-Ville distribution. *In Proceedings, International Conference on, Image Processing*, Vol. 3, Sept. 1996, pp. 141–144.

14. ZHU, Y. M., PEYRIN, F., AND GOUTTE, R., Equivalence between two-dimensional analytic and real signal Wigner distributions. *IEEE Transactions on, Acoustics, Speech, and Signal Processing*, 37(10), 1631–1634, Oct. 1989.

15. ZHU, Y. M., GOUTTE, R., AND AMIEL, M., On the use of wigner distribution for two-dimensional spectral analysis of images. In *Proceedings of the Seventh Workshop on Multidimensional Signal Processing*, Sept. 1991, pp. 2.7–2.7.

16. GRASSIN, S., AND GARELLO, R., Spectral analysis of nonstationary ocean SAR images using the Wigner-Ville transform. Geoscience and Remote Sensing Symposium, IGARSS '96, Remote Sensing for a Sustainable Future, International, Vol. 4, May 1996, pp. 1956–1958.

17. Goswami, J. C. AND Chan A. K., *Fundamentals of Wavelets: Theory, Algorithms, and Applications*. New York: Wiley, 1994, p. 229.

18. PEYRIN, F., RATTON, L., ZEGADI, N., MOUHAMED, S., AND DING, Y., Evaluation of the fractal dimension of an image using the wavelet transform: Comparison with a standard method. In *Proceedings of the IEEE International Symposium on Signal Processing*, 1994, pp. 244–247.

CHAPTER 10

METHODS FOR NEURAL-NETWORK-BASED SEGMENTATION OF MAGNETIC RESONANCE IMAGES

Lia Morra, Silvia Delsanto, and Fabrizio Lamberti

10.1 INTRODUCTION

The introduction of advanced medical imaging techniques, such as magnetic resonance imaging (MRI), computer-assisted tomography (CAT), functional magnetic resonance imaging (fMRI), and positron emission tomography (PET), has dramatically improved the quality of brain pathology diagnosis and treatment. In particular, MRI allows the acquisition of three-dimensional, high-resolution, and highly detailed images of the brain anatomy, with unparalleled soft-tissue contrast with respect to other medical imaging modalities. However, the introduction of such imaging techniques has also increased the quantity and complexity of available image information, thus entailing the adoption of automated tools and techniques for image management (i.e., storage and transmission), analysis, and interpretation. In this chapter, we will focus on the segmentation and classification techniques, which, combined, allow the identification of specific tissue classes and anatomical structures. We will begin by introducing MRI medical applications and successively describe the different MRI segmentation and classification techniques, concentrating in particular on neural-network-based approaches. An example illustrating automatic, single-channel segmentation and classification will conclude the overview.

10.2 APPLICATIONS OF MRI SEGMENTATION

Medical image segmentation has been the subject of extensive research activity in the last years. The segmentation task consists in subdividing an image into a number of homogeneous regions and is closely related to the classification task, which classifies each segmented region as belonging to a specific tissue class or anatomical structure. Medical image segmentation and classification are a daunting task, due to the wide range of possible image types and acquisition protocols, to the inherent complexity of most anatomical structures and to the many sources of artifacts and noise [1, 2]. Image

Handbook of Neural Engineering. Edited by Metin Akay

segmentation is in fact important both as a basis for further processing and independently for diagnosis, treatment, and follow-up of brain pathologies, since it allows the identification and quantitative evaluation of the extension, location, and volume of brain structures and lesions in a noninvasive manner. This can aid in the initial diagnosis as well as treatment and surgical intervention planning (e.g., in radiation treatment planning) of various brain pathologies, such as tumors and multiple sclerosis. Moreover, volumetric analysis is important in the evaluation of various neurological conditions, such as cerebral atrophy, edemas, and brain hematomas, related to several pathologies and neurodegenerative diseases, such as the Alzeihmer and Parkinson syndromes. The measured volume of brain structures and pathologies can thus be correlated with clinical and neurological signs of the disease and used in determining appropriate treatment. Volumetric analysis is also of fundamental importance in patient follow-up, assisting medical doctors in assessing lesion progression or regression and response to therapy.

Magnetic resonance imaging segmentation can also be viewed as a preliminary step for performing 3D volumetric reconstruction. Three-dimensional reconstruction can be of invaluable help to medical doctors, since it enables efficient visualization of tumors and other brain pathologies and/or structures. Furthermore, it allows not only the planning of surgical procedures and in surgery simulation and "rehearsal" but also the assistance of medical experts during actual brain surgery in the operating suite. Finally, volumetric reconstruction can be useful in the analysis of functional parameters, such as the electroencephalographic (EEG) signal; for instance, EEG brain-mapping algorithms allow the correlation of EEG potentials and electrode locations with high-resolution anatomical information, while the availability of segmented MRI images allows the construction of piecewise homogeneous head models which can be used in forming conductivity models to help in simulating and analyzing brain electrical activity [2]. Finally, MRI segmentation constitutes a preprocessing step for many registration algorithms. Basically, image registration consists in transforming the coordinates of each point in one image in order to match it with the physically corresponding point in the other image. Multimodality image registration allows medical doctors to perform the joint analysis of images acquired with different (often complementary) imaging modalities, thus enabling extremely complex clinical examinations which correlate anatomical regions with functional metrics. Monomodality image registration is also useful in studies conducted over a period of time (e.g., treatment evaluation) or on a patient population.

The MRI segmentation techniques have been widely applied in assisting medical experts in brain tumor diagnosis and treatment. Hence, MRI segmentation applications in the oncological field are a notable example, which can help understand better both the advantages that arise from the adoption of such techniques and the main requisites imposed by real clinical applications on segmentation technique accuracy and performance. Current clinical protocols employed by various radiation oncology groups, such as the RTOG (Radiation Treatment Oncology Group, www.rtog.org) and the ECOG (Eastern Cooperative Oncology Group, www.ecog.org), use visual metrics (usually assessed by the radiologist on hard-film copy) in order to estimate tumor volume and reduce operator variability in tumor response rating [3]. A widely adopted criterion for measuring tumor response is based on the measurement of the two largest perpendicular diameters; the radiologist selects the slice with the largest tumor cross section and measures the greatest diameter and its largest perpendicular diameter. An estimate of tumor volume is then calculated by multiplying the lengths of the two diameters (areas are summed in case of multiple lesions). In order to track tumor changes during therapy, tumor volumes are estimated from patient scans taken at different time intervals. According to this protocol, tumor response is classified as follows:

- Complete response is defined as the disappearance of all evidence of active tumor.
- Partial response is defined as at least a 50% decrease in the tumor volume estimate (product of the largest diameters).
- Stable disease is defined as less than 50% decrease or more than 25% increase in tumor volume.
- Progressive disease is defined as more than 25% increase in the tumor volume estimate or appearance of new lesions.

This visual metric suffers from a high variability in the determination of tumor volume estimate; a great interrate variability of the results (16%) has been reported in the published literature. The high variability is due to the difficulties in selecting the best slice and in locating the actual tumor boundary, since both operations are highly operator dependent. Furthermore, this metric is intrinsically approximate and may fail in tracking tumor response over a period of time. In fact, there is no clear evidence that area estimates are actually correlated with tumor volume; for instance, there may be a global decrease in tumor volume, despite local enlargement in one or more slices, or vice versa. The hypothesis on which this protocol is based may sometimes be unrealistic, thus leading in some cases to wrong conclusions.

The MRI segmentation technique may be a precious tool in oncological practice both for diagnosis and treatment planning, as in brain tumor diagnosis, brain surgery, and radiation treatment planning (RTP), and for monitoring, as in tumor response measurement during patient follow-up [3–7]. To support brain tumor diagnosis and RTP, there is a strong need for accurate segmentation techniques. Not only should a segmentation algorithm be able to accurately quantify tumor size and location, but it should also be capable of discriminating the various tissue types within the tumor bed (tumor, necrosis, cyst, and vascular edema) in order to identify active tumor margins. Correct identification of active tumor margins is of paramount importance in RTP, since it allows an accurate selection of the areas to be irradiated. In some cases, selective irradiation, as opposed to whole-brain irradiation, may be of help in limiting therapy countereffects, especially in the long period, while preserving therapy effectiveness [6].

On the contrary, during patient follow-up, it is necessary to specifically track changes in tumor volume by quantifying the relative changes in tumor volume over a period of time to determine tumor response to therapy. In this case, result reproducibility, rather than accuracy, is of paramount importance, and therefore unsupervised segmentation techniques are better suited to this task. However, tumor response measurement is a daunting segmentation task, much more critical than the segmentation of normal volunteers' images. In particular, tumor size, type, stage, and vascularity may hinder the ability of a segmentation technique to differentiate the various tissue types and alter the degree of magnetic resonance contrast enhancement over time. Furthermore, antitumor therapies, such as chemotherapy and radiation treatment, may affect magnetic resonance relaxation parameters of both pathological and normal tissues, potentially altering segmentation results, while surgical resection of the tumor may cause problems related to the presence of edemas and alterations of normal brain anatomy.

Brain tumor diagnosis, treatment and follow-up thus constitutes an excellent example of the wide range of applications of MRI segmentation, illustrating both the potential impact on current clinical procedures and the inherent difficulty in deriving methods capable of obtaining accurate and reproducible results.

10.3 INTRODUCTION TO MRI SEGMENTATION

10.3.1 Overview of MRI Segmentation Techniques

Though the usefulness of segmentation techniques is established, correct segmentation of magnetic resonance images poses many problems. In particular, before examining the different methods employed, a few words on the principal kinds of artifacts are necessary to understand some critical aspects of MRI segmentation. Magnetic resonance images are subject to a number of possible artifacts, besides thermal noise generated by imaging hardware. More specifically, two types of artifacts, that is, gray-level intensity inhomogeneity and the partial-volume artifact, need to be considered, since they can be found, in varying degrees, in all images acquired in this modality and moreover can greatly affect segmentation results. A detailed description of the physics of MRI and of magnetic resonance image characteristics is beyond the scope of this section; for a deeper insight on the fundamentals of MRI, see [8–10].

Thermal or electronic noise is random, mostly Gaussian, white, and additive, and in general tissue dependent, since tissue constituents and their composition vary according to the tissue type. Moreover, the gray-level intensity of pixels associated with voxels of the same tissue class is generally not uniform across different slices of a magnetic resonance data set, even within the same slice; this phenomenon is generally referred to as gray-level intensity inhomogeneity (or simply intensity inhomogeneity). A variety of factors contribute to this artifact, mainly the intrinsic inhomogeneity of biological tissues, the irregularities in the static and radio-frequency (RF) magnetic fields, and nonuniformities in the sensitivity of RF receiver coils. So far, various techniques have been proposed in the literature to correct intensity inhomogeneities [11, 12]. Among these, it is worth mentioning low-pass and homomorphic filtering techniques, which however may corrupt the edges and other high-frequency details of the image. Other techniques are based on the use of phantoms to measure magnetic field profiles of the magnetic resonance scanner; however, in order to reliably correct the effects of inhomogeneities, they require the acquisition of one phantom scan for each subject being imaged and are thus not feasible in everyday practice. The adoption of segmentation techniques able of coping, to some extent, with the presence of such inhomogeneities, thus emerges as a possible alternative to such preprocessing filters [12]. Finally, partial-volume effects arise from the fact that, since each slice being imaged has a finite thickness, the voxel associated to each image pixel may contain more than one tissue type. The gray-level intensity of the related pixel is then determined by the weighted sum of the intensities of the signals generated by each tissue type. Partial-volume artifacts are thus dependent on the slice thickness and result in image blurring (if slice thickness is too high, small brain structures such as nerves may not even be visible in the image). An obvious countermeasure is to acquire images with higher resolution (i.e., smaller slice thickness).

The MRI segmentation techniques can be roughly classified in two main categories: supervised and unsupervised (or automatic) [1]. Supervised segmentation and classification techniques require an experienced radiologist to select a set of pixels or regions of interest for each tissue class in order to train the classifier. The main drawback of these techniques is that the final result depends on the initial choice of the training set, which is operator dependent. Therefore, the results are usually subject to a high interrater (i.e., they depend on the user who selected the training set) and intrarater variability (i.e., the results can vary in time, as the same user selects different training sets), thus limiting the applicability of these algorithms in those cases when result reproducibility is a substantial requisite (e.g., brain tumor patient follow-up). Furthermore, since user input is usually required for each slice to be segmented, this approach can be quite time consuming and consequently impractical in large, multitrial imaging centers. On the contrary, unsupervised segmentation techniques do not require manual selection of a training set. Most of these segmentation techniques are based on clustering algorithms, which subdivide the image in homogenous regions. However, since these regions have no meaning from an anatomical viewpoint, they need to be further classified according to the tissue type (or brain structure). The final classification step can be manual (in this case, the needed manual intervention is limited to assigning a label to each segmented region), automatic, or semiautomatic. Unsupervised segmentation techniques ensure a higher stability and reproducibility of results, since they are independent of the operator. However, some authors report that unsupervised techniques may yield longer execution times than supervised techniques and in some cases may not achieve a meaningful segmentation due to the difficulty in properly initializing clustering algorithms. Investigators have tried to overcome these potential problems by exploiting heuristics based on anatomical properties (knowledge-based approach) [13, 14]. In some cases, semisupervised segmentation techniques, with minimal requisites in terms of operator time and skills, have also been proposed.

Many image processing techniques have been applied to solve various problems related to MRI segmentation (e.g., tissue classification, lesion segmentation). Among those, pattern recognition techniques are generally acknowledged as the most promising methods for performing brain tissue segmentation and/or classification, while other classic image processing techniques, such as thresholding, edge detection methods, and region growing, have been scarcely employed, due to the difficulty in selecting proper thresholds and to their inability to cope with the presence of noise, intensity inhomogeneities, and partial-volume artifacts [1, 15].

Before analyzing the different pattern recognition techniques, a brief discussion about the process of extracting features from magnetic resonance images is needed. Magnetic resonance imaging is an intrinsically multispectral imaging technique, allowing the acquisition of (almost) simultaneous images, with different soft-tissue contrast characteristics, of the same anatomical location (multichannel data sets) by varying acquisition protocols and/or parameter setting. This difference in the contrast is basically due to the fact that different tissue characteristics are responsible for the formation of the resonance image in the different acquisition protocols. Specifically, an image in which contrast is determined predominantly by one of the tissue parameters sensible to MRI, that is, T1 and T2 relaxation time and proton density (PD), is generally referred to as a T1-, T2-, or PD-weighted image. Most published works have relied on this unique characteristic of MRI, exploiting the availability of more information at each pixel site in order to achieve better segmentation results. However, since single-channel images usually yield smaller voxel sizes, they are more suited for precise volumetric analysis of brain structures. Despite this fact, only a few algorithms for single-channel magnetic resonance images have been proposed [16, 18].

Feature extraction is essential in MRI segmentation and classification since it reduces the segmentation task to the grouping of vectors in the feature space. The choice of good features is thus essential in order to ensure an accurate and reliable segmentation [1]. Most published segmentation techniques rely only on gray-level intensities associated with each pixel to perform segmentation, which may be extracted from a single image or from a multichannel image. Useful information may also be extracted by images acquired in modalities different than MRI, provided that the images can be successfully coregistered. To reduce the effects of random noise and enhance segmentation results, many authors have combined gray-level intensities with features extracted from the neighborhood of each pixel (such as the mean intensity) and from a combination of the various channels of the data set [19].

The use of derived features, such as edge detection and texture features, has also been proposed in the literature. However, edge detection is seldom used and is unlikely to achieve good results on magnetic resonance images because of its high sensitivity to noise and its inability to cope with intensity inhomogeneities and other common artifacts. On the contrary, statistical texture features have been exploited in various proposed techniques but are more suited to the classification of image regions than individual pixels (pixel-based classification) since they are necessarily extracted from a large number of pixels [20].

Various pattern recognition methods have been applied in both supervised and unsupervised MRI segmentation techniques. As far as supervised MRI segmentation techniques are concerned, the pattern recognition paradigms can be distinguished in parametric (or statistical) and nonparametric methods. Parametric methods make a hypothesis on the underlying distribution of image features (e.g., gray-level intensities). Usually, Gaussian distributions (either unidimensional or multivariate) are assumed for gray-level intensities of each brain tissue class. The most commonly used parametric method is maximum likelihood (ML): The parameters of the distribution are estimated on the basis of the training set selected by the radiologist, while the remaining pixels are classified as belonging to the most probable class according to the estimated distribution. On the contrary, nonparametric methods such as k-nearest neighbors (kNNs) are not based on any assumption on feature theoretical statistical distribution; on the contrary, they take into account the actual distribution of features, as they are usually based on the distance between points (i.e., pixels) in the feature space. In the k-nearest-neighbor algorithm, the Euclidean distance in the feature space between the pixel to be classified and the training set pixels is considered. The probability of a pixel belonging to a given class is then estimated as the frequency of that class within the pixel's k-nearest neighbors in the feature space [21]. Supervised-learning neural networks (such as multilayer perceptron and radial basis function networks) can be included in the class of nonparametric pattern recognition paradigms. Generally, published results suggest that nonparametric methods achieve higher performances than parametric methods; this may be due to the difficulty in determining the real distribution of features.

Unsupervised pattern recognition techniques, also known as clustering techniques, aim instead at autonomously finding the underlying structure of data in the feature space by grouping into clusters points that are close in the feature space according to some distance measure. From a qualitative point of view, a cluster can be defined as a high-density area in the feature space; various mathematical models have been formulated to achieve this goal. Among clustering techniques, the most widely used in image segmentation are the K-means algorithm, along with its fuzzy equivalent C-means, and unsupervised learning neural networks [22–24]. For instance, the K-means algorithm assigns each input pattern to a given cluster by means of a membership function whose values reflect the degree of similarity between the pattern itself and the cluster prototype (i.e., the centroid). The clustering problem can then be formulated as a minimization problem, with argument equal to the sum of the distances from each input pattern to the relative cluster centroid. The number of clusters is a predefined parameter of the algorithm; a variant of this algorithm, known as ISODATA, allows the dynamic modification of the number of clusters. Another interesting version is the semisupervised fuzzy C-means algorithm, in which an operator "guides" the clustering procedure by selecting a few pixels for each tissue class [25]. In the literature, a number of works have been devoted to the evaluation and comparison of the performances of the various clustering techniques, in both normal and pathological images [5]. Some of the reported results will be analyzed later in this section.

10.3.2 Use of Neural Networks in MRI Segmentation

Among segmentation techniques applied to brain tissue segmentation, a notable position is reserved to neural networks. Neural networks have always been extensively applied to image processing tasks. More details on the various neural network architectures and their applications to image processing may be found in [26, 29]. In general, published results suggest that neural networks can be very effective in accomplishing pixel-based segmentation and classification of magnetic resonance images, achieving high performances, in many cases superior to those of other pattern recognition techniques (such as statistical parametric algorithms or clustering algorithms like K-means and fuzzy C-means) [24]. Some methods are based solely on the gray-level intensity, while others exploit locally calculated features, such as texture: In any case, the vast majority rely on the use of multispectral images. Moreover, the employment of different network architectures has been proposed and compared, suggesting a possible advantage in the use of unsupervised versus supervised training.

Segmentation techniques proposed in the literature generally exploit multispectral images with acquisition channels usually equal to two or three but ranging in some cases up to five [for example, with one inversion recovery and four spin-echo images acquired with different time of echo (TE)] [30]. The most common input features are pixel gray-scale intensity levels in the different images, though features extracted from the combination of the different images and from the pixel neighborhood have also been investigated. Specifically, the mean intensity of the pixel neighborhood permits the reduction of noise, while the use of relative intensity values obtained by the use of two images allows the reduction of additive errors deriving from the spatial nonhomogeneity of the intensities themselves [19, 30, 31]. Contrary to multichannel segmentation, single-channel neural network segmentation has been rare. However, the enhanced resolution offered when comparing with equal acquisition time renders these algorithms clinically significant and advocates their further development. The most significant proposed approach is based on multiresolution analysis: Each pixel is associated with a pattern of features which consists in a scaled family of geometric features. Specifically, this family is derived by the convolution of the image and its derivatives with a Gaussian kernel at different scales, so that each image represents a different level of smoothing with respect to the original; the Gaussian kernel family has been chosen due to its invariance to roto-translations.

Another distinction in the employed neural network architecture depends on whether the applied learning algorithm is supervised or unsupervised. The network architecture most used for supervised segmentation is the multilayer perceptron (MLP). This network is usually trained with a small number of pixels (50/100) for each class selected by a trained radiologist; training set selection is a critical task in both supervised and unsupervised methods, since no established criterion exists to guide the selection itself. The training is generally repeated for each slice, and a study has in fact shown that this architecture is relatively not robust to *interslice* segmentation, that is, segmentation of a slice based on the use of training sets derived from other slices [19]. Though this aspect is highly desirable in supervised techniques, permitting the reduction of operator interaction, it is difficult to achieve, due to the gray-scale inhomogeneity and to the presence of artifacts; attempts have been made to propagate the training set to adjacent slices [1].

The most established unsupervised approaches adopt the self-organizing map (SOM) and Hopfield network architecture. In the latter case, image segmentation can here be seen as the assignment of N pixels, associated to P features, to M classes, in order to minimize the sum of the distances (generally defined as the Euclidean or Mahalanobis distance) from the kth pixel and the centroid of the associated class (clustering). Since the Hopfield network is designed to iteratively minimize an objective function, whose parameters depend on the network weights, this objective function can be mapped on a $M \times N$ Hopfield network: Once the final state is reached, the neuron outputs represent the degree of membership of each pixel to each class [32–36].

Different architectures have been proposed for the SOM maps. In some configurations, the number of neurons, and consequently the number of the classes in which the image is subdivided, is approximately equal to the number of the final classes (usually slightly larger in order to avoid undersegmentation, that is, the inclusion of different tissues in the same class). In other cases, the size of the network is much larger and is used as a preprocessing stage in the segmentation process in order to reduce the dimensionality of the feature space. The use of adaptive resonance theory (ART) networks, which are capable of dynamically changing the number of neurons when the distance between the pattern itself and the existing neurons is greater than a vigilance parameter, does not require a priori knowledge of the number of different tissues (an issue which may be critical in pathological images). However, preliminary results indicate that the obtained segmentation is strongly dependent on the training algorithm parameters and specifically on the vigilance as well as on the order with which the patterns are considered [30].

As already observed, unsupervised segmentation techniques in general, and clustering in particular, do not accomplish pixel classification. Some proposed architectures thus include also a classification stage which is usually subject to supervised training. For example, investigators have proposed a two-stage architecture constituted by a clustering stage (3×3 SOM) followed by a three-layer MLP network for quantitative evaluation of the degree of cerebral atrophy induced by radiotherapy and chemotherapy in medulloblastoma patients [5, 37].

Another study has evaluated the performance of a radial basis function (RBF) network with an unsupervised-trained first layer for clustering followed by a supervised-trained second layer for classification. It is worth remarking that the classification network architectures can be substantially simplified, since the second-layer inputs are constituted by the cluster prototypes instead of the pixel gray levels, thus reducing the sensitivity to noise and gray-level inhomogeneities. The classifier can thus be trained once on a representative ensemble of images, instead of on every image, ultimately yielding an unsupervised segmentation technique.

10.3.3 Result Validation Techniques

Validation and quantitative evaluation of the results obtained by MRI segmentation techniques are essential to their introduction in diagnostic, therapeutic, and surgical simulation clinical protocols. However, segmentation result evaluation still remains an open issue, as the determination of the actual ground truth is not feasible. Two core aspects have to be considered while validating a segmentation algorithm: result accuracy and reproducibility.

The most obvious assessment of result accuracy is obtained by visual inspection by an experienced radiologist, but many other validation strategies have been proposed in the literature. In the following, the most common and promising ones will be briefly introduced [1, 38].

The validation technique closest to the determination of ground truth is the employment of phantom scans to test segmentation algorithms. A phantom is usually subdivided in a few compartments containing known volumes of different substances sensible to MRI (usually paramagnetic liquids or gels doped with paramagnetic agents), and so phantom scans consist of images with a few, highly contrasting classes, representing the various biological tissues, on a homogeneous background. However, phantoms do not exhibit many of the characteristics that make segmentation of human tissues so difficult (i.e., a high geometric complexity and multiple tissue classes). Furthermore, they are incapable of truly mimicking the real distribution of MRI parameters in biological tissues and human RF coil loading. As a consequence, reported accuracy obtained using phantoms is usually very high, especially for large volumes, but can only provide a limited degree of confidence in the reliability of segmentation methods.

A surely more promising validation technique consists of computer-simulated magnetic resonance images; in this case, a reproducible and known ground truth is readily available, as it consists of the anatomical model used to generate the simulated images. Several magnetic resonance signal simulation techniques, along with image reconstruction methods, can be found in the literature. Simulated magnetic resonance images are particularly attractive since they allow the introduction of various sources of noise and artifacts in the generated images. Moreover, such sources can be separately controlled in order to assess the robustness of segmentation methods against the various sources and determine their effects on segmentation results in a quantifiable manner. However, the effectiveness of this validation technique relies on the availability of truly realistic MRI simulation methods.

Finally, a very common approach to MRI segmentation result evaluation requires an experienced radiologist to manually classify MRI study results in order to compare segmentation results with the manual labeling. Since manual labeling is very time consuming, the use of software packages to assist the radiologist in accomplishing this task can be useful in this context to reduce operator intervention. This technique has a major advantage in that it attempts to emulate the radiologist's interpretation, which is the only realistic "valid truth" available for in vivo experiments. Notwithstanding, this technique also has a major drawback, that is, the intrinsic variability of manual segmentation results. To aid in the evaluation and comparison of different segmentation techniques, common databases of manually labeled MRI studies, such as the Internet Brain Segmentation Repository (IBSR), could be very useful for the scientific community.

Accuracy is not the only important parameter in MRI segmentation evaluation; reproducibility is also a key aspect for many clinical applications involving segmentation algorithms. Result reproducibility may be measured in various operating conditions. For instance, in the case of supervised or semisupervised segmentation algorithms, result reproducibility can be quantitatively assessed by having the same radiologist select different training sets on different occasions for the same data set (intrarater variability) or by having different radiologists select different training sets on the same data set (interrater variability). In most cases, the stability of supervised algorithms against operator dependency in the choice of the training set has been evaluated using healthy volunteer scans. This is certainly useful and more indicative of the reliability of the segmentation methods than other validation techniques, such as those based on phantom scans. However, results obtained using MRI studies of normal volunteers may be dramatically different from those of patients, especially after radiation treatment and/or chemotherapy. Once more, this entails the adoption of unsupervised segmentation techniques to eliminate operator dependency on the results.

10.4 SINGLE-CHANNEL UNSUPERVISED SEGMENTATION: CASE STUDY

In this section, we will provide an example of a fully unsupervised neural-network-based approach to MRI segmentation [39]. This example will allow us to concretely illustrate the various steps that constitute a typical unsupervised MRI segmentation technique. In fact, most techniques proposed in the published literature follow the general schema shown in Figure 10.1, which comprises four steps: preprocessing, feature extraction, clustering, and classification. In this case, the preprocessing step consists in extracting the intracranial boundary, but other image enhancing techniques (e.g., noise removal) can also be applied.

The proposed segmentation technique focuses on the use of single-channel images. As previously mentioned, single-channel images usually yield smaller voxel sizes and are therefore useful for performing precise volumetric analysis. Single-channel image segmentation is a difficult task due to the relative lack of information when compared with multispectral images, and only a few authors have dealt with this issue. However, it is worth noting that this approach is quite general and can be easily extended to cover multispectral images and/or other image types by modifying the feature extraction step; in fact, similar approaches have been attempted in the literature to perform segmentation of multichannel images [37].

Finally, in order to perform clustering, a SOM neural network was used; clusters then have to be classified according to the tissue type they represent. This can be done either manually or using a classifier; in this case, a MLP neural network was used.

10.4.1 Preprocessing

Intracranial boundary detection is a preprocessing step that consists in determining the boundary of the intracranial cavity and removing extrameningeal tissues, which are usually irrelevant in the diagnosis and treatment of many brain pathologies. This effectively limits both the number of pixels to be classified, thus reducing the overall processing time, and the number of tissue classes to be considered, simplifying consequently the segmentation task, especially in the case of single-channel images where there is a significant

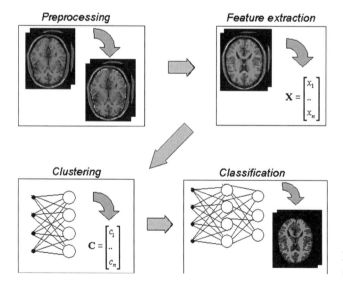

Figure 10.1 Stages of MRI segmentation technique.

overlap in the gray-level intensity distribution of the various tissue classes. To detect the intracranial cavity boundary, a fully automatic technique, proposed in [40], was implemented and adapted to segment T1-weighted images in both the axial and coronal orientations. The procedure is based on three steps.

In the first step, a head mask is determined by removing background noise. First, a rough head mask is generated by histogram thresholding; the threshold is determined by fitting the lower bins of the histogram to a Rayleigh distribution under the assumption that background noise obeys a Rayleigh distribution [41]. Then, the head mask is refined by means of morphological operations.

In the second step, an initial brain mask is generated by using histogram thresholding and morphological operations. Thresholds are again automatically generated by fitting a Gaussian distribution to the central peak of the histogram and then selecting the upper and lower thresholds by half the standard deviation above and below the mean, respectively. These thresholds are usually successful in eliminating most extracranial tissues but may still need manual refinement in some cases. However, a single thresholding step is not sufficient to eliminate all extracranial tissues; therefore a series of morphological operators are applied to the thresholded image. In this step, high-level knowledge on brain anatomy and spatial information generated by the previous head segmentation step are exploited.

Finally, in the third step, the head mask is refined using an active contour model algorithm [42]. In active contour paradigms, the task of segmenting an object from the background is modeled as an energy minimization problem. The energy is calculated as the sum of an external energy, which attracts the contour toward features of interest in the image (such as edges), and an internal energy, which poses smoothness constraints on the final contour. Anisotropic diffusion filtering is used to enhance edges and perform intraregion smoothing in order to obtain better results.

10.4.2 Feature Extraction

Feature selection and extraction comprise a fundamental step in order to ensure accurate and reliable segmentation. Many features have been proposed in the literature and most methods rely on gray-level intensities only in order to classify pixels according to the tissue type. However, in the case of single-channel magnetic resonance images, gray-level intensity is generally not sufficient to obtain satisfactory results. Alternative features have been proposed, such as those extracted by multiscale analysis and image texture (it has to be remarked, though, that statistical texture features, such as those based on the co-occurrence matrix, are not suitable for pixel-based classification and hence are not applicable in this case). In this case, we have employed a nine-dimensional feature vector extracted from each pixel neighborhood. Among various possible combinations, the normalized intensities of the pixel and of its eight nearest neighbors have been selected on an experimental basis. The use of neighbors' gray-level intensity is useful in compensating the effects of random noise while minimizing the loss of resolution. Moreover, introducing the spatial distribution of gray levels in the segmentation process can also be helpful in obtaining more accurate results.

10.4.3 Clustering

Unsupervised clustering is performed by means of a Kohonen SOM [43]. As shown in Figure 10.2, the SOM consists of a single layer of neurons arranged in a bidimensional topology during the learning phase. Each output neuron is associated with a cluster,

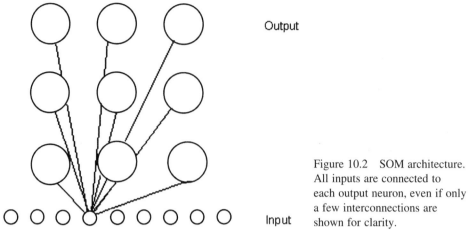

Output

Input

Figure 10.2 SOM architecture. All inputs are connected to each output neuron, even if only a few interconnections are shown for clarity.

and the associated weights represent the cluster centroid (i.e., the prototype) in the feature space. In this case, nine neurons arranged in a 3×3 topology have been used; the number of neurons has been selected in order to allow for sufficient capacity to define correct pixel clusters, without undersegmenting tissue regions, before collapsing to the final number of expected tissue classes. Since the intracranial boundary has been previously extracted in the preprocessing step, the number of tissue classes being considered can be limited to 3 for normal brain parenchyma (i.e., white matter, gray matter, and cerebrospinal fluid). In order to account for intensity inhomogeneities across different slices, the SOM is separately trained for each input slice. After training, each pattern is presented to the network to perform clustering.

Self-organizing maps are trained using a competitive learning algorithm. Each neuron is associated with a vector of weights $\mathbf{w}_i = \{w_{i1}, w_{i2}, \ldots, w_{ij}\}$ which, as previously mentioned, represents the prototype of the cluster associated with that neuron. Each input pattern is associated with the neuron whose prototype is closest to the pattern itself according to some distance measure (usually the Euclidean distance). The training algorithm seeks to minimize an error function, given by the sum of the distances between each input pattern in the training set and the relative winning neuron. In order to do so, a competitive learning rule (also known as the Kohonen learning rule) is employed which iteratively adjusts weight vectors starting from their initial values. At each iteration, all the input patterns in the training set are presented to the network and the winning neuron is recorded. Winning neuron weights are adjusted according to the following rule:

$$\Delta \omega_{ij} = \eta(x_{ij} - w_{ij})$$

Furthermore, the weights of the neighboring neurons are adjusted; in particular, weight updates

$$\Delta \omega_{ij} = \eta f_{\text{neigh}}(x_{ij} - w_{ij})$$

are scaled by a neighborhood function (the most commonly used functions are the Gaussian function and its first derivative, the "mexican hat" function). In this way, the input space topology is preserved in the SOM output space topology (i.e., clusters which are close in the input feature space are associated with neighboring neurons).

As previously mentioned, the SOM is separately trained for each input slice to account for intensity inhomogeneities across different slices. The training set is constituted by 500 randomly selected pixels, where the training set size has been determined experimentally. The training set should not be too small, in order to provide sufficient generalization capabilities, and not too large, in order to preserve the network from the effects of random noise. Moreover, random selection ensures that the choice of the training set is fully unsupervised and operator independent, while the network stability against random initialization has been assessed to guarantee that the final result is substantially independent of the initial choice of the training set.

The training procedure, being iterative, requires initialization of weight vectors. In this case, weights initialization is performed randomly for the first slice, while subsequently final weight vectors are propagated for initialization from each slice to the adjacent. In this way, 3D connectivity of brain structures is exploited to reduce overall computation times and obtain a more accurate segmentation.

Finally, the definition of the training procedure requires the selection of appropriate values for various parameters, namely the neighborhood function and the learning rate. In this case, a Gaussian neighborhood function has been used:

$$f_{neigh}(iter) = \exp\left(\frac{-(x^2 + y^2)}{2\sigma^2}\right)$$

where $x^2 + y^2$ is the relative distance between the winning neuron and the updated neuron and x and y are the coordinate differences in the two dimensions of the output plane. The size of the neighborhood should decrease in time, pursuing network plasticity in the first phase of learning versus selectivity in the final phases.

In this study, the size of the neighborhood is decreased monotonically at each iteration by updating the neighborhood function according to

$$\sigma = 3(0.4^{iter/iter_{max}})$$

where $iter_{max}$ is the predetermined total number of iterations. The decreasing form of the neighborhood ensures that the weight vectors will converge to a final set of anatomically meaningful prototypical vectors [5]. The number of iterations performed is 2000 for the first slice and 300 for subsequent slices. These parameters were selected experimentally in order to ensure reliable convergence of weight vectors.

Finally, an adaptive learning rate η has been used; the initial value is set to 0.01 and is then dynamically adjusted at each iteration. In particular, if the total error significantly increases, the learning rate is decreased in order to avoid network instability, whereas if the error significantly decreases, the learning rate is increased in order to go deeper into the energy landscape.

10.4.4 Classification

In the final classification step, pixel clusters generated by the SOM are mapped into predefined tissue classes by means of a MLP, or feedforward, neural network. The network was trained on one set of manually labeled images, while another set of labeled images was used as a test set to assess the performances of the segmentation procedure on previously unseen cases.

More specifically, the network was separately trained for simulated and real magnetic resonance images to account for the different distributions of gray-level intensities in the two types of input images. Three slices for each data set were selected as representative slices for this purpose, and a preselected number of pixels were randomly selected from each slice to form the training set. More details on the characteristics of the data sets employed for this study will be given in the results section.

It is worth observing that, since classification is performed on the set of prototypical vectors determined by the previous unsupervised clustering procedure, rather than on raw image data, the classification procedure is more robust against noise and has better generalization capabilities, compared to other procedures that directly employ gray-level intensities to perform pixel classification.

As shown in Figure 10.3, the employed MLP has three levels: an input layer, a hidden layer, and an output layer. The input layer consists of nine neurons, one for each component of the feature vectors associated with the pixels in the input image. The vector input is thus constituted by the centroid of the winning neuron associated with each input pixel. The hidden layer and the output layer consist of seven and four neurons, respectively, with sigmoid transfer function. Each neuron in the output layer is associated with a predetermined tissue class (white matter, gray matter, cerebrospinal fluid, and other); each input pattern is therefore associated with the class of the neuron which gives the greatest output; that is, a crisp classification is performed. The initial weights were determined randomly using a random-number generator with Gaussian distribution, zero mean, and variance 0.5.

Finally, the network was trained with error backpropagation using the delta rule for training. Again, the training procedure seeks to minimize an error function which is given by the sum of the distances between the network output and the target (i.e., the desired network output) for each pattern in the training set. When using the delta rule, connection

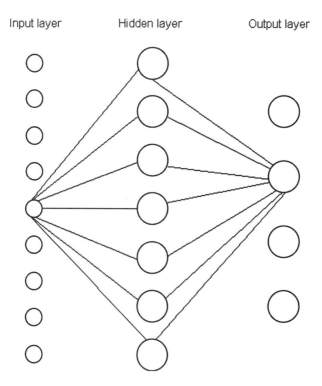

Input layer   Hidden layer   Output layer

Figure 10.3 MLP architecture. All neurons in adjacent layers are fully interconnected, but only a few interconnections are shown for simplicity.

weights are updated at each iteration according to

$$\Delta w_{ji}(t) = -\eta \sum_p \delta_j^p F'(z_j) x_i^p$$

where p denotes the patterns in the training set, $F'(z_j)$ is the first derivative of the transfer function, and δ_j^p is the difference between the actual output and the target of each network layer.

For the output layer, the error is known and can be calculated as

$$\delta_j^p = y_j^p - t_j^p$$

while for hidden layers, the error needs to be backpropagated. In particular, for layers s and $s - 1$, the following relationship applies:

$$\delta_j^{(s-1)_p} = \sum_j \delta_j^{s_p} F^{s'}(z_j^p) w_{ji}^s$$

Appropriate values for the network parameters were determined experimentally; in particular, the learning rate for the network was set to 0.0001, and the network was trained for 5000 epochs.

10.4.5 Results

In this section, we will report the results obtained while testing the segmentation technique discussed in the previous sections on simulated and real magnetic resonance images. The performance of the clustering procedure has been analyzed in greater detail in [39]. In this section we will evaluate the entire segmentation process.

10.4.5.1 *Simulated Magnetic Resonance Images* As discussed in Section 10.2, the use of simulated images facilitates validation of MRI segmentation methods since a reproducible ground truth is known a priori. Furthermore, the effects of various noise sources can be separately investigated [44].

The simulated data sets used in this study were made available by the Brainweb institution (available on-line at http://www.bic.mni.mcgill.ca/brainweb/). A description of the simulation techniques used to generate these data sets can be found in [45, 46]. A number of parameters can be freely set to determine the characteristics of the final images; in this study, all data sets used are T1-weighted images in the axial orientation with a slice thickness of 1.0 mm. In order to assess the effects of random noise and intensity inhomogeneities, 14 data sets with random noise varying in the range from 0 to 9% and level of intensity inhomogeneity varying in the range from 0 to 40% were used. For the purposes of this study, three reference slices were selected for each data set.

The training set was extracted from the reference slices of three preselected data sets and consisted in 2000 randomly chosen pixels for each reference slice. The remaining data sets were instead employed as test sets.

In order to train the classifier, three data sets were chosen to extract the training set, which consisted of 2000 randomly selected pixels for each reference slice, while the others were employed as test sets. In Figure 10.4 a few representative slices are shown, along with the results of the clustering and classification steps. Classification labels are mapped to a color scheme similar to the one used for PET to display color images.

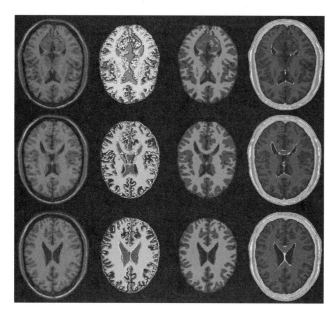

Figure 10.4   Segmentation results for three representative slices. From left to right: simulated T1-weighted magnetic resonance image, gray-scale representation of the clustering step, pseudocolor display of classification results, and ground truth.

In particular, gray- and white-matter tissues are colored red and green, respectively, while cerebrospinal fluid (CSF) is colored blue.

Segmentation results have also been quantitatively compared with the ground truth (i.e., the model used to generate the simulated images). Such quantitative evaluation of results was performed by classifying each reference slice in both the training and test sets and by determining the percentage of correctly classified pixels in every slice. Percentage values have also been separately calculated for each tissue class. Finally, the mean and standard deviation over the entire set of slices of the percentage of correctly classified pixels were calculated. Quantitative results are reported in Table 10.1. The high partial-volume effect at the sharp boundary between the ventricular CSF and the surrounding gray-matter tissue accounts for the relatively low percentage of correctly classified tissue for the CSF class.

In order to deeper evaluate the effects of random noise and gray-level intensity inhomogeneity on final segmentation results, in Figure 10.5 segmentation results are compared against the level of random noise and intensity inhomogeneity, respectively. Numerical results were obtained by varying the percentage of random noise (intensity inhomogeneity) while keeping the value of the other parameter fixed and by calculating the mean percentage of correctly classified pixels over all slices with the same degree of random noise (intensity inhomogeneity).

TABLE 10.1   Mean and Variance of Correctly Classified Pixels Per Slice

	White matter (WM)	Gray matter (GM)	Cerebrospinal fluid (CSF)	Total
Mean	92.77	92.40	77.94	90.41
Standard deviations	3.74	6.40	8.21	2.55

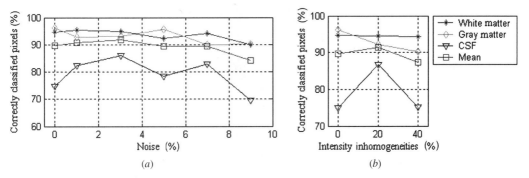

Figure 10.5　Percentage of correctly classified pixels per tissue class with varying level of random noise and intensity inhomogeneity.

10.4.5.2 *Volunteer MR Scans*　Simulated magnetic resonance images are useful in evaluating the performance of segmentation methods but usually cannot encompass all the characteristics of real magnetic resonance images that hinder the capability of segmentation algorithms to accurately segment tissue classes. All the data sets utilized in this study were made available by the IBSR. In this study, 20 volunteer magnetic resonance brain data sets were considered; the data sets and their manual segmentations were provided by the Center for Morphometric Analysis at Massachussetts General Hospital (available online at http://neuro-www.mgh.harvard.edu/cma/ibsr).

All data sets are T1-weighted images in the coronal orientation; in particular, 10 volunteer scans were performed on a 1.5T Siemens Magnetom MR system [repetition time (TR) 40 ms, TE 8 ms, flip angle 50°, field of view (FOV) 30 cm, thk (slice) 3.1 mm, number of excitations (nex) 1], while 10 scans were performed on a 1.5T General Electric Signa MR system [TR 50 ms, TE 9 ms, flip angle 50°, FOV 24 cm, thk 3.0 mm, nex 1). Manual segmentation was performed on the positionally normalized scan by trained investigators using a semiautomated intensity contour mapping algorithm and signal intensity histograms [47]. Once the external border was determined by intensity contour mapping, gray–white matter borders were demarcated using signal intensity histograms. Using this technique, borders are defined as the midpoint between the peaks of the bimodal histogram for a given structure and its adjacent tissue. Other neuroanatomical structures were segmented similarly [48].

For the purposes of this study, 10 volunteer scans were considered and three reference slices per data set were selected. The training set was extracted from three preselected data sets by randomly choosing 1500 pixels for each reference slice, lending great attention to the selection of the training set, in order to ensure that all tissue classes are well represented (i.e., guarantee that the number of training patterns per tissue class does not fall below a certain threshold).

Some representative slices of the volunteer scans, along with the results of the clustering and classification procedures, are shown in Figure 10.6.

As for simulated images, the accuracy of the segmentation algorithm was quantitatively assessed by comparing segmentation results with the ground truth (i.e., the manual segmentation) on a pixel basis; the obtained results are reported in Table 10.2. All the reference slices in the training and test sets were classified and the percentage of correctly classified pixels per slice was determined. The mean and standard deviation over the entire set of slices were then calculated. Since only one manual segmentation was available, it could not be possible to account for inter- and intraobserver variability in result evaluation.

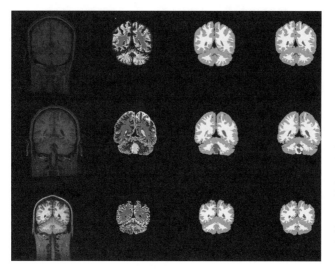

Figure 10.6   Segmentation results for three representative slices. Left to right: raw MRI, gray-scale representation of clustering step, pseudocolor display of classification results, and manual segmentation. All pixels classified as background in the manual segmentation were manually removed in order to enable the comparison of manual and automatic segmentation.

TABLE 10.2   Mean and Variance of Correctly Classified Pixels Per Slice

	White matter (WM)	Gray matter (GM)	Cerebrospinal fluid (CSF)	Total
Mean	83.20	75.09	61.08	77.92
Standard deviation	7.62	11.30	20.26	7.49

However, this source of error, which could possibly cause some variations in the results reported in this section, is common to all validation techniques based on the use of manually labeled studies. As expected, segmentation results are slightly worse when compared to simulated magnetic resonance images due to the greater inherent complexity of real images. Furthermore, the increased slice thickness results in higher partial-volume artifacts, which obviously affect the accuracy of the segmentation algorithm, especially at the boundary between CSF and gray matter.

10.5   FUTURE TRENDS

In this chapter, we have provided an introduction to MRI segmentation techniques, with particular emphasis on a neural-network-based approach. Moreover, we have outlined some of the most interesting applications of segmentation techniques as well as those critical aspects that make MRI segmentation so difficult.

Brain imaging plays a significant role in the diagnosis and treatment of brain pathologies and neurological disorders. Image processing and analysis tools are strongly needed to take full advantage of the advances in imaging modalities and could potentially have a remarkable impact on the efficiency and efficacy of clinical procedures, ultimately leading to a more cost-effective, higher quality health care delivery.

Therefore, it is not surprising that a large amount of past and ongoing research activities have been devoted to MRI segmentation, which is a major task in medical image processing and analysis. However, the introduction of segmentation methods in everyday practice will not be feasible until established techniques, satisfying clinical requirements in terms of result reliability, accuracy, reproducibility, and operator independence, are available.

REFERENCES

1. L. P. CLARKE, R. P. VELTHUIZEN, M. A. CHAMACO, J. J. HEINE, M. VAIDYANATHAN, L. O. HALL, ET AL., MRI segmentation: Methods AND applications. *Magnetic Resonance Imaging*, 13:343–368, 1995.

2. T. HEINONEN, P. DASTIDAR, H. FREY, AND H. ESKOLA, Applications of MR image segmentation. *International Journal of Bioelectromagnetism*, 1:35–46, 1999.

3. L. P. CLARKE, R. P. VELTHUIZEN, M. CLARK, J. GAVIRIA, L. HALL, ET AL., MRI measurement of brain tumor response: Comparison of visual metric and automatic segmentation. *Magnetic Resonance Imaging*, 16:271–279, 1998.

4. M. VAIDYANATHAN, R. VELTHUIZEN, L. P. CLARKE, AND L. O. HALL, Quantization of brain tumor in MRI for treatment planning. *Proc. of the 16th Annual International Conference of the IEEE*, 1:555–556, 1994.

5. M. VAIDYANATHAN, L. P. CLARKE, L. O. HALL, C. HEIDTMAN, R. VELTHUIZEN, ET AL., Monitoring brain tumor response to therapy using MRI segmentation. *Magnetic Resonance Imaging*, 15:323–334, 1997.

6. W. E. REDDICK, M. K. RAYMOND, T. D. ELKIN, T. DAVID, J. O. GLASS, T. E. MERCHANT, ET AL., A hybrid neural network analysis of subtle brain volume differences in children surviving brain tumors. *Magnetic Resonance Imaging*, 16:413–421, 1998.

7. M. VAIDYANATHAN, L. P. CLARKE, L. O. HALL, A. M. BENSAID, R. VELTHUIZEN, S. PHUPHANICH, ET AL., Comparison of supervised MRI segmentation methods for tumor volume determination during therapy. *Magnetic Resonance Imaging*, 13:719–728, 1995.

8. R. ACHARYA, R. WASSERMAN, J. STEVENS, AND C. HINOJOSA, Biomedical imaging modalities: A tutorial. *Computerized Medical Imaging and Graphics*, 19:3–25, 1995.

9. J. W. HENNEL AND J. KLINOWSKI. *Fundamentals of Nuclear Magnetic Resonance*. Longman, Essex, England, 1993.

10. J. B. KNEELAND, W. WEHRLI, AND D. SHAW. *Biomedical Magnetic Resonance Imaging: Principles, Methodology and Applications*. Felix J. BRUCE KNEELAND, New York, 1988.

11. L. Q. ZHOU, Y. M. ZHU, C. BERGOT, A. M. LAVAL-JEANTET, V. BOUSSON, J. D. LAREDO, AND M. LAVAL-JEANTET, A method of radio-frequency inhomogeneity correction for brain tissue segmentation in MRI. *Computerized Medical Imaging and Graphics*, 25:379–389, 2001.

12. R. VELTHUIZEN, L. P. CLARKE, AND H. LIN, Evaluation of non-uniformity corrections for tumor response measurements. *Proc. of the 19th IEEE/EMBS International Conference*, 2:761–762, 1997.

13. M. C. CLARK, L. O. HALL, D. B. GOLDGOF, R. P. VELTHUIZEN, F. R. MURTAGH, AND M. S. SILBIGER, Automatic tumor segmentation using knowledge-based techniques. *IEEE Transactions on Medical Imaging*, 17:187–201, 1998.

14. R. P. VELTHUIZEN, Validity guided clustering for brain tumor segmentation. *Proc. of the 17th Annual Conference of the EMBS*, 1:413–414, 1995.

15. J. C. BEZDEK, L. O. HALL, AND L. P. CLARKE, Review of MR image segmentation techniques using pattern recognition. *Medical Physics*, 20:1033–1048, 1993.

16. J. C. RAJAPAKSE, J. N. GIEDD, C. DECARLI, J. W. SNELL, A. MCLAUGHLIN, Y. C. VAUSS, ET AL., A technique for single-channel MR brain tissue segmentation: Application to a pediatric sample. *Magnetic Resonance Imaging*, 14:1053–1065, 1996.

17. J. C. RAJAPAKSE, J. N. GIEDD, AND J. L. RAPOPORT, Statistical approach to segmentation of single-channel cerebral MR images. *IEEE Transactions on Medical Imaging*, 16:177–186, 1997.

18. S. RUAN, C. JAGGI, J. XUE, J. FADILI, AND D. BLOYET, Brain tissue classification of magnetic resonance images using partial volume modeling. *IEEE Transactions on Medical Imaging*, 19:1179–1187, 2000.

19. A. STOCKER, O. SIPILA, A. VISA, O. SALONEN, AND T. KATILA, Stability study of some neural networks applied to tissue characterization of brain magnetic resonance images. *Proc. of the 13th International Conference on Pattern Recognition*, 3:472–476, 1996.

20. A. PITIOT, A. W. TOGA, N. AYACHE, AND P. THOMPSON, Texture based MRI segmentation with a two-stage hybrid neural classifier. *Proc. of the 2002 International Joint Conference on Neural Networks*, 3:2053–2058, 2002.

21. S. VINITSKI, C. GONZALEZ, C. BURNETT, W. BUCHHEIT, F. MOHAMED, H. ORTEGA, AND S. FARO, 3D segmentation in MRI of brain tumors: Preliminary results. *Proc. of the 17th Annual Conference of the Engineering in Medicine and Biology Society*, 1:481–482, 1995.

22. W. E. PHILLIPS, R. P. VELTHUIZEN, S. PHUPHANICH, L. O. HALL, L. P. CLARKE, AND M. L. SILBIGER, Application of fuzzy C-means segmentation technique for tissue differentiation in MR images of a hemorrhagic glioblastoma multiforme. *Magnetic Resonance Imaging*, 13:277–290, 1995.

23. D. L. PHAM AND J. L. PRINCE, Adaptive fuzzy segmentation of magnetic resonance images. *IEEE Transactions on Medical Imaging*, 18:737–752, 1999.

24. L. O. HALL, A. M. BENSAID, L. P. CLARKE, R. P. VELTHUIZEN, M. S. SILBIGER, AND J. C. BEZDEK, A comparison of neural network and fuzzy clustering techniques in segmenting magnetic resonance images of the brain. *IEEE Transactions on Neural Networks*, 3:672–682, 1992.

25. A. M. BENSAID, L. O. HALL, J. C. BEZDEK, AND L. P. CLARKE, Partially supervised clustering for image segmentation. *Pattern Recognition*, 29:859–871, 1996.

26. M. EGMONT-PETERSEN, D. DE RIDDER, AND H. HANDELS, Image processing with neural networks— A review. *Pattern Recognition*, 35:2279–2301, 2002.

27. J. A. FREEMAN AND D. M. SKAPURA, *Neural Networks: Algorithms, Applications, and Programming Techniques*. Addison-Wesley, Reading, MA, 1991.

28. S. HAYKIN, *Neural Networks: A comprehensive Foundation*, 2nd ed. Prentice-Hall, Englewood Cliffs, NJ, 1999.

29. R. C. GONZALEZ AND R. E. WOODS, *Digital Image Processing*. Prentice-Hall, Upper Saddle River, NJ, 2002.

30. B. ASHJAEI AND H. SOLTANIAN-ZADEH, A comparative analysis of neural network methodologies for segmentation of magnetic resonance images. *International Conference on Image Processing*, 1:257–260, 1996.

31. M. N. AHMED AND A. A. FARAG, Two-stage neural network for volume segmentation of medical images. *International Conference on Neural Networks*, 3:1373–1378, 1997.

32. S. C. AMARTUR, D. PIRAINO, AND Y. TAKEFUJI, Optimization neural networks for the segmentation of magnetic resonance images. *IEEE Transactions on Medical Imaging*, 11:215–220, 1992.

33. R. SAMMOUDA, N. NIKI, AND H. NISHITANI, Hopfield neural network for the multichannel segmentation of magnetic resonance cerebral images. *Pattern Recognition*, 30:921–927, 1997.

34. R. SAMMOUDA, N. NIKI, AND H. NISHITANI, Neural networks based segmentation of magnetic resonance images. *Nuclear Science Symposium and Medical Imaging Conference*, 4:1827–1831, 1995.

35. J.-S. LIN, K.-S. CHENG, AND C.-W. MAO, A modified Hopfield neural network with fuzzy c-means technique for multispectral MR image segmentation. *International Conference on Image Processing*, 1:327–330, 1996.

36. Y. ZHU AND Z. YAN, Computerized tumor boundary detection using a Hopfield neural network. *IEEE Transactions on Medical Imaging*, 16:55–67, 1997.

37. W. E. REDDICK, J. O. GLASS, E. N. COOK, T. D. ELKIN, AND R. J. DEATON, Automated segmentation and classification of multispectral magnetic resonance images of brain using artificial neural networks. *IEEE Transactions on Medical Imaging*, 16:911–918, 1997.

38. J. YANG AND S.-C. HUANG, Method for evaluation of different MRI segmentation approaches. *Nuclear Science Symposium*, 3:2053–2059, 1998.

39. L. MORRA, F. LAMBERTI, AND C. DEMARTINI, A neural network approach to unsupervised segmentation of single-channel MR images. *Proc. of the 1st IEEE/ EMBS Conference on Neural Engineering*, Capri, 1:515–518, 2003.

40. M. S. ATKINS AND B. T. MACKIEWICH, Fully automatic segmentation of the brain in MRI. *IEEE Transactions on Medical Imaging*, 17:98–107, 1998.

41. M. E. BRUMMER, R. M. MERSEREAU, R. L. EISNER, AND R. R. J. LEWINE, Automatic detection of brain contours in MRI data sets. *IEEE Transactions on Medical Imaging*, 12:153–166, 1993.

42. M. KASS, A. WITKIN, AND D. TERZOPOULOS, Snakes: Active contour models. *International Journal of Computer Vision*, 1:321–331, 1988.

43. T. KOHONEN, *Self Organization and Associative Memory*. Springer Verlag, New York, 1989.

44. R. K.-S. KWAN, A. C. EVANS, AND G. B. PIKE, MRI simulation-based evaluation of image-processing and classification method. *IEEE Transactions on Medical Imaging*, 18:1085–1097, 1999.

45. C. A. COCOSCO, V. KOLLOKIAN, R. K.-S. KWAN, AND A. C. EVANS, BrainWeb: Online interface to a 3D MRI simulated brain database. *Proceedings of 3rd International Conference on Functional Mapping of the Human Brain*, vol. 5, no. 4, part 214, S425, Copenhagen, May, 1997.

46. D. L. COLLINS, A. P. ZIJDENBOS, V. KOLLOKIAN, J. G. SLED, N. J. KABANI, C. J. HOLMES, AND A. C. EVANS, Design and construction of a realistic digital brain phantom. *IEEE Transactions on Medical Imaging*, 17:463–468, 1998.

47. D. N. KENNEDY, P. A. FILIPEK, AND V. S. CAVINESS, Anatomic segmentation and volumetric calculations in nuclear magnetic resonance imaging. *IEEE Transactions on Medical Imaging*, 8:1–7, 1989.

48. P. A. FILIPEK, D. N. KENNEDY, AND V. S. CAVINESS, Volumetric analysis of central nervous system neoplasm based on MRI. *Pediatric Neurology*, 7:347–351, 1991.

CHAPTER 11

HIGH-RESOLUTION EEG AND ESTIMATION OF CORTICAL ACTIVITY FOR BRAIN–COMPUTER INTERFACE APPLICATIONS

F. Cincotti, M. Mattiocco, D. Mattia,
F. Babiloni, and L. Astolfi

11.1 INTRODUCTION

In the last decade, high-resolution electroencephalography (EEG) technologies were developed to enhance the spatial information of EEG activity [1, 2]. Furthermore, since the ultimate goal of any EEG recording is to provide useful information about the brain activity, mathematical techniques, known as inverse procedures, have been developed to estimate the cortical activity from the raw EEG recordings. Examples of these inverse procedures are dipole localization, the distributed source, and cortical imaging techniques [1, 3, 4]. Inverse procedures can use linear and nonlinear techniques to localize putative cortical sources from EEG data by using mathematical models of the head as a volume conductor.

Nowadays, it has been shown that the imagery of upper limb movements produced stable EEG patterns on the scalp surface and that such patterns can be used in the buildup of a brain–computer interface (BCI) system between a human and a computer [5–7]. The EEG patterns related to motor imagery are generally located in the centroparietal scalp areas, roughly overlying the primary sensory and motor cortical areas [8–10]. However, several published papers have also underlined the usefulness of collecting brain information related to motor imagery by intracranial recordings in both humans [11–14] and monkeys [15–19]. Such intracortical recordings, often performed on the primary motor cortical areas, provided signals that allow the user to control virtual or physical devices. Despite these successes, the use of intracranial recordings in humans is obviously limited for ethical reasons to those patients that have undergone brain surgery or are severely disabled [11, 13]. However, a question has arosing about the possibility of avoiding such invasive procedures without renouncing the high-quality signals recorded at the cortical level.

More recently, it has been suggested that with the use of modern high-resolution EEG technologies [19–21] it may be possible with estimate the cortical activity associated with the mental imagery of upper limb movements in humans [22, 23]. However, verification of such a statement on a group of normal subjects has not been performed. The scientific question at the base of the present work is whether the estimated cortical activity related

Handbook of Neural Engineering. Edited by Metin Akay

to the mental imagery of the upper limbs provides more useful features with respect to those obtained using scalp EEG recordings. To address this, high-resolution EEG recordings were taken on six healthy subjects during the imaging of upper limb movements. Waveforms from scalp electrodes and those from the estimated cortical activity in particular regions of interest (ROIs) were then compared. Such comparisons returned information about the usefulness of cortical activity for the recognition of mental states with respect to scalp recorded data.

11.2 METHODOLOGY

11.2.1 Data Collection

Six healthy subjects participated in the experiments and were asked to perform the imaging of right-finger movements after a visual stimulus appeared on a screen placed in front of them. The EEG was recorded by a high-resolution EEG cap with 64 electrodes disposed accordingly to an extension of the 10–20 international system. The EEG data sampling frequency was 256 Hz, and the signal was bandpass filtered between 0.1 and 100 Hz before digitization. Electromyographic (EMG) signals were also recorded from the extensor digitoris communis muscle on both arms to check for any small movements of the finger during the mental imagery. Each single EEG trial occurred from 2 s before the arrival of the visual trigger to 1 s after.

11.2.2 Head Model

For all subjects analyzed in this study, sequential magnetic resonance images were acquired and realistic head models were generated. Figure 11.1 shows the realistic head models generated for the experimental subjects analyzed in this study together with the high-resolution electrode array used. Scalp, skull, dura mater, and cortical surfaces of the realistic and averaged head models were obtained with segmentation and countouring algorithms. The surfaces of the realistic head models were then used to build the boundary element model of the head as the volume conductor employed in the study. Conductivity values for scalp, skull, and dura mater were those reported by Oostendorp et al. [24].

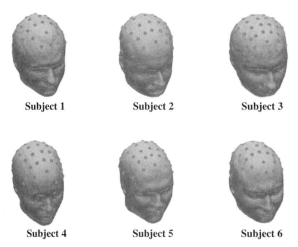

Subject 1 Subject 2 Subject 3

Subject 4 Subject 5 Subject 6

Figure 11.1 Realistic head models generated for the six subjects analyzed in this study, together with their high resolution electrode array used for the EEG recordings.

A cortical surface reconstruction was accomplished for each subject's head with a tessellation of about 5000 triangles on average.

11.2.3   Estimation of Cortical Source Activity

Cortical activity during the mental imagery task was estimated in each subject using the depth-weighted minimum-norm algorithm [25–27]. Such estimation returns a current density estimate for each of the thousand dipoles constituting the modeled cortical source space. Each dipole returns a time-varying amplitude representing the brain activity of a restricted patch of cerebral cortex during the entire task time course. This rather large amount of data can be synthesized by computing the ensemble average of all the dipole magnitudes belonging to the same cortical ROI. Each ROI was defined on each subject's cortical model in accordance with its Brodmann areas (BAs). Such areas are regions of the cerebral cortex whose neurons share the same anatomical (and often also functional) properties. During a decade of neuroimaging studies different BAs were assigned a precise role in the reception and the analysis of sensory and motor commands as well as memory processing. Actually, such areas are largely used in neuroscience as a coordinate system for sharing cortical activation patterns found with different neuroimaging techniques. In the present study, the activity in the following ROI was taken into account: the primary left and right motor areas, related to BA 4, the left and right primary somatosensory and supplementary

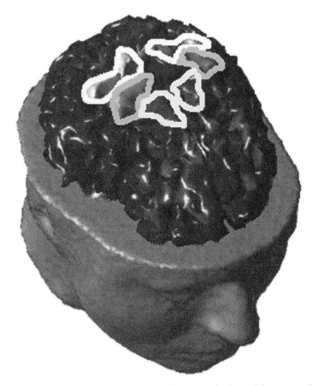

Figure 11.2   The ROIs employed in a particular subject over the realistic cortical and head model reconstructions. Note as each particular ROI is surrounded with a thick line. The lines marked the primary motor, supplementary motor, and somatosensory cortical areas of both hemispheres. In dark gray are represented the other parts of the cortical surface that were used for the estimation of the current density distribution.

motor areas. Figure 11.2 shows the ROIs employed for a particular subject (4) over the realistic cortical and head model reconstructions.

11.2.4 Data Analysis

Artifact correction was performed on each single EEG trial recorded, and the trials contaminated by electrooculogram (EOG) and/or EMG activity at the resting arms were discarded. Artifact-free single trials were then averaged. Each artifact-free single trial was then subjected to the linear inverse procedure, and the associated time-varying cortical distributions were estimated. The procedure of spatial averaging in each ROI explained in the previous paragraph then applied to retrieve the cortical waveforms related to each particular ROI analyzed.

11.3 RESULTS

The average activity observed at the scalp electrodes and those estimated in the cortical ROIs showed consistent spatiotemporal patterns across all subjects of the experimental group analyzed. Such patterns involve the presence of a consistent negative activity over the centroparietal scalp regions, with a variable time delay from the visual trigger onset between 240 and 280 ms and a peak of activity between 290 and 320 ms.

Figure 11.3 shows, for subject 6, the average waveforms gathered from the scalp electrodes C3 and C4 as well as the waveforms obtained from the primary motor areas on the left and right hemispheres (waveforms A4L and A4R, respectively). It is worth noting that the negative deflections of the C3 and C4 waveforms after the visual trigger onset are more balanced with respect to the analogous waveforms obtained in the

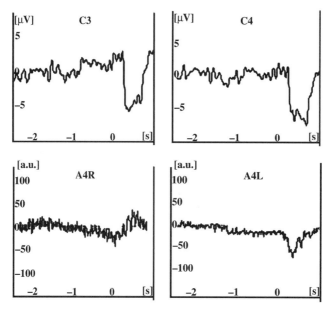

Figure 11.3 Average waveforms for subject 6 gathered from the scalp C3 and C4 leads and the waveforms obtained from the primary motor areas on left and right hemispheres (waveforms represented with the A4L and A4R labels, respectively).

primary motor areas in both hemispheres. In fact, the waveforms related to the primary motor areas of the left hemisphere present more pronounced negative deflections with respect to those observed in the right motor areas. This phenomenon is mainly present on the primary motor area, while the activity from the primary sensory motor areas is present but less unbalanced with respect to those obtained from the motor areas.

Figure 11.4, first row, shows a sequence of time-varying potential distributions generated over the scalp surface in subject 4. A clear negative potential distribution over the centroparietal scalp area is observed during the time period between 260 and 340 ms. Such a potential distribution is similar to the occurrence of the motor potential during normal limb movements [9, 27]. Of particular interest are the peaks of negative activity generated over the vertex and the left lateral scalp areas. It is worth noting that no relevant activity is observed on the scalp areas ipsilateral to the imaged finger movements. The second row of the figure shows the time course of the estimated cortical activity during the mental imagery from 260 to 340 ms after the visual trigger onset. In this figure, the negative cortical values are depicted in red hues while the positive cortical ones are represented with blue hues.

Depth-weighted minimum-norm estimates showed a concentration over the left primary motor areas of the mental activity related to the imaging of the right arm movement during the time period analyzed. No relevant cortical activity is present in the right motor and primary sensory areas, while frontal activity is present. Furthermore, the presence of bilateral negative activity over the cortical regions can be observed relative to the proper supplementary motor area. Also, the cortical patterns are consistent with those recorded from the scalp surface in the same subject (see first row). However, the cortical estimated patterns present higher spatial details than those observed on the scalp surface.

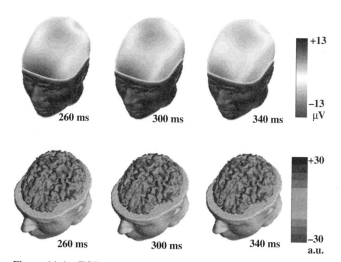

Figure 11.4 TOP row: sequence of the time-varying potential distributions generated over the scalp surface in the subject 4. A clear negative potential distribution over the centro parietal scalp areas is observed during the time period between 260 to 340 ms, after the presentation of the visual trigger starting the imagery of the right middle finger extension. Bottom row: time course of the estimated cortical activity during the mental imagery of the right finger movement from 260 to 340 ms after the visual trigger onset. In this row, the negative cortical values are depicted in red hues, while the positive cortical ones are represented with the blue hues.

11.4 DISCUSSION

The data reported here suggest that it is possible to retrieve the cortical activity related to mental imagery by using sophisticated high-resolution EEG techniques, obtained by solving the linear inverse problem with the use of realistic head models. Of course, the analysis of the distribution of the potential fields associated with the motor imagery in humans have been already described [8, 9]. However, in the context of the BCI, it becomes important if the activity related to the imagination of arm movement can be unbalanced between the two hemispheres. In fact, the greater this imbalance between the activity gathered in scalp electrodes C3 and C4, the easier is the task of recognizing it by a classifier [9]. The relevant finding here is that the group analysis of the cortical waveforms associated with the mental imagery suggested the presence of a more pronounced imbalance between the cortical activity estimated in the primary motor areas of the right and left hemispheres and that gathered from scalp electrodes.

Several EEG or magnetoencephalographic studies have shown that the cortical activity related to the motor imagery can be retrieved using the single equivalent dipole approximation [28–30]. However, it must be noted that the use of a single dipole is a rather crude approximation of the distributed cortical activation occurring during the mental imagery. Furthermore, the dipole localization technique is a nonlinear methodology, that involves the use of iterative algorithms for the search of the minimum of the cost functional associated with the optimal dipole position in the head model. This last aspect makes dipole approximation an unsuitable tool for the on-line processing of the EEG data needed in a BCI device that results in useful feedback to the user [6, 7]. On the other hand, it is worth noting that the cortical estimation methodology illustrated above is suitable for the on-line application needed for the BCI device. In fact, despite the use of sophisticated realistic head models for scalp, skull, dura mater, and cortical surface, estimation of the instantaneous cortical distribution from acquired potential measures required a limited amount of time necessary for a matrix multiplication. Such multiplication occurs between the data vector gathered and the pseudoinverse matrix that is stored off-line before the start of the EEG acquisition process. The complexity of the pseudoinverse matrix is illustrated by the geometric head modeling (obtained with the boundary element or the finite element modeling techniques) as well as by the a priori constraints used for the minimum-norm solutions.

There is a trend in the modern neuroscience field toward using invasive electrode implants to record cortical activity in both animals and humans for the realization of an efficient BCI device [12, 15, 18]. In this chapter we have presented evidence that suggests an alternative methodology for the estimation of such cortical activity in a noninvasive way by using accurate modeling of the principal head structures involved in the transmission of the cortical potential from the brain surface to the scalp electrodes.

ACKNOWLEDGMENTS

This work was supported in part by the European IST Programme FET Project FP6-003758.

REFERENCES

1. NUNEZ, P. L., Neocortical *Dynamics and Human EEG Rhythms*. Oxford University Press, 1995.
2. GEVINS, A., BRICKETT, P., COSTALES, B., LE, J., AND REUTTER, B., Beyond topographic mapping: Towards functional-anatomical imaging with 124–channel EEG and 3-D MRIs. *Brain Topogr.*, 1990, vol. 1, pp. 53–64.
3. DALE, A. M. AND SERENO, M., Improved localization of cortical activity by combining EEG and MEG with

MRI cortical surface reconstruction: A linear approach, *J. Cog. Neurosci.*, 1993, vol. 5, pp. 162–176.

4. SCHERG, M., BAST, T., AND BERG, P., Multiple source analysis of interictal spikes: Goals, requirements, and clinical value. *J. Clini Neurophysiol.*, 1999, vol. 16, pp. 214–224.

5. BIRBAUMER, N., GHANAYIM, N., HINTERBERGER, T., IVERSEN, I., KOTCHOUBEY, B., KÜBLER, A., PETELMOUTER, J., TAUB, E., AND FLOR, H., A spelling device for the paralyzed. *Nature*, 1999, vol. 398, pp. 297–298.

6. PFURTSCHELLER, G. AND NEUPER, C., Motor imagery and direct brain-computer communication. *Proc. IEEE*, 2001, vol. 89, no. 7, pp. 1123–1134.

7. WOLPAW, J. R., BIRBAUMER, N., MCFARLAND, D. J., PFURTSCHELLER, G., AND VAUGHAN, T. M., Brain computer interfaces for communication and control. *Clini Neurophysiol.*, 2002, vol. 113, pp. 767–791.

8. PFURTSCHELLER, G. AND NEUPER, C., Motor imagery activates primary sensorimotor area in man. *Neurosci. Lett.*, 1997, vol. 239, pp. 65–68.

9. NEUPER, C., SCHLÖGL, A., AND PFURTSCHELLER, G., Enhancement of left-right sensorimotor EEG differences during feedback-regulated motor imagery. *J. Clin. Neurophysiol.*, 1999, vol. 16, pp. 373–382.

10. MCFARLAND, D., MINER, L., VAUGHAN, T., AND WOLPAW, J., Mu and beta rhythm topographies during motor imagery and actual movements. *Brain Topog.*, 2000, vol. 12, no. 3, pp. 177–186.

11. KENNEDY, P. R., AND BAKAY, R. A. E., Restoration of neural output from a paralyzed patient by a direct brain connection. *NeuroReport*, 1998, vol. 9, No. 8, pp. 1707–1711.

12. KENNEDY, P. R., BAKAY, R. A. E., MOORE, M. M., ADAMS, K., AND GOLDWAITHE, J., Direct control of a computer from the human central nervous system. *IEEE Trans. Rehabil. Eng.*, 2000, vol. 8, no. 2, pp. 198–202.

13. LEVINE, S. P., HUGGINS, J. E., BEMENT, S. L., KUSHWAHA, R. K., SCHUH, L. A., PASSARO, E. A., ROHDE, M. M., AND ROSS, D. A., Identification of electrocorticogram patterns as a basis for a direct brain interface. *J. Clin. Neurophysiol.*, 1999, vol. 16, pp. 439–447.

14. LEVINE, S. P., HUGGINS, J. E., BEMENT, S. L., KUSHWAHA, R. K., SCHUB, L. A., ROHDE, M. M., PASSARO, E. A., ROSS, D. A., ELISEVICH, K. V., AND SMITH, B. J., A direct brain interface based on event-related potentials. *IEEE Trans. Rehabil. Eng.*, 2000, vol. 8, pp. 180–185.

15. CHAPIN, J. K., MOXON, K. A., MARKOWITZ, R. S., AND NICOLELIS, M. A. L., Real-time control of a robot arm using simultaneously recorded neurons in the motor cortex. *Nature Neurosci.*, 1999, vol. 2, pp. 664–670.

16. NICOLELIS, M. A. L., GHAZANFAR, A. A., STAMBAUGH, C. R., OLIVEIRA, L. M., LAMBACH, M., CHAPIN, J. K., NELSON, R. J., AND KAAS, J. H., Simultaneous encoding of tactile information by three primate cortical areas. *Nature Neurosci.*, 1998, vol. 7, pp. 621–630.

17. TAYLOR, D. M., TILLERY, S. I. H., AND SCHWARTZ, A. B., Direct cortical control of 3D neuroprosthetic devices. *Science*, 2002, vol. 296, pp. 1829–1832.

18. DONOGHUE, J. P., Connecting cortex to machines: Recent advances in brain interfaces. *Nature Neurosci.*, 2002, vol. 5, Suppl. 1, pp. 1085–1088.

19. GEVINS, A., BRICKETT, P., COSTALES, B., LE, J., AND REUTTER, B., Beyond topographic mapping: Towards functional-anatomical imaging with 124-channel EEG and 3-D MRIs. *Brain Topogr.* 1990, vol. 1, pp. 53–64.

20. GEVINS, A., SMITH, M. E., MCEVOY, L., AND YU, D., High resolution EEG mapping of cortical activation related to working memory: Effects of difficulty, type of processing, and practice. *Cereb. Cortex*, 1997, vol. 7, pp. 374–385.

21. BABILONI, F., CINCOTTI, F., CARDUCCI, F., ROSSINI, P. M., AND BABILONI, C., Spatial enhancement of EEG data by surface Laplacian estimation: The use of MRI-based head models. *Clin. Neurophysiol*, 2001, vol. 112, no. 5, pp. 724–727.

22. CINCOTTI, F., MATTIA, D., BABILONI, C., CARDUCCI, F., BIANCHI, L., MILLÁN, J., MOURIÑO, J., SALINARI, S., MARCIANI, M. G., AND BABILONI, F., Classification of EEG mental patterns by using two scalp electrodes and Mahalanobis distance-based classifiers. *Methods Inf. Med.*, 2002, vol. 41, no. 4, pp. 337–341.

23. GRAVE DE PERALTA, R., ET AL., Direct non invasive brain computer interface (BCIS). Presented at the 9th International Conference on Functional Mapping of the Human Brain, June 19–22, 2003, New York. Available on CD-Rom in *NeuroImage*, vol. 19, no. 2, poster 1027.

24. OOSTENDORP, T. F., DELBEKE, J., AND STEGEMAN, D. F., The conductivity of the human skull: Results of in vivo and in vitro measurements. *IEEE Trans. Biomed. Eng.*, 2000, vol. 47, no. 11, pp. 1487–1492.

25. BABILONI, F., BABILONI, C., LOCCHE, L., CINCOTTI, F., ROSSINI, P. M., AND CARDUCCI, F. High resolution EEG: Source estimates of Laplacian-transformed somatosensory-evoked potentials using a realistic subject head model constructed from magnetic resonance images. *Med. Biol. Eng. Comput.*, 2000, vol. 38, pp. 512–519.

26. BABILONI, F., BABILONI, C., CARDUCCI, F., ROMANI, G. L., ROSSINI, P. M., ANGELONE, L. M., AND CINCOTTI, F., Multimodal integration of high resolution EEG and functional magnetic resonance imaging data: A simulation study. *Neuroimage*, 2003, vol. 19, no. 1, pp. 1–15.

27. BABILONI, C., CARDUCCI, F., CINCOTTI, F., ROSSINI, P. M., NEUPER, C., PFURTSCHELLER, G., AND BABILONI, F., Human movement-related potentials vs. desynchronization of EEG alpha rhythm. A high-resolution EEG study. *Neuroimage*, 1999, vol. 10, no. 6, pp. 658–665.

28. SCHNITZLER, A., SALENIUS, S., SALMELIN, R., JOUSMÄKI, V., AND HARI, R., Involvement of primary motor cortex in motor imagery: A neuromagnetic study. *Neuroimage*, 1997, vol. 6, pp. 201–208.

29. HARI, R., FORSS, N., AVIKAINEN, S., KIRVESKARI, E., SALENIUS, S., AND RIZZOLATTI, G., Activation of human primary motor cortex during action observation: A neuromagnetic study. *Proc. Natl. Acad. Sci. USA*, 1998, vol. 95, no. 1, pp. 15061–15065.

30. GREEN, J. B., SORA, E., BIALY, Y., RICAMATO, A., AND THATCHER, R. W., Cortical motor reorganization after paraplegia: An EEG study. *Neurology*, 1999, vol. 53, no. 4, pp. 36–743.

CHAPTER 12

ESTIMATION OF CORTICAL SOURCES RELATED TO SHORT-TERM MEMORY IN HUMANS WITH HIGH-RESOLUTION EEG RECORDINGS AND STATISTICAL PROBABILITY MAPPING

L. Astolfi, D. Mattia, F. Babiloni, and F. Cincotti

12.1 INTRODUCTION

It is well known that high-resolution electroencephalography (HREEG) recording presents a very good temporal resolution (millisecond scale) and a moderate spatial resolution (on the order of 2–3 cm). The spatiotemporal resolution of HREEG is adequate to follow the complex temporal dynamics of brain phenomena [1]. Furthermore, the wide availability of magnetic resonance (MR) images of subjects' heads has made possible the use of realistic models for the head and for the cortical surface in procedures involving the estimation of cortical current activity. It has been demonstrated that the use of realistic head models increases the accuracy of the cortical current reconstruction from HREEG data. Pointlike models of cortical sources (such as the current dipole) are largely used in the analysis of primary evoked potentials/fields or epileptic spikes [2–4]. However, such source models are inefficient at representing extended cortical activations due to cerebral engagement in motor or cognitive tasks [5, 6]. In the distributed source approach, thousands of equivalent current dipoles covering the cortical surface modeled have been used and their strengths estimated by using linear and nonlinear inverse procedures [7–11]. The solution space (i.e., the set of all possible combinations of the cortical dipoles strengths) is generally reduced by using geometric constraints. For example, the dipoles can be disposed along the reconstruction of cortical surface with a direction perpendicular to the local surface. An additional constraint involves forcing the dipoles to explain the recorded data with a minimum or a low amount of energy (minimum-norm solutions [7]). However, statistical characterization of the estimated cortical current densities is also needed to understand if the observed changes in brain activity are due to chance alone or are induced by the

Handbook of Neural Engineering. Edited by Metin Akay
Copyright © 2007 The Institute of Electrical and Electronics Engineers, Inc.

experimental task. However, in the study of cognitive processes with EEG recordings frequency-related instead of temporal analysis is often used [12]. This is because many cognitive processes produced marked variations of EEG spectral contents while temporal variations of the electromagnetic field could be less informative, due to the jitter of the time components of the event-related potentials/fields [13]. Hence, an interest arose in the extension of statistical parametric mapping (SPM) to the analysis of EEG data computed in the frequency domain. Also of interest for a possible extension of SPM techniques to the frequency domain is related to the fact that often the experimental design requires that the subjects perform different tasks. This occurs when a cognitive paradigm such as go/no go or working/not working memory is investigated [14–17]. The comparison between these different tasks in the framework of SPM can be conveniently addressed by the same approach used by the neuroscientific community for functional magnetic resonance imaging (fMRI) studies. In fact, by using the properties of the z-score it is possible to derive a contrast technique that extends the comparisons usually employed by SPM between the task and the baseline periods to the frequency domain.

In this chapter we present a methodology that allows the use of spectral data in the framework of SPM for the analysis of complex behavioral tasks. An application of the proposed approach is provided on data related a short-term memory (STM) and no-short-term memory (NSTM) paradigm recorded by HREEG in a group of healthy subjects. Results obtained with EEG showed active participation of cortical prefrontal and parietal areas during the task phase in which an image had to be held in the memory by the subjects. Such participation was not visible using previously adopted conventional mapping procedures while it was highlighted by the proposed statistical approach.

12.2 MATERIALS AND METHODS

12.2.1 Tasks

For the EEG recordings, the subjects were seated in a comfortable reclining armchair placed in a dimly lit, sound-damped, and electrically shielded room (magnetic insulated room). They kept their forearms resting on armchairs, with right index finger resting between two buttons spaced 6 cm apart. A computer monitor was placed in front of the subjects (about 100 cm). Subjects were exposed to cue visual stimuli on the computer screen. These stimuli consisted of several vertical bar that were first presented (trigger time) and then hidden for a few seconds (delay time) in the STM condition, while in the NSTM condition the vertical bars remained on the screen. Hence, subjects were asked to hold the images in memory (delay period) only during the STM task. In both the STM and NSTM tasks, when the go stimulus appeared on the computer screen, subjects were asked to produce a motor performance in accordance to the image. In particular, subjects had to push the left button if the left bar on the screen was higher than the right vertical bar, or vice versa. A brief on-line feedback on the performance was automatically provided. The timing of the STM task was the following: pretrigger time duration 1 s; warning visual stimulus (trigger time) duration 1 s; cue stimulus duration 5.5–7.5 s; visual go stimulus duration 1 s; and intertrial interval duration 5 s. The control condition in which the spatial bar remained on the screen for the duration of the task was also developed and this was used in the NSTM condition. The experiment included two trial blocks for the NSTM task and two trial blocks for the STM task (each block had 50 trials). The NSTM and STM blocks were intermingled according to a predefined sequence (pseudorandom order). The interblock interval was about 2 min. Subjects were told in advance if the block comprised NSTM or STM trials. Before the recording session, training of about 10 min familiarized subjects with the experimental apparatus and NSTM–STM tasks. Subjects were free to use any memorization strategy, including "visuospatial imagery," somatomotor preparation, and covert verbal coding and rehearsal.

12.2.2   Subjects and EEG Recordings

All subjects involved in the EEG experiments gave informed written consent according to the Helsinki declaration. General recording procedures were approved by the local institutional ethics committees. Nine healthy subjects were considered in this study. Each EEG recording was made using 43 scalp electrodes that were positioned according to an extension of the international 10–20 system. The EEG data were recorded with a 0.3–70-Hz bandpass and linked-earlobe reference. Electrooculographic (0.3–70-Hz bandpass) and surface-rectified electromyographic activity of bilateral extensor digitorum muscles (1–70 Hz bandpass) was also collected. Artifact-free EEG activity from 3.5 s before to 4 s after the onset of the visual go stimulus (zero time) was then considered for the EEG linear inverse procedures. The EEG segments contaminated by blinking, eye movements, mirror movements, and/or other artifacts were rejected off-line. In EEG recordings, frequency bands of interest were theta, alpha 1, alpha 2, and alpha 3 whose band limits were computed according to a standard procedure based on the peak individual alpha frequency (IAF [13]). The IAF was defined as the frequency showing the maximum power density peak within a large frequency range lasting from 4 to 14 Hz. It was calculated at the baseline period (1-s epoch preceding the onset of the warning stimulus, -2000 to -1000 ms). With respect to the IAF, these frequency bands were defined as follows: (i) theta as the IAF $- 7$ Hz to IAF $- 4$ Hz, (ii) alpha 1 or "lower-1 alpha band" as the IAF $- 4$ Hz to IAF $- 2$ Hz, (iii) alpha 2 or "lower-2 alpha band" as IAF $- 2$ Hz to IAF, and (iv) alpha 3 or "upper alpha" as IAF to IAF $+ 2$ Hz. The power spectrum analysis was based on a fast Fourier transform (FFT) using the Welch technique and the Hanning windowing function. To quantify the event-related percentage reduction of alpha oscillations, we followed the widely used procedure for the computation of event-related desynchronization/synchronization (ERD/ERS [12, 18]). The ERD/ERS was computed for theta, alpha 1, alpha 2, and alpha 3 bands based on the individual determination of the IAF.

12.2.3   Cortical Source Estimation

In this study, MR-constructed subjects' realistic head models and an averaged one were used to compute individual cortical current density estimates and to visualize population data, respectively. The averaged head model was constructed using a set of averaged MR images of 152 normal subjects from the Montreal Neurological Institute. Scalp, skull, dura mater, and cortical surfaces of the realistic and averaged head models were obtained with segmentation and countouring algorithms. The surfaces of the realistic head models were then used to build the boundary element model of the head as the volume conductor employed in the present study. A cortical surface reconstruction was accomplished with a tessellation of about 3000 triangles. A current dipole was placed at the center of each triangle. The orientation of each dipole was placed perpendicular to the triangle surface to model the alignment of the cortical neurons with respect to the cerebral cortex surface. Figure 12.1 presents the averaged head model adopted for the visualization of the population data.

When the EEG activity is mainly generated by circumscribed cortical sources (i.e., short-latency evoked potentials/magnetic fields), the location and strength of these

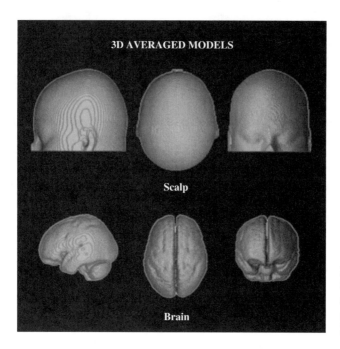

Figure 12.1 The averaged head model as obtained by segmenting the MRIs available from the Montreal Neurologic Institute.

sources can be reliably estimated by the dipole localization technique [19, 20]. In contrast, when EEG activity is generated by extended cortical sources (i.e., event-related potentials/magnetic fields), the underlying cortical sources can be described by using a distributed source model with spherical or realistic head models [7, 21, 22]. With this approach, typically thousands of equivalent current dipoles covering the cortical surface modeled and located at the triangle center were used, and their strength was estimated by using linear and nonlinear inverse procedures [7, 11]. Taking into account the measurement noise $\mathbf{n}$ supposed to be normally distributed, an estimate of the dipole source density magnitudes $\mathbf{x}$ that generated a measured potential $\mathbf{b}$ at a particular time instant t can be obtained by solving the linear system

$$\mathbf{A}\mathbf{x} + \mathbf{n} = \mathbf{b} \qquad (12.1)$$

where $\mathbf{A}$ is a $m \times n$ matrix with number of rows equal to the number of sensors and number of columns equal to the number of modeled sources. We denote by $\mathbf{A}_j$ the potential distribution over the m sensors due to each unitary jth cortical dipole. The collection of all m-dimensional vectors $\mathbf{A}_j$ ($j = 1, \ldots, n$) describes how each dipole generates the potential distribution over the head model, and this collection is called the lead field matrix $\mathbf{A}$. This is a strongly underdetermined linear system in which the number of unknowns, dimension of the vector $\mathbf{x}$, is greater than the number of measurements $\mathbf{b}$ by about one order of magnitude. Estimation of the current density vector $\mathbf{x}$ has been performed by using the weighted minimum-norm solution [21–23].

12.2.4 Spectral Estimation of Current Densities

According to Eq. (12.1), let $\mathbf{x}(t)$ be the source estimates for each time instant t of the recorded event-related potentials/fields $\mathbf{b}(t)$. By computing the Fourier transform of the obtained solution $\mathbf{x}(t)$, we have

$$\mathbf{x}(t) = \mathbf{G} \cdot \mathbf{b}(t) \Rightarrow \mathbf{X}(f) = \mathbf{G} \cdot \mathbf{B}(f) \tag{12.2}$$

where $\mathbf{X}(f)$ and $\mathbf{B}(f)$ are the Fourier transforms of the time-varying sources and potentials, respectively, and $\mathbf{G}$ is the pseudoinverse matrix that maps the recorded potentials $\mathbf{b}$ to the estimated current density vector $\mathbf{x}$. To estimate the spectral contents of the obtained solutions $\mathbf{X}(f)$, the cross-spectral density matrix $\mathbf{xCSD}(f)$ was computed according to the following formula, where the H indicates the Hermitian condition and the apex the transposition:

$$\mathbf{xCSD}(f) = \mathbf{X}(f) \cdot \mathbf{X}(f)^{\mathrm{H}} \tag{12.3}$$

From the positions above it follows that

$$\mathbf{xCSD}(f) = \mathbf{G} \cdot \mathbf{B}(f) \cdot \mathbf{B}(f)^{\mathrm{H}} \cdot \mathbf{G}' = \mathbf{G} \cdot \mathbf{bCSD}(f) \cdot \mathbf{G}' \tag{12.4}$$

where $\mathbf{bCSD}(f)$ is the spectral covariance matrix of the recorded data. The ith entry of the $\mathbf{xCSD}(f)$ matrix $[\mathbf{xCSD}_{ii}(f)]$ is relative to the power spectra of the ith current dipole. After the spectral computations, the logarithmic transform was applied to make the spectral data more Gaussian than the untransformed ones.

12.2.5 Estimation of Cortical Activity in Region of Interest

Estimation of the cortical activity results in a current density estimate for each of the about 3000 dipoles constituting the modeled cortical source space. Each dipole presents a time-varying magnitude representing the spectral power variations generated during the task time course. This rather large amount of data can be synthesized by computing the ensemble average of all the dipole magnitudes belonging to the same cortical region of interest (ROI). Each ROI was defined using the cortical model adopted in accordance with the Brodmann areas (BAs). Such areas are regions of the cerebral cortex whose neurons share the same anatomical (and often also functional) properties. A decade of neuroimaging studies have assigned to different BAs a precise role in the reception and analysis of sensory and motor commands as well as memory processing. Actually, such areas are largely used in neuroscience as a coordinate system for sharing cortical activation patterns found with different neuroimaging techniques. As a result of this anatomical-guided data reduction, we pass from the analysis of about 3000 time series to the evaluation of less than 100 (the BAs located in both cerebral hemispheres). These BA waveforms, related to the increase or decrease of the spectral power of the cortical current density in the investigated frequency band, can be successively averaged across the studied population. The generated grand-average waveforms describe the time behavior of the spectral power increase or decrease of the current density in the population during the task examined.

12.2.6 Statistical Analysis

What is often missing in EEG linear inverse solutions is the level of reliability of the solution. Not all modeled sources have the same degree of sensitivity to measurement noise, so we cannot say whether a source has a high strength because it is the most probable source of that potential distribution or because that source well accounts for the noise superimposed on the potential. Even in the ideal case of absence of noise, some sources seem more inclined to explain a large set of data, just because their geometric properties (i.e., sources positioned on a cortical gyrus rather than deep in a sulcus). A statistical approach to the problem and a measure of the signal-to-noise ratio in the modeled cortical activity are then required. The level of noise in the EEG linear inverse solutions can be addressed by estimating the "projection" of the EEG noise $\mathbf{n}(t)$ onto the cortical surface by means of the computed pseudoinverse operator $\mathbf{G}$; the standard error of the noise on the estimated source strength ξ_j is given by $\langle \mathbf{G}_{j\cdot}\mathbf{n}(t) \rangle = \mathbf{G}_{j\cdot}\mathbf{C}(\mathbf{G}_{j\cdot})'$, where $\mathbf{G}_{j\cdot}$ is the jth row of the pseudoinverse matrix and $\mathbf{C}$ is the EEG noise covariance matrix $(C_{ij} = \langle n_i(t), n_j(t) \rangle)$. This allows us to quantitatively assess the ratio between the estimated cortical activity $\mathbf{x}$ and the amount of noise at the cortical level, quantified through the standard deviation of its estimate. It can be demonstrated that under the hypothesis of a normal estimate for the noise $\mathbf{n}(t)$ obtained with more than 50 time points, the following normally distributed z-score estimator can be obtained for each jth cortical location and for each time point t considered:

$$\mathbf{z}_j(t) = \frac{\mathbf{G}_{j\cdot} \cdot \mathbf{b}(t)}{\sqrt{\mathbf{G}_{j\cdot}\mathbf{C}(\mathbf{G}_{j\cdot})'}} \tag{12.5}$$

where $\mathbf{C}$ is the estimated noise covariance matrix. The uncorrected threshold for the z-score level at 5% is 1.96. Values of z exceeding such a threshold represent levels of estimated cortical activity that are unlikely due to chance alone but are related to the task performed by the experimental subject. However, to avoid the effects of the increase of the type I error due to the multiple z-tests performed, the results will be presented after application of the Bonferroni correction [24].

Here, we extend the SPM approach to the analysis of the power spectra variations during the experimental task. Computation of the z-score level in the spectral case is performed according to the formula

$$\sigma^2_{\text{Noise}_i} = \text{Var}(\text{bCSD}_{ii}(f)) \qquad Z_i(f) = \frac{[G \cdot \text{bCSD}_{ii}(f)G']_i}{\sigma_{\text{Noise}_i}} \tag{12.6}$$

where $\text{Var}(\text{bCSD}_{ii}(f))$ indicates the variance of the estimate of the log-transformed spectral density of the EEG measurements during the baseline period at the considered frequency f and $Z_i(f)$ is the z-score for the ith current dipole for the frequency f. The $Z_i(f)$ was used to construct z-scores that are identical to those introduced in the time domain by Dale and colleagues [25], while the inverse operator G is identical to those computed for the temporal case.

The z-score for each brain region investigated, whether as a result of time- or frequency-domain processing, provides information about the significance of the increase or decrease of the variable considered in the analyzed period with respect to the baseline chosen. Due to the additive properties of the z-estimator, the z-scores obtained for all the dipoles belonging to a particular ROI can be averaged together. Furthermore, such average z-scores for each particular ROI can be averaged across the subject's population by returning a grand-average z-score level for each ROI investigated. These values can be considered as an estimate of the statistical behavior of the variable analyzed (in the time or frequency domain) for the entire population in each considered ROI. In particular, the computation of the population z-score for the ith BA is performed as

$$Z_{i1} = \frac{\sum_{j=1}^{N} Z_{ij1}}{N} \tag{12.7}$$

where z_{ij1} is the average z-score level for the ith ROIs for the jth subject performing task 1.

STM   NSTM

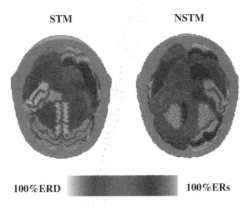

100%ERD   100%ERs

Figure 12.2   The figure presents two maps, obtained by the solution of the linear inverse problem in the spectral domain associated with the gathered EEG data. The head is seen from above, nose down, left ear right. The maps on the figure presents the grand average data of the EEG recorded population relative to the variation of cortical current density power (event-related desynchronization and synchronization; ERD and ERS) in the theta frequency band in the STM and NSTM tasks. Increase of spectral power with respect to the baseline period (ERS) are represented from yellow to red hues, while the decrement of spectral power (ERD) are coded from the dark to the light blue.

12.3 RESULTS

Subjects were exposed to the STM and NSTM tasks while they were under the EEG recording device. All the data in this section are relative to the time-varying spectral changes in different frequency bands during the delay period.

Figure 12.2 presents two estimated cortical distributions obtained by solution of the linear inverse problem in the spectral domain associated with the gathered EEG data. The figure presents the grand-average data of the EEG-recorded population relative to the variation of cortical current density power in the theta frequency band. Increasing values of spectral power with respect to the baseline period (ERS) are represented with yellow to red hues and the decreasing values of spectral power (ERD) with dark to light blue. The left cortical distribution is relative to the STM task while the right one presents data estimated during the NSTM task. Spectral maps are relative to the time period of 1.5 s from the beginning of the delay period. The head is seen from above, nose down, left ear at the right of the visualized head.

It is worth noting that during the STM task there is a large ERS in the theta band in the right and left parietal cortical areas, together a synchronization occurring over the mesial and right premotor cortical areas. Furthermore, a large bilateral frontal reduction of spectral power (ERD) also occurred, as indicated by the blue regions over the cortical map. The NSTM task presents a different pattern of ERD and ERS with respect to the STM task. The grand-average EEG data returned a modest ERS phenomenon in the right parietal area, together with a large bilateral ERD in frontomesial cortical areas. An interesting question is if such variations of spectral power with respect to the baseline period (the ERS and ERD) in the theta band are statistically significant or are only due to noise fluctuations. The answer to this question is presented in Figure 12.3, which shows the distribution of the statistical z-scores associated with variations of the cortical spectral power in each ROI analyzed. In particular, if the power spectrum during the delay period is superior to the power spectrum computed during the baseline for the analyzed

STM   NSTM

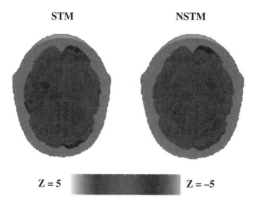

Z = 5   Z = –5

Figure 12.3   The figure shows the distribution of the z-scores associated to the variations of
the cortical spectral power in each one of the ROIs analyzed. A statistical significant decrease
of spectral power in the theta band for is coded in blue, while a statistical significant increase is
coded in yellow.

ROI, a positive z-score is obtained. Furthermore, if the z-score exceeds the threshold for
the statistical significance required, the ROI analyzed was coded with a hue from yellow to
red (in the present case up to a value of $z = 5$; that is, it is associated with a statistical
significance of $p < 10^{-5}$). Instead, a decrease in spectral power in a particular ROI is statistically significant if the associated negative value of z is lower than the required
threshold. In that case, the ROI was coded with a hue from dark to light blue ($z = -5$).
If the spectral power increase or decrease in the analyzed ROI does not reach the threshold
of the statistical significance, the ROI is shown in gray.

Figure 12.3 shows the grand-average significance probability map for the STM task
on the left. It is worth noting the presence of a large statistical significant increase in the
power spectra of the right premotor and the mesial cortical areas, coded in red. Of interest
also is the statistical significant decrease of spectral power in the theta band for the left
frontal areas, coded in blue. The SPM associated with the grand-average data for the
NSTM task in the same frequency band is presented on the right in Figure 12.3. In this
map few cortical areas present variations of spectral power not due to chance alone.
In particular, the left frontal cortical areas show a significant decrease of spectral power
in the theta band during the NSTM task.

12.4   DISCUSSION

The interest in the approach presented here has to do with the extension of the SPM
techniques proposed by Dale and co-workers [25] to the frequency domain and in the possibility to further exploit this approach by applying contrast statistical techniques normally
used in fMRI. It is already accepted that estimation of current densities with the weighted
minimum-norm procedure from EEG recordings can provide an accurate picture of the
underlying cortical activation at a moderate spatial scale [7, 21, 26]. In this context, the
use of SPM provided useful insights on the phenomena under study, going beyond
the analysis of topographic distribution of ERD and ERS at the cortical level.

Animal studies and human neuroimaging studies indicate that tasks requiring STM
activate a functional network linking regions of the prefrontal cortex with posterior association cortices [27–29]. Activation of this network can also be characterized by

measurement of neuroelectric activity recorded on the scalp. Interestingly, recent studies have indicated an enhancement in the power spectra in the theta frequency band of the EEG signal over the midline frontal region during STM tasks [16, 17]. Enhancement of the frontal midline theta rhythm has been noted in a variety of tasks that require sustained, focused attention [16, 17]. Current evidence suggests that this signal is generated in the anterior cingulated cortex [17], an important component of an anterior attentional network critical to the performance of complex cognitive tasks [30–36]. The results we obtained with SPM extended to the frequency domain can be framed in the previously described context. The centromesial activation related to the power spectral increase in theta frequency bands has been obtained, in contrast with the same experimental paradigm but without the memory task. This increase in spectral activity only holds during the delay period, while in the remaining time periods there are no significant differences between memory and NSTM tasks.

In conclusion, the present study highlights the benefit of spectral estimation of cortical current density within the framework of SPM for the study of cortical sources of human brain activity related to STM tasks.

REFERENCES

1. NUNEZ, P. L., 1995, *Neocortical Dynamics and Human EEG Rhythms*, Oxford University Press, New York.

2. TORQUATI, K., PIZZELLA, V., DELLA PENNA, S., FRANCIOTTI, R., BABILONI, C., ROSSINI, P. M., AND ROMANI, G. L., 2002, Comparison between SI and SII responses as a function of stimulus intensity, *Neuroreport* 13(6):813–819.

3. STENBACKA, L., VANNI, S., UUTELA, K., AND HARI R., 2002, Comparison of minimum current estimate and dipole modeling in the analysis of simulated activity in the human visual cortices, *Neuroimage* 16(4):936–943.

4. OTSUBO, H., OCHI, A., ELLIOTT, I., CHUANG, S. H., RUTKA, J. T., JAY, V., AUNG, M., SOBEL, D. F., AND SNEAD, O. C., 2001, MEG predicts epileptic zone in lesional extrahippocampal epilepsy: 12 pediatric surgery cases, *Epilepsia* 42(12):1523–1530.

5. ANOUROVA, I., NIKOULINE, V. V., ILMONIEMI, R. J., HOTTA, J., ARONEN, H. J., AND CARLSON, S., 2001, Evidence for dissociation of spatial and nonspatial auditory information processing, *Neuroimage* 14(6):1268–1277.

6. OKADA, Y. C. AND SALENIUS, S., 1998, Roles of attention, memory, and motor preparation in modulating human brain activity in a spatial working memory task, *Cereb. Cortex* 8(1):80–96.

7. DALE, A. M. AND SERENO, M., 1993, Improved localization of cortical activity by combining EEG anf MEG with MRI cortical surface reconstruction: A linear approach, *J. Cog. Neurosci.* 5:162–176.

8. PHILLIPS, J. W., LEAHY, R., AND MOSHER, J. C., 1997, Imaging neural activity using MEG and EEG, *IEEE Eng. Med. Biol. Mag.* 16:34–41.

9. BAILLET, S. AND GARNERO, L., 1997, A bayesian framework to introducing anatomo-functional priors in the EEG/MEG inverse problem, *IEEE Trans. Biomed. Eng.* 44:374–385.

10. BAILLET, S., GARNERO, L., MARIN, G., AND HUGONIN, P., 1999, Combined MEG and EEG source imaging by minimization of mutual information, *IEEE Trans. Biomed. Eng.* 46:522–534.

11. UUTELA, K., HAMALAINEN, M., AND SOMERSALO, E., 1999, Visualization of magnetoencephalographic data using minimum current estimates, *Neuroimage* 10(2):173–180.

12. PFURTSCHELLER, G. AND LOPES DA SILVA, F. H., 1999, Event-related EEG/MEG synchronization and desynchronization: Basic principles, *Clin. Neurophysiol.* 110(11):1842–1857.

13. KLIMESCH, W., 1999, EEG alpha and theta oscillations reflect cognitive and memory performance: A review and analysis. *Brain Res. Brain Res. Rev.* 29(2/3):169–195.

14. FALLGATTER, A. J. AND STRIK, W. K., 1999, The NoGo-anteriorization as a neurophysiological standard-index for cognitive response control, *Int. J. Psychophysiol.* 32(3):233–238.

15. BOKURA, H., YAMAGUCHI, S., AND KOBAYASHI, S., 2001, Electrophysiological correlates for response inhibition in a Go/NoGo task, *Clin. Neurophysiol.* 112:2224–2232.

16. GEVINS, A., SMITH, M. E., MCEVOY, L., AND YU, D., 1997, High resolution EEG mapping of cortical activation related to working memory: Effects of difficulty, type of processing, and practice, *Cereb. Cortex* 7:374–385.

17. GEVINS, A., SMITH, M. E., LEONG, H., MCEVOY, L., WHITFIELD, S., DU, R., AND RUSH, G., 1998, Monitoring working memory load during computer-based tasks with EEG pattern recognition methods, *Hum. Fact.* 40:79–91.

18. PFURTSCHELLER, G. AND ARANIBAR, A., 1977, Event-related cortical desynchronization detected by

power measurements of scalp EEG, *Electroenceph. Clin. Neurophysiol.* 42:817–826.

19. SCHERG, M., VON CRAMON, D., AND ELTON, M., 1984, Brain-stem auditory-evoked potentials in postcomatose patients after severe closed head trauma, *J. Neurol.* 231(1):1–5.

20. SALMELIN, R., FORSS, N., KNUUTILA, J., AND HARI, R., 1995, Bilateral activation of the human somatomotor cortex by distal hand movements, *Electroenceph. Clin. Neurophysiol.* 95:444–452.

21. GRAVE DE PERALTA, R. AND GONZALEZ ANDINO, S., 1998, Distributed source models: Standard solutions and new developments, in: UHL, C., Ed., *Analysis of Neurophysiological Brain Functioning*, Springer Verlag, New York, pp. 176–201.

22. PASCUAL-MARQUI, R. D., 1995, Reply to comments by Hämäläinen, Ilmoniemi and Nunez, *ISBET Newsletter* 6:16–28.

23. HAMALAINEN, M. AND ILMONIEMI, R., 1984, Interpreting measured magnetic field of the brain: Estimates of the current distributions, Technical report TKK-F-A559, Helsinki University of Technology.

24. ZAR, H., 1984, *Biostatistical Analysis*, Prentice-Hall, New York.

25. DALE, A., LIU, A., FISCHL, B., BUCKNER, R., BELLIVEAU, J. W., LEWINE, J., AND HALGREN, E., 2000, Dynamic statistical parametric mapping: combining fMRI and MEG for high-resolution imaging of cortical activity, *Neuron* 26:55–67.

26. BABILONI, F., CARDUCCI, F., CINCOTTI, F., DEL GRATTA, C., PIZZELLA, V., ROMANI, G. L., ROSSINI, P. M., TECCHIO, F., AND BABILONI, C., 2001, Linear inverse source estimate of combined EEG and MEG data related to voluntary movements, *Human Brain Mapping* 14(4):197–210.

27. BELGER, A., PUCE, A., KRYSTAL, J. H., GORE, J. C., GOLDMAN-RAKIC, P., AND MCCARTHY, G., 1998, Dissociation of mnemonic and perceptual processes spatial and nonspatial working memory using fMRI, *Hum. Brain Mapp.* 6:14–32.

28. CHAFEE, M.V. AND Goldman-Rakic, P. S., 1998, Matching patterns of activity in primate prefrontal

area 8a and parietal area 7ip neurons during a spatial working memory task, *J. Neurophysiol.* 79:2919.

29. FRIEDMAN, H. AND GOLDMAN-RAKIC, P. S., 1994, Coactivation of prefrontal cortex and inferior parietal cortex in working memory tasks revealed by 2DG functional mapping in the Rhesus monkey, *J. Neurosci.* 14:2775–2788.

30. POSNER, M. E. AND PETERSON, S. E., 1990, The attentional system of the human brain, *Annu. Rev. Neurosci.* 13:25–42.

31. GALIN, D., JOHNSTONE, J., AND HERRON, J., 1978, Effects of task difficulty on EEG measures of cerebral engagement, *Neuropsychology* 16:461–472.

32. GEVINS, A. S. AND SCHAFFER, R. E., 1980, A critical review of electroencephalographic EEG correlates of higher cortical functions, *CRC Crit. Rev. Bioeng.* 4:113–164.

33. BABILONI, C., FERRETTI, A., DEL GRATTA, C., CARDUCCI, F., VECCHIO, F., ROMANI, G. L., AND ROSSINI, P. M., 2005, Human cortical responses during one-bit delayed-response tasks: An fMRI study. *Brain. Res. Bull.* 65(5):383–390.

34. BABILONI, C., BABILONI, F., CARDUCCI, F., CINCOTTI, F., DEL PERCIO, C., DELLA PENNA, S., FRANCIOTTI, R., PIGNOTTI, S., PIZZELLA, V., ROSSINI, P. M., SABATINI, E., TORQUATI, K., AND ROMANI, G. L., 2005, Human alpha rhythms during visual delayed choice reaction time tasks: A magnetoencephalography study, *Hum. Brain Mapp.* 24(3):184–192.

35. BABILONI, C., BABILONI, F., CARDUCCI, F., CINCOTTI, F., VECCHIO, F., COLA, B., ROSSI, S., MINIUSSI, C., AND ROSSINI, P. M., 2004, Functional frontoparietal connectivity during short-term memory as revealed by high-resolution EEG coherence analysis, *Behav. Neurosci.* 118(4):687–697.

36. BABILONI, C., BABILONI, F., CARDUCCI, F., CAPPA, S. F., CINCOTTI, F., DEL PERCIO, C., MINIUSSI, C., MORETTI, D. V., ROSSI, S., SOSTA, K., AND ROSSINI, P. M., 2004, Human cortical responses during one-bit short-term memory. A high-resolution EEG study on delayed choice reaction time tasks, *Clin. Neurophysiol.* 115(1):161–170.

CHAPTER 13

EXPLORING SEMANTIC MEMORY AREAS BY FUNCTIONAL MRI

G. Rizzo, P. Vitali, G. Baselli, M. Tettamanti,
P. Scifo, S. Cerutti, D. Perani, and F. Fazio

13.1 INTRODUCTION

Functional magnetic resonance imaging (fMRI) investigates the neural activation in cerebral areas involved in the execution of a task by measuring the related hemodynamic response. Functional activation of a cerebral area causes a local increase in blood flow, leading to an augmented local concentration of oxygenated blood; due to the diamagnetic behavior of oxygenated blood with respect to the paramagnetic deoxygenated blood, the cerebral areas involved in activation present a decrease in the local susceptibility effect and thus an increase in the signal intensity as observed in $T2^*$-weighted MRI [the so-called blood oxygenation-level-dependent (BOLD) effect] [1].

In the 1990s, fMRI has provided invaluable data of cortical activity in a time resolution range that was inconceivable with previous methods based on radioactive tracers with an in-depth anatomical specificity never reached by noninvasive EEG analysis.

In the study of brain function, currently a major issue is in the analysis of cortical activation during the process of speech production involving semantic memory. Several studies have begun to investigate the localization of specific areas in the left posterior temporal cortex which are concerned with the retrieval of specific word categories such as "tools" or "animals" in either fMRI [2] or positron emission tomography (PET) studies [3]. Here the specific role played by the left medial fusiform gyrus (MFG) and left lateral fusiform gyrus (LFG) is investigated [2, 3].

The introduction of suitable protocols in fMRI, with alternate activation and deactivation, permits investigation of the links between the semantic memory regions and language production area [Broca's (BR) area]. For the analysis of these data the methodological approach and the design of appropriate models exploring the correlations between the activities of the addressed regions are crucial. Functional and effective connectivity models [4, 5] provide a general framework which can be applied to group analysis; however, the use of simple parametric models may furnish significant subject-by-subject indexes. In this perspective, the present work will compare these different methods and model structures.

Handbook of Neural Engineering. Edited by Metin Akay
Copyright © 2007 The Institute of Electrical and Electronics Engineers, Inc.

13.2   ORGANIZATION OF SEMANTIC MEMORY IN BRAIN

Memory is usually defined as a sort of "virtual space" where souvenirs are located in specific addresses; within this well-organized system it is possible to plan and execute, intentionally or automatically, research and retrieval procedures of stored information. It is now well recognized that memory is not a single, all-purpose, monolithic system but rather is composed of multiple systems. A first basic division characterizing memory is between short- and long-term memory, the former responsible for holding information for a matter of seconds, the latter able to retain it indefinitely [6].

Data from neuropathology has revealed that both storage systems do not form a unitary structure but can be divided into different subcomponents. Tulving [7] first proposed separation of the long-term system into episodic and semantic memory. Episodic (or autobiographical) memory consists of singular events that a person recalls; it is a neurocognitive (i.e., a thinking) system that enables human beings to remember past personal experiences centerd on the person himself or herself and characterized by a specific spatial and temporal context. Knowledge about the world—all knowledge that is not autobiographical—is referred to by Tulving as semantic memory and includes knowledge of historical events and figures, concepts, and beliefs. It includes the ability to recognize family, friends, and acquaintances. It also includes information learned at school, such as specialized vocabularies and reading, writing, and mathematics. In comparison to episodic memory, it consists of memories that are shared by members of a culture rather than those which are unique to an individual and are tied to a specific time and place.

Clinical observations on patients with brain disorders have revealed the existence of both selective but global (i.e., non-material-specific) impairments of semantic memory with preservation of episodic retrieval [8], thus providing neuropsychological evidence for Tulving's model of a multiple organization of the memory system, and modality-specific semantic impairments [9]. The former condition, which usually follows damage to the left temporal lobe, produces an indiscriminate degradation of the concepts themselves, detectable under a variety of retrieval circumstances, whereas the latter is characterized by an impaired object naming with preserved recognition, limited to a specific sensory modality of stimulus delivery (visual, tactile, or acoustic). This is possible assuming the existence of multiple semantic systems characterized by a modality-specific access (visual, tactile, acoustic, and verbal): For a given modality-specific impairment, the relative semantic system remains fairly intact but disconnected from the others, particularly from the verbal system, thus avoiding name production

With the development of modern techniques of visualization of brain activity in vivo, such as PET and fMRI, cognitive psychologists have become more interested in the neural correlates of the semantic memory and its organization in the brain. Consistent with the clinical literature [10], functional neuroimaging studies of object naming suggest that semantic processing may be critically dependent on the left temporal lobe [11]. Indeed, direct comparisons between brain activities associated with object naming and viewing nonsense objects have revealed those cerebral areas which are more significantly involved in semantic, lexical, and phonological processes (specifically engaged by real object naming and not by viewing nonsense pictures).

There is additional evidence from linguistic fluency tasks that posterior regions of the left temporal lobe mediate lexical and/or semantic processing; in particular, a more predominant involvement of the left prefrontal regions was found for lexical/phonological processing, whereas the ventral part of the left posterior temporal lobe was found to be more activate during semantic elaboration [12].

Disorders of identification and/or naming, which display selectivity for a specific category of entities, such as animals or tools, have been repeatedly described in patients with brain damage and have led to the claim that semantic memory presents a category-specific organization. The most compelling evidence of a cognitive segregation of the semantic information into distinctive categories comes from the double dissociation between the living and nonliving categories, the living domain being more frequently impaired than the nonliving one.

At a neurological level of analysis, a direct method to investigate the correlates of categorical semantic knowledge has been provided by functional neuroimaging of normal subjects. Different locations of brain activation according to the semantic category (usually animals vs. tools) were shown for the first time by Perani and co-workers [13] using a visual picture-matching task. Animal identification was associated with activations in the inferior temporo-occipital visual areas, bilaterally. On the other hand, the identification of nonliving entities (man-made objects) engaged the activation of a predominantly left-hemispheric network involving the left dorsolateral frontal cortex and the left middle temporal gyrus [13, 14]. Similar findings have been successively obtained during the naming of pictures of animals and tools [11, 15]. It has been suggested that these "category-specific" effects reflect the retrieval of category-specific features and attributes [16]. If the main characteristics defining an object are stored in the sensitive and motor areas, activated when we first learned about that object and automatically accessed when it is identified, then the greater activity observed in occipital visual areas for living entities would be consequent to the processing of high-order visual features and attributes defining them; on the contrary, the repeatedly observed activation in the left middle temporal gyrus (close to regions which are active when motion is perceived) and in the left inferior frontal cortex (a region with a motor function) for tools might reflect the retrieval of action knowledge related to tool motion and/or manipulation [17, 18].

All previously reported studies were largely consistent in indicating the posterior part of the left temporal lobe as a major correlate of the semantic processing. The higher spatial resolution of fMRI with respect to PET has led to the demonstration that this region, in particular the fusiform gyrus, is also the site of a category-related difference. The present study was aimed at investigating the functional specialization of the fusiform gyrus in a semantic fluency task and the connections between semantic regions and the speech production system (BR area). In particular, the specific role played by the MFG and the LFG in semantic access was studied.

13.3 METHODOLOGY

13.3.1 Neuropsychological Protocol

The experimental subjects were 10 right-handed male volunteers who gave written informed consent to the study.

During the block-designed fMRI session study, each subject alternated three times between a rest condition where he was instructed to empty his mind and avoid inner speech and one of the following semantic verbal fluency conditions:

Animal category (AN): Subjects were instructed to generate covertly as many common nouns belonging to the animal category as possible throughout the task (30 s times 3).

Tool category (TO): Subjects were instructed to generate covertly as many common nouns belonging to the tool category as possible throughout the task (30 s times 3).

The activation design was the following:

AN: rest−animals1−rest−animals2−rest−animals3

TO: rest−tools1−rest−tools2−rest−tools3

A total of 12 epochs were performed.

Subjects were trained in the different tasks before positioning on the magnet cradle. Instructions were then recalled by the examiner at the beginning of each task epoch by saying aloud a fixed cue (e.g., "animals" for task A, "tools" for task B). After total scan acquisition, subjects were asked to recall the list of words generated during each scan epoch and items were recorded by the examiner. The total number of words generated and the category consistency were thus monitored.

13.3.2 MRI Acquisition

The MRI scans were performed using a 1.5-T scanner (General Electric Medical Systems, Milwaukee, WI) equipped with echo-speed gradient coils and amplifier hardware and a standard quadrature head coil. Before the fMRI session, an anatomical T1 scan acquisition was performed in the same location of the functional images. Field homogeneity was adjusted by means of "global shimming" for each subject to reduce susceptibility artifacts.

The acquisition of the MRI images (T2* weighted) was performed with single-shot gradient echo echo-planar pulse sequence [echo time (TE) = 60 ms, repetition time (TR) = 3000 ms]. Nineteen slices volumes [field of view (FOV) 280 × 280 mm, matrix 64 × 64], 4 mm thick were acquired resulting in a final voxel size of 4.375 × 4.375 × 4 mm. The fMRI acquisition session included the whole activation protocol described above, producing a single data set of 120 volumes (12 activation epochs, 10 scans/epoch lasting 3 s) for each subject.

13.3.3 Image Processing

The first 3 s of each epoch were discarded from analysis in order to eliminate the acoustic stimulation at the beginning of each epoch.

Image processing was performed with SPM99 software (Wellcome Department of Cognitive Neurology, London). First, the image volumes of each subject were realigned to reduce the occurrence of motion artifacts during each scan acquisition. Images were then normalized into the Talairach's stereotactic space [19] to allow intersubject data averaging and comparison across tasks. Stereotactic normalization was also performed for the morphological MRI volume. At this stage images were interpolated into volumes with voxel size of 2 × 2 × 4 mm. After stereotactic normalization, the common stereotactic space for the experimental subjects covered planes from −20 to +56 mm according to the bicommisural plane. All the scans were then spatially smoothed through a Gaussian filter of 10 × 10 × 10 mm to reduce residual anatomical differences among subjects and to improve signal-to-noise ratio.

13.3.4 Analysis of Functional Connectivity

Functional connectivity is used to study the temporal correlations between voxel activity in a sequence of fMRI image volumes and the task the subject is executing in order to verify if this correlation is significant. The analysis is performed using a model (known as the general linear model [20]), in which the activity of each voxel is described as linearly dependent on the experimental conditions (in our case rest condition, AN condition, or TO condition) and estimated with a least-squares method; for each voxel of the image volume sampling the brain, the significance of activity variation during different tasks is then calculated in terms of F-test or t-test values. Starting from this result, a three-dimensional spatial map represented in the Talairach stereotactic space is composed in which a voxel gray value is associated with the statistical significance of that voxel's involvement in an activation task. This map, called the statistical parametric map (SPM), allows a direct spatial identification of cerebral areas involved in a specific activation task, represented by voxels with high gray value.

For functional connectivity analysis, we used implementation of the general linear model for fMRI data devised for SPM99. The experiment was treated as a block design (types of fluency and rest). Comparisons of means were made for each voxel by using the t-statistic, thus generating statistical parametric maps of the t-values SPM{t}. Simple main effects of the different fluencies (animals and tools) in comparison with their time-matched resting scans were calculated. The threshold applied to the statistical maps was $p < 0.05$ (corrected for multiple comparisons).

Functional connectivity was carried out over the entire group (pretending all the MRI data came from the same subject) and subject-by-subject, depending on the subsequent modeling procedure. The central coordinates of BR area, MFG, and LFG were fixed either over the entire group or subject by subject. In subject analyses these coordinates were optimized for each subject by building activation maps based on the correlation of the time series of each voxel with a time series representing the task (i.e., with a dummy signal set to 1 during the stimulus and to 0 elsewhere): correlation was calculated with AN, TO, and both tasks in order to localize the MFG, LFG, and BR area, respectively.

The temporal sequence of activity of BR area, MFG, and LFG was extracted as an average in a spherical volume of 5 mm radius around the region coordinates, thus obtaining three temporal series: br(t), mfg(t), lfg(t). Each temporal sequence was also low-pass filtered in the time domain considering the standard hemodynamic response function and was high-pass filtered at 0.0083 Hz (i.e., half of the RE task switching frequency).

13.3.5 Models and Analysis of Effective Connectivity

Effective connectivity is used to study the influence one brain region has upon another and the task-related changes in the coupling between the two cerebral areas [5]. Here the interest is in the study of influence between semantic regions (MFG and LFG) and BR area during language production, whose interactions can be schematized according to the model presented in Figure 13.1. To build this model, we adopted a hypothesis-led theoretical perspective to constrain the connectivity model to principal anatomic and cognitive elements. Thus, our model include the following: (1) anatomical regions of interest in the left hemisphere which represent the neural substrate involved in the common network for both semantic fluency tasks (AN and TO) as derived by functional connectivity analysis; this neural system is consistent with previous reports in the literature on functional correlates of semantic processing; (2) the cognitive modulations between their connections by the two semantic tasks. In particular, in Figure 13.1, C represents the connection strength between semantic and speech production systems and can be thought of as modulated by the specific semantic task (TO or AN). Direct connections between regions within the model were unidirectional to ensure robust estimates.

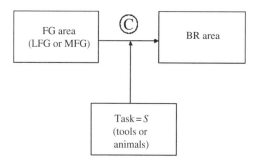

Figure 13.1   Connectivity model for studying influence of semantic areas (LFG and MFG) to BR language production area.

Two different models were derived from that of Figure 13.1 for the effective connectivity analysis:

1. The interaction model shown in Figure 13.2 was used to explore the sequence of AN and TO tasks simultaneously as to the effective connectivity between either MFG or LFG and BR area and the modulation given by the task type. In this model, it is supposed that BR area activity is influenced by the fusiform gyrus area (through the β coefficient, depending on the task S) and directly by the task (through the β_2 coefficient) (Fig. 13.2a). The nonlinear relationship between BR area and the fusiform gyrus expressed by β is linearized by assuming $\beta(S)$ to be the sum of two terms: β_0, not dependent on the task, and β_1, proportional to it (Fig. 13.2b). Thus the model can be described by the equation

$$y(t) = \beta_0 u(t) + \beta_1 u(t)\,s(t) + \beta_2 s(t) + e(t) \qquad (13.1)$$

where $y(t)$ is the signal br(t) normalized [i.e., the mean subtracted and equalized by the Standard deviation (SD)] over the entire AN and TO task sequence; $u(t)$ is the normalized mfg(t) or lfg(t); $s(t)$ is a dummy signal (describing the task) with $s(t) = -1$ during the first half of the acquisition (i.e., 3 AN tasks preceded by 3 REs) and $s(t) = +1$ in the second part of the experiment with TO tasks; $e(t)$ is a random residual noise.

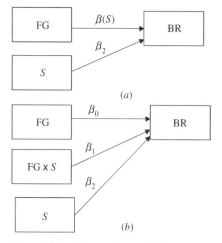

Figure 13.2   Interaction model described by Eq. (12.1).

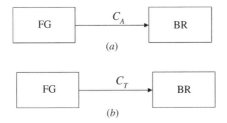

Figure 13.3   Interaction model described by Eqs. (12.2) and (12.3).

2. The AN and TO data were also separately analyzed with simpler parametric models (Fig. 13.3):

$$y_A(t) = C_A u_A(t) + e_A(t) \tag{13.2}$$

and

$$y_T(t) = C_T u_T(t) + e_T(t) \tag{13.3}$$

where $y_A(t)$ and $y_T(t)$ are the br(t) sequence during the AN part (Fig. 13.3a) and the TO part (Fig. 13.3b) of the experiment, respectively; both were normalized within the respective time period. Analogously $u_A(t)$ and $u_T(t)$ are taken and normalized in the AN and TO period and derived from either the mfg(t) or the lfg(t) series. As a consequence of normalization, the regression parameters C_A and C_T directly yield the linear correlation between the input and the output signals.

13.4   RESULTS

13.4.1   Functional Connectivity

The SPM$\{t\}$ maps relative to the group analysis of AN and TO tasks with respect to the corresponding rest conditions are shown in Figures 13.4 and 13.5, respectively. Among the cerebral areas activated during the tasks, it is possible to identify the MFG and BR area regions (Fig. 13.4) and the LFG and BR area regions (Fig. 13.5) as regions of high statistical significance.

These activated areas are visualized in Figure 13.6 on registered morphological MRI images, remapped into the Talairach space, for direct anatomical localization. From this analysis the central coordinates of BR area, MFG, and LFG in the group analysis were derived: -48, 28, 16 (mm in x, y, z directions) for BR area; -24, -48, -16 for MFG; and -46, -56, -12 for LFG.

For a subject-by-subject analysis the coordinates were found for each individual by means of the respective activation maps. In Figure 13.7, an example activation map relevant to the TO task for the detection of the LFG is shown. The map represents the correlation with the TO dummy signal plus the difference of this correlation and that with the AN dummy signal during the AN task. The region of increased activity in the left posterior temporal region can be appreciated by means of this image.

In single-subject analysis, BR area was localized in the same position in seven subjects, while the LFG and MFG presented a more accentuated variation, with respect to the area position estimated by group analysis. These effects were probably due to residual errors in the anatomical spatial normalization of each brain to the Talairach

Animal condition versus Rest condition

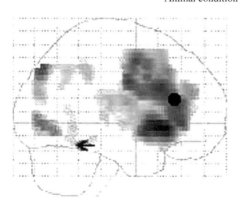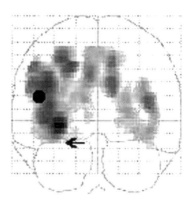

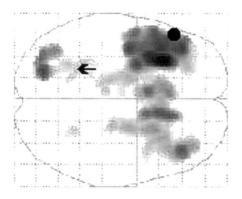

Figure 13.4   *T*-statistical map (corrected at $p < 0.05$) of brain regions activated during covert generation of nouns belonging to animal category. The black arrow indicates the left medial fusiform gyrus (stereotactic coordinates: $x = -24$; $y = -48$; $z = -16$), whereas the black circle represents BR area (stereotactic coordinates: $x = -48$; $y = 28$; $z = 16$). Other foci of cerebral activity were significantly activated during the task, but they are not relevant to the present discussion.

space, but they could also derive from intersubject variability of the functional activation site within the specific specialized cerebral area.

13.4.2   Effective Connectivity

13.4.2.1   Group Analysis   The application of the interaction model [Eq. (12.1)] to the whole sequence, built by attaching the sequences of subjects, gave significant values of β_0 and β_1; parameter values were $\beta_0 = 0.087$, $\beta_1 = -0.039$, and $\beta_2 = -0.002$ with mfg(*t*) as input and $\beta_0 = 0.241$, $\beta_1 = +0.096$, and $\beta_2 = -0.052$ with lfg(*t*) as input. Coefficient β_1 changed significantly according to the different input: the positive sign when the input is lfg(*t*) indicates that the connection from the LFG to BR areas is more active during TO tasks [when the dummy signal $s(t) = +1$] while the link between the MFG and BR area is augmented by the AN task.

Tool condition versus Rest condition

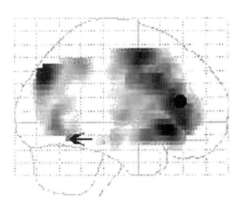

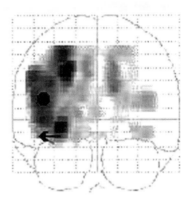

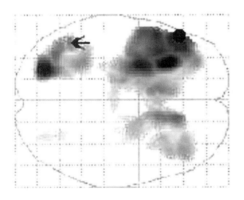

Figure 13.5 *T*-statistical map (corrected at $p < 0.05$) of brain regions activated during covert generation of nouns belonging to tool category. The black arrow indicates the left lateral fusiform gyrus (stereotactic coordinates: $x = -46$; $y = -56$; $z = -12$), whereas the black circle represents BR area (stereotactic coordinates: $x = -48$; $y = 28$; $z = 16$). Other foci of cerebral activity were significantly activated during the task, but they are not relevant to the present discussion.

Similar results were found by the parametric models [Eqs. (12.2) and (12.3)]. In fact, $C_A = 0.162$ and $C_T = 0.040$ with mfg(t) as input, and $C_A = 0.134$ and $C_T = 0.345$ with lfg(t) as input, thus indicating a larger linear correlation between BR area and LFG during TO tasks.

13.4.2.2 Subject-by-Subject Analysis Results relevant to the subject-by-subject application of the interaction model and the linear correlation model of BR area with mfg(t) and lfg(t) input, respectively, are shown in Table 13.1. Here, β_0 and β_2 (not shown) did not present significant changes. With mfg(t) as input β_1 was not able to reproduce the result of the group analysis as in only one subject it displayed a value significantly different from zero, given the degrees of freedom of the least squares estimate. With lfg(t), a slightly better performance was obtained with eight significant figures, seven of which were in the expected direction.

The analysis with the simple correlation model gave more robust indications as with mfg(t) eight significant values (seven in the expected direction) were found as to the

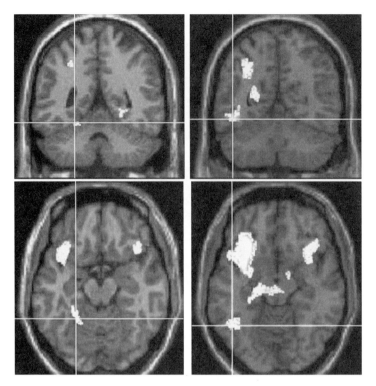

Figure 13.6   Top row: cerebral activity (white areas) in left medial (to left) and left lateral (to right) fusiform gyrus overlaid onto coronal view of standardized T1-weighted MRI brain scan as implemented in SPM99. The two regions of interest are indicated by the crossing point of the white lines. Bottom row: same regions displayed onto transversal view of standardized MRI brain scan.

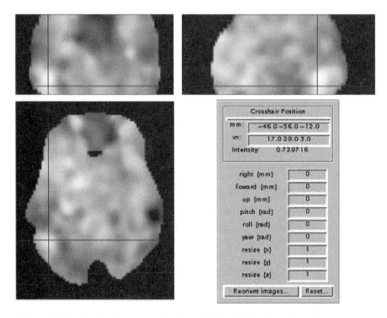

Figure 13.7   Activation map relative to TO task. Activated regions correspond to white color level; note the activation in the lateral posterior temporal region.

TABLE 13.1   Subject by Subject Analysis: Coefficients of the Connectivity Models Shown in Figs 3 and 4

subj	MFG				LFG			
	β_1	C_A	C_T	$C_T - C_A$	β_1	C_A	C_T	$C_T - C_A$
1	0.057	0.779	0.743	− 0.036	**0.037**	0.469	0.698	**0.229**
2	0.063	0.249	0.384	**0.135**	−0.009	0.517	0.618	**0.101**
3	−0.015	0.390	0.116	− **0.274**	**0.155**	0.559	0.816	**0.257**
4	0.004	0.242	0.183	− 0.059	**0.092**	0.680	0.894	**0.214**
5	− **0.177**	0.457	0.015	− **0.442**	**0.080**	0.442	0.748	**0.306**
6	0.036	0.174	0.224	0.050	− **0.176**	0.485	0.046	− **0.439**
7	0.052	0.399	0.363	−0.036	**0.180**	0.560	0.824	**0.264**
8	−0.190	0.205	−0.140	− **0.345**	**0.388**	0.106	0.750	**0.644**
9	−0.044	0.388	0.188	− **0.200**	**0.092**	0.415	0.574	**0.159**
10	−0.014	0.831	0.593	− **0.238**	**0.430**	−0.138	0.853	**0.991**
avg	−0.023	0.411	0.267	−0.145	0.127	0.410	*0.682	*0.273
std	0.092	0.228	0.263	0.184	0.178	0.243	0.245	0.367

*Significant contrast of LFG from MFG.
bold values of β_1 and $(C_T - C_A)$ indicate significant difference from 0.

difference between the two stimuli $(C_T - C_A)$. An even clearer behavior was noticed with lfg(t), as all values were significant and only with subject 6 was performance in the unexpected direction. An analysis of variance (AVOVA) analysis on the 10 subjects indicates a significant contrast of C_T and $(C_T - C_A)$ between the activity of MFG and LFG.

13.5   DISCUSSION

In the past years, functional neuroimaging techniques (PET, fMRI) have been used to investigate the functional specialization of cerebral areas: for a given cognitive function (memory, language, etc.), the main aim was to discover which brain regions were involved in the execution of related mental tasks (learning paradigms, naming tasks, etc.) and thus provide evidence about their functional role.

In the case of language, and in particular with respect to semantic memory, there is a body of evidence that the left inferior frontal gyrus, including BR area, and left posterior temporal regions are implicated in semantic elaboration. Indeed, a large number of neuroimaging studies, using a range of different tasks, have implicated the left inferior frontal gyrus and left ventrotemporal areas in the retrieval of semantic knowledge [15].

The present study confirms that a common activation of the left inferior frontal gyrus (BR area, Ba45) is associated with both living and nonliving semantic fluency, providing further evidence that this area is involved in lexical retrieval based on meaning. Concerning the category effects in the fusiform gyrus, results from the present functional connectivity analysis appear to be in contrast with a series of fMRI investigations by Martin and co-workers which have indicated that activations associated with living entities (animals, faces) cluster in the more lateral aspects of the fusiform gyrus while activations associated with man-made objects cluster in the more medial aspects (see [2] for a review). The advantage of using the effective connectivity analysis with respect to the functional analysis is represented by the ensuing possibility to investigate how distinctive functionally specialized areas interact between them during the execution of cognitive tasks; this refers to the concept of functional integration [21].

A major goal of the present study was to investigate the pattern of functional integration related to lexical retrieval for different semantic categories. The results provided further evidence for the existence of selective patterns of functional integration related to different semantic categories; in particular, animal fluency was associated with increased connectivity between the MFG and BR area, whereas tool fluency showed an enhanced coupling among the lateral posterior fusiform gyrus and BR area. The present, apparently discordant findings, compared to those of Martin and co-workers, could be explained if we speculate that during the semantic fluency task the left Ba45 (BA area) operates a choice between different competing responses belonging to distinct semantic categories.

Two different theories try to account for the role of the left inferior frontal gyrus (Ba45) in language production. The first claims that this area is especially involved in lexical and/or phonological retrieval during attentive verbal processing [22], whereas the second proposes that it may not be implicated in retrieval per se but in the selection of relevant responses and suppression of alternative competing responses [23].

Our study supports this second view. Indeed, we discovered a common activation of the left Ba45 in both fluency tasks, but the effective connectivity analysis has revealed that it is alternatively more connected with the lateral part of the fusiform gyrus during tool production and with its medial part during animal production. If during the semantic fluency tasks the left inferior frontal gyrus Ba45 mediates a choice between competing responses, belonging to distinct semantic categories, we suggest that the left Ba45 actively inhibits the brain areas subserving the opposite semantic category's concepts and that this inhibitor mechanism results in a stronger coupling between the left Ba45 and the inhibited areas. In fact, an enhancement in connectivity between two cerebral regions simply means that the two areas have coherently increased their neural activity but it is not possible to point out if this greater coupling indicates, at the synaptic level, an excitatory or an inhibitory influence of one area upon the other.

The limited sampling rate of the acquisition protocol used in this study prevents the introduction of true causal models to address the questions that are the main goal of effective connectivity; that is, the models cannot usefully introduce correlations between time lags and are constrained to analyze zero-lag correlations at a varying t. So, *correlated* is in this context used as a synonym for *causal*, provided that an a priori knowledge of functional connections with a specific direction can be exploited, as it is in the case of the pathways from the posterior temporal semantic memory areas to the inferior frontal language production area (BR).

On the other hand, the fMRI technique permits us to analyze responses to tasks that can be at least an order-of-magnitude shorter than those that could be considered with nuclear medicine imaging. So, as in the present protocol, a quite rapid switching between activity and rest and between different task types can be performed; consequently, the information content of the data sequences can be considered sufficient for a subject-by-subject analysis provided appropriate models are applied.

The interaction model tested in this work is typical of effective connectivity analysis and, therefore, was conceived for the analysis of ensemble data. The number of parameters estimated by this model is probably too large for consistent performance with single subjects. Conversely, the second simpler correlation model adopts a more appropriate segmentation of the data sequence of the single subject and, probably, for this reason is capable of reproducing the results of the group analysis in single subjects.

The reasons for moving in the direction of an analysis of single subjects are obvious if the aim is an extension of functional imaging of the cerebral cortex from pure physiological studies to clinical analysis. Concerning language production pathologies, the cases of lesions in semantic memory and in language production areas can be considered, as well as specific language disturbances, such as dyslexia.

13.6 FUTURE TRENDS

An appropriate modeling of fMRI protocols relevant to language production mental tasks permits a subject-by-subject identification of the activation of specific working memory areas and of their correlation with language production areas. To this regard, further improvements in data analysis should be obtained by adopting event-related designs in fMRI acquisition protocols, in which stimuli of different types can be intermixed, unlike the block design adopted here, where stimuli of the same type are presented in blocks. This new approach allows the introduction of dynamic connectivity models taking into account hemodynamic responses more accurately.

Additional benefits in the study of complex cerebral functions may also be reached by merging fMRI images to signals describing electrical cortical activity as recorded by multichannel EEG. An EEG-fMRI coupled analysis should permit us to better describe multifocal fast phenomena with optimized spatial and temporal resolutions.

The analysis of effective connectivity we have performed was based on an a priori postulated anatomical model, in which anatomical constraints were derived from previous functional studies and research on monkeys. A possible future development of such a technique may be represented by the adoption of a graph of connectivity comprising not only previously reported regions of interest but also a network of areas derived from the analysis itself of global cerebral activity; all connections between points on the graph will be tested and only significant paths retained. This would allow investigation of the functional integration among brain regions on an individual basis and highlight the organization of the neural substrate of mental activity specific to a single subject. The importance of such an approach is evident if we consider the study of mechanisms of functional recovery after brain damage, which generally involve a cerebral reorganization with distinctive features from patient to patient.

The method presented here and its extension can widen the application field of functional imaging from pure physiological studies to clinical analysis in pathologies involving language impairment.

REFERENCES

1. S. Ogawa, T. M. Lee, A. R. Kay, and D. W. Tank, "Brain magnetic resonance imaging with contrast dependent on blood oxygenation," *Proc. Natl. Acad. Sci. USA* 87, 9868–9872 (1990).

2. A. Martin and L. L. Chao, "Semantic memory and the brain: Structure and processes," *Curr. Opin. Neurobiol.* 11, 194–201 (2001).

3. S. F. Cappa, D. Perani, T. Schnur, M. Tettamanti, and F. Fazio, "The effects of semantic category and knowledge type on lexical-semantic access: A PET study," *Neuroimage* 8, 350–359 (1998).

4. A. R. McIntosh, C. L. Grady, L. G. Ungerleider, J. V. Haxby, S. I. Rapoport, and B. Horwitz, "Network analysis of cortical visual pathways mapped with PET," *J. Neurosci.* 14, 655–666 (1994).

5. C. Buchel and K. J. Friston, "Modulation of connectivity in visual pathways by attention: Cortical interactions evaluated with structural equation modelling and fMRI," *Cereb. Cortex* 7, 768–778 (1997).

6. E. Warrington, "The double dissociation of short- and long-term memory deficits," in L. Cermak, Ed., *Human Memory and Amnesia*, Lawrence Erlbaum, Hillsdale, NJ, 1982.

7. E. Tulving, "Episodic and semantic memory," in E. Tulving and W. Donaldson, Eds., *Organisation of Memory*, Academic Press, New York, 1972.

8. E. Warrington, "The selective impairment of semantic memory," *Q. J. Exper. Psychol.*, 635–657 (1975).

9. T. Shallice, "Impairments of semantic processing: Multiple dissociations," in M. Coltheart, R. Job, and

G. SARTORI, Eds., *The Neuropsychology of Language*, Lawrence Erlbaum, London, 1987.

10. D. CAPLAN, *Language: Structure, Processing, and Disorders*, MIT Press, Cambridge, MA, 1992.

11. A. MARTIN, C. L. WIGGS, L. G. UNGERLEIDER, AND J. V. HUXBY, "Neural correlates of category-specific knowledge," *Nature* 263, 649–652 (1996).

12. C. J. MUMMERY, K. PATTERSON, J. R. HODGES, AND R. J. S. WISE, "Generating a 'tiger' as an animal name or a word beginning with T: Differences in brain activations," *Proc. R. Soc. Lond. B*, 989–995 (1996).

13. D. PERANI, S. F. CAPPA, V. BETTINARDI, S. BRESSI, M. L. GORNO-TEMPINI, M. MATARRESE, AND F. FAZIO, "Different neural networks for the recognition of biological and man-made entities," *NeuroReport* 6, 1637–1641 (1995).

14. D. PERANI, T. SCHNUR, M. TETTAMANTI, M. L. GORNO-TEMPINI, S. F. CAPPA, AND F. FAZIO, "Word and picture matching: A PET study of semantic category effects," *Neuropsychologia* 37, 293–306 (1999).

15. H. DAMASIO, T. J. GRABOWSKI, D. TRANEL, R. D. HITCHWA, AND A. R. DAMASIO, "A neural basis for lexical retrieval," *Nature* 5, 490–505 (1996).

16. L. L. CHAO, J. WEISBERG, AND A. MARTIN, "Experience-dependent modulation of category-related cortical activity," *Cereb. Cortex* 12, 545–551 (2002).

17. M. S. BEAUCHAMP, K. E. LEE, J. V. HAXBY, AND A. MARTIN, "Parallel visual motion processing streams for manipulable objects and human movements," *Neuron* 34, 149–159 (2002).

18. J. A. PHILLIPS, U. NOPPENEY, G. W. BBHUMPHREYS, AND C. J. PRICE, "Can segregation within the semantic system account for category-specific deficits?" *Brain*, 2067–2080 (2002).

19. J. TALAIRACH AND P. TOURNOUX, *Co-planar Stereotactic Atlas of the Human Brain*, Thieme Medical Publishers, New York, 1988.

20. K. J. FRISTON, A. P. HOLMES, K. J. WORSLEY, J. B. POLINE, C. D. FRITH, AND R. S. J. FRACKOWIAK, "Statistical parametric maps in functional imaging: A general linear approach," *Human Brain Mapping* 2, 189–210 (1995).

21. K. J. FRISTON, "Functional integration and inference in the brain. Review," *Prog. Neurobiol.* 68, 113–143 (2002).

22. M. I. POSNER, S. E. PETERSEN, P. T. FOX, AND M. E. RAICHLE, "Localization of cognitive operations in the human brain," *Science* 240, 1627–1631 (1988).

23. S. L. THOMPSON-SCHILL, M. D'ESPOSITO, G. K. AGUIRRE, AND M. J. FARAH, "Role of left inferior prefrontal cortex in retrieval of semantic knowledge: A re-evaluation," *Proc. Natl. Acad. Sci. USA* 94, 14,692–14,797 (1997).

PART II

NEURO-NANOTECHNOLOGY: ARTIFICIAL IMPLANTS AND NEURAL PROTHESES

THIS PART focuses on the design of multielectrode arrays to study how the neurons of humans and animals encode stimuli, the evaluation of functional changes in neural networks after the stroke and spinal cord injuries, and improvements of therapeutic applications using the neural prosthesis.

CHAPTER 14

RESTORATION OF MOVEMENT BY IMPLANTABLE NEURAL MOTOR PROSTHESES

Thomas Sinkjær and Dejan B. Popovic

14.1 INTRODUCTION

Neural engineering seeks to interface directly with the nervous system to obtain sensory or command and control signals, to activate specific outgoing neural signals, or to influence processing within the central nervous system (CNS) [1]. These interfaces are mechanical (e.g., cuff electrodes) or biological (e.g., cultured neuronal electrodes) and range from macroscopic to nanoscopic in size. The class of devices used to restore impaired motor functions by functional electrical stimulation (FES) is often called neural motor prosthesis (Fig. 14.1). A neural prosthesis (NP) is an assistive system that replaces or augments a function which is lost or diminished because of injury or disease of the nervous system. The method most frequently applied in NPs is electrical activation of the appropriate impaired sensorimotor systems, that is, use of FES. Functional electrical stimulation elicits the desired neural activation by delivering a controlled amount of electrical charge patterned as bursts of electrical charge pulses. In principle, it is possible to apply a time-varying magnetic field, thereby inducing electrical currents within the selected parts of the neural pathways; however, this technique is not yet efficient enough for functional activation of the sensorimotor systems. A detailed presentation of most aspects of a NP can be found in [2]. Neural prostheses can be divided into external and implantable systems.

Neural prostheses were suggested for movement control nearly 50 years ago [3]. This original work, termed *functional electrical therapy*, used a surface electrode applied over a single muscle group with the aim of eliminating footdrop and producing carry-over effects. Many NPs with surface electrodes were introduced later and tested in humans with paralysis. In some applications the carry-over effects were substantial [4], yet in most cases the effects were only immediate or short lasting. The life-long use of NPs with surface electrodes is very often not appropriate because of the complexity of daily donning and doffing, cosmesis, insufficient selectivity, fatigue, and overall cosmetic appearance. Thereby, the need for effective implantable NPs became a priority when designing and developing assistive systems for sensorimotor rehabilitation.

For example devices currently in use can improve walking in CNS-injured patients by preventing "footdrop" by FES of the muscles that dorsiflex the foot [5]. Other implantable sensorimotor NPs currently in use can produce grasping movements in a paralyzed hand though FES of forearm muscles, and yet other devices can control bladder function [6].

Handbook of Neural Engineering. Edited by Metin Akay

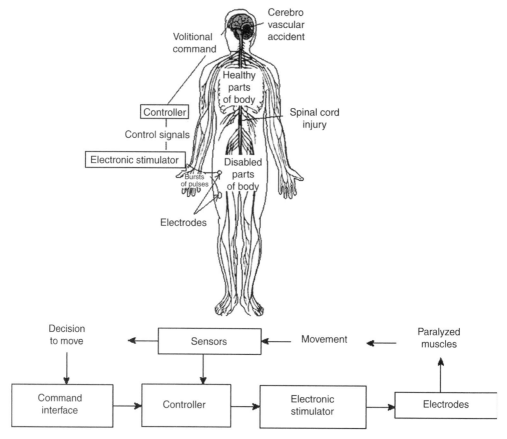

Figure 14.1   General schemes of a motor NP. Components of the NP are controller, electronic stimulator, sensory feedback, and electrodes. The NP acts as a bypass for the diminished or missing transmission of neural signals from and to the CNS. Adapted from [2].

To make neural motor prostheses viable assistive systems for a broad range of patients with injury or disease of the nervous system, we need devices that are easy to implement and operate, that have sufficient complexity to restore important lost motor skills, and that are cosmetically attractive with a long reliable lifespan. Neural engineering research plays an essential role in being able to reach these goals. In this chapter we will focus on implantable neural motor prostheses that interface to the peripheral nerves for sensing and activation and to the CNS for creating command signals.

A peripheral nerve contains thousands of nerve fibers, each of them transmitting information, either from the periphery to the CNS or from the CNS to the periphery. The efferent fibers transmit information to actuators, mainly muscles, whereas afferent fibers transmit sensory information about the state of organs and events such as muscle length, touch, skin temperature, joint angles, nociception, and several other modalities of sensory information. Most of the peripheral nerves contain both afferent and efferent fibers, and the peripheral nerve can thus be seen as a bidirectional information channel. The myelinated nerve fibers range in diameter from approximately 2 to 15 μm. Since nerve fibers are excitable by electric stimuli, this gives the possibility of influencing the neuromuscular system by activating muscles using FES of nerves. Today FES is available as a tool in muscle activation used in picking up objects, in standing and walking, in controlling bladder emptying, and for breathing. Despite substantial progress over nearly four decades of development, many challenges remain to provide a more efficient functionality of FES systems. The forces generated in muscles activated by FES can be graded by varying the stimulus pulses, but the relationship of the force to the stimulus pulse varies in a complex manner which depends on, for example, muscle length, electrode–nerve coupling, and activation history. Several studies have shown that the application of closed-loop control techniques can improve the regulation of muscle activation. Here we review recent advances in

> interfacing the nervous system for stimulation and sensing purposes,
>
> electronic stimulators, and
>
> creating commands signals.

Furthermore, we will outline the clinical applications used today and in the future.

14.2 NEUROELECTRONIC INTERFACE

A neuroelectronic interface connects the electronic world with the living world of nerve cells. The neuroelectronic interface is a large family of devices described based on, among others, the following:

> The anatomic position of the electrode (extraneural, intraneural, epidural, intradural, intraspinal, cortical, deep brain, etc.)
>
> Its purpose (stimulation, recording, or both)
>
> Its material
>
> Its selectivity (in terms of anatomic structures or size of the nerve cells and fibers or direction of the neural activity)
>
> The number of contacts
>
> The level of integration (e.g., with or without integrated electronics, wireless, etc.)

New production techniques, packaging, and interconnections are other important topics to considered when choosing a neuroelectronic interface. Cultured neuron electrodes, such as nerve cells as sensors on a semiconductor or other substrate, are potential new interface methods that ultimately aim at true synaptic connections. Below follows a short description of electrodes for peripheral nerve interfaces. An extensive review can be found in [7].

14.2.1 Intraneural Electrodes

Intraneural electrodes such as the needle microelectrodes [8] and the 25-μm-diameter longitudinal intrafascicular electrode (LIFE) penetrate the nerve fascicle and place their active site in the peripheral nerve endoneurium or, in the case of the dorsal root ganglion, within the neuropil. They inherently have higher recording selectivity and better signal-to-noise ratio as compared to extraneural electrodes. The key development has been to extend their recording stability, biocompatibility, and robustness [7]. Intraneural penetrating electrodes have not yet made their way into clinical FES systems.

14.2.2 Extraneural Electrodes

Extraneural electrodes, such as whole nerve-cuff electrodes, have their active site placed on the surface of the peripheral nerve (the epineurium). The electrode activates or records

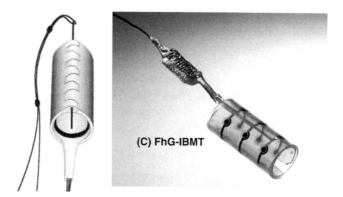

(C) FhG-IBMT

Figure 14.2   Representation of cuff (around nerve) electrodes (left) and newly designed
12-contact circular cuff electrode (with longitudinal slit) for selective stimulation with
four-channel stimulator. To minimize contamination of the neural signal from the surrounding
biological signals, the electrodes must fit tightly around the nerve and be well closed. An electrode
can be closed with the flap covering the longitudinal slip and kept in position by means of the
surgical threads around the cuff (left side), and it can be closed by using a "zipper" method where
the ends of the cuff have little holes and the plastic baton can be pulled through the holes (right side).
The novel self-wrapping polyamide multipolar electrode for selective recording/stimulation of
the whole nerve (C). (Courtesy of T. Stieglitz, Fraunhofer Institute, St. Ingbert, Germany, 2000.)

from a large number of nerve fibers. Extraneural techniques are the least invasive and are
the first to be used in human subjects as an interface to provide feedback from natural
sensors to a fully implantable closed-loop FES system [9, 10]. The advantages of the
electrodes are their stability, durability, and ease of implementation. Figure 14.2
shows a silicon cuff electrode and a polyamide–silicone hybrid cuff electrode. For the
polyamide–silicone hybrid cuff electrode both the shape of the polyamide substrate and
metalization traces are easily realized by standard microfabrication techniques so that arbitrary designs can be formed with high precision and repeatability [11]. Different variations
exist of the cuff electrode. For example, to access nerve fascicles that are located deep
within the nerve trunk, Tyler and Durand [12] designed "spring-loaded" slowly
penetrating intrafascicular electrodes (SPINEs) that insinuated themselves into the nerve.

14.2.3   Sieve Electrodes

Peripheral nerves that have been transected (e.g., when a limb is amputated) can often
regenerate if there is a target to be reached. If such a regenerating nerve is provided
with a biocompatible structure that outgrowing nerve fibers can grow into or through,
an electronic interface can be established. A sieve electrode consists of a matrix of
holes that is positioned at the end of a transected nerve. Ideally the axons of the nerve
will grow through the holes. In principle the holes could be addressable electrode contacts
for recording and/or for electrical stimulation [7]. Sieve devices can be made by microfabrication techniques where the holes are produced through a silicon disk [13]. A
feature of silicon is that the circuitry to process recorded signals and apply multiplexing
can be made as an integral part of the device.

An example where sieves having 30-μm-holes and 30% transparency (ratio of open
space versus closed area of the sieve) yielded to the best nerve regeneration [38% of
functional gastrocnemeus electromyography (EMG) in a rat sciatic model] was presented
by Ceballos et al. [14]. Sieve electrodes have not yet been tested in human applications.

14.2.4 Cultured Neuron Electrodes

In this approach an electronic interface is achieved by growing neural cells onto an electrode structure. The cells are captured in small wells or they are adherent to a substrate having an array of electrode contacts [15]. So-called neurochips are then formed. The idea is then to implant the neurochips into the neural tissue where the neural interface is desired. It is then expected that the cells on the chip will grow collaterals into the host tissue. To guide and/or promote neuronal attachment and axon outgrowth, bioactive substances can be employed.

14.3 ELECTRONIC NERVE STIMULATORS

Coordination of movements by FES requires electronic stimulators that are self-contained with a low power consumption, small, and light and must have the simplest possible interface [e.g., 16–24]. The stimulator should be programmable. The stimulator needs a setup mode—mode of programming when communicating with the host computer. Once the programming is finished, the stimulator should be turned to the autonomous mode. Implantable stimulators for FES may be divided into single- and multichannel devices. Single-channel implants, which have been fabricated, are all radio-frequency (RF) powered and controlled devices. They use relatively few discrete components and have a receiving antenna which is integrated into the circuitry. Two alternative schemes have been considered for multimuscle excitation. It is possible, in principle, to use several one-channel units that are controlled from one controller [19, 23] or to use a single implantable stimulator that will connect with multiple electrodes [20, 25–27]. The single-channel devices have been developed originally for footdrop correction and implanted in many subjects with positive experience, yet the development was continued in a different direction, that is toward multichannel stimulators [26, 28]. The wireless single-channel stimulator, termed BION, has been developed for extensive use in restoring motor functions, and up to now animal experiments show great promise [19]. The BION single-channel wireless stimulator is sealed with glass beads and uses an anodized tantalum and surface-activated iridium electrodes to minimize tissue damage. The diameter of the glass tube is 2 mm, and the length of the whole device is 13 mm. The BION is powered by inductive coupling from an external coil at 2 MHz. A total of 256 units can be driven from a single control based on the Motorola 68HC11 microcomputer. The stimulator delivers charge-balanced monophasic pulses, allowing for the selection of a square of exponential discharge tail.

The problem with all implantable devices without batteries, which require a lot of power to drive the motor systems, is low efficiency of RF transmission. In order to transmit energy, the emitting and receiving antenna must be close and aligned; this is very difficult if a stimulator is injected into a deep muscle. The alternative solution, accepted by most other research teams developing electronic stimulators, is a miniature, implantable, multichannel device that will excite as many muscles as needed. The difficulty is that such a stimulator will be remote from the stimulation points; therefore, connectors and leads have to be used between the stimulator and stimulation points. The use of long leads should be eliminated after the technique of selective nerve stimulation, including potential stimulation of the spinal cord directly or the spinal roots, is perfected. The experiments conducted at Aalborg University with two- and four-channel fully implanted RF-driven stimulators [29] integrated in the cuff electrode suggest that this is a viable technique.

Selective activation of different muscle groups has been achieved in a sitting and walking subject.

The stimulation-telemetry system described by Smith et al. [30] is an example of the state of the art of technology. The device can be configured with the following functions:

1. Up to 32 independent channels of stimulation for the activation of muscles (or sensory feedback), with independent control of stimulus pulse interval, pulse duration, pulse amplitude, interphase delay (for biphasic stimulus waveform), and recharge phase duration (for biphasic stimulus waveform)

2. Up to eight independent telemetry channels for sensors, with independent control of sampling rate and pulse powering parameters of the sensor (power amplitude and duration)

3. Up to eight independent telemetry channels for processed (rectified and integrated) signals measured from muscles or nerves (EMG or ENG), with independent control of the sampling rate and provisions for stimulus artifact blanking and processing control

4. Up to eight independent telemetry channels for unprocessed myoelectric signal (MES) channels, with independent control of sampling rate

5. Up to eight independent telemetry channels for system functions, providing control or sampling of internal system parameters, such as internal voltage levels

Due to the overall timing constraints, implant circuit size, implant capsule size, number of lead wires, circuit and sensor power consumption, and external control and processing requirements, it is not practical to realize a single device having the maximal capabilities outlined above. However, the intent of the stimulator–telemeter system is to provide the means of realizing an optimal implantable device having all the necessary circuitry and packaging to meet the anticipated clinical applications without requiring design or engineering effort beyond that of fabricating the device itself. For example, a basic upper extremity application would require, minimally, eight channels of stimulation for providing palmar and lateral grasp and sensory feedback, along with one joint angle transducer as a command control source. The stimulation–telemetry implant device comprises an electronic circuit that is hermetically sealed in a titanium capsule with feedthrough. A single internal RF coil provides transcutaneous reception of power and bidirectional communication. The lead wires connected to the feedthrough holes extend to the stimulating electrodes, to the implanted sensors, or to the recording electrodes. The later two connections are used as control input. The circuit capsule, coil, and lead-wire exits are conformably coated in epoxy and silicone elastomer to provide physical support for the feedthroughs and RF coil and stress relief to the leads making it suitable for long-term implantation. The functional elements required to realize the system include the following: (1) an RF receiver for recovering power and functional commands transmitted from an external control unit; (2) control logic circuitry to interpret the recovered signals, to execute the command function, and to supervise functional circuit blocks; (3) multichannel stimulation circuitry for generating the stimulus pulses that are sent to the stimulating electrodes; (4) multichannel signal conditioning circuitry which provides amplification, filtering, and processing for the signals to be acquired (MES and sensor signals); (5) data acquisition circuitry for sampling and digitizing these signals; (6) modulation circuitry for telemetering the acquired data through the RF link; (7) power regulation and switching circuitry for selectively powering the included functional blocks of the circuitry, as needed, to minimize the power consumption of the device; and (8) system control circuitry to allow interrogation or configuration of the operation of the device.

One of the technical limitations in FES is the size of the electronic devices and power consumption. Sufficient miniaturization could be reached within a decade according to "Moore's law," predicting that transistor density on integrated circuits will double every 1–1.5 years. However, a complication of increasing the density is susceptibility, especially of digital systems, to electromagnetic interference. In fully implantable neural motor prostheses, the battery energy density and recharging issues will become limited, especially as the FES devices gain functionality and demand more power. The design of the practical RF coupling system for continuous use is at this time not resolved. A method attracting attention is to apply miniature rechargeable batteries that will be charged during no-use periods (e.g., over night) and exploited during the functional tasks. Maximal energy density of implantable batteries is about $1.1 \, \text{Wh/cm}^3$ (lithium ion) [31], and it needs to be improved. This topic remains an open area of research.

14.4 CREATING FES COMMAND SIGNALS FROM BIOPOTENTIALS

The control of current FES systems is mainly based on a variety of hand-operated switches and external transducers [2] which contain no information that is feasible for a feedforward control system or requires too much conscious activity. More recently, the integrated sensors packaged in the biocompatible materials were introduced [32]. In this chapter we are addressing the increased interest in using biopotentials from muscle or brain to control FES systems (for a review see [33]).

14.4.1 Muscle Biopotentials

The use of the EMG recordings to control neuroprostheses is benefiting from both the hardware development dealing with the stimulation artifacts and sophisticated signal processing techniques being applied today in real time due to high-performance digital signal processors. The EMG recordings allow the analysis of the fatigue of the stimulated muscle (e.g., [34]), which is an important feature for FES systems that require prolonged stimulation of muscles such as leg extensors during standing. Recording from multiple muscles around the shoulder is most likely the only method to generate control signals for future neuromotor prostheses controlling reaching and grasping. The implantable sensors that can potentially be applied instead of surface electrodes for EMG recordings offer new horizons, yet these techniques have not been evaluated in practical neuroprosthesis.

14.4.2 Brain Biopotentials

The possible use of brain–computer interface (BCI) technologies through recordings from higher cortical structures has received increasing attention from many researchers as a way to create useful command signals [35]. One possible method is to utilize the focal recordings using EEG instrumentation [36–38]. Another source of control signals is to provide direct access through implantable electrodes to the neural structures that are directly involved in sensorimotor functions (e.g., [39–41]). The implantable systems require insertion of special electrode(s) into the human cortex. The recorded signals can then be transmitted to the external units and processed to extract relevant control signals. For a review on this topic see [43].

Recently FES researchers have begun to explore the BCI technology as a new way to create a command signal for FES systems to control hand grasp [44] and foot control [45]. One of the major challengers in EEG-based FES command signals is the low information transfer rates of maximum 5–25 b/min. [35]. Achievement of greater speed and accuracy depends on improvements in signal processing, translation algorithms, and user training. These improvements depend on increased interdisciplinary cooperation between neuroscientists, engineers, computer programmers, psychologists and rehabilitation specialists and on adoption and widespread application of objective methods for evaluating alternative methods. For a recent review see [35]. For a FES system, a control signal extracted from cortical signals seems attractive, particularly based on our recent understanding of the information coded in a population of cortical cells [46]. The most fundamental finding is that the discharge rate is a broad tuning function covering all directions of movement. Each cell is generally active for all movements. In a population of cells, all cells can be active for most movements, but they may each have different firing related to preferred movement directions and thus slightly different tuning functions. For example, cells in the motor cortex have different preferred directions (detailed by their tuning functions). In a computer-based process, a "population vector" can be calculated and shown to match the movement direction with length corresponding to the movement speed. Animal studies have shown the possibility of recording instantaneous population vectors from many units simultaneously and process them through a computer to extract trajectory information [42]. Tests in animals operating a robotic arm to reach food have demonstrated the potential strengths of such an approach [43]. Ultimately, these methods may be applicable to the control of FES systems in human subjects.

14.5 CLINICAL APPLICATIONS: EXAMPLES

In [2] we discuss clinical areas where FES is already important or where it holds great potential if the adequate technology is developed.

Examples from current research at Aalborg University of the use of interfaces onto the peripheral nervous system to restore motor functions in patients with CNS injuries are given below. See [2] for an extensive review of clinical applications within movement restoration by FES.

Several studies have shown that the application of closed-loop control techniques can improve the regulation of FES muscle activation. Natural sensors such as those found in the skin, muscles, tendons, and joints present an attractive alternative to artificial sensors for FES purposes because they are present throughout the body and contain information useful for feedback control. Moreover, the peripheral sensory apparatus is still viable after brain and spinal cord injuries. Electrical signals can be recorded through one of the neuroelectronic interfaces described above. Here we will limit the focus on using long-term implanted nerve cuff electrodes in the human peripheral nerves. Reliable detection of sensory nerve signals is essential if such signals are to be of use in sensory-based FES neural prosthetics as a replacement for artificial sensors (switches, strain gauges, etc.). For an extensive review on the properties of the signals in long-term implants see [46], and on its application in automatic adjustment in neural prosthetic devices see [48, 49].

14.5.1 Natural Sensory Information Used Drop Foot Prosthesis

Electrical stimulation of the peroneal nerve used for correction of gait has proven to be potentially useful for enhancing dorsiflexion in the swing phase of walking in lower

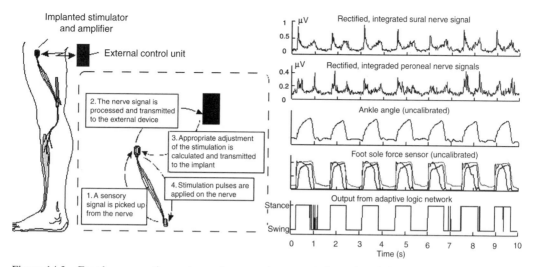

Figure 14.3   Footdrop correction system using natural sensory information picked up either from sural nerve above ankle joint or peroneal nerve above the knee for determination of foot-to-floor contact. The nerve signals were processed by means of an adaptive logic network (ALN), and the output from here was passed through a set of restriction rules (not shown) that decided when to stimulate. The stimulation could then be applied on the peroneal nerve through the same multipolar cuff electrode as used for recording the sensory signal on the Peroneal nerve. (Adapted from Fig. 4 in [33].)

extremities of hemiplegic patients. The stimulation is applied during the swing phase of the affected leg and prevents drop foot. This makes the patient walk faster and more securely. The stimulator is often located distally to the knee on the lateral part of the tibia. The stimulator can be either external or partly implantable. In both cases, the stimulator is triggered by an external heel switch linked to the stimulator through a wire running from the switch under the heel up to the stimulator.

A large clinical impact of an application could be obtained if the external heel switch is replaced with a cuff electrode that records the neural activity from the skin of the foot and uses this as a trigger to turn on and off the electrical stimulation, which enhances the dorsiflexion in the swing phase of walking. The rationale for implanting a cuff electrode on a cutaneous nerve innervating the foot is to remove the external heel switch used in existing systems for footdrop correction and thereby making it possible to use such systems without footwear and preparing it to be a totally implantable system.

By percutaneous lead wires, the nerve cuff electrode was connected in a tripolar configuration to an external amplifier providing an overall gain of approximately 110,000. The output from the system was fed to a microprocessor-controlled stimulator that activated the ankle dorsiflexor muscles through a bipolar surface stimulation of the peroneal nerve. During walking, the nerve signal modulated strongly and gave a clearly detectable response at foot contact and a silent period when the foot was in the air through the swing phase of the walking cycle (Fig. 14.3). After processing the signal, heel contacts could be detected using the afferent nerve signal information alone. One of the most significant problems in correlating the nerve signal activity to the actual heel contact during walking is noise originating from nearby muscle activity.

14.5.2   Natural Sensory Information Used in Hand Prosthesis

A very interesting application of signals from cutaneous receptors is the hand neuroprosthesis. In healthy subjects, the skin receptors play an important role in the control mechanisms when, for example, lifting an object in a precision grip. The fingers of the human hand are subserved by an estimated 200–300 touch-sensing receptors per square centimeter within the skin. Such tactile sensors are required to signal the amount of grasp effort needed to secure an object with sufficient force to prevent slippage but with an economy of effort to avoid undue muscle fatigue. If the initial muscle activity in a hand grasping an object only leaves a minute of safety margin against slips, small frictional slips might elicit brief burst responses in the sensing receptors. These afferent volleys trigger an upgrading of the grip force about 70 ms after the onset of the slip. Healthy subjects are able to control the grip force when holding a given object, independent of the weight and surface texture of the object. This is possible because the cutaneous receptors give information about small slips and skin deformation.

We have implemented an algorithm that makes an FES system able to mimic this function based on the compound information from the cutaneous receptors as recorded by a nerve cuff electrode. Initially, the algorithm was developed in an animal preparation, and later we implemented it in two spinal-cord-injured subjects [48].

Results from a 27-year-old tetraplegic male with a complete C5 spinal cord injury (two years postinjury) are presented here. The patient had no voluntary elbow extension, no wrist function, and no finger function. He used a splint for keeping the wrist stiff. He had partial sensation in the thumb but no sensation in the second to fifth fingers. During general anesthesia, the patient was implanted with a tripolar nerve cuff electrode on the palmar interdigital nerve to the radial side of the index finger branching off the median nerve. The cuff was 2 cm long and had an inner diameter of 2.6 mm. Eight intramuscular wire electrodes were placed in the following muscles: extensor pollicis brevis (EPB), flexor pollicis brevis (FPB), adductor pollicis (AdP), and flexor digitorum longus (FDL). See Figure 14.4.

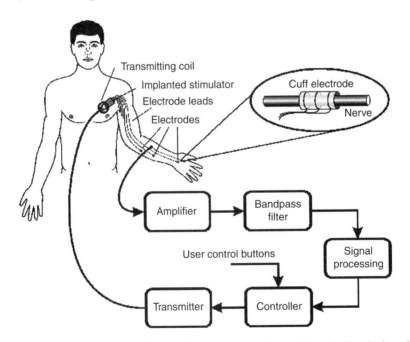

Figure 14.4   Schematic of system for restoration of lateral key grip in spinal-cord-injured volunteer subject, including natural sensory feedback from cutaneous digital nerve. The stimulation system is called "FreeHand" and has been developed at Case Western Reserve University, OH. (Adapted from [10] in Fig. 1.)

The stimulator was controlled by a computer, which also sampled the nerve signals and performed the signal analysis. When stimulation was turned on, the object could be held in a lateral grasp (key grip) by the stimulated thumb. If the object slipped, either because of decreasing muscle force or increasing load, the computer detected this from the processed nerve signal and increased the stimulation intensity with a fixed amount so that the slip was arrested and the object again was held in a firm grip [48].

When comparing the performance of the system to healthy subjects doing the same task, it was shown that the FES-generated force could automatically adjust to levels close to healthy subjects' levels. When extra weight was added, the slipped distance was also comparable to the performance of healthy subjects.

Today this system is developed to an extent where the subject can use it during functional tasks [10]. During an eating session, where the subject has a fork in his or her instrumented hand (grasped between the thumb and index finger), the control system is designed to decrease the stimulation of the finger muscles until the feedback signal from the skin sensors detects a slip between the index finger and the fork. When a slip is detected, the stimulation to the thumb increases automatically proportional to the strength of the sensory feedback, and if no further slips are detected, the controller starts to decrease the stimulation. This continuous tracking of the minimally necessarily needed stimulation means that the hand muscles are only loosely stimulated when the subject decides to rest his or her arm (which typically happens when he or she has a conversation with one or more persons at a dinner table). The stimulation will automatically increase when he or she starts to eat again by placing the fork in the food. A typically eating session will last 20–30 min. A large fraction of this time is dedicated to "noneating" activities. During such times, the stimulation is at a minimum (keeping the fork in the hand with a loose grasp) and thereby preventing the hand muscles from being fatigued. When the feedback is taken away, the subject will typically leave the stimulation on at high stimulation intensity for the full eating session. This will fatigue the stimulated muscles, and the subject will try to eat dinner faster or will rest the muscles at intervals by manually decreasing the stimulation—an effort that requires more attention from the subject than the automatic adjustment of the stimulation intensity.

14.5.3 Natural Sensory Information Used in Control of Neurogenic Bladder

In the case of neurogenic bladder, the elevated intravesicular pressure can force urine to travel back up to the kidneys and produce upper urinary tract infection. Equally common are infections of the lower urinary tract caused by insufficient voiding which leaves a persistent high volume of residual urine. In addition to these problems, overfilling of the bladder can lead to a condition of autonomic dysreflexia, which can also be life threatening.

These problems could be reduced if it would be possible to provide spinal-injured individuals with information about the state of fullness of the bladder. This requires some sensors that can monitor the bladder volume and bladder pressure. An artificial sensor such as a strain gauge device or an ultrasound device attached to the bladder wall might seem like a solution; however, it is difficult to maintain an enduring fixation to the bladder wall. An insertable catheter with integral pressure sensor might be another approach, but the risk of introducing infections from external organisms, especially with regard to chronic devices, is substantial.

Work in an anesthetized cat preparation has shown a good correlation between recorded activity from the S2 sacral root and the intravesical pressure associated with fullness of the bladder [50]. The monotonic increase in the rate of discharge of the recorded, isolated bladder afferent clearly parallels the increase of intravesical pressure. Recently, it has also been demonstrated that a nerve cuff applied around the S2 sacral root or the pelvic nerve innervating the bladder in anesthetized pigs can record activity that correlates with the fullness status of the bladder [51]. But perhaps more importantly, if sensory information from the bladder can be reliably recorded by cuff electrodes placed on suitable nerves without bigger risks, patients with a hyperreflexive bladder (SCI, some incontinence patients) could, when needed, get a closed-loop controlled FES implant that uses this recorded sensory input. Present solutions for such patients are suppression of reflex contractions by drugs and bladder emptying by catheterization, which needs to be done many times during the day and night and is not always pleasant for a patient. In some cases, drugs do not work—then surgical intervention is needed where the detrusor is deafferented by cutting the dorsal sacral roots to prevent reflex contractions, and the bladder can be emptied by use of a sacral root stimulator that has electrodes on the sacral ventral roots. Dorsal rhizotomy increases the bladder capacity, but reflex erection in male patients is lost. To prevent cutting dorsal sacral roots, one could detect fast pressure rises and detrusor activation with nerve cuff recordings from bladder nerves, and the controller could then take appropriate actions, as, for example, inhibit detrusor contractions by stimulating pudendal or penile nerves or block efferent or afferent pelvic nerve transmission to prevent reflex detrusor contractions [51, 52]. In this way continence could be reestablished, low-pressure voiding achieved, and bladder functional capacity increased, and beside medical status improvement, the patients would become more independent as well and could socialize more easily.

14.5.4 Functional Electrical Therapy for Cortical Reorganizing

The adult mammalian nervous system has an ability to reorganize itself in an activity-dependent manner in response to increased or decreased sensory inflow. Manipulating peripheral sensory nerve activity by electrical stimulation, one can modulate in healthy human subjects the magnitude of cortical response and modulate motor pathway excitability, which can produce a mixture of excitation and inhibition at supraspinal levels. Long-term reorganization, as defined by an effect that outlasts the stimulation, of the motor cortex has been reported for repetitive electrical stimulation of peripheral nerves (e.g., [53]). Based on such findings we designed a treatment protocol termed functional electrical therapy (FET)—a new method of treating the more affected arm in humans after a stroke by applying electrical nerve stimulation in an activity-dependent manner to the affected arm. It is an exercise that comprises simultaneous, voluntarily and externally assisted reaching and grasping. Functional electrical stimulation assists the grasp based on control that mimics the patterns typically found in able-bodied subjects. The treatment is intensive exercise of daily functions (e.g., drinking, eating, writing) by the more affected arm/hand. Results from a randomized, blinded, clinical study with 28 patients show that there is a highly significant improvement in performing typical daily activities in all FET-treated patients leading to a better quality of life [54].

The study suggests that systematic electrical stimulation of peripheral nerves in a way that generates lifelike movement timed with the voluntary activity and integrated into a functional scheme leads to faster and greater reorganization of the CNS after stroke and thereby speeds up and promotes the recovery of reaching and grasping in acute stroke subjects. A future challenge is to develop FET systems that can easily and reliably generate lifelike movements for other segments of the body such as elbow and shoulder movements of the arm and ankle and knee and hip movements of the leg.

Future technology will make it possible to interface to the body at a micro- and nanoscale level opening up many new treatments to restore sensorimotor functions in individuals with injuries in the CNS. Howeve, this should not stop us from moving into clinical investigations of FES systems that already have potentials in improving CNS-injured individuals' quality of life.

ACKNOWLEDGEMENTS

The Danish National Research Foundation and The Danish Research Councils are kindly acknowledged for their financial support.

REFERENCES

1. ROBINSON, C. J. Expanding the scope of IEEE transaction on rehabilitation engineering to explicity including neural engineering. *IEEE Trans. Rehabil. Eng.* 8:273–275, 2000.

2. POPOVIC, D. B. AND SINKJÆR, T. Control of Movement for the Physically Disabled: *Control for Rehabilitation Technology*, 2nd ed. Aalborg: Center for Sensory Motor Interaction, Aalborg University, 2003, p. 488.

3. LIBERSON, W. F., HOLMQUEST, H. J., SCOTT, D., AND DOW, A. Functional electrotherapy: Stimulation of the peroneal nerve synchronized with the swing phase of the gait in hemiplegic patients. *Arch. Phys. Med. Rehab.* 42:101–105, 1961.

4. POPOVIC, D. B., POPOVIC, M. B., AND SINKJÆR, T. Neurorehabilitation of upper extremities in humans with sensory-motor impairment. *Neuromod* 5(1): 54–67, 2002.

5. LYONS, G. M., SINKJÆR, T., BURRIDGE, J. H., AND WILCOX D. J. A review of FES-based neural orthosis for the correction of drop foot. *IEEE Trans. Neural Systems & Rehabilitation Engineering* 10(4):206–279, 2002.

6. RIJKHOFF, N. J. M., WIJKSTRA, H., VAN KERREBROECK, P. E. V., AND DEBRUYNE, F. M. J. Urinary bladder control by electrical stimulation: Review of electrical stimulation techniques in spinal cord injury. *Neurourol. Urodynam.* 16:39–53, 1997.

7. YOSHIDA, K., AND RISO, R. Peripheral nerve recording electrodes and techniques. In *Neuroprosthetics: Theory and Practice*, HORCH, K. W., AND DHILLON, G. S. Eds. Imperial College Press, 2004, pp. 638–736.

8. LOEB, G. E., BAK, M. J., SALEMAN, M., AND SCHMIDT, E. M. Parylene as a chronically stable reproducible microelectrode insulator. *IEEE Trans. Biomed. Eng.* 24:121–128, 1977.

9. HAUGLAND, M. K., AND SINKJÆR, T. Cutaneous whole nerve recordings used for correction of footdrop in hemiplegic man. *IEEE Trans. Rehab. Eng.*, 3:307–317, 1995.

10. INMANN, A., HAUGLAND, M., HAASE, J., BIERING-SOERENSEN, F., AND SINKJAER, T. Signals from skin mechanoreceptors used in control of a hand grasp neuroprosthesis. *Neuroreport* 12:2817–2820, 2001.

11. SCHUETTLER, M. AND STIEGLITZ, T. 18 polarhybrid cuff electrodes for stimulation of peripheral nerves. In SINKJAER, T., STRUIJK, J. J. S., AND POPOVIĆ, D. B (eds). *Proceedings of the 5ᵗʰ Annual IESS Conf.* Aalborg, Denmark, 2000, pp. 265–268.

12. TYLER, D. J. AND DURAND, D. M. A slowly penetrating interfascicular nerve electrode for selective activation of peripheral nerves. *IEEE Trans. Rehabil. Eng.* TRE-5(1):51–61, 1997.

13. EDELL, D. J. A peripheral nerve information transducer for amputees: Long-term multichannel recordings from rabbit peripheral nerves. *IEEE Trans. Biomed. Eng.* 33:203–214, 1986.

14. CEBALLOS, D., VALERO-CABRE, A., VALDDERRAMA, E., SCHUTTLER, M., STIEGLITZ, T., AND NAVARRO, X. Morphologic and functional evaluation of peripheral nerve fibers regenerated through polyimide sieve electrodes over long-term implantation. *J. Biomed. Mater. Res.* 60:517–528, 2002.

15. RUTTEN, W. L. Selective electrical interfaces with the nervous system. *Annu. Rev. Biomed. Eng.* 4:407–452, 2002.

16. BIJAK, M., SAUERMAN, S., SCHMUTTERER, C., LANMUELLER, H., AND UNGER, E. A modular PC-based system for easy setup of complex stimulation patterns. In *Proc. Fourth Intern. Conf. IFESS*. Sendai. 1999, pp. 28–32.

17. BRINDLEY, G. S., POLKEY, C. E., AND RUSHTON, D. N. Electrical splinting of the knee in paraplegia. *Paraplegia* 16:428–435, 1978.

18. BUCKETT, J. R., PECKHAM, P. H., THROPE, G. B., BRASWELL, S. D., AND KEITH, M. V. A flexible, portable system for neuromuscular stimulation in the paralyzed upper extremity. *IEEE Trans. Biomed. Eng.* BME-35:897–904, 1988.

19. CAMERON, T., LOEB, G. E., PECK, R. A., SCHULMAN, J. H., STROJNIK, P., AND TROYK P. R. Micromodular implants to provide electrical stimulation of paralyzed

muscles and limbs. *IEEE Trans. Biomed. Eng.* 44(9):781, 1997.

20. DAVIS, R., HOUDAYER, T., ANDREWS, B. J., EMMONS, S., AND PATRICK, J. Paraplegia: Prolonged closed-loop standing with implanted nucleus FES-22 stimulator and Andrews' foot-ankle orthosis. *Stereotact. Funct. Neurosurg.* 69:281–187, 1997.

21. DONALDSON, N. A 24–output implantable stimulator for FES. In *Proc. Second Vienna Int. Workshop Functional Electrostimulation.* Vienna, 1986, pp. 197–200.

22. SMITH, B. T., PECKHAM, P. H., KEITH, M. W., AND ROSCOE, D. D. An externally powered, multichannel, implantable stimulator for versatile control of paralyzed muscle. *IEEE Trans. Biomed. Eng.* BME-37(7): 499–508, 1987.

23. STROJNIK, P., ACIMOVIC, P., VAVKEN, E., SIMIC, V., AND STANIC, U. Treatment of drop foot using an implantable peroneal underknee stimulator. *Scand. J. Rehabil. Med.* 19:37–43, 1987.

24. THROPE, G. B., PECKHAM, P. H., AND CRAGO, P. E. A computer-controlled multichannel stimulation systems for laboratory use in functional neuromuscular stimulation. *IEEE Trans. Biomed. Eng.* BME-31:363–370, 1985.

25. HOLLE, J., FREY, M., GRUBER, H., KERN, H., STOHR, H., AND THOMAS, H. Functional electro-stimulation of paraplegics: Experimental investigations and first clinical experience with an implantable stimulation device. *Orthopaedics* 7:1145–1160, 1984.

26. STROJNIK, P., WHITMOYER, D., AND SCHULMAN, J. An implantable stimulator for all season. In *Advances in External Control of Human Extremities X*, POPOVIC, D. B., Ed. DEMOS Publisher, New York, 1990, pp. 335–344.

27. SMITH, B. T., MULCAHEY, B. J., AND BETZ, R. B. Development of an upper extremity FES system for individuals with C4 tetraplegia. *IEEE Trans. Rehabil. Eng.* TRE-4:264–270, 1996.

28. STROJNIK, P., SCHULMAN, J., LOEB, G., AND TROYK, P. Multichannel FES system with distributed micro-stimulators. In *Proc. Annu. Int. Conf. IEEE EMBS.* 1993, pp. 1352–1353.

29. HAUGLAND, M., LICKEL, A., HAASE, J., AND SINKJÆR, T. Control of FES thumb force using slip information obtained from the cutaneous electroneurogram in quadrip legic man. *IEEE Trans. Rehabil. Eng.* TRE-7(2): 215–227, 1999.

30. SMITH, B. T., TANG, Z., JOHNSEN, M. W., POURMEHDI, S., GAZDIK, M. M., BUCKETT, J. R., AND PECKHAM, P. H. An externally powered multichannel, implantable stimulator-telemeter for control of paralyzed muscles. *IEEE Trans. Biomed. Eng.* BME-45:463–475, 1998.

31. CRAELIUS, W. The bionic man: Restoring mobility. *Science* 295:1018–1021, 2002.

32. HART, R. L., KILGORE, K. L. AND PECKHAM, P. H. A comparison between control methods, for implanted FES hand-grasp systems. *IEEE Trans. Rehab. Eng.* TRE-6:208–218, 1998.

33. SINKJÆR, T., HAUGLAND, M., INMANN, A., HANSEN, M., AND NIELSEN, K. D. Biopotentials as command

and feedback signals in functional electrical stimulation systems. *Med. Eng. Phys.* 25:29–40, 2003.

34. TEPAVAC, D., AND SCHWIRTLICH, L. Detection and pre-diction of FES induced fatigue. *J. Electromyogr. Kinesiol.* 7(1):39–50, 1997.

35. WOLPAW, J. R., BIRBAUMER, N., HEETDERKS, W. J., MCFARLAND, D. J., PECKHAM, P. H., SCHALK, G., DINCHIN, E., QUATRANO, L. A., ROBINSON, C. J., AND VAUGHAN, T. M. Brain-computer interface technology: A review of the first international meeting. *IEEE Trans Rehabil Eng.* 8(2):164–73, 2000.

36. MCFARLAND, D. J., MCCANE, L. M., AND WOLPAW, J. R. EEG-based communication and control: Short-term role of feedback. *IEEE Trans. Rehabil. Eng.* TRE-6(1):7–11, 1998.

37. WOLPAW, J. R., RAMOSER, H., MCFARLAND, D. J., AND PFURTSCHELLER, G. EEG-based communication: Improved accuracy by response verification. *IEEE Trans. Rehabil. Eng.* TRE-6(3):326–333, 1998.

38. VAUGHAN, T. M., MINER, L. A., MCFARLAND, D. J., AND WOLPAW, J. R. EEG-based communications: Analysis of concurrent EMG activity. *Electroencephalo. Clin. Neurophysiol.* 107(6):428–433, 1998.

39. NORMANN, R. A., MAYNARD, E. M., ROUSCHE, P. J., AND WARREN, D. J. A neural interface for a cortical vision prosthesis. *Vision Res.* 39(15):2577–2587, 1999.

40. ROUCHE, P. J., AND NORMAN, R. A. Chronic intracorti-cal microstimulation (ICMS) of cat sensory cortex using the Utah Intracortical Electrode Array. *OIEEE Trans. Rehabil. Eng.* TRE-7:56–58, 1999.

41. KENNEDY, P. R., BAKAY, R. A., MOORE, M. M., ADAMS, K., AND GOLDWAITHE, J. Direct control of a computer from the human central nervous system. *IEEE Trans. Rehabil. Eng.* 8:198–202, 2000.

42. SCHWARTZ, A. B., TAYLOR, D. M., AND TILLERY, S. I. Extraction algorithms for cortical control of arm prosthesics. *Curr. Opin. Neurobiol.* 11(6):701–707, 2001, Review.

43. CHAPIN, J. K., AND MOXON, K. A. *Neural Prostheses for restoration of sensory and motor function*, Boca Raton, FL: CRC Press, 2001, pp. 1–296.

44. LAUER, R., PECKHAM, P. H., AND KILGORE, K. EEG-based control of a hand grasp neuroprosthesis. *Neuroreport* 8:1767, 1999.

45. JUUL, P. R., LADOUCEUR, M., AND NIELSEN, K. D. Coding of lower limb muscle force generation in associ-ated EEG movement related potentials: Preliminary studies toward a feed-forward control of FES-assisted walking. In SINKJAER, T., STRUIJK, J. J. S., AND POPOVIC, D. B. (eds). Proceedings of the 5th Annual IFESS Conf., Aalborg, Denmark, 2000, pp. 335–337.

46. GEORGOPOULOS, A. P. Current issues in directional motor control. *Trends, Neurosci.* 18:506–510, 1995.

47. STRUIJK, J. J., THOMSEN, M., LARSEN, J., AND SINKJÆR, T. Cuff electrodes in long-term recordings of natural sensory information. *IEEE Engineering in Medicine and Biology Magazine* May/June: 91–98, 1999.

48. HAUGLAND, M. K., AND SINKJÆR, T. Control with natural sensors. In *Synthesis of Posture and Movement*

in *Neural Prostheses*, Winters, J. and Crago, P. Eds. New York: Springer, 1999, pp. 617–629.

49. Riso, R. R. Perspectives on the role of natural sensors for cognitive feedback in neuromotor prostheses. *Automedica* 16:329–353, 1998.

50. Hibler, H. J., Janig, W., and Koltzenburg, M. Myelinated primary afferents of the sacral spinal cord responding to slow filling and distention of the cat urinary bladder. *J. Physiol.*, pp. 449–463, 1993.

51. Jezernik, S., Wen, J. G., Rijkhoff, N. J., Djurhuus, J. C., and Sinkjær, T. Analysis of bladder related nerve cuff electrode recordings from preganglionic pelvic nerve and sacral roots in pigs. *J. Urol.*, 163(4): 1304–1314, 2000.

52. Rijkhoff, N. J. M., Wijkstra, H., van Kerrebroeck, P. E. V., and Debruyne, F. M. J. Selective detrusor activation by sacral ventral nerve root stimulation: First results of intraoperative testing in humans during implantation of a Finetech-Brindley system. *World J. Urol.* 16:337–341, 1998.

53. Khaslavskaia, S., Ladouceur, M., and Sinkjær, T. Increase in tibialis anterior motor cortex excitability following repetitive electrical stimulation of the common peroneal nerve. *Exper. Brain Res.* 145(3): 309–315, 2002.

54. Popovic, D., Radulovic, M., Schwirtlich, L., and Jaukovic, N. Automatic vs. hand-controlled walking of paraplegics. *Med. Eng. Phys.* 25:63–73, 2003.

CHAPTER 15

HYBRID OLFACTORY BIOSENSOR USING MULTICHANNEL ELECTROANTENNOGRAM: DESIGN AND APPLICATION

John R. Hetling, Andrew J. Myrick, Kye-Chung Park, and Thomas C. Baker

15.1 INCORPORATING NEURAL CELLS AND TISSUE IN HYBRID DEVICE DESIGNS

The term *biosensor* is used to describe a sensing system comprised in part of living matter or of materials derived from living matter (e.g., enzymes, antibodies, nucleic acids). Generally, the actual sensing element (i.e., front end) in the system is living, and the response of the living component is transduced by a second conventional sensor (e.g., an electrode) and the signal is measured and stored electronically [1]. Living organisms can be quite robust in terms of their ability to maintain homeostasis in a wide range of environments, but isolated cells, tissues, and organs require special consideration to keep alive when isolated from the parent organism. However, it is primarily single cells that are targeted for inclusion in biosensors. A reasonable conclusion is that biosensors are bulky given the significant amount of instrumentation required to keep isolated cells alive and quite fragile under field conditions where a sensor might be used. So what motivates the development of biosensors? The answer includes high sensitivity, high specificity, and speed. The same exquisite sensitivity to individual species of molecules in the surrounding milieu which allows a cell to play its particular role in the physiology of an organism can be exploited to detect those molecules in the parts-per-billion range, and this can often be accomplished on a millisecond time scale.

All cells are sensitive to environmental parameters such as pH, temperature, and toxins (any element or compound becomes a toxin at sufficiently high concentration) as well as basic metabolic compounds (e.g., glucose, oxygen, carbon dioxide). However, many cells are specifically sensitive to certain chemicals for which they express membrane-bound receptors (e.g., glutamate receptors expressed by cortical neurons) or certain types of energy (e.g., light sensitivity of retinal photoreceptor cells, stretch sensitivity of muscle spindle cells). A principal design feature of a sensor is its specificity, or the degree to which its output can be influenced by inputs other than the desired measurand. In other words, if you are building a biosensor to measure glucose, you would not want the

Handbook of Neural Engineering. Edited by Metin Akay
Copyright © 2007 The Institute of Electrical and Electronics Engineers, Inc.

sensor to also be sensitive to oxygen concentration unless you could absolutely control for oxygen concentration in your system. The specificity of membrane-bound receptor proteins is another advantage of cell-based biosensors.

The choice to use a biosensor depends on whether the sensitivity, specificity, and speed offset the fragility, bulky nature, and expense inherent to their use. A primary motivation in biosensor research is to improve durability and reduce size and cost. The specific type of biosensor to be described here is designed to measure airborne chemicals, or odors. The approach is slightly different from most biosensor development in that the living component is the entire olfactory organ of an insect, the antenna. The biological response to an odor is a change in membrane potential of olfactory neurons contained within the antenna, and this biopotential can be recorded with electrodes. The system for recording the electroantennogram (EAG) as well as an EAG analysis approach allowing discrimination between different analytes (airborne volatile compounds) will be described in detail below.

15.2 APPLICATIONS FOR OLFACTORY SENSORS

Olfaction is the sensing of airborne molecules; an olfactory sensor is distinguished from a taste sensor in that it does not need to contact the solid or liquid source of the detected molecule. As many compounds of interest exhibit some degree of volatility, an olfactory biosensor can be thought of as a remote sensor. Potential applications of olfactory biosensors include and surpass applications of artificial sensors, assuming their inherent disadvantages can be overcome (there are theoretical and technical limitations on the size, sensitivity, specificity, and speed of artificial sensors). Industrial uses include quality control of aromatic products (e.g., perfume, wine) and environmental safety (a more refined approach to the canary in the coal mine). Airports routinely use an artificial olfactory biosensor in screening luggage for explosives and drugs (your bag is swabbed with a small white pad, which is then scanned for the illicit substance). Olfactory biosensors are sensitive to a long list of anthropogenic compounds of interest to security and defense, including drugs, explosives, and chemical warfare agents. For many insects, the ability to locate specific species of plants, or even individual plants in a certain state of health or maturity, is important for food or in their reproductive cycle; this is often accomplished by sensing airborne volatiles. Similarly, the body gives off specific odors depending on the state of health (e.g., ketones on the breath of diabetics). Olfactory biosensors may find use in the clinic either for rapid screening or as a high-level diagnostic tool.

15.3 STRATEGIES FOR ARTIFICIAL NOSE DESIGN

The artificial nose, as presently conceived, can be distinguished from other chemical detectors (such as pH or NO electrodes) by the promise of detecting an arbitrary number of different compounds with the same device. Where single-chemical detectors often rely on a semipermeable membrane specific to the molecule to be detected, artificial noses generally consist of an array of semiselective, cross-reactive sensors which demonstrate distributed specificity. There are a number of different artificial nose technologies currently being developed (see Table 15.1) [2–5]. An array of different classes of sensors yields a set of response vectors representing the sensor output, which must then be interpreted by a pattern recognition scheme. This has been done by using pattern recognition methods based on statistical and computational neural network approaches [5].

TABLE 15.1   Signal Transduction Technologies in Use or Under Development for an Artificial Nose [5]. Each Approach Incorporates Multiple Sensor Classes with Differing Sensitivity Spectra Based on Variations in the Sensor Component which Interacts Directly with the Analyte (i.e., Odor)

Signal transduction technology	Principal of operation	Primary limitations
Metal-oxide and MOSFET	Electrical resistance of semiconductor changes when analyte molecules are adsorbed onto surface	High power levels required damaged by sulfur- containing compounds and weak acids Low sensitivity
Conducting polymers	Electrical conductance of film changes with reversible adsorption of analyte molecules	Long response times Drift Poor reproducibility Expensive
Piezoelectric-based (accoustic wave devices)	Analyte molecules adsorbed by thin films on crystals, altering RF resonance frequency	Drift Low sensitivity Poor reproducibility
Fiber-optic/solvatochromic fluorescent dye	Analyte molecules alter polarity of dye's surroundings, affecting fluorescence	Low sensitivity Long response times Limited lifetime (photobleaching)

Note: For example, attachment of various functional groups to the backbone of conducting polymers to create unique polymers with differing affinity for a given analyte. Each class of sensor within a given approach requires individual fabrication; individual sensors then need to be combined into one device. This labor-intensive process imposes a practical limit of about three sensor classes within a given device. This in turn limits discrimination across brad categories of analytes.

Current trends in artificial nose development include the use of temporal features in the detector response and the development of neural networks which can learn new odors on-line, without the need to retrain the entire odor library [5].

However, as noted in Table 15.1, there are three important limitations of artificial nose technology. First, the long response times (tens of seconds to minutes) of most approaches limit them to steady-state measurements, where the steady state may take impractically long times to reach. Steady state is not often attained under many field conditions where an artificial nose might find application (e.g., a turbulent wind stream). Second, the number of sensor classes comprising the array is limited to about three in present designs. This limits the number of compounds which can be distinguished and generally requires advanced knowledge of the compounds to be detected. Third, all artificial nose technologies exhibit low sensitivity as compared to most biological olfactory sensors.

15.4   BIOLOGY OF INSECT OLFACTORY SYSTEM

As stated above, the olfactory biosensor described here uses an insect olfactory organ, or antenna, as the sensing element. A brief description of the anatomy and physiology of this system will be helpful. The principal arrangement of the biological olfactory system is quite well conserved across phyla, from insects to mammals. Olfactory receptor neurons (ORNs) exhibit a response when airborne molecules bind to metabotropic membrane receptors and activate G-protein cascades, providing amplification and eventually leading to membrane potential changes and characteristic trains of action potentials

[6–8]. The antenna is comprised of a cuticle layer, contiguous with the rest of the insect exoskeleton, which contains the dendrites of the ORNs. Contact with the environment is made through a large number of fenestrations or pores in the cuticle; airborne compounds diffuse through the pores and to the dendrites. If the receptor protein expressed by an individual ORN binds to the odorant molecule, a G-protein-mediated transduction cascade, very similar to the phototransduction cascade in photoreceptor cells, results. A single activated G-protein acts on many (hundreds to thousands) substrate molecules, thereby providing a level of molecular amplification to the odor signal; this amplification is one reason a biosensor is inherently very sensitive. If the stimulus is strong enough, that is, enough receptor proteins were bound, threshold is reached and an action potential results.

These sensory neurons synapse onto a variety of interneurons in the antennal lobe (AL), the output of which appears on mitral cells which lead to higher processing structures in the brain [9]. Sensory cells, numbering in the hundreds of thousands, have overlapping, semiselective, yet broad response spectra. The result of the transduction-level coding and the olfactory bulb processing is a system that exhibits a remarkably high sensitivity with broadband detection and discrimination. These are desirable features in any detector system and represent active areas of research in many areas of information technology.

15.5 BIOLOGICAL OLFACTORY SIGNAL PROCESSING

The ORNs, numbering around a few hundred thousand, may be divided into a smaller number of groups defined by the receptor protein expressed by each cell. There are about a hundred individual receptor proteins for the average insect. Each ORN expressing a given receptor protein projects axons to a specific glomerulus in the AL. These glomeruli are the first synaptic relay in the insect olfactory system and are analogous to the olfactory bulb in mammals. Each glomerulus is therefore "tuned" to the sensitivity spectrum of a particular receptor protein. Projection neurons carry the olfactory information from the AL to the mushroom body, where, through a set of feedback mechanisms, the representation of the odor in the neural coding space is refined before being passed on to higher perceptual levels.

This refinement of the neural representation of an odor, well described in a review by Laurent [9], is necessary because organisms are in general capable of discriminating between a number of odors that greatly exceeds the number of different receptor proteins. That is, each odor does not have a "hard-wired" connection to a perceptual center. This is possible because each receptor protein does not bind to only one specific molecule but has a finite sensitivity spectrum. A single receptor protein can bind to a family of compounds which differ slightly in molecular shape and charge distribution; these differences define the parameter space of odorant molecules. The parameter space of odors (i.e., the molecular properties of odorants to which sensory neurons are sensitive) is not precisely defined. However, structure–activity studies (chain elongation, double-bond position, functionality) performed on noctuid olfactory neurons in vivo have been particularly enlightening in understanding that ligand–receptor interactions can behave according to conformational energy and electron distribution models and not merely to space filling [10–12]. In short, some compounds bind more strongly than others, and a plot of binding efficiency versus variation in specific molecular parameters results in the sensitivity spectrum of a given receptor protein. By interpreting the relative responses of several ORNs, each expressing different receptor proteins, a specific odor can be identified as unique.

15.6 PROOF OF CONCEPT FOR MULTICHANNEL EAG: DIFFERENTIAL SENSITIVITY BETWEEN SPECIES

The potential utility of using insect antennae as sensitive and *discriminating* biosensors for known agents in the environment stems from the successful use of the single-antenna EAGs that have been used to monitor the relative concentrations and even plume structures of high concentrations of known pheromone odorants (to which the selected EAG detector was tuned) [13–16]. An EAG response is thought to result from the summed generator (DC) potentials of individual olfactory receptor neurons that have been depolarized by exposure to an odorant, with the depolarizations of those receptor neurons closest to the recording electrode contributing more to the amplitude than those further away [17]. It has long been known, however, that EAG responses to any series of compounds form a graded series and are not compound specific. Even for sex pheromone components, the most specific of behaviorally active odors for insects, anthropogenic analogs differing significantly by chain length or functional group have been shown to be capable of generating significant EAG responses having amplitudes that are only 50–25% reduced from the natural pheromone compound itself [18].

Previous studies have hinted that EAG response spectra to a range of odorants can be at least a little species specific [19–21]. However, the concept of creating a discriminating EAG biosensor system by using an array of antennae that compares response amplitudes across the array had only very recently been suggested and attempted [22]. A discriminating EAG formed by antennae from multiple species would require exploiting only very slight species-to-species differences in the EAG response. Obviously, a single EAG signal from a very broadly tuned antenna responding in graded fashion to a wide variety of odorants cannot be used to discriminate one volatile compound from another.

Park et al. [22] created a discriminating EAG biosensor by utilizing differences in the EAG responses of several antennae from different species of insects monitored simultaneously. Their system made optimal use of the ability of the EAG to respond to sharp peaks in concentration such as those generated by the flux from individual odor strands in an odor plume passing over the antennae [13]; two or three depolarizations (i.e., odor strands) per second were registered. As shown by Baker and Haynes [13], the ability of an insect antenna to respond quickly to peak concentration in the strands in a plume makes it a much more sensitive system than slower responding artificial sensors whose polymers can only take a reading of time-averaged mean concentrations [23, 24].

The proof of the insect antennal array concept came when Park et al. [22] were able to discriminate most of the 20 different compounds they tested by comparing EAG response spectra across only five different species. The 20 compounds were presented as single puffs to individual antennae (Fig. 15.1), and the mean amplitudes of response to each compound over several replicates were calculated and normalized with respect to EAGs from a standard compound. The histograms developed from this process resulted in clear patterns of relative EAG amplitudes across the five species' antennae used. Most of these were discriminable by eye (Fig. 15.2).

Park et al. [22] then used simple clustering algorithms on the EAG amplitude information alone and confirmed what was apparent by eye. The cluster analysis

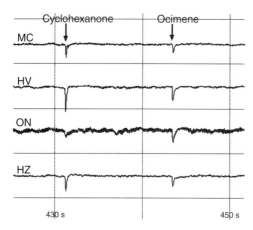

Figure 15.1   Differential EAG response amplitudes when either cyclohexanone or ocimene was puffed over antennae of *Microplitis croceipes* (MC; parasitic wasp), *Heliothis virescens* (HV; moth), *Ostrinia nubilalis* (ON; moth), or *Helicoverpa zea* (HZ; moth). Each horizontal division is 5 mV, and each vertical division is 10 s.

clearly demonstrated that compounds from similar classes (such as medium-chain-length aliphatic alcohols and longer chain length aliphatic alcohols) clustered together (Fig. 15.3). They also showed that compounds within classes were distinguishable based on characteristics such as chain length (Fig. 15.3).

Thus the proof of concept for using a multichannel EAG to discriminate between an arbitrary number of odors had been established. However, several technological issues needed to be addressed to make the EAG biosensor practical. Principal among these were the design of a multichannel electrode array and preamplifier optimized for field condition recording of the EAG and the automation of the odor discrimination analysis. Electronic optimization of the preamplifier will assure that the extreme sensitivity of the antenna can be exploited [i.e., maximize the signal-to-noise (SNR)]. Colocalization of antennae will assure that an odor strand in a turbulent air stream is sensed by each antenna at approximately the same time. Minimizing the size of the preamplifier will minimize perturbations to the air stream. Automation is important in order to take advantage of the inherent speed of the hybrid system and to fulfill many potential applications requiring

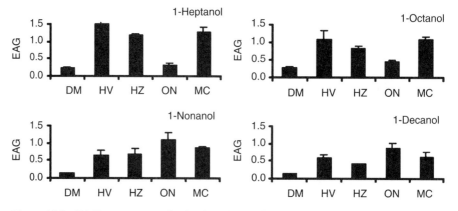

Figure 15.2   EAG response amplitudes from five different insect species' male antennae when presented with four different straight-chain alcohols: DM = *Drosophila melanogaster*, HV = *Heliothis virescens* (moth), HZ = *Helicoverpa zea* (moth), ON = *Ostrinia nubilalis* (moth), and MC = *Microplitis croceipes* (parasitic wasp).

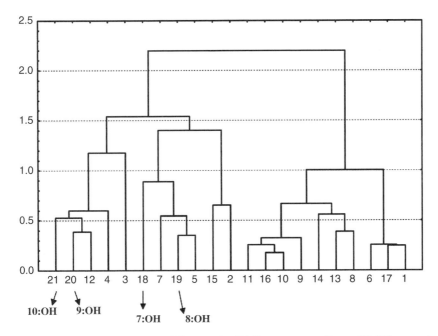

Figure 15.3   Cluster analysis of five-species EAG response profiles to 20 different volatile compounds. Decanol (10:OH) and nonanol (9:OH) cluster together but separate from octanol (8:OH) and heptanol (7:OH). Visually this should be apparent in Figure 15.2. Compounds 8–11 and 13 are all monoterpenes, sitting in a separate cluster of EAG antennal array response patterns than the aliphatic alcohols.

autonomous operation (e.g., roving robots searching for land mines). Early answers to each of these challenges are described in the following sections.

15.7   DESIGN OF MULTICHANNEL EAG SYSTEM

In the design of an EAG acquisition system, it is desirable to maximize the SNR obtained. It is also desirable to eliminate or limit as many sources of interference as possible. Minimization of noise and interference allows maximal information to be extracted from the signal during postacquisition processing. The source of noise is in the physical components that take part in processing the input signal, while interference originates outside the signal path. Sources of interference include 60-Hz power-line radiation, radio-frequency radiation, electronic changes due to temperature drift, mechanical agitation, and power supply ripple. Noise sources are numerous but often include thermal (Johnson) noise, $1/f$ or flicker noise, shot noise, and quantization noise.

A further source of interference arises from inside the "sensor" (i.e., the antenna), which contains not only chemoreceptors but also mechanical receptors. As air is swept across the antenna, mechanoreceptors are stimulated, giving rise to a source of interference in this application that cannot be eliminated. A source of signal noise is likely to be related to the origin of the signal. The EAG voltage changes are caused by ensembles of individual action potentials. The number of neurons involved appears to be on the order of 10^5, probably an order of magnitude less for any individual response (referred to below, by convention, as a *depolarization*). Together, the number of neurons and their firing rate

Antennae

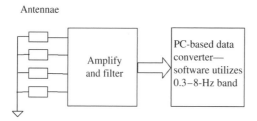

Figure 15.4 EAG acquisition system block diagram.

present a fundamental limit to the amount of information that can be obtained from the signal. Despite the inherent sources of distortion in the sensor, it is still desirable for research purposes to provide the most accurate EAG measurements possible by limiting electrical sources of interference and noise.

Biopotential amplifiers are traditionally located remotely from the signal source, which is connected via shielded cables to a differential amplifier. A long cable has a tendency to pick up interference, or generate interference as a result of mechanical agitation, known as the triboelectric effect. Further, a differential amplifier (two amplifiers per channel) is needed to reduce power line interference. In the EAG application, a single ended active probe can amplify the signal very close to its source, which is also physically small, and send a much higher power signal through a cable for further amplification and filtering. The power gain is high enough to overpower interference sources. For instance, the power gain of the EAG active probe described here, with 500 kΩ input resistance, is 2×10^7. Another advantage to this approach is that only one preamplifier per channel is needed. Further, amplifier bias current is provided through the antenna itself without the need for biasing circuitry.

A good way to begin a design is to first draw a general block diagram (Fig. 15.4). The hardware component to be described here is the "amplify-and-filter" section. The PC-based converter is readily available from many vendors. The amplifier interfaces with the antennae at its inputs and the data converter at its outputs. To begin designing, it is first necessary to characterize both the antennae and the data converter.

Figure 15.5 shows a Thevenin equivalent circuit for a typical insect antenna. The antenna acts as a voltage source, V_A, in series with a source resistance, R_A. The Thevenin equivalent resistance can be expected to vary from approximately 100 to 10 MΩ. This value also varies as the preparation becomes less viable through dehydration and so cannot be easily predicted. Noise that is present across any source of real, finite impedance is termed Johnson noise, which is created by thermal motion of charge carriers. This noise cannot be avoided without reducing the temperature of the source. Johnson noise has equal power density at all frequencies (up to the terahertz range) equal to $4kT$ watts per hertz.

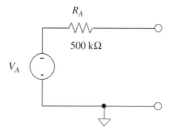

Figure 15.5 Thevenin equivalent circuit for insect antenna.

The root-mean-square (RMS) value of the Johnson noise voltage spectral density across any resistor can be calculated using

$$V_J(f) = \sqrt{4kTR} \qquad \text{V}/\sqrt{\text{Hz}}$$

where $k = 1.38 \times 10^{-23}$ is Boltzmann's constant and T is temperature in kelvin. For example, the noise level of the resistor value $R_A = 500 \text{ k}\Omega$ from 0.1 to 10 Hz at 298 K is

$$V_J = \sqrt{(4)(1.38 \times 10^{-23})(298)(500 \times 10^3)(10 - 0.1)} = 285 \text{ nV RMS}$$

The voltage V_A may also be further characterized. The EAG signal has an approximate "active" level of 400 μV RMS and occupies the frequency range of about 0.1–10 Hz. The maximum signal swing expected is approximately 10 mV peak to peak (p-p). In addition, a large DC offset on the order of 100 mV is typically present on the EAG signal.

A useful comparison between the signal and the noise is the SNR, often expressed in decibels:

$$\text{SNR} = 10 \log_{10}\left(\frac{S}{N}\right) = 20 \log_{10}\left(\frac{V_S}{V_N}\right)$$

where S represents the signal power, N represents the noise power, and V_S and V_N are the signal and noise voltages, respectively. Here, for example,

$$\text{SNR} = 20 \log_{10}\left(\frac{400 \times 10^{-6}}{285 \times 10^{-9}}\right) = 63 \text{ dB}$$

Now the input characteristics of the data converter can be summarized. A simplified analysis can be divided into two parts: the amplifier and the digitizer. Although it is difficult to know the exact analog signal input chain to the analog-to-digital converter (ADC) within a proprietary digitizer design, a good model may be constructed by treating each analog channel as an independent amplifier followed by a first-order low-pass filter (LPF), with negligible noise above the corner frequency of the LPF. Because the floating-point digital output of the digitizer software driver is scaled to the input voltage, the gain of the amplifier is unknown and does not need to be determined. If necessary, it would be possible to determine the gain by measuring the quantization levels and opening the digitizer case to find the ADC part number to determine its input range.

The greatest concern in designing the amplifier is the noise it adds to the input signal. Generally, the input noise voltage V_N and sometimes the input noise current I_N of an amplifier are specified as functions of frequency. Often V_N consists of so-called *flicker noise* and *white-noise* components. For input source impedances that are very high I_N is a concern. In this case, it is expected that the input of the digitizer amplifier will be driven by a low-impedance amplifier output from the "amplify-and-filter" stage. Flicker noise density can be approximated with the equation.

$$V_{Nf}(f) \cong K\sqrt{\frac{1}{f}}$$

White noise, by definition, has a flat spectral density, as in the relation

$$V_{NW}(f) \cong N_W$$

The RMS values of two independent noise sources in series can be represented as a single noise source with an RMS value equal to the sum of their squares. Therefore, the total equivalent input noise density is given by the formula

$$V_N(f) = \sqrt{K^2 \frac{1}{f} + N_W^2}$$

Further, the bandwidth of any real amplifier is limited. A simple way to model this bandwidth limitation is to place a first-order LPF on the output of the amplifier (gain 1). The "output" noise spectral density curve shown in Figure 15.6 is an estimate of the amplifier noise present at the scaled output of the digitizer with the amplifier input grounded. Such a curve is plotted in the figure. Parameters of the input amplifier and multiplexer chain are as follows:

Input referred white-noise spectral density	$250 \text{ nV}/\sqrt{\text{Hz}}$
Input referred flicker noise coefficient	$3.75 \times 10^{-6} \text{ W}$
Amplifier bandwidth	413 kHz
Input impedance	$100 \text{ G}\Omega/100 \text{ pF}$
Bipolar input voltage ranges	$\pm 50 \text{ mV}, \pm 0.5 \text{ V}, \pm 5 \text{ V}, \pm 10 \text{ V}$

Parameters of the digitizer are as follows:

Analog input channels	16
Maximum sample rate (total for all channels)	200,000 samples/s
Sampling resolution	16 bits
Accuracy	± 3 LSBs (least significant bits)

Typically, the signal power and the noise power are compared within a specified bandwidth, where the signal power exists. A noise voltage spectral density may be integrated across a frequency range using the equation

$$V_{\text{NBW}} = \sqrt{\int_{\text{BW}} V_N^2(f)\, df}$$

Fortunately, in this application, quantization noise is not a problem. At a sample rate of 12,500 samples/s, 67 aliased segments of the amplifier white noise integrate from 0.1

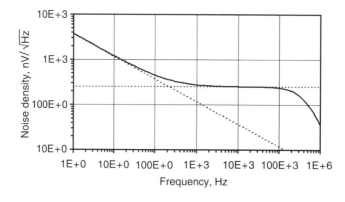

Figure 15.6 Spectral density estimate of digitizer output noise.

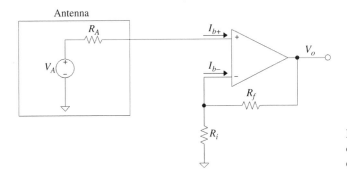

Figure 15.7 Preamplifier connected to antenna equivalent circuit.

to 10 Hz, resulting in 6.4 μV RMS. The $1/f$ noise integrates to 8.0 μV RMS. The total noise is

$$V_N = \sqrt{(6.4 \times 10^{-6})^2 + (8.0 \times 10^{-6})^2} = 10.2 \, \mu\text{V RMS}$$

Without precise knowledge of the source impedance of the antenna, an amplifier with very high input impedance is desirable to measure the EAG voltage accurately. Illustrated in Figure 15.7 is a model of an antenna connected to a noninverting high-input-impedance amplifier. The performance of this amplifier construct is highly dependent on the performance of the operational amplifier (op amp). The performance of an op amp can be evaluated by examining parameters assigned to its nonideal behavior.

Every op amp requires bias current to operate its input amplifiers. This is usually specified on manufacturer's specifications as "input bias current." Further, each individual amplifier has an impedance to ground and an impedance between itself and other amplifier inputs on the same device. These are known as "common-mode input impedance," and "differential-mode input impedance,'" respectively. The degree of power supply ripple and noise feedthrough to the op amp output is usually specified as the power supply rejection ratio (PSRR). Other parameters critical to this application include "input current noise" and "input voltage noise." These will be explained in further detail below.

Before selecting an amplifier, it is a good idea to list the few critical specifications, which will aid in generating a list of candidate devices. Specifications of major concern are listed below. A point should be made that datasheet noise data are representative of a typical amplifier and are only approximate. Thus, consideration should be made for variability between individual parts.

Parts per package	4 in this application
Input bias current	≤ 1 nA
R_{CM}	≥ 1 GΩ
In-band current noise	≤ 40 fA RMS
Voltage noise	Evaluate
PSRR	Maximize

Noise sources in the noninverting amplifier configuration (excluding thermal noise in the antenna) (Fig. 15.7) include the resistors and the amplifier. Amplifier noise is measured and usually published in the op amp specifications. Measured amplifier noise includes input referred noise voltage and input referred noise current. In this application, the noise at the amplifier output can be approximated using the equation

$$V_o \cong \left(1 + \frac{R_f}{R_i}\right) V_{\text{NBW}}$$

This equation applies only because the component values and amplifier specifications have been chosen properly. For more details on op amp noise analysis see Clayton [25] and the application note by Bob Clark [26] listed in the references.

Two particular op amps considered for this design were the AD8608 and the LMC6084. Integration of the noise from 0.1 to 10 Hz results in the following noise specifications (data shown for the LF147, used in a circuit second stage, will become relevant below):

Op amp	PSRR, dB	I_N, fA/$\sqrt{\text{Hz}}$	K, nW	V_{NBW}, nV RMS
AD8608	77	10	91.7	197
LMC6084	72	0.2	285	611
LF147		10	164	353

The SNR at the output of each amplifier has been calculated for an "active" signal input of 400 μV RMS in the preamplifier for $R_i = 10$ kΩ and $R_f = 100$ kΩ and plotted in Figure 15.8 as a function of the input resistance. The SNR is a decreasing function of the input resistance because the available signal power is inversely proportional to the input resistance but the noise power is constant.

A figure of merit for an amplifier is its *noise figure*. The noise figure is expressed in decibels as

$$\text{NF} = 10\log_{10}\left(\frac{S_i/N_i}{S_o/N_o}\right)$$

where S_i/N_i is the signal-to-*thermal*-noise ratio at the input of the amplifier. The noise figure is a measure of how much noise the amplifier adds to the signal over the thermal noise present at the input. The lower the noise figure, the better the amplifier. A perfect amplifier has a noise figure of 0 dB. In this case, the NF can be calculated

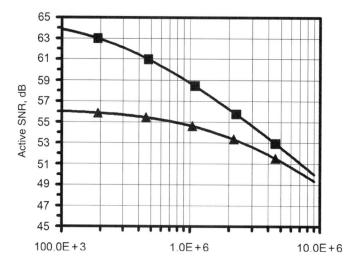

Figure 15.8 Preamplifier SNR as function of antenna resistance: ■, AD8608; ▲, LMC6084.

using the equation

$$\mathrm{NF} \cong 10\log_{10}\left(\frac{V^2_{\mathrm{NBW}}}{4kTR_A\mathrm{BW}}\right)$$

The noise figure of each amplifier considering input current noise and resistor network noise (i.e., not the equation above) as a function of the input resistance is plotted in Figure 15.9a. The noise figures for the AD8608 and LMC6084 are decreasing functions of the input resistance. There are two main contributors to noise for these two amplifiers: V_N and the Johnson noise of the antenna. As the resistance of the antenna increases, so does the Johnson noise voltage, which eventually dominates the noise at the output of the op amp. This means that the op amp does not add much significant power to the noise signal, resulting in a near 0-dB noise figure at 10 MΩ. Conversely, at $R_A = 100\,\mathrm{k}\Omega$, the Johnson noise voltage of the antenna is reduced by a factor of 10, and the amplifier input voltage noise dominates.

It is now appropriate to determine the necessary amount of preamplifier gain. Following the law of superposition, the system noise figure can be calculated by summing the noise sources at the output of the acquisition system and comparing it to the component at the output due to thermal noise at the antenna. The system output noise voltage V_{NS} can be calculated using the equation

$$V_{NS} = \sqrt{(V_{e0}A_2A_1)^2 + (V_{e1}A_2)^2 + V^2_{e2}} = \sqrt{A^2_1A^2_2 4kTR_A\mathrm{BW} + A^2_1A^2_2 V^2_{\mathrm{NBW}} + V^2_{e2}}$$

where

V_{e0} = antenna input thermal noise voltage

V_{e1} = equivalent output noise of preamplifier

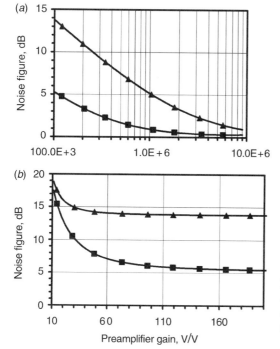

Figure 15.9   (*a*) Preamplifier noise figure as function of antenna resistance and (*b*) system noise figure as a function of pre-amp gain ($R_A = 100\,\mathrm{k}\Omega$): ■, AD8608; ▲, LMC6084.

V_{e2} = equivalent output noise of digitizer

A_1 = preamplifier gain

A_2 = digitizer gain = 1

We have already calculated V_{e2}, the equivalent output noise of the digitizer. This is 10.2 μV. Since the output of the digitizer in the software is equal to the measured input, the gain of the digitizer, A_2, is 1. The system noise figure can be calculated using the formula

$$\text{NF} = 10\log_{10}\left(\frac{V_{NS}^2}{A_1^2 A_2^2 4kTR_A \text{BW}}\right) = 10\log_{10}\left[1 + 2.41 + \left(\frac{80.0}{A_1}\right)^2\right]$$

where the worst-case antenna resistance value $R_A = 100$ kΩ has been selected. The first of the three terms is the thermal noise; the second is the added preamplifier noise, which has been minimized through selection of the amplifier; and the third contribution to the noise figure is the digitizer noise. To approach a perfect amplifier, effort must be made to minimize the second two terms. The second term is minimized by selecting the appropriate preamplifier part. The third term is minimized by selecting the gain. Intuitively, A_1 needs to be increased until the noise due to the digitizer becomes insignificant compared to preamplifier and thermal noise.

When the gain is adjusted by varying R_i, the NF varies as in Figure 15.9b. It can be seen that the noise figure asymptotically approaches its minimum value as the gain is increased. But unfortunately, the DC voltage offset at the input can be expected to be up to 100 mV. The better device, the AD8608, can only operate at up to ±3 V. So the maximum gain possible is approximately 20 V/V. Without the ability to use a high-pass filter (HPF) in this configuration with a reasonable capacitance value, the addition of a second amplification stage is needed. So the first-stage amplifier design is finalized with a gain of 21.

To squeeze out another 7 dB from the system, another gain stage is needed. But to eliminate the large DC offset, it is necessary to use a HPF. A 0.1-Hz first-order HPF will remove the DC offset and has a manageable settling time ($\tau = 1.6$ s) but distorts frequencies of interest in both magnitude and phase from 0.1 to ~1 Hz. Even though component tolerances for a HPF are on the order of ±10%, it is possible to remove most of the distortion in a software-implemented compensation filter [finite impulse response (FIR) described below]. The SNR measurements are subject to variation. Thus, a safe margin should be created in case the digitizer noise is higher or the preamplifier noise is lower than expected. Also, digitizer flicker noise is higher at lower frequencies, but the signal is attenuated more by the HPF. Using a gain of 11 V/V in the second stage results in an overall gain of 231 V/V and a maximum peak-to-peak signal range of 2.31 V, which is just outside the 2-V range of the digitizer. A HPF with a corner frequency of 0.1 Hz and a LPF of 20 Hz is used to help reject 60 Hz as well as an antialiasing filter that removes high-frequency preamplifier noise and other high-frequency interference that could alias into the 0.1–10-Hz band. To deal with all types of interference sources, it is desirable to attenuate as much as possible signal frequencies above $f_s/2$, where f_s is the sampling frequency. For instance, in our laboratory, a large (~10-mV) troublesome, wide-bandwidth 50-kHz interference signal is present that must be highly attenuated to avoid being aliased into the 0.1–10-Hz band when sampled at 12,500 samples/s. To preserve simplicity, we use a first-order filter (Fig. 15.10).

Figure 15.10 Second-stage amplifier schematic; first-order bandpass filter with gain.

The general-purpose junction fie[l]d-effect transister (JFET) quad op amp, the LF147, is used for the second stage. Using the worst-case antenna resistance $R_A = 100\,\text{k}\Omega$, the equations for the two-stage system output noise and noise figure are as follows:

$$V_{NS} = \sqrt{(V_{e0}A_3A_2A_1)^2 + (V_{e1}A_3A_2)^2 + (V_{e2}A_3)^2 + V_{e3}^2} = 55.3\,\mu\text{V}$$

$$\text{NF} = 10\log_{10}\left[1 + \left(\frac{V_{e1}}{V_{e0}A_1}\right)^2 + \left(\frac{V_{e2}}{V_{e0}A_1A_2}\right)^2 + \left(\frac{V_{e3}}{V_{e0}A_3A_2A_1}\right)^2\right]$$

$$= 10\log_{10}(1 + 2.41 + 0.018 + 0.121) = 5.50\,\text{dB}$$

Variable	V_{e0}	A_1	V_{e1}	A_2	V_{e2}	A_3	V_{e3}
Equation	$\sqrt{4kTR_A\text{BW}_1}$		A_1V_{NBW1}		A_2V_{NBW2}		
Value	$0.127\,\mu\text{V}$	21	$4.14\,\mu\text{V}$	11	$3.9\,\mu\text{V}$	1	$10.2\,\mu\text{V}$

There are four terms involved in the above equations: antenna thermal noise, amplifier 1, amplifier 2, and the digitizer. It can be seen that most of the noise is due to the thermal noise and the stage 1 amplifier. For higher values of R_A, the noise performance of stage 1 will increase substantially.

Care is also taken to isolate the preamplifier from the power supply. Much of the power supply noise is removed by a voltage regulator and the PSRR of the amplifier. Fortunately, power supply ripple is $\geq 60\,\text{Hz}$ and will be entirely removed by digital filtering after acquisition. The final EAG amplifier has the block diagram shown in Figure 15.11. The hardware design is complete; however, some signal processing is performed in the software. A block diagram of the signal chain connected to the EAG amplifier output is shown in Figure 15.12.

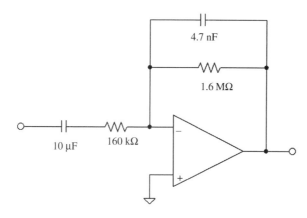

Figure 15.11 EAG amplifier block diagram: R_A, source impedance of antenna; HPF, high-pass filter; LPF, low-pass filter; f_c, cutoff frequency; G, gain.

Because we have more than adequate processing power in the average PC, a very sharp infinite-impulse-response (IIR) Butterworth LPF is employed (characteristics shown in Fig. 15.13). This filter will reduce any 60-Hz signal to well below the in-band amplifier noise level. After decimation, a FIR compensation filter removes distortion introduced by the 0.1-Hz HPF and the 20-Hz LPF in the second-stage amplifier. Nonuniform phase delay in the FIR filter is also compensated for. Compensation is applied from 0.1 to 30 Hz. It is after the compensation filter that the data are considered properly acquired and are recorded to disk for future review. The final design, after assembly on a printed circuit board, is shown in Figure 15.14. The circuit is contained in a slim aluminum case, and four antennae make electrical contact with the four amplifier input electrodes and the common ground electrode via conductive gel (as used with clinical electrocardiogram electrodes). The amplifier is connected via a seven-conductor cable (V^+, V^-, ground, and four single-ended data channels) to the data acquisition card in a laptop computer for digital signal processing (Fig. 15.13) and execution of the algorithm which discriminates between odors based on the four-channel EAG data, as described in the following section.

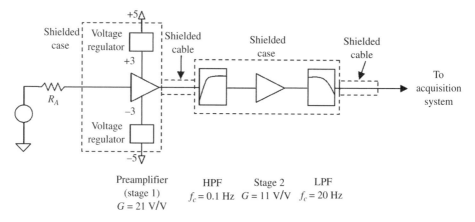

Figure 15.12 Data acquisition and processing system block diagram: FIR, finite-impulse-response filter.

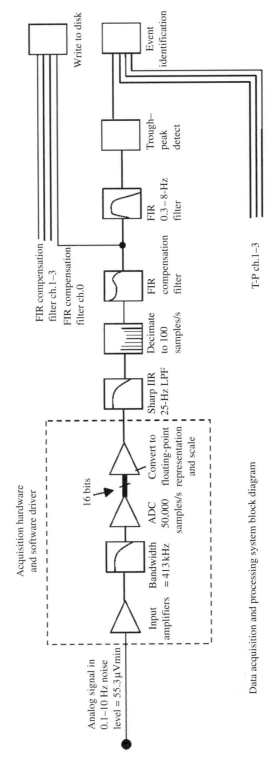

Figure 15.13 A 25-Hz Butterworth LPF, 20th-order frequency response.

259

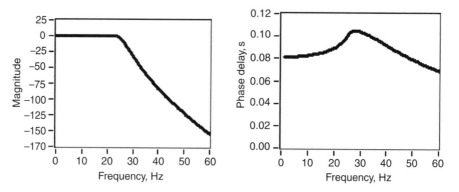

Figure 15.14   Final circuit design executed on custom printed circuit board using surface-mount components to minimize outside dimensions of entire package. The top of the case is removed; shown next to a U.S. dime for size comparison. When assembled, the case is less than 5 mm thick.

15.8   SIGNAL PROCESSING OF MULTICHANNEL EAG: AUTOMATED ODOR DISCRIMINATION

After the data are acquired, they are further processed to discriminate between odors. This is accomplished by identifying and measuring features of each response (depolarization) and then applying a relatively common pattern recognition technique. It should be noted that several strategies could be employed to discriminate odors based on EAG data, and the methods described below are only one solution. Other approaches are being investigated in our laboratory in an effort to maximize robustness, speed, and ease of implementation.

The objective of the signal processing level of the system is to identify EAG depolarizations, measure their amplitude, and group time-correlated depolarizations recorded from the four antennae. The time-correlated depolarizations correspond to a single odor strand passing the array of antennae. The correlated groupings of depolarizations are termed *events*. In order to associate the responses of the antennae with a particular odor, the system must first be trained. Training is accomplished by collecting data while subjecting the antennae to various known odor sources. After training has been completed, a classifier uses the training data to identify the odor that gave rise to the recorded event. The classification is performed on-line within 1 s after the data have been acquired.

The 0.3–8-Hz filter is noncausal and has approximately the frequency response depicted in Figure 15.15. The filter was constructed empirically and assists in peak-to-trough

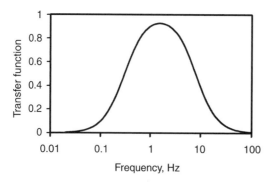

Figure 15.15   The 0.3–8-Hz bandpass filter frequency response.

detection by creating troughs that precede the peaks (purposeful distortion of the signal) and by smoothing the waveform. Most bandpass functions would accomplish these tasks.

A peak detection algorithm finds maxima and minima of polynomials fit to the data; the result is identification of the locations and values of the depolarizations. An example of this step is shown in Figure 15.16. The data are greatly reduced as only the peak-to-nearest-preceding-trough amplitude is retained, along with the time of the peak. These data are then compared to a threshold. Peak–trough events with an amplitude larger than a user-adjustable threshold are kept, while those smaller are considered to be unreliable and ignored.

Recall that events are comprised of the responses of the four antennae to a single strand of odor, which passes transiently across the array. Because the antennae cannot occupy the same point in space, the time of arrival of the odor strand and therefore the time of the four EAG response peaks are not coincident but depend on the air speed and the spacing of the antennae relative to the wind direction (and possibly, to a lesser degree, on species). Event identification is accomplished by associating peak–trough events that are close in time. The user may adjust this correlation time. The largest depolarization within a presumed event is then evaluated by another threshold to retain only events containing significant energy on at least one channel. Figure 15.17 depicts the result of evaluating the four EAG signals for "near-simultaneous" EAG responses which would comprise an event. Only the trough-to-peak amplitudes for the four channels are plotted (filled symbols), along with markers to denote the time of detected events (open circles). Note that in this example the event time window is 16 ms, and the event markers are plotted 16 ms following the last depolarization which is a component of each detected event. That is, depolarizations recorded more than 16 ms apart would not be considered to arise from the same odor strand. The choice of time window is influenced by wind speed and the distance between the most upwind and downwind antennae in the array.

Classification techniques utilize "features" and "classes," where it is hoped to classify an event based on its features. In this application, the features used to describe the odor are the depolarization (peak-to-trough) amplitudes of the four preconditioned EAG signals. Thus, each event is described by four features. Each feature can be considered one dimension in a multidimensional feature space. Odors that can be discriminated have high probability density in different regions of the feature space. Bayes's theorem can be used for classification if the probability distributions of each class are known. Ideally, during training, a four-dimensional histogram could be produced. Unfortunately,

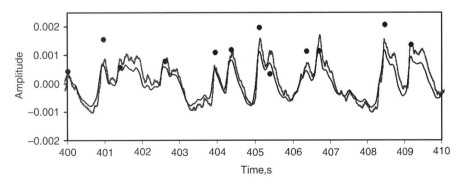

Figure 15.16   Segment from typical one-channel EAG recording. Light trace is the raw signal, dark trace is after filtering. Filled circles indicate times and trough-to-peak magnitudes of responses which were above the user-defined threshold (100 μV in this case).

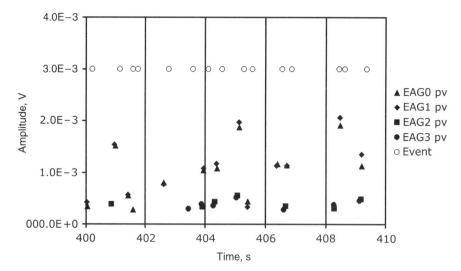

Figure 15.17   Trough-to-peak values (filled symbols) and event time markers (open circles) obtained with a 16-ms correlation time. Events markers are located 16 ms after the last trough-to-peak value contained in that event.

the number of data points required to create an accurate four-dimensional histogram (or up to 16 dimensions) is prohibitive. Thus, alternate classification techniques are available. One such technique, known as the "*k*-nearest-neighbor" technique, was employed here.

The measured features of each event are plotted in four-dimensional space, where each dimension represents one channel (i.e., antenna). The training data are collected first, which involves exposing the sensor to known odors, sequentially, until a requisite number of training events have been collected. In the example presented here, 100 training events were recorded for each odor, which took 1–2 min per odor; any number of odors can be contained in the training set.

The classification technique uses the weighting function $X_{d1\text{-}d10} = 1/(d + 20 \ \mu\text{V})$, where d is the "city block" distance of the point to be classified to training data points in the four-dimensional feature space. The city block distance between two points is the sum of the absolute value of the difference between every dimension. For instance, in just two dimensions, $d = |x_2 - x_1| + |y_2 - y_1|$, which is the distance you would have to travel in most places in Chicago to get from p_1 to p_2. Ten values of X are calculated for each odor in the training data. These 10 values are summed for each training odor, and the unknown point is classified as being the training odor associated with the maximum sum. Presently, the classifier assumes equal prior probabilities, and the training data are not normalized. Also, the coefficient in the weighting function is chosen manually. Automating these aspects of the algorithm is a future goal.

The results of a simple classification experiment are shown in Figure 15.18. Here, the system was trained with three odors. The odor stream was created by placing a drop of one of three volatile liquids on each of three pieces of filter paper that were placed, in turn, about 5 ft upwind in an indoor wind tunnel, giving rise to odor A, odor B, and odor AB. Exhaust from the wind tunnel was vented to the outside, preventing an odor from recirculating in the air stream. Due to turbulent flow, at times the antennae are stimulated by the introduced odor strands, while at others, the array is detecting only components of the room air. As a result, the training data are not pure and represent mixtures of room air and the intended odorant. This presents a problem, because an

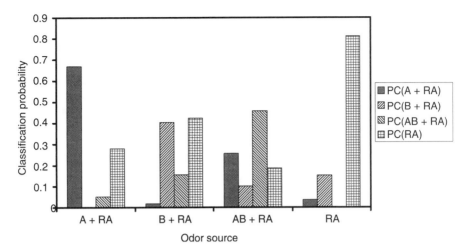

Figure 15.18   Classification results when system was trained on three odors (A, B, and AB) as well as ambient room air that carries odors in wind tunnel (RA). The four-antenna array was then exposed to each of the four conditions until approximately 100 events had been detected for each condition. Every event was classified as A, B, AB, or RA.

unknown portion of the events in the training set contain events due to room air only. The classification scheme described here is a forced-choice algorithm, and with three training sets, any event will be classified as one of those three odors, even if the stimulus is actually a component of the background room air. By adding a fourth training set comprised of responses to room air (RA) only, it is more probable that RA will be correctly classified. The results of Figure 15.17 are typical when this technique is applied. It can be seen that odor A is correctly classified 95% of the time, odor B is correctly classified 83% of the time, and odor AB is correctly classified 64% of the time. Odor AB was a mixture of odor A and odor B, and while the pure mixture should give rise to a novel EAG response, the presumed classification mistakes occurring when odor AB was introduced were components odor A or odor B, as opposed to RA.

It is important to remember that each classification event takes less than 1 s. Unfortunately, even if the odors are highly distinguishable from RA, it is still possible for RA to be incorrectly classified as one of the training odors. One way to address this problem is to remove the clean-air events from the known odor training sets by reclassifying each training event as either clean air or the intended odor and then using these "verified" training data to classify unknown events. This represents just one possible refinement to this interesting pattern recognition problem.

15.9   FUTURE TRENDS

We have described the advantages, applications, technology, and methodology for incorporating a living olfactory organ into a hybrid-device biosensor. *Fast*, *sensitive*, and *discriminating* are all words that can be used to describe this sensor, but there is much yet to be done. Work is ongoing to refine the algorithm used to classify odors in order to increase accuracy. This includes the evaluation of features of the EAG signal other than peak amplitude and more robust pattern recognition strategies. Automating selection of thresholds and filter parameters and providing statistics describing the confidence of a given classification are also being pursued. One fundamental challenge is the absence of a "gold standard" for odor strand composition under most field and laboratory conditions; that is, it is very difficult to know what the true odor is in a packet of air that elicits a response from the four-channel EAG biosensor. One method employed by Park et al. [22] was to bifurcate the air stream, passing half to an antenna and the other half to the flame ionization detector (FID) of a gas chromatograph. The arrival of a packet of air to each sensor (antenna and FID) was time locked, allowing unambiguous identification of the odor eliciting each antenna response. However, this arrangement is cumbersome and not easily translated to other settings.

Another aspect under development is the inclusion of wind direction and wind speed sensors. For an array of static sensors, for example, spaced 10 m apart along the edge of a field, wind information and triangulation can be used to locate the distant source of an odor. Incorporated into a single sensor, wind information could be used to guide a robotic mobile autonomous odor seeker (MAOS) in situations where approaching the odor source would present a high risk to people or trained dogs, for example, locating land mines or toxic chemical spills. Using Global Positioning System (GPS) technology, telemetry, and cameras, such a roving sensor could be deployed in military reconnaissance and search-and-rescue missions or to gather information near natural disaster areas. The extreme sensitivity could provide unexpected diagnostic benefits. The applications of olfactory biosensors are potentially quite broad, and it will likely be some time before artificial olfactory sensors can compare in performance. Perhaps the greatest limitation on this approach is the short-lived nature of the antennae (less than 1 h when removed from the insect); increasing the useful life of the hybrid sensor is also being pursued. Potential strategies include leaving the antenna attached to the insect, which increases the lifespan to several days, or using a supported olfactory receptor neuron culture system. The bioengineering challenges of this last approach are significant.

REFERENCES

1. J. J. PANCRAZIO, J. P. WHELAN, D. A. BORHODER, W. MA, AND D. A. STENGER, Development and application of cell-based biosensors. *Ann. Biomed. Eng.* 27:697–711, 1999.
2. J. WHITE, T. A. DICKINSON, D. R. WALT, AND J. S. KAUER, An olfactory neuronal network for vapor recognition in an artificial nose. *Biol. Cybernet.* 78(4):245–251, 1998.
3. S. E. STITZEL, L. J. COWEN, K. J. ALBERT, AND D. R. WALT, Array-to-array transfer of an artificial nose classifier. *Anal. Chem.* 73(21):5266–5271, 2001.
4. K. J. ALBERT, M. L. MYRICK, S. B. BROWN, D. L. JAMES, F. P. MILANOVICH, AND D. R. WALT, Field-deployable sniffer for 2,4-dinitrotoluene detection. *Environ. Sci. Technol.* 35(15):3193–3200, 2001.
5. T. A. DICKINSON, J. WHITE, J. S. KAUER, AND D. R. WALT, Current trends in "artificial-nose" technology. *TIBTECH* 16:250–258, 1998.
6. V. TORRE, J. F. ASHMORE, T. D. LAMB, AND A. MENINI, Trasduction and adaptation in sensory receptor cells. Review. *J. Neurosci.* 15(12):7757–7768, 1995.
7. P. A. ANDERSON AND B. W. ACHE, Voltage- and current-clamp recordings of the receptor potential in olfactory receptor cells *in situ*. *Brain Res.* 338(2):273–280, 1985.
8. T. V. GETCHELL, F. L. MARGOLIS, AND M. L. GETCHELL, Perireceptor and receptor events in vertebrate olfaction. Review. *Prog. Neurobiol.* 23(4):317–345, 1984.
9. G. LAURENT, Olfactory network dynamics and the coding of multidimensional signals. *Nature* 3:884–895, 2002.
10. T. LILJEFORS, B. THELIN, AND J. N. C. VAN DER PERS, Structure-activity relationships between stimulus molecules and response of a pheromone receptor cell in turnip moth, *Agrotis segetum:* Modifications of the acetate group. *J. Chem. Ecol.* 10:1661–1675, 1984.
11. T. LILJEFORS, B. THELIN, J. N. C. VAN DER PERS, AND C. LÖFSTEDT, Chain-elongated analogues of a pheromone component of the turnip moth, *Agrotis segetum.* A structure-activity study using molecular mechanics. *J. Chem. Soc. Perkin Trans II*, 1957–1962, 1985.
12. T. LILJEFORS, M. BENGTSSON, AND B. S. HANSSON, Effects of double-bond configuration on interaction between a moth sex pheromone component and its

receptor: A receptor-interaction model based on molecular mechanics. *J. Chem. Ecol.* 13: 2023–2040, 1987.

13. T. C. BAKER AND K. F. HAYNES, Field and laboratory electroantennographic measurements of pheromone plume structure correlated with oriental fruit moth behaviour. *Physiol. Entymol.* 14:1–12, 1989.

14. A. E. SAUER, G. KARG, U. T. KOCH, J. J. DE KRAMER, AND R. MILLI, A portable EAG system for the measurement of pheromone concentrations in the field. *Chem. Senses* 17:543–553, 1992.

15. G. KARG AND A. E. SAUER, Spatial distribution of pheromone in vineyards treated for mating disruption of the grape vine moth *Lobesia botrana* measured with electroantennograms. *J. Chem. Ecol.* 21:1299–1314, 1995.

16. J. N. C. VAN DER PERS AND A. K. MINKS, A portable electroantennogram sensor for routine measurements of pheromone concentrations in greenhouses. *Ent. Exp. Appl.* 87:209–215, 1998.

17. K.-E. KAISSLING, Insect olfaction. In *Handbook of Sensory Physiology*, Vol. 4, L. M. BEIDLER, Ed. Springer Verlag, Berlin, pp. 351–431, 1971.

18. W. L. ROELOFS AND A. COMEAU, Sex pheromone perception: Electroantennogram responses of the red-banded leaf roller moth. *J. Insect Physiol.* 17:1969–1982, 1971.

19. B. H. SMITH AND R. MENZEL, The use of electroantennogram recordings to quantify odourant discrimination in the honey bee, *Apis mellifera. J. Insect Physiol.* 35:369–375, 1989.

20. J. H. VISSER, P. G. M. PIRON, AND J. HARDIE, The aphids' peripheral perception of plant volatiles. *Ent. Exp. Appl.* 80:35–38, 1996.

21. J. H. VISSER AND P. G. M. PIRON, Olfactory antennal responses to plant volatiles in apterous virginoparae of the vetch aphid *Megoura viciae. Ent. Exp. Apll.* 77:37–46, 1997.

22. K. C. PARK, S. A. OCHIENG, J. ZHU, AND T. C. BAKER, Odor discrimination using insect electroantennogram responses from an insect antennal array. *Chem. Senses* 27:343–352, 2002.

23. D. R. WALT, T. DICKINSON, J. WHITE, J. KAURER, S. JOHNSON, H. ENGELHARDT, J. SUTTER, AND P. JURS, Optical sensor arrays for odor recognition. *Biosens. Bioelectron.* 13:697–699, 1998.

24. N. KASAI, I. SUGIMOTO, M. NAKAMURA, AND M. KATOH, Odorant detection capability of QCR sensors coated with plasma deposited organic films. *Biosens. Bioelectron.* 14:533–539, 1999.

25. G. B. CLAYTON, *Operational Amplifiers*, 2nd ed. Butterworth-Heinemann, Oxford, 1979.

26. B. CLARK, Analog Devices Application Note 253, "Find Op Amp Noise with Spreadsheet," http://www.analog.com/UploadedFiles/Application_Notes/353070850AN253.pdf.

CHAPTER 16

RECONFIGURABLE RETINA-LIKE PREPROCESSING PLATFORM FOR CORTICAL VISUAL NEUROPROSTHESES

Samuel Romero, Francisco J. Pelayo, Christian A. Morillas,
Antonio Martínez, and Eduardo Fernández

16.1 INTRODUCTION

Most of the sensory information we perceive comes from our eyes. We rely on this sense to perform almost every daily task, such as driving, walking, grasping objects, and so on. Blindness deprives millions of individuals of receiving an enormous amount of precious information on color, distances, shapes, motion, textures, and so on. A number of research groups around the world aim to reach a remedy and restore vision to blind people. The developments of intracortical arrays of electrodes pave the way for direct electrical stimulation of the visual areas of the brain. This chapter presents an approach to feed these electrodes with visual information and to automatically translate image sequences into the neural language of the primary visual cortex.

16.2 HUMAN VISUAL PATHWAY AND BLINDNESS

The human visual system is a complex structure composed of several processing stages that lead from the reception of light at the eye to the production of visual sensations at the brain. This chain of processing steps is known as the "visual pathway." This path includes the eyeball with the retina, the optic nerve, the lateral geniculate nucleus, and the cortical primary and higher visual areas in the brain.

The human eye acts as a camera, regulating the amount of light and focusing to get a sharp projection of the image on the retina. The retina transduces light into electrochemical neural signals but also performs additional processing and information compression of the images before sending them to the visual centers of the brain. The retina, as described first by Ramón y Cajal, is organized into layers of specialized neurons.

The first layer corresponds to the photoreceptors (rods and cones), responsible for translating luminance and color into electrochemical potentials. Several layers of retinal horizontal, bipolar and ganglion cells, among others, collect the output of photoreceptors. The output of neighboring groups of photoreceptors, called the receptive field, is

Handbook of Neural Engineering. Edited by Metin Akay
Copyright © 2007 The Institute of Electrical and Electronics Engineers, Inc.

connected to a ganglion cell. The number of photoreceptors converging into a ganglion cell varies from just one at the fovea to a large number at the periphery of the retina. This leads to high resolution of the perceived image at its center. Most of these cells respond to the contrast of activity between the center and the surround of its receptive field.

So the retina not only functions in the simple transduction from light and color into neural signals but also performs different spatiotemporal locally computed functions and information compression before sending the signals to the brain through the optic nerve. This processing includes color contrast enhancing, motion detection, and edge highlighting. The retina cells convey the signals obtained at about 125 million photoreceptors into about one million fibers of the optic nerve.

Finally, the nerve converges with the brain at the lateral geniculate nucleus and then reaches the primary visual area located at the occipital region of the brain cortex. This is the starting point for further high-level processing in other areas, such as the recognition of shapes, faces, evaluation of motion, and so on.

A detailed explanation of the organization of the visual pathways can be found, for example, in [1, 2].

There are a huge number of causes of dysfunction in sight, provoked by illnesses or by accidents. The number of totally blind individuals rises to about 45 million people worldwide. The amount grows to 180 million persons with some other vision impairments [3].

16.3 APPROACHES FOR SIGHT RESTORATION: CURRENT RESEARCH ON VISUAL PROSTHESES

At the writing of this chapter, a number of research groups worldwide are facing the problem of blindness. The variety of proposed solutions corresponds with the multiplicity of causes of this illness.

Not every proposed solution is suitable for each patient. Blindness is the result of damage or malfunction of at least one of the stages comprised by the visual pathway, ranging from the cornea to the visual areas of the occipital lobe of the brain.

Some of these prosthetic approaches, as mentioned in [4], are the following:

- *Retinal Implants*: Inserting a device over (epiretinal) or under (subretinal) the retina that would provide electrical stimulation of the retina cells. These prostheses require a good functioning of, at least, the optic nerve.

- *Optic Nerve Implants*: Again requires functional optic nerve and is still in the early developmental stage. Difficult surgical access and very limited image resolution with the available electrodes are two important drawbacks of this approach.

- *Cortical Implants*: Direct electrical stimulation of the visual areas of the brain, by external planar electrodes or by penetrating intracortical electrodes. This is the only solution for patients with unrecoverable retinas or optic nerves.

Some other attempts are biological studies on gene therapy [4] or translation of visual information into some other good working sense, such as the "Seeing with Sound" project [5].

The decision on which approach is the most adequate is controversial, especially due to the still primitive knowledge about the coding of information by the brain and to the difficulties of ensuring long-term functioning with useful image resolution.

16.4 ELECTRICAL STIMULATION OF BRAIN

With some precedents, Luigi Galvani carried out the first neural electrical stimulation experiment in 1781. He attached the leg of a frog to an electrical machine, and the injection of electrical current caused violent muscular contractions. After a number of other discoveries on bioelectromagnetism, some researchers, such as Hubel [1] and Brindley and Lewin [6], contributed to the development of electrical stimulation and registering of cortical visual neurons, helping to extend the application of electricity not only to motor tissues but also to sensory perception.

The success of artificial cochleas encouraged research groups worldwide to use these devices for visual prostheses, although the amount of information handled by the eyes is far more complex and extensive. This way, the first experiments on direct electrical stimulation of the visual cortex led to the perception of phosphenes by blind individuals. William Dobelle [7] created a prosthesis that stimulates the cortex with a set of planar, wired surface electrodes. This is the only known experience with long-term implantation of cortical prostheses in humans. Although it produced some rudimentary vision, it presented important disadvantages: low spatial resolution between phosphenes and the need for high currents with undesirable effects.

The Utah Electrode Array, another prosthetic device developed by the research team led by Richard Normann [4], is able to simultaneously register or stimulate 100 electrodes in a small surface. In this case, the electrodes are penetrating the cortex, so less current is needed to elicit phosphene perception, making this option safer. However, this device has not been tested in chronic human implants.

In [4], Norman exposes as disadvantages of the cortical approach the absence of retinal processing, a poor visuotopic organization, and problems in representing multiple features in the primary visual cortex (color, ocular dominance, motion, lines).

In this chapter, we describe an architecture for a system that complements the cortical approach, aiming to overcome some of the above-mentioned problems. Such a system makes the cortical approach one of the most attractive for sight restoration.

The Utah Electrode Array is showing to be a promising solution for delivering electrical signals to the cortex of the brain, but a higher level of abstraction is also needed to decide how, when, and where these signals should be delivered to the proper electrode. This way, the approach is extended beyond just a problem of biocompatible interfaces between electrical devices and the cerebral tissue, extending the problem to an information and coding level. This device is intended to provide the translation between the visual expression of the world in terms of color and brightness and the neural language of spikes for complex stimuli, beyond the phosphene.

16.5 COMPUTATIONAL MODEL OF RETINA AND SYSTEM REQUIREMENTS

From a computational point of view, the retina is regarded in our model as a set of parametric filters, each one highlighting a specific feature of the image, such as color contrast, illumination intensity, and so on. This set of filters corresponds to the representation of different functionalities of the cell layers of the retina, as described in [8, 9] (see Fig. 16.1).

Most of the filters in our retina model operate with different color channels as its input. The opponent-color theory is based on the investigations by Jameson and

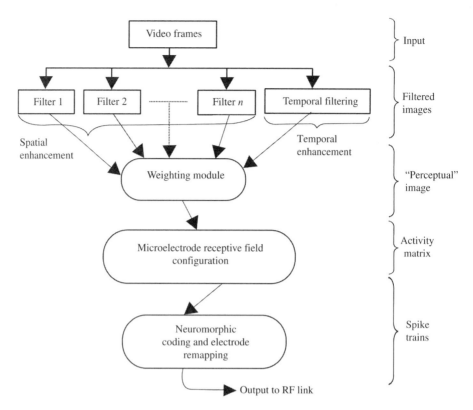

Figure 16.1 Schematic organization of retina-like processing model.

Hurvich (10). They proposed a model in which the opponent mechanism receives input from a linear combination of the three classes of cone receptor, channels L (red), M (green), and S (blue) in our model. This model combination can be mathematically expressed as a set of difference-of-Gaussian (DoG) filters. The most important filters are the red–green and yellow–blue processes [11].

The output resulting from the use of every filter on the original image is then combined into a single "saliency map" or perceptual image with enhanced features. This way, no explicit color channels are maintained, but their relative contribution is taken into account. The contribution of every filter to the weighted sum must also be configurable. With the resulting camera resolution image, an assignment of groups of pixels to its corresponding electrode must be performed, achieving what we call an *activity matrix*. The group of pixels assigned to one electrode resembles the concept of the receptive field.

This first representation of the activity of the retina is encoded in amplitude. However, the neural tissue of the visual cortex must receive signals in the form of trains of spikes whose frequency should vary according to the intensity of the stimulus. So another stage for intensity to frequency is needed. This stage is called *neuromorphic coding*. This last step must include a possible remapping of the electrodes to achieve a good visuotopic organization.

Finally, the signals for every electrode are sent to the microelectrode array through a radio-frequency (RF) link.

16.6 ARCHITECTURE

16.6.1 Sensor and Photoreceptors

When our choice is the cortical approach, which is suitable for patients with damage to the eye, the retina or even the optic nerve, a device must perform the functions of that segment of the visual pathway.

The entry point is a light sensor that will translate light level and color intensities into electronic format to allow digital automated processing. Any charge-coupled devide (CCD) or complementary metal–oxide–semiconductor (CMOS) sensor with its corresponding optics would be fine for the case. However, the system must be able to face scenes with high contrast and lighting levels. In these cases, the linear response of these devices leads to saturation and the consequent loss of information at the first stage of the prosthesis. A better result is obtained when employing logarithmic response cameras, which are able to register very dark and bright objects in the same frame, saving that information for later processing, as the human eye does. So the first point is that a system conceived to send visual information to the brain cortex should start by using a logarithmic sensor.

The output of these sensors offers three color planes for every image frame. These channels would correspond to every kind of cone, sensible to long, medium, and short wavelengths, here referred to as red, green, and blue channels. The intensity channel can be computed as the normalized sum of every color channel for every pixel of the image.

Some studies have been carried out to determine the minimum resolution in terms of pixels (or phosphenes or electrode tips in the array) in order to have a functional prosthesis which would allow us to navigate or read big characters [12]. If we want to make good use of the limited number of electrodes in current arrays, it is not adequate to dedicate specific electrodes for different channels, but we must put as much useful information as possible in each electrode.

16.6.2 Filtering and Receptive Fields

The second stage is the set of color-opponent filters resembling the bipolar cell receptive fields, along with a temporal filter with a contrast gain control mechanism that highlights motion information. Taking into account the available data on the different channels at a human retina [11, 13], we can consider an adequate basic model of the retina consisting of the following filters:

- *ML versus S*: Yellow versus blue color channels, computed by using a DoG
- *LS versus M*: Red versus green, again with a DoG
- *LMS*: Achromatic luminance filter, computed by a Laplacian
- Temporal bandpass filter

To compute the color-opponent channels, a Gaussian weighting function or "mask" is employed. The mask is centered on the pixel that is currently under processing, and it covers the pixels included in its neighborhood, for a given radius. As a temporal filter, we can consider the one based on the retina ganglion cell model presented in [14], which implements a nonlinear temporal contrast gain control scheme (e.g., the one proposed in [15]). This model is indeed a complete approximation of the spatiotemporal receptive field features of biological retinal ganglion cells, as described below. Thus it can be used as the only visual processing module in the future visual prosthesis if we want to better approximate the function of a biological retina. However, as shown in Figure 16.1, we combine the output of the temporal filter with those produced by the spatial-only ones. In this way, temporal changes as those produced by motion onsets are highlighted in the resulting feature map (the "perceptual image"), but we avoid having the output vanish for static visual stimulus as it occurs in an actual retina.

According to the description in [15], the temporal filtering module computes the temporal-enhanced (TE) images as the product, for each point $\rho = (x, y)$ in the image, of a modulation factor $g(\rho, t)$ by the convolution of the stimulus s and a kernel function $K(\rho, t)$:

$$\text{TE}(\rho, t) = g(\rho, t)[K(\rho, t) * s(\rho, t - \delta(t))] \tag{16.1}$$

where $\delta(t) > 0$ is the response latency.

The model assumes space–time separability in such a way that the kernel function $K(\rho, t)$ can be decomposed into different spatial (K_S) and temporal (K_t) functions as follows [14, 15]:

$$K(\rho, t) = K_S(\rho)K_t(t) \tag{16.2}$$

$$K_S(\rho) = \frac{g_+}{2\pi\sigma_+^2} e^{-\rho^2/(2\sigma_+^2)} - \frac{g_-}{2\pi\sigma_-^2 e^{-\rho^2/(2\sigma_-^2)}} \tag{16.3}$$

$$K_t(t) = \delta(t) - \alpha H(t)e^{-\alpha t} \tag{16.4}$$

The spatial kernel is modeled as a DoG, with g_+ and g_- determining the relative weights of center and surround, respectively; σ_+ and $\beta\sigma_+$ (with $\beta > 1$) are their diameters. In the temporal function, Eq. (16.4), H denotes the Heaviside step function and α^{-1} is the decay time constant of the response.

The modulation factor in Eq. (16.1) implements a contrast gain control feedback loop, since it is obtained as a nonlinear function of a low-pass temporal filtered version of TE [14, 15].

Once all the filters are applied and combined into a single perceptual image, the system proceeds to a reduction in spatial resolution, grouping the values of neighboring pixels to obtain the activity level at each electrode. The size of the receptive field varies depending on the location at the retina. Usually, the value for the electrode is the average of the pixels on its receptive field.

16.6.3 Neural Pulse Coding and Visuotopic Remapping

The activity matrix representing remarkable perceptual features of the input image is encoded in amplitude. Before sending this information to the cortex, a translation into the neural language expected by the brain is required.

Neurons represent activity values in the form of spike trains, in which the amount of spikes and its frequency encode the level of activity incoming to the neuron. A known model of this cell is the *leaky integrate-and-fire spiking neuron*. Further references on these models can be found in [16].

In the digital implementation we have proposed for this model [8], the activity level incoming to the neuron is accumulated until a threshold is reached. In that moment, a spike is issued and the neuron potential returns to a resting value. Additionally, the accumulated potential drops slowly due to a leakage, so that in the absence of inputs the activity of the neuron decays. For simplicity in digital implementation, we have adopted a linear decay for the leakage term.

The expected response is shown in Figure 16.2, in which high-sustained stimuli give out a longer spike train and low stimuli may not get the neuron to reach its threshold and no spike is issued to the corresponding electrode.

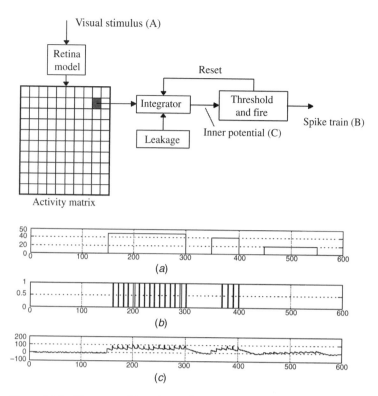

Figure 16.2 Neuromorphic pulse coding. Schematic model of integrate-and-fire neuron (upper); simulation example for different stimuli.

Each time a spike is produced, the corresponding electrode address is sent over to the array. This communication scheme is known as *address-event representation* (AER) [17, 18]. In our case, no arbitration mechanism is employed, as the addresses are sent serially, multiplexed in time. This representation has been tested on bioinspired very large scale integrated (VLSI) vision systems, such as those described in [19, 20].

This working model for one electrode can be extended to any number of electrodes, either serially (computing the electrodes one by one) or in parallel. Either method is right if it is able to deliver signals with a temporal resolution of 1 ms or better, as this time is lower than the time a neuron needs to be able to fire again after a spike.

In any case, the correspondence between points stimulated at the retina and the neurons that respond to it on the cortex does not follow a visuotopic organization. It was believed that the projection of an image on the retina would elicit electrical activity on the brain with a similar shape. At a macroscopical level, close points stimulated in the retina give activity in close points in the brain cortex. But when we deal with implants such as the Utah Electrode Array, in which the distance between electrodes corresponds to the width of cortical columns, this mapping loses its organization, and a strong remapping is required [21], mainly due to the irregularities of the surface of the cortex.

This remapping can be achieved with a look-up table of the correspondence between the stimulated point and the perceived position in the prosthesis. The contents of this table must be determined for every implanted patient on an acute clinical process in which individual electrodes would be activated in pairs and the patient describes the relative position of the phosphenes. Then, the table stores the electrode that would give the right spatial position with respect to the reference electrode, thus providing a translation from retina pulse addresses to electrode addresses.

16.6.4 Wireless Link

The train of events is sent to the electrode matrix implanted on the cerebral cortex. Commonly, percutaneous connectors have been employed to gain electrical access to the prosthesis. However, this connection is not desirable for chronic implants due to the risks of infections and a possible psychological rejection.

Ideally, the bond to send data and energy to the prosthesis should be made through a wireless RF link. This link would send energy through induction coils and the data following a protocol able to indicate the electrode to which an event must be sent and some other information related to the waveform. This information should be transmitted in real time so that even serial scanning of the electrodes would result in simultaneous perception in terms of biological response times.

This RF link would be useful not only for stimulation but also for registering the activity on the visual cortex and would relate that activity recordings to controlled stimuli, which will help in understanding the encoding of complex images on the brain.

Efforts on the development of RF links for electrode arrays are being carried out by several researchers [22–24].

16.6.5 Electrode Stimulation

The final interface with the cortical visual area of the brain is an array of penetrating microelectrodes. Previous approaches, such as Dobelle's prosthesis, used planar platinum electrodes. However, it has been demonstrated that the current needed to elicit the perception of phosphenes through a single electrode is 10 times smaller when using penetrating electrodes than with planar surface contacts. This fact reduces power consumption, heat dissipation, and some undesirable effects of having relatively high currents injected on the brain.

The Utah Electrode Array, developed by the research team of Richard Normann, is one of the best options for a cortical implant. It is a silicon base with 100 penetrating electrodes. A detailed description of the array has been made in [4].

These kinds of devices, in which a limited bandwidth channel must deliver information to a large number of electrodes, require demultiplexation and a number of digital-to-analog converters (DACs) to stimulate every point in time.

There are some other devices, such as the electrode arrays fabricated at Michigan [22]. The number of electrodes in the array is only 64, but a wireless communication system is integrated on it.

16.7 CORTIVIS PROJECT

The architecture described here is being applied to a prosthetic device under development in a European project. This initiative, named CORTIVIS [25] (Cortical Visual Neuroprosthesis for the Blind, ref. QLK6-CT2001-00279), has as its main goal to produce a wireless, portable device to stimulate the visual cortex by using an array of penetrating implanted microelectrodes. Its main architecture, depicted in Figure 16.3, is conceptually similar to those adopted by other initiatives [26, 27]. One of the advantages of the system

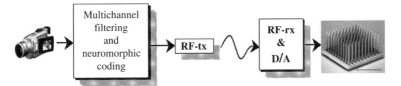

Figure 16.3   Main blocks of CORTIVIS Project prosthesis, including a camera, a retina-like filtering and encoding block, a radio frequency transmitter (tx), the corresponding receiver (rx) and D/A converter, and the microelectric array to interface the neural tissue of the visual cortex.

being developed by the CORTIVIS consortium is that every stage of the prosthesis is reconfigurable in every parameter. The goal of CORTIVIS is to obtain:

- A tunning workbench to adjust and refine retina models in acute and chronic implants, for both stimulation and registering
- A wireless reconfigurable device with a configuration that can be altered even after implantation, avoiding further surgery
- A higher computational layer able to automatically translate a continuous stream of video into a set of spike events and electrode addresses in order to feed the wireless implanted array of microelectrodes
- Additional knowledge on suitability of patients, implant adjustment and training procedures, long-term chronic implant studies, and so on

To ease and accelerate the development, we are progressively moving from a full software implementation model to a fully hardware real-time portable chip. We are implementing the retina model on field-programmable gate array (FPGA) circuits, which makes the development of prototypes less expensive and faster, allowing reprogramming of the circuit. The FPGA circuits allow the implementation and testing of complex digital designs without the fabrication process and costs of other technologies such as a custom ASIC chip (application-specific integrated circuit).

At the current development stage, the software model is able to accept as input either continuous video streams from a web cam (optionally, these streams can be saved), video files, or static image files. The gain of each color plane, corresponding to the photoreceptor, can be adjusted.

This software allows building and testing very different retina models, as the user is able to add new filters and vary multiple parameters.

The filters (Gaussians and DoGs, Laplacian of Gaussians, and so on) can be defined over the incoming color and intensity channels or even over the output of previously designed filters. This way, a chain of filters can be specified.

The system also allows any mathematical expression taking already defined filters as operands. This high degree of flexibility makes this platform a useful tool to test and compare retina models.

With respect to the definition of receptive fields, the user can select fixed and variable size and shape to determine which neighboring pixels contribute to a specific electrode, as can be seen in Figure 16.4. The options offered are:

- Circular (fixed or variable radius)
- Elliptical
- Square
- Rectangular

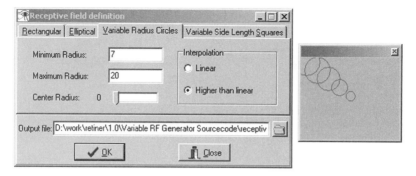

Figure 16.4 CORTIVIS tool to select sizes and shapes of ganglion receptive fields.

This tool is useful to model a variable spatial resolution, as it occurs in the human retina, in which the fovea has high resolution in contrast with the periphery.

The last stage in the model translates the array of activity values produced by previous stages into a stream of pulses. These pulses represent spike events to be applied to the microelectrode array for stimulation. The software is designed to work with different

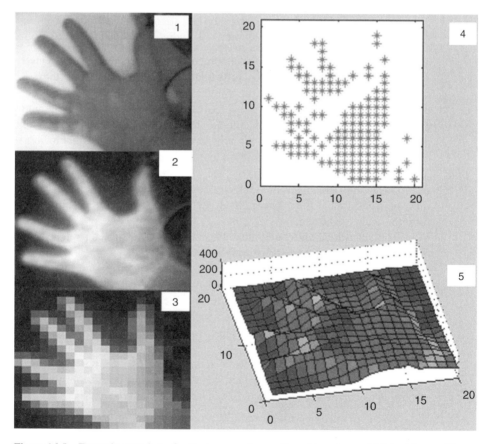

Figure 16.5 Example snapshot of retinal processing on live video stream. Original frame (1), retina-like filtering (2), mapping to receptive field of every electrode (3), activity matrix (4), and spiking for each electrode (5).

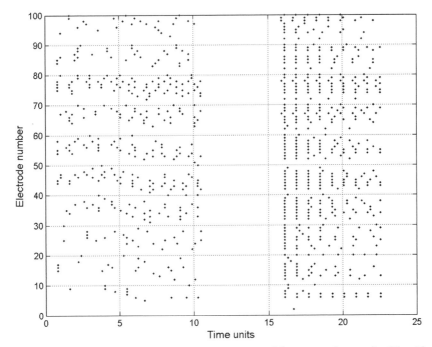

Figure 16.6   Example spike event sequence generated from natural scene for 10×10 microelectrode array. First, a complex image is registered, then the camera's objective is covered, so there is no activity, and last, the sudden exposure to light tends to produce an initial stimulus-locked synchronized response at most of the output channels.

array dimensions. A leaky integrate-and-fire spiking neuron model has been implemented to determine the production of events at the prosthesis. The user can select different values for this model:

- Firing threshold
- Resting potential
- Leakage factor

The effects of changing these parameters over the firing of electrodes can be displayed in real time even with live image capture, and a detailed register of the spiking is also available once the processing is over.

Thus, this software platform is a test bench for the refinement of retina models and a prototype for later implementation on a single-chip portable device. Currently, the visuotopic remapping module is not included, as it is under development. Two examples show the processing performed by the system in Figures 16.5 and 16.6.

16.8   FUTURE TRENDS

Cortical prostheses for the profoundly blind are in the early stage of development. Intense research is been carried out to achieve interfaces with the brain that can be employed for long-term stimulation. We know few cases of human implantation of such devices in which some form of rudimentary vision is produced, but it has not been shown to be fully functional and better than classical aids to the blind, such as canes or guide dogs.

In the future the level of study should be raised not only to electrically interface with the neural tissue but also to automatically feed this interfaces, going beyond the production of single phosphenes for simple stimuli.

For the research groups working on this, the development of a full prosthetic device, from a camera to the implanted stimulator, must be the objective. Ideally, this device will be wireless and portable, have low power consumption, be reconfigurable without surgery, and have high quantity (resolution) and quality (including color) of information.

Some other issues remain open for researchers, such as providing with stereovision not only by duplicating the prosthesis but also by connecting the information flowing through each one to obtain information on perspective and distances, including automatic local gain control on camera sensors to allow the use of these devices indoors and outdoors, out of controlled laboratory environments; automatically compensating the motion of the eyeball to allow the patient to move it, by changing the camera orientation through neural motor signals; enhancing the resolution achieved by electrode arrays [22]; studying the effects and solutions for long-term usage of electrical stimulators of the brain; and developing faster and more comfortable procedures for the tunning and visuotopic remapping of the prostheses.

This is an encouraging research field for neuroscientists, engineers, and computer scientists in which significant advances still need to be made.

ACKNOWLEDGMENTS

The developments described in this chapter have been achieved due to the support of the CORTIVIS Project (Cortical Visual Neuroprosthesis for the Blind), ref. QLK6-CT2001-00279, funded by the European Community.

REFERENCES

1. D. H. HUBEL, *Eye, Brain and Vision*, Scientific American Library, New York, 1995.
2. H. KOLB ET AL., Webvision, http://retina.umh.es/webvision/index.html.
3. WORLD HEALTH ORGANIZATION, Elimination of Avoidable Blindness. Report by the Secretariat, http://www.who.int/gb/EB_WHA/PDF/WHA56/ea5626.pdf, 56th World Health Assembly, April 2003.
4. R. A. NORMANN ET AL., A neural interface for a cortical vision prosthesis, *Vision Res.*, 39, 2577–2587 (1999).
5. P. MEIJER, The vOICe. Wearable computing for the blind, http://www.seeingwithsound.com.
6. G. S. BRINDLEY AND W. S. LEWIN, The sensations produced by electrical stimulation of the visual cortex, *J. Physiol.* 196, 479–493 (1968).
7. W. H. DOBELLE AND W. G. MLADEJOVSKY, Phosphenes produced by electrical stimulation of human occipital cortex, and their application to the development of a prosthesis for the blind, *J. Physiol.* 243, 553–576 (1974).
8. F. J. PELAYO ET AL., Cortical visual neuro-prosthesis for the blind: Retina-like software/hardware preprocessor, in *Proceedings of the first International IEEE EMBS Conference on Neural Engineering*, M. AKAY (ed.), Capri, Italy, March 20–22, 2003, IEEE, paper 2.4.5–1.
9. F. J. PELAYO ET AL., Retina-like processing and coding platform for cortical neuro-stimulation, paper presented at the 25th Annual International Conference of the IEEE EMBS, Cancun, Mexico, September 17–21, 2003.
10. JAMERSON AND L. M. HURVICH, Some quantitative aspects of an opponent-color theory. I. Chromatic responses and spectral saturation, J. Opt. Soc. Amer. 45(7), 1955.
11. E. ZRENNER ET AL., in L. SPILLMAN and J. S. WERNER, Eds., *Visual Perception: The Neurophysiological Foundations*, Academic Press, San Diego, 1990, p. 163.
12. K. CHA ET AL., Mobility performance with a pixelized vision system, *Vision Res.* 32, 1367–1372 (1992).
13. R. SEKULER AND R. BLAKE, *Perception*, 3rd ed., McGraw-Hill, New York, 1994, pp. 193–214.
14. S. D. WILKE ET AL., Population coding of motion patterns in the early visual system, *J. COMP. Physiol. A* 187, 549–558 (2001).
15. M. J. BERRY ET AL., Anticipation of moving stimuli by the retina, *Nature* 398, 334–338 (1999).

16. W. GERSTNER AND W. KISTLER, *Spiking Neuron Models. Single Neurons, Populations, Plasticity*, Cambridge University Press, Cambridge, 2002.

17. M. MAHOWALD, VLSI analogs of neural visual processing: A synthesis of form and function, Ph.D. Thesis, Computational and Neural Systems, California Institute of Technology (1992).

18. A. MORTARA AND E. A. VITTOZ, A communication architecture tailored for analog VLSI networks: Intrinsic performance and limitations, *IEEE Trans. Neural Networks* 5(3), May (1994).

19. K. A. BOAHEN, Point-to-point connectivity between neuromorphic chips using Address-Events, *IEEE Trans. Circuits Syst. II* 47(5), 416–434 (2000).

20. P. VENIER ET AL., An integrated cortical layer for orientation enhancement, *IEEE J. Solid State Circuits*, 32(2): 177–186. (1997).

21. R. A. NORMANN ET AL., Representations and dynamics of representations of simple visual stimuli by ensembles of neurons in cat visual cortex studied with a microelectrode array, in *Proceedings of the First International IEEE EMBS Conference on Neural Engineering*, M. AKAY, (ed.), Capri, Italy, March 20–22, 2003, IEEE, paper 2.0.1–1.

22. M. GHOVANLOO ET AL., Toward a button-sized 1024-site wireless cortical microstimulating array, in *Proceedings of the First International IEEE EMBS Conference on Neural Engineering*, M. AKAY, (ed.), Capri, Italy, March 20–22, 2003, IEEE, paper 2.3.4–3.

23. G. J. SUANING AND N. H. LOVELL, CMOS neurostimulation system with 100 electrodes and radio frequency telemetry, in *Proceedings of the IEEE EMBS Conference on Biomedical Research on the 3rd Millennium*, Victoria, Australia, February 22–23, 1999.

24. M. R. SHAH ET AL., A transcutaneous power and data link for neuroprosthetic applications, in *Proceedings of the 15th Annu. Int. Conf. IEEE EMBS* San Diego, pp. 1357–1358, October 28–31, 1993.

25. http://cortivis.umh.es.

26. INTRACORTICAL VISUAL PROSTHESIS, Prosthesis Research Laboratory, Illinois Institute of Technology, http://www.iit.edu/~npr/intro.html.

27. J. F. HARVEY ET AL., Visual cortex stimulation prototype based on mixed-signal technology devices, Int paper presented at the International Functional Electrical Stimulation Society Conference (IFESS'99), Sandai (Japan), August 1999.

CHAPTER 17

BIOMIMETIC INTEGRATION OF NEURAL AND ACOUSTIC SIGNAL PROCESSING

Rolf Müller and Herbert Peremans

17.1 INTRODUCTION

Despite an ever-increasing speed of technological achievements, biological systems still outperform technical systems in many ways, but not in every respect. In fact, for many isolated, well-defined tasks, man-made technology has already outperformed biological systems. For example, manufacturing robots can replace and even outperform humans in repetitive tasks comprising a fixed sequence of actions. Such robots cannot, however, reproduce the versatility of human skill in performing tasks which require adopting a strategy on the fly. Part of the explanation why biological systems have retained superiority in these more involved tasks may lie in nature's ability to create well-integrated systems comprising many components, each of which contains evolutionarily embedded knowledge about the particular tasks it performs and the control loops to which it belongs. In order to be successful, neural engineering strategies should take this into account and avoid the fallacy of thinking of nervous systems as isolated collections of neurons. In doing so, it can readily make connections to trends in technology which shift intelligence into peripheral parts. For instance, the interfaces through which many computer peripherals are attached have been made progressively smarter, which makes accessing the peripherals simpler (e.g., hard-drive interfaces, plug-and-play protocols). This chapter reviews the design process for building a piece of biomimetic, intelligent periphery. The biological system it seeks to reproduce—at a functional level—is the biosonar system found in bats. Biosonar outperforms manmade sonar technology in its ability to support versatile, fully autonomous navigation as well as a variety of other tasks in often demanding natural environments as a sufficient far sense. In order to provide a robotic platform capable of reproducing these skills, the biomimetic system couples sensing with actuation and integrates peripheral with neural processing by feedback loops spanning all stages.

17.2 BIOMIMETIC SONAR HEADS: STATE OF THE ART

The biosonar system of bats uses intertwined acoustic and neural signal processing with various feedback control loops spanning both domains to achieve an unparalleled performance. Although the ways in which the animals achieve this performance remain largely

Handbook of Neural Engineering. Edited by Metin Akay

unknown, some basic system specifications have been determined and are given in Table 17.1. The animals have high power ultrasonic emission systems combined with very sensitive reception. The sites of emission (mouth or nostrils) and reception (ears) are often surrounded with elaborate baffle shapes (auricles, "nose leafs" and "horse shoes"). Many species of bats have, to a varying extent, control over both orientation and shape of these structures. These shapes and behaviors are most likely serving sophisticated, adaptive signal processing schemes which operate in the acoustic domain on the outgoing and incoming wave fields. Similarly, numerous specializations exist in both the peripheral [1] and the central stages of the auditory system [2].

Biosonar systems have long been recognized as exceptional examples of powerful yet parsimonious natural perception/sensorimotor systems by engineers. Consequently, there is some prior art in terms of robotic models, which is briefly reviewed here and compared to the known specifications of biosonar systems in Table 17.1.

While each of the various systems which have been developed over the last 10 years has been successful in reproducing some of the interesting aspects to biosonar function, so far none of them has integrated all known functional features of a bat head. In terms of size, most robotic sonar systems have been out of scale by a factor of at least 4–5 [3–5] or evenmore [6]. Being much larger than natural biosonar systems renders these models incapable of duplicating the diffraction effects around the head which facilitate sonar sensing in their biological counterparts. Those model systems which have been approximately to scale [7] have achieved this at the expense of crucial functional features such as ear mobility. In terms of transducers, all existing systems have been using standard, commercially available technology [3–6], either narrow-band piezoelectric

TABLE 17.1   Specifications of a Current State-of-the-Art Robotic Models (S1 [8], S2 [5], S3 [9], S4 [16], S5 [11], S4, and S5 are shown in Fig. 1.1), the CIRCE Prototype (Under Development) and The Bat Sonar System

	S1	S2	S3	S4	S5	CIRCE	Bats
Scale[1]	$\sim$10	$\sim$1–2	$\sim$4	$\sim$4–5	$\sim$4–5	$\sim$1	1.0
N rot. DF for ears[2]	0	0	1	2	2	2	2
Variable chirp[3]	No	No	No	Yes	Yes	Yes	Yes
Pinnae	No	Yes[4]	No	No	No	Yes	Yes
Var. directivitywidth[5]	No	No	No	No	No	Yes	Yes
F-dep. directivity-axis[6]	No	No	No	No	No	Yes	Yes
Max. BW[7] [kHz]	$\ll$10	$\ll$10	$\sim$100	$\sim$100	$\sim$100	160	160
SNR[8] [dB]	?	?	$\sim$30	$\sim$60	>60	$\gg$60	?
Output SPL[9] [dB]	?	?	?	?	$\sim$110	$\sim$120	$\sim$130
Number of IHC[10]	1	Few	1	Software	Software	$\sim$1000	700–2000
Number of SGC[11]	0	Few	0	Software	Software	$\sim$10,000	13,000–55,000

[1]A natural bat head is assumed here to have a diameter of 4 cm.
[2]Number of rotational degrees of freedom.
[3]Bandwidth and instantaneous frequency as a function of time.
[4]Applies to one version of the system.
[5]Directivity pattern can be made wider or narrower.
[6]Central axis of the directivity pattern depends on frequency [26].
[7]Total non-zero frequency band (union of all species in case of bats).
[8]Signal-to-noise ratio as level difference between an echo from a plane in 1 m distance and the noise floor.
[9]Sound pressure level in 1 m distance straight ahead.
[10]Inner hair cells, see text (data from [25]).
[11]Spiral ganglion cells, see text (data from [25]).

(a)   (b)

Figure 17.1   Examples of current state-of-the-art robotic models for biomimetic heads:
(a) system built by the Institute of Perception, Action, and Behaviour, University of
Edinburgh [4, 8] (system S4 in Table 17.1); (b) system built by the Biosonar Lab, University
of Tübingen [5] (system S5 in Table 17.1).

transducers [6, 7] or comparatively more wide-band capacitive transducers [3–5].
However, no transducer is known to be commercially available and hence none of these
systems is capable of covering the full band over which the biosonar chirps emitted by
different bat species extend (~20–200 kHz). Ear mobility with either one or two
rotational degrees of freedom has been realized in several systems [3–5] but is lacking
in others, where the transducers are statically mounted [6, 7]. The same is true for neuromimetic signal representations, which have been used in [5, 7], but not in any of the other
systems. Even where present, the generated spike codes have not come close to being a
quantitatively correct reproduction of what propagates in the auditory nerve of bats.
Only one of the static systems is equipped with pinnae [7]. For a technical reproduction
of the mobile, non–rigidly deforming, beam-shaping antennae present in nature, no
prior art exists.

It is evident from the specifications summarized in Table 17.1 that there is a wide
gap between the current robotic models (Fig. 17.1) and natural sonar heads and that realizing a more realistic functional reproduction of the biological system requires several challenges to be addressed:

Biomimetic Antennae Shapes   In bats, sonar emitters (mouth or nose) and receivers
(ears) are surrounded by often elaborate baffle shapes. These shapes are hypothesized to serve as beam-forming aperture tapers. Both, the beam-forming mechanisms and the resulting beam properties need to be understood in order to be able
to mimic the acoustic signal processing that is performed by biosonar systems
before the signals are transformed into a neural representation.

Transducer Technology   Bats are able to produce high-energy sonar pulses apparently in a very efficient manner. This results in good signal-to-noise ratios even
for otherwise unfavorable target ranges and target strengths. On the receiver side,
bats can rely on the superb sensitivity of the mammalian hearing system. Coming
close to these specifications requires careful design of electromechanical

transducers systems which uses an integrated approach for designing the transducers proper and matched driver electronics.

Actuation   Many bat species possess several degrees of freedom in orienting their outer ears (pinna) as well as controlling the shape of the baffle structures surrounding the sites of sound emission. This enables them to perform adaptive sensing in a configuration space which is made up not only by the position and orientation of their sonar heads but also by the orientation and shape parameters for the sensors.

Neuromorphic Signal Processing   Ultrasonic signal processing poses a special challenge for neural systems because the high bandwidth of the signals has to be adapted to the comparatively low rates at which neural signals (action potentials, or "spikes") can be generated. The principal mechanism to achieve this is a filter bank representation, where each channel codes for the portion of the signal bandwidth which falls into the passband of a bandpass filter, which is much narrower than the entire signal bandwidth. However, the number of primary receptors (inner hair cells) and nerve fibers making up the auditory nerve in bats (approximately 700–2200 and 15,000–55,000, respectively; see [9]) are still in reach of what digital neuromimetic hardware can match quantitatively.

17.3   BIOMIMETIC ANTENNA SHAPES

The approximately 1000 species of bats [10] harbor a large variety of baffle shapes associated with ultrasonic emission and reception. For designing biomimetic aperture tapers it is essential to understand what effects these shapes have on the system properties, in particular the directivity patterns and what the underlying mechanisms are. For these purposes, geometric data on the biological shapes have been collected and used as a basis for numerical simulations of the diffraction phenomena.

Samples of ears and other facial structures were obtained from bat carcasses belonging to the following species: *Eptesicus fuscus* (big brown bat), *Myotis nattereri* (Natterer's bat), *Nyctalus noctula* (noctule bat), *Plecotus auritus* (brown long-eared bat), *Pipistrellus nathusii* (Nathusius' pipistrelle), *Pipistrellus pipistrellus* (pipistrelle), *Rhinolophus ferrumequinum* (greater horseshoe bat), *Rhinolophus rouxi* (Rufous horseshoe bat), and *Tadarida* sp. (free-tailed bat). The material was either fresh or had been kept in a freezer for various amounts of time. The samples were mounted in lifelike postures using photographs of live animals and subjected to fanbeam scanning with X rays using a microcomputer-tomography system (SkyScan 1072) equipped with a 1024 × 1024-pixel, 12-bit charge-coupled device (CCD) chip. In the obtained shadow images the tissue–air boundaries as well as inhomogeneities in the tissue are clearly visible (Fig. 17.2*a*).

Shadow images were obtained with an angular resolution of 0.9° over an arc spanning slightly more than 180°. Transverse cross sections were estimated using the Feldkamp cone-beam reconstruction algorithm (implementation provided by Skyscan). Depending on the magnification used, cross-sectional pixel size ranged between 10 and 19 μm. Image postprocessing was performed using the Visualization Toolkit (VTK) [11]: Images were segmented based on the gray-scale value with the threshold being manually adjusted to the properties (brightness, signal-to-noise ratio) of the individual data sets. Where necessary, manual image postprocessing was carried out to remove noise and reconstruction artifacts. From the stacks of segmented cross sections (Fig. 17.2*b*) trigonal mesh

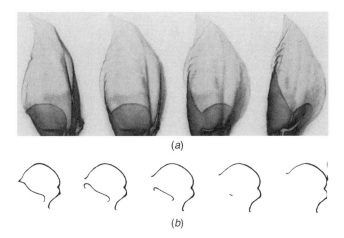

(b)

Figure 17.2   Examples of tomography data: (*a*) shadow images; (*b*) segmented cross sections. In the cross-sectional images, ear tissue is shown in black, the surrounding air in white. The images show a right-ear sample taken from a horseshoe bat (*R. ferrumequinum*); the cross sections shown in (*b*) are reconstructed from the shadow image data set to which the examples shown in (*a*) belong.

representations of the ear surfaces were estimated using image blurring with a 2D Gaussian filter mask and the marching cubes algorithm [12]. The meshes were decimated [13] and smoothed with a windowed sinc function interpolation kernel [14]. Renderings of the resulting ear meshes (Fig 17.3) show some typical features of bat ears, for instance, appendages partially occluding the frontal access to the concha and washboard patterns on the inner side of the pinna.

The acoustic wave diffraction phenomena caused by the pinna can be predicted numerically using a finite-element simulation model. Because of numerical error considerations, elements with aspect ratios close to 1 as well as equal sizes between elements are desirable. Meshes with these properties can be generated directly froma stack of reconstructed computed tomography (CT) cross sections, where the voxel centers define a rectangular grid already. Elements for the air volume surrounding the pinna can therefore be created by checking a simple rule for the voxels on the element corners falling inside or outside the pinna. For example, a new air volume element is added if all eight corners are outside the pinna, that is, the gray values of the respective pixels exceed a chosen threshold. Smoothing with a low-pass convolution mask can be applied to the stack of images prior to meshing in order to reduce the bias of such a conservative rule. This meshing procedure generates an estimate of the surrounding air volume with staircase boundaries (Figs. 17.4*a,b*).

If the element size is chosen not to exceed the resolution requirements of the simulation (e.g., 10–20 elements per wavelength as a rule of thumb), the discontinuous nature of the approximation should not affect the numerical results. Because of these resolution requirements and the three-dimensional nature of the problem (the pinna are not symmetric), the number of elements increases rapidly with the space to be filled. For example, for a frequency of 50 kHz with ∼6.8 mm wavelength, an element edge length of ∼0.5 mm is appropriate. This corresponds to a density of 8000 elements/cm^3, or 8×10^9 elements/m^3. Therefore, the use of finite-element simulations is restricted to the immediate vicinity of the pinna surface (Figs. 17.4*c,d*). The CFS++ program, currently under development at the Department of Sensor Technology of the University of Erlangen Nürnberg, was used to carry out the finite-element simulations. Beyond the

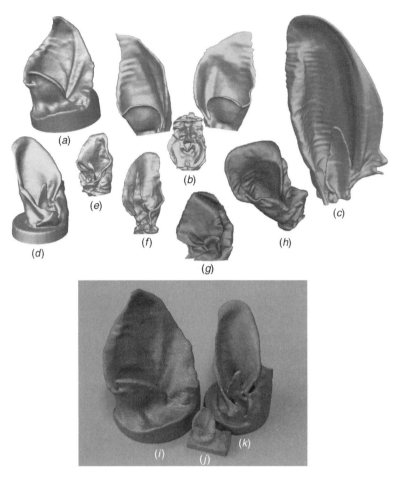

Figure 17.3   Example of pinna shapes estimated from CT data sets: (*a*) *R. rouxi* (Rufous horseshoe bat); (*b*) *R. ferrumequinum* (greater horseshoe bat), shown are the two ears and the appendages ("horseshoe") around the nostrils; (*c*) *P. auritus* (brown long-eared bat); (*d*) *E. fuscus* (big brown bat); (*e*) *P. nathusii* (Nathusius' pipistrelle); (*f*) *M. nattereri* (Natterer's bat); (*g*) *N. noctula* (noctule bat); (*h*) *Tadarida* sp. Physical prototypes made from stainless steel grains in a bronze matrix for two of these shapes: (*i*) *R. rouxi* (s.a.), magnified 4 times; (*j*) *R. rouxi* natural size (for comparison); (*k*) *E. fuscus* (s.d.).

region immediately bordering the pinna, the finite-element results need to be forward propagated into free space in order to estimate far-field system properties such as the directivity pattern.

   Mesh representations of the pinna shapes offer the advantage that a fairly wide range of experimental shape manipulations can be carried out with relative ease. Experimental manipulations can be used to assess the effect of structural pinna features on the system-properties and to understand the mechanisms that link structural features to system properties. Appendages such as a tragus or an antitragus can be removed by devising a Boolean combination of implicit functions to cover exclusively the parts of the pinna to be removed. The Boolean combination of implicit functions can then be used as a clipping mask to remove the appendages in question, as shown in Figures 17.5*a,b,c*.

   Washboard patterns on the inner pinna surface can be found to varying extent in several bat species (Fig. 17.3). At present, it is not clear yet whether these structures

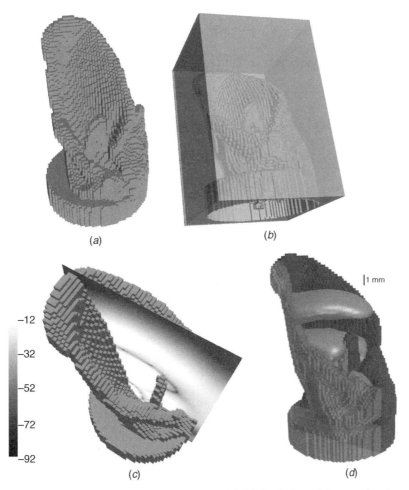

Figure 17.4   Example of finite-element sound field simulations: (*a*) approximation of ear surface using cubic elements; (*b*) air volume surrounding ear meshed by cubic elements. (*c*,*d*) Example visualizations of numerical wave field predictions for excitation by 50-kHz sine wave in ear canal: (*c*) cut through wave field (gray level encodes power relative to excitation on decibel scale); (*d*) field contours for power level − 32 dB below source level in ear canal. The ear sample shown is from *E. fuscus*; (Fig. 17.3*d*).

serve an acoustic purpose, either exclusively or in addition to some other function. However, once a mesh representation of the pinna surface is obtained, acoustic effects of these washboard patterns can be tested readily by using a low-pass filter to smooth the ear surface and hence remove the washboard patterns (Figs. 17.5*c*,*d*).

Besides studying the properties of natural pinna shapes, interesting baffle shapes can be searched via an optimization process with starts from simple primitives [15]. The latter approach provides an opportunity to optimize a shape for performance of a specific sonar task, contrary to natural shapes that can be expected to represent compromises between the possibly conflicting requirements of all the sonar tasks which the animal has to perform. However, both approaches can be combined by using meshes representing natural shapes as a starting point for the optimization process. In this way, it can be determined if the antenna performance of a natural pinna shape in a particular task is being limited by

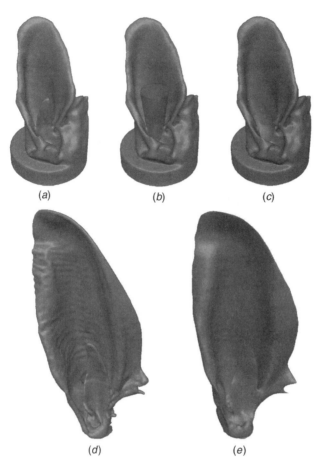

Figure 17.5   Example of experimental pinna manipulations. $(a-c)$ Removal of ear appendages via "Boolean surgery": (a) mesh representation of original pinna; (b) pinna with superimposed Boolean combination of implicit functions (a cone and two planes); (c) pinna with tragus removed by clipping with implicit function; the ear sample is from *E. fuscus* (Fig. 17.3*d*). (d,e) Example of washboard surface smoothing via low-pass filtering: (d) mesh representation of original pinna; (e) pinna smoothed with windowed sinc function interpolation kernel; the ear sample is from *P. auritus* (Fig. 17.3*c*).

other constraints and how this compromise differs from the optimum that could be reached starting from this natural configuration.

It is a reasonable precaution to compare numerical predictions to experimental results even if the numerical methodology can be considered mature, because comparing a few reference cases allows us to confirm that the choices made, for example, in terms of sampling are valid. Physical prototypes can be generated from polygonal mesh representations via various rapid prototyping methods such as selective laser sintering (Figs. 17.3*i, j, k*). Since the ratio of wavelength and object dimensions is the main factor which determines scattering and diffraction, the prototypes can be scaled up in order to facilitate the measuring process. Magnified versions of the pinna shapes allow the usage of standard (e.g., $\frac{1}{8}$-in.) measuring microphones and lower frequencies. Rapid prototyping methods allow the use of various materials; however, since the impedance mismatch between air and any biological tissue is about four orders of magnitude, the material of the prototype is not a significant factor for acoustic measurements.

17.4   ULTRASONIC TRANSDUCER TECHNOLOGY

Faithfully mimicking the sonar system of different bat species requires the ultrasonic transducer to meet a demanding set of specifications. The transducer design goals of the Chiroptera Inspired Robotic Cephaloid (CIRCE) system are given in Table 17.2.

Commercially available broadband ultrasonic transducers (e.g., [16]) are too large to fit on a biomimetic bat head with realistic dimensions and furthermore lack adequate bandwidth. Hence, new transducer technology is required to meet these specifications.

A literature search complemented with measurements and finite-element simulations was conducted to compare different candidate transducer technologies. For capacitive transducers, conventional designs with taut metallic polymer over a counter electrode were investigated as well as an array comprising several capacitive micromachined transducers (CMUTs). For piezoelectric transducers, both bending mode resonators [i.e., piezoelectric ceramic (PZT) ceramic and polyvinylidene fluoride (PVDF) foil] and thickness mode resonators [i.e., multilayer stack actuator and electromechanical film (EMFi)] were evaluated in more detail. Based on a thorough assessment of the pros and cons of the various approaches, detailed in [17], EMFi-based transducers were selected because they matched best the specifications given in Table 17.2.

The EMFi is a polypropylene film; its thickness for sensor and actuator applications is typically 30–70 μm. The film has a cellular structure which results from its manufacturing process [18]. Charge is injected into the polypropylene film by a corona method making use of a strong electric field with $E_{polarization} \simeq 10\,kV/cm$. The resulting buildup of internal charge at the surfaces of the cavities turns the latter into macroscopic dipoles (see Fig. 17.6a) which retain their dipole moments after the polymer cools to room temperature. The cellular structure and the macroscopic dipoles result in relatively high piezo constants $d_{33} = 130 - 450\,pC/N$.

A simple prototype comprising a patch of EMFi (10 × 10 mm) mounted on a copper plate (2 mm thick) (see Fig. 17.6b) as backing was used for investigating the properties of the EMFi material. The polymer was fixed on one side using conductive adhesive so that the piezo material can oscillate in the thickness mode. Actuated with an AC voltage of up to 600 V peak to peak (pp), the deflection of the polymer was measured as a function of frequency (see Fig. 17.6c). The measurement shows a pronounced dependence of the amplitude of the displacement on the applied driving voltage. Higher sound pressure levels (SPLs) are therefore achieved at higher driving voltages. As a result of discharges occurring at the boundary of the polymer, the maximal SPL is, however, limited by a maximum driving voltage of 1000 Vpp. In addition, the relatively constant displacement of the polymer surface in the frequency range between 20 and 200 kHz together with the reciprocal nature of the transduction mechanism indicates that EMFi can also be used as broadband receiver material.

TABLE 17.2   Specifications for Ultrasonic Transducers

	Emitter	Receiver
Size	15 × 15 mm	<10 × 10 mm
Frequency band	20–200 kHz	20–200 kHz
Sound pressure level (SPL)	80–100 dB at 1 m	
Sensitivity		>1 mV/Pa
Dynamic range		>80 dB
Equivalent SPL at noise floor		<40 dB

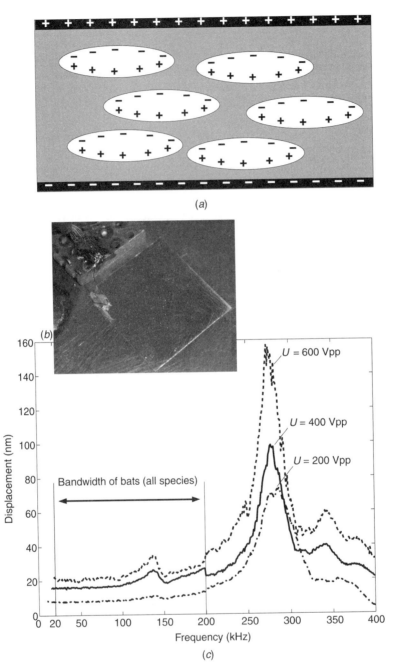

Figure 17.6   The EMFi transducer technology: (*a*) presence of permanent dipoles in polymer film gives EMFi its electrical properties; (*b*) test setup for evaluation of EMFi transducer technology; (*c*) frequency response for displacement of EMFi transmitter surface.

The properties exhibited by the EMFi material which are of direct relevance for sonar applications are summarized in Table 17.3. Because of their very good mechanical properties, EMFi-based receivers can be placed directly in the biomimetic bat head's ears without impairing their freedom of movement. Furthermore, the polymer's flexibility allows changing the shape of the transmitter surface during operation. This makes possible

TABLE 17.3 Properties of Electromechanical Film

	Pros	Cons
Mechanical behavior:	Light, thin and flexible	Sensitive surface
	Easy to process	Temperature-sensitive, $T_{max} < 90°C$
Acoustic behavior:	$\sim$ Constant deflection for [20–200] kHz	SPL <80 dB at 1 m, used as thickness vibrator
Electrical behavior:	Low power consumption	High actuation voltage, [200–1000] Vpp

the use of adaptive beam-shaping/steering techniques that are of great importance to active sensing applications. The relatively low temperature at which the material loses its electret properties complicates contacting the material somewhat, forcing the use of conductive glue instead of soldering, but has no important impact on its use as the operating temperature range can be kept below T_{max} in this application. The results summarized in Table 17.3 support the conclusion that only the transduction efficiency of the material, both as a sensor and an actuator, may prove to fall short of the requirements. Therefore further testing is currently under way in order to determine precisely to what extent optimized prototypes of sonar systems with both EMFi-based receivers and transmitters can comply with the specifications given above.

17.5 ACTUATION

The actuation subsystem of the biomimetic sonar head needs to provide the actuation mechanisms required for duplicating the essential degrees of freedom of a real bat head. In particular, independent panning and tilting of each of the two ear structures as well as shape deformation in a tightly constrained space has to be achieved. This is possible only through innovative ways for mechanically driving the biomimetic bat head.

The mechanical design of the biomimetic sonar head is to a large extent determined by the actuator technology. An overall comparison of the different candidate microactuator technologies—electromagnetic, piezoelectric, electrostatic, pneumatic, shape memory alloy, and hydraulic small-size actuators—has been conducted. It strongly indicates that the most suitable actuators for the ear rotations of the biomimetic sonar head are electromagnetic motors. For ear shape changes, electroactive polymer (EAP) actuators are currently considered the most promising technology. Electromagnetic rotary motors offer the advantage of moderate driving voltages (5–10V). Furthermore, the use of harmonic gears makes it possible to attain the required speeds and torques in conjunction with a high repeatability while at the same time staying within the geometric constraints (size motors $\sim$15 mm). Currently the most suitable electromagnetic motor that is commercially available is the EC6 motor of the Maxonmotor Company. This is a BLDC-type motor that can be combined with a microharmonic drive to provide the necessary reduction and accuracy. The whole group (motor, reduction, and 100-pulse encoder) has a size of $\oslash = 8 \times 35$ mm (for comparison this is about the size of an AAA battery) and provides an accuracy of 10 in. and 10 mNm torque. This will be adequate to meet the requirements on speed, small size, and acceleration.

An overview of state-of-the-art mechanical systems (e.g., sonar heads, vision systems with requirements similar to our design, and micromanipulation systems), detailed in [19], has led to the conclusion that parallel architectures like the Agile Eye [20] and differential architectures are the most promising configurations. The adopted design is a differential architecture, illustrated in Figures 17.7a,b. The choice has been based on considering compactness, number of joints, complexity, control issues, and requirements as to manufacturing precision. As illustrated in Figures 17.7a,b, the differential configuration combines the movements of two gears to move a third gear with the required two rotational degrees of freedom. If both motors have the same sense of rotation, the ear makes a pan movement (Fig. 17.7b). If both motors have opposite rotation senses, a tilt movement is executed (Fig. 17.7a). The differential, apart from being very compact,

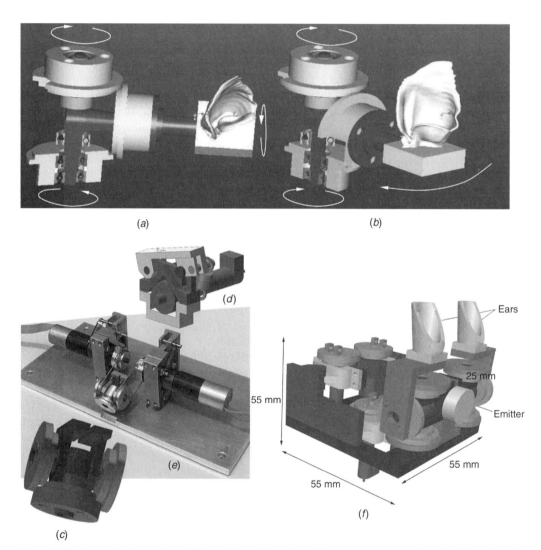

Figure 17.7 Sonar head actuation: Differential configuration of two rotary motors allows both (a) tilt and (b) pan movements of each ear in a small footprint. For tilt movements, the driving wheels are moved in opposite directions; for pan movements, they are moved in synchrony. (c–e) Details of the actuation system: (c) differential pulley; (d) motor suspension with provisions to pretension the cable; (e) test setup. The dimensions of the complete structure are indicated in the design of the biomimetic bat head shown in (f). In (a), (b), and (f), the ears and the emitter are not represented by their final shapes, which are still being worked on at the time of writing.

has the added bonus of sharing the much larger inertia of the tilt movement between the two motors. However, the absence of end-effector position information (only the motors are fitted with encoders) combined with the occurrence of backlash makes it difficult for a gear-based differential system to guarantee the required accuracy. Hence, a new differential mechanism was designed where forces and motion are transmitted through cables and pulleys. In this setup, a miniature stainless steel cable is woven through the differential pulleys to obtain a backlash-free motion of the third pulley. The price paid for this gain in accuracy is increased complexity: The cables have to be pretensioned. This complicates the motor suspension mechanism, as is clearly seen in Figures 17.7c,d,e. The complete biomimetic head, shown in Figure 17.7f, provides independent pan and tilt capabilities for both ears, that is, one differential pulley and two motors/suspensions for each ear, while still fitting inside a $55 \times 55 \times 55$-mm cube. This arrangement keeps the volume filled by the parts which are expected to interact significantly with the acoustic wave fields, similar to the head sizes of the larger bat species.

17.6 NEUROMORPHIC SIGNAL PROCESSING

Very little is known about the functional significance of many features of the mammalian hearing system in the context of biosonar. Furthermore, data from bats are unavailable for choosing the parameters of detailed neuronal models. Therefore, a parsimonious, efficient model was adopted which selectively reproduces functionally significant features of the neural code and generates a quantitatively correct representation of the code in the auditory nerve with respect to these features. The model (Fig. 17.8a) consists of linear

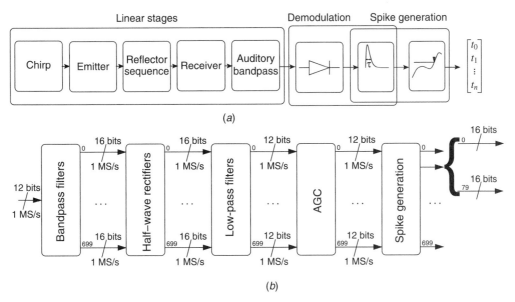

Figure 17.8 Neuromimetic signal processing system. (a) Block diagram of effective signal processing model in context of entire biosonar system description; the low-pass filtering stage (with time constant τ) can be thought of as being shared between the demodulation and the spike generation stage. (b) Summary of data flow specifications for digital implementation; the numbers given are for a single echo data stream from one ear. The complete neuromorphic system will duplicate the processing stages and data streams between them in order to support a binaural system.

bandpass filtering followed by a demodulation operation implemented as a half-wave rectification and subsequent low-pass filtering. These stages correspond to well established models of periphery of the mammalian hearing system [21]. This is followed by a simple spike generation, where spikes are caused by the signal amplitude exceeding a threshold. Together with the low-pass filtering in the demodulation stage, the thresholding operation forms an integrate-and-fire neuron model, which in turn can be regarded as an approximation of the Hodgkin–Huxley equations [22]. Since there are no data on the refractory behavior of the neurons making up the auditory nerve in bats, only the first spike triggered by an echo for every bandpass channel and firing threshold will be considered. This restricts the spike code generated by the model to a subset of the information available to natural biosonar systems, but it avoids dependencies on speculative re-creations of poorly understood system properties. The model provides for an optional automatic gain control step implemented as a normalization prior to spike generation. This stage can be viewed to mimic the natural facilities for automatic gain control present at several stages and levels of the auditory system. This normalization prior to thresholding offers the additional advantage that the spike time representation is a vector of constant length (one spike time per threshold). Therefore, all echoes share a code space of the same dimensionality, which allows for a straightforward analysis of the code statistical features and their information content with regard to certain tasks [23].

The digital implementation of the model is an extension of previous work on digital cochlea modeling of human hearing [24, 25]. The extension poses a worthy challenge because the ultrasonic frequency range of biosonar requires the system to cope with much higher data rates. In addition, a quantitatively correct representation of the auditory nerve is attempted in order to study code properties on a neural population level. For this purpose, the digital model of the auditory periphery should not only generate the spike code but also generate statistics of second and maybe higher order across bandpass channels and firing thresholds on the fly. The later capability is important in order to assess what influence different natural habitats have on the statistical dependence of spiking times in the auditory nerve.

The key advantages of a digital model are a flexible design process aided by advanced synthesis tools and the availability of mature, million-gate reprogrammable field-programmable gate array (FPGA) technology [25]. For an experimental system, where the model in conjunction with the statistical tools to characterize it is the major research objective, these advantages offset the disadvantage of greater power consumption.

The Xilinx Virtex II FPGA was chosen as a hardware platform for implementation of the model. The FPGAs are integrated into a host PC system via a peripheral component interconnect (PCI) board. Digital-to-analog conversion for the creation of arbitrary waveform, biomimetic sonar pulses and analog-to-digital conversion for echo acquisition are performed via commercial multifunction data acquisition boards (National Instruments NI 611X series). Data transfer between data acquisition and FPGA boards is performed under direct memory access (DMA) control to the main memory of the host PC. The incoming echo waveform data are digitized for each receiver (ear) at a rate of 1 megasample/ps and with 12-bit resolution (Fig. 17.8*b*). In order to maintain numerical stability, 16-bit resolution is used for the auditory bandpass filtering stages. After low-pass filtering, the resolution is decreased back to 12 bits for normalization and thresholding. Finally, spike times are reported with 16-bit word length allowing to address spike times in $2^16 = 65,536$ sampling intervals, that is, over a range of 65.5 ms corresponding to approximately 11 m in target range. Since the FPGAs can run at clock speeds much higher than the input data rate, the computations on the different bandpass channels can be multiplexed. This enables us to model at least the lower number of bandpass channels (inner hair cells) and firing thresholds (spiral ganglion cells) found in bats ([9], Table 17.1). Because the system introduced here provides a biomimetic representation of the input and matches the numbers of its natural paragon, it provides a unique experimental platform to study quantitative aspects of the spike representation formed in the auditory nerve. This allows us to investigate, for example, how neural sample sizes affect the estimator performance of the brain.

17.7   FUTURE TRENDS

The primary objective for the biomimetic system described here is to serve as a platform for experiments intended to provide hypotheses for biological function. The use of a robotic system suitable for operation in the real, physical world ensures that the explanatory power of any such hypothesis is automatically proven by virtue of a working implementation. The obtained results can then be used in experimental work performed with animals with the goal of testing the applicability of hypotheses. The investigation of echo features generated by active pinna movements is but one example of the kind of research that would benefit greatly from the availability of a small, actuated, biomimetic bat head. Irrespective of whether they mirror the actual functional significance to biological systems, principles discovered by considering the properties of such biomimetic systems can be used to augment technology. The shapes of ultrasonic transducers and indeed all technical antennae do not display the amount of sophistication seen in the pinna, nose leafs, and horseshoes of bats around the world. However, the continuing spread of devices communicating with electromagnetic wavelengths not too far from the wavelengths employed by biosonar systems opens large application areas for smart antennas. Future designs of such antennas could benefit from functional principles discovered in the study of acoustic signal processing performed by structural features on the heads of bats.

ACKNOWLEDGMENTS

This work was supported by the European Union, IST Program, Life-like Perception Systems Initiative (IST-2001-35144). The original work reviewed here is the collective effort of a six-partner consortium of the CIRCE (Chiroptera Inspired Robotic Cephaloid) project. The project partners are Universiteit Antwerpen, Friedrich-Alexander-Universität Erlangen-Nürnberg, Eberhard-Karls-Universitäat Tübingen, Katholieke Universiteit Leuven, University of Bath, University of Edinburgh, University of Southern Denmark. More information on the CIRCE project can be found at the project's website: http://www.circe-project.org. Pinna samples were provided by the Zoological Institutes of the Universities of Erlangen, Munich and Tübingen. Professors Joris Dirckx and Stefan Gea, Department of Physics, BIMEF, Universiteit Antwerpen, provided access to their MicroCT facilities as well as invaluable help in data acquisition.

REFERENCES

1. N. SUGA, J. A. SIMMONS, AND P. H. S. JEN. Peripheral specialization for fine analysis of Doppler-shifted echoes in the auditory system of the 'cf-fm' bat *Pteronotus parnellii. J. Exp. Biol.*, 63:161–192, 1975.

2. N. SUGA. Biosonar and neural computation in bats. *Sci. Am.*, 262(6):60–68, June 1990.

3. R. KUC. Biomimetic sonar recognizes objects using binaural information. *J. Acoust. Soc. Amer.*, 102(2): 689–696, August 1997.

4. H. PEREMANS, V. A. WALKER, AND J. C. T. HALLAM. 3d object localisation with a binaural sonarhead, inspirations from biology. In *IEEE International Conference*

on Robotics and Automation, Vol. 4. IEEE, New York, 1998, pp. 2795–2800.

5. R. MÜLLER. A synthetic biosonar-observer. In N. ELSNER and G. W. KREUTZBERG, Eds, *Proceedings of the 4th Meeting of the German Neuroscience Society 2001; 28th Göttingen Neurobiology Conference*, Vol. II. Thieme, Stuttgart, 2001, p. 1040.

6. R. KUC. Three dimensional tracking using qualitative sonar. *Roboti. Autonomous Syst.*, 11:213–219, 1993.

7. T. HORIUCHI AND K. HYNNA. Spike-based modeling of the ILD system in the echolocating bat. *Neural Networks*, 14:755–762, 2001.

8. H. PEREMANS, R. MÜLLER, J. M. CARMENA, AND J. C. T. HALLAM. A biomimetic platform to study perception in bats. In *Proc. of the SPIE Conf. on Sensor Fusion and Decentralized Control in Robotic Systems III*. Boston, November 2000, pp. 168–179.

9. M. VATER. Cochlear physiology and anatomy in bats. In P. E. NACHTIGALL AND P. W. B. MOORE, Eds, *Animal Sonar Processes and Performance*. Plenum Press, New York, 1988, pp. 225–242.

10. R. M. NOWAK. *Walker's Mammals of the World*, 5th ed. Johns Hopkins University Press, Baltimore, MD, 1991.

11. W. SCHROEDER, K. MARTIN, AND B. LORENSEN. *The Visualization Toolkit an Object Oriented Approach to 3D Graphics*, 3rd ed. Kitware, 2003.

12. W. E. LORENSEN AND H. E. CLINE. Marching cubes: A high resolution 3D surface construction algorithm. *Computer Graphics (Proc. SIGGRAPH 87)*, 21: 163–169, 1987.

13. W. J. SCHROEDER, J. A. ZARGE, AND W. E. LORENSEN. Decimation of triangle meshes. *Computer Graphics*, 26: 65–70, 1992.

14. G. TAUBIN, T. ZHANG, AND G. H. GOLUB. Optimal surface smoothing as filter design. In B. BUXTON AND R. CIPOLLA, Eds, *Computer Vision—ECCV '96. 4th European Conference on Computer Proceedings*, Vol. 1. Springer-Verlag, 1996, 283–292.

15. J. M. CARMENA, N. KÄMPCHEN, D. KIM, AND J. C. T. HALLAM. Artificial ears for a biomimetic sonarhead: From multiple reflectors to surfaces. *Artificial Life*, 7(2):147–69, 2001.

16. C. BIBER, S. ELLIN, E. SHENK, AND J. STEMPECK. The polaroid ultrasonic ranging system. Paper presented at the 67th Convention of the Audio Engineering Society, New York, October 1980.

17. A. STREICHER AND R. LERCH. Overview of high-potential transducer technologies, http://www.circe-project.org/results.htm, December 2002.

18. M. PAAJANEN, J. LEKKALA, AND K. KIRJAVAINEN. Electromechanical film (emfi)—A new multipurpose electret material. *Sensors and Actuators A*, 84: 95–102, 2000.

19. N. SALUSTIANO, F. SCHILLEBEECKX, H. BRUYNINCKX, AND D. REYNAERTS. Discussion of existing sonar heads suggestions for improvements, http://www.circe-project.org/results.htm, December 2002.

20. C. M. GOSSELIN, E. ST. PIERRE, AND M. GAGNÉ. On the development of the agile eye. *IEE Robot. Automation Mag.*, pp. 29–37, December 1996.

21. T. DAU AND D. PÜSCHEL. A quantitative model of the "effective" signal processing in the auditory system. I. model structure. *J. Acoust. Soc. Am.*, 99:3615–3622, 1996.

22. W. M. KISTLER, W. GERSTNER, AND J. L. VAN HEMMEN. Reduction of the Hodgkin-Huxley equations to a single-variable threshold model. *Neural Comput.* 9:1015–1045, 1997.

23. R. MÜLLER. A computational theory for the classification of natural biosonar targets based on a spike code. *Network: Comput. Neural Syst.*, 14:595–612, August 2003.

24. S. JONES, R. MEDDIS, S. C. LIM, AND A. R. TEMPLE. Toward a digital neuromorphic pitch extraction system. *IEEE Trans. Neural Networks*, 11(4):978–987, July 2000.

25. L. QIANG AND S. JONES. Overview of feasible neuromorphic implementations of reception subsystem, http://www.circe-project.org/results.htm, February 2003.

26. J. M. WOTTON, R. L. JENISON, AND D. J. HARTLEY. The combination of echolocation emission and ear reception enhances directional spectral cues of the big brown bat, *Eptesicus fuscus. J. Acoust. Soc. Am.*, 101:1723–1733, 1997.

CHAPTER 18

RETINAL IMAGE AND PHOSPHENE IMAGE: AN ANALOGY

*Luke E. Hallum, Spencer C. Chen, Gregg J. Suaning,
and Nigel H. Lovell*

In passing conversation (at parties and in the tearoom), a thumbnail description of the prosthetic vision problem often elicits the response from the informed layperson, "You need to implement the function of the photoreceptors!" An excellent first approximation—we agree that the task at hand, to some extent, is to mimic the neural processing implemented by the visual system (although the analogy goes much deeper than the retina—to the visual cortex and beyond). The aim of this chapter, however, is to ask a question that is, perhaps, easily overlooked: What can be gleaned from the optics, as opposed to the neural processing, of the eye in our work toward designing a prosthetic retina? That is, to what extent are the retinal and phosphene images analogous?

18.1 PROSTHETIC VISION PROBLEM

It has been demonstrated that stimulation via an electrode array implanted at the retina, optic nerve, or visual cortex may evoke the perception of spots of light—so-called phosphenes—in the visual field of a profoundly blind person [1–3]. Consequently, there is work afoot to provide a prosthesis in a similar vein to the cochlear implant. Notionally, the implanted array would be coupled to a photosensor (e.g., a digital video camera worn on spectacles) and processor, which acquire successive scenes and actuate electrodes accordingly. Thus, the problem at hand is to represent scenes via relatively few phosphenes—few by comparison to the abundance of neurons which form the substrate for early vision.

Indeed, the operative word is "few." Given the state of the art of interfacing electrodes and visual tissue, any prosthesis realized in the foreseeable future will comprise tens of electrodes, or a couple hundred at best. While this may make a profound difference to quality of life, it is a meager approximation of the native system. For the sake of comparison, take the cochlear implant: First made available as a therapeutic device in 1982, the implant typically involves an array of 22 electrodes which interface to spiral ganglia [4]. Under normal, biological circumstances, the front line of the auditory sensory epithelium—the hair cells—number some 16,000. The few data available from stimulation trials in human retina demonstrate irregularly shaped phosphenes subtending a visual arc on the order of

---

Handbook of Neural Engineering. Edited by Metin Akay

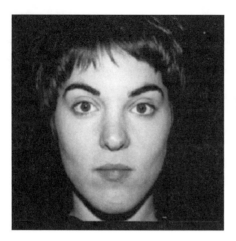 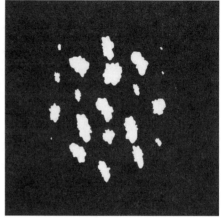

Figure 18.1   Prosthetic vision problem. The aim is to represent the scene (left) by way of relatively few phosphenes (white blobs, right). Note that here the sizes of high-contrast phosphenes are modulated; a device may ultimately effect the modulation of phosphene intensity or perhaps even both size and intensity. With further research into electrode fabrication, implantation, and interfacing to neural tissue, one may envisage the evolution of this picture: more phosphenes, smaller phosphenes, less irregularly shaped phosphenes, and a more ordered arrangement of phosphenes. (Left image source [6].)

$1°$ [5]. Phosphenes are typically disordered across the visual field and separated by several degrees. Figure 18.1 depicts the problem.

How was the phosphene image in Figure 18.1 formed? Note that the image comprises 23 phosphenes (2 of which are not activated and therefore only 21 are visible). First, the 23 phosphene locations are mapped on to the scene. Then, at each location in the scene, a spatial filter is applied which yields a scalar output. This scalar indicates the level of activation of the phosphene in question, and the phosphene's size is set accordingly.[1] Prior to setting phosphene size, the scalar output is quantized for the following reason: Since phosphenes are actuated by discrete implanted electronics, the number of phosphene sizes that can be rendered in the visual field is likely finite. With a little consideration, the analogy with early vision neurons (such as retinal ganglion cells) and their receptive fields will become readily obvious to the reader; we may think of each spatial filter as an effective artifical receptive field.

The filtering process may be conceptualized in a second way—as the operation of a single spatial filter that varies as it is shifted about the scene. A shift-variant filter like this one could be readily designed so that, when it conincided with a phosphene location, it implemented the desired artificial receptive field. Via this second conceptualization, the analogy with the eye's optics becomes apparent (keep in mind that a ray of light originating from a point on the eye's optical axis—a paraxial ray—is affected differently by the eye's optics than is an oblique ray en route to more eccentric areas of the retina (e.g., [7]). Our (shift-variant) filtered scene is then sampled at the pertinent phosphene locations. Figure 18.2 frames the problem.

Two approaches, of intuitive appeal, to forming the phosphene image come to mind: impulsive sampling and regional averaging. The latter, for each phosphene, averages the

---

[1]Once the size of a phosphene is determined, the phosphene profile is generated. The profile (a synthesis parameter) has perceptual consequences, but given the scope of this chapter, it is the analysis side (determination of phosphene size) to which we will restrict our attention.

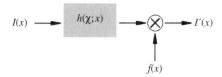

Figure 18.2   In one dimension, formation of the phosphene image. Our original scene, $I(x)$, is to be sampled so as to form the phosphene image, $I'(x)$. The locations of phosphenes in the visual field of the implantee are used to form the sampling function $f(x)$; the sampling function comprises an array of impulses for multiplication with the filtered scene. Prior to sampling, we filter via $h(\chi; x)$—a linear filter that varies as it is shifted about the scene.

intensity of the scene in the region local to the phosphene, and the phosphene is actuated accordingly. That is, $h(\chi; x)$ (as per Fig. 18.2) is a window of uniform intensity. The former simply samples the scene at phosphene locations. That is, filtering, $h(\chi; x)$, is done away with. These intuitive approaches form the basis for early attempts at simulating prosthetic vision in the literature [8, 9]. We will return to impulsive sampling and regional averaging in Section 18.5 and examine their efficacy.

The crux of this chapter is to draw an analogy between the formation of the retinal image (sometimes referred to as the receptoral image) and that of the phosphene image. Both images are discrete signals which differ, perhaps most notably, in the spatial scale by several orders of magnitude. Human vision is indeed remarkable; we are presented with such a seemingly clear "picture" of the world. This would seem at times to belie the fact that a discrete (retinal) image lies at the heart of vision. Can we use to our end in the prosthetic vision problem an understanding of the retinal image discretization process?

18.2 FORMATION OF RETINAL IMAGE

En route to the retina, a ray of light passes the eye's optical media: the cornea, aqueous and vitreous humors, and the lens. The cornea has the most marked dioptric effect—indeed, two-thirds of the bending of light occurs at the air–cornea interface—followed by the lens. The curvature of the lens, and therefore its dioptric effect, is under voluntary control; the contraction of the intraocular ciliary muscle, resulting in increased curvature of the surface of the lens, is referred to as *accommodation*. By this means, rays emanating from points of interest can be converged to a focal point on the retina. Here, photoreceptors form a mosaic that samples incident light intensity, thus forming the retinal image [10].

The dioptrics work in conjunction with the pupil (serving as the aperture) and iris (acting as the aperture stop). The diameter of the natural pupil ranges between approximately 2 and 8 mm [11], allowing for a 16-fold change in area and therefore a 16-fold change in the amount of light admitted to the eye. Two antagonistic muscles, controlled via reflex, determine pupil size: the sphincter and dilator pupillae. The former is "hardwired" to the sympathetic nervous system and the latter to the parasympathetic nervous system [12, pp. 207–227]. Therefore stress, for example, will likely cause pupillary dilation, and drugs that act on the parasympathetic system (*muscarinic agonists*) cause pupillary constriction.

Since light is a field, the finite aperture that is the pupil makes for diffraction patterns impressed on the retina. In many applications, a geometric description (light as ray) is sufficiently accurate. However, in the eye the scales of the fringes of the diffraction patterns formed are of the same order as the photoreceptor diameter [10, p. 32]. Therefore, diffraction effects are present in the retinal image, and a deeper understanding of the formation of the retinal image requires a physical description (light as a field).

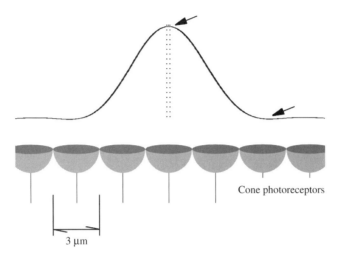

Figure 18.3   Diffraction pattern formed on retina for 2-mm pupil. A narrow ray of light (geometric pattern shown via dotted line) is in fact spread, forming an Airy intensity pattern. An Airy pattern comprises a central lobe (containing 84% of incident light energy) surrounded by progressively faint concentric rings (only one of which is shown). The intensity pattern excites photoreceptors (at the fovea, spaced by approximately 3 μm) which collectively comprise the retinal image. Arrows show the positions of the first minimum of the Airy pattern.

A diffraction pattern is shown in Figure 18.3: A point of light in object space, which we may think of as a detail in the scene, is in fact spread so as to activate numerous photoreceptors. This phenomenon is referred to as *point spread* and described mathematically by the *point spread function* [7].

18.3   SPATIAL SENSITIVITY OF RETINAL IMAGE

Since a point source of light is, in fact, spread by diffraction, it follows that diffraction poses the upper limit on the retina's sensitivity to spatial detail [13, p. 472]. For example, consider a pair of car headlights viewed at nighttime. Each light forms an Airy pattern in the retinal image. As the distance at which the headlights are viewed increases (imagine that we are in a second car that overtakes and accelerates away from the first), the Airy patterns approach each other on the retina. Once the patterns overlap to the extent that they cannot be distinguished, the spatial sensitivity of the retina has been surpassed: The two points cannot be resolved and the car headlights appear as one. Consider now the convergence of two broad Airy patterns on the retina as opposed to two narrow patterns; the broad patterns will form a homogeneous-looking lump sooner than the narrow patterns.

Our headlight example is, in fact, a lay description of Rayleigh's criterion for resolution: If two points are to be resolved, they need subtend θ_c radians of visual arc, where

$$\theta_c = \frac{1.22\lambda}{D}$$

where λ is the wavelength of light involved and D is the pupil diameter. This so-called *angular limit of resolution* is the separation for which the global maximum of one Airy pattern coincides with the first zero of the other (see arrows, Fig. 18.3), allowing for discrimination of the two maxima.

As described by the previous equation the pupil diameter and angular limit of resolution, θ_c, are inversely related. Accordingly, it would seem that the 8-mm pupil is favorable for acuity tasks such as reading and other near work: That a larger pupil makes for a more spatially sensitive retinal image. This is, however, not the case. For pupil sizes above, typically, 3 mm, aberrations in the eye's optical media make for more point spread [14]. That is, for pupil diameters beyond 3 mm, a diffraction-limited description of the eye's optics is no longer accurate. To fix ideas, it is useful to envisage the two competing mechanisms at work: The diffraction effect of pupil dilation serves to narrow the point spread function, but pupil dilation incorporates the effects of aberrations, which serve to further spread the point spread function. Clearly, the task at hand for the eye, in the case of near work, is to find some "happy medium" in the interest of retinal image quality.

18.4 PUPILLARY NEAR REFLEX

When an object is brought near for focusing—an index finger, for example, moved along the midline, from arm's length to near the nose—the *near reflex* is elicited in the viewer. Much of the literature reports the reflex as a three-way *synkinesis*:[2] The eyes converge, lenses accommodate, and the pupils constrict. The reflex makes for good viewing of the near object (finger), but not only because of accommodation (rays emanating from the near finger require more bending if they are to be brought to a focal point on the retina) and convergence (which improves depth perception). The reflex also mitigates the deleterious effect of increased point spread on the sensitivity of the retinal image, constricting to a diameter so as to exclude the effect of the eye's aberrated optical media. The assumed diameter is typically between 2 and 3 mm [14]—a diameter proven in numerous studies to suit acuity tasks [e.g., 11]. In assuming this diameter, the eye has found the aforementioned happy medium.

It is theorized that the density of foveal cones and the diffraction-limited optics of the eye have coevolved. That is, were cone packing more or less dense, the quality of the retinal image would be adversely affected. The converse of this—if the eye's diffraction-limited optics were to effect a filter of less or more spread, retinal image quality would be adversely affected—holds implications for the prosthetic vision problem. Specifically, we now have some clues as to how to design filter $h(\chi; x)$ as per Figure 18.2.

To summarize so far: Impulsively sampling the scene at each phosphene location and averaging the scene over the apertures of each phosphene have intuitive appeal in approaching the prosthetic vision problem. Eye optics, however, over and above impulsive sampling and averaging, implement point spread, which makes for a better quality retinal image. If the pupil were to constrict further, what would be the effect on the retinal image? What if the pupil had constricted less? These questions are brought to bear in the following section, where the connection is made with the prosthetic vision problem: When acuity is paramount, the filter $h(\chi; x)$ needs to take a specific form.

18.5 NUMERICAL EXPERIMENT

To demonstrate the effect of the eye's optics on retinal sensitivity, consider the following short numerical experiment. We array 100 simulated photoreceptors in one dimension,

[2]A synkinesis involves a voluntary movement effecting an involuntary movement, such as the movement of an unpatched eye in sympathy with its patched fellow.

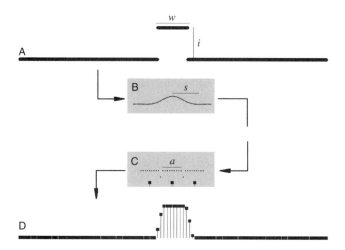

Figure 18.4 Implementation of numerical experiment. Each "continuous," one-dimensional stimulus image (example shown in A) comprises 1000 pixels. The stimulus is first analyzed by the filter shown in B—an Airy pattern simulating the blurring effect of the eye's (diffraction-limited) optics—forming the "passed image." Each receptor (the 100 receptors are shown in D, equally spaced by 10 pixels) then responds according to the passed image. Specifically, as shown in C, the average intensity in the neighborhood of a receptor determines the activation level; this simulates averaging over the aperture of cone inner segments [17]. Finally, the simulated retinal image is quantized to some number of levels, Q, and normalized—a crude simulation of the light-adaptive retina. Lowercase letters indicate experiment variables: w, stimulus width; i, stimulus intensity; o, stimulus offset from central receptor; s, point spread of simulated optics; a, aperture of receptor averaging.

equally spaced. To this array of receptors we present a randomly generated battery of (one-dimensional) stimulus images. Each image is analyzed by our receptor array; taken together, receptor outputs form the simulated retinal image. What is of interest is the number of unique retinal images our array generates in response to the battery, indicating its fitness to reconstruct the (unexpected) visual world—its sensitivity to detail. The implementation is depicted in Figure 18.4.

This experiment borrows from Snyder, Laughlin, and Stavenga [15]. Those authors used images of circles of randomly generated sizes and intensities as an input battery to an array of receptors. Therefore, the spatial-frequency content of stimulus images spanned the spectrum, and the battery may be thought of as epitomizing the unexpected. Circles also simulate the tendency of matter to cohere: An aspect of the physical world to which the visual system is attuned [16].

Accordingly, our present one-dimensional images are rectangular windows of various height, width, and horizontal offset. Since sensitivity is our present interest, parameter w varied uniformly between 1 and 10 (the spacing of receptors); o varied between -5 and 5; and i varied between 0.001 and 1.0 in increments of 0.001. Thus, our present battery may be thought of as the "epitome of the high-frequency unexpected." We set quantization levels $Q = 16$ so as to simulate implanted electronics capable of delivering 16 different stimulation levels.

We can use this numerical setup to explore the two intuitive approaches to the prosthetic vision problem, as mentioned in Section 18.1—impulsive sampling and averaging. For the former, consider the setup where there exists no filter in B (as per Fig. 18.4, $s \rightarrow 0$)—that is, the passed image is a replica of the stimulus image, simulating diffraction-free optics (an unlikely event)—and the aperture of receptors is minimal ($a = 1$; i.e., receptors do not average but simply sample the passed image). From a battery of 100 stimulus images, our receptor array can only construct two unique images. Increasing the apertures to $a = 9$ examines the second intuitive approach—averaging. Our receptor array can now construct 17 unique images. In the case of impulsive sampling, the simulated retinal image is aliasing heavily. For the purpose of this chapter, it is convenient to think of aliasing as changes in the original signal (in this case, the one-dimensional image; in the case of Fig. 18.1, the scene) that produce no change in the sampled signal (in this case, the simulated retinal image; in the case of Fig. 18.1, the phosphene image). The effect of averaging is to dealias the simulated retinal image to some extent, thus making for the reconstruction of more unique images.

Returning to our experiment, supposing we implement, as filter at B, an Airy pattern with point spread $s = 20$ (i.e., a point spread of twice the receptor spacing, which, as shown in Fig. 18.3, simulates fringes formed by the 2-mm pupil on a mosaic of receptors spaced by 3 μm). In this case, 35 unique images are reconstructed. What if the point spread of the Airy pattern is doubled, simulating a pupil that constricts further into the diffraction-limited region (beyond the "happy medium"—a concept introduced in the previous section)? The number of unique images falls to 24; our receptor mosaic has become less spatially sensitive.

It follows from the above results that neither impulsive sampling nor averaging make for an optimally spatially sensitive phosphene image. With reference again to Figures 18.1 and 18.2, the above results suggest that, if the scene were filtered with $h(\chi; x)$ equal to an Airy pattern, the spatial sensitivity of the phosphene image would be increased.

18.6 INTO THE PERIPHERY

In the previous section, we showed that spatial filtering is of particular importance to the prosthetic vision problem. Specifically, we demonstrated that a filter combining spatial averaging and an Airy pattern of particular spread make for a more sensitive simulated retinal image and may likely make for a more sensitive phosphene image. Our analogy between retinal and phosphene images thus far has concentrated on central vision: diffraction-limited optics and light incident on the all-cone retina where receptors are tightly packed. We now turn our attention to some aspects of peripheral vision.

As mentioned in Section 18.1, paraxial rays of light are affected differently by the eye's optics as compared with oblique rays. To illustrate, in this section we examine the effect of optical aberrations on off-axis light entering the eye under natural viewing conditions. In the peripheral retina, the density of receptors declines [10]. Therefore, increasingly, the eye is met with the challenge of analyzing visual stimuli that it cannot resolve. Nonetheless, the peripheral retina remains an excellent detector, if not resolver, of spatial detail and movement [18]. It so happens that the blurring effect of optical aberrations on oblique rays serves to limit, but not eliminate, aliasing from peripheral vision [19]. While neural processing (such as motion sensitivity of ganglion cells [20]) plays an important role in detection, it is nonetheless instructive to examine the eccentric point spread function and its relationship to receptor density. Again, we do so with a mind to informing the design of our shift-variant filter $h(\chi; x)$ as per Figure 18.2 for implementation in a prosthesis.

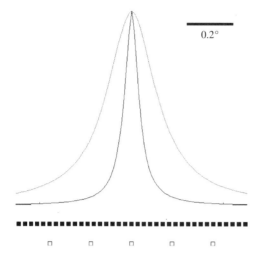

Figure 18.5   Point spread and spacing of retinal receptors. The faint line depicts the point spread function for light incident on the retina at 40° eccentricity; the solid line for 5°. Squares depict cone photoreceptor density at 40° eccentricity. (Density data from [23].)

In the 4-mm pupil of subjects allowed free accommodation, Navaro and colleagues [7] measured aerial images[3] of a laser shone through the pupil at various oblique angles. Since light forming the aerial image has passed the eye's optics twice, it is related to the point spread function via a deconvolution (effectively removing the blurring effect of the light's second pass). The modulation transfer function, describing the effect of the optics at various spatial frequencies, is related to the Fourier transform of the aerial image. The retreival of the point spread function from the aerial image is somewhat problematic.[4] However, to a first approximation, we can retreive the point spread function using Fourier theory.[5] Navarro and colleagues fitted the following biexponential to the modulation transfer function:

$$\text{MTF}(f) = (1 - C)\,\exp\,(-Af) + C\,\exp\,(-Bf)$$

Here, A, B, and C are fitting parameters, fitted at each eccentricity of interest (the fovea, 5°, 10°, 20°, 30°, 40°, 50°, 60°), and f is frequency in cycles per degree.[6]

From the equation for the modulation transfer function, we can derive the expression for the point spread function using (inverse) Fourier theory:

$$\text{PSF}(x) = \frac{A}{1 + [2\pi(x/A)]^2} - \frac{AC}{1 + [2\pi(x/A)]^2} + \frac{BC}{1 + [2\pi(x/B)]^2}$$

where x is degrees of visual arc. The derived functions for 5° and 40° are shown in Figure 18.5.

Following the eye's biological lead, filtering [$h(\chi; x)$ as per Fig. 18.2] that accounts for local phosphene density could be readily implemented. For example, the top, rightmost phosphene in Figure 18.1 has fewer near neighbors than do the more central phosphenes. Accordingly, in forming the retinal image, filtering at this phosphene location could have

---

[3]The aerial image is taken by a charge-coupled display (CCD) camera placed in front of the eye which images light reflected from the *ocular fundus*—the posterior portion of the eye's interior.

[4]The aerial image contains no phase information.

[5]Navarro et al. [7] and Pearson [21] indicate that there's likely little phase information for the 4-mm pupil.

[6]It is assumed that the point spread function is approximately circularly symmetric.

implemented a broad artificial receptive field like that depicted in Figure 18.5. Doing so would better incorporate information from the top, right-hand region of the original scene (and therefore limit aliasing). In a previous paper, the present authors suggested artificial receptive field profiles geometrically transformed in accordance with the local geometry of the phosphene mosaic [22]. So, too, we can readily envisage a scenario where a prosthesis primarily serves as a detector of movement in the scene: for example, an implantee using a prosthesis in conjunction with a cane when crossing the road.

18.7 DISORDER IN CONE MOSAIC

Hitherto, we have paid little attention to the mosaic of photoreceptors per se—the neural substrate that effectively forms the sampling function. It so happens that photoreceptors on the retina are not arranged in a regular mosaic, with regular center-to-center spacing. Rather, photoreceptors are slightly disordered [28]; it is thought that photoreceptor disorder makes for better quality spatial vision.

Photoreceptor disorder exchanges aliasing for broadband noise in the retinal image. That is, frequencies from the original signal for sampling (in our case the scene) that are beyond the Nyquist limit will not appear in the subsampled image (the phosphene image) as spurious low frequencies; instead, the subsampled image is affected by broadband noise. The visual system is more robust when faced with broadband noise as compared with aliasing, which effectively introduces "blind spots." It is beyond the scope of this chapter to provide a full elaboration, using Fourier theory as reasoning, as to the mechanism of disordered sampling; the interested reader may refer to [24]. Figure 18.6 demonstrates the phenomenon on a small scale in space.

In engineering circles, the reconstruction of signals sampled in a disordered fashion is known as Yen interpolation [25]. The reconstructed signal we yield from disordered sampling is equivalent to that of the ordered if (1) the average density of the disordered sampling function is at least equal to that of the ordered and (2) the positions at which disordered samples were taken is known to the reconstruction process. For this reason, the notion that photoreceptor mosaic disorder is a "feature" of the visual system was met with opposition [26]; for this to be true, visual centers downstream of the retina, where the elaboration of the visual world begins, would need know the locations of

Figure 18.6 Effect of disordered sampling. In both images, a stub of the underlying stimulus is shown at the top: a cosinusoid that varies in the x direction. On the left, the stimulus is sampled by a regular, rectangular mosaic. The frequency of the cosinusoid is twice that of sampling; aliasing makes for samples of uniform intensity. On the right, jitter has been added to each sample location. The result is to mitigate aliasing in exchange for broadband noise in the sampled image. (After [24].)

photoreceptors; otherwise, visual information would be lost. This sounds like a reasonably complicated developmental process.

At first glance, the disorder of phosphenes seems to have a deleterious effect on the phosphene image (see Fig. 18.1). However, when we consider the following characteristics of the phosphene image, it stands to reason that disorder in the retinal mosaic may inform design and make for better quality prosthetic vision: (1) phosphenes are loosely packed in the visual space and therefore there is ample room for the deleterious effect of aliasing and (2) the positions of phosphenes (i.e., of disordered sample locations) are known to the observer of the phosphene image (who effectively serves as the reconstruction process). Therefore, should the design criterion of many, closely packed phosphenes, likely arranged in a regular hexagonal mosaic [27], be closely followed by ensuring that phosphene locations are slightly disordered? If so, this holds implications for the manufacture of implantable arrays. This question, among others, will be borne out by continuing psychophysics experiments.

18.8 CONCLUSION

A concise, holistic description of human vision is elusive. So, too, a definitive solution to the prosthetic vision problem—that is, how to render phosphenes in the visual field so as to best convey vision—is unlikely. Work toward the former is a composite of numerous fields of endeavor. So, too, work toward the latter will benefit from points of view physiological, anatomical, physical, engineeringwise, and psychological. In this chapter, we have presented some aspects of the formation of the retinal image—dealiasing by lens–cornea, point spread as a function of eccentricity, and disorder in the photoreceptor mosaic—and suggested means of their informing a solution to the prosthetic vision problem. Such approaches will be borne out, in simulations and in human trials, by psychophysical experiments—an ongoing endeavor of our laboratory, as is the development of ideas and preliminary experiments put forward in this chapter.

ACKNOWLEDGMENT

The authors thank Dr. Paul Curmi of Physics, UNSW, for valuable insights into physical optics.

REFERENCES

1. HUMAYUN, M. S., WEILAND, J. D., FUJII, G. Y., GREENBERG, R., WILLIAMSON, R., LITTLE, J., MECH, B., CIMMARUSTI, V., VAN BOEMEL, G., DAGNELIE, G., AND DE JUAN, E. (2003). Visual perception in a blind subject with a chronic microelectronic retinal prosthesis. *Vision Res.*, 43, 2573–2581.

2. VERAART, C., WANET-DEFALQUE, M. C., GERARD, B., VANLIERDE, A., AND DELBEKE, J. (2003). Pattern recognition with the optic nerve visual prosthesis. *Artif. Organs*, 27, 996–1004.

3. BRINDLEY, G. S. AND LEWIN, W. S. (1968). The sensations produced by electrical stimulation of the visual cortex. *J. Physiol. (Lond.)*, 196, 479–493.

4. RUBINSTEIN, J. T. AND MILLER, C. A. (1999). How do cochlear prostheses work? *Curr. Opin. Neurobiol.*, 9, 399–404.

5. RIZZO, III, J. F., WYATT, J., LOEWENSTEIN, J., KELLY, S., AND SHIRE, D. (2003). Perceptual efficacy of electrical stimulation of human retina with a microelectrode array during short-term surgical trials. *Invest. Ophthalmol. Vis. Sci.*, 44, 5362–5369.

6. HANCOCK, P. (2003). Psychological image collection at Stirling, University of Stirling Psychology Department. www: http://pics.psych.stir.ac.uk; accessed July 1, 2003.

7. NAVARRO, R. N., ARTAL, P., AND WILLIAMS, D. R. (1993). Modulation transfer function of the human eye

as a function of retinal eccentricity. *J. Opt. Soc. Am. A*, 10, 201–212.

8. CHA, K., HORCH, K. W., AND NORMANN, R. A. (1992). Mobility performance with a pixelized vision system. *Vision Res.*, 32, 1367–1372.

9. THOMPSON, R. W., BARNETT, JR., G. D., HUMAYUN, M. S., AND DAGNELIE, G. (2003). Facial recognition using simulated prosthetic pixelized vision. *Invest. Ophthalmol. Vis. Sci.*, 44, 5035–5042.

10. PIRENNE, M. H. (1967). *Vision and the Eye*. Chapman and Hall: London.

11. CAMPBELL, F. W. AND GREGORY, A. H. (1961). Effect of size of pupil on visual acuity. *Nature*, 187, 1121–1123.

12. REMINGTON, L. A. (1998). *Clinical Anatomy of the Visual System*. Butterworth-Heinemann: Boston.

13. HECHT, E. (2002). *Optics*. Addison Wesley: San Francisco.

14. ATCHISON, D. A. AND SMITH, G. (2000). *Optics of the Human Eye*. Butterworth-Heinemann: Oxford.

15. SNYDER, A. W., LAUGHLIN, S. B., AND STAVENGA, D. G. (1977). Information capacity of eyes. *Vision Res.*, 17, 1163–1175.

16. MARR, D. (1982). *Vision: A Computational Investigation into the Human Representation and Processing of Visual Information*. W. H. Freeman: New York.

17. MILLER, W. H. AND BERNARD, G. D. (1983). Averaging over the foveal receptor aperture curtails aliasing. *Vision Res.*, 23, 1365–1369.

18. THIBOS, L. N., CHENEY, F. E., AND WALSH, D. J. (1987). Retinal limits to the detection and resolution of gratings. *J. Opt. Soc. Am. A*, 4, 1524–1529.

19. WILLIAMS, D. R., ARTAL, P., NAVARRO, R., MCMAHON, M. J., AND BRAINARD, D. H. (1996). Off-axis optical quality and retinal sampling in the human eye. *Vision Res.*, 36, 1103–1114.

20. DEARWORTH, J. R. AND GRANDA, A. M. (2002). Multiplied functions unify shapes of ganglion-cell receptive fields in retina of turtle. *J. Vision*, 2, 204–217, http://journalofvision.org/2/3/1/, doi:10.1167/2.3.1.

21. PEARSON, D. E. (1975). *Transmission and Display of Pictorial Information*. Wiley: London.

22. HALLUM, L. E., SUANING, G. J., TAUBMAN, D. S., AND LOVELL, N. H. (2004). Towards photosensor movement-adaptive image analysis in an electronic retinal prosthesis. In *Proceedings of the 26th Annual International Conference of the IEEE/Engineering in Medicine and Biology Society*, Sept. 2004, 1–5, San Francisco.

23. CURCIO, C. A., SLOAN, K. R., KALINA, R. E., AND HENDRICKSON, A. E. (1990). Human photoreceptor topography. *J. Compar. Neurol.*, 292, 497–523.

24. YELLOTT, J. I. (1983). Spectral consequences of photoreceptor sampling in the rhesus retina. *Science*, 221, 382–385.

25. YEN, J. L. (1956). On nonuniform sampling of bandwidth-limited signals. *IRE Trans. Circuit Theory*, 3, 251–257.

26. HIRSCH, J. AND HYLTON, R. (1984). Quality of the primate photoreceptor lattice and limits of spatial vision. *Vision Res.*, 24, 347–355.

27. HALLUM, L. E., SUANING, G. J., AND LOVELL, N. H. (2004). Contribution to the theory of prosthetic vision. *ASAIO J.*, 50, 392–396.

28. HIRSCH, J. AND MILLER, W. H. (1987). Does cone positional disorder limit resolution? *J. Opt. Soc. Am. A*, 4, 1481–1492.

CHAPTER 19

BRAIN-IMPLANTABLE BIOMIMETIC ELECTRONICS AS NEURAL PROSTHESES TO RESTORE LOST COGNITIVE FUNCTION

Theodore W. Berger, Ashish Ahuja, Spiros H. Courellis, Gopal Erinjippurath, Ghassan Gholmieh, John J. Granacki, Min Chi Hsaio, Jeff LaCoss, Vasilis Z. Marmarelis, Patrick Nasiatka, Vijay Srinivasan, Dong Song, Armand R. Tanguay, Jr., and Jack Wills

19.1 INTRODUCTION

One of the frontiers in the biomedical sciences is developing prostheses for the central nervous system (CNS) to replace higher thought processes that have been lost due to damage or disease. Prosthetic systems that interact with the CNS are currently being developed by several groups [1], though virtually all other CNS prostheses focus on sensory or motor system dysfunction and not on restoring cognitive loss resulting from damage to central brain regions. Systems designed to compensate for loss of sensory input attempt to replace the transduction of physical energy from the environment into electrical stimulation of sensory nerve fibers (e.g., cochlear implant or artificial retina) or sensory cortex [2–4]. Systems designed to compensate for loss of motor control do so through functional electrical stimulation (FES), in which preprogrammed stimulation protocols are used to activate muscular movement [5, 6], or by "decoding" premotor/motor cortical commands for control of robotic systems [7–9].

The type of neural prosthesis that performs or assists a cognitive function is qualitatively different than the cochlear implant, artificial retina, or FES. We consider here a prosthetic device that functions in a biomimetic manner to replace information transmission between cortical brain regions [10, 11]. In such a prosthesis, damaged CNS neurons would be replaced with a biomimetic system comprised of silicon neurons. The replacement silicon neurons would have functional properties specific to those of the damaged neurons and would both receive as inputs and send as outputs electrical activity to regions of the brain with which the damaged region previously communicated (Fig. 19.1). Thus, the class of prosthesis being proposed is one that would replace the computational function of damaged brain and restore the transmission of that computational result to other

Handbook of Neural Engineering. Edited by Metin Akay
Copyright © 2007 The Institute of Electrical and Electronics Engineers, Inc.

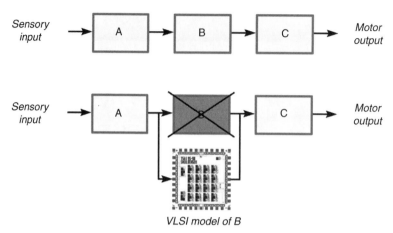

Figure 19.1   Schematic diagram for general case of replacing damaged central brain region with VLSI (very large scale integrated) system implementation of biomimetic model and connecting inputs of VLSI-based model to afferents of damaged region and outputs of VLSI-based model to efferents of damaged region.

regions of the nervous system. Such a new generation of neural prostheses would have a profound impact on the quality of life throughout society, as it would offer a biomedical remedy for the cognitive and memory loss that accompanies Alzheimer's disease, the speech and language deficits that result from stroke, and the impaired ability to execute skilled movements following trauma to brain regions responsible for motor control.

19.2   THE SYSTEM: HIPPOCAMPUS

We are in the process of developing such a cognitive prosthesis for the hippocampus, a region of the brain involved in the formation of new long-term memories. The hippocampus lies beneath the phylogenetically more recent neocortex and is comprised of several different subsystems that form a closed feedback loop (Fig. 19.2), with input from the neocortex entering via the entorhinal cortex, propagating through the intrinsic subregions of hippocampus, and then returning to neocortex. The intrinsic pathways consist of a cascade of excitatory connections organized roughly transverse to the longitudinal axis of the hippocampus. As such, the hippocampus can be conceived of as a set of interconnected, parallel circuits [12, 13]. The significance of this organizational feature is that, after removing the hippocampus from the brain, transverse "slices" (approximately 500 μm thick) of the structure may be maintained in vitro that preserve a substantial portion of the intrinsic circuitry, and thus allow detailed experimental study of its principal neurons in their open-loop condition [14, 15].

The hippocampus is responsible for what have been termed long-term "declarative" or "recognition" memories [16, 17]: the formation of mnemonic labels that identify a unifying collection of features (e.g., those comprising a person's face) and to form relations between multiple collections of features (e.g., associating the visual features of a face with the auditory features of the name for that face). In lower species not having verbal capacity, an analogous hippocampal function is evidenced by an ability, for example, to learn and remember a sequence of environmentally cued motor movements [18–22]. Major inputs to the hippocampus arise from virtually all other cortical brain regions and transmit to the hippocampus high-level features extracted by each of

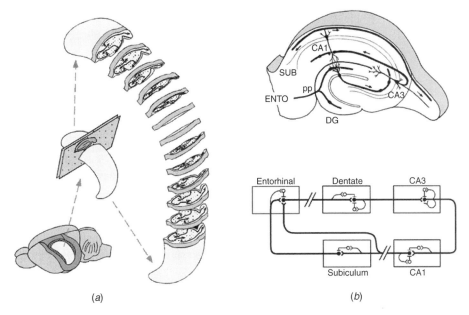

(a)   (b)

Figure 19.2   (*a*) Diagrammatic representation of rat brain (lower left) showing relative location of hippocampal formation on left side of brain (white); diagrammatic representation of left hippocampus after isolation from brain (center) and slices of hippocampus for sections transverse to longitudinal axis. (*b*) Diagrammatic representation of transverse slice hippocampus illustrating its intrinsic organization: fibers from entorhinal cortex (ENTO) project through perforant path (pp) to dentate gyrus (DG); granule cells of dentate gyrus project to CA3 region, which in turn projects to CA1 region; CA1 cells project to subiculum (SUB), which in intact brain then projects back to entorhinal cortex. In a slice preparation, return connections from CA1 and the subiculum are transected, creating an open-loop condition for experimental study of hippocampal neurons.

the sensory systems subserved by these cortical areas. Thus, the hippocampus processes both unimodal and multimodal features for virtually all classes of sensory input and modifies these neural representations so that they can be associated (e.g., forming a link between a face and a name) and stored in long-term memory in a manner that allows appropriate additional associations with previously learned information (the same face may have context-dependent names, e.g., first-name basis in an informal, social setting vs. position title in a formal, business setting) and that minimizes interference (the same name may be associated with several faces). After processing by the hippocampal system, new representations for important patterns are transmitted back to other cortical regions for long-term storage; thus, long-term memories are not stored in the hippocampus, but propagation of neural representations through its intrinsic circuitry and the transformations in those representations consequent to that propagation are required for a reencoding essential for long-term memory.

It is well established that damage to the hippocampus results in a loss of ability to form new long-term memories [16, 17]. It is the degeneration and malformation of hippocampal neurons that are the underlying cause of the memory disorders associated with Alzheimer's disease. Similarly, hippocampal pyramidal cells, particularly those in region CA1, are highly susceptible to even brief periods of anoxia, such as those that accompany stroke. Even blunt head trauma has been shown to be associated with preferential loss of hippocampal neurons in the hilus of the dentate gyrus. Finally, there is a long history of association between hippocampal dysfunction (particularly in region CA3) and epileptiform activity. Thus, there are is a wide array of neural damage and neurode-generative disease conditions for which a hippocampal prosthesis would be clinically relevant.

19.3 GENERAL STRATEGY AND SYSTEM REQUIREMENTS

Our strategy for achieving a hippocampal neural prosthesis is based on several system requirements for replacing any CNS region with a biomimetic device that interacts bidir-ectionally with the rest of the undamaged brain (i.e., sensing and communicating input to the biomimetic device from afferents of the damaged region; communicating and electri-cally stimulating outputs from the biomimetic device to efferents of the damaged region) [10, 11]. First and foremost is the nature of the biomimetic model that constitutes the core of the prosthetic system. Information in the hippocampus and all other parts of the brain is coded in terms of variation in the sequence of all-or-none, point process (spike) events, or temporal pattern (for multiple neurons, variation in the spatiotemporal pattern). The essen-tial signal processing capability of a neuron is derived from its capacity to change an input sequence of interspike intervals into a different, output sequence of interspike intervals. In all brain areas, the resulting input/output transformations are strongly nonlinear, due to the nonlinear dynamics inherent in the cellular/molecular mechanisms comprising neurons and their synaptic connections [15, 23–28]. As a consequence, the output of virtually all neurons in the brain is highly dependent on temporal properties of the input. The spatiotem-poral patterns of activity expressed by neurons in the hippocampus are the only "features" that the neocortex has to work in constructing representations for long-term memory. Thus, identifying the nonlinear input/output properties of hippocampal neurons, and the compo-site input/output transformations of hippocampal circuitry, is essential to mimicking the short-term to long-term memory reencoding process of the hippocampal system. It is this fundamental functionality of hippocampal neurons that must be captured by any mathematical model designed to replace damaged hippocampal tissue.

Attempting to accomplish this modeling goal with compartmental neuron models [29, 30] based on cable theory is simply not feasible. The number of parameters required to represent complex dendritic structures and the number and variety of ligand- and voltage-dependent conductances common to hippocampal neurons is simply too large to include in a multineuron network model that is sufficiently compact for a microchip or even a multichip module. Although simplifications of compartmental neuron models are an option, the trade-offs between complexity of the model and representation of neuron and network dynamics are not yet fully understood [25]. For this reason, we are using a nonlinear systems analytic approach to modeling hippocampal neurons [14, 15, 31–34]. In this approach, neurons and circuits or networks to be modeled are first charac-terized experimentally using a "broadband" stimulus, for example, a series of impulses (typically 1000–2000 total) in which the interimpulse intervals vary according to a random (Poisson) process. Because the distribution of interimpulse intervals is exponen-tial, the mean frequency can remain relatively low (2 Hz), and thus be physiological, yet the range of intervals can be wide (10–5000 ms). Such a stimulation protocol ensures that the majority, if not all, of synaptic and cellular mechanisms are activated and, as a conse-quence, contribute to neuron, circuit, or network output which is measured (e.g., electro-physiologically). The modeling effort then becomes focused on estimating linear and nonlinear components of the mapping of the known input to the experimentally measured output. Given accurate estimation methods [35–37], the result is a model that is "compact" (many fewer terms than a compartmental neuron network model), predictive for virtually any temporal pattern, and incorporates at least the majority of known and unknown mechanisms (thus, not requiring modification and optimization for each new discovery in the future).

A biomimetic model of a hippocampal network of neurons must be reduced to a hardware implementation in microcircuitry (VLSI system) for at least three reasons. First, for a neural prosthetic device used clinically, it will be necessary to simulate multiple neurons and neural circuits in parallel; hardware implementations provide the most efficient means for realizing parallel processing. Second, by definition the device in question will be required to interact with the intact brain in realtime if it is to substitute for lost neural function and at least partially reinstate normal levels of cognition and behavior. Again, the rapid operational rates of modern VLSI technology provide the best means for achieving the goal of real-time signal processing and sufficient computational speed to support on-going interaction of a patient with the environment. Finally, there is the practical consideration of integration of the prosthetic system with the patient, that is, it must be miniaturized sufficiently that it easily can be carried "on board" with the patient.

A biomimetic device for central brain regions must interact bidirectionally with the undamaged brain to support cognitive function and influence behavior. As discussed above, one of the indisputable characteristics of the mammalian brain is that information (a recognized object, object relations, a motor target or plan, etc.) is represented in the activity of multiple neurons, that is, population or "ensemble" coding. Anatomic connections between multiple neurons in one brain region and multiple neurons in a second brain region are not random—there is typically an identifiable topography in which neurons from one subregion of a given brain region project primarily to a localized subset of neurons in the target structure. Moreover, virtually all brain areas have region-specific cytoarchitectures, for example, varying degrees of cellular and/or dendritic layering and geometries of curvature, within which topographical connections are embedded. As a consequence of these considerations, bidirectional communication between a biomimetic device and the brain must be accomplished with one or more multisite electrode arrays, with the spatial distribution and density of recording/stimulating electrodes designed to match (be "conformal" with) the cytoarchitecture and topography density of the brain regions providing the inputs and outputs of the device.

Issues of power and data transmission can be substantial given that, at present, it is not known how many neuron models or what level of network complexity are required to achieve a clinically significant improvement in quality of life when attempting to restore cognitive brain function. In-depth treatment of these issues must await results of preclinical studies in behaving animal models (which we have scheduled as the next step in the development of a hippocampal prosthesis). Issues of long-term biocompatibility also are relevant to chronic use of implanted recording/stimulating electrodes used to interface with the brain. Although we will not discuss those issues here, we have dealt with them elsewhere and continue to research this fundamental problem [38].

19.4 PROOF OF CONCEPT: REPLACEMENT OF CA3 REGION OF HIPPOCAMPAL SLICE WITH BIOMIMETIC DEVICE

We have developed a multistage plan for achieving a neural prosthesis for the hippocampus. The first stage will be described in this chapter and involves a "proof of concept" in which we develop a replacement biomimetic model of the CA3 subregion of hippocampal in vitro slice. We have chosen to realize our first-generation prosthesis in the context of a hippocampal slice for several reasons. Among them is that the 500-μm

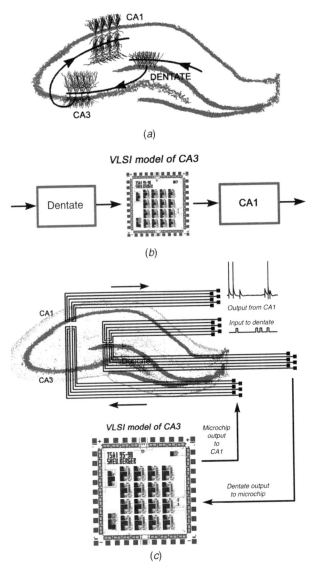

Figure 19.3 Strategy for replacing CA3 region of hippocampus with VLSI model of its nonlinear dynamics and interfacing VLSI biomimetic device with the remaining, active slice through conformal, multisite electrode array, thus restoring whole-circuit dynamics. (*a*) Diagrammatic representation of trisynaptic circuit of hippocampus. (*b*) Conceptual representation of replacing CA3 field with VLSI-based model. (*c*) Hippocampal slice in which CA3 field has been removed. Overlaid is an integrated system in which impulse stimulation from an external source is used to activate dentate granule cells and is delivered through one component of a multisite electrode array. A second component of the electrode array senses the responses of dentate granule cells and transmits the responses to the VLSI-based model. The VLSI device performs the same nonlinear input/output transformations as biological CA3 neurons and transmits the output through the multisite electrode array to the dendrites of CA1 neurons, thus activating the last component of the trisynaptic pathway.

thickness of the slice allows us essentially to reduce the problems of modeling the three-dimensional function of the hippocampus, and of interfacing with its complex, three-dimensional structure, to a more tractable two dimensions. This allows us to develop the initial stages of experimental strategies, modeling methodologies, hardware designs, and interfacing technologies within the context of a more simplified and controlled set of conditions.

The major intrinsic circuitry of the hippocampus consists of an excitatory cascade of the dentate, CA3, and CA1 subregions (dentate $\rightarrow$ CA3 $\rightarrow$ CA1; Fig. 19.3a) and is maintained in a slice preparation. Our stage 1 prosthesis demonstration consists of (i) surgically eliminating dentate input to the CA3 subregion, (ii) replacing the biological CA3 with a VLSI-based model of the nonlinear dynamics of CA3 (Figs. 19.3b,c), and (iii) through a specially designed multisite electrode array, transmitting dentate output to the VLSI model and VLSI model output to the inputs of CA1 (Fig. 19.3c). The definition of a successful implementation of the prosthesis is the propagation of spatiotemporal patterns of activity from dentate $\rightarrow$ VLSI model $\rightarrow$ CA1 reproduces that observed experimentally in the biological dentate $\rightarrow$ CA3 $\rightarrow$ CA1 circuit.

19.5 EXPERIMENTAL CHARACTERIZATION OF NONLINEAR PROPERTIES OF THE HIPPOCAMPAL TRISYNAPTIC PATHWAY

The first step in achieving the defined proof of concept is to experimentally characterize the nonlinear input/output properties both of field CA3 and of the entire trisynaptic pathway, that is, the combined nonlinearities due to propagation through the dentate $\rightarrow$ CA3 $\rightarrow$ CA1 subfields. The model of CA3 will be used to develop the biomimetic replacement device to substitute for CA3 dynamics after the biological CA3 has been removed from the slice. The input/output properties of the trisynaptic pathway will be used to evaluate the extent to which hippocampal circuit dynamics have been restored after substituting the biomimetic device for field CA3.

Completing such an experimental characterization requires a stable in vitro slice preparation in which trisynaptic responses (simultaneous recordings from dentate granule cells, CA3 pyramidal neurons, and CA1 pyramidal neurons) to random impulse train stimulation of perforant path afferents to the dentate gyrus can be recorded reproducibly, thus providing the data required for the modeling effort. After experimenting with slices prepared from several different points along the septotemporal axis of the hippocampus, varying the concentration of the chloride channel blocker picrotoxin, which was used to facilitate transsynaptic propagation through the trisynaptic circuit, and systematically sampling from different subregions of the dentate, CA3, and CA1 fields, we successfully defined an optimal preparation for characterizing hippocampal nonlinearities. Studies have been conducted using both conventional glass electrode, extracellular recording (three different electrodes in the three fields) and silicon-based multisite electrode arrays (see Section 19.8).

It is important to note that input in the form of random impulse train stimulation is applied only to the perforant path afferents to dentate. The nonlinearities of the dentate then determine the actual input to CA3; in turn, the nonlinearities of CA3 determine the input to CA1. In this stage 1 prosthesis development, field potentials are used as the measure of output from each of the three hippocampal regions: population spikes of dentate granule cells, population spikes of CA3 pyramidal cells, and population EPSPs (excitatory postsynaptic potentials) of CA1 pyramidal cells. This is important because it means that for both fields CA3 and CA1, not only does the continuous input of the random train vary in terms of interimpulse interval but also, because of nonlinearities "upstream," the input also varies in terms of the number of active afferents. As will become clear below, this places a major constraint on modeling CA3 input/output properties; that is, the model must be capable of predicting CA3 output as a function of both the temporal pattern and the amplitude of dentate input.

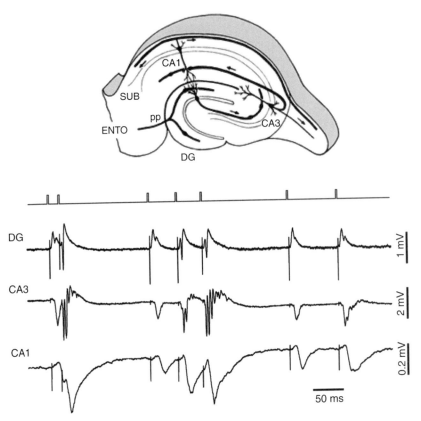

Figure 19.4   Top: diagrammatic representation of hippocampal slice (from Andersen). First trace: segment of series of impulses with randomly varying interimpulse intervals delivered in one in vitro hippocampal slice experiment to perforant path (pp) fibers originating from entorhinal cortex (ENTO) and terminating in dentate gyrus (DG). Second trace: Field potential responses recorded from DG. The narrow, biphasic deflections preceding the field potential responses are stimulation artifacts and thus correspond to the occurrence of stimulation impulses in the first trace. The large negative-going deflection in each response is the "population spike." Amplitude of the population spike correlates positively with the number of granule cells reaching threshold and generating an action potential. It is the amplitude of the population spike that is used as the measure of output from both DG and CA3. Third trace: Field potential responses recorded from CA3. The stimulation artifacts are much smaller in amplitude. Note the occurrence of multiple population spikes, which is presumed to be due to excitatory, recurrent collaterals unique to the CA3 field. Bottom trace: Field potential responses recorded from CA1. These responses are recorded from the dendritic region of CA1 and thus reflect population EPSP responses of CA1 neurons rather than population action potentials ("population spikes"). Note the progressively longer delay in the onset of response following the stimulation artifact for fields DG, CA3, and CA1, reflecting the propagation of activity through the hippocampal trisynaptic circuit.

A small segment of one random train and the evoked dentate, CA3, and CA1 field potential responses is shown in Figure 19.4. The top trace represents the impulse stimulations used to activate perforant path afferents to dentate. The strong nonlinearities of the dentate are evident in the second trace; The amplitude of the negative-going population spike varies considerably as a function of interimpulse interval. Likewise, CA3 population spikes and CA1 population EPSPs also exhibit strong variation in amplitude as a function of the temporal intervals of the train. It is the dependence of CA3 output on interimpulse interval and dentate population amplitude that must be captured by the model.

19.6 NONLINEAR DYNAMIC MODELING OF CA3 INPUT/OUTPUT PROPERTIES

Replacement of the CA3 hippocampal region requires that a sufficiently accurate quantitative representation of the CA3 functional mapping be developed, leading to a scalable implementation with reduced complexity. Considering that such quantitative representation should be developed solely relying on the input/output data sets generated experimentally, the Volterra–Poisson modeling approach was chosen and appropriately adapted to our problem. The Volterra modeling approach is a mathematically rigorous, scalable method [39, 40] with predictive capabilities that models nonlinear dynamics with an arbitrarily chosen level of accuracy, based on input/output data, and has been successfully applied in modeling biological systems [41]. The nonlinear dynamic input/output characteristics of the modeled system are quantitatively captured by the Volterra kernels. Volterra kernels are system descriptors that remain invariant irrespective of the type or the power of the stimulus.

The experimental data sets used to estimate the Volterra–Poisson model of CA3 were population spike sequences recorded at the granule cell layer (input) and the corresponding population spike sequences recorded at the pyramidal cell layer of CA3 (output). These population spike sequences were evoked by fixed-amplitude, Poisson-distributed (2 Hz mean rate) random impulse train (RIT) stimulation applied at the perforant path. The model estimation was carried out using only the amplitude and the interspike intervals of the population spikes in the input and output sequences (Fig. 19.5). Given the sequence of events in the input and the output data sets, quantitative determination of the nonlinear dynamic properties of CA3 was achieved by computing the Volterra–Poisson kernels. One of the key advantages of Volterra–Poisson kernels is their invariance with respect to the variability of the amplitude of the spikes in the stimulus sequence.

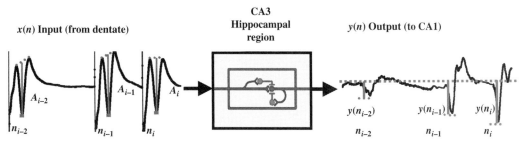

Figure 19.5 The nonlinear dynamic model of the CA3 hippocampal region receives input $x(n)$ from dentate, a Poisson point process with events (electrical impulses) of variable amplitude (A_i) and randomly varying interimpulse intervals and generates a series $y(n)$ of variable-amplitude point process events synchronous to the input.

The computation of Wiener–Poisson kernels [32, 42] and Volterra–Poisson kernels [37] has been addressed in the literature before, when the amplitude of the impulses in the input sequence was fixed. However, in our case, besides variability of the interimpulse interval there was variability in the amplitude of the impulses at the input sequence. As a result, the equation representing the single-input/single-output, third-order Volterra–Poisson model employed to capture the CA3 nonlinear dynamic properties was adapted as follows:

$$y(n_i) = A_i k_1 + A_i \sum_{n_i - \mu < n_j < n_i} A_j k_2(n_i - n_j)$$

$$+ A_i \sum_{n_i - \mu < n_{j_1} < n_i} \sum_{n_i - \mu < n_{j_2} < n_i} A_{j_1} A_{j_2} k_3(n_i - n_{j_1}, n_i - n_{j_2}) \qquad (19.1)$$

where A_i, A_j represent the varying amplitudes of the population spikes recorded at the granule cell layer (input); $y(n_i)$ represents the amplitude of the population spikes recorded at CA3 (output); k_1, k_2, and k_3 are the first-, second-, and third-order kernels, respectively; n_i is the time of occurrence of the current impulse in the input/output sequence; and n_j is the time of occurrence of the jth impulse prior to the present impulse within the kernel's memory window μ.

The estimation of the kernels was facilitated by expanding them on the orthonormal basis of the associated Laguerre functions $\mathscr{L}_l(n)$} [35] as follows:

$$k_2(n_i - n_j) = \sum_{l=0}^{L-1} a_l \mathscr{L}_l(n_i - n_j)$$

$$k_3(n_i - n_{j_1}, n_i - n_{j_2}) = \sum_{l_1=0}^{L-1} \sum_{l_2=0}^{L-1} b_{l_1 l_2} \mathscr{L}_{l_1}(n_i - n_{j_1}) \mathscr{L}_{l_2}(n_i - n_{j_2}) \qquad (19.2)$$

Combining Eq. (19.1) and (19.2), we obtain

$$y(n_i) = A_i k_1 + A_i \sum_{l=0}^{L-1} a_l \left(\sum_{n_i - \mu < n_j < n_i} A_{j_2} \mathscr{L}_l(n_i - n_j) \right)$$

$$+ A_i \sum_{l_1=0}^{L-1} \sum_{l_2=0}^{L-1} b_{l_1 l_2} \left(\sum_{n_i - \mu' < n_{j_1} < n_i} \sum_{n_i - \mu' < n_{j_2} < n_i} A_{j_1} A_{j_2} \mathscr{L}_{l_1}(n_i - n_{j_1}) \mathscr{L}_{l_2}(n_i - n_{j_2}) \right) \qquad (19.3)$$

which can be solved via least squares to yield estimates of the expansion coefficients a_l, $b_{l_1 l_2}$. The computed expansion coefficients are used to reconstruct the kernel estimates and are the quantities entered in the field-programmable gate array (FPGA) along with k_1 to implement the model on the chip for the functional replacement of CA3.

The accuracy of the model and the reliability of the kernels were evaluated using the prediction normalized mean-square error (NMSE), mathematically defined by the equation

$$\text{NMSE} = \frac{\sum_i [y_{\text{model}}(n_i) - y_{\text{data}}(n_i)]^2}{\sum_i [y_{\text{data}}(n_i)]^2} \qquad (19.4)$$

where y_{model} is the model response obtained by substituting the computed kernels into Eq. SC.1 and y_{data} is the sequence of amplitudes of the population spikes recorded at CA3. Small NMSE values indicate that the kernels reliably capture the nonlinear dynamics of the system they model. Larger NMSE values suggest that higher order terms are needed or that the data are noisy.

Data collected from several hippocampal slices were analyzed and the associated Volterra–Poisson kernels were computed. A third-order model was selected as it provided a consistent improvement in NMSE between 4 and 9% compared to a second-order model. The class of computed kernels with consistently low NMSE in the neighborhood of 6% exhibited behavior similar to the representative case shown in Figure. 19.6. In particular, the second-order kernels exhibited a fast depressive phase in the beginning followed by a brief facilitatory phase and a slow, shallow depressive phase before returning to the zero line. The third-order kernels exhibited a fast facilitatory phase in the beginning followed by slower and shallower depressive and facilitatory phases before returning to the zero plane.

One of the most important properties of a Volterra–Poisson model is its predictive capability to arbitrary input patterns. It is a property that is necessary for the CA3 model to function as a CA3 replacement, since it is not bound to a specific input sequence. Each

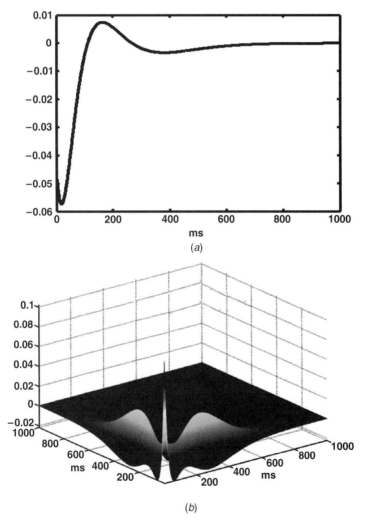

Figure 19.6   Representative second-order kernel (*a*) and third-order kernel (*b*) of CA3 nonlinear dynamic mapping.

value predicted by the Volterra–Poisson model is the resultant of the nonlinear functional terms of the corresponding Volterra–Poisson series. The nth term of the series contributes to the computation of the predicted output with the effect of nth-order interactions among the input impulses weighed by the nth-order kernel across its memory window. Thus, the second-order term represents the effect on the predicted output of the interaction between

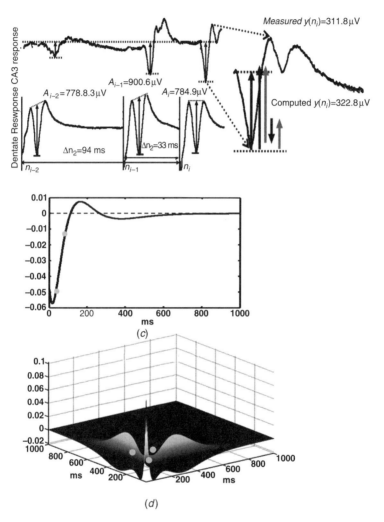

Figure 19.7 Example of model prediction. (*a*) *Bottom waveform (measured CA3 input)*: three consecutive dentate population spikes and corresponding population spike amplitudes (gray); *top waveform (measured CA3 output)*: three consecutive CA3 population spikes and corresponding population spike amplitudes (black arrows). (*b*) Detail of third population spike of top waveform in (*a*) recorded at CA3, its measured amplitude (positive, black arrow starting the peak of the population spike), and its model predicted counterpart (positive, black arrow to the right of the measured amplitude), which is sum of component attributed to first-order interactions (rightmost, positive, black arrow), component attributed to second-order interactions (negative, gray arrow to left of first-order term), and component attributed to third-order interactions (positive, gray arrow to left of second-order term). (*c*) Second-order kernel with gray dots marking its weight to contribution of second-order interactions of input population spike amplitudes shown on bottom waveform in (*a*). (*d*) Third-order kernel with gray dots marking its weight to contribution of third-order interactions among input population spike amplitudes shown on bottom waveform in (*a*).

pairs of input impulses weighed by the second-order kernel across the memory window of about 1000 ms. Similarly, the third-order term represents the effect on the predicted output of the interaction among triplets of input impulses weighed by the third-order kernel across the memory window of the kernel. An example of how each term of the Volterra–Poisson model of the CA3 contributes to the computation of predicted output of the model is shown in Figure. 19.7.

The amplitude of the third spike of the CA3 response (Fig. 19.7a, top waveform) is predicted using the computed Volterra–Poisson model of CA3. The amplitudes of the three population spikes (Fig. 19.7a, bottom waveform) measured at the dentate (CA3 input) are processed by the model of Eq. (19.1) to produce the model response for the third CA3 spike amplitude. In this case, Eq. (19.1) gives

$$
\begin{aligned}
y(n_i) = {}& A_i k_1 + \text{first–order term (rightmost positive gray arrow)} \\
& + A_i[A_{i-1} k_2(\Delta n_1) + A_{i-2} k_2(\Delta n_2)] \\
& + \text{second–order term (right negative black arrow)} \\
& + A_i[A_{i-1}A_{i-1} k_3(\Delta n_1, \Delta n_1) + A_{i-2}A_{i-2} k_3(\Delta n_2, \Delta n_2) + 2A_{i-1}A_{i-2} k_3(\Delta n_1, \Delta n_2)] \\
& + \text{third–order term (gray arrow left of 2nd-order term)} \qquad (19.5)
\end{aligned}
$$

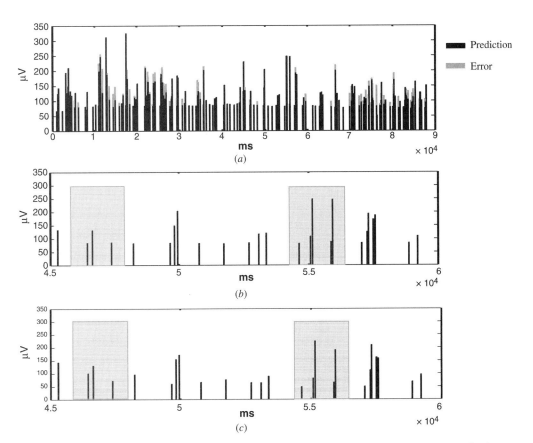

Figure 19.8 The CA3 model prediction (NMSE = 8.21%). (a) Predicted population spike amplitudes (black) and their differences from corresponding measured population spike amplitudes (gray). (b) Segment of CA3 model prediction. (c) Corresponding segment of measured CA3 population spike amplitudes. The shaded rectangles highlight two areas for comparison between model-predicted values and recorded values.

The term of each order in Eq. (19.5) contributes to the value predicted by the model with the effect of input interactions of the same order weighed by the corresponding kernel.

The predictive capability of the Volterra–Poisson model enabled us to evaluate quantitatively the accuracy of the model representing the CA3 functional mapping and the reliability of the computed kernels that reflect the nonlinear dynamic input/output properties of CA3. The measure we used for this evaluation was the prediction NMSE defined by Eq. (19.4). The small NMSE values (on the order of 6%) observed during the analysis for the selected class of models indicate that the kernels computed from the experimental data sets reliably captured the CA3 nonlinear dynamics and that the prediction accuracy of the resulting CA3 model is high. An example of model prediction using the Volterra–Poisson model developed for CA3 using the slice experimental data is shown in Figure 19.8.

19.7 MICROCIRCUITRY IMPLEMENTATION OF CA3 INPUT/OUTPUT MODEL

19.7.1 System Overview

The goal of the hardware implementation phase of the hippocampal prosthesis project is to build a highly efficient, low-power system-on-a-chip for replacing damaged portions of the hippocampus. The system, as depicted in Figure 19.9 accepts analog signals from the biological tissue, buffers and amplifies the signals, and then performs an analog-to-digital conversion (ADC). The sequence of processing is controlled by a finite-state machine (FSM), and automatic gain control (AGC) circuitry adjusts the amplitude of the input signal. Once the signals are in binary form, other circuitry identifies and calculates the amplitude of the population spike ("spike detect" in Fig. 19.9) and transmits the result to the circuitry that performs the nonlinear modeling function ("response generator" in Fig. 19.9). Each digital nonlinear response prediction is converted to a biphasic representation compatible with electrical stimulation of neuronal tissue ("output waveform

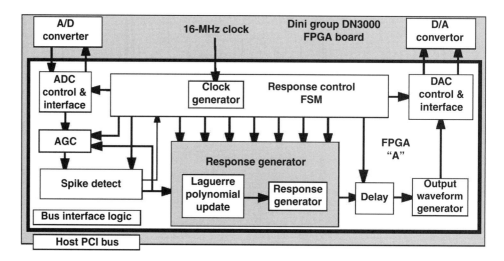

Figure 19.9   Top-level system diagram showing major functional blocks and signal flow, including relationship of real-time spike detection, response generator, and output waveform generator, which are all in digital domain.

generator" in Fig. 19.9). The result is converted from digital to analog (digital-to-analog conversion, DAC) and is transmitted to the biological tissue through a conformal multisite electrode array (described in Section 19.8).

19.7.2   Real-Time Population Spike Amplitude Extraction

Real-time population spike amplitude identification and measurement are performed in the digital domain. Digital circuitry implemented on an FPGA is used to determine the population spike amplitude by a process of filtering, differentiation, and intelligent integration. The algorithm for real-time population spike extraction was written in C language. The C programs were converted to VHSIC Hardware Description Language (VHDL) code. The VHDL description can be used to program the FPGA and to synthesize the VLSI implementation.

By using the FPGA, we have an opportunity to validate our algorithm and the VHDL in an experimental setting before we proceed with fabrication of an integrated circuit. This provides a significant level of risk reduction when we design and fabricate the integrated circuit. The FPGA we selected is configured as a plug-in card for a conventional Intel/AMD-style personal computer. The computer is used to compile the VHDL and then to download the VHDL into the FPGA. The computer also allows data logging while the FPGA is operating. In this fashion, we recorded all data samples from our ADC as well as the population spike amplitudes generated by the FPGA. This allows us to verify the accuracy and precision of the population spike identification and amplitude measurement.

We designed and built a chassis for the interface electronics that includes an instrumentation amplifier providing very high input impedance, a variable-gain amplifier (used to set the operating range of our analog-to-digital converter), a high-pass filter at 0.5 Hz, a low-pass filter at 30 kHz, and a level shifter. The filtered signal is then applied to the ADC. The converter is a 16-bit device with an input range from 0 to 5 V. The converter is clocked at 10 kHz, controlled by the FPGA. Communications between the interface electronics and the FPGA are handled by low-voltage differential signaling (LVDS).

The population spike identification and amplitude measurement algorithm consists of filtering, differentiation, and integration of signals above threshold. Spectral analysis of excitatory postsynaptic potential (EPSP) waveforms showed little energy above 500 Hz, so it was chosen as the upper cutoff, frequency. We then worked to optimize the filter design to use the smallest number of gates. The optimum choice was to use a single-section, infinite-impulse-response (IIR) low-pass filter, with a 3-dB frequency of 452 Hz. This filter is described by the recursive equation

$$y_{i+1} = y_i + \frac{x_i - y_i}{2^2}$$

and shows that the filter can be implemented with one subtraction, one addition, and a right shift of 2 bits (which is just a wiring connection). Differentiation is performed using first-order differences

$$\frac{d}{dt} x_i \approx x_i - x_{i-1}$$

The integration algorithm consists of accumulating the absolute value of the derivatives as long as they exceed a predetermined threshold.

In Figure 19.10, we show the dentate population spike (detector input signal) and the filtered difference values used to compute the size of the spike. The integration begins and ends when the difference value crosses software-settable thresholds (shown here as $y = 0$).

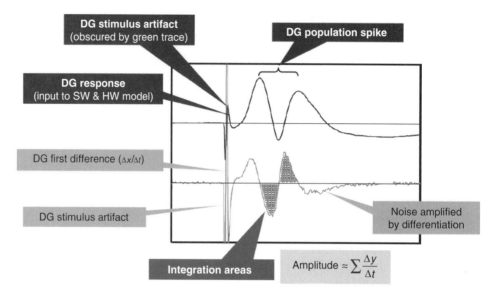

Figure 19.10   Input waveforms. Dentate gyrus response and DG population spike and results of intermediate processing: first difference and integration areas that are gated by stimulus artifact.

We have determined threshold values that deliver good accuracy but are relatively insensitive to the noise amplified by the differencing operation. Future implementations of this logic will retain the capability to adjust threshold values for greater experimental flexibility.

The population spike amplitude measurement procedures operate with very high accuracy. Figure 19.11 depicts the comparison of spike data from three experiments. The first two panels show near-perfect agreement between the floating-point software models and fixed-integer software models computed using hardware. The third panel shows a higher NMSE due to the loss of a single sample in the logging file: Because the values do not correlate in time, the point appears as an off-axis error. The detected spike agrees with the software model when the sample logging is padded by one time step.

We have completed initial design and construction of the interface electronics chassis as described earlier in this section. The interface electronics receive analog population spike

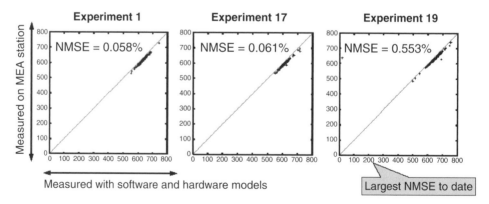

Figure 19.11   Results of three separate experiments showing close agreement of actual experimental measurements to software and hardware models. The higher error in the third panel was caused by an input sample that was not detected correctly.

waveforms from the existing experimental apparatus, which is either a multielement array built by Multi Channel Systems [57] or a multi-microelectrode plate systems built by Gunter Gros. The complete system has been repeatedly tested with live slices of hippocampal tissue to verify the algorithms and hardware implementation.

19.7.3 Hardware Implementation of Nonlinear Modeling

We have previously explored several different hybrid analog/digital designs for hardware implementation of hippocampal nonlinear input/output properties [43–47]. Due to difficulties in controlling offset currents and the lack of scalability of capacitors, however, we have opted for totally digital designs [48, 49]. As described in the previous section, nonlinearities of the hippocampal system are to be expressed by Volterra kernel functions. Each kernel function is expressed as a summation of basis functions which are chosen to be generalized Laguerre functions. By appropriate choice of the Laguerre parameter α, we obtain an excellent fit to experimental data with a small set of basis functions.

Laguerre functions can be evaluated either directly by summing a power series or indirectly by a recursive tableau. We have written computer programs implementing both methods to evaluate complexity. The recursive form of the Laguerre calculations is shown below:

$$\mathscr{L}_i(n) = \begin{cases} \sqrt{\alpha^n(1-\alpha)} & i = 0,\ n \geq 0 \\ \sqrt{\alpha}\,\mathscr{L}_{i-1}(0) & i > 0,\ n = 0 \\ \sqrt{\alpha}\,[\mathscr{L}_i(n-1) + \mathscr{L}_{i-1}] - \mathscr{L}_{i-1}(n-1) & i > 0,\ n > 0 \end{cases}$$

Recursion is desirable for hardware implementation for several reasons. First, only one time step of memory is required. We can overwrite each memory location as "stale" results are no longer needed. Second, the operating speed of the circuit can be greatly reduced: Only values required for the current time step need be calculated. Finally, the calculation never changes: Hardware required for direct computation is much less complicated than a full programmable processor. The three cases of the calculation (initialization or time-step zero, polynomial order zero, and all others) are handled by multiplexing the boundary values into the circuit. The tableau in Figure 19.12 shows the calculation of three orders of Laguerre polynomial over three time steps.

A major concern in calculation is numerical accuracy. The general requirement that implants use minimal power forces calculation with fixed-point integer arithmetic. Numerical experimentation determined that 20-bit fixed-point calculation is adequate for the Laguerre recurrence. We use signed integers and round all calculations to the closest integer. It is not necessary to use a full 20-bit multiplier; by factoring the recurrence

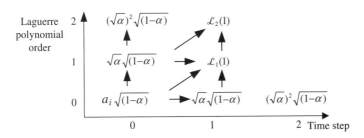

Figure 19.12 Tableau showing calculation of three orders of Laguerre polynomial over three time steps and how values are reused across time steps.

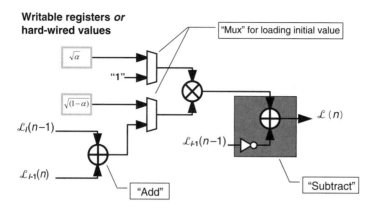

Figure 19.13 "Register transfer" level diagram of recursive function used to compute Laguerre polynomials. Note that only two adders and one multiplier are required in this "compact" implementation.

relation, we find we can subtract quantities of nearly equal magnitude obtaining a much smaller magnitude which significantly reduces the multiplier size.

It is possible to compute the Laguerre polynomials one time only and store the values for all orders and all important time lags. The remainder of the calculations could be based on table lookup. Typical values are orders $0, 1, \ldots, 8$ and time lags $0, 1000$. As a consequence, a new Laguerre recursion can be initiated each time a pulse arrives, and (instead of saving all values) the recursion relation can be used to obtain the current values of the Laguerre functions. Thus, we plan to have 10 (or fewer) recursions running simultaneously, with each recursion requiring storage of only approximately 16 numbers (the current values of the Laguerre functions of order $0-7$ as well as the values of the Laguerre functions at the previous time step). Each recursion will terminate when the Laguerre functions approach zero. Figure 19.13 depicts the block diagram of the recursive hardware showing the "compactness" of this implementation.

The logic for the CA3 response calculation can be implemented in approximately 1000 gates, a small fraction of an FPGA on a commercial board available from Dini Group in San Diego, California. To ensure real-time response, on every sample, the state of the Laguerre polynomials must be updated, the presence of a CA3 population spike detected, and a response generated if a spike is detected. Figure 19.14 shows a

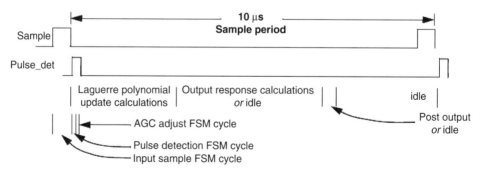

Figure 19.14 "Timing diagram" showing time sequence within 10-μs sample period of calculations performed after spike is detected.

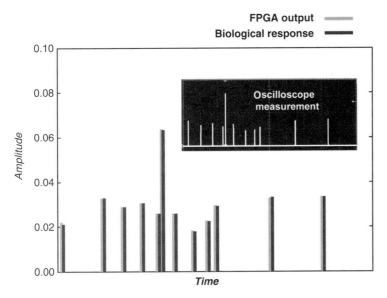

Figure 19.15   Comparison of FPGA-based nonlinear model predictions of hippocampal CA3 output with biological CA3 output as function of dentate population spike input for segment of random impulse train stimulation of perforant path. Predicted (light gray bars) and electrophysiologically recorded (dark bars) CA3 population spike amplitudes are shown for a segment of random interval train stimulation. Recorded dentate population spike amplitudes for impulses within the random interval train were used as model input.

"timing diagram" illustrating how these functions are related in time. In addition, a C program was written to demonstrate the computation when performed with integer arithmetic. Integer arithmetic is highly desirable for hardware implementations as it is very compact when compared to floating-point processors. The results of the C program were compared to values derived by the mathematical modeling group to ensure accuracy and consistency.

The C program was also used as a platform for studying the numerical accuracy required for the recurrent computation of the Laguerre polynomials used as the basis function in the mathematical model. It was determined that 16-bit signed arithmetic would produce stable and accurate results over a recursion of 1000 iterations. Figure 19.15 shows, for several inputs of a random impulse train applied to the perforant path, results of the FPGA-based CA3 model prediction compared to biological CA3 responses recorded from a hippocampal slice. It is evident that the correspondence between the hardware model and the biological system is high.

The elements were modeled using FPGA design tools which provide an integrated design management interface for the user. An example of this design method is shown in Figure 19.16, which depicts the main state machine controlling the update and response calculations. Once specified, all elements were described with the VHDL and simulated. The results of the simulation were compared to the results of the C program executions to verify the circuit would function correctly. Once the simulation was verified, the VHDL description was used to generate a configuration file for the FPGA. The configured FPGA was evaluated in several ways to determine proper operation. A data file captured during a brain-slice experiment was used to generate an input waveform.

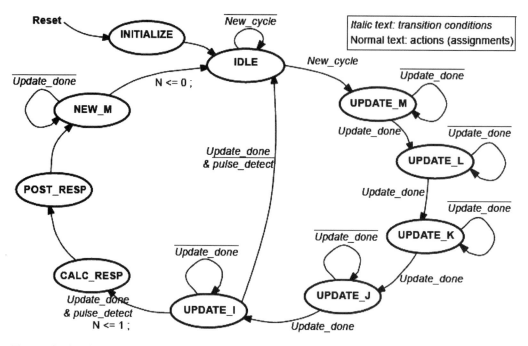

Figure 19.16   Finite-state machine that controls all update and response calculation sequencing.

19.7.4   Output Waveform Generation

Finally, the output amplitude as calculated from the Laguerre nonlinear model is divided into two roughly equivalent pulses in time and amplitude. These two digital pulses, one positive and one negative, are input to the DAC, which produces a biphasic analog pulse to stimulate the tissue.

19.8   CONFORMAL, MULTISITE ELECTRODE ARRAY INTERFACE

In this section, we describe in detail the design, fabrication, and testing of novel planar *conformal* multielectrode arrays (cMEAs) that comprise topographically mapped, spatially-distributed electrodes that conform to the local cytoarchitecture of the brain regions stimulated and recorded in a given experiment. These conformally mapped arrays allow the spatiotemporal relationships of neuronal activity to be recorded without the need for external electrodes. Such site-specific electrical simulation and recording enable efficient data collection for complete nonlinear systems modeling of the hippocampal trisynaptic pathway, as described above.

Two specific types of cMEAs have been implemented: (1) a multisite stimulation/recording MEA capable of sampling the entire trisynaptic pathway, including the dentate gyrus, CA1, and CA3 regions of rat hippocampus, and (2) a similar stimulation/recording MEA capable of stimulating at the input to the dentate, recording the output from dentate, interfacing with an external electronic or computer system that is capable of simulating the nonlinear input/output characteristics of the CA3 hippocampal region, stimulating at the input to CA1, recording the output from CA1, and interfacing once again with an external electronic or computer system. The CA3 replacement demonstration experiments to date have been based on both computer and FPGA interfaces, pointing toward full integration with an application-specific integrated circuit (ASIC) implemented in a silicon-based VLSI circuit.

Electrode pad layouts were devised to provide full coverage of the three critical regions of rat hippocampal slices for use in extracting nonlinear dynamical models and also to provide an interface between the neural tissue and the FPGA/VLSI hardware during acute hippocampal slice testing. In order to find the optimal electrode pad placement, we conducted numerous experiments with conventional sharp glass electrodes designed to locate regions of high neuronal density, both for stimulation and recording.

Based on the exact coordinates provided by these external probe experiments, we designed two sets of cMEAs, one for full coverage of the trisynaptic pathway (so-called trisynaptic cMEAs) and one for the CA3 replacement experiments (so-called CA3 replacement cMEAs). The latter arrays incorporate optimally placed sets of recording pads in the CA3 region and also have optimally placed stimulating and recording electrodes in both dentrate and CA1. As such, these cMEAs have allowed for direct interfacing with the FPGA/VLSI hardware that implements the functional nonlinear dynamical model of CA3.

The trisynaptic cMEA design, as shown in Figure 19.17, includes nine sets of seven linearly spaced 28-μm-diameter pads with 50 μm center-to-center spacing, each set spanning one of the key input/output regions of the dentate, CA3, and CA1 regions of rat hippocampus, thereby allowing for a complete diagnostic assessment of the nonlinear dynamics of the trisynaptic hippocampal circuitry.

The CA3 replacement cMEA design, as shown in Figure 19.18, includes two different circular pad sizes: (1) 28-μm-diameter pads with 50 μm center-to-center spacing are grouped in series to form sets of stimulating pads in dentate gyrus (three at a time) and CA1 (two at a time) and (2) 36-μm-diameter pads also with 50 μm center-to-center spacing are used for recording in dentate, CA3, and CA1. By grouping sets of stimulating pads in series, we are able to achieve significantly larger pad surface areas and correspondingly larger total stimulating currents than are achievable with single pads, while still maintaining essential conformality to cytoarchitecturally relevant features. These stimulating pads have been placed only in dentate and CA1 in order to interface with the FPGA/VLSI hardware that replaces the CA3 region entirely and to thereby support the

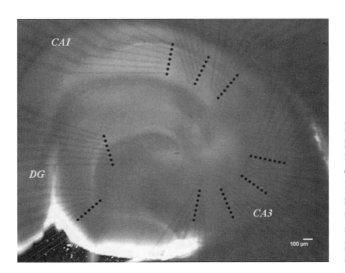

Figure 19.17 Optical photomicrograph of cMEA incorporating trisynaptic electrode pad layout proximity coupled to rat hippocampal slice. The nine sets of seven linearly spaced 28-μm-diameter pads conform to DG, CA3, and CA1 regions.

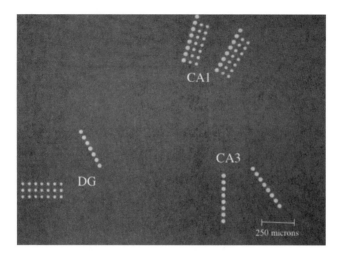

Figure 19.18 Optical photomicrograph of central region of new conformal multielectrode array specifically designed for CA3 replacement demonstration. Stimulating and recording electrodes are arranged in both the DG (lower left) and CA1 (upper middle) regions, while two sets of recording electrodes are provided in the CA3 (lower right) region for prereplacement characterization. The array pin-out is designed to be compatible with the Multi Channel Systems MEA-60 apparatus.

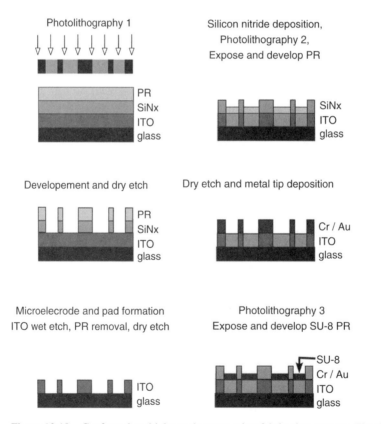

Figure 19.19 Conformal multielectrode array microfabrication process. The three-photomask lithographic fabrication procedure allows for submicrometer resolution to be achieved and allows for the design of many different electrode pad layouts based on the photomask design.

envisioned CA3 replacement experiment protocol. These cMEAs have provided a durable electrical interface between the neural tissue and the FPGA/VLSI hardware in initial CA3 replacement experiments, as described in more detail in other sections.

The general microfabrication procedure employed for these cMEAs is as follows (see Fig. 19.19). A set of indium tin oxide (ITO) electrical leads is defined on an ITO-coated glass substrate [50] of dimensions 49 mm × 49 mm × 1.1 mm by the first photomask exposure and subsequent acid-bath etching. A silicon nitride insulation layer is then deposited by means of plasma-enhanced chemical vapor deposition (PECVD). This layer electrically insulates individual ITO lines from each other and from the saline solution that the cMEA will be exposed to during acute slice testing. A second photomask is then used to define the vias that will form the *conformal* metal (Cr/Au) electrode tips. The Cr/Au metal layer is deposited by electron beam evaporation. After deposition of a second insulating layer composed of SU-8 photoresist, a final photo-lithographic exposure is performed to pattern vias through the insulation layer to the previously deposited gold electrodes. The thick (1.5-µm) epoxy-based SU-8 photoresist provides for decreased shunt capacitance of the cMEA as a whole, thereby enabling higher amplitude neural recordings. The array pinout includes 60 signal channels and is designed to be compatible with the Multi Channel Systems MEA-60 apparatus.

The development of such high-density cMEAs enables not only the CA3 replacement experiment described herein as a specific example but also the high-precision extraction of nonlinear dynamical models from other brain regions. The combination of high-density electrode arrays with conformal mapping also anticipates the development of functional cortical neural prosthetic devices. By way of comparison, it should be noted that many research groups have employed conventional MEAs (based on geometrically regular grid patterns) for purposes of neurophysiological characterization [51–55]. Currently, several such conventional arrays are commercially available (e.g., Multi Channel Systems [56] and the Center for Network Neuroscience at the University of North Texas [57]). These arrays typically feature large pad sizes (on the order of 150–250 µm) as well as electrode placements that allow for large area coverage but not for conformal topological mapping with high-density electrode placements in the region(s) of interest. The electrode array pads we have employed herein are considerably smaller in size, allowing for higher density placement, and have been accurately placed so that they conform to the specific cytoarchitecture of the brain region of interest (e.g., hippocampus).

19.9 SYSTEM INTEGRATION: RESTORATION OF HIPPOCAMPAL TRISYNAPTIC CIRCUIT DYNAMICS WITH CA3 PROSTHESIS

Recall that the goal of this multidisciplinary effort is to achieve the stage 1 prosthetic system for the hippocampal slice by integrating the components described above. More specifically, we seek to functionally replace the biological CA3 subregion of the hippocampal slice with an FPGA/VLSI-based model of the nonlinear dynamics of CA3, such that the propagation of temporal patterns of activity from dentate → VLSI model → CA1 reproduces that observed experimentally in the biological dentate → CA3 → CA1 circuit. We have successfully performed the first integration of an FPGA-based nonlinear model of the CA3 hippocampal region with the hippocampal slice by interconnecting the FPGA and the living slice through a conformal, planar, multisite electrode array. At the time of this review, we have succeeded in supporting real-time communication of

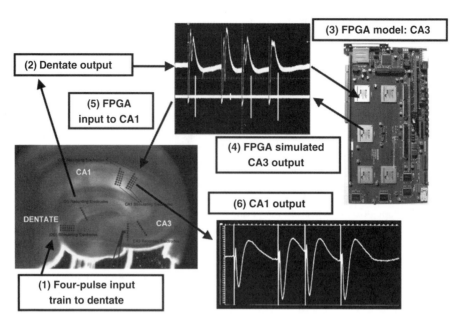

Figure 19.20 System integration of FPGA-based nonlinear model of CA3 hippocampal region with hippocampal slice in which output from dentate gyrus (mossy fibers in the hilus) has been transected, thus eliminating normal propagation of activity from dentate → CA3 → CA1. The FPGA-based model bidirectionally communicates with the living slice through a conformal, planar, multisite electrode array. This system supports real-time communication of stimulation-induced responses of dentate granule cells from the slice to the FPGA, generation of CA3-like outputs from the FPGA device, FPGA-triggered stimulation of CA1, and recording of electrophysiological output from CA1.

stimulation-induced responses of dentate granule cells from the slice to the FPGA, generation of CA3-like outputs from the FPGA device, FPGA-triggered stimulation of CA1, and recording of electrophysiological output from CA1.

The preparation we are using to test our stage 1 prosthetic system is a hippocampal slice with severed CA3 afferents (Fig. 19.20; the transection cannot be seen visually). Electrical stimulation of inputs to the dentate gyrus still evokes excitatory responses from dentate granule cells in this preparation, but excitation cannot propagate through the remainder of the trisynaptic pathway, that is, to CA3 and from CA3 to CA1. To reinstate the function of CA3 and thus reinstate propagation through the trisynaptic, we have interconnected our FPGA model of CA3 through the specially designed, conformal multisite electrode array (black dots in the slice photomicrograph) described in the previous section, which allows recording electrical activity from and stimulating electrical activity in the slice.

A demonstration of the integration of all components of our hippocampal slice prosthesis is shown in Figure 19.20. In this example, we have used as inputs to a transected slice only a subset of possible stimulation patterns and, more specifically, four-impulse trains with variable interimpulse intervals. Although limited in terms of number of impulses, the interimpulse intervals were chosen to evoke large-magnitude nonlinearities in the responses of dentate granule cells. Thus, stimulation of inputs to the dentate with a variable-interval four-impulse train (Fig. 19.20, panel 1) generates variable-amplitude output from dentate granule cells (panel 2). The output is recorded by the multisite array and transmitted to the FPGA model of CA3 (panel 3). The FPGA model A/D converts the extracellular population waveform, identifies the population spike component of that waveform, and calculates its amplitude. Based on the amplitude and history of interimpulse intervals, the FPGA-based nonlinear model of CA3 computes the appropriate CA3 output in terms of biphasic stimulation pulses (panel 4), the magnitudes of which are equivalent to the population spike amplitudes that would have been generated in the CA3 by that particular output. The biphasic output pulses generated by the FPGA are transmitted through the multisite array and used to electrically stimulate inputs to CA1 (panel 5). The resulting variable-amplitude output (population EPSP) is recorded from CA1 (panel 6), demonstrating the functional reinstatement of the dentate $\rightarrow$ CA3 $\rightarrow$ CA1 circuit. These above results demonstrate that the various components we have proposed for our prosthesis can be developed and effectively integrated into a working system.

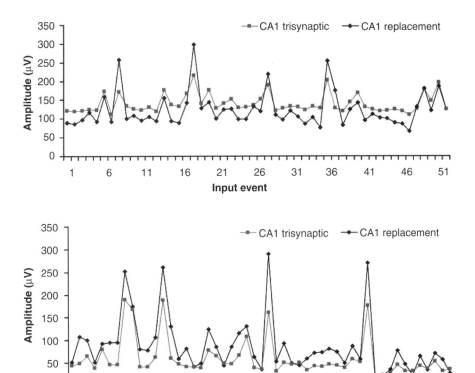

Figure 19.21   Comparison of hippocampal CA1 output in response to FPGA model of CA3 with output of CA1 in response to biological CA3. Examples of results from two random impulse train experiments are shown in the two panels, respectively: amplitudes of population EPSPs recorded from the CA1 region are shown as a function of 50 impulses chosen from among 2400 impulses of the random trains (1200 administered before transecting inputs to CA3; 1200 administered after transection). Time intervals between impulses are not represented; only "input event" number (sequence of sample impulses) is shown to "collapse" the *x* axis. Data for the intact slice (CA1 trisynaptic) are shown by gray boxes; data for the "hybrid" slice with the substituted FPGA model of CA3 (CA1 replacement) are shown by black diamonds. For what is a wide range of intervals captured in this 50-impulse sequence and what is a three- to fivefold difference in population EPSP amplitude, note that CA1 output from the hybrid slice matches extremely well the CA1 output from the intact slice.

We are currently in the process of fully evaluating the output generated in the CA1 region by our CA3 prosthesis system, that is, comparing CA1 output in response to the FPGA model of CA3 with the output of CA1 in response to the biological CA3. For these tests, we are using random impulse train stimulation of the perforant path (rather than four-impulse trains with variable interimpulse intervals as described for Fig. 19.20). Examples of results from two random impulse train experiments are shown in Figure 19.21. Each panel illustrates results from one experiment: Amplitudes of population EPSPs recorded from the CA1 region are shown as a function of 50 impulses chosen from among 2400 impulses of the random trains (1200 administered before transecting inputs to CA3; 1200 administered after transection). Time intervals between impulses are not represented in the figures; only "input event" number (sequence of sample impulses) is shown to "collapse" the x axis. Data for the intact slice (CA1 trisynaptic) are shown in gray boxes; data for the "hybrid" slice with the substituted FPGA model of CA3 (CA1 replacement) are shown in black diamonds. For what is a wide range of intervals captured in this 50-impulse sequence and what is a three- to fivefold difference in population EPSP amplitude, CA1 output from the "hybrid" slice matches extremely well the CA1 output from the intact slice. Further quantification studies are in progress.

ACKNOWLEDGMENTS

This research was supported by the Brain Restoration Foundation, DARPA DSO (HAND Program), the NSF (ERC in Biomimetic Microelectronics Systems), NSF (BITS Program), the ONR (Adaptive Neural Systems Program), the ONR (AINS Program), and the NIH/NIBIB (USC Biomedical Simulations Resource).

REFERENCES

1. BERGER, T. W., AND GLANZMAN, D. L., *Toward Replacement Parts for the Brain: Implantable Biomimetic Electronics as the Next Era in Neural Prostheses.* Cambridge, MA: MIT Press, 2005.

2. HUMAYUN, M. S., DE JUAN, JR., E., WEILAND, J. D., DAGNELIE, G., KATONA, S., GREENBERG, R. J., AND SUZUKI, S. Pattern electrical stimulation of the human retina. *Vision Res.*, 1999; 39(15): 2569–2576.

3. LOEB, G. E. Cochlear prosthetics. *Ann. Rev. Neurosci.* 1990; 13: 357–371.

4. NORMANN, R. A., WARREN, D. J., AMMERMULLER, J., FERNANDEZ, E., AND GUILLORY, S. High-resolution spatio-temporal mapping of visual pathways using multi-electrode arrays. *Vision Res.* 2001; 41: 1261–1275.

5. LOEB, G. E., PECK, R. A., MOORE, W. H., AND HOOD, K. BION™ system for distributed neural prosthetic interfaces. *Med. Eng. Phys.* 2001; 23: 9–18.

6. MAURITZ, K. H. AND PECKHAM, H. P. Restoration of grasping functions in quadriplegic patients by Functional Electrical Stimulation (FES). *Int. J. Rehab. Res.* 1987; 10: 57–61.

7. DONOGHUE, J. P. Connecting cortex to machines: Recent advances in brain interfaces. *Nature Neurosci.* 2002; 5 Suppl: 1085–1088.

8. NICOLELIS, M. A. Brain-machine interfaces to restore motor function and probe neural circuits. *Nature Rev. Neurosci.* 2003; 4: 417–422.

9. SHENOY, K. V., MEEKER, D., CAO, S., KURESHI, S. A., PESARAN, B., BUNEO, C. A., BATISTA, A. P., MITRA, P. P., BURDICK, J. W., AND ANDERSEN, R. A. Neural prosthetic control signals from plan activity. *Neuroreport* 2003; 14: 591–596.

10. BERGER, T. W., BAUDRY, M., BRINTON, R. D., LIAW, J.-S., MARMARELIS, V. Z., PARK, Y., SHEU, B. J., AND TANGUAY, JR., A. R. Brain-implantable biomimetic electronics as the next era in neural prosthetics. *Proc. IEEE* 2001; 89: 993–1012.

11. BERGER, T. W., BRINTON, R. B., MARMARELIS, V. Z., SHEU, B. J., AND TANGUAY, JR., A. R. VLSI implementations of biologically realistic hippocampal neural network models. In BERGER, T. W., AND GLANZMAN, D. L. (Eds.), *Toward Replacement Parts for the Brain: Implantable Biomimetic Electronics as the Next Era in Neural Prosthetics.* Cambridge, MA: MIT Press, in press.

12. ANDERSEN, P., BLISS, T. V. P., AND SKREDE, K. K. Lamellar organization of hippocampal excitatory pathways. *Exp. Brain Res.* 1971; 13: 222–238.

13. AMARAL, D. G., AND WITTER, M. P. The three-dimensional organization of the hippocampal formation: A review of anatomical data. *Neuroscience* 1989; 31: 571–591.

14. BERGER, T. W., HARTY, T. P., XIE, X., BARRIONUEVO, G., AND SCLABASSI, R. J. Modeling of neuronal networks through experimental decomposition. *Proc. IEEE 34th Mid Symp. Cir. Sys.* 1992; 91–97.

15. BERGER, T. W., CHAUVET, G., AND SCLABASSI, R. J. A biologically based model of functional properties of the hippocampus. *Neural Networks* 1994; 7: 1031–1064.

16. MILNER, B. Memory and the medial temporal regions of the brain. In PRIBRAM, K. H., AND BROADBENT, D. E. (Eds.), *Biology of Memory*. New York: Academic, 1970, pp. 29–50.

17. SQUIRE, L. R. AND Zola S. M. Episodic memory, semantic memory, and amnesia. *Hippocampus* 1998; 8: 205–211.

18. BERGER, T. W. AND BASSETT, J. L. System properties of the hippocampus. In GORMEZANO, I. AND WASSERMAN, E. A. (Eds.), *Learning and Memory: The Biological Substrates*. Hillsdale, NJ: Lawrence Erlbaum, 1992, pp. 275–320.

19a. EICHENBAUM, H. The hippocampus and mechanisms of declarative memory. *Behav. Brain Res.* 1999; 3: 123–133.

19b. SHAPIRO, M. L. AND EICHENBAUM, H. Hippocampus as a memory map: Synaptic plasticity and memory encoding by hippocampal neurons. *Hippocampus* 1999; 9: 365–384.

20. O'KEEFE, J. AND NADEL, L. *The Hippocampus as a Cognitive Map*. Oxford University Press, 1978.

21. DEADWYLER, S. A., BUNN, T., AND HAMPSON, R. E. Hippocampal ensemble activity during spatial delayed-nonmatch-to-sample performance in rats. *J. Neurosci.* 1996; 16: 354–372.

22. HAMPSON, R. E., SIMERAL, J. D., AND DEADWYLER, S. A. Distribution of spatial and nonspatial information in dorsal hippocampus. *Nature* 1999; 402: 610–614.

23. MAGEE, J., HOFFMAN, D., COLBERT, C., AND JOHNSTON, D. Electrical and calcium signaling in dendrites of hippocampal pyramidal neurons. *Ann. Rev. Physiol.* 1998; 60: 327–346.

24. DALAL, S. S., MARMARELIS, V. Z. AND BERGER, T. W. A nonlinear positive feedback model of glutamatergic synaptic transmission in dentate gyrus. *Proc. 4th Joint Symp. Neural Computation* 1997; 7: 68–75.

25. SONG, D., WANG, Z., MARMARELIS, V. Z., AND BERGER, T. W. Non-parametric interpretation and validation of parametric models of short-term plasticity. *Proc. IEEE EMBS Conf.* 2003; 1901–1904.

26. WANG, Z., XIE, X., SONG, D., AND BERGER, T. W. Probabilistic transformation of temporal information at individual synapses. *Proc. IEEE EMBS Conf.* 2003; 1909–1912.

27. GHOLMIEH, G., COURELLIS, S. H., SONG, D., WANG, Z., MARMARELIS, V. Z., AND BERGER, T. W. Characterization of short-term plasticity of the dentate gyrus-CA3 system using nonlinear systems analysis. *Proc. IEEE EMBS Conf.* 2003; 1929–1932.

28. DIMOKA, A., COURELLIS, S. H., SONG, D., MARMARELIS, V. Z., AND BERGER, T. W. Identification of lateral and medial perforant path using single- and dual-input random impulse train stimulation. *Proc. IEEE EMBS Conf.* 2003; 1933–1936.

29. WANG, Z., SONG, D., AND BERGER, T. W. Contribution of NMDA receptor channels to the expression of LTP in the hippocampal dentate gyrus. *Hippocampus* 2002; 12: 680–688.

30. SONG, D., WANG, Z., AND BERGER, T. W. Contribution of T-type VDCCs to TEA-induced long-term synaptic modification in hippocampal CA1 and dentate gyrus. *Hippocampus* 2002; 12: 689–697.

31. SCLABASSI, R. J., KRIEGER, D. N., AND BERGER, T. W. A systems theoretic approach to the study of CNS function. *Ann. Biomed. Eng.* 1988; 16: 17–34.

32. SCLABASSI, R. J., ERIKSSON, J. L., PORT, R., ROBINSON, G., AND BERGER, T. W. Nonlinear systems analysis of the hippocampal perforant path-dentate projection I. Theoretical and interpretational considerations. *J. Neurophysiol.* 1988; 60: 1066–1076.

33. SCLABASSI, R. J., ERIKSSON, J. L., PORT, R., ROBINSON, G., AND BERGER, T.W. Nonlinear systems analysis of the hippocampal perforant path-dentate projection II. Effects of random train stimulation. *J. Neurophysiol.* 1988; 60: 1077–1094.

34. SCLABASSI, R. J., ERIKSSON, J. L., PORT, R., ROBINSON, G., AND BERGER, T. W. Nonlinear systems analysis of the hippocampal perforant path-dentate projection III. Comparison of random train and paired impulse analyses. *J. Neurophysiol.* 1988; 60: 1095–1109.

35. MARMARELIS, V. Z. Identification of nonlinear biological systems using Laguerre expansions of kernels. *Ann. Biomed. Eng.* 1993; 21: 573–589.

36. ALATARIS, K., BERGER, T. W., AND MARMARELIS, V. Z. A novel network for nonlinear modeling of neural systems with arbitrary point-process inputs. *Neural Networks* 2000; 13: 255–266.

37. GHOLMIEH, G., COURELLIS, S. H., MARMARELIS, V. Z., AND BERGER, T. W. A novel method for modeling short-term synaptic plasticity. *J. Neurosci. Methods* 2002; 21(2): 111–127.

38. BRINTON, R. B., SOUSSOU, W., BAUDRY, M., THOMPSON, M., AND BERGER, T. W. The biotic/abiotic interface: Achievements and foreseeable challenges. In BERGER, T. W., AND GLANZMAN, D. L. (Eds.), *Toward Replacement Parts for the Brain: Implantable Biomimetic Electronics as the Next Era in Neural Prosthetics*. Cambridge, MA: MIT Press, in press.

39. WIENER, N. *Nonlinear Problems in Random Theory*. New York: The Technology Press of MIT and Wiley, 1958.

40. VOLTERRA, V. *Theory of Functions and of Integral and Integro-differential Equations.* New York: Dover Publications, 1930.

41. MARMARELIS, P. Z. AND MARMARELIS, V. Z. *Analysis of Physiological Systems: The White-Noise Approach.* New York: Plenum Press, 1978. Russian translation: Mir Press, Moscow, 1981. Chinese translation: Academy of Sciences Press, Beijing, 1990.

42. KRAUSZ, H. I. AND FRIESEN, W. O. Identification of nonlinear systems using random impulse train inputs. *Biol. Cybernet.* 1975; 19: 217–230.

43. SHEU, B. J., BERGER, T. W., WU, T. H., AND TSAI, R. H. VLSI neural network implementation of a hippocampal model. *Proc IEEE Int. Symp. Circuits Syst.* 1995; 3: 1664–1667.

44. BERGER, T. W., SHEU, B. J., AND TSAI, R. H.-J. Analog VLSI implementation of biological neural network models. *Proc. IEEE Int. Conf. Neural Networks* 1996; 12–17.

45. TSAI, R., SHEU, B., AND BERGER, T. W. Design of a programmable pulse-coded neural processor for the hippocampus. *Proc. IEEE Int. Conf. Neural Networks* 1998; 1: 784–789.

46. PARK, Y., LIAW, J.-S., SHEU, B. J., AND BERGER, T. W. Compact VLSI neural network circuit with high-capacity dynamic synapses. *Proc. IEEE Int. Conf. Neural Networks* 2000; 4: 214–218.

47. TSAI, R. H., SHEU, B. J., AND BERGER, T. W. A VLSI neural network processor based on a model of the hippocampus. *Analog Integr. Circuits Signal Process.*, 1998; 15: 201–213.

48. BERGER, T. W., GRANACKI, J. J., MARMARELIS, V. Z., SHEU, B. J., AND TANGUAY, Jr., A. R. Brain-implantable biomimetic electronics as neural prosthetics. *Proc. 1st Int. IEEE EMBS Conf. Neural Eng.* 2003; 108–111.

49. BERGER, T. W., GRANACKI, J. J., MARMARELIS, V. Z., SHEU, B. J., AND TANGUAY, Jr., A. R. Brain-implantable biomimetic electronics as neural prosthetics. *Proc. IEEE EMBS Conf.* 2003; 1956–1959.

50. Delta Technologies, Stillwater, Maine; NPL-49/11, 15 to 25 Ω/square sheet resistivity.

51. NOVAK, J. L., AND WHEELER, B. C. Multisite hippocampal slice recording and stimulation using a 32 element microelectrode array. *J. Neurosci. Meth.* 1988; 23(2): 149–159.

52. GROSS, G. W., AND SCHWALM, F. U. A closed flow chamber for long-term multichannel recording and optical monitoring. *J. Neurosci. Meth.* 1994; 52(1): 73–85.

53. GROSS, G. W., WEN, W. Y., AND LIN, J. W. Transparent indium-tin oxide electrode patterns for extracellular, multisite recording in neuronal cultures. *J. Neurosci. Meth.* 1985; 15: 243–252.

54. STOPPINI, L., DUPORT, S., AND CORREGES, P. New extracellular multirecording system for electrophysiological studies: Applications to hippocampal organotypic cultures. *J. Neurosci. Meth.* 1997; 72: 23–33.

55. BORKHOLDER, D. A., BAO, J., MALUF, N. I., PERL, E. R., AND KOVACS, G. T. A. Microelectrode arrays for stimulation of neural slice preparations. *J. Neurosci. Meth.* 1997; 77: 61–66.

56. Multi Channel Systems GmbH, Reutlingen, Germany, http://www.multichannelsystems.com/.

57. Center for Network Neuroscience, Department of Biological Sciences, University of North Texas, Denton, Texas, http://www.cnns.org/.

CHAPTER 20

ADVANCES IN RETINAL NEUROPROSTHETICS

Nigel H. Lovell, Luke E. Hallum, Spencer C. Chen, Socrates Dokos,
Philip Byrnes-Preston, Rylie Green, Laura Poole-Warren,
Torsten Lehmann, and Gregg J. Suaning

20.1 INTRODUCTION

It is indeed a lofty goal to engineer a vision prosthesis that would provide patterned vision to sufferers of profound blindness. At the time of writing, no commercial device exists that performs this function at even the most rudimentary level. There is however considerable research activity in the area. One of the most investigated approaches has been an electrical neurostimulation device comprising multiple individually addressable electrodes. Certain aspects of the technology are proven due to the long-term success of recipients of cochlear implants—devices that are used for hearing restoration in people suffering from profound deafness and severe hearing impairment. However, in terms of complexity of design, processing, and difficulties in neural interfacing and implantation, there is an order-of-magnitude increase in difficulty. Despite the inherent complexity of the task, progress in this research area has been rapid and exciting.

In this chapter, we will present a review of the state of art with regards vision neuroprostheses, with an emphasis on retinal prostheses. Through an examination of the neurobiology and biophysics at the electrode–tissue interface and by drawing heavily on biologically inspired designs, we discuss critical areas that require concentrated research and development. Briefly covered are topics relating to application-specific integrated circuits (ASICs), appropriate image processing for prosthetic vision, the psychophysics of prosthetic vision, and models of parallel stimulation at multiple electrode sites. As a means of highlighting the need for greater understanding of parallel current injection in the retina, we present modeling results from a hexagonal electrode array. This array has been designed to maximize packing density and to form guard (current return paths) to reduce leakage currents when stimulating simultaneously using parallel current sources. In two companion chapters, we present, first, a detailed discussion of design criteria for such devices with a focus on electrode array fabrication, ASIC design, and more detailed modeling of parallel stimulation using hexagonal electrode constructs [1]. Second, we conceptualize the prosthetic vision problem with regards the optics of the visual pathway and the relationship of this to the formation of a retinal image [2].

In broad terms, the retinal neuroprosthesis field has spawned from experiences gained in neurostimulation of the heart and of the cochlea. However, just as the stimulation requirements, when moving from myocardium to cochlea, were magnitudes more

Handbook of Neural Engineering. Edited by Metin Akay
Copyright © 2007 The Institute of Electrical and Electronics Engineers, Inc.

sophisticated in terms of both encoding and stimulation rates, so too are there magnitudes of difference when moving from the cochlea to the retina. Many research efforts in the vision prosthetics area have simply adapted existing neurostimulation designs employed in other organ systems. However, we advocate that significantly different approaches will be necessary in order to realize a truly efficacious retinal implant. Some of these approaches are detailed in the latter sections of this chapter.

20.2   RETINAL NEUROPROSTHESIS REVIEW

In terms of vision prosthetics, an approach that has gained much favor in recent years, and indeed looks likely to realize a device analogous to the cochlear implant, is that of a neuro-stimulator and electrode array implanted within the ocular anatomy. We review the field of retinal prosthesis from a bioengineering standpoint.

In general terms, a vision prosthesis (or bionic eye as it is commonly called) will capture an image (usually by way of an external camera), process this image into a sequence of commands, and broadcast these commands to an implantable device using a radio-frequency (RF) or laser source. Electronics in the implantable device will decode the broadcast signal and use the incipient energy of the transmitter (as recovered by the implant electronics) to inject charge into various locations of the neural retina via an electrode array implanted epiretinally (see Fig. 20.1). A variant of this approach is where

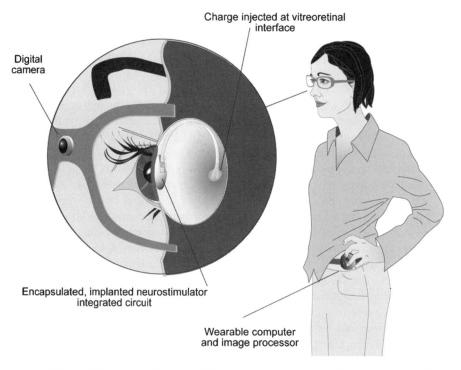

Figure 20.1   A Schematic depiction of vision prosthesis system based on epiretinal implant. An image acquired using a digital camera located on a glass frame is sent to a belt-worn processing unit. Signals from this unit are sent to an RF coil located on the lens of the glasses. An ASIC in the implanted part decodes the RF signals and uses the RF energy as a power source to direct stimulus currents to various predetermined electrodes implanted epiretinally.

the phototransduction is performed in situ by a subretinal implant. Typically such implants have much larger number of electrodes (thousands as opposed to tens or hundreds) but in many instances are unproven in terms of biocompatibility and capacity to inject sufficient charge to elicit localized visual percepts.

In a retinal prosthesis, electrical stimulation of surviving neurons is performed by an implantable device in order to elicit perception—that is, prosthetic vision. There have been a number of excellent reviews of vision prosthesis approaches, including that of Zrenner of University Eye Hospital, Tübingen [3], and Humayun's group of the University of Southern California [4–6], who addressed the issue from an ophthalmologic standpoint. Humayun also reviewed the area with an emphasis on psychophysical phenomena [7]. Hetling and Baig-Silva [8] reviewed approaches but focused on the functional interface with the retina.

20.2.1   Phosphene Phenomenon

It was the pioneering effort of Brindley and Lewin in 1968 [9] who instigated the current, concerted effort by the scientific community to provide an electronic vision prosthesis. However, the recorded phenomenon of the phosphene, the perception of light brought on by electrical stimulation, and therefore the cornerstone of all electronic vision prostheses date back to 1755. At this time, LeRoy, in investigating the application of electricity to "heal various illnesses," noted that an electrical pulse delivered to the eye produced the sensation of light [10]. In 1929, Foerster went further, applying stimulation to the occipital cortex and reporting on the perception of spots of light. In so doing, he coined the term *phosphene* [11].

Despite Foerster's cortical approach, the Australian inventor and radio engineer Graham Tassicker patented a retinal stimulator in 1953 based on the Becquerel effect (the voltage produced by the unequal illumination of two electrodes placed in an electrolyte) and the photosensitive material selenium [12]. His device was implanted in a female subject, suffering from a large central scotoma, between the choroid and sclera. After an apparent event-free recovery, the subject reported seeing uniform white light in the region of the implant and an increased confidence in mobility. Tassicker's paper indicated that further findings would be reported. These, however, were not forthcoming to the best of the authors' knowledge, and Tassicker's efforts became obscured over time, the retinal prosthesis remaining in its infancy.

In the late 1960s, Brindley and Lewin devised a helmet comprising 80 transmitters which communicated to 80 receivers, implanted at the cortical surface, by way of transcutaneous electrical induction [9]. Their subject was a 52-year-old female volunteer suffering from bilateral glaucoma whose visual impairment comprised limited perception of light in one eye and the detection of only some movement in its fellow. "Like a star in the sky"—the subject's description of a phosphene—heralded the success of phosphene vision; the simultaneous perception of some 35 phosphenes could be elicited by the apparatus, 16 of which were within 10° of the subject's point of regard. Furthermore, underlying the complex interactions in the cortex, the phosphenes moved with eye movements. A subsequent patient was able to read Braille characters using the device alone.

Subsequent to Brindley's experiments, the question arose as to where to place the implantable stimulation apparatus: at the brain itself, the optic nerve, or indeed the retina? The former attracted much early attention, and in the last 10 years focus has markedly shifted to the latter. As Brindley himself admitted, for a retinally placed prosthesis, surgical approaches are less invasive, obviating the risks of intracranial neurosurgery in otherwise healthy patients. Further, the spatial mapping of the retina, maintained between its outer and inner neuronal layers [13], is simpler and more functionally accessible than the retinotopic mapping of the visual cortex; focal stimulation has been proven in humans and animal models [14–17]; and it has been found that the inner layer retinal neurons survive in the face of outer retinal degeneration [18]. That said, the changes in the neural retina (neural deafferentation and remodeling) and in degenerative conditions such as age-related macular degeneration (AMD) and retinitis pigmentosa (RP) result in substantial loss of neural tissue [19]. These changes are progressive with the loss of the sensory retina effectively causing deafferentation of the neural retina. As a response there can be large-scale reorganization of the retina with substantial gliosis. The question as to whether degeneration of this nature invalidates a retinal approach is vexatious. While the degeneration certainly makes the interfacing to the neural retina more difficult, there is enough evidence to demonstrate that crude phosphene vision is achievable in humans with advanced retinal degeneration [14, 15, 18, 20, 21]. What is also unknown is whether a neurostimulation device chronically implanted will have a rescue effect on the neural retina and possibly arrest further degeneration.

20.2.2 Electronic Vision Prosthesis

The retinal neuroprosthesis approach proposes electrical stimulation at the retina, making use of an intact optic nerve, to restore vision for sufferers of photoreceptor degeneration, such as AMD and RP. However, for patients suffering from diseases such as glaucoma, which affects the retinal ganglion cells (RGCs) and hence the optic nerve, an electronic vision prosthesis would need to directly stimulate more central sites of the human visual system (HVS), such as the geniculocalcarine tract [22], the superior colliculus [23], or the visual cortex itself. Out of these three, by way of the pioneering work of Brindley, Donaldson, Marg, Dobelle, and Henderson [9, 22, 24–26], the latter has shown the most promise. The Dobelle group has attracted much coverage from the popular media for their implanted array of cortical surface electrodes [27]. The patient is provided phosphenized vision by way of 64 cortical surface electrodes, the phosphenes being highly disarrayed in the visual field. With training, the subject, despite making little use of the device overall, is reported to be able to see Snellen letters and count fingers, although follow-up scientific reports of this work have not appeared in recent literature.

Surface stimulation of the cortex, however, requires charge injections substantially greater than those required at the retina if it is to elicit visual response ([28], cf. [15]). Further to the possibly deleterious effects hereof, the electrotonicity of the cortex, given these relatively large charge injections, indicates that electrode densities high enough to elicit useful multifocal vision are not achievable. Several groups are in pursuit of intracortical approaches in answer to the shortcomings of surface stimulation [29–31]. Notable recent progress has been made by Bradley et al. [32] with the use of 152 iridium electrodes layered with an activated iridium oxide film, implanted in the primary visual cortex of a rhesus monkey. Open questions remain: How does one deliver meaningful perception in the face of the complex retinotopic mapping of the visual cortex? How does one mitigate the risk of intracortical neurosurgery for largely developmental work?

Alternately, a stimulator at the optic nerve would draw on less invasive surgical techniques and on procedures in related electroneurophysiological fields, such as functional electrical stimulation of afferent nerve fibers for muscle control using "cuff" electrodes [33]. Furthermore, sieve electrodes for intrafascicular recording and stimulation such as those being developed by Stieglitz and co-workers [34], in combination with promising results in optic nerve regeneration [35], may eventuate in an optic nerve stimulator with profound restorative capabilities. Belgian researchers, led by Veraart, by way of a percutaneous connector and a "self-sizing spiral-cuff" electrode with four contacts [36, 37], have chronically tested and modeled [37] an implant in a totally blind RP sufferer. The device is reported to deliver phosphenized perception, with phosphenes typically occurring in clusters of up to 30, subtending between 8 and 42 min of arc. While the "optic nerve" approach exhibits shortcomings similar to the visual cortical approach in terms of retinotopic mapping, given the small number of electrodes, the manner in which stimulation parameters have been demonstrated to modulate phosphene location, size, and luminosity is most encouraging for the future of optic nerve stimulation. Indeed after extensive training, the subject has demonstrated the ability to recognize patterns.

20.2.3 Electronic Retinal Vision Prosthesis

The approach to retinal prosthesis is twofold: subretinal and epiretinal. The former proposes a single device implanted in the subretinal space [38, 39], between the outer retina and the retinal pigment epithelium, powered by ambient light incident on an array of photodiodes integral to the device. Hypothetically, the device stimulates remaining intact neurons of the degenerate retina—predominantly, surviving bipolar cells. Herein lies an important advantage of the subretinal approach: Hypothetically any intact middle layers of the retina, and their ability to encode information, are utilized. The whole implantation of a single device would make use of the eye's existing optics and its motility and obviate the need for external, supporting electronic equipment, factors which may prove advantageous.

The epiretinal approach proposes a two-unit device [40–42] wherein an extraocular and intraocular device communicate by way of transcutaneous RF telemetry or transcorneal laser. Here, the extraocular device comprises a camera and microelectronic circuitry for encoding and transmitting stimulation patterns. The intra-ocular device receives the transmission and, by way of an electrode array leading to the vitreoretinal interface, provides controllable charge injection to intact neurons at the inner retina. Whereas the subretinal approach proposes the transduction of light and retinal stimulation in situ (at each site there is situated a photodiode and a stimulating electrode), the epiretinal approach offers configurability. Analog and digital electronics that lie functionally intermediate to the image capture and the stimulating electrodes mean that optimal stimulation parameters may be devised and then implemented without the patient undergoing further surgery (for further discussion, see [43]).

One of the issues with both subretinal and epiretinal approaches is the implantation surgery. As a possible alternate approach, our group demonstrated the efficacy of transretinal stimulation in the instance of a damaged R-membrane in the sheep [44]. Subsequently, Japanese researchers [45] have reported on successful transchoroidal stimulation in rats. However, the ability to localize phosphenes in human subjects is as yet unproven with this approach.

20.2.4 Subretinal Prosthesis

Two principal groups, one American and one German, propose a subretinal prosthesis: the American commercial entity Optobionics and a collaboration of numerous German universities, led by Eberhart Zrenner of University Eye Hospital, Tübingen. Both "subretinal" groups face the problem of photodiodes being insufficient light transducers when implanted in the subretinal space. That is, for little incident light, small photodiodes generate small charge injection which has ultimately proven insufficient in eliciting retinal neural responses. The early work (1997, 1999) of Chow et al. reported that this elicitation could, in fact, be achieved [46, 47]. These findings were inconsistent with those of Zrenner and co-workers [48], who now argue for the need for an additional power source that is gated by the photodiode response to light.

In the acute study of Chow et al. [46], relatively large, external photodiodes (with photosensitive areas of up to 5 mm by 5 mm) were coupled to bipolar electrodes implanted subretinally in the rabbit. The photodiodes were illuminated by a flash stimulator. Cortical potentials were observed; however, the experimental setup—the flash stimulator and requisite size of the photodiodes—effectively foreshadowed problems with the design's scalability. The success of a scaled device was reported two years later [47] but was subsequently discounted [49] as the rabbit retina was shown to be indeed sensitive to infrared light. Further animal studies demonstrated that, over the course of long-term implantation, the device indeed stayed fixed in the subretinal space, and, despite its inability to stimulate the visual system, was electrically functional [50]—important findings for the subretinal approach. Postexplantation histological studies, however, showed photoreceptor loss at the site of the device's implantation (electroretinograms with the device in situ revealed a 10–15% decline). Corrosion of electrodes was observed, and accordingly electrical functionality declined markedly after five months of implantation. Chow and co-workers hypothesize that increasing the ratio of photodiode area to electrode area will overcome the device's subthreshold charge injection and that the development of a chip with fenestrations, thus allowing fluid communication between the choroid and the retina, will overcome adverse effects on the retina [39].

Nonetheless, Optobionics has proceeded to human trials [51], having reported months-long implantation in 10 subjects causing no significant "migration, degradation, rejection, infection, inflammation, erosion, or retinal detachment." Here, the device, 2 mm in diameter and 25 μm thick, consists of 3500 photodiodes and was implanted $20°$ superior and temporal to the macula. Improved vision across, surprisingly, the whole visual field in patients up to 18 months postoperative has been reported. This improvement, as Optobionics concedes, is due to some secondary effect of the surgical procedure or a possible generalized neurotrophic-type rescue effect caused by the presence of the device rather than the device's functioning as specified. No subject in the study was profoundly blind and a number of subjects also underwent cataract surgery at the same time.

Gekeler and co-workers have proposed a 7600-photodiode device, likewise for implantation wholly within the subretinal space [38]. Achieving sufficient charge injection for Zrenner and co-workers, too, has proven problematic, despite their having shown the chronic acceptance of the device in animal models [52]. This has evolved into a two-unit device by way of an external unit for the provision of power [3]. At this point, one would expect that not only power but also data relating to stimulation strategies would be transmitted between the two devices, allowing the image processing strategy some additional configurability, though it is still unlikely to be as flexible as an epiretinal approach.

The German team has also developed an in vitro "sandwich" technique for hypothetical proof of their concept. Here, a 60-photodiode device is positioned subretinally and an array of recording electrodes epiretinally [53]. Light at 30 klx, 100 times that of recommended office space illumination, incident on the device has been found to generate charge injection levels capable of eliciting a retinal neural response. This approach has also proven that spatial mapping persists between outer and inner layers of the retina. Of further interest is the group's foray into optical imaging [17]. Here, microelectrode arrays were acutely implanted in cats both sub- and epiretinally. Currents of between 4 and 78 μA drove individual electrodes, and retinotopic excitation was observed at the visual cortex. The results indicate that these electrode sizes (750 μm for epiretinal; 330 μm for subretinal) could make for multifocal artificial vision with a resolution of better than $2°$ of arc. The group has devoted much effort to developing ab externo and ab interno surgical means of accessing the subretinal space [38, 54].

For the subretinal approach, the device, being hypothetically wholly contained within the eye and fixed in the subretinal space, is coupled to eye movements. If an implant recipient's eye motility functions normally, this would serve as markedly advantageous. For the normally sighted, in regarding a scene, the eye constantly saccades in order to obtain a fixed visual image independent of head movement; in the case of an externally mounted camera serving a retinal prosthesis, head movements would be the only means of the recipient's changing their point of regard. Indeed, in the psychophysical studies of Cha et al., adverse effects of head movements on balance and therefore on scene cognition were reported [55]. However, a device implanted within the subretinal space that seeks to mimic photoreceptor function would require some "hard-coded" stimulation strategy, there existing limited means for adjusting this strategy in situ. As a comparison, rudimentary analog electronic signal processing exists in most digital cameras, despite its being, at best, a poor approximation of the retina's function (e.g., see Joseph's approach to designing a complementary metal–oxide–semiconductor, or CMOS, pixel with logarithmic dynamic range [56]). This processing, to a large extent, is dependent on the initial electromechanical design and supporting electronics of the photodetector array—that is, it is decided at the design stage and apart from features such as dynamic range adjustment and pixel binning—remain relatively inflexible in comparison with software-based image processing approaches, as would be the case for an epiretinal device.

20.2.5 Epiretinal Prosthesis

The epiretinal approach, as compared with the subretinal, has accrued more data surrounding human trials and accordingly faces problems "downstream" in the evolution of a clinical device. Charge injection at the vitreoretinal interface remains little understood, as does electrically interfacing to visual mechanisms that underlie the site or sites of stimulation. In acute human trials, Rizzo and co-workers report varied success [57]. Eliciting the perception of reproducible geometric patterns remains problematic. Rizzo et al. posit that interference between so-called direct and indirect stimulation (indirect being that by way of synaptic interface with more outer layer retinal neurons) of ganglion cells contributes to smaller than expected perceived phosphenes. Modeling electrode behavior with both near and distant return electrodes at the vitreoretinal interface [58] and creative means of minimizing the invasivity of requisite surgery by way of polymeric, nanofabricated electrode arrays [59] are also under pursuit by Rizzo, Wyatt, and co-workers.

Findings by Rizzo et al. regarding biological variability and the approach of Eckmiller et al. [43] make for interesting reconciliation. Eckmiller proposes that the implant recipient is provided a user interface to a so-called learning retina encoder and can tune stimulation parameters so as to achieve perceived stimulus that well approximates visual stimulus.

Hornig and Eckmiller have simulated charge injection by very small, epiretinally positioned electrodes at idealized, small regions of the inner retina [60]. Current steering, a technique applied to functional electrical stimulation in its use of cuff electrodes in stimulating nerve fascicles [61], involves manipulating parallel bipolar stimulations so as to shift current density maxima. Said simulations indicate the efficacy of this approach in increasing information channels by increasing the number of effective sites of stimulation. It follows that central to the work by Hornig and co-workers is an eight-channel CMOS stimulator with on-chip memory, each channel being capable of delivering four different stimulation profiles [62]. Acute trials in humans of a flexible, epiretinal electrode are reported [63], subsequent to chronic experimentation in primates [64]. Here, a novel foil electrode was mounted at the vitreoretinal interface in conformal contact with ganglion cell layers and connected to supporting external electronics. Stimulation amplitudes of approximately 100 μA elicited perception. An interesting by-product of this study is the inner layer's apparent robustness in the face of repeated contact with electrodes.

Regarding charge injection, Greenberg et al. have applied computer modeling to the vitreoretinal interface [65]. At any point on the inner layer of the retina, axons from ganglion cells at farther eccentricities overlie ganglion cell somas. It is thus reasonable to assume that charge injected should innervate said axons, leading to more diffuse perceptions. Results, however, in in vivo human studies are to the contrary [15]. In seeking to explain this focal perception, Greenberg et al. find that excitation thresholds are indeed lower, but not significantly so, at or near a retinal ganglion cell soma, as compared with sites farther along its axon, for extracellular stimulation by disk electrode. Regardless, it is Humayun and co-worker's empirical data that lend much weight to the epiretinal approach over other approaches with a number of reports from human trials [66] in which chronic trials of a 16-channel epiretinal implant were conducted using a modified cochlear implant neurostimulator (Advanced Bionics). Prior to this, the six-month-long implantation of an electrically inactive platinum electrode array, affixed by way of retinal tacks, into dogs suggested feasible surgical approaches and biocompatibility [40].

The authors' team has developed a 100-channel RF stimulator for and intraocular situation [42], delivering data and power by an external, portable Linux environment [67]. Our electrode and stimulation approaches are proven to evoke cortical potentials in animal models [68]. The requisite ophthalmic surgery proves feasible on acute implantation of a nonfunctioning prototype [69]. The requisite intimate contact of epiretinal electrodes so as to achieve the elicitation of multifocal perception is subject to on-going work, as is the investigation of transretinal stimulation, which would obviate the need for substantial ocular invasion, allowing the placement of the implanted electrode array and its associated electronics in the retrobulbar space [44]. In the case that an implant recipient is only partially blind, this approach leaves open the possibility that light incident on the partially surviving retina is not obscured.

If low electrode densities prevail, as indeed seems likely in clinical trials and early generations for both epiretinal and subretinal devices, it follows that supporting platforms become all the more important. Lower electrode densities, in that individual elements of the retina likely cannot be selectively stimulated, make for less use of the HVS's innate processing—processing that will therefore befall external hardware and software should prosthetic vision be rendered optimally useful. Trialing early devices will call for the assessment of various stimulation strategies—strategies that will need to be reprogrammed without additional surgery.

While many ASICs exist for neurostimulation, only a few have been reported for electrical stimulation of the neural retinal [42, 62, 70], all of which propose RF communication to an intraocularly implanted device. Our 100-channel device [42], at its present iteration, is divided into two functional blocks of 50 electrodes each—a "stimulating" block and a "reference" block. One or more electrodes from the stimulating block are driven with respect to one or more reference electrodes; polarity is reversible in that blocks can be switched. A single current source is multiplexed to the blocks, allowing the delivery of parallel stimuli, limited only by the capacity of the current source (1860 μA). Balancing the current division across electrodes, however, is problematic, requiring prior knowledge of the electrode–tissue impedances: These cannot be assumed equal since, in practice, there exist inevitable inequalities in contact with neural tissue, insulation, and microscopic electrode surface characteristics (all of which may change over the course of device implantation). Our ASIC is bidirectional, featuring means of measuring, with moderate-accuracy electrode–tissue impedances (for a discussion on this topic, see [71]). In an updated approach by Lui et al. [72], multiple current sources are postulated along the same lines as that of Hornig [62]. However, in vitro or in vivo testing of multiple current injection sites has not been reported. This will be discussed in more detail in subsequent sections.

20.3 BIOLOGY INSPIRES DESIGN

It is readily apparent from the review of work presented that there is much research and development activity in the vision prosthesis area. It is also equally apparent that many skills and research findings need to coalesce in order to create a final solution. In the following sections, we briefly detail our design approach and then discuss some of the necessary experiments and developments that we believe will be necessary to realize a functioning device.

Interfacing electrodes with the neural retina is likely the biggest challenge in creating a vision prosthesis. Without a functional low-impedance electrode–tissue interface, it is not possible to depolarize the neural retina with a focal injection of charge within biologically safe limits. One of our research aims is to modify our epiretinal electrode for improved stability of stimulation at the neural tissue interface. This involves the development of bioactive electrodes that are surface modified or have drug-releasing capabilities for biological tissue in-growth and which will provide an environment conducive to cellular attachment at the microelectrode sites. This approach to neural prostheses provides biological factors, such as neurotrophins, to direct neuron growth onto the electrodes and extracellular matrix proteins to encourage adhesion. By developing more intimate contact between neural cells and electrodes, a more effective method of charge transfer to the excitable tissue is theoretically possible. Studies of neural in-growth in cortical electrodes have shown that this approach effectively stabilizes the tissue–electrode interface, dramatically improves the signal-to-noise ratio (SNR), and enhances long-term performance of the neural implant [73–76].

As implied in the introductory remarks, a vision prosthesis will likely require a generational advancement in current technology in order to be effective. The primary issues are the number of electrodes that can be mechanically packaged and surgical implanted in close proximity to neural tissue and a means of electrically communicating with same. Based on current neurostimulator designs, we would multiplex a single constant current source, scanning electrodes from top to bottom, to create a rasterized image in much the same way a television screen forms a picture. We have demonstrated that this is possible for typical scenes with 10×10 electrodes, communicating using a 2.5-MHz RF link. However, should we wish to increase the number of electrodes, it would not be possible to refresh all electrodes fast enough so as not to create a flicker effect. In a related fashion, we would be constrained by the bandwidth of the communications link.

We postulate that answers may lie in the underlying biology of the eye, whereby parallel signals are processed from over 100 million photoreceptors, synapsing with over 1 million RGCs. In order to adopt a parallel processing scheme, however, a number of electrode design and stimulus paradigm issues must be overcome. Before investigating these issues, a further examination of comparative retinal biology is valuable.

In 1940, L. Fejes Toth showed that the hexagonal lattice is indeed the densest packing generally, periodic or otherwise [77]. Mother Nature however predated Toth. By way of example, consider the near-perfect hexagonal arrangement of unit eyes comprising the compound eye of *Drosophila* (fly), as shown in Figure 20.2. Each fly retina is composed of approximately 750 identical unit eyes, or ommatidia. In a self-assembling manufacturing

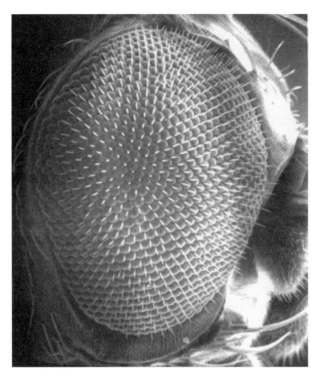

Figure 20.2 Scanning electron micrograph of adult *Drosophila* (fly) eye. Notice the near-perfect rows of ommatidia, or unit eyes, arrayed hexagonally. (Courtesy of Jamie C. Rusconi, University at Albany, State University of New York).

process the *Drosophila* eye undergoes developmentally regulated programmed cell death (apoptosis). The complex patterning stems from cell–cell interactions and is vital for sculpting tissues and deleting unnecessary cells and structures. As a final step in patterned self-assembly, unpatterned interommatidial cells are organized into an interweaving hexagonal lattice of nine secondary and tertiary pigment cells (Fig. 20.3). Indeed, approximately one-third of interommatidial cells are removed by apotosois.

20.4 PARALLEL CONCURRENT STIMULATION IN RETINA

The biology informs us of an optimal hexagonal electrode packing. It also belies the concept of parallelism. From the tightly packed retinal pigment epithelium to the myelinated fibers comprising the optic nerve, each nerve impulse is communicated in a way that does not produce crosstalk with its neighbors (except by design in the intermediate processing layers of the retina). Engineering an electrode array to achieve such selective communication in the presence of a conducting physiological saline medium is a major challenge. Fortunately, the hexagonal packing also allows the natural capture of an injected current by the surrounding six return electrodes (much like the surrounding interommatidial lattice in *Drosophila*) connected in parallel to form a guard ring.

To investigate this phenomenon, simulations were performed using a simplified 2D computational model, and the distribution of voltage arising from multiple parallel current sources connected to an electrode array was visualized. The model was formulated using a hexagonal lattice of 98 cylindrical electrodes of radius 0.2 mm and height 0.2 mm immersed in a thin volume of physiological saline measuring $20 \times 20 \times 0.2$ mm. Note

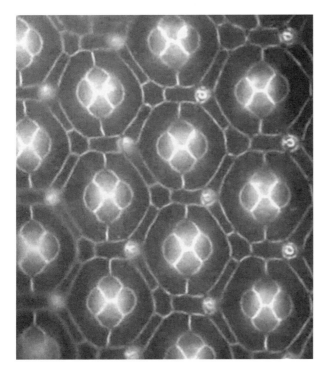

Figure 20.3   Fluorescent images of *Drosophila* (fly) retina 42 h after pupal development stained with an antibody to *Drosophila* β-catenin. Each ommatidial core is surrounded by an interommatidial lattice of secondary and tertiary pigment cells. (Courtesy of Jamie C. Rusconi, University at Albany, State University of New York.)

that this geometric arrangement is symmetric in the third z-dimension, reducing the problem to solving a 2D system. The Poisson governing equation for the voltage distribution, V, throughout the saline region was

$$-\nabla \cdot (\sigma \, \nabla V) = I \tag{20.1}$$

where σ is the conductivity of the medium and I is the volume current density injected into the medium at the given location [78]. For physiological saline, σ was taken to be 1 S/m [79]. Using cylindrical electrodes and constant σ, Eq. (20.1) reduces to the 2D partial differential system

$$\frac{\partial^2 V(x,y)}{\partial x^2} + \frac{\partial^2 V(x,y)}{\partial y^2} = \frac{-i_s^n}{\sigma \pi r^2 h} \qquad \text{within the } n\text{th stimulus electrode} \tag{20.2}$$

$$\frac{\partial^2 V(x,y)}{\partial x^2} + \frac{\partial^2 V(x,y)}{\partial y^2} = \frac{-i_r^m}{\sigma \pi r^2 h} \qquad \text{within the } m\text{th return electrode} \tag{20.3}$$

$$\frac{\partial^2 V(x,y)}{\partial x^2} + \frac{\partial^2 V(x,y)}{\partial y^2} = 0 \qquad \text{outside all electrodes} \tag{20.4}$$

$$V(x_c, y_c) = 0 \qquad \text{at all return electrode center coordinates } (x_c, y_c) \tag{20.5}$$

where i_s^n is the absolute value of current injected into the nth stimulus electrode, i_r^m is the absolute value of current returned from the mth return electrode, and r, h are the radius and height of the electrode cylinders. In general, the i_s^n currents injected are given, whereas the return currents i_r^m are unknown when there is more than one return electrode. These return currents can be determined from the additional constraint given by Eq. (20.5). Zero-flux boundary conditions $\partial V/\partial x = 0$, $\partial V/\partial y = 0$ were imposed on the x and y edges of the domain, respectively.

Equations (20.1)–(20.5) were solved using a custom spectral collocation method implemented in MATLAB (MathWorks). In brief, the 2D domain was segmented into a number of smaller rectangular regions enclosing each circular electrode as well as the spaces between electrodes and between electrodes and the domain boundary. Within each segment, a nonuniform $x–y$ mesh was generated using Chebyshev node spacing [80]. This resulted in a more dense packing of nodes near each segment boundary compared with the centers of each segment. First-order derivative continuity in voltage was enforced at each segment boundary. The resulting system of differentiation matrices generated was then inverted to solve for the voltage at each node. Spectral differentiation allowed for more accurate derivatives with fewer nodes than the more traditional finite-difference approach [81]. For the simulations presented here, 7 nodes were placed in each x, y direction within all electrodes, 9 nodes in the interelectrode spaces, and 23 nodes between the outermost electrodes and the boundary of the domain. That is, a total of 219×331 nodes in x, y respectively. This modest number of nodes allowed the models to be solved within 10 min using a Pentium 4 desktop workstation.

Figure 20.4 models the situation whereby all 14 current sources are injecting 1 mA each through an electrode (not drawn) at the center of each hexagon. Surrounding each injection point are six return electrodes all connected to zero potential. There is an obvious concentration of current achieved by the guard electrodes. Figure 20.5 demonstrates the situation of bipolar stimulation with one electrode per source only (that to the immediate right of the injecting electrode) acting as a return electrode. Looking at the individual currents returned in each guard, with the six distributed guards, the average current in each "hexagon" is 0.99 ± 0.06 mA [mean $\pm$ standard deviation (SD)]. In the case of the single guard, the average current is 1.00 ± 0.22 mA. The much higher standard deviation reflects the imbalance of currents as a result of leakage from neighboring current sources. The greatest return current is the bottom left return

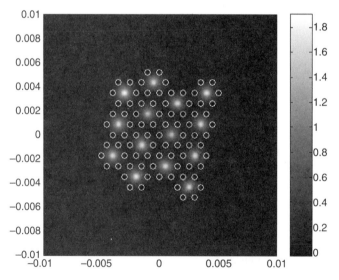

Figure 20.4 Simulated 2D voltage distribution derived from 14 parallel current sources, injecting 1 mA each into a 20 mm $\times$ 20 mm $\times$ 2 mm physiological saline region ($\sigma = 1.0$ S/m) by way of 400-μm-diameter cylindrical electrodes. Each injection electrode is surrounded by six guard electrodes arranged in a hexagonal pattern (depicted as open circles), each guard tied to zero potential. The vertical colorbar indicates the simulated potential in volts. The guard electrodes effectively focus current in the region of each hexagon and reduce current leakage in other regions.

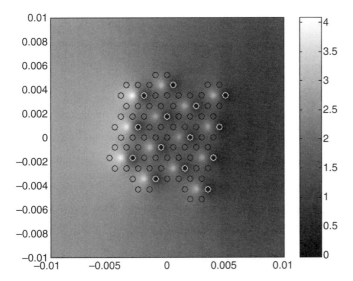

Figure 20.5 Simulated 2D voltage distribution as per Figure 20.4 but with only one return electrode (immediately to the right of each current-injecting electrode). The remaining five guard electrodes (depicted as open black circles) are effectively floating. With the higher current density between injecting and return electrodes, the overall voltage in the medium increases, as does the leakage of current between adjacent return electrodes (even in this instance where current injection was identical at 1 mA for each current source).

electrode, returning 1.39 mA, while the return which is second from the top on the right only received 0.68 mA. These current imbalances are obviously of considerable importance when trying to accurately inject and recover charge from various regions of the neural retina.

20.5 CRITICAL AREAS FOR RESEARCH AND DEVELOPMENT

In this section a number of fundamental research tasks, which are in the critical path for realizing a functioning retinal neuroprosthesis, are outlined. For the purpose of this discussion the differentiator for a fundamental research need has been based on the fact that it relates intimately to the underlying retinal neurophysiology and is not merely an adaptation of similar neurostimulation approaches in other target tissue.

20.5.1 Application-Specific Integrated Circuit Design

Connected to each electrode within the hexagonally arranged electrode array is a requisite electronic circuit. These circuits serve to deliver precise stimuli in accordance with instructions from an external controller. The external controller acquires a digital representation of the visual scene and subsequently generates a phosphene representation of this scene by way of processing strategies. Each phosphene is mapped to an electrode interfaced with the surviving neurons of the epiretinal surface. Once the phosphene representation of the visual scene is established, instructions are conveyed by way of wireless telemetry [82] to the implanted electronics for subsequent delivery of electrical stimulations.

Critical flicker fusion (CFF) in sighted individuals is such that a flickering light incident upon the retina appears continuous and of constant intensity. In the intact retina CFF is intensity dependent, in accordance with the Ferry–Porter law. The association of a "fusion frequency" with retinal prostheses has been debated for some time and indeed has been explored by Humayun et al. [20], who found the fusion frequency in the diseased retina to be of the order of two to three times that of a healthy retina stimulated by light of moderate intensity.

In the degenerated retina, intraretinal processing is dysfunctional and what mechanisms that may remain are likely bypassed via the way in which the bipolar and retinal ganglion cells may be preferentially stimulated over other surviving neurons [65, 83]. The apparent result of this is a reduction in the duration of the perceived phosphenes and thus an increase in the fusion frequency.

Implications in visual prosthesis include the need to deliver stimuli at a rate substantially higher than would be necessary for stimulation of the intact retina with light. Further exacerbating the difficulties for the neuroprosthesis designer is the fact that as the quantity of stimulation sites increase so as to improve image and information resolution, the complexity of maintaining sufficient stimulation to avoid flicker increases accordingly.

From the foregoing, it is apparent that stimulation must take place at multiple sites simultaneously. The authors have published, and further describe in an adjacent chapter [1], a novel addressing and multiplexing scheme that substantially decreases the requisite data (instruction) throughput to facilitate simultaneous stimulations with focused charge distributions [84].

In employing this methodology, we illustrate how as many as 14 electrodes of 98 within an array may be stimulated simultaneously. Furthermore, this methodology is scalable to arrays of much larger dimensions. We note, however, that there exists a limit to the quantity of electrodes that may be stimulated at any one time owing to power supply considerations. There further exist rate-limiting constraints from residual charge recovery following stimulation (e.g., via shorting of electrodes after stimulation). Finally there are critical matters associated with the charge distribution and exchange between electrodes stimulated in unison (the topic of a separate section within this chapter and further explored in an adjacent chapter [1]). These factors combine to establish practical boundaries that will dictate the form of the next generation of retinal prostheses.

20.5.2 Biohybrid High-Density Electrode Arrays

Platinum is one of the most commonly used materials for biomedical electrode applications. It has shown excellent biological performance over almost three decades in sensory stimulating electrodes such as in the cochlear implant and is considered biocompatible in such applications. Examining the electrode design and function in applications such as the cochlear implant is instructive, and many of the issues are similar to those in the vision implant.

In the cochlea, there are on the order of 30,000 spiral ganglion cells (SGCs), the equivalent sensory cells in the hearing pathway to the RGCs in the visual pathway. With the loss of hair cells (equivalent to the photoreceptors), the cochlear implant's electrode array stimulates these cells with an electric impulse which is then converted to an ionic current that propagates the signal to the hearing centers in the brain via the cochlear nerve.

Of the tens of thousands of functional SGCs theoretically available for specific stimulation, the cochlear implant electrode stimulates approximately 20 nerve bundles, each likely to be made up of many hundreds of cell bodies and associated axons. Speech recognition is good in the state-of-the-art cochlear devices, although there is a dramatic decrease in speech recognition in people with cochlear implants when the noise approaches similar levels to the signal. The combination of large electrode size (relative to SGC size), lack of direct electrode-to-cell contact, and limited electrode numbers result in limitations in the quality of hearing and definition of complex sound such as music or speech in a crowded room.

In contrast with the hearing system the visual system has over a million RGCs forming a network of axons that converge in the optic nerve. The ideal number and size of electrodes for functional visual perception are not known; however, it does appear that current electrode arrays for epiretinal prostheses are a limiting component of the device design. Design criteria for future devices should address the need for higher density, encapsulated electrodes, a greater number of channels for stimulation, bioactive materials for close contact with neurons, and maintaining neural cell viability and function and finally should consider increasing the surface area of electrodes via porous or surface-roughened electrodes.

The latter is a strategy for reducing electrode impedance as well as enhancing the surface area available for cell contact, thereby enhancing adhesion [75, 76, 85]. It is proposed that in-growth of the neurons into the electrode surface could adequately reduce the current magnitude required for stimulation and consequently reduce diffusion of the current to adjacent neurons, effectively increasing spatial selectivity. In addition, formation of this intimate connection may prevent or minimize the growth of the nonfunctional and nonconductive encapsulation layer [75], while the use of softer and more compliant materials for electrode coating has the potential to reduce inflammation caused by strain mismatch at the tissue–electrode interface [86]. These stringent criteria are unlikely to be attainable by platinum alone, and there is a need for novel approaches to material design and electrode fabrication for these high-demand applications.

In terms of bioactivity, there are three broad approaches that have been proposed. These can be summarized as surface modification or coating of existing electrode materials [75, 76], development of novel bioactive electrode materials [87], and incorporation of cultured cells which are then encouraged to extend neurites to make the synaptic connections with neurons in the host tissue [88]. For all approaches, bioactivity may be conferred by controlled release or presentation of neurotrophic factors, by modification with extracellular matrix proteins or more specifically neural cell adhesion sequences derived from these proteins, and by release of other active agents, including essential nutrients, anti-inflammatory drugs, or even chemical signaling agents.

Of critical importance is the electrode material or surface coating which must be conductive, biocompatible in an active and passive state, and easily manufactured to produce the uniform layer necessary for effective charge transfer. More challenging is the requirement that the material must be capable of retaining the aforementioned biological factors in matrices over an extended period of time with constantly varying electrical stimulation. Additionally it must possess the ability to be modified in vivo to release certain factors while maintaining others. These constraints have prompted research into novel conducting polymers, carbon–polymer-based composites, and enzyme-coated metals and glassy carbons [87]. Thus, in the near future we expect the emergence of a new generation of smarter vision prostheses with a variety of biological functions.

20.5.3 Validated Models of Electrical Stimulation in Retina

Very little has been published on mathematical models of retinal tissue stimulation (either from the viewpoint of passive resistive and capacitive tissue models or models that incorporate Hodgkin–Huxley kinetics of membrane processes). There is an obvious need for greater understanding of the underlying activation processes in the neural retina in order to optimize both ASIC and electrode design. Indeed we will model the electrical charge distribution under circumstances of multiple current injection sites. Passive and active properties of retinal tissue will be considered.

It is important to be able to predict the effects of electrode geometry and material composition, electrode proximity, and electrical parameters on the biological environment. Significant peripheral nerve models exist for a variety of electrode interfaces, configurations and states of activity. Some elementary models of the retina are also in existence [89–91]. One model by Hornig and Eckmiller models all cells of the inner retina, including the horizontal, bipolar, amacrine, and retinal ganglion cells [60]. This model will prove useful if the amacrine cells are used for in-growth and attachment to the electrode surface. An alternate model by Greenberg et al. explores the stimulation of RGCs by extracellular fields with active ion channels and realistic cell morphology derived directly from a neuronal tracing with three membrane states individually applied: a linear passive model, a Hodgkin–Huxley model with passive dendrites, and a model composed of all active compartments with five nonlinear ion channels [90]. With considerable enhancement, these models could be used as a basis for modeling the neural interface.

20.5.4 Image Processing and Psychophysics of Prosthetic Vision

There in only a small handful of literature on the psychophysics of prosthetic vision [55, 92–96]. Should work in the above-mentioned critical areas make for the realization of a device in coming years analogous to the cochlear implant, the question remains, "How should phosphenes be rendered in the visual field of the implantee so as to best convey vision?" In cochlear prosthetics, improvement in recent years on standardized tests such as monosyllabic word recognition is one-third attributed to improved speech processing strategies [97]. Clearly, there is work to be done on the prosthetic vision problem—the psychophysics and the image processing.

Due to the fact that human trials of retinal prosthetics are few to date, the prosthetic vision problem is broached by way of Simulations. Again, this is not unlike research in the "cochlear" domain: Simulated, low-channel signals are played to the normally hearing on loudspeakers. These simulations follow work by Cha and colleagues, where sighted subjects were presented simple acuity and mobility tasks while viewing the world through, effectively, a mask of pinholes. More recent simulations have made use of technology—immersive displays, head tracking, and more computationally expensive image processing algorithms [98]. The development of simulations that track the eye so as to stabilize the phosphene image on the retina, and therefore better approximate prosthetic vision, is a topic on-going in the authors' laboratory and presumably in other labs. We have hypothesized the notion of movement-adaptive filtering [99]; likely developments also include moves away from linear, spatiotemporal strategies toward connectionist networks, error minimization algorithms, and "intelligent" algorithms such as region-of-interest approaches [100].

20.6 CONCLUSIONS

From the conspectus of work detailed in this review of retinal neuroprosthestics, it is apparent that significant advances have been achieved, particularly in the last half decade. In spite of this, however, there is still considerable ground to be covered. Detailed understanding of neural activation in parallel concurrent stimulation is notably lacking, with a paucity of literature on the effects of current leakage in general and models of the neural retina specifically. Up until recent times, there has been no major need for such research, but as we strive for denser electrode grids with larger quantities of electrodes, the need becomes more pressing. This is by no means the only issue that needs concerted research; there are still major advances to be made in our understanding of phosphene generation, the complexity of the neural–tissue interface, and the long-term biocompatibility of visual prostheses. Until these and other issues are more fully understood, restoration of vision to the profoundly visually impaired in a manner that the everyday person would consider as "acceptable" remains a "science fiction". However, the "science fact" is that sufficient evidence exists to suggest that the provision of some degree of visual perception will occur in the near to medium future.

REFERENCES

1. G. J. SUANING, L. E. HALLUM, S. CHEN, P. Byrnes-Preston, S. DOKOS, T. LEHMANN, AND N. H. LOVELL, "Design approaches in retinal neuroprosthetics," in *Neural Engineering*, M. AKAY, Ed., Wiley, Hoboken, NJ, 2005.

2. L. E. HALLUM, N. H. LOVELL, AND G. J. SUANING, "Retinal image and phosphene image: An analogy," in *Neural Engineering*, M. AKAY, Ed., Wiley, Hoboken, NJ, 2005.

3. E. ZRENNER, "Will retinal implants restore vision?" *Science*, vol. 295, pp. 1022–1025, 2002.

4. R. R. LAKHANPAL, D. YANAI, J. D. WEILAND, G. Y. FUJII, S. CAFFEY, R. J. GREENBERG, E. DE JUAN, Jr., AND M. S. HUMAYUN, "Advances in the development of visual prostheses," *Curr. Opin. Ophthalmol.*, vol. 14, pp. 122–127, 2003.

5. J. WEILAND and M. HUMAYAN, "Past, present and future of artificial vision," *Artif. Organs*, vol. 27, pp. 961–962, 2003.

6. E. MARGALIT, M. MAIA, R. J. GREENBERG, G. Y. FUJII, G. TORRES, D. V. PIYATHAISERE, T. M. O'Hearne, W. LIU, G. LAZZI, G. DAGNELIE, D. A. SCRIBNER, E. DE JUAN, Jr., AND M. S. HUMAYUN, "Retinal prosthesis for the blind," *Surv. Ophthalmol.*, vol. 47, pp. 335–356, 2002.

7. M. S. HUMAYUN, "Intraocular retinal prosthesis," *Trans. Am. Ophthalmol. Soc.*, vol. 99, pp. 271–300, 2001.

8. J. R. HETLING AND M. S. BAIG-SILVA, "Neural prostheses for vision: Designing a functional interface with retinal neurons," *Neurol. Res.*, vol. 26, pp. 21–34, 2004.

9. G. S. BRINDLEY AND W. S. LEWIN, "The sensations produced by electrical stimulation of the visual cortex," *J. PHYSIOL. (Lond.)*, vol. 196, pp. 479–493, 1968.

10. C. LEROY, "Ou l'on rend compte de quelques tentatives que l'on a faites pour guerir plusieurs maladies par l'electricite," *Mem. Acad. Roy. Sci. Paris*, pp. 60–98, 1755.

11. O. FOERSTER, "Beitrage zur pathophsiologie der sehbahn und der sehsphare," *J. Psychol. Neurol. Lpz.*, vol. 39, pp. 463–485, 1929.

12. G. E. TASSICKER, "Preliminary report on a retinal stimulator," *Br. J. Physiol. Opt.*, vol. 13, pp. 102–105, 1956.

13. H. WASSLE AND B. B. BOYCOTT, "Functional architecture of the mammalian retina," *Physiol. Rev.*, vol. 71, pp. 447–480, 1991.

14. J. D. WEILAND, D. YANAI, M. MAHADEVAPPA, R. WILLIAMSON, B. V. MECH, G. Y. FUJII, J. LITTLE, R. J. GREENBERG, E. DE JUAN, Jr., AND M. S. HUMAYUN, "Electrical stimulation of retina in blind humans," presented at 25th International Conference of the IEEE Engineering in Medicine and Biology Society, 2003.

15. M. S. HUMAYUN, E. DE JUAN, Jr., G. DAGNELIE, R. J. GREENBERG, R. H. PROPST, AND D. H. PHILLIPS, "Visual perception elicited by electrical stimulation of retina in blind humans," *Arch. Ophthalmol.*, vol. 114, pp. 40–46, 1996.

16. M. S. HUMAYUN, R. PROPST, E. DE JUAN, Jr., K. McCORMICK, AND D. HICKINGBOTHAM, "Bipolar surface electrical stimulation of the vertebrate retina," *Arch. Ophthalmol.*, vol. 112, pp. 110–116, 1994.

17. U. T. EYSEL, P. WALTER, F. GEKELER, H. SCHWAHN, E. ZRENNER, H. G. SACHS, V.-P. GABEL, AND Z. F. KISVÁRDAY, "Optical imaging reveals 2-dimensional patterns of cortical activation after local retinal stimulation with sub- and epiretinal visual prostheses," presented at Invest. Ophthalmol. Vis. Sci., 2002.

18. A. Santos, M. S. Humayun, E. De Juan, Jr., R. J. Greenburg, M. J. Marsh, I. B. Klock, and A. H. Milam, "Preservation of the inner retina in retinitis pigmentosa. A morphometric analysis," *Arch. Ophthalmol.*, vol. 115, pp. 511–515, 1997.

19. R. E. Marc, B. W. Jones, C. B. Watt, and E. Strettoi, "Neural remodeling in retinal degenerations," *Prog. Retin. Eye Res.*, vol. 22, pp. 607–655, 2003.

20. M. S. Humayun, E. De Juan, Jr., J. D. Weiland, G. Dagnelie, S. Katona, R. Greenberg, and S. Suzuki, "Patterned electrical stimulation of the human retina," *Vision Res.*, vol. 39, pp. 2569–2576, 1999.

21. J. I. Loewenstein, S. R. Montezuma, and J. F. Rizzo, 3rd, "Outer retinal degeneration: An electronic retinal prosthesis as a treatment strategy," *Arch. Ophthalmol.*, vol. 122, pp. 587–596, 2004.

22. E. Marg and G. Dierssen, "Reported visual percepts from stimulation of the human brain with microelectrodes during therapeutic surgery," *Confinia Neurol.*, vol. 26, pp. 57–75, 1965.

23. B. S. Nashold, Jr., "Phosphenes resulting from stimulation of the midbrain in man," *Arch. Ophthalmol.*, vol. 84, pp. 433–435, 1970.

24. P. E. K. Donaldson, "Experimental visual prosthesis," *Proc. IEE*, vol. 120, pp. 381–298, 1973.

25. W. H. Dobelle, M. G. Mladejovsky, and J. P. Girvin, "Artifical vision for the blind: Electrical stimulation of visual cortex offers hope for a functional prosthesis," *Science*, vol. 183, pp. 440–444, 1974.

26. D. C. Henderson, J. R. Evans, and W. H. Dobelle, "The relationship between stimulus parameters and phosphene threshold/brightness, during stimulation of human visual cortex," *Trans. Am. Soc. Artif. Intern. Organs*, vol. 25, pp. 367–371, 1979.

27. W. H. Dobelle, "Artificial vision for the blind by connecting a television camera to the visual cortex," *Am. Soc. Artif. Intern. Organs*, vol. 46, pp. 3–9, 2000.

28. V. Chowdhury, J. W. Morley, and M. T. Coroneo, "The neurophysiologic response to surface stimulation of the visual cortex in the cat by a bionic eye device," presented at Invest. Ophthalmol. Vis. Sci., 2003.

29. E. M. Schmidt, M. J. Bak, F. T. Hambrecht, C. V. Kufta, D. K. O'Rourke, and P. Vallabhanath, "Feasibility of a visual prosthesis for the blind based on intracortical microstimulation of the visual cortex," *Brain*, vol. 119 (Pt. 2), pp. 507–522, 1996.

30. R. A. Normann, E. M. Maynard, P. J. Rousche, and D. J. Warren, "A neural interface for a cortical vision prosthesis," *Vision Res.*, vol. 39, pp. 2577–2587, 1999.

31. P. R. Troyk, W. Agnew, M. Bak, J. Berg, D. Bradley, L. Bullara, S. Cogan, R. Erickson, C. Kufta, D. McCreery, E. Schmidt, and V. Towle, "Multichannel cortical stimulation for restoration of vision," presented at Proceedings of the 24th Annual International Conference of the IEEE Engineering in Medicine and Biology Society, 2002.

32. D. C. Bradley, P. R. Troyk, J. A. Berg, M. Bak, S. Cogan, R. Erickson, C. Kufta, M. Mascaro, D. McCreery, E. M. Schmidt, V. Towle, and H. Xu, "Visuotopic mapping through a multichannel stimulating implant in primate v1," *J. Neurophysiol.*, 93(3): 1659–1670, 2000.

33. M. Schuettler, T. Stieglitz, R. Keller, J.-U. Meyer, and R. R. Riso, "A hybrid cuff electrode for functional electrostimulation of large peripheral nerves," *Med. Biol. Eng. Computing*, vol. 37, pp. 792–793, 1999.

34. T. Stieglitz, X. Navarro, S. Calvet, C. Blau, and J.-U. Meyer, "Interfacing regenerating peripheral nerves with a micromachined polyimide sieve electrode," presented at Proceedings of the 18th Annual International Conference of the IEEE Engineering in Medicine and Biology Society, 1997.

35. D. Fischer, P. Heiduschka, and S. Thanos, "Lens-injury-stimulated axonal regeneration throughout the optic pathway of adult rats," *Exper. Neurol.*, vol. 172, pp. 257–272, 2001.

36. C. Veraart, C. Raftopoulos, J. T. Mortimer, J. Delbeke, and D. Pins, "Visual sensations produced by optic nerve stimulation using an implanted self-sizing spiral cuff electrode," *Brain Res.*, vol. 813, pp. 181–186, 1998.

37. J. Delbeke, M. Oozeer, and C. Veraart, "Position, size and luminosity of phosphenes generated by direct optic nerve stimulation," *Vision Res*, vol. 43, pp. 1091–1102, 2003.

38. F. Gekeler, H. Schwahn, A. Stett, K. Kohler, and E. Zrenner, "Subretinal microphotodiodes to replace photoreceptor function: A review of the current state," in *Vision, Sensations et Environnement*, M. Doly, M.-T. Droy, and Y. Christen, Eds., Irvinn, Paris, 2001, pp. 77–95.

39. A. Y. Chow, M. T. Pardue, V. Y. Chow, G. A. Peyman, C. Liang, J. I. Perlman, and N. S. Peachey, "Implantation of silicon chip microphotodiode arrays into the cat subretinal space," *IEEE Trans. Rehabil. Eng.*, vol. 9, pp. 86–95, 2001.

40. A. B. Majji, M. S. Humayun, J. D. Weiland, S. Suzuki, S. A. D'Anna, and E. De Juan, Jr., "Long-term histological and electrophysiological results of an inactive epiretinal electrode array implantation in dogs," *Invest. Ophthalmol. Vis. Sci.*, vol. 40, pp. 2073–2081, 1999.

41. V. Ortmann, M. Fuchs, and R. E. Eckmiller, "Electrical stimulator chip for retinal implants," presented at *Invest. Ophthalmol. Vis. Sci.*, 2002.

42. G. J. Suaning and N. H. Lovell, "Cmos neurostimulation asic with 100 channels, scaleable output and bi-directional radio frequency telemetry," *IEEE Trans. Biomed. Eng.*, vol. 48, pp. 248–260, 2001.

43. R. Eckmiller, M. Becker, and R. Hunermann, "Towards a learning retina implant with epiretinal contacts," presented at IEEE International Conference on Systems, Man, and Cybernetics, 1999.

44. G. J. Suaning, N. H. Lovell, and Y. Kerdraon, "Trans-retinal electrical stimulation using a neuroprosthesis: The effects of damage to the r-membrane," presented at 24th Annual International Conference of the IEEE Engineering in Medicine and Biology Society, 2002.

45. H. Kanda, T. Morimoto, T. Fujikado, Y. Tano, Y. Fukuda, and H. Sawai, "Electrophysiological studies of the feasibility of suprachoroidal-transretinal stimulation for artificial vision in normal and rcs rats," *Invest. Ophthalmol. Vis. Sci.*, vol. 45, pp. 560–566, 2004.

46. A. Y. Chow and V. Y. Chow, "Subretinal electrical stimulation of the rabbit retina," *Neurosci. Lett.*, vol. 225, pp. 13–16, 1997.

47. N. S. Peachey, A. Y. Chow, M. T. Pardue, J. I. Perlman, and V. Y. Chow, "Response characteristics of subretinal microphotodiode-based implant mediated cortical potentials," J. G. Hollyfield, R. E. Anderson, and M. M. Lauail (eds.), Plenum New York, *Retinal Degen. Dis. Exper. Therapy*, pp. 471–478, 1999.

48. E. Zrenner, A. Stett, S. Weiss, R. B. Aramant, E. Guenther, K. Kohler, K. D. Miliczek, M. J. Seiler, and H. Haemmerle, "Can subretinal microphotodiodes successfully replace degenerated photoreceptors?," *Vision Res.*, vol. 39, pp. 2555–2567, 1999.

49. N. S. Peachey, M. T. Pardue, S. L. Ball, J. R. Hetling, V. Y. Chow, and A. Y. Chow, "Unexpected sensitivity of the mammalian retina to infrared light," presented at Invest. Ophthalmol. Vis. Sci., 2000.

50. A. Y. Chow, V. Y. Chow, K. H. Packo, J. S. Pollack, G. A. Peyman, and R. Schuchard, "The artificial silicon retina microchip for the treatment of vision loss from retinitis pigmentosa," *Arch. Ophthalmol.*, vol. 122, pp. 460–469, 2004.

51. A. Y. Chow, K. H. Packo, J. S. Pollack, and R. A. Schuchard, "Subretinal artificial silicon retina microchip implantation in retinisis pigmentosa patients: Long term follow-up," presented at Invest. Ophthalmol. Vis. Sci., 2003.

52. H. N. Schwahn, F. Gekeler, K. Kohler, K. Kobuch, H. G. Sachs, F. Schulmeyer, W. Jakob, V. P. Gabel, and E. Zrenner, "Studies on the feasibility of a subretinal visual prosthesis: Data from yucatan micropig and rabbit," *Graefes Arch. Clin. Exp. Ophthalmol.*, vol. 239, pp. 961–967, 2001.

53. A. Stett, W. Barth, S. Weiss, H. Haemmerle, and E. Zrenner, "Electrical multisite stimulation of the isolated chicken retina," *Vision Res.*, vol. 40, pp. 1785–1795, 2000.

54. F. Gekeler, K. Kobuch, H. N. Schwahn, A. Stett, K. Shinoda, and E. Zrenner, "Subretinal electrical stimulation of the rabbit retina with acutely implanted electrode arrays," *Graefes Arch. Clin. Exp. Ophthalmol.*, vol. 242, pp. 587–596, 2004.

55. K. Cha, K. W. Horch, and R. A. Normann, "Mobility performance with a pixelized vision system," *Vision Res.*, vol. 32, pp. 1367–1372, 1992.

56. D. Joseph and S. Collins, "Modeling, calibration, and correction of nonlinear illumination-dependent fixed pattern noise in logarithmic cmos image sensors," *IEEE Trans. Instrum. Measur.*, vol. 51, pp. 1996–1001, 2002.

57. J. F. Rizzo, R. J. Jensen, J. Loewenstein, and J. Wyatt, "Unexpectedly small percepts evoked by epi-retinal electrical stimulation in blind humans," presented at Invest. Ophthalmol Vis. Sci., 2003.

58. K. L. Roach, L. Theogarajan, and J. Wyatt, "An electrical model of electrode behavior in retinal prostheses," presented at Invest. Ophthalmol. Vis. Sci., 2002.

59. D. B. Shire, J. L. Wyatt, and J. F. Rizzo, "Progress toward an inflatable neural prosthesis," presented at Invest. Ophthalmol. Vis. Sci., 2001.

60. R. Hornig and R. Eckmiller, "Optimizing stimulus parameters by modeling multi-electrode electrical stimulation for retina implants," presented at International Joint Conference on Neural Networks, 2001.

61. J. H. Meier, W. L. Rutten, A. E. Zoutman, H. B. Boom, and P. Bergveld, "Simulation of multipolar fiber selective neural stimulation using intrafascicular electrodes," *IEEE Trans. Biomed. Eng.*, vol. 39, pp. 122–134, 1992.

62. R. Hornig and R. Eckmiller, "Retina implant stimulator with on-chip memory for complex stimulus profiles," presented at Invest. Ophthalmol. Vis. Sci., 2002.

63. R. E. Eckmiller, R. Hornig, V. Ortmann, and H. Gerding, "Test technology for acute clinical trials of retina implants," presented at Invest. Ophthalmol. Vis. Sci., 2002.

64. H. Gerding, R. E. Eckmiller, R. Hornig, V. Ortmann, A. Kolck, and S. Taneri, "Safety assessment and acute clinical tests of epiretinal retina implants," presented at Invest. Ophthalmol. Vis. Sci., 2002.

65. R. J. Greenberg, T. J. Velte, M. S. Humayun, G. N. Scarlatis, and E. De Juan, Jr., "A computational model of electrical stimulation of the retinal ganglion cell," *IEEE Trans. Biomed. Eng.*, vol. 46, pp. 505–514, 1999.

66. M. S. Humayun, J. D. Weiland, G. Y. Fujii, R. Greenberg, R. Williamson, J. Little, B. Mech, V. Cimmarusti, G. Van Boemel, and G. Dagnelie, "Visual perception in a blind subject with a chronic microelectronic retinal prosthesis," *Vision Res.*, vol. 43, pp. 2573–2581, 2003.

67. G. J. Suaning, L. E. Hallum, S. C. Chen, P. J. Preston, and N. H. Lovell, "Phosphene vision: Development of a portable visual prosthesis system for the blind," presented at 25th International Conference of the IEEE Engineering in Medicine and Biology Society, 2003.

68. G. J. Suaning, N. H. Lovell, and Y. A. Kerdraon, "Physiological response in ovis aries resulting from electrical stimuli delivered by an implantable vision prosthesis," presented at 23rd International Conference of the IEEE Engineering in Medicine and Biology Society, 2001.

69. Y. Kerdraon, J. Downie, G. J. Suaning, M. Capon, M. Coroneo, and N. H. Lovell, "Surgical implantation of a vision prosthesis model into the ovine eye," *Clin. Exp. Ophthalmol.*, vol. 30(1), pp. 36–40, 2002.

70. W. Liu, K. Vichienchom, M. Clements, S. C. DeMarco, C. Hughes, E. McGucken, M. S. Humayun, E. De Juan, Jr., J. D. Weiland, and R. Greenberg, "A neuro-stimulus chip with telemetry unit for retinal prosthetic device," *IEEE J. Solid-State Circuits*, vol. 35, pp. 1487–1497, 2000.

71. G. J. SUANING, N. H. LOVELL, AND C. Y. KWOK, "Fabrication of platinum spherical electrodes in an intraocular prosthesis using high-energy electrical discharge," *Sensors and Actuators A: Physical*, vol. 108, pp. 155–161, 2003.

72. W. LIU, M. SIVAPRAKASAM, P. R. SINGH, R. BASHIRULLAH, AND G. WANG, "Electronic visual prosthesis," *Artif. Organs*, vol. 27, pp. 986–995, 2003.

73. P. KENNEDY AND R. BAKAY, "Restoration of neural output from a paralyzed patient by a direct brain connection," *NeuroReport*, vol. 9, pp. 1707–1711, 1998.

74. P. R. KENNEDY, "The cone electrode: A long-term electrode that records from neurites grown onto its recording surface," *J. Neurosci. Methods*, vol. 29, pp. 181–193, 1989.

75. X. CUI, V. A. LEE, Y. RAPHAEL, J. A. WILER, J. F. HETKE, D. J. ANDERSON, AND D. C. MARTIN, "Surface modification of neural recording electrodes with conducting polymer/biomolecule blends," *J. Biomed. Mater. Res.*, vol. 56, pp. 261–272, 2001.

76. Y. ZHONG, X. YU, R. GILBERT, AND R. V. BELLAMKONDA, "Stabilizing electrode-host interfaces: A tissue engineering approach," *J. Rehabil. Res. Dev.*, vol. 38, pp. 627–632, 2001.

77. L. F. TOTH, "Uber einen geometrischen satz," *Math Z*, vol. 46, pp. 79–83, 1940.

78. R. PLONSEY, "Volume conductor theory," in *The Biomedical Engineering Handbook*, J. D. BRONZINO, Ed., CRC Press, Boca Raton, FL, 1995, pp. 119–125.

79. B. ROTH, "The electrical properties of tissues," in *The Biomedical Engineering Handbook*, J. D. BRONZINO, Ed., CRC Press, Boca Raton, FL, 1995, pp. 126–138.

80. J. A. C. WEIDEMAN AND S. C. REDDY, "A matlab differentiation matrix suite," *ACM Trans. Math. Software*, vol. 26, pp. 465–519, 2000.

81. L. N. TREFETHEN, *Spectral Methods in Matlab*, SIAM, Philadelphia, 2000.

82. G. J. SUANING, L. E. HALLUM, S. C. CHEN, P. J. PRESTON, AND N. H. LOVELL, "Phosphene vision: Development of a portable visual prosthesis system for the blind," presented at 25th Annual International Conference of the IEEE-EMBS, Cancun, Mexico, 2003.

83. F. RATTAY, S. RESATZ, P. LUTTER, K. MINASSIAN, B. JILGE, AND M. DIMITRIJEVIC, "Mechanisms of electrical stimulation with neural prosthesis," *Neuromodulation*, vol. 6, pp. 42–56, 2003.

84. G. J. SUANING, L. E. HALLUM, P. J. PRESTON, AND N. H. LOVELL, "An efficient multiplexing method for addressing large numbers of electrodes in a visual neuroprosthesis," presented at 26th Annual International Conference of the IEEE-EMBS, San Francisco, CA, 2004.

85. X. CUI AND D. C. MARTIN, "Fuzzy gold electrodes for lowering impedance and improving adhesion with electrodeposited conducting polymer films," *Sensors and Actuators A: Physical*, vol. 103, pp. 384–394, 2003.

86. A. WIDGE, M. JEFFRIES-EL, C. F. LAGENAUR, V. W. WEEDN, AND Y. MATSUOKA, 'Conductive polymer 'molecular wires' for neuro-robotic interfaces," presented at IEEE International Conference on Robotics and Automation, 2004.

87. S. ZHANG, G. WRIGHT, AND Y. YANG, "Materials and techniques for electrochemical biosensor design and construction," *Biosensors Bioelectron.*, vol. 15, pp. 273–282, 2000.

88. W. M. GRILL, J. W. McDONALD, P. H. PECKHAM, W. HEETDERKS, J. KOCSIS, AND M. WEINRICH, "At the interface: Convergence of neural regeneration and neural prostheses for restoration of function," *J. Rehabil. Res. Devel.*, vol. 38, pp. 633–639, 2001.

89. F. RATTAY, S. RESATZ, P. LUTTER, K. MINASSIAN, B. JILGE, AND M. R. DIMITRIJEVIC, "Mechanisms of electrical stimulation with neural prostheses," *Neuromodulation*, vol. 6, pp. 42–56, 2003.

90. R. J. GREENBERG, T. J. VELTE, M. S. HUMAYUN, G. N. SCARLATIS, AND E. DE JUAN, JR., "A computational model of electrical stimulation of the retinal ganglion cell," *IEEE Trans. Biomed. Eng.*, vol. 46, pp. 505–514, 1999.

91. F. RATTAY, "Analysis of models for extracellular fiber stimulation," *IEEE Trans. Biomed. Eng.*, vol. 36, pp. 676–682, 1989.

92. K. CHA, K. W. HORCH, R. A. NORMANN, AND D. K. BOMAN, "Reading speed with a pixelized vision system," *J. Opt. Soc. Am.*, vol. 9, pp. 673–677, 1992.

93. K. CHA, K. HORCH, AND R. A. NORMANN, "Simulation of a phosphene-based visual field: Visual acuity in a pixelized vision system," *Ann. Biomed. Eng.*, vol. 20, pp. 439–449, 1992.

94. R. W. THOMPSON, G. D. BARNETT, M. HUMAYAN, AND G. DAGNELIE, "Facial recognition using simulated prosthetic pixelised vision," *Invest. Ophthalmol. Visual Sci.*, vol. 44, pp. 5035–5042, 2003.

95. J. S. HAYES, V. T. YIN, D. PIYATHAISERE, J. D. WEILAND, M. S. HUMAYUN, AND G. DAGNELIE, "Visually guided performance of simple tasks using simulated prosthetic vision," *Artif. Organs*, vol. 27, pp. 1016–1028, 2003.

96. L. E. HALLUM, G. J. SUANING, D. S. TAUBMAN, AND N. H. LOVELL, "Simulated prosthetic visual fixation, saccade, and smooth pursuit," *Vision Res.*, vol. 45, no. 6, 775–788, 2005.

97. J. T. RUBINSTEIN AND C. A. MILLER, "How do cochlear prostheses work?" *Curr. Opin. Neurobiol.*, vol. 9, pp. 399–404, 1999.

98. S. C. CHEN, N. H. LOVELL, AND G. J. SUANING, "Effect of prosthetic vision acuity by filtering schemes, filter cut-off frequency and phosphene matrix: A virtual reality simulation," presented at 26th Annual International Conference of the IEEE-EMBS, San Francisco, CA, 2004.

99. L. E. HALLUM, G. J. SUANING, D. S. TAUBMAN, AND N. H. LOVELL, "Towards photosensor movement-adaptive image analysis in an electronic retinal prosthesis," presented at 26th Annual International Conference of the IEEE-EMBS, San Francisco, CA, 2004.

100. J. R. BOYLE, A. J. MAEDER, AND W. W. BOLES, "Challenges in digital imaging for artificial human vision," presented at Proceedings of the SPIE, 2001.

CHAPTER 21

TOWARDS A CULTURED NEURAL PROBE: PATTERNING OF NETWORKS AND THEIR ELECTRICAL ACTIVITY

W. L. C. Rutten, T. G. Ruardij, E. Marani, and B. H. Roelofsen

21.1 INTRODUCTION

Efficient and selective electrical stimulation and recording of neural activity in peripheral, spinal, or central neural pathways require microelectrode arrays (MEAs) at micrometer or nanometer scale. At present, wire arrays in brain, flexible linear arrays in the cochlea, and cuff arrays around nerve trunks are in experimental and/or clinical use. Two- and three-dimensional brushlike microarrays and "sieves," with around a hundred electrode sites, have been proposed, fabricated in microtechnology, and/or tested in a number of laboratories.

As there are no "blueprints" for the exact positions of fibers in a peripheral nerve or motor neurons in a ventral root region, an insertable multielectrode has to be designed in a redundant way. Even then, the efficiency of a multielectrode will be less than 100%, as not every electrode will contact one neural axon or soma.

Therefore, "cultured probe" devices are being developed based on cell-cultured planar MEAs (Figs. 21.1 and 21.2). They may enhance efficiency and selectivity because neural cells have been grown over and around each electrode site as electrode-specific local networks. If, after implantation, collateral sprouts branch from a motor fiber (ventral horn area) and if they can each be guided and contacted to one specific "host" island network, a one-to-one (i.e., very selective and efficient) stimulatory interface will result.

The islands perform the function of a biofriendly surrounding for the sprouts, producing neuroattractive proteins to guide the sprouts toward them.

A number of aspects relevant to the successful development of a cultured probe have been studied in our group intensively during recent years. Among them are the electrical behavior of the neuron–electrode contact [1–3], the capability to record and stimulate neuron activity of cells positioned on top of MEA electrodes or closely beside them [4], and the chemical surface modification of flat substrates into neurophilic and neurophobic regions [5, 6]. In summary, these studies resulted in understanding the neuron–electrode contact and the wave shapes of recorded action potentials, the discovery of a stimulation current "window," and the discovery of chemical coatings suited for patterning of host islands.

Handbook of Neural Engineering. Edited by Metin Akay
Copyright © 2007 The Institute of Electrical and Electronics Engineers, Inc.

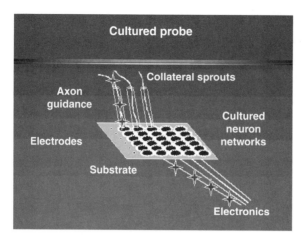

Figure 21.1 Schematic impression of "cultured probe"-type of neural information transducer/ prosthetic device. The black circular "stand-alone-islands" are cultured neuronal networks, separated from each other, at and around each electrode, acting as biofriendly hosts/attractors for the sprouts. The primary goal is selective stimulation of neuronal sprouts (motor fibers). The series of stars stand for information flow from one electronic stimulus channel to one neural fiber via one electrode site.

As stated, the local island networks serve only to provide a biofriendly surrounding for the sprouts. However, it is well established that (large) developing cortical neuron networks will start to exhibit spontaneous activity after about one week in vitro. It has been shown that the patterns resemble those seen in vivo [7]. Recording of the firing activity of

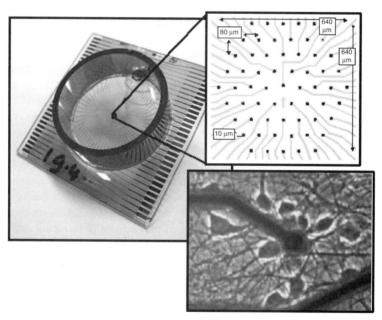

Figure 21.2 Multielectrode array device with culture dish (diameter 30 mm) on top. The MEA is based on a glass substrate (5 × 5 cm); it has 61 embedded electrodes. The 61 electrode tips (tip diameter 10 μm) lie in the center of the chamber (magnified in right-top figure). The figure below shows a detail of one tip surrounded by a 4-DIV (days in vitro) old developing cortical cell culture.

individual neurons has become established, since the pioneering work of Gross and Pine [8, 9], as a useful technique, as is also shown by Jimbo et al. [10] in a study of activity-dependent plasticity at the synaptic level.

The question is whether smaller networks, such as our islands, will become spontaneously active also. Is there a minimal network size?

In this chapter we report the latest progress regarding two of the aforementioned aspects: (1) enhanced neurophilic/neurophobic contrast and (2) spontaneous and synchronized activity of patterned host islands.

21.1.1 Enhanced Neurophobic/Neurophilic Contrast

Essential properties of the probe are that cells must adhere to the surface of the substrate and develop into networks of somata and outgrowing neurites/axonal fascicles within a predefined area around the electrodes. The goal should also be to keep the networks flat and confined as long as possible, so as to delay the natural tendency of neurons to aggregate or island networks to become connected by "neural cables." For that, chemical modification of the substrate and electrode areas with various coatings, such as neurophilic polyethyleneimine (PEI) and neurophobic fluorocarbon (FC) monolayers is applied to promote or inhibit adhesion of cells.

Then, one of the key problems in long-term neuronal patterning studies on MEAs is the biofouling of the background material with cell-adhesive proteins, which in turn promote the random overgrowth with neuronal tissue [11]. Polyethylene oxide (PEO)–coated surfaces are known for their ability to inhibit the adsorption of proteins and are promising alternatives as neurophobic background surfaces. The methods to fabricate PEO-coated surfaces can be divided in two different subgroups: covalent bonding of relatively short polyethylene oxide chains (PEO; also termed polyethylene glycol, PEG) and adsorption of polyethylene oxide–polypropylene oxide (PEO–PPO) block copolymers onto hydrophobic materials [12]. The advantages of covalent coupling of PEO chains to surfaces are the initial stability of the layer and the prohibited displacement of PEO by proteins in solution. However, the physiological environment in time could dissociate the chemical bonds between the PEO chains and the underlying substratum. Another important point is the fact that the chemistry involved should be transferable onto MEAs, which are usually vulnerable to more laborious chemistry. For instance, cleaning of MEAs with aggressive acids before chemical modification is prohibited because the conductive metal leads would be dissolved. Therefore, it is relevant to test more simple modification routes as potential methods to be used on MEAs. Thus, adsorbed layers of PEO–PPO block copolymers on hydrophobic surfaces were investigated as potential neurophobic background surfaces in patterning studies over a time period of 30 days (commercially available PEO–PPO–PEO triblock copolymers, called Synperonics F108 and F127, were tried).

Background adhesion results are shown over a period of 30 days for neurophobic PI- and FC-coated surfaces, each supplemented by F108 or F127 Synperonics.

21.1.2 Spontaneous Activity of Patterned Islands

It is well known that large unpatterned ("random") cultured networks of cortical neurons start to fire spontaneously after about a week in vitro. The implications of such firing for the proper functioning as host has to be considered. First, the cultured probe is meant for stimulation purposes. In that case, electrodes deliver stimuli directly to the sprouts (antidromic propagation) and the activity of the network itself is of no importance (antidromic stimulation of sprouts by natural synaptic connections between axonal sprout and network is impossible). Second, if one would consider, for future two-way interface applications, the use of the cultured probe for recording purposes, the spontaneous activity of networks may seriously distort the information from the sprouts.

For smaller networks, such as the patterned islands of the cultured probe, it is not known yet what the minimum network size and cell density conditions are for the occurrence of spontaneous activity. It may be that small-diameter islands show no activity at all.

Therefore, it is very interesting to investigate whether the patterned islands of a cultured probe indeed show spontaneous activity. Part of this chapter is devoted to that topic, to answer questions such as: Do islands also become spontaneously active after a week? With the same characteristics as unpatterned (random) networks? Is there a minimum diameter above which networks become spontaneously active? If stand-alone networks become interconnected, do firing patterns synchronize?

21.2 MATERIALS AND METHODS

21.2.1 Cortical Neuron Isolation/Culturing and MEA Fabrication

Dissociated trypsin/ethylenediaminetetraacetic acid (EDTA) cortical neurons (1-day postnatal rats) were seeded onto patterned structures with a plating density of 5000 living cells/mm^2. Cells were allowed to adhere onto the surfaces during a time period of 4 h. Samples were rinsed with NaCl (0.9%) solution to remove nonadherent cells. Neurons were cultured in chemically defined R12 medium [Dulbecco's modified Eagle's medium (DMEM)/Ham's F12, Gibco] without serum. The cultures were stored in a CO_2 incubator with a constant temperature of 37°C and a constant CO_2 level of 5%. The culture medium was refreshed half, 3 times a week.

In short, MEAs were fabricated from 5 × 5-cm glass plates with gold-deposited tracks leading to 61 hexagonal ordered electrodes. The MEAs were isolated with a sandwich layer of $SiO_2-Si_3N_4-SiO_2$ (ONO) using a plasma-enhanced chemical vapor deposition (PECVD) process. Electrode tips were deinsulated with a SF_6 reactive ion etching (RIE) technique and platinized to reduce the electrode impedance down to 200 kΩ at 1 kHz.

Neurophilic islands were created by PEI microstamping of circular patterns.

21.2.2 Recording of Action Potentials

Electrode signals were amplified, filtered between 0.3 and 6 kHz (first order), and captured by a 16-channel 12-bit National Instruments PCI-6023E data acquisition PC card. The input range as well as the sampling frequency was software controlled by a Labview program. The real-time data processing software reduced the data stream by rejection of data which did not contain bioelectrical activity. Artifact rejection was severe: If activity is measured at the same time in different channels, the waveforms are rejected. In each channel, the root-mean-square (RMS) noise level was constantly monitored and determines the setting of a level detector to detect spike activity. The threshold was set at 6 times the noise level (typically 7 μV RMS). Each time bin of 10 ms with recorded activity was stored and analyzed with MATLAB computer software. Before further processing, wave shapes were classified to distinguish multiunit form single-unit activity [13].

21.2.3 Fabrication of Neurophobic Background Materials

1. *Polyimide* Polyimide (PI, Probimide 7510®, Arch Chemicals N.V., Zwijndrecht, Belgium) was spin coated (4000 rpm, 30 s) onto 25-cm^2 glass plates (Glaverbel, Belgium). The PI was diluted in *n*-methyl pyrolidon (1 : 1 v/v), dried on a hot plate (120°C, 5 min), exposed to UV light, and baked (300°C, 90 min). Then plates were cut into square pieces of approximately 2.6 cm^2.

2. *Plasma-Coated Fluorocarbon (FC)* In a RIE system, spin-coated PI samples were treated with an etching CHF$_3$/O$_2$ plasma (25 sccm CHF$_3$, 5 sccm O$_2$, 150 mTorr and 2.1 × 10^{-1} W/cm^2) for 20 s, a depositing CHF$_3$ plasma (25 sccm CHF$_3$, 150 mTorr, and 2.1 × 10^{-1} W/cm^2) for 40 s, and a final depositing CHF$_3$ plasma treatment at 1.2 × 10^{-1} W/cm^2 for 8 min.

3. *Synperonics F108 and F127* The triblock copolymers Synperonics F108 (EO$_{127}$–PO$_{48}$–EO$_{127}$; ICI, Holland BV, Rozenburg) and F127 (EO$_{95}$–PO$_{62}$–EO$_{95}$; ICI, Holland BV, Rozenburg) were dissolved in 0.1 M phosphate-buffered saline (1% w/w) and adsorbed onto PI- and plasma-coated FC samples over a time period of 24 h. Subsequently, samples were rinsed twice with sterile water (Aqua Purificata, Bufa BV, Uitgeest, The Netherlands). Surfaces were dried by aspiration of residual water with a glass pipette connected to a vacuum pump.

21.2.4 Preparation of Polydimethylsiloxane (PDMS) Microstamps

Sylgard 184 silicone (Mavom bv, The Netherlands) was mixed with the curing agent in a 10 : 1 ratio. Air bubbles were removed from the mixture by evacuation with a water jet pump. Collapse of air bubbles was promoted by following a cycle of evacuation and pressure release for 6 times. A metal ring with an inner diameter of 4.3 mm (height 0.8 mm) was placed around the central area of a PI mould containing three different regions of 20 microwells (12 μm deep) with diameters of 50, 100, and 150 μm at the bottom. The spacing distance between the wells was fixed at 90 μm for all three regions. The ring was filled with the mixed silicone, covered with a 76-mm microscope slide, and crosslinked at room temperature in the mold for four days. Finally, stamps were carefully removed from the mold and stored in plastic tubes until use.

21.2.5 Quantification of Background Adhesion

Microphotographs were taken on two separate subsections of each pattern after 1, 4, 8, 15, and 30 days with a digital photo camera (AxioCam HR, Carl Zeiss, Germany) attached to an inverted phase contrast microscope (Nikon Diaphot-TMD, Tokyo, Japan). The images were first corrected for nonuniform illumination. Then, they were converted into black-and-white images by a threshold operation: white pixels for the cell areas, black for the background material.

The calculated fraction *F* of white pixels in the image now represented a quantitative measure for the adhesion on the background material only:

$$F = \frac{\text{number of white pixels}}{\text{number of white pixels} + \text{black pixels}}$$

21.3   RESULTS

21.3.1   Improved Neurophobic/Neurophilic Contrast

Figure 21.3 shows the results of cell adhesion in the neurophobic (called "background") areas. It can be clearly observed that adhesion in the neurophobic regions decreases with time for PI only (first column) but remains equal or varies a bit for all other

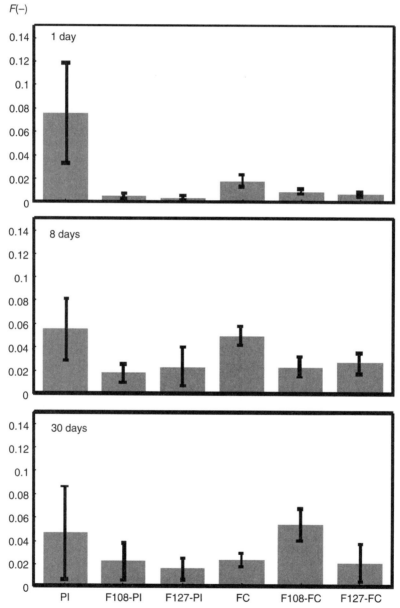

Figure 21.3   Quantified background adhesion fraction F (see Section 21.2.4) of neuronal tissue on six different background surfaces (neurophobic part) after 1, 8, and 30 days. Mean $\pm$ S.D. ($n = 6$). PI: polyimide coating only; F108-PI: superonics F108 coating on top of PI layer; F127-PI: superonics F127 coating on top of PI layer; FC: fluorocarbon coating only; F108-FC: superonics F108 coating on top of FC layer; F127-FC: superonics F127 coating on top of FC layer.

coating conditions. Except for the F108-FC coating, which shows an increase of cells in the neurophobic parts. The conclusion from Figure 21.3 is that FC coating is better than PI coating, also after 30 DIV. Additional coating of PI or FC layers with triblock-copolymers Synperonics F108 or F127 repels cells even better, as well for PI as for FC, but not after 30 days in the F108-FC case. Best results are obtained for the two coatings on the PI basic layer, with only a few percent of neurophobic layer area taken by cells after 30 days; see Figure 21.3 (full report is in [12]).

21.3.2 Spontaneous Activity of Patterned Islands

On one MEA, with unconnected islands, spike rates in a particular island were observed after 22 DIV. Neurons fired at a rate of 8 spikes/s at 22 DIV and 4–5 spikes/s at 28 DIV.

A second, more systematic search for spontaneous activity in patterned but con-nected islands was done on 2 MEAs, each with 60 stamped circles of PEI, diameters 50, 100, and 150 μm (20 each). Activity started at normal age, 8 DIV, and could last until the last observation day, 35 DIV. Almost no activity was seen for the 50-μm-diameter islands. Five percent of the 100-μm-diameter islands showed spontaneous activity and 10% of the 150-μm-diameter islands. Spike rates were much higher than in the uncon-nected island, ranging from averages (± 1 SD) of 17 (12), 302 (199), 229 (192), 95 (—), and 248 (—) at 8, 12, 14, 19, and 35 DIV, respectively (number of active electrodes per islands were 5, 3, 4, 1, and 1, respectively).

The experimental results are summarized in Table 21.1. On 150-μm PEI circles, the neuronal connectivity N_C between neuronal clusters (defined as the average number of surrounding electrodes, connected to a single PEI-coated island through neurite fas-cicles) was only slightly higher for N_C-active than for N_C-nonactive electrodes. However, on 100-μm circles the connectivity was significantly higher on the active electrodes.

An example of connected clusters is given in Figure 21.4. It shows the result of out-growing neuronal tissue as observed after 30 days on a microprinted pattern of PEI circles onto FC. Cells initially migrated toward the center of the circles (not shown) and allowed free space for the outgrowth of neurites around the aggregates and within the borders of the PEI circles. The tissue on the two circles were interconnected by a bundle of neurites and/or axons.

TABLE 21.1 Number of Neurons N Adhering on Microprinted PEI Circles with Diameters of 50, 100, and 150 μm After 1 Day

	50 μm	100 μm	150 μm
N (—), 1 day	16.4 ± 2.7 ($n = 20$)	45.6 ± 5.0 ($n = 20$)	81.2 ± 15.1 ($n = 20$)
P_{AC-EL} (%), 8 days	0.7 ± 1.9 ($n = 7$)	2.8 ± 3.9 ($n = 7$)	10.7 ± 12.7 ($n = 7$)
P_{COV-EL} (%), 8 days	70.0 ± 23.3 ($n = 6$)	93.6 ± 8.0 ($n = 6$)	93.6 ± 9.0 ($n = 6$)
N_C (—), active, 8 days	— ($n = 0$)	6.0 ± 0.0 ($n = 3$)	3.5 ± 2.6 ($n = 7$)
N_C (—), nonactive, 8 days	2.0 ± 2.1 ($n = 36$)	2.9 ± 2.4 ($n = 34$)	2.7 ± 2.5 ($n = 31$)
P_{AC-EL} (%), 15 days	0.8 ± 2.0 ($n = 6$)	5.0 ± 7.7 ($n = 6$)	10.0 ± 13.8 ($n = 6$)
P_{COV-EL} (%), 15 days	64.2 ± 23.8 ($n = 6$)	93.3 ± 8.8 ($n = 6$)	95.8 ± 5.9 ($n = 6$)
N_C (—), active, 15 days	— ($n = 0$)	6.0 ($n = 1$)	3.3 ± 2.2 ($n = 4$)
N_C (—), nonactive, 15 days	2.0 ± 2.0 ($n = 36$)	2.7 ± 2.4 ($n = 35$)	2.4 ± 2.3 ($n = 33$)

Abbreviations: P_{AC-EL}: percentage of corresponding electrically active electrodes on MEAs after 8 and 15 days; P_{COV-EL}: percentage of electrodes covered with neuronal cells on MEAs after 8 and 15 days; N_C: number of neuronal connections with surrounding electrodes for electrically active electrodes and nonactive electrodes after 8 and 15 days.

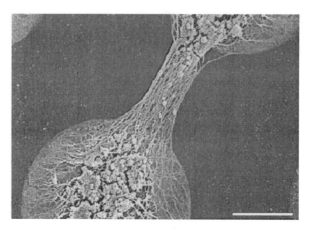

Figure 21.4   Scanning electron micrograph of cortical neuronal tissue (30 days) present on pattern of PEI circles microprinted onto FC. Diameter of circular islands is 150 μm. Spacing distance between circles is 90 μm. Scaling bar = 50 μm. The micrograph shows a connected pair of islands in detail.

21.3.3   Synchronized Activity of Patterned Islands

Figure 21.5 gives a typical example of two connected islands at 12 DIV. They fired synchronously during periods of typically 120 s, with regular silent intervals between these 120-s epochs. Figure 21.6 presents the spike rates of these islands, first as a spike

Electrode 34, div 12, 550.0 spikes per 60 s, file: k5mldl2gv24
Electrode 40, div 12,163.4 spikes per 60 s, file: k5mldl2gv24

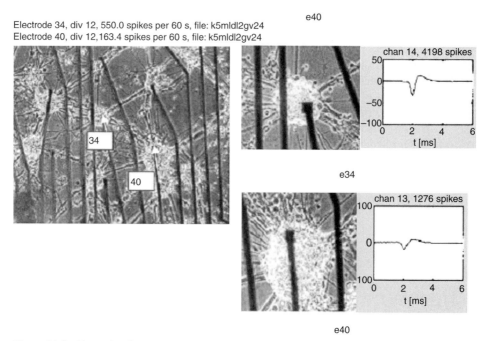

Figure 21.5   Example of spontaneous bioelectrical activity from cortical neurons seeded onto PEI-microprinted MEA (12 DIV). Two islands at electrodes 34 and 40 were spontaneously active after a week. Electrode separation is 190 μm. Islands became interconnected later and showed synchronized behavior (Fig. 21.6). Details on the right show the two clusters and the derived averaged spike waveform (top: electrode 40, 4198 spikes, bottom: electrode 34, 1276 spikes).

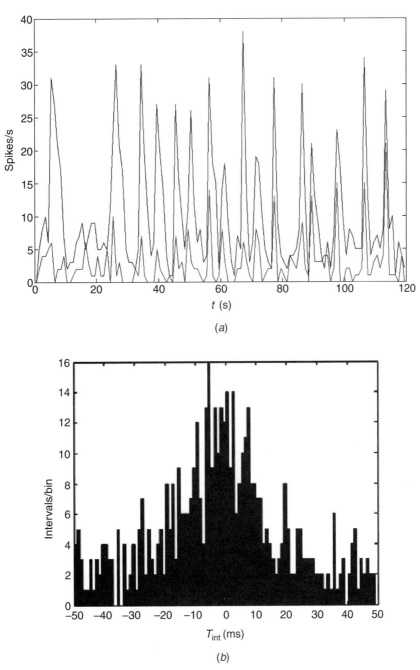

Figure 21.6   Analysis of spike activity of two connected islands at electrodes 34 and 40 of Figure 21.5. (*a*) Spikes per second during 120 s of bursting period. Vertical scale: spikes/second. The two curves are for electrode 34 (upper, dark gray curve) and 40 (light gray). The two curves clearly resemble the same oscillating pattern of bursts, pointing at synchronization between the two networks. Spearman correlation coefficient $\rho = 0.61$. (*b*) Interelectrode (electrodes 34 and 40) interval time (T_{int}) histogram, bin width 1 ms. This representation of the data also clearly illustrates synchronization between the firing patterns of the two island networks.

rate plot over a time interval of 120 s, then as a interelectrode interval histogram. It is clearly observed that the two clusters fire in a highly synchronized way with periodic bursting. In the peaks, spike rates go up to 30 spikes/s in one channel and about 20 spikes/s in the other.

21.4 DISCUSSION

Activity in patterned MEAs will be briefly compared to unpatterned (random) network activity, as reported in [14]. In the latter, activity starts around one week after seeding, that is, 7 DIV. Spike rate, summed over all 61 electrodes, develops gradually from zero to 32 spikes/s between 9 and 42 DIV. This implies a maximum of about 0.5 spikes/s per electrode at 42 DIV.

However, in time, and per electrode, spike rate may differ considerably; for example, 22 spikes/s was observed maximally in one electrode (lasting a few days). Comparing unpatterned networks with patterned islands, activity seems to start at the same age, 7–8 DIV. The variability in the random networks and absence of statistics so far in the unconnected island case make it hard to draw a comparison. A cautious observation may be that the spike rate in unconnected islands is in the same range as in the random network, that is, 4–8 spikes/s at 22–28 DIV.

It is however clear that the connected clusters fire at a very high spike rate, much higher than observed in the unpatterned network [14].

Another comparison can be made with regard to the probability that spontaneous firing develops in unpatterned large networks and in patterned ones. In unpatterned networks it was found [14] that 16 out of 24 experiments [each consisting of 7 cultured MEAs) were selected (on several grounds, but not presence/absence of activity)] for longitudinal measurements. Of this set of 112 cultures (i.e., 112 MEAs with 60 electrodes each), 47 (40%) exhibited spontaneous activity at 7 and 8 DIV, of which 16 had four or more active electrode sites. This means that 60% had no active electrodes at all, 26% ($n = 31$) showed activity at less than four electrodes, and 14% ($n = 16$) at four or more electrodes. So, one can estimate conservatively that the average probability that an electrode is active is on the order of 1% (under the assumption that the electrodes behave "independently," so we do not consider the bursting periods, in which many of the available electrodes show activity). Clearly, this percentage is of the same order or lower as found for the patterned islands, as we saw in Table 21.1, that is, 2.8% and 10.7% for the 100- and 150-μm islands, respectively, at 8 DIV. However, as the number of elements in each sample is only 7 and the standard deviations are considerable (Table 21.1), these figures are only indicative.

The start of activity at diameter 100 μm and onward in circular islands probably indicates a minimum number of neurons needed to develop spontaneous activity. However, also other variables will play an essential role in that process, such as number and density of synaptic connections and density of cells.

Control of local density was done by Chang et al. [15] on hippocampal neurons showing that alternating line patterning (40-μm-wide lines) of substrates gave control over local density of neurons, 100–500 cells/mm^2, and did enhance the activity compared to randomly plated networks with about the same density.

The data give some indication that small islands, diameter 50 μm, develop hardly any spontaneous activity. This may be advantageous for the use of a cultured probe as a recording device. As stated in the introduction, for the stimulatory use of a cultured probe, spontaneous activity is unimportant as long as axonal sprouts grow from the host tissue toward the implant. Other types of cultured probes may do the reverse, that is, "send out" axons to the host tissue. In that case, spontaneous activity of the intermediate cultured networks is undesirable.

REFERENCES

1. J. R. BUITENWEG, W. L. C. RUTTEN, AND E. MARANI, "Geometry based finite-element modeling of the electrical contact between a culture neuron and a microelectrode," *IEEE Trans. Biomed. Eng.* 50, 501–510, 2003.

2. J. R. BUITENWEG, W. L. C. RUTTEN, E. MARANI, S. K. L. POLMAN, AND J. URSUM, "Extracellular detection of active membrane currents in the neuron-electrode interface," *J. Neurosci. Methods* 115, 211–221, 2002.

3. J. R. BUITENWEG, W. L. C. RUTTEN, AND E. MARANI, "Extracellular stimulation window explained by a geometry-based model of the neuron-electrode contact," *IEEE Trans. Biomed. Eng.* 49, 1591–1600, 2002.

4. J. R. BUITENWEG, W. L. C. RUTTEN, AND E. MARANI, "Modeled channel distributions explain extracellular recordings from cultured neurons sealed to microelectrodes," *IEEE Trans. Biomed. Eng.* 49, 1580–1591, 2002.

5. T. G. RUARDIJ, M. H. GOEDBLOED, AND W. L. C. RUTTEN, "Adhesion and patterning of cortical neurons on polyethyleneimine and fluorocarbon-coated surfaces," *IEEE Trans. Biomed. Eng.* 47(12), 1593–1599, 2000.

6. T. G. RUARDIJ, M. H. GOEDBLOED, AND W. L. C. RUTTEN, "Long-term adhesion and survival of dissociated cortical neurons on miniaturized chemical patterns," *Med. Biol. Eng. Comp.* 41, 227–232, 2003.

7. M. A. CORNER AND G. J. A. RAMAKERS, "Spontaneous firing as an epigenetic factor in brain development—Physiological consequences of chronic tetrodotoxin and picrotoxin exposure in cultured rat neocortex neurons," *Dev. Brain Res.* 65, 57–64, 1992.

8. G. GROSS, "Simultaneous single unit recording in vitro with a photoetched laser deinsulated gold multielectrode surface," *IEEE Trans. Biomed. Eng.* 26, 273–278, 1979.

9. J. PINE "Recording action potentials from cultured neurons with extracellular microcircuit electrodes," *J. Neurosci. Meth.* 2, 19–31, 1980.

10. Y. JIMBO, T. TATENO, AND H. P. C. ROBINSON, "Simultaneous induction of pathway-specific potentiation and depression in networks of cortical neurons," *Biophys. J.* 76, 670–678, 1999.

11. J. M. COREY, B. C. WHEELER, AND G. J. BREWER, "Micrometer resolution silane-based patterning of hippocampal neurons: Critical variables in photoresist and laser ablation processes for substrate fabrication," *IEEE Trans. Biomed. Eng.* 43, 944–955, 1996.

12. T. G. RUARDIJ, M. A. F. VAN DEN BOOGAART, AND W. L. C. RUTTEN, "Adhesion and growth of electrically-active cortical neurons on polyethyleneimine patterns microprinted on PEO–PPO–PEO triblockcopolymer-coated hydrophobic surfaces," *IEEE Trans. Nanobiosci.* 1(1), 1–8, 2002.

13. G. W. VAN STAVEREN, J. R. BUITENWEG, T. HEIDA, AND W. L. C. RUTTEN, "Wave shape classification of spontaneous neuronal activity in cortical cultures on micro-electrode arrays," in *Proceedings Second Joint IEEE-EMBS/BMES Conference*, IEEE, Houston, TX, 2002.

14. J. VAN PELT, P. S. WOLTERS, M. A. CORNER, W. L. C. RUTTEN, AND G. J. A. RAMAKERS, "Long-term characterisation of firing dynamics of spontaneous bursts in cultured neural networks," *IEEE Trans. Biomed. Eng.* 51, 2051–2062, 2004.

15. J. C. CHANG, G. J. BREWER, AND B. C. WHEELER, "Modulation of neural network activity by patterning," *Biosensors Bioelectron.* 16(7/8), 527–533, 2001.

CHAPTER 22

SPIKE SUPERPOSITION RESOLUTION IN MULTICHANNEL EXTRACELLULAR NEURAL RECORDINGS: A NOVEL APPROACH

Karim Oweiss and David Anderson

22.1 INTRODUCTION

Since the late 1950s, techniques for recording single-unit activity have dominated the analysis of brain activity at the microscale due to perfection in isolating and characterizing individual neurons' physiological and anatomical characteristics [1–3]. It was not until the early 1980s that techniques for recording multiple units became feasible [4]. The advent of microprobe fabrication technology greatly accelerated the ability to simultaneously record from ensembles of neurons with minimal tissue damage [5–7]. Because of the ability to integrate a large number of recording electrodes on a single probe, the technology provided neuroscientists with tools to monitor the spontaneous and stimulus-driven behavior of large populations of cells [8]. This has dramatically increased the ability to understand the general computational principles of neuronal populations acting in concert and how they respond individually and collectively to independent stimuli [9].

In a typical neurophysiological experiment using a microelectrode array, the activity of a population of neurons is recorded in response to a certain stimulus. The outcome is multiple time series expressing the extracellular voltage potentials resulting from the population response to the stimulus parameter vector. Each time series contains the activity of an unknown number of neurons in the form of spike trains. Spike sorting has to take place to segregate the response from each single unit. Extraction of the temporal relations among multiple spike trains is subsequently performed to assess functional interdependency among local neural circuit elements and characterize their dynamic properties over long periods of time. A good review for spike-sorting methods is provided in [10].

There remain, however, many challenges that encumber multiunit multielectrode ensemble recording data analysis. Nonstationarity of the spike waveform shape, variability in the total number of neurons in the recorded population among replicas of the same experimental conditions, and the multivariate nature of the data are common problems faced by many investigators. Variability in spike waveform shape requires an astounding amount of human intervention that scales up exponentially with the number of neurons

Handbook of Neural Engineering. Edited by Metin Akay

recorded due to the lack of automated spike sorting with negligible margin of error [11–13]. In the simplest case, neurons fire independently from each other. In the more complex case, two or more neurons can fire simultaneously. The probability of more than two neural sources firing simultaneously in a spontaneous activity is usually negligible. However, in stimulus-driven activity or in some brain regions with increased synchrony, this coincidence probability can be significant [14]. In either case, the result is a complex waveform that often does not resemble any of the simple spike waveforms elicited by the individual neural sources. Such waveforms constitute a significant challenge as they cause a well-known phenomenon of decreased cross correlation at zero offset [15].

To date, the most commonly used techniques to resolve spike superposition is template matching [16]. It relies on comparing the putative superposition waveforms to a dense grid of possible combinations of each pair of templates with all possible delays in the range of −1 to +1 ms between the two templates. While this may seem feasible in single-channel recording and a considerably small number of spike templates, it is clearly impractical and cumbersome when a large population is recorded by an electrode array with high channel count. Possibilities for economical resolution of overlap improve greatly for arrays of electrodes because unique spatial distribution information is added to the time series that characterizes the spike waveform, thereby enabling an additional degree of freedom to resolve the ambiguity.

The algorithm reported in this chapter is a new approach to resolve spike overlap based on an augmented representation of the observation space to incorporate the spectral and spatial information of the spike waveforms. It also comprises a natural extension to the robust spike-sorting technique initially proposed in [17] and detailed in [18]. The approach relies on an elegant intermix between the theory of multiscale decomposition by means of the discrete wavelet transform (DWT) and classical eigendecomposition commonly used in array signal processing theory.

22.2 THEORY

22.2.1 Mathematical Preliminaries

It is assumed that the activity of P neural sources is recorded within the discrete time interval $[t_1, \ldots, t_N]$ by an M-electrode array. The result is a matrix $\mathbf{Y} \in \mathfrak{R}^{M \times N}$ that can be expressed as

$$\mathbf{Y} = \mathbf{AS} + \mathbf{Z} \tag{22.1}$$

where $\mathbf{S} \in \mathfrak{R}^{P \times N}$ denotes the signal matrix, $\mathbf{A} \in \mathfrak{R}^{M \times P}$ denotes the mixing matrix whose columns express the array response to each source, and $\mathbf{Z} \in \mathfrak{R}^{M \times N}$ denotes a zero-mean additive noise component with arbitrary spatial and temporal covariance modeling the background neural activity and the thermal and electrical noises from the conditioning electronics. The objective is to recover the matrix $\mathbf{S}$ and determine the parameter P to be able to segregate the individual signal components in the rows of $\mathbf{S}$.

One way to recover the signal matrix $\mathbf{S}$ from the observed mixture stems from linear systems theory, which relies on prior knowledge of the mixing matrix $\mathbf{A}$ [19]. One approach is to use the *best linear unbiased estimator* (BLUE) using the *pseudoinverse* of $\mathbf{A}$, which is expressed as

$$\tilde{\mathbf{S}} = (\mathbf{A}^\mathsf{T} \mathbf{R}_Z^{-1} \mathbf{A})^{-1} \mathbf{A}^\mathsf{T} \mathbf{R}_Z^{-1} \mathbf{Y} \tag{22.2}$$

where $\mathbf{R}_Z \in \mathfrak{R}^{M \times M}$ denotes the noise spatial covariance matrix. The term that premultiplies the observation matrix $\mathbf{Y}$ is used to cancel out all the interference and correlated noise components. If $\mathbf{R}_Z$ is diagonal (spatially uncorrelated noise), then Eq. (22.2) amounts to applying appropriate weights to every channel to maximize the signal-to-noise ratio (SNR). If the noise is spatially correlated, then the inverse of the noise covariance acts as a spatial filter for all nonprincipal components in $\mathbf{Y}$. However, the signal estimate using this method is contingent upon knowledge of the noise kernel $\mathbf{R}_Z$ and the mixing matrix $\mathbf{A}$. Besides, this method has some limitations as the number of *nonprincipal* sources becomes large because the cancellation process increases the noise on channels where sources are not present. Other approaches for recovering the signal matrix use independent-component analysis (ICA), which relies on the statistical independence between sources [20]. This approach does not require prior knowledge of $\mathbf{A}$ or the noise kernel $\mathbf{R}_Z$ and may have a substantial amount of success provided the independence assumption holds. However, it requires precise knowledge of P—the number of sources present in the complex waveform—and *at most* one of the sources is required to be Gaussian distributed. In highly synchronized neuronal activity, there is no guarantee that these assumptions always hold.

An important observation is that one needs only to use columns of $\mathbf{A}$ that express the array response to the principal neural sources making up the complex spike waveform. This may become feasible if a reliable estimation of the *signal subspace* is feasible. There is a vast amount of literature dealing with this problem and we will not attempt to review them. It is only sufficient to mention that estimating the signal subspace becomes very complex when the sources have correlated waveform shapes, which is typically the case for neuronal spikes. An alternative in such a case is to project the complex waveform onto a *nested* set of subspaces using an orthogonal basis set, hoping that some sources may have a "large" correlation with a subset of these bases and therefore may not be strongly present in the corresponding subspaces may lead to a *sparser* representation of these sources in the corresponding subbands, while they may be very close to zero in other subbands where the correction is "small". The sparseness of neural spikes in time helps this approach to a large extent, because it implies that the probability of compound events in those subspaces will be low but not zero. If knowledge of this subset is known for single sources from their simple spike waveforms, then $\mathbf{A}$ can be effectively reduced to fewer columns, thus increasing the SNR. Estimation of the $\mathbf{A}$ matrix for simple spike waveforms has been the subject of previous work [21, 22]. We briefly overview in the next section how the single-source case is characterized before we embark on the complex multisource case.

22.2.2 Single-Source Characterization

When the spike waveform is a result of a single neuron firing, the observation model in (22.1) takes the form

$$\mathbf{Y} = \mathbf{as}^{\mathrm{T}} + \mathbf{Z} \tag{22.3}$$

where now the matrix $\mathbf{A}$ becomes the $M \times 1$ column vector $\mathbf{a}$ and the simple spike is expressed by the $N \times 1$ column vector $\mathbf{s}$. The covariance of $\mathbf{Y}$ is expressed as

$$\mathbf{R}_Y = E\{\mathbf{Y}\mathbf{Y}^{\mathrm{T}}\} = \mathbf{as}^{\mathrm{T}}\mathbf{sa}^{\mathrm{T}} + \mathbf{R}_Z \tag{22.4}$$

assuming the noise to be uncorrelated with the spike waveform. Using singular-value decomposition (SVD), $\mathbf{R}_Y$ is spectrally factored as

$$\mathbf{R}_Y = \sum_{i=1}^{P} \delta_i \mathbf{u}_i \mathbf{u}_i^{\mathrm{T}} + \sum_{i=P+1}^{M} \zeta_i \mathbf{u}_i \mathbf{u}_i^{\mathrm{T}} \tag{22.5}$$

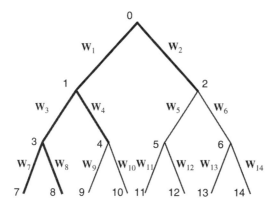

Figure 22.1 Discrete wavelet packet transform binary tree representation for three resolution levels. Node indexing starts at node 0 (time domain) and increases linearly following a top–bottom left–right order. The total number of nodes in the DWT tree (thick lines) is $2^L + 1$ and in the DWPT tree is $2^{L+1} - 1$. **W** is an orthonormal transformation matrix.

where δ_i denotes the ith eigenvalue corresponding to the ith signal source, ζ_i denotes the eigenvalue corresponding to the ith noise source, and the $M \times 1$ vectors $\mathbf{u}_i$ are the eigenvectors spanning the column space of **Y**. In the single-source case, the parameter $P = 1$, and therefore it is sufficient to examine the dominant eigenvalue/eigenvector pair $\{\delta_1, \mathbf{u}_1\}$ to characterize the source in terms of the spike's temporal energy content expressed by δ_1 and spatial distribution expressed by $\mathbf{u}_1$. In the multisource case, the parameter P is unknown, and it becomes nontrivial to determine the number of dominant modes in Eq. (22.5) due to the mutual correlation of spike waveforms from distinct neurons [18].

22.2.3 Orthogonal Transformation

As discussed previously, projecting the simple spike waveform **s** onto a nested set of subspaces using an orthogonal basis set yields variable degrees of correlation that can be used to identify a subset of bases that best characterize the spike waveform. This can be conveniently achieved using orthogonal wavelet bases due to their excellent properties in approximating a wide variety of signals, especially transient ones that are spike-*alike* [23]. For the lack of space, we omit the mathematical details of the theory of wavelet decomposition and the interested reader is referred to [24].

The overcomplete representation of signal projection using the two-band discrete wavelet packet transform (DWPT) offers an augmented set of subspaces from which one obtains a *dictionary* of bases from which to choose. The binary tree representation illustrated in Figure 22.1 is a conventional way of representing a multilevel DWPT. For an L-level DWPT, there is a total of $J = 2^{L+1} - 1$ nodes in the tree. To identify a characteristic *best-basis* tree for a given signal, one needs to "prune" the tree based on an objective function to obtain an *admissible* binary tree. An *admissible* binary tree is a binary tree where each node has either zero or two children.[1] In [24], Mallat showed that the number B_L of bases in a full DWPT binary tree of depth L leading to admissible binary trees is bounded by $2^{2^{L-1}} \leq B_L \leq 2^{\frac{5}{4}2^{L-1}}$. High-amplitude wavelet coefficients in a certain node indicate the presence of the corresponding basis in the signal and measure its contribution. On

[1]For a Q-band DWPT, there are Q children nodes for a given parent node [27].

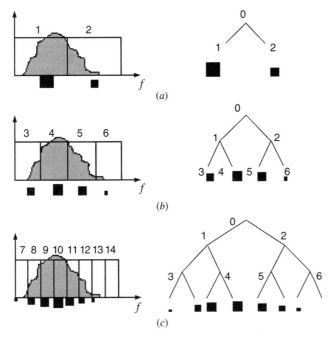

Figure 22.2 Hypothetical spectral representation of an arbitrary signal superimposed on frequency tiling obtained by DWPT at every level. Black squares indicate the relative correlation of the signal with the corresponding wavelet basis and illustrated also on the binary tree. The size of each square is directly proportional to the dominant eigenvalue in that subband. [Node numbers are not shown (on the right) for the last scale level in (c) for clarity].

the other hand, they evaluate the content of the signal inside the related frequency subband, as illustrated in Figure 22.2. Better frequency resolutions are obtained as we proceed from tree top to bottom at the expense of time-domain resolution and vice versa.

To obtain a best-basis binary tree for a given signal, multiple criteria have been proposed in the literature depending on the application [25, 26]. Obviously in our case, one would choose the bases that have the "highest" correlation with the spike waveform since these will best represent the spike waveform. The definition of the highest correlation becomes nontrivial in the presence of strongly correlated neural noise. This is because most of the neural noise consists of spike waveforms from similar neurons. The only information in the noise-free term $\mathbf{as}^\mathrm{T}$ in Eq. (22.3) that is minimally shared by the noise components in $\mathbf{Z}$ is the column vector $\mathbf{a}$, which expresses the array response to the single source. Accordingly, this vector can be considered a *footprint* characterizing the source energy *at* the array.

To be more precise, denote by $\mathbf{Y}_j \in \Re^{M \times N}$ the orthogonal transformation of $\mathbf{Y}$ (row wise) in the jth node of a DWPT binary tree up to L levels. By linearity of the transform, we can express $\mathbf{Y}_j$ as

$$\mathbf{Y}_j = \mathbf{as}_j^\mathrm{T} + \mathbf{Z}_j \qquad (22.6)$$

Note that the column vector $\mathbf{a}$ is independent of the node index j. This is a valid assumption because there is a very negligible probability that the array responds differently to every time sample in the spike waveform, which implies that $\mathbf{a}$ is "stationary in time" over the length of the spike interval. Similar to the analysis in Eqs. (22.4) and (22.5), the covariance of $\mathbf{Y}_j$ can be spectrally factored with SVD as

$$\mathbf{R}_Y^j = \sum_{i=1}^{P} \delta_i^j \mathbf{u}_i^j {\mathbf{u}_i^j}^{\mathrm{T}} + \sum_{i=P+1}^{M} \zeta_i^j \mathbf{u}_i^j {\mathbf{u}_i^j}^{\mathrm{T}} \qquad j = 0, 1, \mathrm{K}, J \qquad (22.7)$$

The eigenvalue/eigenvector pair $\{\delta_1^0, \mathbf{u}_1^0\}$ is equivalent to $\{\delta_1, \mathbf{u}_1\}$ because node 0 is the time-domain signal and therefore $\mathbf{u}_1^0$ is always *similar* to $\mathbf{a}$ up to a scale factor.

For a single source indexed by p, the best-basis selection process can be formulated using a top-down search algorithm that searches the dictionary of bases, starting from node 0, and selects the nodes for which the dominant mode $\{\delta_i^j, \mathbf{u}_i^j\}$ has an eigenvector $\mathbf{u}_i^j$ that is *invariant*, that is, resembles the column vector $\mathbf{a}$. Specifically, the invariance property can be formulated in terms of a distance metric between $\mathbf{u}_1^0$ and $\mathbf{u}_i^j$ in a *Euclidean* norm sense [18]. A cost is thus defined for splitting a given parent node to obtain its children. Specifically, the cost of each node in the DWPT binary tree of source p is compared to the cost of its children and a decision is made to split the parent node if

$$\mathrm{Cost}(j, p) = \min_{j \in \Im_p} \left\| \mathbf{u}_p^{\mathrm{Parent}} - \mathbf{u}_p^{\mathrm{Child}} \right\|^2 \qquad (22.8)$$

where $\Im_p\{j\}$ denotes the set of best bases satisfying the inclusion criterion for nodes with indices less than j (i.e., higher in the tree). The outcome is a characteristic binary tree representing source p along with the feature set $\{\delta_i^j, \mathbf{u}_i^j\}$ for all $j \in \Im_p\{J\}$.

Figures 22.3 and 22.4 demonstrate the applicability of this approach to two different cases of simple noisy spike waveforms extracted from our experimental data archives along with their "cleaned" versions using a wavelet-based denoising algorithm [28]. These spikes were recorded on a subarray of four channels (labeled in different colors in Figs 22.3a and 22.4a). The characteristic features of each spike waveform are shown in (b)–(d) of each figure. These are the characteristic best-basis wavelet packet tree $\Im_1\{J\}$, the dominant eigenvalue profile across best tree nodes $\{\delta_1^j\}, j \in \Im_1\{J\}$, and the principal eigenvector $\{\mathbf{u}_1^j\}, j \in \Im_1\{J\}$.

22.2.4 Multisource Characterization

When the observation matrix contains a complex spike resulting from two or more sources (i.e., $P > 1$), the resolution problem has two levels of complexity. First, estimating P and, second, characterizing each source with features similar to the ones illustrated in Figures 22.3 and 22.4. Consider the specific case of $P = 2$ for simplicity, so we have two sources: source A and source B. We start with the easier scenario in which the two sources elicit spike waveforms with different spatial distribution across channels. This means each source appears on a *principal* channel with high SNR. It is important to note that it is not necessary for each source to be "detectable" on other non-principal channels. Denote the principal channel for A by m_A and for B by m_B. Accordingly, the case considered is one for which $m_A \neq m_B$. The mixing matrix $\mathbf{A}$ thus has two linearly independent columns $[\mathbf{a}_1 \ \mathbf{a}_2]$. It is required to apply the best-basis approach described in Eqs. (22.7) and (22.8) to obtain a characteristic best-basis tree and feature set for each of the two sources. This case is relatively simple, because we only need to examine the two dominant modes $\{\delta_m^j, \mathbf{u}_m^j\}, m = 1, 2$ for all j, to identify the two best-basis sets $\Im_A\{J\}$, and $\Im_B\{J\}$. This situation is illustrated in Figure 22.5 for two different overlap scenarios: when the spikes partially overlap (Fig. 22.5c) and when they fully overlap (Fig. 22.5d). The overlap reference point is

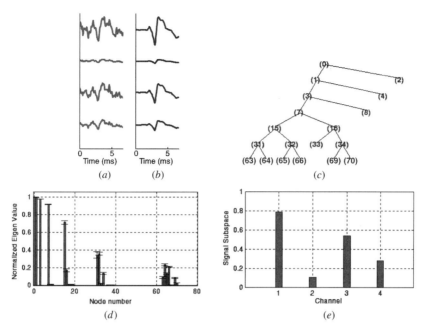

Figure 22.3 (*a*) Single-source waveform along a four-channel array before and after denoising [28]. (*b*) Characteristic best-basis wavelet packet tree $\Im_1\{J\}$. (*c*) Sample mean of dominant eigenvalue profile $\{\delta_1^j\}$ for $\Im_1\{J\}$ across 200 realizations of spike occurrences from the same single unit. (*d*) Sample mean of dominant eigenvector $\{\mathbf{u}_1^j\}$ for $\Im_1\{J\}$.

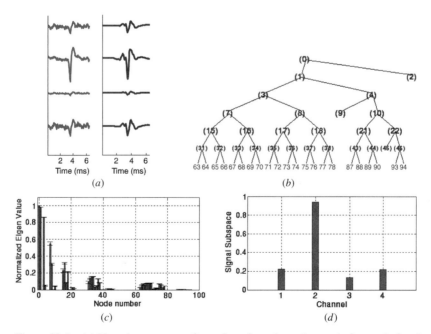

Figure 22.4 (*a*) Neural source waveform along four-channel array before and after denoising. (*b*) Characteristic best-basis wavelet packet tree $\Im_1\{J\}$. (*c*) Sample mean of dominant eigenvalue profile $\{\delta_1^j\}$ in $\Im_1\{J\}$ for 200 realizations of spike occurrences of the same single unit in (*a*). (*d*) Sample mean of dominant eigenvector $\{\mathbf{u}_1^j\}$ across $\Im_1\{J\}$.

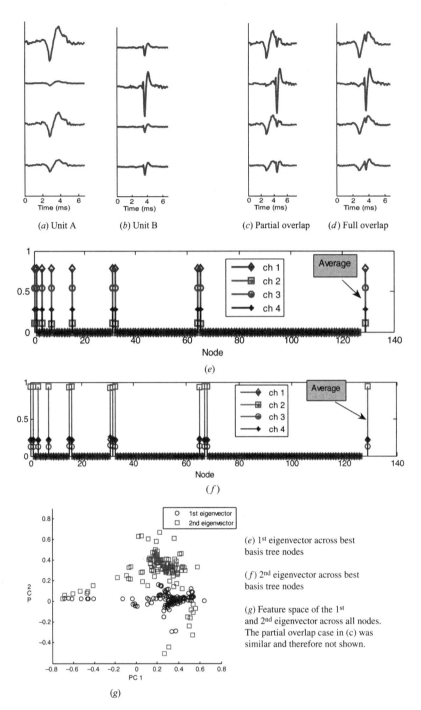

(a) Unit A   (b) Unit B   (c) Partial overlap   (d) Full overlap

(e)

(f)

(e) 1st eigenvector across best basis tree nodes

(f) 2nd eigenvector across best basis tree nodes

(g) Feature space of the 1st and 2nd eigenvector across all nodes. The partial overlap case in (c) was similar and therefore not shown.

(g)

Figure 22.5 (a) Source A across four channels. (b) Source B across four channels. (c) Complex waveform from partially overlapping sources A and B. (d) Fully overlapped sources A and B. (e) First eigenvalue profile for the fully coherent waveform case. (f) First eigenvector profile for fully coherent waveform case. Partial-overlap case (not shown) was very similar. (g) Second eigenvalue profile for fully coherent waveform case. (h) Second eigenvector profile for fully coherent waveform case. Partial-overlap case (not shown) was very similar.

taken at the zero crossing of the first source spike waveform. The features for the two dominant modes $\{\delta_m^j, \mathbf{u}_m^j\}$, $m = 1, 2$, are illustrated in Figures 22.5e–5h for the full-overlap case. The partial-overlap case was very similar and therefore is not shown.

The second scenario we consider is more complex. The two sources now elicit spike waveforms with identical spatial distribution across channels. This means that $m_A = m_B$. Accordingly, the two columns $[\mathbf{a}_1 \; \mathbf{a}_2]$ of the mixing matrix $\mathbf{A}$ may be linearly dependent and $\mathbf{A}$ will be rank deficient. Therefore, examining the second dominant mode in the wavelet tree of the complex waveform will not resolve the overlap due to the so-called *array ambiguity* [29]. Nevertheless, the advantage of the orthogonal transformation plays an important role in this context. Recall from Section 22.2.3 that the dominant mode of the best-basis set $\Im_1\{J\}$ is always associated with the wavelet bases that have the highest correlation with the spike waveform. In the case of more than one source, each of the simple spike waveforms $\mathbf{S} = [\mathbf{s}_A^T \; \mathbf{s}_B^T]^T$ forming the complex spike will have distinct correlation with the wavelet basis at any given node j. Accordingly, the best-basis subset $\Im_1\{J\}$ obtained from the dominant-mode analysis can be further split into two subsets: The first, $\Im_1^A\{J\} \subset \Im_1\{J\}$, will include those nodes in which the correlation of the source A waveform with the wavelet basis $\mathbf{w}_j$ "dominates" that of source B. This can be expressed using the following criterion: For all $j \in \Im_1\{J\}$,

$$
\begin{aligned}
\text{Decide } j &\in \Im_1^A\{J\} \qquad \text{if } \langle \mathbf{s}_A, \mathbf{w}_j \rangle > \langle \mathbf{s}_B, \mathbf{w}_j \rangle \\
\text{Decide } j &\in \Im_1^B\{J\} \qquad \text{if } \langle \mathbf{s}_B, \mathbf{w}_j \rangle > \langle \mathbf{s}_A, \mathbf{w}_j \rangle
\end{aligned}
\tag{22.9}
$$

where $\langle \cdot \rangle$ denotes a dot product. The second subset, $\Im_1^B\{J\} \subset \Im_1\{J\}$ on the other hand, will contain those nodes for which the source B spike waveform has a higher correlation with $\mathbf{w}_j$ than that of source A. Note that it is easy to see that both sets are disjoint, that is, $\Im_1^A\{J\} \cap \Im_1^B\{J\} = \{\}$. Figure 22.6 demonstrates an example of this complex scenario for noise-free spike templates for clarity.

The generalization of the approach to more than two sources making up the complex spike is straightforward. For an arbitrary number of sources within the complex spike window, it can be easily shown that in an arbitrary node j that belongs to the set $\Im_1\{J\}$ the sources can be rank ordered depending on the strength of their correlation with the wavelet basis $\mathbf{w}_j$. It is thus evident that the larger the cardinality of the set $\Im_1\{J\}$, more *resolution power* can be achieved. This can be achieved by either implementing a Q-band wavelet transform [27] or increasing the number of decomposition levels L. It is also clear that increasing the number of channels in the array contributes to the robustness of the cost function evaluation as it reduces the variance of the Euclidean distance metric.

Generally speaking, the probability of more than three neurons spiking simultaneously may be neglected. In our approach, we do not assume prior knowledge of the number of sources contributing to the complex spike. We have proposed elsewhere a criterion to determine the number of sources within the analysis window [30]. This is formulated in terms of a *sphericity test* to determine the equality of the smallest eigenvalues (presumably the noise eigenvalues) or equivalently how *spherical* the noise subspace is [31]. The test takes the form of a likelihood ratio in an arbitrary node j that calculates the ratio of the geometric mean to the arithmetic mean of the noise eigenvalues as

$$
\gamma_j = \frac{\left(\prod_{n=i}^{M} \delta_m^j \right)^{1/(M-i+1)}}{[1/(M-i+1)] \sum_{n=i}^{M}} \qquad i = 1, \ldots, M-1
\tag{22.10}
$$

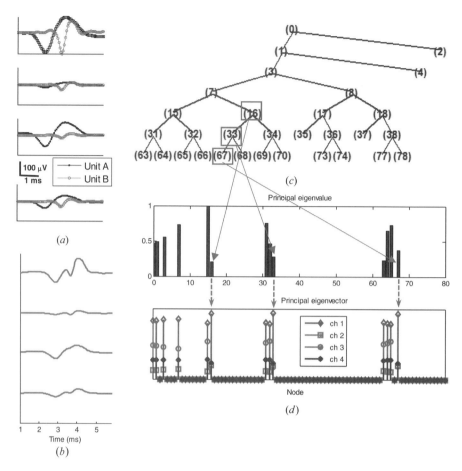

Figure 22.6 (*a*) Noise-free spike waveforms from two sources with near-identical spatial distribution [source A was scaled on channel 3 to enable distinction of features in (*c*)]. (*b*) Complex spike waveform across four channels resulting from superimposing both templates in (*a*). (*c*) *Top*: Dominant-mode best-basis tree $\Im_1\{J\}$. *Bottom*: Dominant eigenvectors of set $\Im_1\{J\}$ illustrating two subsets $\Im_1^A\{J\}$ and $\Im_1^B\{J\}$. Nodes 8, 18, 33, and 68 show stronger responses to source B and therefore are included in $\Im_1^B\{J\}$. Others respond stronger to source A.

This test can be used to determine how many sources $P_j \leq P$ have the strongest representation in the *j*th node and accordingly decide how many sources to consider within the first mode if the array is ambiguous or how many eigen modes to consider if it is not. In practice, an augmented matrix $\overline{\mathbf{A}}$ can be formed by concatenating all the eigenvectors from all the sphericity tests in (22.10) for $j \in \Im_1\{J\}$ in the columns of $\overline{A}$ as follows

$$\overline{\mathbf{A}} = [\mathbf{u}_1^0 \, \mathbf{u}_2^0 \cdots \mathbf{u}_{P_0}^0 \, \mathbf{u}_1^1 \, \mathbf{u}_2^1 \cdots \mathbf{u}_{P_1}^1 \cdots \mathbf{u}_1^J \, \mathbf{u}_2^J \cdots \mathbf{u}_{P_J}^J] \qquad (22.11)$$

where $\mathbf{u}_i^j$ denotes the *i*th dominant eigenvector in the *j*th node. Then, the matrix $\overline{A}$ is projected onto a feature space determined by its first two principal components to allow visualization of the number of clusters corresponding to the number of sources in the complex spike. It is clear that the algorithm can resolve up to *M* different sources within the complex event window.

22.3 PERFORMANCE

Besides the results presented in Figures 22.3–22.6, we demonstrate in this section through simulations the capability of the approach in resolving multiple spikes with variable degrees of overlap. We let ρ denote the ratio of temporal overlap between the spikes such that $0 < |\rho| < 1$. When spike waveforms fully overlap in time (i.e., their zero crossings coincide), $\rho = 0$, and when they do not overlap, $|\rho| = 1$. The absolute value implies we will not distinguish between the case when spike 1 precedes or trails spike 2; that is, the order of the spikes is not important, and in either case ρ will be the same. Three overlap scenarios were simulated as illustrated in Figure 22.7a corresponding to ρ values of -0.5, 0, and $+0.5$ overlap. Accordingly, source A was fixed within the window, while source B was added to A at three different time instants corresponding to two partial overlaps [left ($\rho = -0.5$) and right ($\rho = 0.5$)] and one full overlap ($\rho = 0$) in the middle panel. Figure 22.7b illustrates the feature space obtained by projecting the matrix $\overline{A}$ onto the subspace of its first two principal components. It is clear that the clusters indicate the presence of two sources and are independent of the lag within the window for the data in Figure 22.7a. This performance strengthens the validity of the proposed approach for precisely resolving the number of sources within the complex event as well as its robustness to shifts within the analysis window. The latter property is essential to cope with nonstationarity of the neural signal environment.

22.4 FUTURE TRENDS

Spike superposition resolution is a classical problem that has received significant attention in neuroscience research for a long time. The complexity of the problem dictated an intrinsic level of complexity of the approach. All the methods proposed to date are innately classical in the sense that the underlying theory is governed by exhaustive search approaches. In contrast, the methodology we described was capable of resolving the overlap with no constraints on the time window in which the superposition occurs or the amount of overlap between spike waveforms. Furthermore, the methodology is capable of determining the number of sources involved in a detected event (without actually separating the waveforms). In addition, it segregates the individual waveforms and identifies the exact spectral and spatial characteristics of the corresponding sources.

With the advent of continuous advances in microfabrication of high-density electronic interfaces to the nervous system, it is clear that the resulting neural yield renders exhaustive search techniques very inefficient and time consuming. Nevertheless, the advent of this technology incorporates additional information in the space domain that can be efficiently utilized to provide *smarter* techniques for resolving complex overlaps. The information provided by sampling the space surrounding the microelectrode arrays coupled with the powerful theory of wavelets enabled us to integrate the temporal, spectral, and spatial information to resolve substantially complex situations of spike overlap. It is noteworthy that the choice of the wavelet basis plays an important role in improving the resolution power of the technique.

The outcome of the proposed technique will potentially aid studies focusing on neurons in small volumes that are highly synchronized and/or are participating in local circuits. The bias inflicted on correlation calculations due to nearly simultaneous events can be largely removed by fully identifying each event recorded by an array

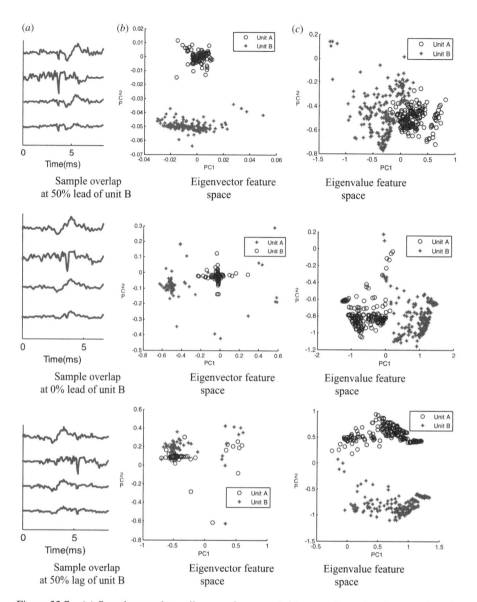

Figure 22.7   (*a*) Sample complex spike waveform created by superimposing two templates in Figures 22.5*a,b* and (*b*) feature space obtained by projecting matrix **A** onto subspace of first two principal components: *top*: 50% lead; *middle*: 0% lag; *bottom*: 50% lag.

whether it is a single spike or an overlapped complex. The methodology is part of a suite of algorithms that facilitate development of a fully automated, near-optimal system to enhance the signal processing technology of implanted devices for recording and stimulating neural cells at the microscale. In experiments where information is being extracted from cortical neuron signals for adaptive control of external devices such as neuroprosthetic arms, it is well known that multiple single-neuron recordings outperform unresolved clusters of neurons. Closely spaced arrays and the signal processing methods that accompany them may well be worth the added complexity.

REFERENCES

1. R. C. GESTELAND, B. HOWLAND, J. Y. LETTVIN, AND W. H. PITTS, "Comments on microelectrodes," *Proc. IRE*, 47:1856–1862, 1959.

2. C. A. TERZUOLO AND T. ARAKI, "An analysis of intra-versus extra-cellular potential changes associated with activity of single spinal motor neurons," *Ann. NY Acad. Sci.*, 94:547–558, 1963.

3. M. VERSEANO AND K. NEGISHI, "Neuronal activity in cortical and thalamic networks," *J. Gen. Physiol.*, 43: 177–195, 1960.

4. M. A. NICOLELIS, A. A. GHAZANFAR, B. M. FAGGIN, S. VOTAW, AND L. M. OLIVEIRA, "Reconstructing the engram: Simultaneous, multisite, many single neuron recordings," *Neuron*, 18:529–537, 1997.

5. K. FRANK AND M. C. BECKER, "Microelectrodes for recording and stimulation," in *Physical Techniques in Biological Research*, Vol. 5, W. L. NASTUK, Ed., New York: Academic, 1964.

6. K. D. WISE, D. J. ANDERSON, J. F. HETKE, D. R. KIPKE, AND K. NAJAFI, "Wireless implantable microsystems: High-density electronic interfaces to the nervous system," *Proc. IEEE*, 92:76–97, 2004.

7. R. NORMANN, E. M. MAYNARD, P. J. ROUSCHE, AND D. J. WARREN, "A neural interface for a cortical vision prosthesis," *Vision Res.*, 39(15):2577–2587, 1999.

8. G. BUZSÁKI, "Large-scale recording of neuronal ensembles," *Nature Neurosci.*, 7:446–451, 2004

9. J. K. CHAPIN, "Using multi-neuron population recordings for neural prosthetics," *Nat. Neurosci.*, 7:452–455, 2004.

10. M. S. LEWICKI, "A review of methods for spike sorting: The detection and classification of neural action potentials," *Network Comput.: Neural Syst.*, 1:53–78, 1998.

11. M. S. FEE, P. P. MITRA, AND D. KLEINFELD, "Variability of extracellular spike waveforms of cortical neurons," *J. Neurophysiol.*, 76:3823–3833, 1996.

12. M. S. FEE, P. P. MITRA, AND D. KLEINFELD, "Automatic sorting of multiple unit neuronal signals in the presence of anisotropic and non-Gaussian variability," *J. Neurosci. Methods*, 69:175–188, 1996.

13. K. HARRIS, D. HENZE, J. CSICSVARI, H. HIRASE, AND G. BUZSÁKI, "Accuracy of Tetrode spike separation as determined by simultaneous intracellular and extracellular measurements," *J. Neurophysiol.*, 84:401–414, 2000.

14. G. GERSTEIN, "Cross-correlation measures of unresolved multi-neuron recordings," *J. Neurosci. Methods*, 100(1/2):41–51, 2000

15. I. GAD ET AL., "Failure in identification of overlapping spikes from multiple neuron activity causes artificial correlations," *J. Neurosci. Methods*, 107:1–13, 2002.

16. E. M. SCHMIDT, "Computer separation of multi-unit neuroelectric data: A Review," *J. Neurosci. Methods*, 12:95–111, 1984.

17. K. G. OWEISS AND D. J. ANDERSON, "Spike sorting: A novel amplitude and shift invariant technique," *Journal of Neurocomputing*, 44–46 (C):1133–1139, 2002.

18. K. OWEISS AND D. ANDERSON, "Tracking signal subspace invariance for blind separation and classification of nonorthogonal sources in correlated noise," EURASIP Journal on Advances in Signal Processing, 2007, article ID 37485, 20pp., 2007.

19. T. KAILATH, Linear Systems, Englewood Cliffs, NJ: Prentice-Hall, 1985.

20. K. OWEISS AND D. ANDERSON, "Independent component analysis of multichannel neural recordings at the micro-scale," *Proc. of the 2nd Joint EMBS/BMES Conference*, 1:2012–2013, Houston, TX, 2002.

21. K. G. OWEISS AND D. J. ANDERSON, "MASSIT: Multiresolution analysis of signal subspace invariance technique: A novel algorithm for blind source separation," *Proc. IEEE 35th ASSC*, 1:819–823, 2001.

22. K. G. OWEISS AND D. J. ANDERSON, "A new technique for blind source separation using subband subspace analysis in correlated multichannel signal environments," *Proc. IEEE-ICASSP*, 1:2813–2816, May 2001.

23. M. AKAY, Ed. *Time Frequency and Wavelets in Biomedical Signal Processing*, New York: Wiley-IEEE Press, 1997.

24. S. MALLAT, *A Wavelet Tour of Signal Processing*, 2nd ed., Academic Press, 1999.

25. R. COIFMAN AND M. WICKERHAUSER, "Entropy based algorithms for best basis selection," *IEEE Trans. IT*, 38:713–718, 1992.

26. H. KRIM ET AL., "Best basis algorithm for signal enhancement," *Proc. ICASSP*, 1:1561–1564, 1995.

27. D. LEPORINI AND J.-C. PESQUET, "High-order properties of M-band wavelet packet decompositions," Paper presented at the IEEE Signal Processing Workshop on Higher-Order Statistics, 1997.

28. K. G. OWEISS AND D. J. ANDERSON, "A new approach to array denoising," *Proc. IEEE 34th Asilomar Conf. SSC*, 1403–1407, 2000.

29. H. VAN TREES, *Optimum Array Processing*, 1st ed., New York: Wiley, 2002.

30. K. G. OWEISS, "Source detection in correlated multichannel signal and noise fields," *Proc. IEEE-ICASSP*, 5:257–260, 2003.

31. D. LAWLEY, "Tests of significance for the latent roots of covariance and correlation matrices," *Biometrika*, 43:128–136, 1956.

CHAPTER 23

TOWARD A BUTTON-SIZED 1024-SITE WIRELESS CORTICAL MICROSTIMULATING ARRAY

Maysam Ghovanloo and Khalil Najafi

23.1 INTRODUCTION

Neuroscientists have shown that there is a topographic representation of each sensory modality over the appropriate region of sensory cortex. This suggests that by implanting an array of electrodes in an individual with a sensory deficit and stimulating specific sites on sensory cortex we can restore a limited sensation [1]. A cortical neural prosthesis must apply electrical stimulations in a well-controlled temporal–spatial pattern similar to the natural cognitive neural activity in order to be effective [2]. This approach has been highly successful in auditory cortex with tens of stimulating sites [3]. Nevertheless, it is significantly more challenging in visual restoration and needs hundreds of stimulating sites at least to afford a minimum functional resolution [4–10].

The University of Michigan (UM) micromachined (microelectro-mechanical-systems or MEMs) silicon microelectrode array technology has been developed during the past two decades with more than 50 passive and active recording and stimulating designs. It can provide an appropriate medium for the above applications by taking advantage of the features available from thin-film, integrated circuits (ICs) and micromachining technologies on silicon wafers. These advantages include precise definition of probe shanks and site geometries with a high level of consistency, which helps in reproducing biological experiments as well as integration of active circuitry on the probes, which eases many of the interconnection problems and shrinks the device size [11]. These micromachined silicon probes are now being widely distributed throughout the world to the medical community. They are being utilized by neuroscientists in an increasing number of experiments to advance both theoretical and clinical aspects of this promising technology [12]. In the clinical phase, though, when these microprobes are considered to be used in humans, safety plays a major role. Safety issues cover a wide range from short- and long-term tissue reactions to the device packaging. However, they can be categorized into patient and device safety. There is one requirement that significantly assists both categories: wireless operation of the implantable microelectronic devices. Elimination of the wires that otherwise would breach through the skin will reduce the risk of infection and patient discomfort. On the device side, the untethered implantable device is physically more robust and easier to be hermetically packaged [10].

In our laboratory, research on developing a wireless intracortical stimulating microprobe started by designing a wireless interface chip, called Interestim-1 [13], for

Handbook of Neural Engineering. Edited by Metin Akay
Copyright © 2007 The Institute of Electrical and Electronics Engineers, Inc.

STIM-2, a hardwired active 64-site multishank microstimulator probe [14], and gained experience from prior research on several peripheral-nerve wireless microstimulators [15–17]. Interestim-1 mounts on a flat micromachined platform that supports the receiver coil and vertical STIM-2 probes, as shown in Figure 23.1. It is designed to generate dual regulated ± 5-V supplies capable of providing the microstimulating probes with up to 50 mW of power. It also regenerates a 4-MHz internal clock and extracts data bit streams of up to 100 kbps from the amplitude shift keying (ASK) 4-MHz radio-frequency (RF) carrier. The internal clock is stepped down by a user-selectable ratio to be synchronized with the recovered data and sent to the STIM-2 probes along with synchronizing strobe pulses [13].

Interestim-2, a four-channel stand-alone wireless stimulating chip, was designed by transferring all active circuitry, including the stimulation and site selection circuits from the active probes to a wireless microstimulating chip, while still using the same microassembly structure of Figure 23.1 [18]. This new circuit allocation has two advantages. First, the wireless chip can stimulate tissue through passive micromachined probes (or even microwires), which are more prevalent and have simpler and lower cost processing. Second, the back-end height of the cortical implant, which was constrained by the vertical active probe circuitry, can be reduced. This is critical in providing enough clearance for the implant in the very limited space between the cortical surface and the skull. A more recent version of Interestim-2 is developed in a foundry-based standard complementary metal–oxide–semiconductor process, which has a modular architecture and addresses 32 stimulating sites per module [19].

When the number of stimulating sites increases, interconnections between the wireless microstimulating chip and passive probes in Interestim-2 architecture become so elaborate that integration of some sort of site selection circuitry on the probe in order to reduce the number of interconnects is inevitable [20]. Reducing the number of interconnects becomes even more crucial when the goal is a 1024-site stimulating system, because it is shown that interconnects are the major cause of failure in cortical implants with large number of stimulating or recording sites [10, 21]. To solve the interconnection problem, all the active circuitry, including wireless interfacing, site selection, and stimulation circuits, were transferred to the probe back end in a modular architecture, which was called Interestim-3. Every Interestim-3 probe is a stand-alone module that addresses 16 stimulating sites and only needs a pair of connections to a receiver LC-tank circuit, which is the only hybrid component in the Interestim-3 wireless cortical microstimulating

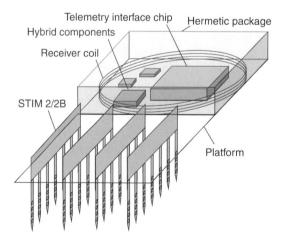

Figure 23.1 Wireless microstimulator array with hybrid coil and interface chip mounted on micromachined platform [13, 18].

system. The rest of this chapter is dedicated to a description of the Interestim-3 architecture. Section 23.2 also includes a brief depiction of some circuit blocks down to the transistor level. The microassembly of the planar micromachined Interestim-3 probes into a three-dimensional (3D) microstructure is shown in Section 23.3. Section 23.4 shows some of the measured results, followed by concluding remarks in Section 23.5.

23.2   SYSTEM OVERVIEW

The Interestim-3 block diagram is shown in Figure 23.2a [22]. Each individual probe, as a module, has all the necessary components to operate as a 16-site stand-alone wireless microstimulator such as RF power conditioning, wireless receiver, digital controller, and site-manipulating circuits. The only inputs to this probe, called coil 1 and coil 2, are the

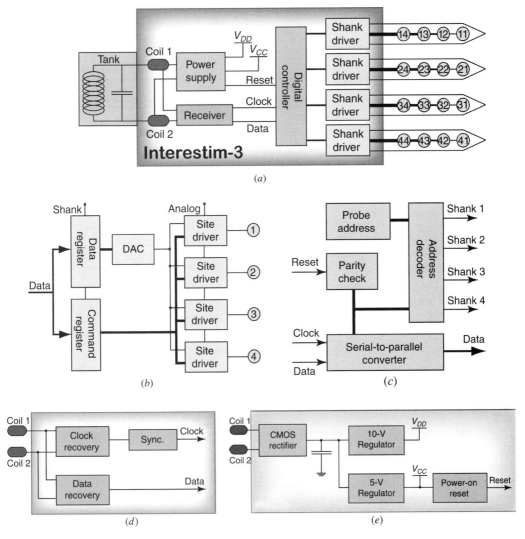

Figure 23.2   (a) Overall block diagram of Interestim-3. (b) Shank driver block diagram. (c) Shank digital controller block diagram. (d) Receiver block diagram. (e) Power supply block diagram [14].

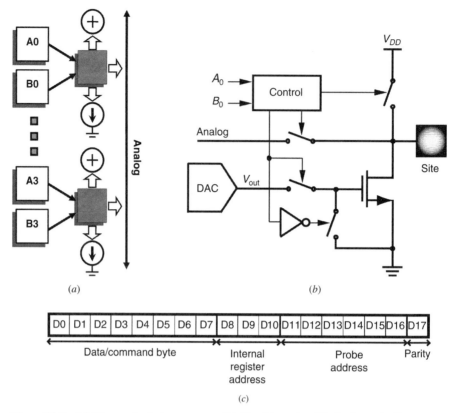

(a)   (b)

| D0 | D1 | D2 | D3 | D4 | D5 | D6 | D7 | D8 | D9 | D10 | D11 | D12 | D13 | D14 | D15 | D16 | D17 |

Data/command byte   Internal   Probe   Parity
                                     register   address
                                     address

(c)

Figure 23.3 (a) Site manipulation strategy. (b) Site driver simplified schematic diagram. (c) Interestim-3 data frame protocol.

nodes of an implantable miniature hybrid LC-tank circuit, which are shared between all probes that are connected in parallel to form a 3D microstimulating array (see Fig. 23.8 below).

Each Interestim-3 probe has four shanks and each shank has 4 stimulating sites on it, which are controlled by an individual shank driver block. These 16 sites are arranged in a 4×4, 1-mm^2 square array with 250 μm center-to-center spacing in the X and Y directions (see Fig. 23.7 below). A more detailed block diagram of the shank driver is shown in Figure 23.2b. There are two 8-bit registers in every shank driver. One of them is the command register that keeps the electrical configuration of its four associated sites by assigning 2 bits to each site. These 2 bits (A, B), as shown in Figure 23.3a and summarized in Table 23.1, define whether each site is at high impedance (inactive), involved in a

TABLE 23.1 Interestim-3 Site Status and Functions

State	Bit A	Bit B	Function
0	0	0	High impedance
1	0	1	V_{DD}
2	1	0	Current sink
3	1	1	Common analog

biphasic-bipolar stimulation, or connected to the common analog line (CAL). The CAL is simply an electrical connection that passes through all site driver blocks for charge balancing. Every stimulating site has an individual site driver circuit, shown in Figure 23.3*b*, with a single nMOS nonlinear current sink [18]. In the first phase of a biphasic-bipolar stimulation between two sites, one site is connected to V_{DD} (state 1), while the other one is connected to the nMOS current sink (state 2). The amplitude information of the resulting stimulus current that passes through the excitable neural tissue is stored in the data register, which is the second 8-bit register in each shank driver block and controls the nMOS current sink gate voltage by means of a voltage-mode 7-bit digital-to-analog converter (DAC). In the second phase of stimulation, the site states are switched by a new command and, therefore, the direction of stimulus current reverses in the tissue. This simple current-steering strategy works as long as the implant circuitry is totally isolated from the tissue, which is the case in bipolar stimulation with an electrically floated biomedical implant. For monopolar stimulation, where each site needs source and sink current versus a common ground (usually the same as the circuit ground) a dual-supply stimulating circuit would be needed [14, 23].

The digital controller block is shown in Figure 23.2*c*. The serial data bit stream out of the wireless receiver block is initially converted to a parallel data frame by an 18-bit shift register (Fig. 23.3*c*). The shift register content is latched once it is filled by a data frame and then a parity-checking circuit is activated to look for an even parity. In case of a correct parity, the data frame gets decoded. Otherwise, a reset line is activated and the entire probe resets without taking any further steps.

Each Interestim-3 data frame, shown in Figure 23.3*c*, has 1 byte of either stimulation amplitude data or site configuration command (D0–D7), 3 bits for internal register addressing (D8–D10), 6 bits for probe addressing (D11–D16), and 1 bit for parity checking (D17). Each Interestim-3 probe has a 6-bit hard-wired user-selectable address, which is initially set to 0 when all of the links are shorted to ground. Cutting each link with laser changes the associated address bit from 0 to 1. Therefore, up to 64 (2^6) probes can be individually addressed in order to achieve a 1024-site 3D microstimulator array. After decoding the data frame, if the data frame address matches with the probe address, the 3-bit internal register address is used to send the configuration command or amplitude data byte to one of the eight internal registers in each probe (two in every shank driver, as shown in Fig. 23.2*b*). Using separate addresses for data and command registers in every shank driver can save the amount of data transfer across the limited-bandwidth wireless link in those situations where the stimulation amplitude is constant and only the active site configuration or timing of stimulation pulses would change.

High data rate is one of the key requirements of a wireless 1024-site stimulating microsystem. In broadband wireless communications such as the IEEE 802.11 standard for wireless local-area networks (LANs), data rates as high as 54 Mbps have been achieved at the expense of increasing the carrier frequency to 5.8 GHz, resulting in a data rate–carrier frequency ratio of only 0.93%. In other words, each data bit is carried by 107.4 carrier cycles. However, the maximum carrier frequency for biomedical implant applications is limited to a few tens of megahertz due to the coupled coil self-resonant frequency, more power loss in power transfer circuitry at higher frequency, and excessive power dissipation in the tissue, which increases as the square of the carrier frequency [24]. Therefore, the goal is to transfer each data bit with a minimum number of carrier cycles to maximize the data rate–frequency ratio and minimize the amount of power consumption.

So far, the ASK modulation scheme is commonly used in biomedical implants because of its fairly simple modulation and demodulation circuitry [10, 13, 15–18].

This method, however, faces major limitations for high-bandwidth data transfer, because high-bandwidth ASK demodulator circuits need high-order filters with sharp cutoff frequencies, whose large capacitors cannot be easily integrated in this low-frequency range of RF applications. A remedy that is proposed in the so-called suspended carrier modulation boosts the modulation index up to 100% to achieve high data rates with low-order integrated filters at the expense of an average 50% reduction in the transferred power [10]. Therefore, the frequency shift keying (FSK) data modulation technique was utilized for wirelessly operating Interestim-3, which simply means sending binary data with two frequencies f_0 and f_1, as shown in Figure 23.4. Frequency shift keying is more robust than ASK in noisy environments and against coil misalignments due to motion artifacts, which is a common problem in inductively operated biomedical implants. Yet, another excellent characteristic of FSK modulation for wireless implants is that, unlike ASK, the transmitted power is always constant at its maximum level irrespective of f_0 and f_1 or the data contents.

The Interestim-3 receiver block consists of data recovery, clock recovery, and synchronization circuits, as shown in Figure 23.2d. The received sinusoidal carrier across the receiver LC-tank circuit is used in this block to recover the clock and serial data bit stream for the rest of the chip. The clock recovery block is a cross-coupled differential pair which squares up the sinusoidal FSK carrier into a similar square waveform called CK_{in}. In order to achieve data rates above 1 Mbps, while keeping the carrier frequency below 10 MHz, a particular FSK modulation protocol was utilized. In this protocol, a logic 1 bit is transmitted by a single cycle of the carrier at f_1, a logic 0 bit is transmitted by two cycles of the carrier at f_0, and f_0 is chosen twice as f_1, as shown in Figure 23.4 ($f_0 = 2f_1$). The transmitter frequency switches at a small fraction of a cycle and only at negative-going zero crossings. This technique is generally known as phase-coherent FSK and leads to a consistent data transfer rate of f_1 bits per second. As a result, if we consider the average carrier frequency to be $(f_0 + f_1)/2$, then the data rate–carrier frequency ratio can be as high as 67%. It is also useful to notice that any odd number of consecutive f_0 cycles in this protocol is an indication of data transfer error [25].

The FSK demodulation in data recovery block is based on measuring the period of each received carrier cycle, which is the information-carrying entity in FSK modulation scheme. If the period is higher than a certain value, a logic 1 bit is detected and otherwise

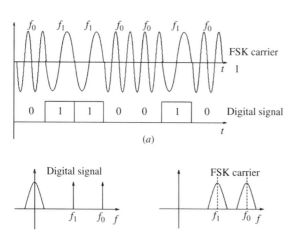

Figure 23.4 The FSK data transfer protocol [25]: (a) time domain; (b) frequency domain.

a logic 0 bit is received. A simple method for time measurement in analog circuits is charging a capacitor with a constant current source and monitoring its voltage. Charging and discharging of this capacitor should be synchronized with the FSK carrier signal. If the capacitor voltage is higher than a certain value, a logic 1 bit is detected and otherwise a logic 0 bit is received. In order to make the data recovery block more robust against noise and process parameter variation, the circuit was designed fully differential by charging two unequal capacitors ($C_L > C_H$) with different currents ($I_H > I_L$) and comparing their voltages with a hysteresis comparator, as shown in Figure 23.5a [25]. The output

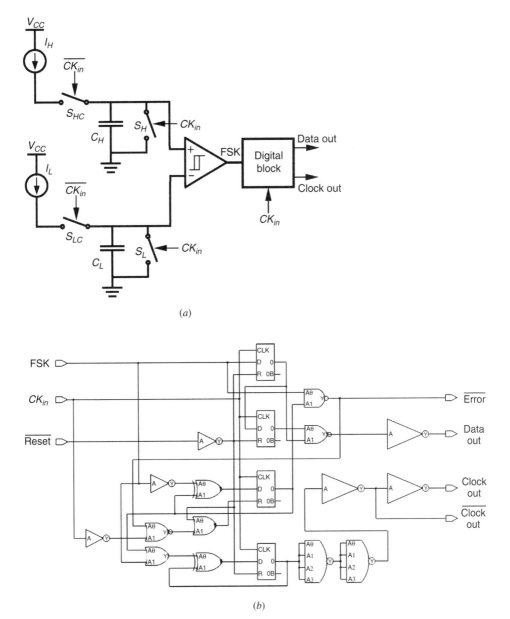

Figure 23.5 (a) Digital block of data recovery circuit. (b) Simulation waveforms with $f_0 = 5$ MHz and $f_1 = 2.5$ MHz [25].

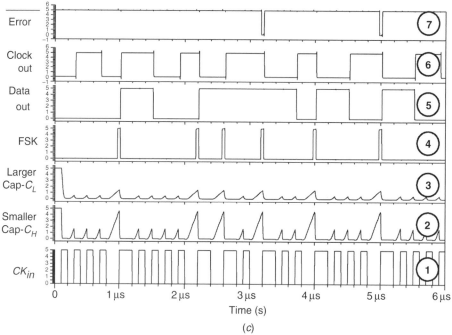

Figure 23.5 *Continued.*

of the hysteresis comparator is a series of square pulses which discriminate between the carrier long and short cycles. These pulses along with CK_{in} are fed into the digital block of Figure 23.5b to generate a serial data bit stream (Data-Out) as well as a constant-frequency clock (Clock-Out). We have also developed two other demodulator circuits based on the same FSK protocol, which are discussed in [26–28].

The performance of the FSK demodulator circuit is simulated in Figure 23.5c. When CK_{in} (trace 1) is low, S_L and S_H switches are open and S_{LC} and S_{HC} are closed. Therefore, the current sources I_L and I_H linearly charge C_L and C_H up to V_L (3) and V_H (2), respectively. During a logic 1 long cycle, V_{L1} and V_{H1} are twice V_{L0} and V_{H0} when a logic 0 short cycle is being received, because in our FSK protocol in Figure 23.4a, f_0 is twice f_1. A hysteresis comparator, with hysteresis window width (W_{hyst}) set somewhere between $V_{H1}-V_{L1}$ and $V_{H0}-V_{L0}$, compares the capacitor voltages. The comparator output, called FSK (4), switches to high during a logic 1 long cycle but not during a logic 0 short cycle. The S_L and S_H switches discharge the capacitors in a fraction of the FSK carrier second half cycle, when CK_{in} is high. Meanwhile, S_{LC} and S_{HC} switches open to reduce power consumption. The train of pulses at the output of the hysteresis comparator (4), which discriminates between short and long carrier pulses, is applied to the digital block, shown in Figure 23.5b, which turns them into a serial data bit stream (5) and a constant-frequency clock (6). This circuit is also capable of detecting any odd number of successive short pulses, which is an indication of error due to our FSK modulation protocol (7).

Supplying a 1024-site wireless microstimulator in an efficient way is another challenge in this design, especially because everything except the receiver coil should be integrated for size reduction. The distributed power supply scheme which is adopted here by including a fully integrated power supply block on every probe has several advantages over a central power supply for the entire 1024-site wireless stimulator system:

1. Each Interestim-3 probe is a self-sufficient planar stimulating array and can be used either individually or in parallel with other similar probes in a 3D array.

2. The output current capacity of the power supply scales by the number of probes that are used in the 3D stimulating microsystem without any unutilized capacity.

3. Less leakage current flows into the substrate due to lower current levels through each power supply rectifier block [29].

4. Better heat dissipation is achieved by distributing the high-current components, such as rectifiers and regulators, across all probes in a larger area.

Figure 23.2*e* shows the Interestim-3 power supply block diagram, which generates regulated 5- and 10-V supplies for the probe internal circuitry and stimulating site drivers, respectively. A CMOS full-wave rectifier [29, 30], shown in Figure 23.6*a*, uses

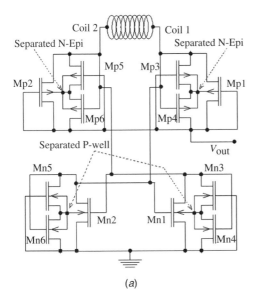

(*a*)

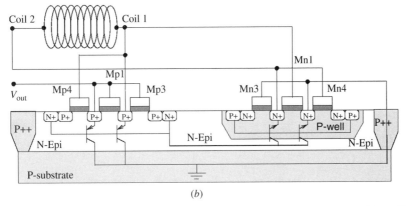

(*b*)

Figure 23.6 (*a*) Full-wave CMOS rectifier bridge. (*b*) Cross section of CMOS rectifier bridge with its parasitic transistors in UM-BiCMOS process [29].

diode-connected MOS transistors (Mp1, Mp2, Mn1, Mn2), which are controlled by the receiver coil voltage (coil 1 and coil 2). Since the coil voltages are higher than the other internal nodes, the rectifier should be implemented in separated N-Epi and P-well regions to be always isolated from the rest of the chip by reverse-biased diodes. To eliminate latch-up and substrate leakage problems, two auxiliary MOS transistors are added to each rectifying MOS to dynamically control the potentials of these separated regions (Mp3–Mp6, Mn3–Mn6). The effect of the auxiliary transistor switching is that the N-Epi potential always follows max(Coil, V_{out}) and the P-well potential follows min(Coil, GND). Therefore, none of the parasitic vertical bipolar junction transistors (BJTs) that are shown in cross section in Figure 23.6b can turn on. The power supply block also includes a 400-pF capacitive low-pass filter for ripple rejection and a power-on reset block, which activates the reset line at startup and releases it after 15 μs when all transient voltages are suppressed.

Figure 23.7 shows photomicrographs of two versions of the Interestim-3 probes, fabricated in the 3-μm, one-metal, two-poly, N-Epi, UM-bipolar CMOS (BiCMOS) process. Having only one metal layer in this MEMS-oriented, large-feature-size process has made the 21.2-mm^2 full-custom IC layout more challenging and larger than what is achievable in today's state-of-the-art submicrometer processes. An order-of-magnitude reduction in the back-end area is expected by implementing the probe in a 0.5-μm, three-metal process. To reduce the back-end height of the probe from 5 to only 1.5 mm above the cortical surface in the current process, probe 2 in Figure 23.7b, was designed with a

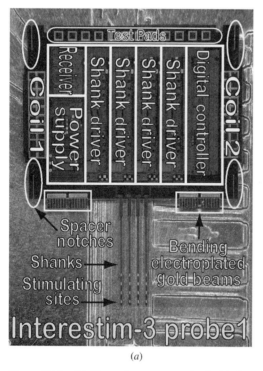

(a)   (b)

Figure 23.7 (a) Floor plan of Interestim-3, 16-site wireless addressable microstimulating probe, with micromachining features such as spacer notches and bending electroplated gold beams for 3D microassembly [22]. (b) Second version of Interestim-3 with bending back end to reduce height of probe back end from 5 to 1.5 mm [20].

bending back end. The shank drivers are located at the middle of the probe, surrounded by digital controller, receiver, and power supply blocks, to have direct access to the four sites on each shank. A functional probe only needs coil 1 and coil 2, the two large input pads on either side of the probe. Twelve test pads are added on the upper edge in addition to several internal cutting links and test points to test each Interestim-3 block individually as well as the entire probe.

23.3 THREE-DIMENSIONAL ARRAY MICROASSEMBLY

Interestim-3 is a planar probe that is designed with all the features required for typical microassembly of micromachined probes on a platform, as shown in Figure 23.7 and described in [21]. The body of each probe is defined by deep boron diffusion followed by standard CMOS processing steps in a dissolved-wafer micromachining process, which is described in [11, 14]. Two wide electroplated gold beams, each with seven $100 \ \mu m \times 300 \ \mu m$ fingers, are added to each side of the probe to provide probe–platform interconnects. All of the gold beams on each side of the probe are electrically connected to the large pad on that side, which represent coil 1 and coil 2 input nodes in the Interestim-3 block diagram of Figure 23.2.

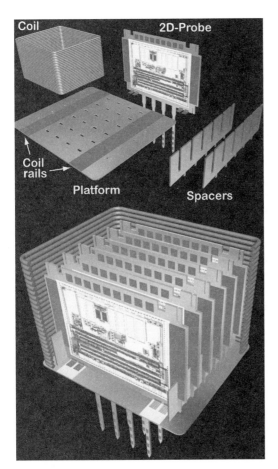

Figure 23.8 Conceptual drawing of Interestim-3 microassembly components with assembled 3D probe [22].

Bai mentions in [21, p. 284] that, "the most time consuming step in the 3-D microassembly process is that of placing spacers. The greater the number of probes the more difficult this is because the probes inserted into the platform are not initially parallel to each other and tend to stick back to back before the spacer is put in." In order to solve this problem, two new sets of spacer notches were added to the lower edges of the probe back end (see Fig. 23.7) to insert probes after the two spacers are installed on the platform. This procedure is easier than keeping 16 or more probes in parallel and orthogonal to the platform while inserting two vertical spacers. When spacers are installed first, each probe should be placed between only two vertical spacers and it will be kept upright parallel to previously installed probes by the spacers. After inserting and ultrasonically bonding the last probe gold beams to the platform, a cylindrical coil, which is the only hybrid component of this structure, is mounted on the platform while encompassing the entire back-end volume of the 3D array. In acute experiments, the coil and the entire back end are potted by epoxy or silicone to fix all the components before implantation. For chronic usage, however, the probe back end should be hermetically sealed using MEMS packaging techniques. A rendering of an Interestim-3 3D array and its building blocks are shown in Figure 23.8 [22].

23.4   MEASUREMENT RESULTS

The entire Interestim-3 probe has not been wirelessly operated yet. However, since the individual blocks can be isolated from the rest of the chip by cutting links, we were able to achieve promising results from individual block measurements. In the setup, a graphical user interface (GUI) in the LabView environment runs on an IBM PC and generates a serial data bit stream at a user-defined rate using a high-speed digital input–output card (National Instruments DIO-6534). This signal is used as an external modulating source for a function generator (Agilent 33250A), which generates an FSK-modulated sinusoidal waveform based on the FSK protocol that was described in Section 23.2 (Figure 23.4). The FSK sine wave is passed through a wide-band isolation transformer (North Hills 0904) and applied to the probes at wafer and die levels.

In Figure 23.9a the stimulation amplitude register value is swept from 0 to 127, while monitoring the 7-bit DAC output voltage and the nMOS current sink. The DAC output is the gate voltage for the single transistor current sink in Figure 23.3b and controls the nonlinear stimulus current that is shown on the right axis in Figure 23.9a. The nMOS nonlinearity offers finer resolution over the smaller stimulus currents and coarser resolution over the larger stimulus currents. This is generally what is needed in a neurophysiological setup, where smaller stimulus currents are used to interact with only one or a small group of neurons. In contrast, large stimulus currents are used to stimulate large populations or bundles of neurons, where small variations of the stimulating current do not make much of a difference. Figure 23.9b shows sample bipolar-biphasic stimulation waveforms between two sites, A and B, with a 50-kΩ resistive load connected between them, resembling the tissue. Every biphasic pulse is a result of four consecutive commands as described in Section 23.2 [18]. In a more practical in vitro experiment, when the probe shanks are immersed in saline, up to 10% of the stimulus current can leak into the grounded, deep boron diffused, probe substrate, which is loosely isolated from the surrounding saline solution by its superficial native oxide. To cancel out the charge imbalance caused by this leakage

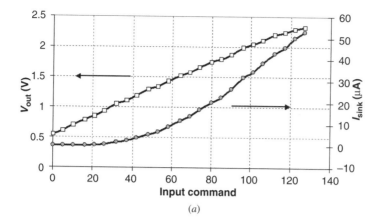

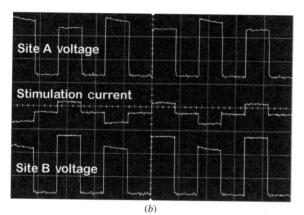

Figure 23.9   (*a*) Output characteristics of voltage-mode 7-bit DAC and nonlinear nMOS current sink versus digital input commands. (*b*) Sample bipolar-biphasic stimulation waveforms. From top: site A voltage, stimulation current passing through 50-kΩ load, site B voltage [18].

current, all the sites can be shorted to ground through the CAL after several stimulation pulses. A better solution is to coat the back side of each probe with Parylene-C in order to isolate the grounded substrate from tissue or saline.

The receiver block measurements are shown in Figure 23.10. A comparison between these waveforms and the simulated traces (1–5) in Figure 23.5*c* shows that the circuit is working precisely as expected. The power consumption of this block was 0.55 mW at 200 kbps [25].

Figure 23.11 shows the measurement results on the power supply block. Figure 23.11*a* shows the CMOS full-wave rectifier waveforms at 100 kHz. It can be seen that the output voltage (V_{out}) is a full-wave rectified sine wave and the N-Epi potential follows max (coil 1, V_{out}) waveforms, as expected from auxiliary transistor switching (see Fig. 23.6). Figure 23.11*b* shows the regulator input (V_{in}) and outputs (5 and 10 V) when the rectifier is receiving a 4-MHz carrier. Each regulator output is supplying 5 mA in this measurement. The internal current consumption of the regulator is measured as 340 μA at 15 V DC input [29, 30].

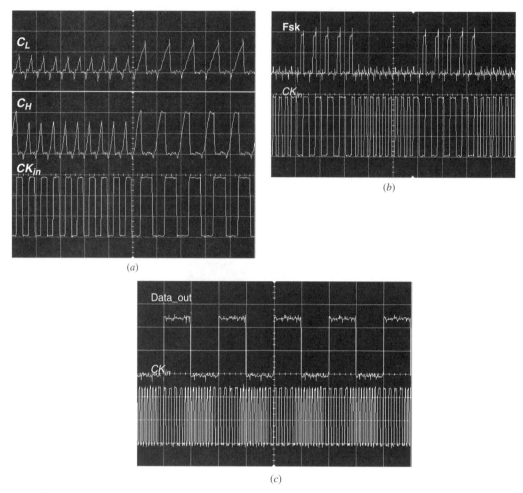

(a)

(b)

(c)

Figure 23.10   Fully differential FSK test chip measured waveforms with f_1 and f_0 equal to 1 and 2 MHz, respectively: (a) larger and smaller capacitor voltages (1 μs/div); (b) comparator output pulses (2 μs/div); (c) demodulated serial data bit stream (5 μs/div) [25].

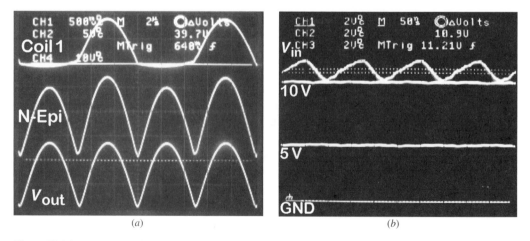

(a)

(b)

Figure 23.11   (a) Measured CMOS rectifier waveforms at 100 kHz from top: coil input, separated N-Epi, and output voltages. (b) Full-wave rectified carrier input to regulators with 8-MHz ripple along with regulated 5 and 10 outputs, each supplying 5 mA [29].

TABLE 23.2   Interestim-3 Specifications

Parameter	Value
Process technology	UM-BiCMOS 3-μm
Probe size	8.1 mm $\times$ 6.5 mm
Back-end size	6.5 mm $\times$ 5.0 mm
Shank size	3.1 mm $\times$ 60 μm
Shank and site spacing	250 μm
Stimulation site area	1810 μm^2
Circuit area	21.2 mm^2
Inputs nodes	Coil 1 and coil 2
Number of sites (outputs)	16
Maximum stimulation current	100 μA
Carrier frequency (f_0 and f_1)	2 and 4 MHz
Maximum internal clock frequency	2 MHz
Maximum simulated data rate	2 Mbps
Maximum measured data rate	625 kbps
Data rate–carrier frequency ratio	67%
Supply voltage	5 and 10 V
Maximum supply current	5 mA
Number of transistors	4200
Estimated power consumption	15 mW

23.5   CONCLUSION

The goal is to develop a 1024-site intracortical wireless stimulating 3D array for visual or auditory prostheses based on the University of Michigan micromachined silicon microprobe technology. The research was started with a wireless interface chip for the existing hard-wired stimulating probes (Interestim-1) and was carried on by a four-channel stand-alone wireless stimulator chip (Interestim-2). Interestim-3 is a stand-alone 16-site wireless active microstimulating probe with a modular architecture. A 6-bit user-selectable internal address allows using up to 64 probes in parallel with only two connections per probe to a mutual receiver LC-tank circuit to achieve a 1024-site wireless microstimulating 3D array. Interestim-3 is fabricated in a 3-μm, one-metal, two-poly, N-Epi, UM-BiCMOS process with micromachining features for 3D microassembly. It has four 60 μm $\times$ 3.1 mm shanks with 250 μm center to center spacing. Each shank has four 1810-μm^2 sites that are 250 μm spaced. Interestim-3 circuitry has 4200 transistors that occupy 65% of the 32.5-mm^2 probe back end. A summary of Interestim-3 specifications is provided in Table 23.2.

ACKNOWLEDGMENTS

The authors thank Professor K. D. Wise and Dr. W. J. Heetderks for their guidance, M. Gulari for fabricating the probes, and R. G. Gordenker for equipment support. This work was supported by the National Institutes of Health, under contract number NIH-NINDS-N01-NS-9-2304. This work made use of the WIMS Engineering Research Center shared facilities supported by the National Science Foundation under Award Number EEC-0096866.

REFERENCES

1. P. K. CAMPBELL, K. E. JONES, R. J. HUBER, K. W. HORCH, AND R. A. NORMANN, "A silicon-based, three-dimensional neural interface: Manufacturing process for an intracortical electrode array," *IEEE Trans. Biomed. Eng.*, vol. 38, no. 8, pp. 758–767, Aug. 1991.

2. W. H. DOBELLE AND M. G. MLADEJOWSKY, "Phosphenes produced by electrical stimulation of human occipital cortex and their application to the development of a prosthesis for the blind," *J. Physiol.*, vol. 243, pp. 553–576, 1974.

3. J. P. RAUSCHECKER AND R. V. SHANNON, "Sending sound to the brain," *Science*, vol. 295, pp. 1025–1029, Feb. 2002.

4. G. S. BRINDLY AND W. S. LEWIN, "The sensation produced by electrical stimulation of visual cortex," *J. Physiol.*, vol. 196, pp. 479–493, 1968.

5. M. BAK, J. P. GIRVIN, F. T. HAMBRECHT, C. V. KUFTA, G. E. LOEB, AND E. M. SCHMIDT, "Visual sensations produced by microstimulation of the human occipital cortex," *Med. Biol. Eng. Comp.*, vol. 28, pp. 257–259, May 1990.

6. R. A. NORMANN, "Sight restoration for individuals with profound blindness," Available: http://www.bioen.utah.edu/cni/projects/blindness.htm, cited Dec. 2003.

7. E. ZRENNER, "Will retinal implants restore vision?" *Science*, vol. 295, pp. 1022–1025, Feb. 2002.

8. J. WEINER, "Sight seeing," *USC Health Magazine*, winter 2003, Available: http://www.usc.edu/hsc/info/pr/hmm/win03/sight.html, cited Dec. 2003.

9. K. CHA, K. HORCH, AND R. A. NORMANN, "Simulation of a phosphene-based visual field: Visual acuity in a pixelized vision system," *Ann. Biomed. Eng.*, vol. 20, no. 4, pp. 439–449, 1992.

10. P. R. TROYK, Reports and presentations, NIH contract # N01-NS-7-2365, 2001.

11. K. NAJAFI, K. D. WISE, AND T. MOCHIZUKI, "A high yield IC-compatible multichannel recording array," *IEEE Trans. Electron Devices*, pp. 1206–1211, July 1985.

12. Center for Neural Communication Technology (CNCT), University of Michigan, http://www.umich.engin.edu/facility/cnct/.

13. M. GHOVANLOO, K. BEACH, K. D. WISE, AND K. NAJAFI, "A BiCMOS wireless interface chip for micromachined stimulating microprobes," in *Proc. IEEE-EMBS Special Topic Conf. on Microtechnologies in Med. and Biol.*, pp. 277–282, Madison, WI, May 2002.

14. C. KIM AND K. D. WISE, "A 64-site multishank CMOS low-profile neural stimulating probe," *IEEE J. Solid-State Circuits Systems*, vol. 31, no. 9, pp. 1230–1238, Sept. 1996.

15. B. ZIAIE, M. D. NARDIN, A. R. COGHLAN, AND K. NAJAFI, "A single-channel implantable microstimulator for functional neuromuscular stimulation," *IEEE Trans. Biomed. Eng.*, vol. 44, no. 10, pp. 909–920, Oct. 1997.

16. J. A. VON ARX AND K. NAJAFI, "On-chip coils with integrated cores for remote inductive powering of integrated microsystems," Int. Conf. Solid-State Sensors Actuators (Transducers 97), pp. 999–1002, June 1997.

17. J. A. VON ARX, "A single ship, fully integrated, telemetry powered system for peripheral nerve stimulation," Ph.D. dissertation, Dept. Elec. Eng. Comp. Science, University of Michigan, Ann Arbor, 1998.

18. M. GHOVANLOO AND K. NAJAFI, "A BiCMOS wireless stimulator chip for micromachined stimulating microprobes," in *The IEEE 2nd Joint EMBS-BMES Conference Proceedings*, pp. 2113–2114, Oct. 2002.

19. M. GHOVANLOO AND K. NAJAFI, "A modular 32-site wireless neural stimulation microsystem," paper presented at the IEEE Solid-State Circuits Conference, Feb. 2004.

20. M. D. GINGERICH, "Multi-dimensional microelectrode arrays with on-chip CMOS circuitry for neural stimulation and recording," Ph.D. dissertation, Dept. Elec. Eng. Comp. Science, University of Michigan, Ann Arbor, 2002.

21. Q. BAI, K. D. WISE, AND D. J. ANDERSON, "A high-yield microassembly structure for three-dimensional micro-electrode arrays," *IEEE Trans. Biomed. Eng.*, vol. 47, no. 3, pp. 281–289, Mar. 2000.

22. M. GHOVANLOO, K. D. WISE, AND K. NAJAFI, "Towards a button-sized 1024-site wireless cortical microstimulating array," in *1st International IEEE/EMBS Conference on Neural Engineering Proc.*, pp. 138–141, Mar. 2003.

23. K. E. JONES AND R. A. NORMANN, "An advanced demultiplexing system for physiological stimulation," *IEEE Trans. Biomed. Eng.*, vol. 44, no. 12, pp. 1210–1220, Dec. 1997.

24. C. POLK AND E. POSTOW, *Handbook of Biological Effects of Electromagnetic Fields*, CRC Press, Boca Raton, FL, 1986.

25. M. GHOVANLOO AND K. NAJAFI, "A high data transfer rate frequency shift keying demodulator chip for the wireless biomedical implants," *IEEE 45th Midwest Symp. Circuits Syst. Proc.*, vol. 3, pp. 433–436, Aug. 2002.

26. M. GHOVANLOO and K. NAJAFI, "A fully digital frequency shift keying demodulator chip for the wireless biomedical implants," in *IEEE Southwest Symposium on Mixed-Signal Design Proc.*, pp. 223–227, Feb. 2003.

27. M. GHOVANLOO AND K. NAJAFI, "A high data-rate frequency shift keying demodulator chip for the wireless biomedical implants," *IEEE Int. Symp. Circuits Syst. Proc.*, vol. 5, pp. 45–48, May 2003.

28. M. GHOVANLOO AND K. NAJAFI, "A wideband frequency shift keying wireless link for inductively powered biomedical implants," *IEEE Trans. Circuits Sys. I*, vol. 51, no. 12, pp. 2374–2383, Dec. 2004.

29. M. GHOVANLOO AND K. NAJAFI, "Fully integrated power-supply design for wireless biomedical implants," in *Proc. IEEE-EMBS Special Topic Conf. on Microtechnologies in Med. and Biol.*, pp. 414–419, May 2002.

30. M. GHOVANLOO AND K. NAJAFI, "Fully integrated wideband high-current rectifiers for inductively powered biomedical implants," *IEEE J. Solid-State - Circuits*, vol. 39, no. 11, pp. 1976–1984, Nov. 2004.

CHAPTER 24

PRACTICAL CONSIDERATIONS IN RETINAL NEUROPROSTHESIS DESIGN

Gregg J. Suaning, Luke E. Hallum, Spencer Chen,
Philip Preston, Socrates Dokos, and Nigel H. Lovell

24.1 INTRODUCTION

In our collective anxiety toward providing a useful visual prosthesis for the blind, we may be overlooking some important clues that exist within the natural world and indeed within mammalian biology itself that may lead to substantial benefits for the future of prosthetic vision. In terms of electrical and mechanical design, retinal neuroprostheses have largely evolved from their auditory forbearers, such as the cochlear implant, which itself evolved from the cardiac pacemaker. As a result of this evolutionary process, substantial similarities exist between current retinal neuroprostheses and the devices from which they were derived. This chapter explores the virtues and drawbacks of this evolutionary approach toward neuroprosthesis design and combines our knowledge of neurostimulation and modelling approaches focused on a deeper understanding of current injection to suggest and justify particular attributes of visual neuroprosthesis design.

Specific areas that are addressed include the geometry of electrode arrays used for epiretinal stimulation, anticipated electrode numbers, and the relationship between numbers and rates of stimulation; approaches toward electrode fabrication; the psychophysics of electrical stimulation in the retina; implantation issues; implant electronics; and models of current injection given the aforementioned electrode and electronic designs. The multidisciplinary skills needed to address these issues are representative of the broad skill base necessary to design and construct a functional retinal neuroprosthesis.

24.2 ELECTRODE ORGANIZATION AND PHOSPHENIZED VISION

Cha and colleagues [1–3] pioneered the simulation of prosthetic vision. In short, subjects viewed the world through a mask of pinholes. Pinholes (and therefore phosphenes) were arranged in a regular, rectangular fashion in the visual field—a so-called Manhattan geometry in reference to the ordered street arrangements of the Borough of Manhattan in New York City. The density of the Manhattan arrangements of phosphenes and the

Handbook of Neural Engineering. Edited by Metin Akay
Copyright © 2007 The Institute of Electrical and Electronics Engineers, Inc.

Figure 24.1 Populated electrode array containing 100 platinum spheres of 450 μm diameter within cast polydimethylsiloxane (silicone elastomer) substrate. Electrodes are placed in a square grid with a 10-cent Australian coin shown for reference.

overall number of phosphenes were varied, and the effects on subjects' mobility, reading ability, and other tasks were assessed.

The Manhattan geometry has intuitive appeal—television and computer displays, for example, are based on a regular, rectangular arrangement of pixels. Further, implantable arrays being manufactured at the time featured electrodes arranged in a rectangular pattern [4]. Figure 24.1 illustrates an early square electrode array manufactured in our laboratory and used for epiretinal implantation in animals.

Had Cha et al. varied the geometry of phosphene layout, as opposed to simply density and quantity, improved results would have likely been demonstrated. As an alternative to the Manhattan geometry, consider a hexagonal mosaic. The two-dimensional (2D) variant of the seventeenth-century conundrum known as Kepler's conjecture involves the dense packing of circles on a plane. That the densest packing is, in fact, a hexagonal mosaic was not demonstrated until three centuries later by Féjes-Toth [5].

Scribner et al. (2001) [6, 7] reported the manufacture of an electrode wherein microchannels carrying conductors were arranged in a hexagonal mosaic; they likely realized that the most efficient means of arranging their circular conduits on a glass substrate was by way of a hexagonal mosaic. Hallum et al. (2004) [8] suggested hexagonal arrangements of electrodes be adopted for larger scales, citing the benefits of hexagonal arrangements on concentrating charge injected into the retina which should lead to a more dense packing of discrete phosphenes in the visual field. Indeed, simulations of prosthetic vision at the authors' laboratory indicate hexagonal packing effects improved acuity [9, 10]. Further, it so happens that the hexagonal arrangement of electrodes makes for more efficient multiplexing in implantable microelectronics [11].

24.3 ELECTRODE QUANTITIES AND STIMULATION RATES

The cardiac pacemaker requires only a single stimulation channel (electrode) with the return path for the neurostimulation via the pacemaker capsule itself. For the cochlear implant, substantial benefit can be achieved with as few as 16 stimulation channels and associated electrodes. In contrast, little visual information can be conveyed with any number of phosphenes fewer than 64 [3], with hundreds or indeed thousands of phosphenes necessary to approximate the rich sense of vision that sighted people enjoy.

The mean stimulation frequency for each of the above examples differs markedly. Assuming a single source of stimuli, said source must supply approximately two stimulus pulses per second for the cardiac pacemaker so as to maintain a heart rate of 120 beats per minute and up to approximately 15,000 pulses per second for the cochlear implant in order to take advantage of modern speech processing techniques [12].

In the case of a visual prosthesis stimulating the retina, the optimal stimulation rate is not yet established and is likely to be the topic of intense debate well into the future. The practical lower limit of the stimulation rate is the visual prosthesis variant of the so-called critical flicker fusion (CFF) frequency, below which phosphenes conveyed by way of a visual prosthesis will appear pulsatile. The CFF is further dependent upon the site of stimulation, for example, the retinal neurons owing to inhibitory effects that may be elicited by way of stimulation of neurons other than the retinal ganglion cells (RGCs). The authors posit that frequency modulation above the CFF will play a substantial role conveying safe and effective phosphene information in image processing and stimulation strategies.

In anticipation of such implementation of frequency modulation techniques, any visual prosthesis for the retina should possess stimulation rate capabilities well above the CFF with an upper limit of the signal-carrying capacity of individual optic nerve fibres which is of the order of 150–250 signals per second [13]. In instances where large numbers of electrodes are employed, the stimulation rate is a linear multiple of the electrode quantity.

The CFF of the intact retina is dependent upon light intensity (Ferry–Porter law). In an illustrative example of a television powered by 50-Hz electricity, no flicker is observed at a screen refresh rate of 25 Hz (interleaved wherein every second line of pixels is updated with each pass at 50 Hz) at moderate brightness. A different situation applies in electrical stimulation applied to the diseased retina. In these instances, intraretinal processing is dysfunctional as a result of the disease, and what mechanisms that remain are bypassed by the way in which the bipolar and RGCs are preferentially stimulated over other surviving neurons [14, 15]. The apparent result of this is a reduction in the duration of the perceived phosphenes. In testing on human subjects, flicker fusion occurs at two to three times the frequency in electrical stimulation when compared with visual flicker fusion of normal eyes [16].

Ignoring the differential CFF thresholds for electrical versus light stimulation, to avoid the perception of flicker, a single stimulation source must be capable of stimulating once every 20 ms. For pulse widths of the order of 1 ms each (2 ms with a charge recovery phase), as few as 10 electrodes may be driven in series from a single source. For electrical stimulation the number of electrodes that may be driven in series is likely to be fewer still.

To achieve and facilitate going beyond the frequency of stimulation necessary for even the most rudimentary conveyance of vision through a neuroprosthesis, parallelization of stimuli is the only practical means of delivery.

Consider a simplified example of two separate electrode pairs through which simultaneous stimulation is occurring. Assume that electrons traveling between the first pair of electrodes are colored red, and electrons traveling between the second pair are colored blue. Some red electrons passing between the first pair will be diverted toward the second pair in the event that any path of less than infinite impedance exists between the first and second pairs through the tissue or body fluids within which the electrodes are harbored. The same applies to the blue electrons. If the objective is to physiologically stimulate excitable tissue between the respective electrode pairs, additional or different sites may be stimulated as a result of the red electrons passing toward the blue and vice versa. This problem is exacerbated when the currents delivered between electrode pairs differ from one another.

Thus, there exist two factors: a need for high quantities of electrodes to facilitate the conveyance of complex patterns and images and a need for high stimulation rates through each of the stimulation channels to avoid flicker and take advantage of any psychophysical benefits that may be frequency encoded above the flicker frequency.

24.4 ELECTRODE FABRICATION

As the center-to-center spacing of electrodes fabricated in the hexagonal geometry need not differ from that of the Manhattan geometry, the advantages of the hexagonal geometry are achieved without cost to the technology used in their fabrication. However, fabrication of organized arrangements of multiple electrode arrays is, in general, a complex subject in itself.

It is universally accepted that more phosphenes will contribute to improved performance of a visual prosthesis. As electrode quantities increase with the requirement for additional stimulation sites, traditional hand fabrication methods such as those used in the manufacture of exquisite cochlear arrays are becoming, or are already, obsolete in terms of visual prosthesis. Thus strategies must be derived to achieve increased numbers of phosphenes.

Polyimide-based arrays, capable of automated fabrication using traditional photo lithographic techniques, offer the possibility of packing dense patterns of electrodes upon a flexible substrate. Figure 24.2 illustrates the authors' first attempt at such an array, initiated prior to the realization of the advantages to hexagonal geometry.

Other centers have developed complex designs that perform well in acute in vivo trials [17]. However, questions over water uptake and subsequent delamination of metallic conductors plague the polyimide approach.

When considering regulatory approval processes, researchers are inherently reluctant to propose unproven or new materials in the fabrication of implantable devices. An illustrative example of this is benzocyclobutene (BCB), a material likened to polyimide but with substantially lower water uptake. Despite promising in vivo results [18], materials that require biocompatibility verification are all but universally avoided in the development of a device intended for immediate or near-term clinical application. On the one

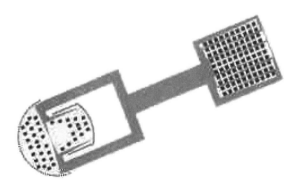

Figure 24.2   Early attempt at manufacture of polyimide-based electrode with rectangular-arrayed electrodes illustrated on right. Connection pads are shown on the left.

hand, this stifles progress on the pursuit of new materials but, on the other hand, creates a degree of conservatism that is consistent with maximization of safety when it comes to trial of new devices.

It is a common misconception that there is a specific and concise list or directory of "approved" biomaterials. While it is true that some materials are more widely accepted as being more biocompatible over others—silicone, alumina oxide ceramic, titanium, and platinum, to name a few—regulatory bodies almost universally require characterization of materials on a "target tissue" basis. For example, materials appropriate for intramuscular implantation may not be appropriate for intraocular use and vice versa. This said, the overriding factor is the track record of a material that is used in determining its forward potential for compatibility with another particular tissue type.

Consistent with this conservative approach, the epiretinal devices in clinical trial at the time of writing employ electrode arrays consisting of platinum and silicone rubber only [19]. The electrode array design itself is substantially influenced by the auditory brainstem implant that has been used in clinical application for more than a decade.

For the next generation of epiretinal devices, additional stimulation sites will be employed. In order to remain consistent with the aforementioned design conservatism, electrodes that will be used in the foreseeable future are likely to take a more advanced form of the auditory brainstem implant but maintain the same material composition.

Difficulties in deposition and electroplating of platinum have led the authors' research team to devise a methodology of laser ablation that allows for the use of the traditional biomaterials of platinum and silicone rubber in the fabrication of complex arrays [20]. This technique is a substantially simplified variant to methods introduced originally by Mortimer [21]. In this technique a Nd:YAG laser is used to ablate platinum foil that is adhered to a spun layer of silicone rubber atop a rigid substrate. Following ablation and excess foil removal (Fig. 24.3), a second layer of silicone rubber is deposited by way of spin coating, effectively locking the foil into position between the two layers of silicone rubber. The final steps include opening of electrode and interconnect contacts (either by ablation or plasma etching) prior to removal of the finished device from the rigid

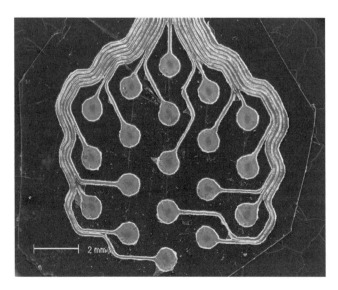

Figure 24.3 A 19-channel prototype electrode fabricated using laser ablation methodology. Subsequent spin coating of a thin layer of silicone rubber will lock each of the conductors in place while at the same time filling any imperfections in the layer beneath.

substrate. Work is underway to extend this technique to devise multilayer devices for increased electrode numbers.

24.5 PSYCHOPHYSICS OF ELECTRICAL STIMULATION

In order to evoke the most meaningful visual percepts, the sequence and parameters of electrical impulses to be delivered to retinal tissue remain open questions. Indeed, the most substantial advancements in cochlear neuroprosthesis in recent years have involved new stimulation strategies, not improved electronics or mechanical implant design [22]. With retinal prosthesis having recently begun clinical trials, it is therefore timely to consider stimulation strategies.

For the epiretinal approach, increased phosphene quantities correspond with increased stimulation sites. However, it should be noted that a one-to-one relationship between electrode quantities and phosphenes is not necessarily the case in all areas of vision prosthetics. Researchers in Belgium have shown that discrete phosphenes exceeding 100 in quantity can be elicited with a four-electrode cuff about the optic nerve [23]. In addition, Rizzo and colleagues [24] showed that a single electrode may elicit the perception of a cluster of phosphenes. Therefore, it we are to arrange phosphenes effectively in the visual field, it is crucial that electrodes comprising an implanted array are likewise driven efficaciously.

Electrical stimulation provides very simplistic input to the visual pathway when compared with a constant influx of ambient light from the visual scene. In other words, the rich stream of visual data carried by the optic nerve in response to ambient light is fundamentally different from the data carried in response to electrical stimuli.

The information conveyed by the optic nerve is arranged in rich spatial (the nerve comprises some 1.2 million fibers) and temporal [conduction velocities differ from fiber to fiber en route to the lateral geniculate nucleus (LGN)] patterns. To draw an analogy with physical optics, we may think of the information as "coherent"—it is readily interpreted by visual areas downstream of the retina.

In stark contrast, when electrical stimulus is applied to the RGCs (assuming these are the cells directly stimulated in response to electrical stimuli [25]), only one action potential may be elicited from any one cell and indeed action potentials from all cells physiologically excited by the stimulation event are prone to be elicited almost simultaneously. It is unlikely that such signals are readily interpreted by any rule-based methodology within the cognitive centers and thus any interpretation of said signals may be as a result of mere statistical chance. In other words, because the signals resulting from electrical stimulus likely constitute "incoherent" information within a rule-based interpretation, the interpretation of such signals as being a phosphene is arguably unlikely unless some features of the data stream mimic features consistent with fundamental rules or these rules are in some way overruled or overwhelmed.

There exists growing evidence that even though very small (tens-of-micrometer-diameter) electrodes can indeed yield physiological excitation of RGCs [26], psychophysical phosphene perception from electrodes of similar size may not eventuate. Studies employing small, epiretinal electrodes have not elicited phosphenes despite attempts being made to obtain them [27]. The reasons for this remain unclear, but if indeed action potentials are being evoked, these signals do not equate to perception of a phosphene. Stimuli from larger (hundreds-of-micrometer-diameter) electrodes consistently evoke psychophysical perceptions of phosphenes [19].

The absence of observed phosphenes could be a simple case of insufficient coupling between electrode and tissue. Indeed Grumet's work [26] was conducted in vitro whereas the psychophysical testing has the requisite of being performed in vivo, thus introducing a fundamental variation in the methodologies, but there may be alternative explanations such as the importance of eye movements and the physical size of the area of excitation that contribute to the psychophysics of perception.

For purposes of illustration only, let us explore a rudimentary rule-based interpretation methodology. This methodology is not meant to strictly adhere to the known neurophysiology within the retinal layers but serves more as an insight into how electrode and cell geometries can interact. Assume that within the complex processes facilitating the psychophysical recognition of a phosphene a form of "weighted voting" occurs at the cognitive centers such that the combination of excitatory or inhibitory signals from the RGCs determine whether or not a phosphene is to be perceived as such or simply ignored as being extraneous or incoherent. In reference to the hexagon arrangement of the RGCs in Figure 24.4, we refer the reader to Figure 15 of Dacey [28], which shows RGCs approximating a hexagonal mosaic—implying that mother nature had discovered this efficiency long before Kepler and Féjes-Toth. Let us further assume that at the site of excitation all RGCs with sufficiently large voltage gradients across their membranes elicit action potentials indiscriminately but upon arrival at their destination the first to arrive is interpreted as the center of the hexagon (activated) and the remainder of the cells within the hexagon considered inhibited.

As the size of the electrode increases, it logically follows that increasing numbers of RGCs (hexagon clusters thereof) contribute to the physiological data being sent to the cognitive centers. As this quantity increases, the ratio of excited to inhibited cells asymptotically approaches a ratio of 1 : 2.

With the overtly simplified configuration of Figure 24.5, as the radius of the electrode (and thus the stimulated area) increases, RGCs counted at the cognitive centers as being "activated" sum according to the formula

$$\text{RGCA} = 6b \qquad b > 0 \tag{24.1}$$

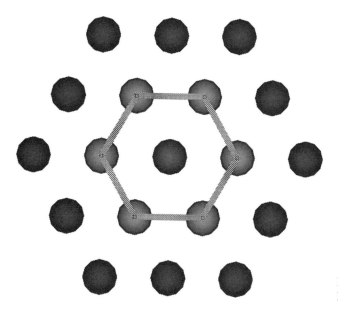

Figure 24.4   Layout of RGCs.

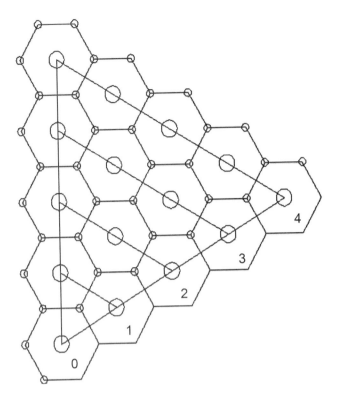

Figure 24.5 One-sixth of a hexagonal mosaic. When arrayed six times about the center marked 0, the complete mosaic is formed.

where b is the "band" of hexagonal clusters as labeled in Figure 24.5. At the same time, RGCs counted as being "inhibited" are added according to the formula

$$\text{RGCI} = 6(2b + 1) \tag{24.2}$$

For small areas of excitation, the ratio of activated to inhibited RGCs is substantially less than 1 : 2. This progression of the ratio of excited to inhibited cells as a function of the electrode radius is shown in Figure 24.6. If the aforementioned RGC-weighted voting occurs, phosphene perceptions from small electrodes are substantially less likely to occur.

While the authors do not suggest such a simplified rule-based methodology exists in this form, this example, specifically the nexus between the electrode radius and its relationship to the size of the hexagonal RGC clusters, illustrates the plausibility of cognitive processing precluding perceptions from being elicited from small electrodes even though action potentials are elicited from the RGCs either directly or indirectly by way of bipolar or other surviving retinal neurons.

Microsaccades may also have implications toward the geometric size of the electrodes, particularly where the edge of the perceived phosphene is concerned. For a sighted individual viewing an object with ambient light, microsaccades are essential for continued viewing of an object [29]. For small electrodes (tens of micrometers in diameter), an elicited phosphene can be said to be comprised primarily of its edge alone whereas phosphenes derived from larger electrodes would have large central regions that may not be as sensitive to the absence of microsaccades in terms of perception, as small movements would simply move the perceived image to a different location within the same phosphene.

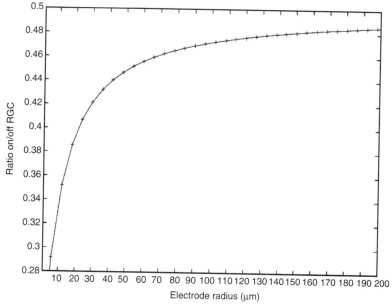

Figure 24.6   Ratio of excited to inhibited RGCs plotted as function of electrode radius.

In some of the epiretinal devices being proposed [30, 31] a camera acquires an image from the visual scene. This image is subsequently processed and used to determine the electrical stimulation instructions that will be used to convey a phosphene representation of the original image. Movement of a sighted eye affects a corresponding movement of the image relative to the retina. This is not the case in the camera/phosphene interaction as any movement of the eye will not affect the phosphene image upon the retina unless coordination of eye movements is made with the image acquired by the camera.

The virtues of accurate eye tracking may not be limited to that of image movement coordination alone. Huang and Tsay are credited with being the first to propose the super resolution restoration concept wherein several low-resolution images acquired from a rudimentary detector are combined to derive an image of substantially higher spatial resolution by way of the spatial aliasing effect [32]. Simply put, multiple images, each acquired with a shift of the detector by a known or determinable proportion of the detector diameter, may be combined to significantly improve resolution. The vision prosthesis analogue of the detector in this context is the resolved phosphene upon the topographically mapped retina. In development of an image processing strategy, ideally, it will not only be the captured image that is conveyed on a frame-by-frame basis, but additional information describing the spatial changes from the previous captured image will be further incorporated into an overall perception of the underlying scene. With knowledge of the degree of movement between images, the implant recipient may learn to combine said images in a way similar to that proposed by Huang and Tsay, thereby improving cognitive resolution without increasing stimulation site numbers. Indeed a rudimentary form of this movement–image correlation has been established in patients fitted with visual neuroprostheses—head movements give rise to multiple images in concert with vestibular feedback that may be combined to affect pattern recognition with clear benefit [23, 33].

24.6   IMPLANTATION AND PROSTHESIS DESIGN

Majji and co-workers have consistently demonstrated an ability to fixate electrode arrays comprised of platinum contacts within a silicone elastomer substrate using retinal tacks [34]. Others pursuing the subretinal approach have proposed and indeed demonstrated implantation of electrode arrays in the subretinal space [35, 36]. While not yet routine operations, both approaches show promise for long-term implantation of prosthetic devices interfaced with retinal tissue.

Beyond the promising ability to interface with the retina, the need arises for implantation and support of other components of a stimulating device, the electronics in particular. Epiretinal device proposals invariably possess some form of power supply, whether it be by way of rectification of radio-frequency instructions [37–39] or by way of laser light [40]. Chow et al.'s elegant but otherwise unproven subretinal photodiode devices incorporate both stimulator and electrodes in one device [41]. However, both clinical testing and the laws of physics indicate that these subretinal devices fail in terms of their ability to evoke physiological excitation of retinal neurons consistent with vision. Zrenner and co-workers have recognized that such subretinal devices require supplementary power supplies in order to be made efficacious and, as a result, have employed such features in their devices [42]. Thus, at least some portion of any visual prosthesis must deal with power. Further, without some form of intrinsic image processing (e.g., as is present in subretinal devices), a receiver of some nature is also required so as to receive stimulation instructions. Finally, control and delivery of electrical current are integral requirements of a stimulating retinal prosthesis. Thus, researchers are presented with a problem adjunct to their electrode-positioning efforts of where to fixate the receiver, stimulator, and power supply.

The only functioning retinal device being used in chronic clinical trial at the time of writing is a modified cochlear implant undergoing evaluation by Weiland and colleages at Doheny Eye Institute at the University of Southern California [43]. Their approach to implantation of the whole of the device is to place the receiver/stimulator in a location consistent with cochlear devices, with a flexible cable routed to and around the eye prior to transscleral insertion into the globe. While efficacious for low electrode quantities owing to a cable of moderate diameter (millimeters), extrapolation of this approach to larger quantities of electrodes will undoubtedly be limited by the capacity of the globe to accommodate the transscleral cable of increasing signal quantities. Were these signals to be multiplexed, an additional requirement of demultiplexing would be introduced within the globe, thereby defeating much of the advantage gained by placing the receiver/stimulator elsewhere.

Proposed subretinal devices by Schwahn and co-workers [36] draw upon a similar approach wherein transscleral cables feed implanted electrodes in the subretinal space. Here again the technology is limited to the physical size of the cable and the globe's capacity to tolerate it. However, given the proximity of the electronics to the globe, the cable itself need not be comparatively as robust as the cable used by Weiland and colleagues.

Other proposals such as those of the authors' laboratories [44] include the incorporation of the whole of the implanted electronics within the globe itself. The basic advantage of this approach is the avoidance of a breach in the scleral wall that is a potential source of complications. This further facilitates an increase in electrode quantities as the transcleral cable is eliminated and thus the mechanical requirements of the cable linking the electronics to the electrodes are substantially relaxed. This approach, however, increases the difficulties associated with implantation owing to the increased size and volume of the intraocular components involved. This said, fixation of the stimulating electronics to the pars plana has been demonstrated to be a plausible methodology [45], as is the widely used method of phacoemulsification [46] conceivably facilitating placement of at least some of the stimulating body within the lens capsule, both lending weight to the all-inclusive, intraocular approach.

24.7 IMPLANT ELECTRONICS

The importance of implant electronics is often diminished by some researchers who may not appreciate the complexities involved with the delivery of electrical stimuli. It is well known that inappropriate stimuli can lead to damage to both electrodes and the tissue with which they interface.

There exist two basic caveats in this area. First, the net charge delivered to the site of stimulation must be neutral. Second, the charge injection must stop prior to gas formation from electrolysis of water and hydrogen reduction reactions. If either of these occur, the probability of tissue damage or electrode dissolution is increased.

The electrochemical reactions that can take place include (assuming a platinum electrode) electrolysis of water that can cause gas production and changes in pH at the electrode site; oxidation of saline; oxidation of platinum that is likely to lead to electrode dissolution; oxidation of organics (e.g., glucose) that potentially leads to toxic by-products; and hydrogen reduction reactions. In order to avoid some of these reactions, Lilly et al. [47] proposed a balanced, biphasic pulse which, theoretically, reverses the reactions through the addition of a second phase of equal charge but opposite polarity. This, however, does not facilitate unlimited injection of charge.

Figure 24.7 shows a typical voltage waveform resulting from biphasic, constant current injection from platinum electrodes in a saline electrolyte. While the current waveform may itself be symmetric through careful design and control, the voltage waveform may be substantially asymmetric due to the capacitive and resistive coupling with the tissue and surrounding media. This alone indicates that the chemical processes are not entirely reversible. Provided, however, that gas formation is not initiated (as in the electrolysis of water and hydrogen reduction reactions described above), the reactions are in fact reversible.

Avoidance of gas formation has been investigated in detail by Robblee and Rose [48]. Detailed analysis of the physics and chemistry involved in the avoidance of gas formation is beyond the scope of this text, but as a general summary, a limit on the charge per unit area of electrode surface may be derived as a function of the electrode material, the electrolyte composition, and the stimulus waveform parameters. By reversing the charge prior to reaching the charge density limit, it is generally accepted that long-term delivery of electrical stimuli is safe. It is noted that the so-called safe limit for platinum electrodes is generally accepted as being of the order of 100 $\mu C/cm^2$ with several variants of this quantity described within the literature, illustrating the "black art" behind the charge density issue.

Even though safe charge injection limits are maintained and biphasic waveforms are applied, slow kinetics within the oxidative–reductive reactions that store electrical charge on electrodes allow for a small amount of charge to remain [49]. In some applications, subsequent shorting together of all of the electrodes dissipates residual charge to the same electrical potential prior to the next delivery of stimulus [50]. The second method employs capacitors in series with each electrode which prevent the passage of net direct

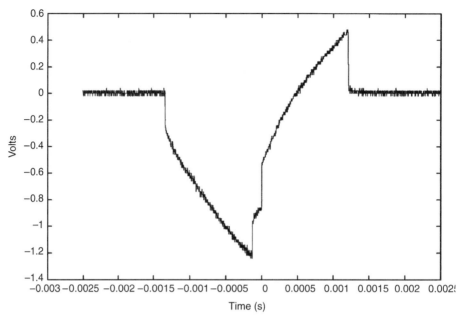

Figure 24.7 Typical voltage waveform resulting from balanced, biphasic, constant current stimulus pulse. In this instance the electrodes are platinum spheres, 400 μm in diameter, immersed within physiological saline. Stimulus parameters are 0.96 mA current, 1.1 ms pulse width, 330 μs interphase gap.

current (DC) into the tissue, thus injecting no net charge. It is noted that the former requires time, thereby affecting the stimulation rate (see Fig. 24.8 for further details) and the latter requires electronic components that collectively become impractically large for devices with a large number of electrodes.

The charge balance issue is again raised with the introduction of parallel stimulation. Crosstalk between stimulating electrode pairs and their return path (guard) counterparts is a source of DC flow. Such net DC has been found to produce significant damage to neuronal tissue with as little as 1 μA [51]. Safe DC flows are generally accepted as being of the order of 100 nA.

A distant, common return electrode, as is present in the modified cochlear implant—cum visual prosthesis—in clinical trials by Humayun and colleagues [19] affords an elegant solution to the crosstalk problem described above. However, substantial limitations exist in the employment of such a methodology in order for high numbers of electrodes to be employed and stimulated in unison. First, unless the current from each of the simultaneous stimulus sites is uniform, a difference in the electrical potential between adjacent stimulating electrodes is established, thereby promoting current flow from one stimulating electrode toward another. Second, given that the collective current flow to/from the stimulating electrodes is additive, the current-carrying capacity (and hence the surface area) of the distant return electrode must be appropriately large. Further, the additive effects of the simultaneous current may indeed cause unwanted muscle contractions or other deleterious effects at the site of the common return electrode. Finally, to facilitate the electrode being distant to the desired localized stimuli, it is requisite that an electrical conductor breach the scleral wall.

In terms of current distribution and localization, a particular advantage is achieved with the hexagonal electrode arrangement—that is, the ability to "guard" a stimulating

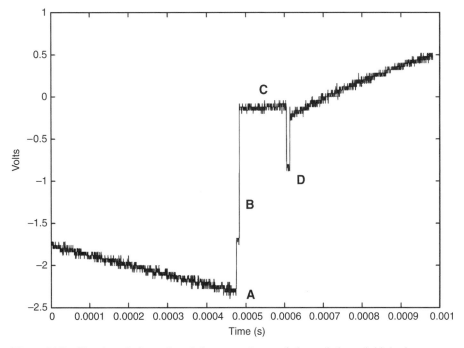

Figure 24.8 Shorting of electrode pair between phases of charge-balanced, biphasic stimulation pulse. Cathodic phase concludes at A with shorting of electrodes to ground at B. Shorting is held during C and subsequently released at D. Note the abrupt return to a nonzero voltage at D prior to the resistive voltage rise at the onset of the anodic phase resulting from the purely resistive properties of the electrolyte—only some of the residual charge has been dissipated during C. Note also that charge dissipation shorting normally occurs after the completion of both phases as opposed to interphase. Interphase shorting is shown here as an illustration of the residual charge phenomenon.

electrode with six surrounding electrodes. The aim of this approach is to capture the charge within a given hexagon and preclude crosstalk between neighboring pairs of hexagons from occurring. By introduction of a multiplexing methodology that provides for the center of each hexagon to be redefined, any electrode within the array is capable of being a stimulating electrode, with each of its immediate neighbors serving as "guards." In order to better understand the requirements for parallel stimulation and how this relates to electrode design, we present in the next section a number of numerical simulations employing different electrode placements and configurations.

24.8   MODELS OF ELECTRODES AND CURRENT INJECTION

For the purposes of the simulations in this section we use the approach that is described in more detail in [52]. In brief, we solve the Poisson governing equation $-\nabla \cdot (\sigma \nabla V) = I$ for the voltage distribution V throughout a $20 \times 20 \times 0.2$-mm region of physiological saline using a custom spectral collocation method [53, 54] implemented in MATLAB (MathWorks). The conductivity of the medium σ was taken to be 1 S/m [55] and I was the volume current density injected into the medium at a number of predetermined locations determined by our interest in parallel current injection from hexagonally config-ured electrode arrays. For the sake of simulation ease, cylindrical electrodes of 0.2 mm and height 0.2 mm are used as this geometric arrangement reduces the problem to solving a 2D system.

Figure 24.9a describes the situation whereby 14 parallel current sources are con-nected to 14 separate active electrodes at the center of each of 14 hexagons in a 98-electrode array. Owing to the hexagon cluster layout, each of the specified stimulating electrodes is surrounded by six equally spaced guard electrodes that may be used as the electrical return path for stimulus delivered to the center of each hexagon cluster. Thus, surrounding each injection point are six return electrodes all connected to zero potential. Redefining the central electrode of each hexagon (Fig. 24.9b) introduces "edge effects" as six guard electrodes do not exist at the edges of the array but in the worst case only three electrodes serve as guards.

In this instance the active electrode has moved down and to the right but there are seven unique locations to which an active electrode can be translated. This redefinition is a very useful feature that arises out of the electrode multiplexing and addressing that are performed by the application-specific integrated circuit (ASIC) [11].

Through modeling we can visualize the guarding afforded by these electrode con-figurations (Fig. 24.10). In this instance 1 mA is injected through the central electrodes of each of the 14 hexagonal clusters. There is an obvious concentration of current achieved by the guard electrodes (Fig. 24.10a). On a translation of the central injecting electrodes downward and to the right (Fig. 24.10b)—with the resultant absence of some guard electrodes surrounding injection sites—there is a generalized depolarization of the

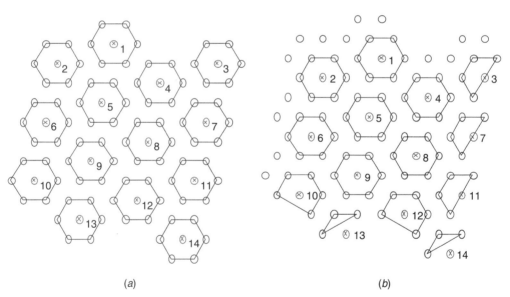

(a)   (b)

Figure 24.9   Hexagonal electrode array lattice comprising 98 electrodes and 14 individual current sources that may be activated in parallel. (a) Active electrode is in a position such that all surrounding guard electrodes completely encompass the active (shown pictorially by the solid lines connecting each guard ring). (b) One of the other six scenarios whereby the active electrode position has been translated downward and to the right. In this configuration not all active electrodes are completely surrounded by six guard electrodes. While this could be fixed by including extra electrodes at the periphery, this would necessitate a larger electrode array with additional interconnects and support electronics on the ASIC.

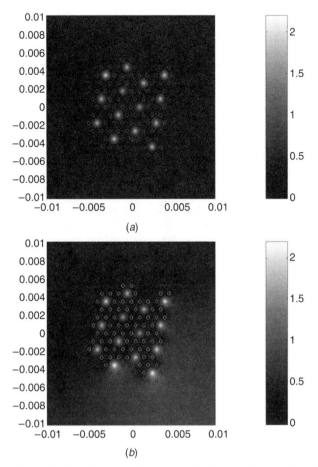

Figure 24.10   Simulated 2D voltage distribution derived from 14 parallel current sources, injecting 1 mA each into 20 mm × 20 mm × 2 mm physiological saline region ($\sigma = 1.0$ S/m) by way of 400-μm-diameter cylindrical electrodes. The X and Y axes are distances in millimeters. (*a*) Each injection electrode is surrounded by six guard electrodes arranged in a hexagonal pattern (depicted as open circles), each guard tied to zero potential. The vertical colorbar indicates the simulated potential in volts. The guard electrodes effectively focus current in the region of each hexagon and reduce return current interaction with other regions. (*b*) The central electrode for each current source has been translated downward and to the right. In this configuration, a number of the current sources are not totally surrounded by guards. These electrodes (depicted as open black circles) are effectively at a floating potential. Note the generalized depolarization of the tissue in the lower right of the simulated saline region due to the lack of return electrodes effectively anchoring the tissue at zero potential.

tissue. However, on examination of the crosstalk between surrounding injection sites and their corresponding guards, there is only a slight degree of current exchange. In the configuration of Figures 24.9*a* and 24.10*a*, the sum of the individual currents returned in each guard ranges from 0.83 mA in hexagon 8 to 1.07 mA in hexagons 2 and 9. The average current in each "hexagon" is 0.99 ± 0.06 mA (mean ± SD).

In the case of the translated configuration as in Figures 24.9 and 24.10*b*, the average summed current in each hexagon is 0.98 ± 0.09 mA. The slightly higher variance in the currents is associated with larger leakage in the peripheral hexagons with minimum return currents of 0.81 mA in hexagon 13 and 1.12 mA in hexagon 12.

These current imbalances are obviously of considerable importance when trying to accurately inject and recover charge from various electrodes interfaced to the neural retina. In general terms, the guards effectively attenuate crosstalk between hexagons, even in the case where the central injecting electrode is not entirely guarded by six return electrodes. Naturally more detailed simulations and experimental data are needed to fully characterize the effectiveness of the hexagonal guarding approach. This includes data where current injections are not equal across all hexagons. It will also be necessary to compare the usefulness of this method of bipolar stimulation versus monopolar stimulation using a large distant return as the indifferent electrode. While the latter approach has some possible advantages associated with inherently reducing crosstalk between adjacent electrodes, it also has a number of practical barriers, including the need to place a separate return electrode exterior to the globe in the case of an intraocular implant.

Owing to the nature of electrical stimuli and the complexities of the electrode–tissue interface, no single stimulation profile entirely eliminates differential charge from occurring between electrodes within the array. Therefore, supplementary means are required after stimulation occurs so as to return the electrodes to equal potential. This charge balance must occur in a regional sense. With the residual charge imbalances on the electrodes clearly being substantially smaller when using the hexagon–guard arrangement over other methods, shorting between stimulation events need only be brief in order to achieve the necessary outcomes. This eliminates the need for high numbers of capacitors in series with the electrodes.

24.9 CONCLUSION

In a progression through various practical aspects of vision prosthesis design and by introducing a number of conjectural ideas, we have presented our approaches toward the realization of a functioning epiretinal neuroprosthesis. A particular focus has been on a hexagonal electrode array construct and how this design impacts electrode size, electrode fabrication, interfacing, and implant electronics. Rudimentary modeling of the electrode array has highlighted the inherent advantages of the proposed design, indicating that hexagonal electrode array geometries have advantages in terms of parallelization of current sources and associated minimization of leakage currents between adjacent sources. On-going testing, both on the bench and in vivo, will further inform the design process.

REFERENCES

1. K. CHA, K. W. HORCH, R. A. NORMANN, AND D. K. BOMAN, "Reading speed with a pixelized vision system," *J. Opt. Soc. Am.*, vol. 9, pp. 673–677, 1992.

2. K. CHA, K. W. HORCH, AND R. A. NORMANN, "Mobility performance with a pixelized vision system," *Vision Res.*, vol. 32, pp. 1367–1372, 1992.

3. K. CHA, K. HORCH, AND R. A. NORMANN, "Simulation of a phosphene-based visual field: Visual acuity in a pixelized vision system," *Ann. of Biomed. Eng.*, vol. 20, pp. 439–449, 1992.

4. P. K. CAMPBELL, K. E. JONES, AND R. A. NORMANN, "A 100 electrode intracortical array: Structural variability," *Biomed. Sci. Instrum.*, vol. 26, pp. 161–165, 1990.

5. L. FÉJES-TOTH, "Uber einen geometrischen satz," *Math. Z.*, vol. 46, pp. 79–83, 1940.

6. D. SCRIBNER, M. HUMAYUN, B. JUSTUS, C. MERRITT, R. KLEIN, J. G. HOWARD, M. PECKERAR, F. PERKINS, L. JOHNSON, W. BASSETT, P. SKEATH, E. MARGALIT, E. KAH-GUAN AU, J. WEILAND, E. DE JUAN, J. FINCH,

R. GRAHAM, C. TRAUTFIELD, AND S. TAYLOR, "Intra-ocular retinal prosthesis test device," Engineering in Medicine and Biology Society, *Proceedings of the 23rd Annual International Conference of the IEEE.* October 25–28, 2001, Istanbul, Turkey, vol. 4: 3430–3435.

7. D. SCRIBNER, L. JOHNSON, R. KLEIN, W. BASSETT, J. G. HOWARD, P. SKEATH, L. WASSERMAN, B. WRIGHT, F. PERKINS, AND A. PECKERAR, "A retinal prosthesis device based on an 80 × 40 hybrid microelectronic-microwire glass array," presented at Custom Integrated Circuits Conference, 2003.

8. L. E. HALLUM, G. J. SUANING, AND N. H. LOVELL, "Contribution to the theory of prosthetic vision," *ASAIO* J., vol. 50, no. 4, pp. 392–396, 2004.

9. S. C. CHEN, N. H. LOVELL, AND G. J. SUANING, "Effect of prosthetic vision acuity by filtering schemes, filter cut-off frequency and phosphene matrix: A virtual reality simulation," presented at 26th Annual International Conference of the IEEE-EMBS, San Francisco, CA, 2004.

10. S. C. CHEN, N. H. LOVELL, AND G. J. SUANING, "Visual acuity measurement of prosthetic vision: A virtual-reality simulation study," *J. Neural Eng.*, vol. 2, pp. 135–145, 2005.

11. G. J. SUANING, L. E. HALLUM, P. J. PRESTON, AND N. H. LOVELL, "An efficient multiplexing method for addressing large numbers of electrodes in a visual neuroprosthesis," presented at 26th Annual International Conference of the IEEE-EMBS, San Francisco, CA, 2004.

12. C. Q. HUANG, R. K. SHEPHERD, P. M. CARTER, P. M. SELIGMAN, AND B. TABOR, "Electrical stimulation of the auditory nerve: Direct current measurement in vivo," *IEEE Trans. Biomed. Eng.*, vol. 46, pp. 461–470, 1999.

13. B. O'BRIEN, T. ISAYAMA, R. RICHARDSON, AND D. BERSON, "Intrinsic physiological properties of cat retinal ganglion cells," *J. Physiol.*, vol. 538, pp. 787–802, 2002.

14. R. J. GREENBERG, T. J. VELTE, M. S. HUMAYUN, G. N. SCARLATIS, AND E. DE JUAN JR., "A computational model of electrical stimulation of the retinal ganglion cell," *IEEE Trans. Biomed. Eng.*, vol. 46, pp. 505–514, 1999.

15. F. RATTAY, S. RESATZ, P. LUTTER, K. MINASSIAN, B. JILGE, AND M. DIMITRIJEVIC, "Mechanisms of electrical stimulation with neural prosthesis," *Neuromodulation*, vol. 6, pp. 42–56, 2003.

16. M. S. HUMAYUN, E. DE JUAN, JR., J. D. WEILAND, G. DAGNELIE, S. KATONA, R. GREENBERG, AND S. SUZUKI, "Patterned electrical stimulation of the human retina," *Vision Res.*, vol. 39, pp. 2569–2576, 1999.

17. T. STIEGLITZ, H. H. RUF, M. GROSS, M. SCHUETTLER, AND J.-U. MEYER, "A biohybrid system to interface peripheral nerves after traumatic lesions: Design of a high channel sieve electrode," *Biosensors Bioelectron.*, vol. 17, pp. 685–696, 2002.

18. R. S. CLEMENT, A. SINGH, B. OLSON, K. LEE, AND J. HE, "Neural recordings from a benzocyclobutene (bcb) based intra-cortical neural implant in an acute animal model," presented at 25th Annual International Conference of the IEEE Engineering in Medicine and Biology Society, Coucun, Mexico, September 17–21, 2003.

19. M. S. HUMAYUN, J. D. WEILAND, G. Y. FUJII, R. GREENBERG, R. WILLIAMSON, J. LITTLE, B. MECH, V. CIMMARUSTI, G. VAN BOEMEL, AND G. DAGNELIE, "Visual perception in a blind subject with a chronic microelectronic retinal prosthesis," *Vision Res.*, vol. 43, pp. 2573–2581, 2003.

20. M. SCHUETTLER, S. STIESS, B. V. KING, AND G. J. SUANING, "Fabrication of implantable microelectrode arrays by laser-cutting of silicone rubber and platinum foil," *J. Neural Eng.*, vol. 2, pp. 159–161, 2005.

21. J. T. MORTIMER, "Electrodes for functional electrical stimulation," 2000.

22. J. T. RUBINSTEIN AND C. A. MILLER, "How do cochlear prostheses work?" *Curr. Opin. Neurobiol.*, vol. 9, pp. 399–404, 1999.

23. C. VERAART, M. WANET-DEFALGUE, B. GERARD, A. VANLIERDE, AND J. DELBEKE, "Pattern recognition with the optic nerve visual prosthesis," *Artif. Organs*, vol. 27, pp. 996–1002, 2003.

24. J. RIZZO III, J. WYATT, J. LOEWENSTEIN, S. KELLY, AND D. SHIRE, "Perceptual efficacy of electrical stimulation of human retina with a microelectrode array during short-term surgical trials," *Invest. Ophthalmol. Vis. Sci.*, vol. 44, pp. 5362–5369, 2003.

25. R. J. GREENBERG, "Analysis of electrical stimulation of the vertebrate retina—Work toward a retinal prosthesis," Johns Hopkins University, Ph.D., 1998.

26. A. E. GRUMET, "Electric stimulation parameters for an epi-retinal prosthesis," Massachusetts Institute of Technology, Ph.D., 1999.

27. J. F. RIZZO, R. J. JENSEN, J. LOEWENSTEIN, AND J. WYATT, "Unexpectedly small percepts evoked by epi-retinal electrical stimulation in blind humans," presented at Invest Ophthalmol. Vis. Sci., 2003.

28. D. DACEY, "The mosaic of midget ganglion cells in the human retina," *J. Neurosci.*, vol. 13, pp. 5334–5355, 1993.

29. E. G. HECKENMUELLER, "Stabilization of the retinal image: A review of method, effects and theory," *Psychol. Bull.*, vol. 63, pp. 157–169, 1965.

30. G. J. SUANING, L. E. HALLUM, S. C. CHEN, P. J. PRESTON, AND N. H. LOVELL, "Phosphene vision: Development of a portable visual prosthesis system for the blind," presented at 25rd International Conference of the IEEE Engineering in Medicine and Biology Society, 2003.

31. J. WYATT AND J. RIZZO, "Ocular implants for the blind," *IEEE Spectrum*, pp. 47–53, May 1996.

32. T. S. HUANG AND R. Y. TSAY, "Multiple frame image restoration and registration," in *Advances in Computer Vision and Image Processing*, vol. 1, T. S. HUANG,

Ed., Greenwich: *Journal of Artificial Intelligence*, 1984, pp. 317–339.

33. J. S. HAYES, V. T. YIN, D. PIYATHAISERE, J. D. WEILAND, M. S. HUMAYUN, AND G. DAGNELIE, "Visually guided performance of simple tasks using simulated prosthetic vision," *Artif. Organs*, vol. 27, pp. 1016–1028, 2003.

34. A. B. MAJJI, M. S. HUMAYUN, J. D. WEILAND, S. SUZUKI, S. A. D'ANNA, AND E. DE JUAN, JR., "Long-term histological and electrophysiological results of an inactive epiretinal electrode array implantation in dogs," *Invest. Ophthalmol. Vis. Sci.*, vol. 40, pp. 2073–2081, 1999.

35. F. GEKELER, K. KOBUCH, H. N. SCHWAHN, A. STETT, K. SHINODA, AND E. ZRENNER, "Subretinal electrical stimulation of the rabbit retina with acutely implanted electrode arrays," *Graefes Arch. Clin. Exp. Ophthalmol.*, vol. 242, pp. 587–596, 2004.

36. H. N. SCHWAHN, F. GEKELER, K. KOHLER, K. KOBUCH, H. G. SACHS, F. SCHULMEYER, W. JAKOB, V. P. GABEL, AND E. ZRENNER, "Studies on the feasibility of a subretinal visual prosthesis: Data from yucatan micropig and rabbit," *Graefes Archi. Clini. Exp. Ophthalmol.*, vol. 239, pp. 961–967, 2001.

37. G. J. SUANING AND N. H. LOVELL, "Cmos neuro-stimulation asic with 100 channels, scaleable output, and bidirectional radio-frequency telemetry," *IEEE Trans. Bio-Med. Eng.*, vol. 48, pp. 248–260, 2001.

38. T. STIEGLITZ, W. HABERER, C. LAU, AND M. GOERTZ, "Development of an inductively coupled epiretinal vision prosthesis," presented at 26th Annual International Conference of the IEEE-EMBS, San Francisco, CA, 2004.

39. W. LIU, M. SIVAPRAKASAM, P. R. SINGH, R. BASHIRULLAH, AND G. WANG, "Electronic visual prosthesis," *Artif. Organs*, vol. 27, pp. 986–995, 2003.

40. J. RIZZO AND J. WYATT, "Prospects for a visual prosthesis," *Neuroscientist*, vol. 3, pp. 251–262, 1997.

41. A. Y. CHOW, V. Y. CHOW, K. H. PACKO, J. S. POLLACK, G. A. PEYMAN, AND R. SCHUCHARD, "The artificial silicon retina microchip for the treatment of vision loss from retinitis pigmentosa," *Arch. Ophthalmol.*, vol. 122, pp. 460–469, 2004.

42. E. ZRENNER, A. STETT, S. WEISS, R. B. ARAMANT, E. GUENTHER, K. KOHLER, K. D. MILICZEK, M. J. SEILER, AND H. HAEMMERLE, "Can subretinal microphotodiodes successfully replace degenerated photoreceptors?" *Vision Res.*, vol. 39, pp. 2555–2567, 1999.

43. J. D. WEILAND, D. YANAI, M. MAHADEVAPPA, R. WILLIAMSON, B. V. MECH, G. Y. FUJII, J. LITTLE, R. J. GREENBERG, E. DE JUAN, JR., AND M. S. HUMAYUN, "Visual task performance in blind humans with retinal prothetic implants," presented at 26th Annual International Conference of the IEEE EMBS, San Francisco, CA, 2004.

44. Y. KERDRAON, J. DOWNIE, G. J. SUANING, M. CAPON, M. CORONEO, AND N. H. LOVELL, "Development and surgical implantation of a vision prosthesis model into the ovine eye," *Clin. Exp. Ophthalmol.*, vol. 30, pp. 36–40, 2002.

45. Y. KERDRAON, J. DOWNIE, G. J. SUANING, M. CAPON, M. CORONEO, AND N. H. LOVELL, "Surgical implantation of a vision prosthesis model into the ovine eye," *Clin. Exp. Ophthalmol.*, vol. 30, pp. 36–40, 2002.

46. C. KELMAN, "Symposium: Phacoemulsification. History of emulsification and aspiration of senile cataracts," *Trans. Am. Acad. Ophthalmol. Otolaryngol*, vol. 78, pp. 5–13, 1974.

47. J. C. LILLY, J. R. HUGHES, E. C. ALVORD JR., AND T. W. GALKIN, "Brief, noninjurious electric waveform for stimulation of the brain," *Science*, vol. 121, pp. 468–469, 1955.

48. L. S. ROBBLEE AND T. L. ROSE, "Electrochemical guidelines for selection of protocols and electrode materials for neural stimulation," in *Neural Prostheses Fundamental Studies*, W. F. AGNEW AND D. B. MCCREERY, Eds., Englewood Cliffs, NJ: Prentice-Hall, 1990, pp. 26–66.

49. P. M. CARTER, R. K. SHEPERD, AND J. F. PATRICK, "Safety studies for a prototype nucleus 22 channel implant at high stimulation rates," presented at 3rd In. Congress on Cochlear Implants, Paris, France, 1995.

50. J. F. PATRICK, P. M. SELIGMAN, D. K. MONEY, AND J. A. KUZMA, 'Cochlear Prostheses,' Edinburgh: Churchill-Livingstone, 1990, pp. 99–124.

51. R. K. SHEPHERD, J. MATSHUSHIMA, R. E. MILLARD, AND G. M. CLARK, "Cochlear pathology following chronic electrical stimulation using non-charge balanced stimuli," *Acta Otolaryngol.*, vol. 111, pp. 848–860, 1991.

52. N. H. LOVELL, L. E. HALLUM, S. CHEN, S. DOKOS, R. STAPLES, L. POOLE-WARREN, AND G. J. SUANING, "Advances in retinal neuroprosthetics," *in Handbook of Neural Engineering*, M. AKAY, Ed., Hoboken, NJ: Wiley, pp. 337–356, 2007.

53. L. N. TREFETHEN, *Spectral Methods in Matlab*, Philadelphia: SIAM, 2000.

54. R. PLONSEY, "Volume conductor theory," in *The Biomedical Engineering Handbook*, J. D. BRONZINO, Ed., Boca Raton, FL: CRC Press, 1995, pp. 119–125.

55. B. ROTH, "The electrical properties of tissues," in *The Biomedical Engineering Handbook*, J. D. BRONZINO, Ed., Boca Raton, FL: CRC Press, 1995, pp. 126–138.

NEUROROBOTICS AND NEURAL REHABILITATION ENGINEERING

THIS **PART** focuses on the recent developments in the areas of the biorobotic system, biosonar head, limb kinematics, and robot-assisted activity to improve the treatment of elderly subjects at the hospital and home. It also focuses on the interactions of the neuron chip, neural information processing, perception and neural dynamics, learning memory and behavior, biological neural networks, and neural control.

Handbook of Neural Engineering. Edited by Metin Akay
Copyright © 2007 The Institute of Electrical and Electronics Engineers, Inc.

CHAPTER 25

INTERFACING NEURAL AND ARTIFICIAL SYSTEMS: FROM NEUROENGINEERING TO NEUROROBOTICS

P. Dario, C. Laschi, A. Menciassi, E. Guglielmelli,
M. C. Carrozza, and S. Micera

25.1 INTRODUCTION

For many scientists and engineers, including robotics researchers, nature and biology have always represented an ideal source of inspiration for the development of artificial machines and, more recently, of biomimetic components and robotic platforms. In fact, for millennia, the dream of developing truly effective biomimetic machines remained remote for humans. Today, the progress of mechatronics and robotics technology in such components as sensors, actuators, processors, control, and human–machine interfaces is making possible the development of biomimetic platforms usable as tools for neuroscientists, neurophysiologists, physiologists, and psychologists, for validating biological models, and for carrying on experiments that may be difficult or impossible using human beings or animals.

The interaction between biological sciences and robotics is twofold [1], as illustrated in Figure 25.1. On the one hand, biology provides knowledge on the human system needed to build humanoid robots (or humanlike components); on the other hand, anthropomorphic robots represent a helpful platform for experimental validation of theories and hypotheses formulated by scientists [2]. The first line of research has been pursued in robotics in order to develop robots that can better interact with real-world environments by imitating the solutions implemented by nature. This approach has received further impetus with the growth of "embodied" artificial intelligence (AI), based on the consideration that intelligence can only be developed along with a physical body [3]. The second line of research began quite recently to be defined, and in this chapter we refer to it as "neurorobotics." According to this approach, robots or robotic parts can be built by implementing a model of the corresponding biological system under observation; focused experiments on such robotic systems help verify/falsify the starting hypothesis on the biological model and even formulate new ones. For instance, the idea that biologically inspired robots can be used as experimental platforms to test the inspiring models has been applied to confirm some scientific hypotheses in ethology [4].

Handbook of Neural Engineering. Edited by Metin Akay

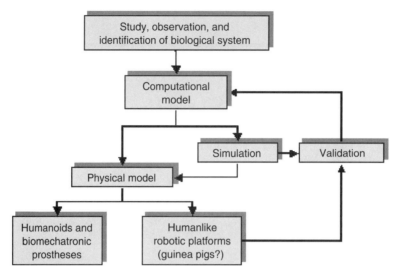

Figure 25.1   Scheme of twofold relation between biological science and robotics. In the study of a biological system, a computational model is formulated on the mechanisms/functions under observation; the model can be validated by a computer simulation or a physical model can be built, directly from the computational model or from the simulation. This physical model consists of a robotic artifact that can be used for validating the starting model (neurorobot) or for development of biologically inspired robots and body parts.

In the following, we discuss two case studies of the implementation of this concept of the twofold interaction between neuroscience and robotics. The first is aimed at developing a robotic manipulation platform which mimics human mechanisms of perception and action and can implement neurophysiological models of sensorimotor coordination and learning for their experimental validation. A close interaction between roboticists and neuroscientists is the starting point for the realization of the robotic components.

The second case study consists of the design and development of a cybernetic hand prosthesis, intended as a humanlike robotic hand with a neural interface allowing the patient to wear, feel, and control it in a natural way by his or her nervous system. In particular, the mechanical structure of the prosthesis, its sensorization, and its control scheme together with the first experimental results are described.

25.2 NEUROROBOTIC PLATFORM FOR EXPERIMENTAL VALIDATION OF SENSORIMOTOR COORDINATION AND LEARNING IN GRASPING

Development of the anthropomorphic robotic platform for manipulation presented here is motivated by the need for validating some hypotheses on the mechanisms of human learning formulated in neuroscience in the context of the sensorimotor coordination functions that are needed in grasping actions. Taking into account such neuroscientific findings, which hypothesize a stepwise progressive learning process in humans [5], development of the robotic components is aimed at mimicking as much as possible the corresponding biological systems.

In neurorobotics, the goal of reproducing humanlike features and replicating biological models has the priority of achieving good robot performance; solutions that are more similar to those adopted by nature are preferred even when more "artificial" robotic solutions can provide better performances, in terms of, for instance, action accuracy and speed or sensory sensitiveness and resolution, in order to follow as closely as possible the model to be validated.

The robotic platform presented here is composed of sensors and actuators replicating some levels of anthropomorphism in the physical structure and/or functionality [6], and their specifications were defined starting from:

1. On the neuroscience side, the definition of the functionality that the robotic platform should possess, in terms of sensory modalities, features to be perceived, motor capabilities, control schemes, and ideal tasks to be performed;

2. on the robotics side, the best available anthropomorphic robotic components, described in terms of sensors, actuators, control schemes, and technology for new improvements; this description includes demonstrations and tests on available devices, when relevant and feasible.

The basic arrangement of the robotic humanlike manipulation system consists of a robotic arm with a hand and a head. The hand is equipped with tactile sensors and the head supports a binocular vision system. Software modules implement basic humanlike sensory data processing. The platform is conceived so as to be modular and to be rearranged as largely and easily as possible and thus to provide different tools for different experiments.

25.2.1 Robotic Limbs

The robotic arm is the 8-degree-of-freedom (DOF) "Dexter" arm [7] whose physical structure is somehow anthropomorphic (see Fig. 25.2a), with the link arrangement reproducing the human body from trunk to wrist, so as to mimic human movements. The mechanical structure is light and flexible, due to steel cable transmission: the six distal motors are located on the first link, which represents the trunk. This solution increases the anthropomorphism of the distal part, even though reducing accuracy, with respect to traditional robotic solutions. The kinematic structure redundancy allows dexterous manipulation. The arm can provide proprioceptive data for all the eight joints and it can be controlled by a traditional PID (proportional–integrative–derivative) control scheme as well as by compliant and impedance control schemes in either the Cartesian or the joint space [8].

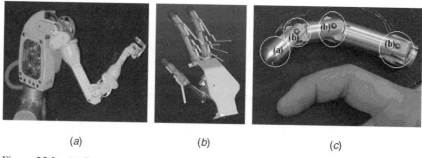

(a)   (b)   (c)

Figure 25.2 (a) Dexter robot arm (Scienzia Machinale srl, Pisa). (b) View of biomechatronic hand as designed. (c) Fabricated finger integrating three-component force sensor in fingertip and three-joint angle sensors.

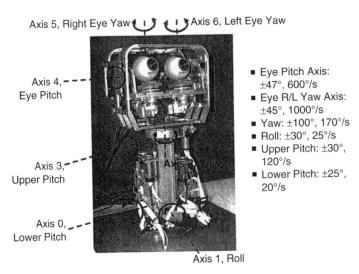

Axis 5, Right Eye Yaw   Axis 6, Left Eye Yaw

Axis 4,
Eye Pitch

- Eye Pitch Axis:
  ±47°, 600°/s
- Eye R/L Yaw Axis:
  ±45°, 1000°/s
- Yaw: ±100°, 170°/s
- Roll: ±30°, 25°/s
- Upper Pitch: ±30°,
  120°/s
- Lower Pitch: ±25°,
  20°/s

Axis 3,
Upper Pitch

Axis 0,
Lower Pitch

Axis 1, Roll

Figure 25.3   View of robotic head (Okino Industries Ltd., Japan) with indications of 6 DOFs and corresponding ranges of motion and speeds.

A robotic hand is under development according to a biomechatronic design, that is, by integrating sensors, actuators, and smart mechanisms in order to obtain a structure and a functional behavior comparable as much as possible to those of the human hand. The design is based on the experience we gained in developing prosthetic hands for restoring grasping ability and cosmetic appearance in amputees [9, 10].

The hand design is based on the concept of underactuation, meaning that even if only four motors are present, the hand has 9 DOFs, some active and some passive. The fingers can self-adapt to the object shape by using the passive DOFs and a differential mechanism integrated in the transmission system. The flexion/extension movements can be achieved independently by the three fingers, and the thumb can also abduct and adduct. Finally, the robotic hand can also be considered anthropomorphic in terms of mass and size (see Figs 25.2b,c).

The robotic humanlike head was designed in order to provide both neck and eyes movements to the cameras (see Fig. 25.3). The neck movements are implemented by a 4-DOF mechanism that reproduces the humanlike movements of lateral flexion, rotation, and ventral and dorsal flexions at different heights. The eye movements are reproduced by a common tilt and by two different pans (3 DOFs). The ranges of motion, the speeds, and the acceleration of the seven movements are very close to those of humans and thus allow to reproduce accurately the human eye movements of saccades and smooth pursuing.

25.2.2   Artificial Perception

The perceptual part of the platform includes visual and haptic (proprioceptive and tactile) sensory systems.

The *proprioception* on the robotic hand is based on eight joint position sensors (Hall effect sensor based) embedded in all the joints of each finger (see Fig. 25.2c) and an incremental encoder at each motor. As in humans the Golgi tendon organs provide information on the tendon strain, and three strain-gauge-based sensors measure the tension on the cables controlling the fingers flexion. Finally, contact of the hand and the

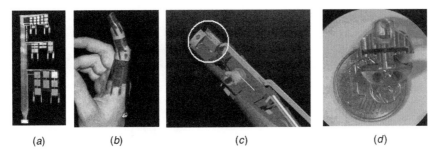

(a)   (b)   (c)   (d)

Figure 25.4   (a) Prototype of finger on–off touch sensor and (b) example of its application on human finger. The mechanical structure of three-component force sensors integrating strain gauges sensors (c) and model of fingertip (d).

hand-held object with the environment is detected by means of an accelerometer integrated in the palm.

The *somatosensory* information is essentially tactile information.

On–off touch sensors are placed on the internal and lateral surface of the phalanges and on the palm of the hand in order to sense the contact with the whole hand and to sense the contact area. More sensitive on–off touch sensors are placed on the hand dorsum in order to detect accidental contact with the environment. As for the sensor design, the on–off sensors are deposited on a flexible matrix in order to allow uniform sensor distribution and easy arrangement of the sensors on the phalanges (see Figs. 25.4a,b).

Three-component force sensors are mounted at the fingertips to measure the force vector at the contact point. A first version is a sensor based on a cross disposition of the strain gauges and located at the base of the fingertip so as to make the whole fingertip a three-component force sensor (see Figs 25.4c,d). In a second version, the quantitative information (intensity and direction) about the force is provided by an array of microfabricated force sensors, as those depicted in Figure 25.5.

Each microsensor cell has been designed to measure a maximum force of 7 N and to measure normal and tangential working forces in the range of 200–300 mN [11–13]. The theoretical values of the gauge factor and sensitivity are 162.8 and

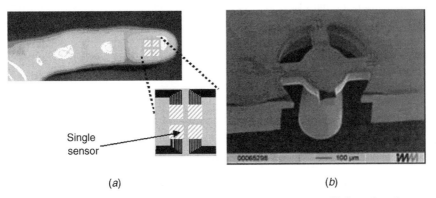

Single sensor

(a)   (b)

Figure 25.5   (a) Fingertip sensorization with flexible sensor array. (b) Scanning electron micrograph of cross section of three-component force sensor microfabricated in silicon (in collaboration with IMM Mainz, D).

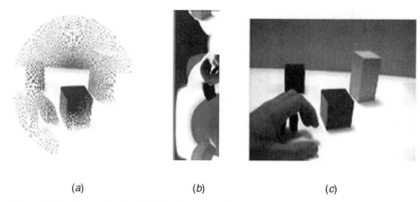

(a)   (b)   (c)

Figure 25.6   Example of retinalike image: (a) in retinalike image, pixel resolution is higher in center and degrades toward periphery, still providing meaningful indication of image content; (b) log-polar transformation, as an imitation of image projection on brain cortex, gives actual amount of data; (c) standard image of same scene provided as reference.

6.1×10^{-3} mV/kPa, respectively, for the normal force, while they are 63.8 and 0.013 mV/kPa, respectively, for the tangential force.

The silicon microsensor is based on the piezoresistive effect and it exploits four piezoresistors that are ion implanted in a silicon membrane and that are used as independent strain gauges. The sensing element consists of a circular membrane with four tethers and a cylindrical mesa located in the center of the membrane. The particular 3D mechanical structure is obtained starting from a silicon on insulator (SOI) wafer and exploiting a technological process of 52 steps that treat the wafer from both sides by applying the advanced silicon etching (ASE) technology [14].

As regards the vision system, a retinalike vision system is integrated in the robotic head. The system is based on a space-variant image whose space resolution is higher in the center (fovea) and degrades toward the periphery, thus imitating the structure of the human retina (see Fig. 25.6) [15, 16].

The main advantage of retinalike cameras is that they drastically reduce the amount of data to be processed while still providing significant information on the scene. As the amount of data is much smaller, the computational cost and consequently time are also reduced significantly. This allows us to include some visual processing in the control loop of the head and eye movements. On the other hand, retinalike vision is inherently active and very fast, and accurate camera movements are needed in order to focus the points of interest and to allow the extraction of good visual data. Therefore, it is crucial that the retinalike sensors are mounted on a robotic head able to move them quickly and precisely.

25.3   CYBERNETIC HAND PROSTHESIS

25.3.1   Three-Fingered Anthropomorphic Hand

A three-fingered anthropomorphic robotic hand has been developed as a human prosthesis and to be ultimately interfaced with the amputee's nervous system. The hand incorporates the tactile and joint angle sensors described in the previous section. Four motors are employed, one to activate the adduction/abduction of the thumb and the others for the opening and closing of the three fingers. Emphasis is on a device that is lightweight, reliable, cosmetic, energy efficient, and highly functional and that will ultimately be commercially viable. The development of the new hand is based on the experience acquired with a previous prototype [the so-called RTR (formerly Centre of Research in Technologies for Rehabilitation) II hand [17], in which the solution proposed by Shigeo Hirose in his Soft Gripper design [18] was applied to the case of the two fingers and the thumb].

Underactuated mechanisms allow obtaining self-adaptive grasping capabilities due to a large number of DOFs controlled with a limited number of actuators and differential mechanisms. This approach allows reproducing most of the grasping behaviors of the human hand without augmenting the mechanical and control complexity. This feature is particularly important in prosthetics, when only a few control signals are usually available from the electromyography (EMG) control interface, and therefore it is difficult for the amputee to control in a natural way more than two actuators [19].

The hand is based on a tendon transmission system. The tension of the tendons generates a flexing torque around each joint by means of small pulleys and allows the flexion movement; this transmission structure works in the same way as the *flexor digitorum profundus* in the human hand. The extension movement is realized by torsion springs. The adduction and abduction movements of the thumb are realized by means of a four-bar link mechanism.

An adaptive grasp system has been designed based on compression springs: Both finger wires are connected to a linear slider through two compression springs (see Fig. 25.7).

During a general grasp, index and middle fingers may not come in contact with the grasped object at the same time; one of the fingers and the thumb will usually come contact with the object first. When the first finger (e.g., the middle finger) comes in contact with the object, the relative spring starts to compress, the slider is now free to continue its

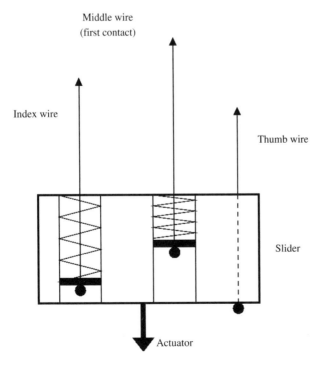

Figure 25.7 Adaptive grasp mechanism scheme.

motion, and the second finger (e.g., index finger) can flex, reaching the object. When high forces are required, compression springs behave as a rigid link and the entire force is transmitted from the slider to the fingers; this is the main advantage of using compression springs instead of extension springs.

The hand is equipped with an advanced artificial sensory system which is the core of the hand control system and has a twofold function: First, it provides input signals for the low-level control loop of the grasping phase, thus enabling local and autonomous control of the grasp without requiring the user's attention and reaction to incipient slippage. Second, it generates sensory signals to be transmitted to the amputee through an appropriate neural interface (such as the implantable one described in [20, 21]). The biomimetic sensors are the same as those used in the robotic hand described in the previous section.

25.3.2 Control of Prosthesis

An efficient and natural control of a sensorized multi-DOF prosthetic hand requires addressing the following problems: (1) the development of algorithms for extracting information on the user's intentions by processing the natural efferent neural signals, called *high-level pattern recognition module* (HLPRM); (2) the development of algorithms for closed-loop control of the artificial prosthesis according to the commands coming from the HLPRM and the sensory information obtained from the biomimetic sensors embedded in the prosthesis, called *low-level controller* (LLC); and (3) the development of a strategy for the stimulation of the afferent nerves in order to provide some sensory feedback to the user. The development of the HLPRM is a typical pattern recognition problem. The HLPRM should be able to correctly identify what is the task t^* the user would like to perform among the different possible tasks. In order to achieve this result the HLPRM is composed of the following subsystems:

1. A system to increase the signal-to-noise ratio of the efferent neural signals recorded by the regeneration-type neural electrodes [22, 23]. It is worth noting that in this case we must consider as "noise" not only the interference due to the thermal or electrical noise but also the presence of other neural signals which are not related to the task we would like to discriminate. For this reason we consider specific algorithms, such as "blind deconvolution" algorithms, principal-component analysis, or cross-time–frequency representations, in order to decompose the neural signals and "erase" the noise components.

2. A system for the extraction of different time and frequency parameters ("neural features") from the neural signals. These features are used for pattern recognition.

3. A system for the discrimination of the task desired. Different architectures have been designed and tested. In particular, the applicability of soft-computing techniques (neural networks, fuzzy systems, genetic algorithms) is analyzed. Different architectures based on statistical and soft-computing algorithms have been preliminarily investigated [24, 25].

These systems have been tested using some neural signals recorded by microneurography (see Fig. 25.8).

The LLC should control by closed loop the actuation of the prosthesis according to the task t^* selected by the HLPRM (and the information provided by the biomimetic sensors). Even in this case several algorithms based on soft-computing techniques will

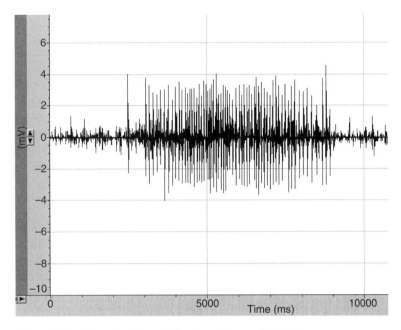

Figure 25.8   Example of neural signal used to test HLPRM.

be implemented. In preliminary experiments, the LLC was implemented as a fuzzy logic controller (FLC) in order to drive the prosthetic hand according to the information extracted by the high-level controller (HLC) in a closed-loop way. An FLC does not require an accurate model of the plant, and this can be very useful in motor drive applications where the mechanical load can be nonlinear or partially unknown or time varying. In these preliminary experiments the HLC was simply a random function selecting among the different possible choices. If the command chosen by the HLC was "open the hand," the system read the present position of the hand and provided the necessary commands to the motor in order to reach the desired position (which was selected to be the maximum opening of the hand). If the command was "grasp an object," the system acquired force information from the hand by communicating with a microcontroller, through the parallel port, and provided the command to motors depending on the desired grip force. The hand was then closed until the contact was reached.

After this event, the thumb was stopped and the index and middle fingers were closed by the FLC commands until the difference between the actual and desired force levels was less than a tolerance threshold. The inputs to the FLC were the position error (or the force error, depending on the implemented action) and the change in position error (force); the output was the variation of motor voltage. The control increased or decreased the motor voltage according to the position (force) error amplitude. The rule base implements a PID controller as described in [25].

The FLC was compared with a classical PID controller in the implementation of the different tasks (opening and closing the hand). The FLC outperformed the PID controller in both tasks, obtaining rise times compatible with real-time applications.

Finally, in order to implement sensory feedback, afferent neural signals should be characterized [26, 27].

25.4 CONCLUSIONS

This chapter has briefly analyzed the status and perspectives of interactions between neuroscience and robotics and presented two case studies.

The first case is the application of robotics to the study of neuroscientific models. The hypothesis is that a robot can ultimately become a sort of artificial "guinea pig" for neuroscience experimental investigation. The authors developed an anthropomorphic robotic platform for the validation of neurophysiological models of learning sensorimotor coordination in human grasping. The interaction between roboticists and neuroscientists allowed us to formulate the requirements for the robotic platform, especially in terms of tactile sensing, grasping capability, and vision processing, so as to replicate the features of interest for implementing and testing the proposed model for learning and sensorimotor coordination.

The second case study concerns the development of a new prosthesis, intended to be interfaced with the human nervous system. Preliminary results on the sensory system are quite promising, both for the implementation of the low-level control loop of the grasping phase and for the generation of sensory signals to be transmitted to the user through an appropriate neural interface. This system is the current base for the realization of a new prosthesis which could be felt by the amputee as the lost natural limb delivering a natural sensory feedback (by means of the stimulation of some specific afferent nerves) and could be controlled in a very natural way by processing the efferent neural signals coming from the central nervous system (thus increasing the responsiveness and functionality of the current EMG-based control prosthesis).

Considerations and results presented in this chapter are quite preliminary, but they clearly indicate that there is a very promising future (and lots of research to be done) for the new field of neurorobotics.

ACKNOWLEDGMENTS

Part of the work was supported by the European Union within the IST-FET PALOMA and IST-FET CYBERHAND Projects. The authors wish to acknowledge the fruitful collaboration with and contributions provided by the partners of these projects. The development of the biomechatronic hand was partly supported by the INAIL RTR Centre (Research Centre in Rehabilitation Bioengineering of the INAIL Prostheses Centre). The microfabrication of tactile sensors was supported by the EU "Improving Human Potential" (IHP) Programme in collaboration with the Institut fuer Mikrotechnik Mainz (IMM), Germany.

REFERENCES

1. P. DARIO, E. GUGLIELMELLI, AND C. LASCHI, "Humanoids and personal robots: Design and experiments," *J. Robot. Syst.*, 18(12):673–690, 2001.
2. M. KAWATO, "Robotics as a tool for neuroscience: Cerebellar internal models for robotics and cognition", in M. H. HOLLERBACH AND D. E. KODITSCHEK (Eds.), *Robotics Research, The Ninth International Symposium*, Springer-Verlag, London, 2000, pp. 321–328.
3. R. BROOKS, *Cambrian Intelligence*, MIT Press, Cambridge, MA, 1999.
4. D. LAMBRINOS, M. MARIS, H. KOBAYASHI, T. LABHART, R. PFEIFER, AND R. WEHNER, "An autonomous agent navigating with a polarized light compass", *Adaptive Behav.*, 6:175–206, 1997.
5. Y. BURNOD, P. BARADUC, A. BATTAGLIA-MAYER, E. GUIGON, E. KOECHLIN, S. FERRAINA, F. LACQUANITI, AND R. CAMINITI, "Parieto-frontal coding of reaching: An integrated framework." *Experi. Brain Res.*, 129(3):325–346, 1999.
6. P. DARIO, M. C. CARROZZA, E. GUGLIELMELLI, C. LASCHI, A. MENCIASSI, S. MICERA, AND

F. VECCHI, "Robotics as a 'future and emerging technology': Biomimetics, cybernetics and neuro-robotics in European projects," *IEEE Robot. Automation Mag.*, 12(2):29–45, 2005.

7. B. ALLOTTA, L. BOSIO, S. CHIAVERINI, AND E. GUGLIELMELLI, "A redundant arm for the URMAD mobile unit," paper presented at the 6th International Conference on Advanced Robotics, Tokyo, Japan, 1993, pp. 341–346.

8. L. ZOLLO, B. SICILIANO, C. LASCHI, G. TETI, AND P. DARIO, "An experimental study on compliance control for a redundant personal robot arm," *Robot. Autonomous Syst.*, 44(2):101–129, 2003.

9. M. C. CARROZZA, B. MASSA, S. MICERA, R. LAZZARINI, M. ZECCA, AND P. DARIO, "The development of a novel prosthetic hand–Ongoing research and preliminary results," *IEEE/ASME Trans. Mechatron.*, 7(2):108–114, June 2002.

10. M. C. CARROZZA, P. DARIO, F. VECCHI, S. ROCCELLA, M. ZECCA, AND F. SEBASTIANI, "The cyberhand: On the design of a cybernetic prosthetic hand intended to be interfaced to the peripheral nervous system", *IEEE/RSJ Int. Conf. Int. Rob. Syst.*, 3:2642–2647, 2003.

11. L. BECCAI, S. ROCCELLA, A. ARENA, A. MENCIASSI, M. C. CARROZZA, AND P. DARIO, "Silicon-based three axial force sensor for prosthetic applications," Paper presented at the 7th National Conference on Sensors and Microsystems, Bologna, February 4–6, 2002.

12. L. BECCAI, S. ROCCELLA, A. ARENA, F. VALVO, AND P. VALDASTRI, A. MENCIASSI, M. C. CARROZZA, AND P. DARIO, "Design and fabrication of a hybrid silicon three-axial force sensor for biomechanical applications," *Sensors and Actuators A*, 120(2):370–382, 2005.

13. P. VALDASTRI, S. ROCCELLA, L. BECCAI, E. CATTIN, A. MENCIASSI, M. C. CARROZZA, AND P. DARIO, "Characterization of a novel hybrid silicon three-axial force sensor," *Sensors and Actuators A*, 123/124C: 249–257, 2005.

14. S. A. MCAULEY, H. ASHRAF, L. ATABO, A. CHAMBERS, S. HALL, J. HOPKINS, AND G. NICHOLLS, "Silicon micromachining using a high-density plasma source," *J. Phys. D: Appl. Phys.*, 34:2769–2774, 2001.

15. G. SANDINI AND P. DARIO, "Active vision based on a space-variant Sensor," in H. MIURA AND S. ARIMOTO (Eds), *Robotics Research*, vol. 5, MIT Press, Cambridge, MA, 1990, pp. 75–84.

16. G. SANDINI AND M. TISTARELLI, "Vision and space variant sensing," in H. WECHSLER (Ed.), *Neural Network for Perception*, Academic Press, San Diego, 1992, pp. 398–425.

17. F. PANERAI, G. METTA AND G. SANDINI, "Visuo-inertial Stabilization in Space-variant Binocular Systems", *Robot. Autonomous Syst., Special Issue on Biomimetic Robotics*, 30(1/2):195–214, 2000.

18. B. MASSA, S. ROCCELLA, M. C. CARROZZA, AND P. DARIO, "Design and development of an underactuated prosthetic hand," *IEEE International Conference on Robotics and Automation*, Washington, D. C., May 11–15, 4:3374–3379, 2002.

19. S. HIROSE and Y. UMETAMI, "The development of soft gripper for the versatile robot hand," *Mechanism and Machine Theory*, 13:351–359, 1977.

20. M. ZECCA, S. MICERA, M. C. CARROZZA, AND P. DARIO, "On the control of multi functional prosthetic hands by processing the electromyographic signal," *Crit. Rev. Biomed. Eng.*, 30(4/6):459–485, 2002.

21. S. HIROSE AND S. MA, "Coupled tendon-driven multijoint manipulator," paper presented at IEEE Conf. on Robotics and Automation, 1991, pp. 1268–1275.

22. P. DARIO, P. GARZELLA, M. TORO, S. MICERA, M. ALAVI, J.-U. MEYER, E. VALDERRAMA, L. SEBASTIANI, B. GHELARDUCCI, C. MAZZONI, AND P. PASTACALDI, "Neural interfaces for regenerated nerve stimulation and recording", *IEEE Trans. Rehab. Eng.*, 6(4):353–363, 1998.

23. D. CEBALLOS, A. VALERO-CABRE, E. VALDERRAMA, M. SCHUTTLER, T. STIEGLITZ, AND X. NAVARRO, "Morphologic and functional evaluation of peripheral nerve fibers regenerated through polyimide sieve electrodes over long-term implantation," *J. Biomed. Mater. Res.*, 60(4):517–528, 2002.

24. S. MICERA, A. M. SABATINI, P. DARIO, AND B. ROSSI, "A hybrid approach for EMG pattern analysis for classification of arm movements," *Med. Eng. Phys.*, 21:303–311, 1999.

25. C. FRESCHI, A. DI GIGLIO, S. MICERA, A. M. SABATINI AND P. DARIO, "Hybrid control of sensorised hand prosthesis: Preliminary work," paper presented at EUREL Conf: Pisa, September 21–24, 1999.

26. S. MICERA, W. JENSEN, F. SEPULVEDA, R. R. RISO, AND T. SINKJAER, "Neuro-fuzzy extraction of angular information from muscle afferents for ankle control during standing in paraplegic subjects: An animal model," *IEEE Trans. Biomed. Eng.*, 48(7):787–794, 2001.

27. E. CAVALLARO, S. MICERA, P. DARIO, W. JENSEN, AND T. SINKJAER, "On the intersubject generalization ability in extracting kinematic information from afferent nervous signals," *IEEE Trans. Biomed. Eng.*, 50(9):1063–1073, 2003.

CHAPTER 26

NEUROCONTROLLER FOR ROBOT ARMS BASED ON BIOLOGICALLY INSPIRED VISUOMOTOR COORDINATION NEURAL MODELS

E. Guglielmelli, G. Asuni, F. Leoni, A. Starita, and P. Dario

26.1 INTRODUCTION

Children from a very early age attempt to reach objects which capture their attention. It can be observed how, by making endogenous movements and correlating the resulting arm and hand spatial locations, they allow an autoassociation to be created between visual and proprioceptive sensing. During the growing phase children improve their capability to reach objects more quickly and accurately by an appropriate trajectory. The result of this learning process consists in the production of a set of motor programs that become linked, with the aim of the selected desired task. Furthermore, these motor programs can autonomously adapt themselves during the execution of the task according to the current perception of the environment. This scheme has been identified from a psychological point of view by Piaget [1] as the circular reaction to explain visuomotor associations. Neurocontrollers for robotic artifacts can be devised taking inspiration from such understanding in neuroscience. In this chapter a case study of the design and the development of a general, biologically inspired, artificial neural controller able to manipulate sensorial data coming from visual and proprioceptive artificial sensors and to control two different robotic arms for the execution of simple reaching tasks is presented.

26.2 BIOLOGICAL FRAMEWORK OF MOTOR CONTROL

In humans, sensorimotor behaviors are generated by neural interactions occurring in different parts of the brain and connected to different sensorimotor structures, not fully known yet. Experimental data that relate electrical activity in the brain to limb movements are fairly abundant providing the essential data on which a top-down modeling approach can be based. However, these models are still far from providing accurate models describing in quantitative terms the internal functionality of the sensorimotor cortex areas involved in motor control. The generation and control of limb movements come out as the result of information processing in a complex hierarchy of motor centers within the

Handbook of Neural Engineering. Edited by Metin Akay

nervous system, including the motor cortex, the brainstem, and the spinal cord [2]. Neurophysiological studies that provide the essential data on which a top-down modeling of these phenomena is based could be used as a reference also in robotics, which seeks to design more and more robust and adaptable systems. For instance, the challenges posed by the development of humanoid robots cannot be faced with a "standard" engineering robotic approach to the problem of control [3]. Humanoid robots are required to have a complex mechanical structure including a lot of active degrees of freedom actuating their artificial limbs, and they are also required to interact and to communicate with humans in their living space in a natural, possibly anthropomorphic manner [4, 5]. To achieve these functions humanoid robots are endowed with a lot of different sensors (e.g., tactile, force, vision) enabling them to perceive the surrounding environment. The amount of concurrent data incoming from the sensors may be so large that it is necessary to develop complex processing subsystems able to handle and integrate such data successfully providing appropriate inputs to the control systems. Biological control systems successfully solve the problem of translating sensory data into appropriate motor behaviors by using optimized selection and fusion techniques. The objective of this work is to take inspiration from biology for the development of novel visuomotor coordination control schemes able to capture in an autonomous way all the mechanical, kinematic, and dynamic parameters of real robots. From a functional point of view, the voluntary execution of an arm movement toward a visual target involves the selection of suitable kinematic and dynamic parameters. Kinematic parameters are related to the perception of the target in the extrapersonal space, the elaboration of a path that the hand should follow (starting from the current position) to reach the target, and, consequently, the planning of the limb joint trajectories in time. Dynamic parameters concern the selection of the appropriate muscle contractions required in order to produce the right joint torques so that they are able to follow the imposed trajectory. These parameters could vary dynamically according to unpredictable changes that may appear during the execution of the movement. Examples of these changes are external payloads or mechanical constraints such as obstacles and impairments or may be "soft" in the sense that they could be represented also by handled tools that may change their references as to the hand in an unpredictable way (e.g., a hammer slipping in the hand). Moreover, humans are able to reach a target with an unknown tool which is grasped by the hand without any preliminary learning phase. The human reaching movements show straight paths and bell-shaped velocity profiles of the hand, as observed by several researchers [6]. To explain the reasons for this fact a number of hypotheses have been made; for example Flash and Hogan [7] described this behavior in terms of the minimization of joint acceleration changes (jerk model), and others explained it observing that the straight path followed by the hand corresponds to the minimum distance between the hand and the target. An important finding, also useful from a modeling point of view, is that such a motor behavior is due to the arm dynamics and is not related to some particular neural state [8]. At the level of the central nervous system (CNS) it is well known that sensory and motor areas show topographic organization of the body area to which they are related. Muscle surfaces and receptors have a corresponding point on the cortical surface and the resolution of sensory acuity and motor dexterity of body segments is directly proportional to the size of the cortical areas to which they correspond. Besides, adjacent muscles are mapped onto adjacent cortical sites. A sketch of a plausible connecting model of the cortical area involved in motor functionality is drawn in Figure 26.1.

Movement planning and control in the brain exploits complex pathways involving many cortical areas such as association, motor, somatosensory, and visual areas. The aim of these areas and pathways is to create a chain of coordinate transformations from

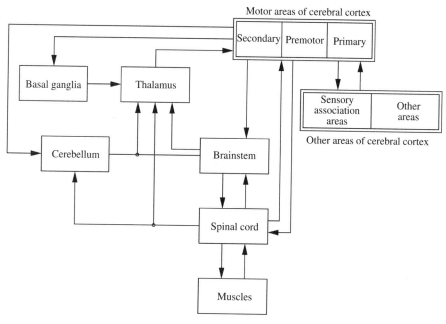

Figure 26.1   Rough diagram and connections of cortical areas devoted to generation of motor commands.

the outworld frames to the internal ones providing multimodal sensorial integration tools for proprioceptive and exteroceptive sensitivity. Moreover, Tanji and Evarts [9] provided findings about specific functions of cortical neurons such as the coding of movement kinematic properties of neurons in the primary motor cortex (MI) and the correspondence between the direction of the arm movements and the distribution and entity of the discharge of the neurons [10]. It is also shown that genetic information by itself does not permit the formation a correct model of connections between cortical areas able to provide a functional scheme for a proper controller [11]. Therefore, taking into account that many physical parameters are time dependent, such as muscle length, muscle strength, and body mass, the continuous adaptation of the connectivity patterns with other physiological parameters is required. To date, however, there is still no general consensus about how biological neural networks actually generate voluntary movement. There are experimental evidences that motor neurons encode both hand paths (movement direction—kinematics) and muscle activity in single- and multijoint movements (dynamics) [12]. Also, there are experimental findings showing that the emerging kinematic movement properties are not dependent on the dynamics of the muscle system [13]. To explain this different behavior it was suggested that the motor cortical activity could encode both visual and motor aspects over time allowing visuomotor transformation [14]. In this way visual feedback can provide kinematics data which can be related to the actuation properties of the muscles through the learning. The result of the learning may appear as a coding of the muscle dynamics. There is also evidence that the nonlinear properties of the muscles related to their viscoelastic nature allow the development of a simple feedforward scheme control requiring a few parameters to perform the whole hand trajectory [15]. This scheme could take advantage of the direction-coding properties of the motor neurons that, in this case, are required to activate only a limited number of motor parameters. These speculations may have a proper match with the results of Mussa-Ivaldi et al. concerning the hypothesis that the CNS may coordinate complex movements "through the vectorial superposition of a few motor primitives stored within the neural circuits in the spinal cord" [16, p. 7534]. In conclusion, even though complete understanding of how humans can effectively develop and apply sensorimotor coordination schemes is still not at hand, neuroscientists are providing more and more advanced hypotheses and accurate models which represent a powerful source of inspiration for designing innovative supervision and control schemes for robotic artifacts. At the same time, such robots, where a bioinspired control scheme has been instantiated, could represent a candidate physical model with an application potential as a tool made available to neuroscientists for validating their hypotheses and models. In the following a specific case study on the development of a novel neurocontroller for robotic manipulators based on neurophysiological models is presented in detail, including simulation and experimental results.

26.3 MODELING AND ROBOTIC APPROACHES

Robotics, which aims at designing more and more robust and adaptable systems working in the real, unstructured world, is looking with increasing interest at biomimetic and bioinspired solutions directly derived from the observation of nature. From this perspective, neurophysiological models can offer a different point of view allowing the development of biologically inspired models to control in a "natural way" anthropomorphic robotic biologically inspired sensors and actuators. This approach can potentially offer a deeper insight in the comprehension of the biology itself providing platforms to evaluate the models of brain functions. From another point of view, psychological approaches can help in the comprehension of the neuropsychological mechanism behind these abilities. For example, the circular reaction mechanism proposed by Piaget [1] represents a well-known traditional psychological approach to the problem of visuomotor coordination. In particular we are interested in understanding what mechanisms are established and allow a person to learn stable transformations between two separate sensorial systems, such as the visual and motor sensorial systems. It could be noted that during visuomotor tasks the ocular movements tend to bring the projection of the image of the hand onto the retina (visual target) such that it corresponds at the level of ocular fovea in order to improve the vision of the hand. If these movements do not conform to a correct foveation, the actual ocular position causes a visual error. Grossberg and Kuperstein [17] described a model showing how these error signals can be used by the cerebellum to learn those values of the parameters needed to control a correct foveation. Burnod and co-workers [18] proposed a distributed neural network architecture showing how motor commands for reaching tasks occur at the same time in different cortical areas. The proposed architecture shows similar results to those found in an experimental way by Georgopoulos [10] regarding the coding of the direction of movement in 3D space exhibited by a neuronal population in primates. The DIRECT (direction-to-rotation effector control transform) model proposed in [19] includes descriptions of the role of sensorimotor cortex in the learning of visuomotor coordination correlating proprioceptive and visual feedback to the generation of joint movements. The model shows successful reaching of targets located in the workspace of the simulated 3-degree-of-freedom (DOF) arm taking into account the redundancy of the arm. The DIRECT model represents a very interesting biologically inspired motor control behavior emphasizing some peculiar humanlike characteristics, such as the successful reaching of targets despite the different operative conditions from those taken into account during the learning phase (normal reaching). Typical modification of operative conditions for which a robust behavior can be obtained by the

DIRECT model are adding a tool of variable length at the end of last link, the execution with a clamped joint, with distortions of the vision due to the use of a prism and blind reaching. All of these limitations are executed without a specific learning phase; the system learns to move the joints in "normal" conditions (i.e., with correct visual input and allowing free motion of all the joints); then it is able to successfully perform the reaching task in the "modified" conditions. Inverse kinematics is one of the crucial problems in developing robot controllers. Traditionally, solutions to the inverse kinematics problem are obtained by different techniques based on mathematical computational models, such as inverse transform or iterative methods. Sometimes, these methods may suffer from drawbacks, especially when the number of degrees of freedom increases: Inverse transform does not always guarantee a closed-form solution, while iterative methods may not converge and may be computationally expensive. Neural network approaches, which provide robustness and adaptability, represent an alternative solution to the inverse kinematics problem, especially when the number of DOFs to be controlled is high, for example, in the case of redundant and/or modular reconfigurable kinematic chains, or the external spatial reference system is not easy to be modeled, such as in visuo-motor coordination tasks. Two main approaches have been proposed for the use of neural networks to solve the problem of inverse kinematics. The first, based on mathematical models, considers the artificial neural networks as a tool to solve nonlinear mappings without, or in some cases with a limited or partial, knowledge of the robotic structure [20, 21]. The second approach builds the mapping between motor commands and sensory input on the basis of repeated sensorimotor loops; the final map is then used to generate appropriate motor commands to drive the robot toward the target sensory input. These latter methods make use of self-organizing maps [22] to build the internal mapping: Their self-organization and topological preservation features make them well suited for capturing mechanical constraints; moreover, they show a good capability of processing artificial sensory signals, according to the analogous mapping features of the somatosensory cortex by which they are inspired. These abilities could be useful in closed-loop control systems which have to deal with somatosensory signals coming from anthropomorphic artificial sensors, such as proprioceptive and tactile, in order to allow the generation of proper sensorial inputs toward the motor areas [23]. In the past, visuomotor coordination problems have been extensively and quite successfully approached using Kohonen's maps [24, 25]. Nevertheless, this type of network may suffer from the necessity of a priori knowledge of the probability distribution of the input domain in order to choose a proper cardinality and structure of the net, so as to avoid over- or underfitting problems. Furthermore, they are not suitable to approach dynamic environments or continuous learning [26]. In general, approaches involving self-organizing techniques assign a subarea of the input domain (e.g., joint space) to a single neuron, according to a specified resolution. This methodology divides the whole space in a set of equally probabilistic subspaces, disregarding the correlations between the task space and the sensory domain in which they are performed. Growing neural gas (GNG), proposed by Bernd Fritzke [27], is an incremental neural model able to learn the topological relation of input patterns. Unlike other methods, this model avoids the need to prespecify the network size. On the contrary, from a minimal network size, a growth process takes place which is continued until a condition is satisfied. The objective of the reported case study is to provide a computational model that embodies a solution of the inverse kinematic problem capturing the kinematic proprieties of real robots by learning and improving some aspect of the DIRECT model. The proposed neural model will be evaluated by two sets of experimental trials: first, a comparative analysis of the simulation results obtained considering the same planar manipulator used by the DIRECT model will be presented and discussed; second, results in the application to two different real robotic arms will be reported, thus demonstrating the feasibility and the portability of the proposed model, including its learning capabilities.

26.4   STRUCTURE AND FUNCTIONALITY OF PROPOSED NEURAL MODEL

Looking at the overall block diagram of the DIRECT model in Figure 26.2 and having in mind its transposition into a control scheme for a robotic arm, we can outline at least two main "weak" characteristics. The first concerns the method adopted to realize the integration of the proprioceptive and visual sensorial modalities. This integration is realized by the spatiomotor present position vector (PPVsm) combining the current proprioceptive perception of the arm posture (motor present position map) and the current perception of the hand (spatial present position vector) on the retinic frame. Then, the desired motion direction is computed by the spatial direction vector (DV) stage as the difference between the spatial target position vector (TPV) and the PPVsm; consequently, the dimension of the output of the PPVsm must be equal to the dimension of the TPV.

Indeed, objects and, in general, real images are represented in a topographic fashion onto the retina and in the brain at the level of the temporal and parietal cortex association areas. Here, other sensorial modalities, such as proprioception and touch, are integrated together; in these areas we find a distributed and integrated perception of the same "scene" acquired by all the sensors available [23]. Our approach consists in using self-organizing growing neural maps for realizing the mapping of proprioception and vision perception and of their integration. The second novelty consists in the use of the motor present position map (PPMm) and of the position direction map (PDMms). The dimensions of these maps are initially decided based on a priori and empirical assumptions. Also in this case we will release this constraint using the same type of neural maps described before. Hence, we will not distinguish between maps or vectors, thus simplifying the model. Besides, we will use neural maps able to grow in accordance with the complexity of the input space to which they are connected (e.g., visual, proprioception, associative, and motor domains). These neural maps, called GNG, and their equations are in accordance with those introduced by Fritzke [27]. As mentioned, GNG is an incremental neural model capable of learning the topological relation of input patterns by means of "Hebbian" learning, similar to other self-organizing networks. Unlike other methods, however, this model avoids the need to previously specify the network size. On the contrary, from an initial minimal network size, a growth process takes place which is continued until an ending condition is fulfilled. Also, learning parameters are constant over time, in contrast to other models which heavily rely on decaying parameters. In the original algorithm proposed by Fritzke [27] the insertion of a new cell is driven by a constant parameter. More precisely, one new cell is added everytime that λ input signals are presented, so that the total number of cells that compose the network results in about $\lfloor |\mathcal{A}|/\lambda \rfloor$, where $|\mathcal{A}|$ is the cardinality of the training set. A different approach to control the growing process in such networks is described in [28]. In this work the insertion occurs only when the cumulated local error is greater than a proper threshold. This algorithm has been used by the author to learn an inverse kinematic model in the context of sensorimotor coordination. Winter et al. [29] used the same approach with some modifications in the case of function estimation. The modification is a heuristic function that moves the neurons toward regions in the input space where the error is large. As a result, the

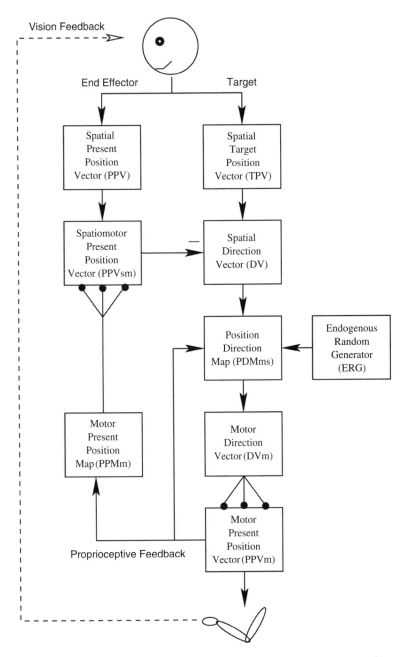

Figure 26.2   Overall block diagram of DIRECT model. (Redrawn from [19].)

neurons are concentrated in regions where the function that should be approximated is more complicated. In our approach, we use the λ parameter as the insertion frequency counter. The goal is to add cells proportionally to the magnitude of the quantization error of the net over the time. The error is greater and more frequently new units are added to the net. In this case, when the input patterns remain quasi-stationary (e.g., if the input patterns do not differ from each other more than a given threshold), we expect that the system could feature a linear reduction of the quantization error. The error decrement is assured by the "insertion law." The new unit is connected through weighted connections between the two units with the largest errors belonging to the same neighbor. The new connections are weighted by a portion of the original connection between the oldest units, guaranteeing that the new local error is less than the previous one. In this way the insertion of cells could be virtually never stopped: A low quantization error means that a new cell will be inserted after a long, long time. However, we do not insert new units if the quantization error is below a specified threshold. Furthermore, this mechanism allows us to deal with unbound training sets. If at a certain time a never-seen-before input pattern is generated, then a consistent quantization error will be calculated and so the insertion procedure starts again. In detail

$$\lambda = \frac{K}{\overline{\text{Err\_qnt}}} \tag{26.1}$$

where K is an appropriate constant and $\overline{\text{Err\_qnt}}$ is the mean quantization error computed by the formula

$$\overline{\text{Err\_qnt}} = \frac{\sum_{\xi \in \mathcal{A}} \| \xi - \mathbf{w}_{s_1} \|^2}{|\mathcal{A}|} \tag{26.2}$$

where $\mathcal{A}$ is the training set, ξ is the pattern in input, and $\mathbf{w}_{s1}$ is the reference vector associated to winner unit s_1.

Using these maps to control different kinematic chains, that is, different robots, we will expect to not modify neither the architecture (number of maps, connections schemes, etc.) of the neural controller nor the learning procedures nor the learning equations. The only change we expect from robot to robot is the different way the growing process of the map will be performed according to the kinematic and mechanical parameters of the controlled robots. The output of these maps is given by the equation

$$\Phi(\xi) = \frac{\mu \mathbf{w}_{s_1} + \sum_{i \in N_{s_1}} (\nu \mathbf{w}_i)/|N_{s_1}|}{\| \xi - \mathbf{w}_{s_1} \|^2} \tag{26.3}$$

where ξ is the input pattern to the map, μ and ν are appropriate constants, $\mathbf{w}_{s1}$ is the weight vector associated with the winner unit, and N_{s1} is the set of direct topological neighbors of the winner unit.

Finally, Figure 26.3 illustrates the overall picture of the neural model that we propose. The vision map (VM) receives an appropriate visual coding concerning the target and the end-effector positions. Then, according to the Eq. (26.3), a winner population (the winner unit and its neighbors) representing the target and a winner population for the end effector will be activated.

Defining c_e as the coding for the end-effector position and c_t for the target, the VM output is defined as

$$\mathbf{v} = [\Phi(c_e), \Phi(c_t) - \Phi(c_e)] \tag{26.4}$$

where $\Phi(c_t) - \Phi(c_e)$ is the neural representation in the VM of the desired direction vector that starts from the end-effector position and points to the target position . The output of the proprioceptive map (PM) is computed in the same way and it produces an appropriate proprioceptive end-effector representation. Then, the sensorimotor map (SMM) receives as input both the vision and proprioceptive representation of the end effector and the direction vector, producing as output the set of weights belonging to the winner unit for that input according to Eq. (26.3). Regarding generation and learning of the arm movements, we use almost the same computational method proposed by the DIRECT model. In this

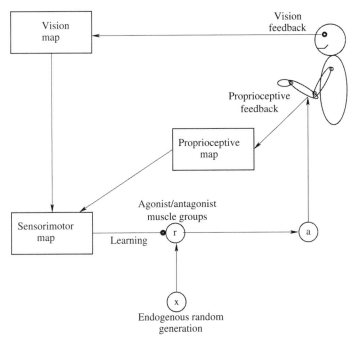

Figure 26.3 The proposed neural model. During the learning phase a random activation of the muscles is generated by the endogenous random generator (ERG) similarly to the approach proposed by the DIRECT model [19].

respect, r is a population of units devoted to the generation of the joint movements; their number is equal to the DOF of the manipulator multiplied by 2 according to the use of a pair of muscle (agonist and antagonist) for each joint. Each unit of **r** receives connections from all the SMM units. Only the weights of the winner unit and those of its neighbors are allowed to learn in order to move the end effector toward the target. The adaptive weights between SMM cell k and the unit i are modified according to the following learning equation (Grossberg's [30] "outstar" law):

$$\frac{d}{dt}z_{ki} = \gamma c_k(-\delta z_{ki} + r_i) \tag{26.5}$$

where γ is a small learning rate parameter and δ is a decay rate parameter, while the output of the units of r is given by the equation

$$r_i = x_i + \sum_k c_k z_{ki} \qquad i = 1, \dots, n \tag{26.6}$$

where x_i is the ERG activation (only present during the learning phase) and z_{ki} is the adaptive weights from cell k in SMM to unit i of **r**; c_k is 1 if the cell k in SMM is activated and 0 if it is not.

The a_i cell has the task of calculating the joint rotation displacement according to the activation of the r units that are connected them. The updating rule used is

$$\frac{d}{dt}a_i = \epsilon(r_i^E - r_i^I)g(r_i^E, r_i^I, \psi_i) \tag{26.7}$$

where a_i is the angle command for joint i, r_i^E and r_i^I are the corresponding excitatory and inhibitory r cell activities, respectively, and ϵ is an integration rate parameter, while the g function is defined as

$$g(e, i, \psi_i) = \begin{cases} \psi_{\max} - \psi_i & \text{if} \quad (e - i) \geq 0 \\ \psi_i & \text{if} \quad (e - i) < 0 \end{cases} \tag{26.8}$$

where ψ_i is the angle of the joint corresponding to a_i and $\psi_{\max}$ is the maximum angle of this joint.

This rule regards the "length" of the muscles [2] as well as their viscoelastic characteristic. Learning in our model is achieved through autonomously generated repetition of an action–perception cycle, which produces the associative information needed to learn coordination visuomotor during a motor babbling phase; in fact, the model, by r cell activities, endogenously generates movement commands that activate the correlate visual, proprioceptive, and motor information used to learn its internal coordinate transformations.

26.5 EXPERIMENTAL RESULTS

The implemented system was tested and evaluated through experimental trials aiming at verifying the learning capabilities and motor coordination functionality. The experimental scenario includes:

1. A simulator of a pan-tilt vision system
2. A 2D robot arm simulator
3. The 6-DOF PUMA 560 (Unimation, Sewickley, PA) and the 8-DOF DEXTER (SM Scienzia Machinale srl, Pisa, Italy) robot arms.

26.5.1 Simulator of Pan-Tilt Vision System

The vision system is required to focus both the target and the end effector by providing the vision feedback as the input for the VM. According to the results described in [31] the human vision system has a double functionality. The first consists in recognition tasks and the other is involved in visuomotor tasks. In particular, inside the inferior temporal cortex the middle temporal (MT) area contains a column organization of neurons devoted to represent the binocular disparity. Starting from these results, we simulate a pan-tilt vision system composed of two cameras able to move independently. The task of the vision system is to focus both well-defined relevant points of the end effector and the target (avoiding ambiguities) and after that to provide as its output the angular positions of the pan-tilt system of each camera. This is in accordance with the well-known triangulation method used to compute the distance between such vision systems and a target. To better simulate the encountered problems using a real vision system, we add a Gaussian noise to the sensed positions simulating shadows or environmental illumination changes.

26.5.2 Two-Dimensional Robot Simulator

The 2D robot simulator consists of a cinematic chain of three links as Figure 26.4 shows.

The lengths of the links and the joint ranges are $L_1 = 280$ mm, $L_2 = 280$ mm, $L_3 = 160$ mm and $30° \leq \theta_1 \leq 240°$, $30° \leq \theta_2 \leq 180°$, $30° \leq \theta_3 \leq 190°$, respectively.

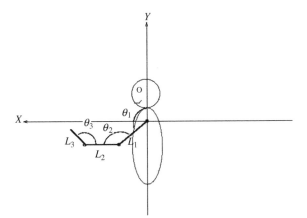

Figure 26.4   Planar 2D simulated robot arm with three links.

The performance of the proposed model is fully comparable with those pointed out by Bullock et al. [19] in all the same experimental scenarios (normal reaching, clamped joint, visual shift, and tool and blind reaching). Two graphs, showing the behavior of the error distance and the trajectory of the joints during a normal reaching trial, have been plotted using simulation output. The monotonic trend shown by the graph in Figure 26.5a is a general behavior found in all the trials.

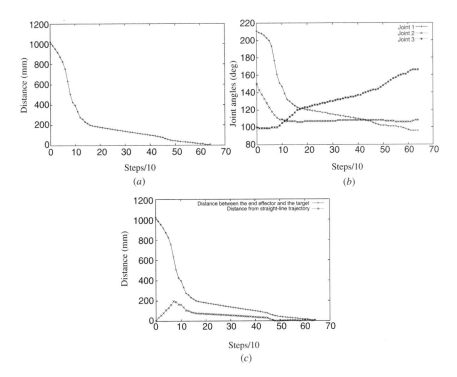

Figure 26.5   Two-dimensional simulator: (a) distance in normal reaching task; (b) joint trajectories in normal reaching task (3 DOFs); (c) comparison between graph of distance between end effector and target and graph of distance from straight-line trajectory during normal reaching task.

TABLE 26.1 Cardinalty of Maps: 2D Simulator

	Vision map	Proprioceptive map	Sensorimotor map
Proposed model	2,369	1,333	3,357
DIRECT model	—	15,625	10,290

We can underline how the monotonic trend and the absence of notable oscillations ensure a linear movement (straight-line trajectory) of the end effector toward the target, which corresponds to a minimal waste of energy according to the findings in [32]. Figure 26.5c shows the graph of the distance between the end effector and the target in comparison with the graph of the distance from the straight-line trajectory which links the initial and the target positions.

In Table 26.1 we report a comparison between the cardinality of the DIRECT maps and the proposed neural model.

Note that the VM is not present in the DIRECT model and the SMM corresponds, with some differences, to the DIRECT PDM. Looking at Table 26.1 it is evident that the number of network units created when the growing approach is used is reduced with respect to the case of a priori cardinality imposed by the DIRECT model.

26.5.3 PUMA 560 and DEXTER Robot Arms

The PUMA 560 robot arm is an industrial robot with 6 DOFs equipped with motors mounted on the joint rotation axes, while the DEXTER arm is an anthropomorphic 8-DOF robot arm in which the joints are cable actuated by motors located on the first link of the robot (for a complete description of the DEXTER arm see [33]). Due to the strong differences in the mechanical transmission systems in the elasticity of the actuation systems and in their internal friction the two robots clearly need different control strategies. Applying the same proposed model described here we expect different map cardinalities in order to reflect the differences between the two robots. Figures 26.6a,b show the performance of the system in terms of the graph of the error distance from the target and the joint trajectories in a reaching with a tool task involving the PUMA arm. Figure 26.6c shows the trend of the distance between the end effector and the target in comparison with the trend of the distance from the straight-line trajectory.

Moreover, Figure 26.7a highlights a monotonic trend behavior during a normal reaching task with the DEXTER manipulator and Figure 26.7c shows the graph of the distance between the end effector and the target in comparison with the trend of the distance from the straight-line trajectory. Table 26.2 shows the number of units created for the two robots.

Focusing on Table 26.2 it is possible to note a number of units in the case of the DEXTER robot greater than in the case of the PUMA robot. The difference consists in the wider workspace of the DEXTER arm with respect to that of the PUMA arm. Comparing also Tables 26.1 and 26.2 we can point out that the model always requires less units for the learning in the 3D space than in the case of the DIRECT model, where only the 2D case was present (3 DOFs). Figures 26.8a,b,c show some pictures of the PUMA robot during attempts to reach targets using a tool, while Figures 26.8d,e,f show some pictures of the DEXTER robot during the attempts to reach targets in normal reaching tasks. As illustrated in Figures 26.8c,f, the final reached positions differ from the initial ones in terms of

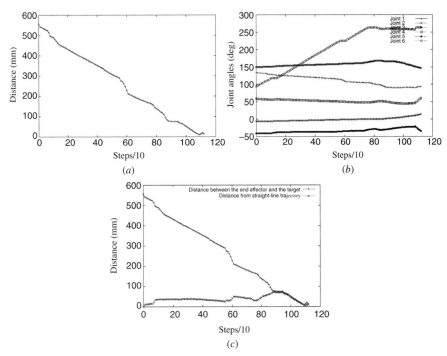

Figure 26.6   PUMA manipulator: (*a*) distance in reaching with tool; (*b*) joint trajectories in reaching with tool (6 DOFs); (*c*) comparison between graph of distance between end effector and target and graph of distance from straight-line trajectory during reaching with tool.

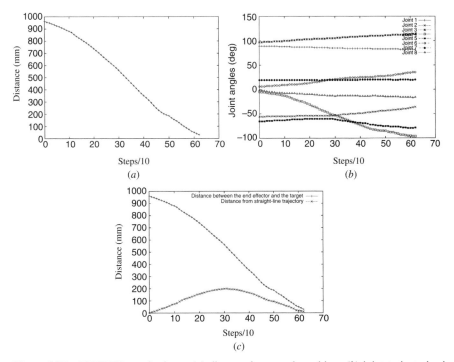

Figure 26.7   DEXTER manipulator: (*a*) distance in normal reaching; (*b*) joint trajectories in normal reaching (8 DOFs); (*c*) comparison between graph of distance between end effector and target and graph of distance from straight-line trajectory during normal reaching task.

TABLE 26.2   Cardinalty of Maps: Real Robot

	Vision map	Proprioceptive map	Sensorimotor map
PUMA 560 (6 DOFs)	10,970	6,063	9,708
DEXTER (8 DOFs)	15,004	6,580	11,018

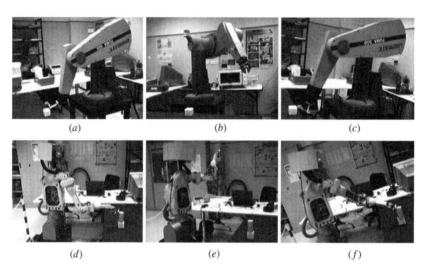

Figure 26.8   Real robot. (*a*) Required final position, (*b*) starting point posture, and (*c*) final reached position with PUMA manipulator (Unimation, PA). (*d*) Required final position, (*e*) starting point posture, and (*f*) final reached position with DEXTER manipulator (SM Scienzia Machinale srl, Pisa, Sewickley, Italy).

orientation (not considered in this work) due to the redundancy of the manipulators with respect to the reaching task in 3D space.

26.6   CONCLUSION

In this chapter, the current state of the art in the field of sensorimotor coordination models inspired from neuroscience findings in primates and humans has been analyzed. A case study of the design and development of a novel neurocontroller, inspired by the DIRECT model, has been proposed by the authors. The main improvement introduced with respect to the DIRECT model is the integration of the visual and proprioceptive information using a GNG approach. This has permitted us to simplify the model architecture adopting the same computational structure in manipulating different sensory modalities; it has also permitted a drastic reduction in the number of required units compared to the number of nodes obtained by using the DIRECT model. These neural maps are able to grow, preserving topological relations, in accordance with the kinematic proprieties of the robot arm during the motor babbling phase. Experimental results obtained by simulation and trials on two different types of robotic arms show the effectiveness of the proposed model, pointing out the importance of using growing neural network topological preservation paradigms in developing sensorimotor coordination in robotic systems. Particularly, the cardinalities of each map have been demonstrated to be lower, both in

3D space and 2D space, when compared to those proposed by the DIRECT model considering the 2D case only. The neural model has been shown to control real different complex robotic systems with a comparable performance producing no modifications neither in the model nor in the learning equations. Future developments will be aimed at using the same model to learn and perform other, more complex robotic tasks, such as explorative complex procedures for manipulative tasks which require the integration of tactile and force feedback. Finally, the model will be refined so to take into account hand orientation constraints and dynamic aspects, such as nonlinear muscle models and direct control of the joint torques involving feedforward control schemes.

ACKNOWLEDGMENTS

Part of the work presented in this chapter has been supported by the European Commission in the Fifth Framework Programme, within the IST (Information Society Technology) FET (Future and Emerging Technologies) PALOMA (Progressive and adaptive learning for object manipulation: a biologically inspired multi-network architecture) Project, under contract No. IST-2001-33073. The authors wish to acknowledge the fruitful collaboration with the project partners.

REFERENCES

1. PIAGET, J.: 1976, *The Grasp of Consciousness: Action and Concept in the Young Child*, Cambridge, MA: Harvard University Press.
2. KANDEL, E. R., SCHWARTZ, J. H., AND JESSELL, T. M.: 2000, *Principles of Neural Science*, 4th ed., New York: McGraw-Hill.
3. MATARIC, M. J., ZORDAN, V. B., AND WILLIAMSON, M. M.: 1999, Making complex articulated agents dance, *Autonomous Agents and Multi-Agent Systems* 2(1), 23–44.
4. MIWA, H., OKUCHI, T., TAKANOBU, H., AND TAKANISHI, A.: 2002, Development of a new human-like head robot we-4, in *Proceedings of IROS 2002, 2002 IEEE/RSJ International Conference on Intelligent Robots and Systems*, vol. 3, IEEE Press EPFL, Lausanne, Switzerland, pp. 2443–2448.
5. BROOKS, R., BREAZEAL, C., MARJANOVIC, M., SCASSELLATI, B., AND WILLIAMSON, M.: 1999, The Cog Project: Building a humanoid robot, in *Computation for Metaphors, Analogy and Agents*, C. NEHAMIV (Ed.), Lecture Notes in Artificial Intelligence, 1562. New York, Springer, 52–57.
6. GROSSBERG, S. AND PAINE, R. W.: 2000, A neural model of cortico-cerebellar interactions during attentive imitation and predictive learning of sequential handwriting movements, *Neural Networks* 13(8/9), 999.
7. FLASH, T.: 1984, The coordination of arm movements: An experimentally confirmed mathematical model, Technical Report AIM-786, MIT.
8. MASSONE, L. L. E. AND MYERS, J. D.: 1996, The role of plant properties in arm trajectory formation: A neural network study. *IEEE Trans. Sys. Man Cybernetics* 26(5), 719–732.
9. TANJI, J. AND EVARTS, E. V.: 1976, Anticipatory activity of motor cortex neurons in relation to direction of an intended movement, *Neurophysiology* 39(5), 1062–1068.
10. GEORGOPOULOS, A.: 1996, On the translation of directional motor cortical commands to activation of muscles via spinal interneuronal systems, *Cognitive Brain Research* 3, 151–155.
11. JACOBSON, M.: 1969, Development of specific neuronal connections, *Science* 163(867):543–547.
12. SCOTT, S. H.: 1997, Comparison of onset time and magnitude of activity for proximal arm muscles and motor cortical cells before reaching movements, *Journal of Neurophysiology* 77, 1016–1022.
13. BIZZI, E. AND MUSSA-IVALDI, F.: 2000, *Toward a Neurobiology of Coordinate Transformations, The Cognitive Neuroscience*, 2nd ed., Cambridge, MA: MIT Press.
14. JOHNSON, M., MASON, C., AND EBNER, T.: 2001, Central processes for the multiparametric control of arm movements in primates, *Current Opinion in Neurobiology* 11, 684–688.
15. KARNIEL, A. AND INBAR, G. F.: 1997, A model for learning human reaching movements, Biological Cybernetics 77(3):173–183.
16. MUSSA-IVALDI, F. A., GISZTER, S. F., AND BIZZI, E.: 1994, Linear combinations of primitives in vertebrate motor control, *Proceedings of the National Academy of Sciences USA* 91(16), 7534–7538.

17. GROSSBERG, S. AND KUPERSTEIN, M.: 1989, *Neural Dynamics of Adaptive Sensory-Motor Control*, Expanded Edition, Elansford, NY, Pergamon Press.

18. BURNOD, Y., BARADUC, P., BATTAGLIA-MAYER, A., GUIGON, E., KOECHLIN, E., FARRAINA, S., LAQUANITI, F., AND CAMINITI, R.: 1999, Parieto-frontal coding of reaching: An integrated framework, *Experimental Brain Research* 129(3), 325–346.

19. BULLOCK, D., GROSSBERG, S., AND GUENTHER, F. H.: 1993, A self-organizing neural model of motor equivalent reaching and tool use by a multijoint arm, *Journal of Cognitive Neuroscience* 5(4), 408–435.

20. LIN, S. AND GOLDENBERG, A. A.: 2001, Neural network control of mobile manipulators, *IEEE Transactions on Neural Networks* 12(5), 1121–1133.

21. PATINO, H. D., CARELLI, R., AND KUCHEN, B.: 2002, Neural networks for advanced control of robot manipulators, *IEEE Transactions on Neural Networks* 13(2), 343–354.

22. KOHONEN, T.: 1997, *Self-Organizing Maps*, 2nd ed., Berlin, Springer-Verlag.

23. LEONI, F., GUERRINI, M., LASCHI, C., TADDEUCCI, D., DARIO, P., AND STARITA, A.: 1998, Implementing robotic grasping tasks using a biological approach, in *Proceedings of the International Conference on Robotics and Automation*, IEEE Press, Leuven, Belgium, pp. 16–20.

24. KUPERSTEIN, M. AND RUBINSTEIN, J.: 1989, Implementation of an adaptive neural controller for sensory-motor coordination, *IEEE Control Systems Magazine* 9(3), 25–30.

25. WALTER, J. A. AND SHULTEN, K. J.: 1993, Implementation of self-organizing neural networks for visuo-motor control of an industrial robot, *IEEE Transactions on Neural Networks* 4(1), 86–95.

26. MARSLAND, S., SHAPIRO, J., AND NEHMZOW, U.: 2002, A self-organizing networks that grows when required, *Neural Networks* 15(8/9), 1041–1058.

27. FRITZKE, B.: 1994, Growing cell structures a self-organizing network for unsupervised and supervised learning, *Neural Networks* 7(9), 1441–1460.

28. CARLEVARINO, A., MARTINOTTI, R., METTA, G., AND SANDINI, G.: 2000, An incremental growing neural network and its application to robot control, paper presented at Intl. Joint Conference on Neural Networks, Como, Italy, pp. 24–27.

29. WINTER, M., METTA, G., AND SANDINI, G.: 2000, Neural-gas for function approximation: A heuristic for minimizing the local estimation error, paper presented at International Joint Conference on Neural Network (IJCNN 2000), Como, Italy, pp. 24–27.

30. GROSSBERG, S.: 1974, Classical and instrumental learning by neural networks, *Progress in Theoretical Biology* 3, 51.

31. GOODALE, M. A. AND D. M. A.: 1992, Separate visual pathways for perception and action, *Trends in Neurosciences* 15, 20–25.

32. JORDAN, M. I. AND WOLPERT, MILMER, D. A. 1999, Computational motor control, in M. GAZZANIGA, Ed., *The Cognitive Neurosciences*, 2nd ed. Cambridge, MA: MIT Press.

33. ZOLLO, L., SICILIANO, B., LASCHI, C., TETI, G., AND DARIO, P.: 2002, Compliant control for a cable-actuated anthropomorphic robot arm: An experimental validation of different solutions, paper presented at IEEE International Conference on Robotics and Automation, Washington, DC, pp. 1836–1841.

CHAPTER 27

MUSCLE SYNERGIES FOR MOTOR CONTROL

Andrea d'Avella and Matthew Tresch

27.1 INTRODUCTION

Biological motor control is an extraordinary achievement of evolution, yet it is still a largely unexplained phenomenon. Movements that we perform efficiently and effortlessly require controlling and coordinating a large number of interacting degrees of freedom. Even a simple movement like reaching for an object involves activating tens of different muscles, with specific amplitudes and timings, to generate the appropriate torque profiles at the shoulder, elbow, and wrist joints. Given the complexity of the dynamics of an articulated mechanical system like the arm, specifying the proper torque profiles and selecting the suitable muscle activation patterns are a challenging computational problem. How the central nervous system (CNS) solves this problem is a central question in the study of the neural control of movement [1] and a possible source of inspiration for neurorobotic design.

A long-standing hypothesis is that the CNS simplifies the control of movement through a hierarchical and modular architecture [2–6]. At the lowest level of the hierarchy, muscle recruitment might be controlled by a few functional units, thereby reducing the dimensionality of the output space. Higher levels in the hierarchy might recruit and flexibly combine these output modules to control a variety of different movements.

In this chapter we will consider a specific model for the organization of a set of output modules. According to this model, groups of muscles are recruited coherently in space or time, that is, as *muscle synergies*. We will review here a number of methods to identify muscle synergies from a collection of muscle activity patterns observed over a variety of different movements. Using simulations, we will characterize the performance of these methods. Finally, we will show the results of the analysis of experimental data.

27.2 MUSCLE SYNERGIES

One way in which the CNS might simplify the task of controlling many degrees of freedom is by organizing muscle synergies. The basic idea (Fig. 27.1a) is that of reducing the dimensionality of the output space by grouping control signals into a set of output modules. The output signals for a biological controller are muscle activations and an output module corresponds to a muscle synergy—the coherent activation in space or in time of a group of muscles. Such a modular architecture would produce an adequate controller if the combinations of the output modules can generate all the output patterns

Handbook of Neural Engineering. Edited by Metin Akay

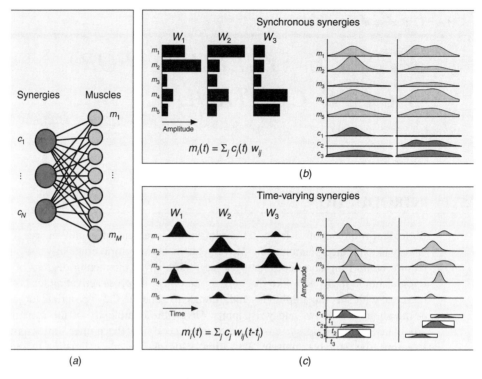

Figure 27.1 Synergy-based models for generation of muscle patterns. (*a*) A set of independent modules (*synergies*) recruits different combinations of output channels (*muscles*). The actual value of each output channel (m_i) is given by the sum of the contributions of each module for that channel, each scaled by a coefficient (c_j). (*b*) Simulated example of two different patterns with five muscles constructed by the combinations of three synchronous synergies. The profiles below each pattern correspond to the time course of the amplitude coefficients [$c_j(t)$] for each synergy. (*c*) Example of combination of three time-varying synergies to construct two distinct muscle patterns. The height of the rectangles below the muscle patterns is proportional to the amplitude coefficient (c_j) of each synergy; the horizontal position of the rectangles corresponds to the time interval spanned by each synergy resulting from the shifting of the synergy onset time by the onset delay (t_j) of each synergy. The profile within each rectangle shows the average time course of the muscles for the corresponding synergy.

necessary to control the movement repertoire of the system under consideration. Since the patterns associated with different movements might share a substantial amount of structure and since the movement repertoire might be limited, it seems realistic that a small number of modules might act as a basis for movement generation. Output patterns might share structure because different movements are constructed by similar components and because they are generated by the same biomechanical apparatus. The movement repertoire of a species might be limited because only a subset of all the movements that its biomechanical apparatus can produce might be actually behaviorally relevant. Finally, a modular architecture would represent an efficient implementation of a controller because it uses the same modules for different movements and has just enough modules for the movement repertoire that it has to control.

We introduce here two different models of muscle synergies to capture different types of structure in the muscle patterns that might be shared across movements: spatial structure—across muscles—and spatiotemporal structure—across both muscles and time.

27.2.1 Synchronous Synergies

A synchronous synergy represents an output module that activates a group of muscles with a fixed balance in their activation levels. We model the generation of a muscle pattern by the recruitment of N synchronous synergies as the linear combination of N vectors $\{\mathbf{w}_i\}$, $i = 1, \ldots, N$,

$$\mathbf{m}(t) = \sum_{i=1}^{N} c_i(t)\mathbf{w}_i \qquad (27.1)$$

where $\mathbf{m}(t)$ is a M-dimensional vector representing the activations of M muscles at time t; $c_i(t)$ a coefficient scaling the amplitude of the ith synergy at time t, and $\mathbf{w}_i$ the ith muscle synergy. Since muscle activation is a nonnegative quantity,[1] it is convenient to constrain both synergies and amplitude coefficients to nonnegative values.

Given a set of synergies, different muscle patterns are generated by choosing different temporal profiles for the amplitude coefficients of each synergy. Figure 27.1*b* illustrates this model with an example of the generation of two different patterns with the same three synergies. By construction, the synchronous synergies capture the underlying relationship, at each moment in time, among the amplitude of the various muscles across all the patterns.

If we sample the muscle patterns at discrete time intervals, the relationship between the muscle activations defined by this model can also be expressed as the existence of a factorization of the matrix of the muscle patterns as a product of a synergy matrix times a matrix of activation coefficients:

$$\mathbf{M} = \mathbf{W}\mathbf{C} \qquad (27.2)$$

where $\mathbf{M}$ has M rows and K columns (K total number of samples), $\mathbf{W}$ has M rows and N columns (N number of synergies), and $\mathbf{C}$ has N columns and K rows.

27.2.2 Time-Varying Synergies

A time-varying synergy represents an output module that activates a group of muscles with a specific and fixed spatiotemporal profile. The activation time course can be different across muscles and thus a time-varying synergy can represent an asynchronous as well as a synchronous activation profile. As with synchronous synergies, we model the generation of a muscle pattern by the recruitment of N time-varying synergies as the linear combination of the elements of each synergy. Since the muscle activation described by a time-varying synergy spans a certain time interval, we can shift in time the onset of each synergy before summing their elements at each point in time:

$$\mathbf{m}(t) = \sum_{i=1}^{N} c_i\mathbf{w}_i(t - t_i) \qquad (27.3)$$

where $\mathbf{w}_i(\tau)$ is the M-dimensional vector representing the activations of M muscles in the ith synergy at the time τ after the synergy onset, t_i is the onset delay for the ith synergy, and all other symbols are as in Eq. (27.1). Thus the generative model for time-varying synergies includes both an amplitude parameter and a temporal parameter. Comparing Eqs. (27.1) and (27.3), we notice that the temporal dependence in the muscle patterns

---

[1]In fact, muscles can only generate active force in one direction, shortening their contractile elements. Analogously, the neural drive to the muscles, represented by action potentials traveling along the axons of motoneurons from the spinal cord to the neuromuscular junction, can be quantified as a nonnegative firing rate.

derives from the scaling coefficients $[c_i(t)]$ in the case of synchronous synergies while it derives from the time course of the synergies $[\mathbf{w}_i(t - t_i)]$ in the case of time-varying synergies. Hence any regularity in the temporal structure of the muscle patterns has to be captured by a regularity in the scaling coefficients in the synchronous synergies model whereas it is captured by the structure of the synergies in the time-varying model. Moreover, given a set of synergies, to describe one muscle pattern, the synchronous model requires as many parameters $[c_i(t)]$ as the number of synergies times the number of time points while in the time-varying model the number of parameters (c_i and t_i) is equal to twice the number of synergies.

By shifting in time and scaling in amplitude a given set of time-varying synergies, we can generate a variety of muscle patterns with different time courses in the activation of each muscle. Figure 27.1c shows an example of two muscle patterns generated by combining the same three time-varying synergies with different recruitment parameters.

In discrete time, the ith time-varying synergy of duration T can be expressed as a matrix $\mathbf{W}^i$ with M rows and J columns, each representing the synergy activation vector at time t_j, $0 \leq t_j < T$ ($j = 1, \ldots, J$). We use the matrix $\mathbf{W} = [\mathbf{W}^1 \ \mathbf{W}^2 \cdots \mathbf{W}^N]$ with M rows and $J \times N$ columns to compactly represent a set of N synergies. If we introduce a time-shifting matrix $\mathbf{\Theta}_i[k, K]$ to align, by matrix multiplication of $\mathbf{W}$ with $\mathbf{\Theta}_i$, the first sample of the ith synergy with the kth sample ($2 - J \leq k \leq K$) of a muscle pattern (K samples long) and to truncate the synergy's samples shifted before the beginning or after the end of the pattern, we can rewrite Eq. (27.3) as

$$\mathbf{M} = \mathbf{W}\left(\sum_{i=1}^{N} c_i\mathbf{\Theta}_i[k_i, K]\right) = \mathbf{WH} \qquad (27.4)$$

27.3 METHODS TO IDENTIFY SYNERGIES

When trying to reverse engineer the design of a biological controller, we can tackle the problem using two complementary approaches. We can extract information about the controller's design from the regularities of its behavior and its observable output or we can gain information on the controller's architecture from knowledge of its internal structure. The latter approach may lead to stronger inferences on the controller's design than the former approach, especially when it allows testing causal models through the study of the effect of external interventions on the controller's components. This *structural* approach, however, requires accessing the physical substrate that implements the controller and understanding its mechanisms. For a biological system this means gaining access to the neural circuitry and decoding the neural representation of the relevant control variables and algorithms. Both these requirements are often quite difficult to meet in practice, and they often restrict the range and the naturalness of conditions in which the system can be studied. The *behavioral* approach, on the contrary, does not require direct access to the controller's implementation, only to the controller's output. This approach not only simplifies the testing of a large range of conditions but also directly benefits from this range. In fact, any inference based on the existence of invariant characteristics will be stronger if the set of conditions over which the invariance is observed is large.

The approach we advocate in the study of the neural control of movement is to characterize the regularities in the motor output by identifying a set of muscle synergies that can generate the muscle patterns observed during natural motor behaviors. The existence of a small set of generators for the entire muscle pattern repertoire of a species supports the hypothesis of a modular architecture of the controller. Moreover, the identification of muscle synergies from the regularities of the observed behavior can be instrumental in guiding a more systematic structural approach.

27.3.1 Decomposition Methods for Synchronous Synergies

One simple approach to identify synchronous synergies is to study the pairwise correlations among muscle activations. If two muscles are recruited with a fixed balance, we expect their activations to be correlated, and thus the observation of pairwise correlations indicates a synergistic activation. More generally, the recruitment of *one* synergy will generate muscle activities that lie on a one-dimensional subspace of the muscle activation space. The single vector that generates this subspace corresponds to a synchronous synergy. The problem with this correlation-based approach is that it will only work if the synergies are recruited one at a time, that is, only if one of the scaling coefficients $[c_i(t)$ in Eq. (27.1)] are nonzero at any one time. When different synergies activate the same muscles and several synergies are recruited simultaneously, the underlying correlations may be masked by the synergy combinations.

A more general approach for the identification of synchronous synergies consists in determining the generators of the subspace spanned by the synergy combinations. Any method that can factorize the muscle pattern matrix $\mathbf{M}$ as a product of a matrix $\mathbf{W}$ made by N synergy vectors times a coefficient matrix $\mathbf{C}$ [see Eq. (27.2)] provides a set of generators. Different methods can identify different generators since any invertible transformation of the generators will still span the same subspace. Classical methods like principal-component analysis (PCA) [7], factor analysis (FA) [8], or more recent methods like independent-component analysis (ICA) [9] differ on the assumptions about the nature of the noise in the pattern generation process, and they can all be used to obtain a linear decomposition. In case of nonnegative synergies and combination coefficients, we can use one of the recently introduced nonnegative matrix factorization (NMF) methods [10–12].

The NMF methods use an iterative algorithm to find a factorization of the muscle pattern matrix that minimizes a given cost function. We briefly present here the algorithm introduced by Lee and Seung [12] to minimize the cost function given by the reconstruction error E:

$$E^2 = \mathrm{trace}\{(\mathbf{M} - \mathbf{WC})^{\mathrm{T}}(\mathbf{M} - \mathbf{WC})\} \tag{27.5}$$

that is, the Euclidean distance between observed data and reconstructed data. The algorithm starts by initializing synergies ($\mathbf{W}$) and coefficients ($\mathbf{C}$) with random nonnegative values and proceeds by iterating two steps until convergence:

1. Given the synergies $\mathbf{W}$, update the coefficients $\mathbf{C}$ decreasing the error E, according to the rule

$$C_{ij} = C_{ij} \frac{(\mathbf{W}^{\mathrm{T}}\mathbf{M})_{ij}}{(\mathbf{W}^{\mathrm{T}}\mathbf{WC})_{ij}} \tag{27.6}$$

2. Given the coefficients $\mathbf{C}$, update the synergies $\mathbf{W}$ decreasing the error E, according to the rule

$$W_{ij} = W_{ij} \frac{(\mathbf{MC}^{\mathrm{T}})_{ij}}{(\mathbf{WCC}^{\mathrm{T}})_{ij}} \tag{27.7}$$

These multiplicative update rules allow for faster convergence than a gradient descent rule,[2] they automatically constraint synergies and coefficients to nonnegative values, and they do not require adjusting any convergence parameter.

Since the number of synergies N is a free parameter of the model, we need to determine its optimal value. Such a process of model order selection can be approached using many different approaches. One approach is to examine the amount of reconstruction error $E(N)$ as a function of the number of synergies N used in the extraction algorithm. The reconstruction error will decrease as the number of synergies increases and, ideally, in the absence of noise the actual number of synergies used to generate the data N^* could be easily determined as the smallest N for which $E(N) = 0$. In presence of noise, the reconstruction error for $N < N^*$ synergies has two components: one due to the noise and one due to the fact that we are reconstructing the data with less synergies than the number used to generate the data. Since for $N \geq N^*$ only the noise contributes to the reconstruction error, we expect a change in slope in the $E(N)$ curve at N^*. Thus, in principle we can use a change in slope of the $E(N)$ curve as a criterion for the choice of the optimal number of synergies.

27.3.2 Time-Varying Synergies Decomposition

The construction of muscle patterns by combinations of time-varying synergies is achieved using two operations: scaling in amplitude and shifting in time each synergy [see Eq. (27.3)]. Because of the time shift operation, the time-varying synergies used to construct the patterns are not the linear generators of a subspace in the muscle activation space as with synchronous synergies.[3] We cannot therefore use the same factorization methods described above to decompose the muscle pattern matrix.

We have recently introduced a novel algorithm to factorize a muscle pattern matrix into a nonnegative matrix representing a set of time-varying synergies and a matrix representing a set of recruitment parameters [see Eq. (27.4)] [13, 14]. Similar to the algorithm presented above for synchronous synergies, this algorithm uses the reconstruction error as a cost function and a multiplicative rule for updating scaling coefficients and synergies. Because of the additional time shift operation, at each iteration the time-varying algorithm updates not only synergies and amplitude scaling coefficients but also the onset delay times. After initializing N time-varying synergies with random nonnegative values, the algorithm is applied to a set of S muscle patterns, each pattern composed by K_s samples, and it iterates through the following steps until convergence:

(i) For each pattern $\mathbf{M}_s$, given the synergies $\mathbf{W}$, find N synergy onset times t_{is} using the following matching procedure:

(a) Set $\mathbf{R}^{(0)} = \mathbf{M}_s$ and $n = 0$; compute $\mathbf{\Psi}[i, j] = \mathbf{W}\,\mathbf{\Theta}_i[j, K_s]$, that is, the synergy $\mathbf{W}^i$ with the onset shifted at the sample j, for every $i \in [1, N]$ and every $j \in [2 - J, K_s]$; and normalize each $\mathbf{\Psi}[i, j]$ using the norm $\|\mathbf{\Psi}\|^2 = \mathrm{trace}(\mathbf{\Psi}^{\mathsf{T}}\mathbf{\Psi})$.

---

[2] $\mathbf{W} \leftarrow \mathbf{W} - \mu\nabla_{\mathbf{W}}E^2$, where μ is an adjustable learning rate parameter.

[3] If we consider all possible M-muscle, K-sample patterns, they form a space of $M \times K$ dimensions. The patterns generated by the combinations of N synchronous synergies constitute an $(N \times K)$-dimensional subspace of the muscle patterns space; that is, at each point in time we have an independent N-dimensional space. Once a J-sample long time-varying synergy is shifted at a particular onset time, it can be seen as an $(M \times K)$-dimensional vector. Such a vector will have at most $M \times J$ nonzero elements and the $K + J - 1$ vectors obtained with all possible onset time shifts of the same synergy will be in general linearly independent. Thus a single time-varying synergy corresponds to a set of $K + J - 1$ basis vectors.

(b) Compute the scalar products between $\mathbf{R}^{(n)}$ and $\mathbf{\Psi}[i, j]$,

$$\Phi_{ij}^{(n)} = \text{trace}(\mathbf{M}_s^{\mathrm{T}}\mathbf{\Psi}[i,j]) \tag{27.8}$$

(c) Select the synergy (i^n) and the onset sample (j^n) with the maximum scalar product and set the onset delay coefficient (t_{is}) for the selected synergy accordingly.

(d) Set $n \leftarrow n + 1$: If $n < N$, update the residual $\mathbf{R}^{(n+1)} = \mathbf{R}^{(n)} - \mathbf{\Psi}[i^n, j^n]$ and set $\mathbf{\Psi}[i^n, j] = 0$ for every j and return to (b); stop otherwise.

(ii) For each pattern $\mathbf{M}_s$, given the synergies $\mathbf{W}$ and the onset times t_{is}, update the scaling coefficients c_{is} using

$$c_{is} \leftarrow c_{is} \frac{\text{trace}(M_s^{\mathrm{T}}W\Theta_i(t_{is}))}{\text{trace}(H_s^{\mathrm{T}}W^{\mathrm{T}}W\Theta_i(t_{is}))} \tag{27.9}$$

(iii) Compute $\mathbf{H}$ [see Eq. (27.4)] and update the synergies using

$$W_{ij} \leftarrow W_{ij} \frac{\lfloor MH^{\mathrm{T}}\rfloor_{ij}}{[WHH^{\mathrm{T}}]_{ij}} \tag{27.10}$$

In step (i) the algorithm determines the onset delays (t_i) for each synergy and for each pattern using a nested matching procedure based on the cross correlation between the synergies and the data. In step (ii) the nonnegative scaling coefficients for each synergy and for each pattern (c_{is}) are updated using a multiplicative rule similar to the one used in the NMF algorithm [Eq. (27.6)] but taking into account the synergies' time shifting (through the matrix Θ_i). Finally in step (iii) the nonnegative time-varying synergies (W_{ij}) are updated with a multiplicative rule again similar to the corresponding NMF rule [Eq. (27.7)] but modified for the time shifting.

27.4 SIMULATIONS

We tested the extraction algorithms presented above on simulated data. For both synchronous and time-varying algorithms we constructed a data set combining four randomly generated synergies composed of random coefficients. We then extracted a set of four synergies from the simulated data set and tested the capability of the algorithms to recover the synergies and coefficients used to generate the data. We also characterized the performance of the algorithms when the simulated data were corrupted by noise and when the number of extracted synergies was different from the number of generating synergies.

27.4.1 Synchronous Synergies

Simulated data were generated by randomly selecting a set of four synergies, each synergy defining the activation levels of 12 muscles. These four synergies were weighted by activation coefficients drawn from an exponential distribution, and a predicted muscle activation pattern was determined according to Eq. (27.2). Gaussian noise was then added to each muscle, with the noise variance for each muscle set to be a fraction (0, 5, 10, 15, 20, 50%) of the variance of that muscle's activation. Data sets consisting of 1000 data points were used.

The NMF synergy extraction method described above was able to correctly identify the synergies which generated the simulated data set, even in the presence of relatively

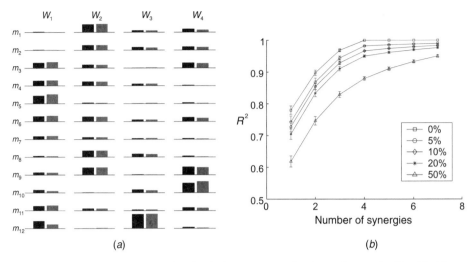

Figure 27.2   Synchronous synergies simulation. (*a*) Example of set of synergies identified by NMF algorithm (in grey) from 12-dimensional (m_{1-12}) data set generated by four muscle synergies (shown in block) corrupted by Gaussian noise with variance equal to 20% of each muscle's individual variance. (*b*) Amount of explained variance (R^2) as function of number of synergies used in synergy extraction method for differing amounts of noise (separate lines). For all but the highest amount of noise, the slope of each curve changes abruptly at four synergies, correctly identifying the number of synergies used to generate the data set.

high levels of corrupting noise. In Figure 27.2*a*, we show the synergies identified using the NMF synergy extraction algorithm applied to a simulated data set corrupted by noise levels equal to 20% of the variance of individual muscle activations. As can be seen in the figure, the identified synergies were very similar to the generating synergies. Table 27.1 summarizes the similarity between synergies identified by NMF and those actually used to generate simulated data sets in the presence of different amounts of corrupting noise. As can be seen in the table, even with substantial amounts of noise, the extraction algorithm identified the correct synergies well, with a mean similarity greater than 0.9 (an average similarity of 0.84 in the worst case). The correlation between identified and generating activation coefficients was also greater than 0.9 for all noise levels but

TABLE 27.1   Comparison of Extracted and Generating Synergies for Four Synchronous Synergies*

Noise level (%)	R^2	s_{mean}	s_{min}	s_{max}	r^c_{mean}	r^c_{min}	r^c_{max}
0	1.000	0.991	0.978	0.998	0.983	0.967	0.996
5	0.982	0.989	0.973	0.998	0.962	0.938	0.978
10	0.967	0.986	0.969	0.997	0.939	0.915	0.960
20	0.950	0.977	0.945	0.995	0.913	0.869	0.945
50	0.879	0.917	0.844	0.966	0.771	0.677	0.845

*Simulated data sets were generated from a combination of four randomly generated non-negative synergies. Each synergy was weighted by a non-negative activation coefficient and corrupted by Gaussian Noise. The noise level was manipulated to be a fraction of the variance of each muscle's final activation coefficient. Similarity between synergies was calculated as the scalar product between each synergy. Similarity between activation coefficients was taken as the correlation between the activation coefficients. Mean, maximum, and minimum values for each similarity measure are also shown.

the highest: With noise equal to 50% of the variance of each recorded muscle, the mean correlation coefficient fell to 0.77. This low correlation with high noise likely reflects the fact that activation coefficients for synchronous synergies can be used by the algorithm to fit the noise in individual data samples. In contrast, the synergy is held constant across different data samples and so might be expected to be more robust to noise. These results suggest that the NMF algorithm is capable of extracting the correct set of synergies and activation coefficients used to generate a simulated data set, even in the presence of substantial noise.

We also examined the results of the extraction algorithm using different numbers of synergies. In all cases for this analysis, the number of synergies used to generate the data was maintained at 4, while the number of synergies used in the extraction algorithm was varied. As seen in Figure 27.2b, as the number of synergies used in the extraction was increased, the average R^2 value also increased. When four synergies were used in the extraction, the same number used to generate the data, the slope of the R^2 curve changed abruptly for the first four levels of noise. As mentioned above, in these cases this change in slope in the R^2 plot could be used to determine the correct number of synergies to use in the extraction algorithm. In the case with the largest noise (50% of variance), however, there was no discernible change in the R^2 slope for any particular number of synergies. Thus, only at these high levels of corrupting noise was it impossible to determine the correct number of synergies in these simultated data sets.

27.4.2 Time-Varying Synergies

We tested the time-varying extraction algorithm using two different types of generating synergies. We generated a set of four synergies, each composed of 20 samples for 12 muscles, either directly sampling from a uniform distribution in the interval [0 1] or using Gaussian functions with centers and widths randomly chosen from uniform distributions and with exponentially distributed amplitudes. We constructed a simulated data set composed of 25 sections, each 40 samples long, by combining the four synergies with random exponentially distributed amplitude coefficients and random uniformly distributed onset delays. Finally, we added to each data sample, independently for each muscle, Gaussian noise with a variance equal to a fraction of the data variance for that muscle ranging from 0 to 50%.

When we extracted a set of four synergies from the simulated data set we found that even in the presence of a substantial amount of noise the algorithm was able to recover both synergies and coefficients used in the data generation. Figure 27.3a shows an example of generating and extracted synergies obtained in one simulation run with noise added to the data (20% of data variance). The extracted synergies (dashed line in Fig. 27.3a) were essentially identical to the generating synergies (solid line). In fact, the mean similarity between the best matching pairs of synergies in the two sets was 0.99.[4] Table 27.2 summarizes the results of the comparison between generating and extracted synergies in 10 simulation runs with different amount of noise for either Gaussian or uniform random synergies. For each level of noise added to the simulated data we evaluated a number of indicators for the quality of the reconstruction of synergies and coefficients averaging the results of the 10 runs. This table shows that even when 50% of noise is added to the generated data, lowering the R^2 to 86%, the extracted synergies

---

[4]We quantified the degree of similarity between two time-varying synergies by computing the maximum of the scalar products of one synergy with a time-shifted version of the other synergy over all possible relative delays. We compared two synergy sets by matching the synergies in pairs, starting from the pair with the highest similarity, removing that pair, and continuing by matching the remaining synergies.

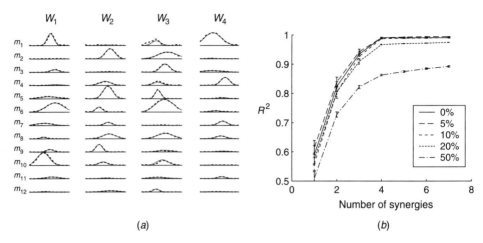

(a)   (b)

Figure 27.3   Time-varying synergy simulation (Gaussian synergies). (*a*) Example of comparison of four generating synergies (solid line) and extracted synergies (broken line) in case of data generated by combinations of Gaussian synergies with 20% noise added. (*b*) The R^2 for reconstruction of data generated by four synergies with variable number of extracted synergies for different amounts of added noise (separate lines).

are very similar to the generating ones (in the worst case the similarity is on average above 0.9 with Gaussian synergies and above 0.98 with uniform synergies), the respective amplitude coefficients are highly correlated (average correlation coefficient above 0.9 in the worst case with Gaussian synergies and above 0.96 with uniform synergies), and the

TABLE 27.2   Comparison of Extracted and Generating Synergies for Four Time-Varying Synergies

Noise level (%)	R^2	s_{mean}	s_{min}	s_{max}	r^c_{mean}	r^c_{min}	r^c_{max}	ΔT_{mean}	ΔT_{max}	ΔT_{min}
				Gaussian Synergies						
0	0.991	0.997	0.993	0.999	0.955	0.907	0.999	1.215	2.202	0.225
5	0.990	0.992	0.975	0.999	0.985	0.963	0.999	0.976	1.585	0.457
10	0.988	0.997	0.994	0.999	0.995	0.987	0.999	0.883	1.480	0.280
20	0.967	0.987	0.963	0.998	0.992	0.980	0.998	1.575	2.582	0.794
50	0.863	0.962	0.908	0.993	0.964	0.906	0.994	3.186	5.074	1.698
				Uniform Synergies						
0	0.996	0.998	0.996	1.000	0.986	0.955	0.998	0.823	1.652	0.293
5	0.990	0.998	0.995	0.999	0.988	0.977	0.995	1.003	1.761	0.448
10	0.989	0.999	0.998	0.999	0.994	0.986	0.999	0.816	1.281	0.311
20	0.964	0.996	0.992	0.998	0.987	0.964	0.997	1.370	2.365	0.646
50	0.840	0.987	0.981	0.991	0.979	0.964	0.990	2.242	3.248	1.369

Four synergies were extracted from data generated with four gaussian or uniform synergies. Averages over 10 simulation runs were computed for: The fraction of data variation explained by the extracted synergies (R^2), the mean similarity between the four pairs of generating and extracted synergies (s_{mean}), the similarity between the least similar pair (s_{min}), the similarity between the most similar pair (s_{max}), the mean correlation coefficient between the generating and extracted amplitude coefficients for the four synergies (r^c_{mean}), the correlation coefficient for the pair of generating and extracting synergies with the minimum correlation (r^c_{min}), the correlation coefficient for the pair with the maximum correlation (r^c_{max}), the mean time difference between the onset each pair of generating and extracted synergies (ΔT_{mean}), the time difference for the pair with the maximum difference (ΔT_{max}) and for the pair with the minimum difference (ΔT_{min}).

difference in the onset delay coefficient is small (5 time samples at most with Gaussian synergies and 3 at most with uniform synergies). These results clearly indicate that the extraction algorithm is capable of recovering the synergies and the coefficients used to generate the data even in the presence of substantial amounts of noise.

In the case of the extraction of a set of synergies with a number of elements different from the number of synergies used to generate the data, we found that when up to 20% of noise was added to the data, the slope of the curve representing R^2 as a function of the number of extracted synergies showed a sharp change in slope at four, corresponding to the number of generating synergies. Only with the highest level of noise (noise variance equal to 50% of the data variance) was the inflection on the R^2 curve not clearly distinguishable.

27.5 SYNERGY EXTRACTION FROM MUSCLE ACTIVITY PATTERNS

We applied the two extraction algorithms discussed above to the study of the modular organization of the motor output in the vertebrate motor system. First, we considered withdrawal reflexes of frogs with the spinal cord surgically isolated from the rest of the CNS. As in most vertebrates, the isolated frog spinal cord can generate a rich variety of reflexive behaviors. Thus the spinal frog is an ideal preparation to investigate the role of the spinal cord circuitry in the construction of the motor output. The aim of this study was to characterize the spatial structure of the muscle patterns and we used the synchronous extraction algorithm for this analysis. We then examined a natural behavior of the intact frog, defensive kicking. Since the muscle patterns observed in intact frogs appeared to have specific temporal characteristics, in this second study we investigated the spatiotemporal structure of the patters using the time-varying extraction algorithm.

27.5.1 Synchronous Synergies Extracted from Spinal Frogs

Withdrawal reflexes were evoked in the spinalized frog from scratching different regions of the hind-limb skin surface [10]. Stimuli were applied systematically at cutaneous sites ranging from the back of the calf to the back of the foot to the front of the foot and the front of the knee. This variation in stimulation site evoked a range of different muscle activation patterns, recorded in the EMG activity of a set of 9 hind-limb muscles.

We examined whether this range of EMG activity could be explained as the combination of a small number of muscle synergies using the synchronous synergy extraction method (NMF) described above. Figure 27.4 shows the R^2 value as a function of differing numbers of synergies used in the extraction algorithm for data obtained from one frog. As can be seen in the figure, there was a sharp change in the slope of the curve at four synergies, at which point the extraction algorithm explained more than 90% of the variance in the recorded data set. This observation suggested that this behavior might have been produced through the combination of four muscle synergies. Figure 27.5a shows the four synergies identified by the extraction algorithm, showing the balance of muscles for each of the nine muscles recorded. These synergies were very similar across different frogs, with a mean similarity greater than 0.9 between the extracted synergies. Figure 27.5b shows, for two frogs, the mean activation coefficient for each of the four extracted synergies used to reconstruct the muscle activation patterns measured

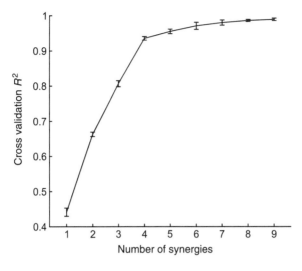

Figure 27.4 Amount of explained variance as function of number of synergies extracted using NMF for set of muscle activation patterns observed during withdrawal reflexes in spinalized frog. There was a clear change in slope at four synergies.

across different regions of the skin surface. For all synergies, there was a very similar variation in the activation levels of each synergy across the skin surface, especially for the first three synergies. Taken together, these results suggest that withdrawal reflexes in the spinalized frog might be well described as the flexible combination of a small number of muscle synergies.

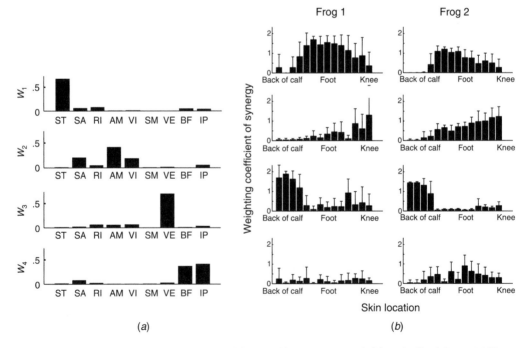

(a)   (b)

Figure 27.5 Synchronous synergies extracted from EMG patterns recorded in spinalized frogs. (a) Four synergies extracted from EMG patterns using NMF with four synergies. (b) Variation across skin surface of activation coefficients for four synergies extracted in two different frogs. The x axis indicates the surface of the hind limb moving from the back of the calf, down the heel to the back of the calf, up the front of the calf, and to the knee [10]. Reproduced with permission.

27.5.2   Time-Varying Synergies Extracted from Intact Frogs

We investigated the spatiotemporal structure of a large set of muscle patterns observed during kicking in intact frogs [14]. We recorded the EMG activity simultaneously from 13 hind-limb muscles during a total of 239 kicks elicited by cutaneous stimulation of the foot in four frogs. Depending on the location of the stimulation, the frogs performed kicks in different directions: At the point of maximum extension of the limb the ankle reached a range of different positions (medial, caudal, or lateral of the starting position).

We extracted sets of time-varying synergies with a number of elements ranging from 1 to 10 (Fig. 27.6) from the rectified, filtered (20-Hz low-pass), and integrated (10-ms interval) EMGs. Each synergy was chosen to be 30 samples long for a total duration of 300 ms. Based on the change in slope of the R^2 curve,[5] we selected three as the optimal number of synergies. Each of the three extracted synergies (Fig. 27.7) showed specific spatiotemporal characteristics. Extensor muscles were the most active muscles in the first two synergies (W_1 and W_2 in Fig. 27.7) while flexor muscles were the most active muscles in the third synergy (W_3). The first synergy showed a short and almost synchronous burst of activation of hip extensor muscles (RI and SM). Knee extensor muscles (VI and VE) were recruited in the second synergy with short and slightly delayed bursts. The bursts of the flexor muscles in the third synergy had substantially longer durations than the bursts in the first two synergies.

Muscle patterns corresponding to kicks in different directions were reconstructed by scaling in amplitude and shifting in time the three extracted synergies. In particular, the amplitude and the onset time of the two extension synergies were modulated systematically with the direction of kick. For example, in a medially directed kick (Fig. 27.8*a*, i), involving mainly hip extension, the muscle pattern (thin line and shaded area) was reconstructed (thick line) by the combination of the first and third synergies. In contrast, the muscle pattern of a laterally directed kick (Fig. 27.8*a*, iii), involving mainly knee extension, was reconstructed by the activation of the second and third

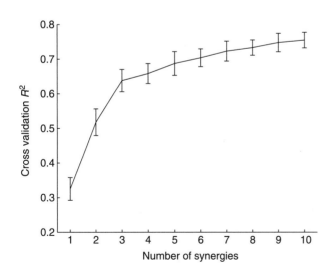

Figure 27.6   Fraction of frog's kicking data variation explained as function of number of synergies extracted by time-varying algorithm [14]. Adapted with permission.

---

[5]We computed the R^2 of the reconstruction of 20% of the kicks, randomly selected, using the synergies extracted from the remaining 80%. Such a cross-validation procedure provides a better estimate of the capability of the model to explain data not used to the model estimation, that is, its generalization performance.

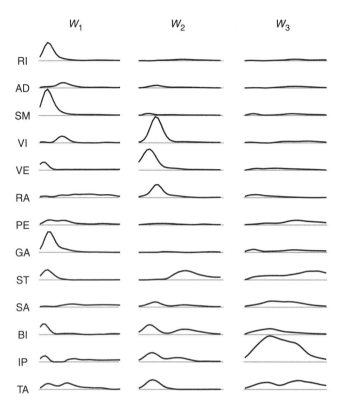

Figure 27.7   Three time-varying synergies extracted from entire set of EMG patterns recorded in four intact frogs during kicking in different directions [14]. Each synergy has a duration of 300 ms. Adapted with permission.

synergies. Interestingly, the pattern for a caudal kick, obtained by a combination of hip and knee extension, was reconstructed by a simultaneous activation of the first two synergies. When we examined the entire set of kicks, we found that there was a gradual and systematic modulation of the ratio of the activation amplitude of the first two synergies (as in the examples of Fig. 27.8b) as well as a gradual shifting of their onset time [14]. Thus the different synergies appear to have specific functional roles in the control of limb movements. The fact that combinations of the same few elements, properly modulated, could reconstruct muscle patterns observed during movements in different directions supports the idea of a modular organization of the lowest level in the control hierarchy of the vertebrate motor system.

27.6   MUSCLE SYNERGIES FOR MOTOR CONTROL

We have introduced a model for the generation of muscle patterns through the combination of muscle synergies. Muscle synergies—coherent activations in space or time of a group of muscles—might be organized by the CNS to simplify the task of mapping a motor goal into the muscle pattern that can achieve it, that is, the process of motor control. Such a modular organization might in fact provide the CNS with a low-dimensional representation of the motor output and allow us to perform high-level planning and optimization on a reduced set of variables. The set of synergies would

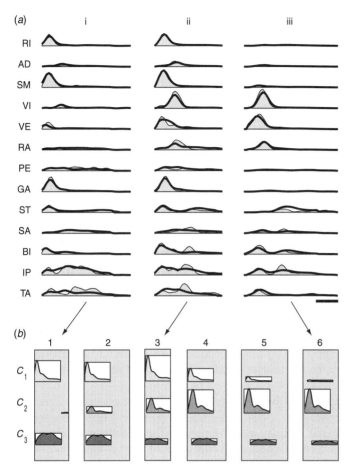

Figure 27.8   Reconstruction of EMG patterns during frog kicking as combinations of three time-varying synergies [14]. (*a*) Three different EMG patterns (thin lines and shaded area) are reconstructed by scaling in amplitude, shifting in time, and combining the three synergies of Figure 27.7 (thick line). (*b*) Amplitude and onset delay coefficients for three patterns shown in (*a*) and three more patterns are indicated by the height and the horizontal position of three rectangles. Going from patterns 1 to 6 the direction of the kicks gradually shifts from medial to caudal to lateral. Reproduced with permission.

thereby constitute a basic vocabulary whose elements can be composed to generate many useful motor behaviors. The structure of the synergies should then capture the invariant characteristics of the muscle patterns shared in the control of different movements. We thus considered two different muscle synergy models that capture different types of structure in the muscle patterns: synchronous synergies expressing spatial relationships among muscles' activations and time-varying synergies expressing spatiotemporal relationships. We presented two algorithms for the identification of a set of muscle synergies from the regularities in the motor patterns observed in a variety of conditions. We have shown through simulations that these algorithms are capable of recovering the synergies and the combination coefficients used in the generation of the data even when the data are corrupted by a substantial amount of noise. Finally, we have applied the extraction algorithms to the analysis of experimental data obtained from spinalized and intact frogs. We have found that the muscle patterns observed during the performance of natural motor behaviors can be reconstructed by the combinations of a small number of synergies and that the synergies are systematically modulated to control different movements. These results support the idea of a modular organization of the output of the motor controller in the vertebrate CNS and provide a framework to better guide a direct physiological investigation of the controller's implementation.

27.7 FUTURE TRENDS

It will be interesting to extend the approach we presented to the study of a larger range of behaviors and vertebrate systems. In particular, the muscle patterns used by primates to control arm movements have been studied extensively, but the synergy decomposition approach might provide novel insights into the processes underlying the pattern generation. To study vertebrate systems with a very rich repertoire of movements and a high degree of sophistication in their control, we might need to make the model more sophisticated as well. One natural extension of the time-varying synergy model would be to allow for time scaling, that is, adding a third parameter for each synergy corresponding to the total duration of the muscle activation time course described by the synergy.

Biomechanical simulations could be employed to address the issue of whether a synergy-based controller is capable of generating the muscle patterns necessary for a natural movement repertoire. The use of a realistic biomechanical model of a vertebrate limb might both allow the testing of the synergy model on a realistic system and provide insights on the biomechanical characteristics and requirements for the synergy structure.

Finally, an approach based on combinations of muscle synergies could be used to implement a controller in a system using FES of peripheral nerves for motor recovery in patients with spinal cord injuries. Mimicking the hypothetical organization of the neural controller, an artificial controller could optimize a set of synergies and learn movement control though a low-dimensional synergy representation. Beyond the computational benefits of this approach, the input signal for such a controller might naturally match a synergy-based cortical representation of movement that could be obtained with a brain–machine interface system.

ACKNOWLEDGMENTS

We thank E. Bizzi for constant support and P. Saltiel for many helpful discussions. This work was supported by National Institutes of Heath grants NIH NS09343 and NIH NS39865.

REFERENCES

1. N. BERNSTEIN, *The Co-ordination and Regulation of Movement*, Oxford: Pergamon, 1967.
2. T. G. BROWN, "The intrinsic factors in the act of progression in the mammal," *Proceedings of the Royal Society of London. Series B: Biological Sciences*, vol. 84, pp. 308–319, 1911.
3. W. A. LEE, "Neuromotor synergies as a basis for coordinated intentional action," *Journal of Motor Behavior*, vol. 16, pp. 135–170, 1984.
4. P. S. G. STEIN AND J. L. SMITH, "Neural and biomechanical control strategies for different strategies of vertebrate hindlimb motor tasks," in *Neurons,*

Networks, and Motor Behavior, P. S. G. STEIN, Ed., Cambridge, MA: MIT Press, 1997.

5. E. BIZZI, A. D'AVELLA, P. SALTIEL, and M. TRESCH, "Modular organization of spinal motor systems," *Neuroscientist*, vol. 8, pp. 437–442, 2002.

6. M. C. TRESCH, P. SALTIEL, A. D'AVELLA, AND E. BIZZI, "Coordination and localization in spinal motor systems," *Brain Research and Brain Research Review*, vol. 40, pp. 66–79, 2002.

7. I. T. JOLLIFFE, *Principal Component Analysis*, 2nd ed., New York: Springer, 2002.

8. A. BASILEVSKY, *Statistical Factor Analysis and Related Methods: Theory and Applications*, New York: Wiley, 1994.

9. A. HYVARINEN, J. KARHUNEN, AND E. OJA, *Independent Component Analysis*, New York: Wiley, 2001.

10. M. C. TRESCH, P. SALTIEL, AND E. BIZZI, "The construction of movement by the spinal cord," *Nature Neuroscience*, vol. 2, pp. 162–167, 1999.

11. D. D. LEE AND H. S. SEUNG, "Learning the parts of objects by nonnegative matrix factorization," *Nature*, vol. 401, pp. 788–791, 1999.

12. D. D. LEE AND H. S. SEUNG, "Algorithms for nonnegative matrix factorization," in *Advances in Neural Information Processing Systems*, vol. 13, T. K. LEEN, T. G. DIETTERICH, AND V. TRESP, Eds., Cambridge, MA: MIT Press, 2001, pp. 556–562.

13. A. D'AVELLA AND M. C. TRESCH, "Modularity in the motor system: Decomposition of muscle patterns as combinations of time-varying synergies," in *Advances in Neural Information Processing Systems*, vol. 14, T. G. DIETTERICH, S. BECKER, AND Z. GHAHRAMANI, Eds., Cambridge, MA: MIT Press, 2002, pp. 141–148.

14. A. D'AVELLA, P. SALTIEL, AND E. BIZZI, "Combinations of muscle synergies in the construction of a natural motor behavior," *Nature Neuroscience*, vol. 6, pp. 300–308, 2003.

CHAPTER 28

ROBOTS WITH NEURAL BUILDING BLOCKS

Henrik Hautop Lund and Jacob Nielsen

28.1 INTRODUCTION

In this chapter, we will show how artificial neural networks can be constructed in a physical manner and how a user may be able to utilize this new concept to construct robotic systems. To this end, intelligent building blocks (I-BLOCKS) are used. I-BLOCKS are a collection of physical building blocks that each contains processing and communication capabilities. By attaching a number of building blocks together, it is possible to make a system that obtains specific behaviors. Differences in physical structures imply differences in the overall processing and communication. These differences can be explored, and so a user may be allowed to "program by building" [1].

In this work, we can view each building block as a neuron with synapses in the connectors. When connecting building blocks together, the user will be connecting "neurons" together to form an artificial neural network. Imagine the building blocks being cubes and then connectors being in each of the six sides of the cube. The connectors could be either one way (input or output) or both ways (input and output). When two cubes (neurons) are attached together, they will form a connection (one-way or two-way connection, depending on the physical implementation of the building blocks) and information can be passed from one building block (neuron) to another. If processing in the individual building block is (on some abstract level) modeling the functionality of a neuron and the communication on the connections is modeling the functionality of synapses, we can create artificial neural networks through building physical structures with the building blocks.

It is interesting and challenging to construct such physical artificial neural networks because it allows us to investigate:

- The 3D property of the neural networks
- The relationship between morphology and control
- New educational principles
- New entertainment tools

Indeed, the strong relationship between morphology and control is explicit in these intelligent artifacts. The physical structure has a defining role for the overall behavior of the constructed artifact (e.g., robot). In contrast, much research in traditional artificial intelligence has suggested that the physical aspects can be abstracted away and that it is

Handbook of Neural Engineering. Edited by Metin Akay

possible to work on a purely abstract level (e.g., a symbol system) in order to understand intelligence. The intelligent artifacts constructed from the neural building blocks are exemplifying our opposing view that, at least on some levels, the physical aspects play a crucial role in defining the "intelligence" of the overall system. In the present study, this is shown with the morphology having a defining role for the overall behavior, whereas in other studies it may be possible also to include other physical aspects such as material, elasticity, and energy use, for example, as exemplified with our humanoid robots, Viki, which won the RoboCup Humanoids Free Style World Championship in 2002 [2].

Apart from the interesting aspect of allowing us to study some physical aspects of neural networks, the implementation of neural networks in I-BLOCKS also allows us to develop a new methodology in edutainment. With a neural network implementation in I-BLOCKS it becomes possible to investigate an edutainment scenario where the user first builds a construction of I-BLOCKS and afterward trains the construction to express a behavior. We can imagine both supervised learning and unsupervised learning methods for the training of the I-BLOCKS. Hence, in such a scenario, the user is simply building and training—a simple process which may allow many different users to be able to work with the system.

28.2 I-BLOCKS

We constructed a number of hardware building blocks with individual processing power and communication capabilities. In order to exemplify in a clear manner how these function as building blocks, we chose to implement them in LEGO DUPLO[1] housing. Other housing can easily be used, and the LEGO DUPLO housing simply functions as an example. Most building blocks are in 4×2 DUPLO bricks (and a few in 2×2 DUPLO and in 4×2 DUPLO of double height). Hence, they are not cubes, and connections are put in the center of each 2×2 area on the top and on the bottom of the DUPLO bricks. So each building block has two connectors on the top (one on the left and one on the right) and two connectors on the bottom (see Figs. 28.1 and 28.2).

Each building block contains a PIC microcontroller, in this case a PIC16F876 40-pin, 8-bit complementary metal–oxide–semiconductor Flash microcontroller (see Fig. 28.2). Further, in the case of rectangular LEGO DUPLO, each brick contains four serial two-way connections, two connections on the top and two connections on the bottom of each building block. We developed a number of standard electronic building blocks that allow processing and serial communication and some specialized building blocks that include sensors or actuators. The standard building block only needs few components (e.g., a PIC, a few diodes, some capacitors and resistors, four connections, a crystal, and a latch), and therefore, in the future, it can be produced fairly inexpensively.

A series of such building blocks were designed. All include the standard functionality of being able to process and communicate, but some include additional hardware, so among others, there exist input and output building blocks as indicated in Table 28.1. Further, there exists a rechargeable battery building block and a battery building block for standard 9-V batteries. Also, some standard building blocks contain small backup batteries that allow a construction to function for a period of time (up to 5 min if no actuator building blocks are included) when detached from the centralized battery building block. It is possible to construct a number of other building blocks not included

[1]LEGO and DUPLO are trademarks of LEGO System A/S.

Figure 28.1   Neural building block implemented in LEGO DUPLO. In this implementation, there are two connectors on the top and two on the bottom of each neural building block. On each stud, there is connection for power transfer. (Copyright © H. H. Lund, 2002.)

in the list of currently available building blocks, for example, building blocks with digital compass, sonar, and accelerometer.

   With these building blocks, it is possible to construct a huge variety of physical objects with various functionalities. The processing in the physical construction is distributed among all the building blocks, and there is no central control opposed to traditional computerized systems (e.g., traditional robots). The distribution of control is obtained by allowing processing within each individual building block.

28.3   NEURAL NETWORK BUILDING BLOCKS

The building blocks become neural building blocks by the specific implementation of artificial neural network processing in the PIC microcontroller of each building block. In the simplest form, the individual processor reads the input (activation) on each input channel (connector), sums up the activation, applies a function (e.g., sigmoid or threshold) and

Figure 28.2   Neural building block implemented in LEGO DUPLO that contains PIC microcontroller (center), crystal (upper right), and communication channels (center of two round connectors). (Copyright © H. H. Lund, 2002.)

TABLE 28.1 Classification of Different Hardware Types of Building Blocks and their Corresponding, Possible Functionality

General type	Specific type	Functionality
Standard building blocks	Standard	Processing & communication
	Back-up battery	As standard, but functions with own battery
Input building blocks	LDR sensor	Light level
	Two LDR sensors	Light level, light direction
	Two microphones	Sound level, direction of sound
	IR sensor	Infra red light level and patterns
Output building blocks	Touch sensor	Pressed or not
	Potentiometer	Turn level
	Double motor brick	Turn $+/-$ 45°
	Motor brick	Axe turn 360°
	Eight LEDs	Light in patterns
	Sound generator	Sounds
	IR emitter	Wireless communication
	Display	Displays messages

sends the result to the output channels (connectors). So, the individual building block works as a simple artificial neuron, and the connection of a number of building blocks can work as a traditional artificial neural network with input, processing, and output.

It is also possible to implement spiking neural networks in the neural building blocks. In this case, in a neural building block, action potentials build up toward an activation threshold, and when they reach the threshold, the neural building block may be able to fire the action potential to other connected neural building blocks. Afterward the neural building block enters into a repolarization phase and after-hyperpolarization phase (see further details below).

Initially, in order to exemplify the potential of neural building blocks, let us concentrate on the simplest form with a traditional artificial neural network. For the individual building blocks, the activation function is chosen to be a sigmoid function

$$\text{Output} = \frac{1}{1 + e^{-\text{sum(input)}}}$$

Each building block operates with activation in the range of 0–255, so 0–127 is inhibition and 127–255 is excitation (by subtracting 127 from the activation value). So sum(input) becomes

$$\text{Sum(input)} = \text{bias} + \sum_{i=1}^{N} \frac{x_i - 127}{2N}$$

where N is number of inputs. The number 2 in the denominator represents a general weight of $\frac{1}{2}$ (times number of inputs to the neuron). This weighting can be changed in an adaptive manner in learning neural building blocks.

The bias of each building block can be set by attaching a "set level" building block with touch sensors or a potentiometer that allows a user to set a specific value, which is written into the neural building block before the user again detaches the set level building block (Fig. 28.3).

Figure 28.3   Example of intelligent building blocks implemented in LEGO DUPLO. Left: sensor building block that contains two microphones. Right: motor building block that contains servomotor that allows the top element to turn. (Copyright © H. H. Lund, 2002.)

28.3.1   Results

Initially, a number of basic tests were performed to verify that feedforward inhibition and excitation and recurrent inhibition and excitation could be performed. With a positive result on these basic tests, a number of tests regarding system behavior were performed. These tests included performing simple math, creating sound melodies, performing as a light-searching mobile robot, performing as an exploring mobile robot, and performing as a curtain controller. Here, only the exploration robot will be described (due to lack of space).

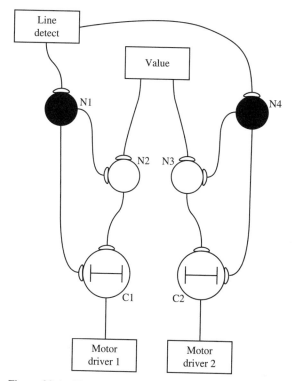

Figure 28.4   Neuronal circuit for exploring mobile robot. White neurons (N2 and N3) are excitatory neurons and black neurons (N1 and N4) are inhibitory neurons. C1 and C2 are combinational neurons in the sense that they integrate the inhibition and excitation. (Copyright © H. H. Lund and J. Nielsen, 2002.)

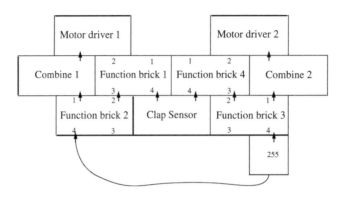

Figure 28.5 Building blocks to make neuronal circuit for exploring mobile robot from Figure 28.4. (Copyright © H. H. Lund and J. Nielsen, 2002.)

The exploration task consists of driving around in an area on the floor and covering the area without crossing the boundaries of the area. The area could be confined with black stripes on a white floor, but in our case, we allow the user to interactively confine the area by clapping with his or her hands when the boundary is met.

The neuronal structure in Figure 28.4 shows how a solution may look. The two excitatory neurons, N2 and N3, are fed by a value, which makes the vehicle go forward whenever these neurons are producing an action potential. The line detect sensor is directly connected to two inhibitory neurons, N1 and N4, that can inhibit the two excitatory neurons. The inhibitory neurons are to be considered two feedforward networks of different size and with inhibitory neurons at the end. This results in two different time constants for the inhibitory neurons that make them send their action potential at a given duration no matter how long the input signal from the line detector is. To start, both inhibitory neurons inhibit the excitatory neurons while controlling the combinational neurons that thereby make the vehicle go backward until the one inhibitory neuron stops inhibiting. Now the vehicle turns until the other inhibitory neuron stops sending its action potential, and then the vehicle starts moving forward again.

The brick structure shown in Figure 28.5 does the same thing as described for the neuronal circuit. Two of the "function building blocks" have been programmed to be inhibitory neurons. The timers of these inhibitory neurons have been set to 3 and 5 s,

Figure 28.6 Robot built from neural building blocks for exploration task. (Copyright © H. H. Lund and J. Nielsen, 2002.)

which makes the vehicle back up for 3 s and turn for 2 s whenever the clap sensor is activated. The excitatory neurons are fed with a value of 255 to ensure that they are activated as long as they receive no inhibitory input. The actual structure that has been built can be seen in Figure 28.6. It is connected to LEGO motors and LEGO wheels and thereby becomes a mobile robot for the exploration task.

28.4 SPIKING NEURAL NETWORK BUILDING BLOCKS

It is also possible to implement spiking neural network control in the building blocks so that each block represents a neuron that fires through the connections when a threshold is reached. After firing, there will be a refractory period. The physical stimuli intensity is frequency modulated via the rate of the action potentials. However, the generator potential has to be higher than the activation threshold for action potentials to be generated at all.

Another issue in the neurosensors is adaptation, which is related to the generator potential of the sensor. With no adaptation, the generator potential remains the same when a constant stimulus is applied. With adaptation the generator potential declines over time either slowly or rapidly depending on the type of sensor nerve. Adaptation is used in the central nervous system to avoid "sensory overload" and it allows less important or unchanging environmental stimuli to be partially ignored. Rapid adaptation is also important in sensory systems, where the rate of change is important. When a change occurs in the sensor stimulus, the phasic response will occur again, reproducing the generator potential curves, just at other potential values.

We used Hebbian learning [4], which is an adaptation rule where a synapse is strengthened if pre- and postsynaptic neurons are active at the same time. When speaking of spiking neural networks, the meaning of "the same time" must however be specified. Due to the time it takes for a neuron to build up toward its threshold level, one has to introduce a time window to find out which presynaptic neurons take part in firing this neuron [3].

Hebbian learning is an unsupervised learning method, and when formulated as a spike-learning method, the process of learning is driven by the temporal correlations between presynaptic spike arrival and postsynaptic firing.

28.4.1 Implementation Issues

The implementation of spiking neurons in the intelligent artifacts uses hardware (a PIC16F876 processor) where the action potentials emitted are digital pulses. In order to support both excitatory and inhibitory signals, these are coded to be of different lengths. The generation of action potentials is based on the input of the neurons via a leaky integrator, which sums up the amount of input while decaying at a certain rate:

$$A(t) = \alpha \times A(t-1) + I(t) \qquad 0 < \alpha < 1$$

where $A(t)$ is the activation function, $I(t)$ is the input, and α is the decay rate. The activation function is a simple threshold function which sends an action potential when the summed input reaches above a given threshold.

An implementation of absolute and relative refractory periods has also been investigated, where the absolute refractory period decides the maximum action potential frequency of a neuron while the relative refractory period demands higher input in order for the neuron to generate action potentials. The current implementation has an absolute refractory period of about 10 ms, which reflects the refractory period of biological neurons well.

Another thing that has been investigated and implemented is adaptation in the sensors. Here the adaptation is made so that when adapting to high levels of sensory input the steady-state level of the adaptation will be higher than when adapting to lower level sensory input. In this way sensory inputs of different values can still be distinguished although adaptation has taken place. The current values for adaptation have been preset to take around 2 s for adapting from a change from total darkness into daylight. The curves for adaptation are exponential.

Given the limitations of only four communication channels per intelligent building block, the threshold values have been preprogrammed at a fairly high level so that input from only one building block is enough to generate output action potentials given that the input frequency is high enough to allow for temporal summation.

Hebbian learning has been implemented in the CPU blocks to stimulate connections at a certain learning rate. This rate has been set fairly high (connection strengths updated every 2 s) in order to speed up the learning process. Besides, whenever an action potential is fired, the input activity of the last 10 ms is traced back to the actual input port to determine which connections to strengthen.

28.4.2 Experiments

In order to verify that the spiking neural building blocks can be used to make robotic systems, we performed an experiment where the structure has to learn to control a mobile robot to perform a light-seeking task. Learning happened by allowing a user to show light (e.g., with a flashlight) to the robot construction over a period of approximately 3–5 min while the building blocks would perform sensor adaptation and Hebbian learning.

The scenario for the testing of the building blocks has been selected so as to demonstrate some of the capabilities of the intelligent artifacts implemented as spiking neurons. The testing takes place within a 150×150-cm no-walls-area where a mobile robot is to trace and approach a light source using a structure of assembled spiking neural building blocks. The light source is placed in the middle of the field 10 cm away from the borderline furthest away from a camera.

The blocks used for solving the light-tracing task are the battery block, the light dependant resistor (LDR) sensor block, the CPU block, and the LEGO motor controller block. The battery block supplies the final structure through the individual power-sharing capabilities of the blocks. The LDR sensor block is a light-sensitive sensor block that uses the above-mentioned theory to modulate light levels into frequency-based pulse trains while adapting to light levels over time. It should be mentioned that even though there are two LDR sensors on this block, only input from the one is used to generate the output spike trains. The CPU block implements the spiking neural model described above, including stimulation of connection weights due to output resultant input over time using the rules of Hebbian learning. Finally, the LEGO motor controller block converts the input spike trains into motor pulse wave modulation (PWM) signals used to control standard LEGO motors.

The actual assembly of the blocks models a Braitenberg vehicle [5], where the left light sensor ends up controlling the right motor and vice versa. Two schematics of the connection can be seen in Figure 28.7, where the first is a neural connection diagram that shows how the sensory input ends up controlling the motor due to the spike trains traveling through the structure. The second part of Figure 28.7 shows how all the blocks are connected to form the control structure for the vehicle.

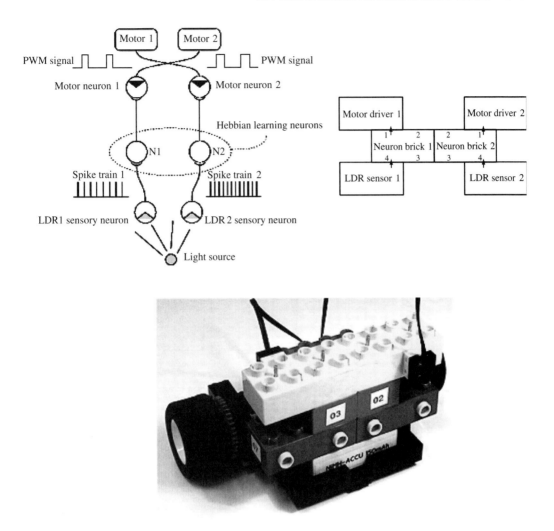

Figure 28.7   Built structure in two schematic views and as photo.

28.4.3   Results

The results of the tests are described here considering both the built structure and the functionality of the single block. Figure 28.8 shows results of eight test runs in the same scenario. In each test run, the vehicle has been placed at an angle less than or equal to 45° toward the light. Each test of Figure 28.8 will be described in the following starting with the top leftmost figure. The vehicle has been fully trained before starting the test session, which means that it is just about unable to generate enough spikes for the motors to drive under the present ambient lighting conditions. This means that it has to have a brighter light source within its field of vision in order to drive the motors forward.

In the first test the vehicle starts at angle zero to the light source driving straight toward the goal, and due to little differences in the lightning conditions as well as the motor forces, the vehicle slides off to the right slowly until the right light sensor goes outside of the light cone slowing down the left motor and finally making the vehicle

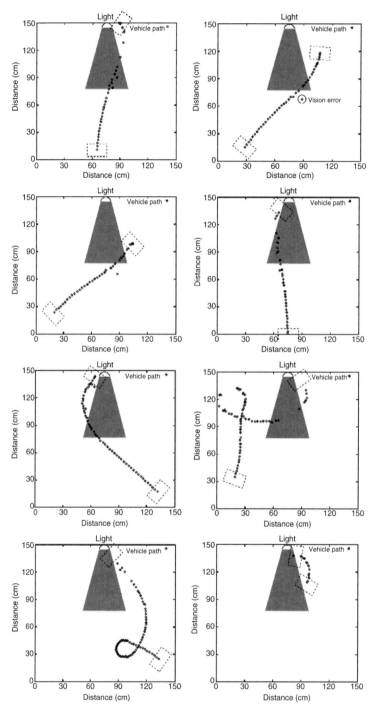

Figure 28.8   Eight different runs in test scenario with results recorded by camera and processed with vision system.

reach its destination. Because of the adaptation in the light sensors, the vehicle reacts fast over a short period time, and through a longer period of time the reaction flattens out. Because of this, in the next test the vehicle does not reach its goal. Here the vehicle is started at an angle approximately 45° to the light source, and because the vehicle is a little to the right as in the previous test, it ends up crossing the light cone so far away from the light source that the edge between the light cone and the surroundings is just blurry enough to make the vehicle drive through it without adjusting the vehicle direction enough to keep one light sensor inside the light cone. As can be seen, the result in the next figure is almost identical to this result. In the rest of the figures the vehicle reaches the goal, but in two of the cases it has to go through a loop to get there. The loops are possible because of the before-mentioned different powers of the motors, meaning that one motor might just be able to drive forward under the ambient lighting conditions while the other motor is standing still. These results in loops are made even though the light source is not within the line of sight of either sensor. It can also be seen that it is the vehicle's right motor that drives under ambient lighting conditions because both loops are going counter clockwise.

The next series of test results, shown in Figure 28.9, displays the output of the LDR sensor block with the built-in adaptive spiking neuron. Figures 28.9a, b, c show how the sensor adapts to normal office ambient light brightness. In Figure 28.9a the block has been kept under total darkness for a while and then suddenly put into the light, resulting in a maximum frequency spike train. After a short time (Fig. 28.9b), the sensor has adapted further to the brightness of the ambient light, and finally, in Figure 28.9c the adaptation has reached its steady state, resulting in a somewhat lower frequency spike train than those in (a) and (b). Now, in Figure 28.9d a bright flashlight is being held in front of the light sensor resulting in a maximum frequency spike train which immediately starts

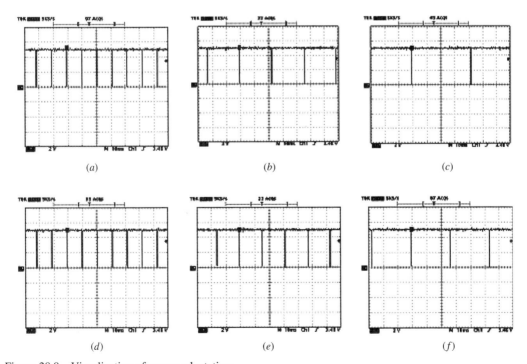

      (a)   (b)   (c)

      (d)   (e)   (f)

Figure 28.9   Visualization of sensor adaptation.

adapting toward its steady state in Figure 28.9*f*. It should be noted that the steady states of the ambient light (Fig. 28.9*c*) and the bright flashlight (Fig. 28.9*f*) are not the same and that brighter light still results in a higher spike frequency, although adaptation has been introduced.

The last measurement taken is the learning capability of the CPU block. In these measurements the CPU block gets its input from the LDR block just as in the vehicle described above. The outputs of both of these blocks are displayed in the picture sequence of Figure 28.10. Here the upper channel of the oscilloscope pictures shows the output of the sensory neuron and the lower shows the output of the learning neuron. The learning starts at point zero (Fig. 28.10*a*), where the sensor is kept under normal office lighting conditions. To start the learning a bright light source is introduced to the light sensor, and this in turn generates a fast series of spikes which just overcomes the threshold of the learning neuron, making it send out spikes, although at a lower frequency (Fig. 28.10*b*). During this process the connection to the light sensor is strengthened, which, after the light is turned off and on several times, results in the learning neuron sending out spikes at increasingly higher frequencies (Figs. 28.10*c*, *d*). However, in Figure 28.10*e* it can be seen that the

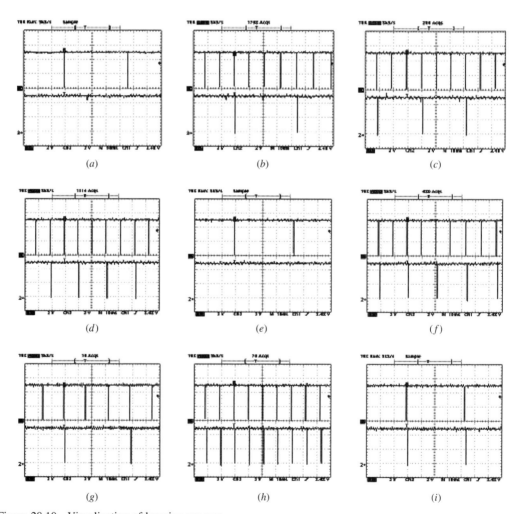

Figure 28.10 Visualization of learning process.

learning has not been going on long enough for the learning neuron to send out spikes when the sensory neuron is exposed just to ambient light so the learning process continues. In Figure 28.10*g* the learning neuron starts to react to light brightness just a little above ambient conditions, and being exposed to this kind of light for some time it starts to reflect the frequencies of the light sensor neuron with just a slight delay. This can be seen in Figures 28.10*h,i*.

28.5 DISCUSSION AND CONCLUSION

In this chapter, we have described how it is possible to develop physical neural networks with neural building blocks. The topology of the neural network is implied directly from the morphology of the robot constructed by the user, and so the functionality of the robot and the neural network is decided by the physical structure designed by the user. Different physical structures will result in different behaviors.

The first example shown here is simple but illustrative. It shows that the simple implementation of neural building blocks is enough to obtain a system that allows the user to develop robots similar to those known from most entertainment scenarios (e.g., following, searching, exploring, avoiding tasks).

Learning was not shown with the first example, but one implementation is shown with the second example. Indeed, it may be easy to implement learning, since feedback can be achieved through the input building blocks described in Table 28.1. For instance, the simple touch sensors may be used to provide good/bad feedback, while potentiometers can give further detailed feedback for supervised learning. The learning can then take place as backpropagation of error or with a similar supervised learning method. Also, unsupervised learning such as Hebbian learning is a possibility, as shown with the second example in this chapter. In these cases, the learning may act upon the weights of the synapses in the connectors of the neural building blocks.

In general, the neural building blocks allow us to explore the topology of neural networks in a novel manner and allows users to develop their own neural network by constructing a physical artifact with functionality. Indeed, the user will be "programming by building" [1]. This means that programming of a specific behavior simply consists of building physical structures known to express that specific behavior. To allow everyday users to develop functionality of artifacts, it is important to make the process of functionality creation accessible. We believe that for everyday users it is important to find simpler ways for creating technological artifacts. Especially, it is wrong to believe that everyday users should be able to program prebuilt physical robotic structures in a traditional programming language. The programming in a traditional programming language demands that the everyday user learn both syntax and semantics of the programming language, and this is a long and tedious process. Hence, such an approach will exclude most everyday users from becoming creative with the new technology. If we want to achieve ambient intelligence with integration of technology into our environment, so that people can freely and interactively utilize it, we need to find another approach. Therefore, we have suggested moving away from programming the artifacts with traditional programming languages and instead provide methods that allow people to program by building without the need for any a priori knowledge about programming languages. Indeed, we even suggest completely removing the traditional host computer (e.g., a personal computer) from the creative process. The I-BLOCKS are the tool that supports investigation of this innovative way of manipulating conceptual structures. Interaction with the surrounding environment happens through I-BLOCKS that obtain sensory input or produces actuation output. So the overall behavior of an "intelligent artifact" created by the user with the I-BLOCKS depends on the physical shape of the creation, the processing in the I-BLOCKS, and the interaction between the creation and the surrounding environment (e.g., the users themselves).

Further, the general research focus allows us to explore and understand the relationship between morphology and control. The work presented here should make it evident that the morphology plays a crucial role in defining the overall behavior of the system. On longer term, we are interested in researching how such building blocks may be expanded to have the possibility to do self-reconfiguration. This option is explored in the HYDRA project (http://www.hydra-robot.com).

ACKNOWLEDGMENTS

H. Caspersen made some hardware development. Also thanks to K. Nielsen, L. Pagliarini, O. Miglino, P. Marti, and the HYDRA consortium. The views presented here are those of the author and do not necessarily present the views of any sponsor.

REFERENCES

1. H. H. LUND, Intelligent artefacts, in *Proc. of 8th Int. Symposium on Artificial Life and Robotics*. Oita, Japan: ISAROB, 2003.
2. H. H. LUND, L. PAGLIARINI, L. PARAMONOV, AND M. W. JOERGENSEN, Embodied AI in humanoids, in *Proc. of 8th Int. Symposium on Artificial Life and Robotics*. Oita, Japan: ISAROB, 2003.
3. W. GERSTNER AND W. M. KISTLER, *Spiking Neuron Models*. Cambridge University Press, 2002.
4. D. O. HEBB, *The Organization of Behavior*. Wiley, New York, 1949.
5. V. BRAITENBERG, *Vehicles: Experiments in Synthetic Psychology*. Cambridge, MA: MIT Press, 1984.

CHAPTER 29

DECODING SENSORY STIMULI FROM POPULATIONS OF NEURONS: METHODS FOR LONG-TERM LONGITUDINAL STUDIES

Guglielmo Foffani, Banu Tutunculer, Steven C. Leiser, and Karen A. Moxon

29.1   INTRODUCTION

Information in the cortex is coded not simply by small numbers of highly tuned neurons but through the combined actions of large populations of widely tuned neurons. Traditionally, however, investigations into the nature of neural codes have utilized data from serially recorded single neurons averaged over repeated trials [1] and therefore were unable to demonstrate the coding of brain information at the population level. During the past 30 years, there has been a remarkable development of techniques for simultaneous (parallel) neuronal population recording [2–5]. Randomly sampled neuronal populations can be used to demonstrate a surprisingly rich encoding of information in the brain, in both the sensory [6–10] and motor [11–15] systems.

   We are particularly interested in studying the plasticity of neural systems in response to behavioral modification or injury. Because of the stability and large information yield of multineuron recording techniques, it has now become possible to measure changes in the function of populations of neurons to study specific issues related to plasticity in intact animals. However, two main problems arise when one attempts to perform these types of long-term longitudinal studies. The first problem is long-term population recording and the second is population analysis with large data sets.

   To address these problems, the first part of this chapter will review our established techniques for chronically recording from large populations of neurons through electrode arrays implanted in sensorimotor regions of the brain, including cortex, thalamus, and brainstem structures (Section 29.2; Fig. 29.1). These techniques are well suited for measuring long-term (i.e., from days to months) functional changes in neuronal ensembles, and we extend them to longitudinal studies that examine the plasticity of the system in response to injury. The second part of the chapter will describe the efficient data analysis method we developed for decoding stimulus location using populations of

Handbook of Neural Engineering. Edited by Metin Akay

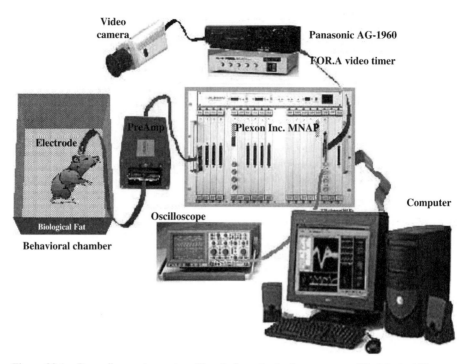

Figure 29.1 Recording system setup. Signals from the brain are preamplified (gain 100, bandpass 150 Hz–15 kHz) and sent to a multineuron acquisition processor (MNAP; Plexon, Inc.), which performs further conditioning (amplification and filtering). This system enables on-line neuronal spike sorting and simultaneous recordings from every microwire electrode.

single neuron (Section 29.3). Long-term longitudinal studies generate large amounts of data, for which linear or nonlinear multivariate techniques can easily become computationally prohibitive. Our approach overcomes this limitation, maximizing both computational efficiency and the amount of information that can be extracted from the neural code. The last part of the chapter will provide directions for future work (Section 29.4).

29.2 RECORDING POPULATIONS OF SINGLE NEURONS

29.2.1 Extracellular Electrophysiology In Vivo

Chronic single-neuron electrophysiology provides a stable technique for monitoring the activity of neurons, not only on the single-unit level but also on a population level in both anesthetized and awake animals [4, 16]. In its most simple form, extracellular recordings can be performed by placing the exposed tip of a single insulated wire near a neuron in the brain. This wire acts as an electrode and can pick up voltage fluctuations due to action potentials from neurons near the electrode (generally within 50 μm). Recording this signal provides a means to quantitatively assess when a neuron fires an action potential, often called a "spike" due to its appearance when viewed on an oscilloscope. By using multiple-wire electrodes in this manner, it is possible to record from a large number of neurons (100 or more) simultaneously.

The action potential results when the cell's membrane reaches a threshold potential for transiently opening voltage-gated sodium channels, allowing positively charged sodium ions to flow down their concentration gradient into the cell. This creates a fluctuation in voltage between the immediate area (position of electrode) and distant locations (position of the referencing wire). The spike represents this transient change in the membrane potential, which is generally several times the amplitude of the background signals recorded from the electrode. It should be noted that the potential difference between the electrode and the reference is directly related to the input impedance of the electrode. Because larger input impedances result in greater voltage output, a better signal-to-noise ratio is reached with high-impedance electrodes.

In order to improve the signal-to-noise ratio, proper signal conditioning (i.e., amplification and filtering) should occur as close to the source (animal's head) as possible. However, since the goal is to be able to record from a freely moving animal, the connection at the animal's head must be as small as possible. Therefore, the current being recorded from each electrode that is chronically implanted is amplified using small junction field-effect transistors (JFETs) that can fit directly on the animal's head (i.e., headsets). The output of the headset is then carried over relatively short wires (0.5–1 m) into a preamplifying box, which supplies preamplifier gain and filtering for spike detection. The signal is typically bandpass filtered around the frequency content of an action potential (e.g., 150 Hz–15 KHz). The output of the preamplifier sends the neural signals to an acquisition computer for further amplification and spike analysis. Spike analysis includes discriminating individual cells from each other and from the background noise [17] and monitoring the times at which each cell fired an action potential for further analyses.

29.2.2 Choice of Microelectrode Arrays

There are several different types of multiple-site recording electrodes designed to effectively increase the yield of recorded neurons [16]. In most instances these electrodes are easy to use. However, they are designed to work optimally under certain conditions, and this must be considered when selecting the appropriate electrode for the application. Moreover, successful recordings are dependent on proper handling of the electrode. Generally, special care must be used when the electrodes are implanted. Successful use of these electrodes requires skill acquired through practice.

Multiple-wire electrode arrays have been in use in varied configurations since the 1960s. It is only recently that sufficient hardware and software have been developed so that large numbers (>32) of neurons can routinely be recorded simultaneously from these electrodes. The characteristics of the individual wires used remains the same. The wire must have a small diameter while still maintaining adequate stiffness and tip shape to penetrate the dura and traverse the tissue with minimal bending and mechanical disturbance to surrounding tissue. The wire must be sufficiently insulated with a material that is biologically inert in order to minimize the reactivity with the tissue. If the impedance of the electrodes is relatively low, with minimal noise, it is likely to record multiple neurons per wire, thus greatly increasing the yield of these electrode arrays. These electrodes have generally been produced within the laboratory by paying particular attention to the tips.

In addition to microwire arrays, several laboratories have employed standard thin-film technologies from the semiconductor industry to manufacture multisite recording probes on a silicon substrate [18–24]. However, the yield for single-unit recordings from silicon probes has been very low [25] compared with microwires [26]. These results were initially thought to suggest that multiple recording sites along the shaft of the electrode would not be feasible for chronic recording because these recording sites would be in contact with neural tissue that was severely damaged during the insertion of the electrode and would, therefore, not support chronic recordings. The solution for silicon thin-film electrodes has been to insert it into the neural tissue immediately before recording [27].

We developed a novel thin-film electrode based on inert ceramic [28] and demonstrated that these ceramic-based thin-film microelectrodes can record single-neuron action potentials for several weeks. These microelectrode arrays are about the size of a single microwire (approximately 50–100 µm) and have multiple recording sites precisely spaced along its length, generating more recording sites for the same volume of electrode compared to microwires. Our initial studies showed that the yield from ceramic-based multisite electrodes (CBMSEs) is comparable to the recording capabilities of microwires. We have been able to consistently record from these thin-film microelectrode arrays for several weeks. This discounted the conclusion that multisite thin-film electrodes were inadequate for chronic recording and opened the possibility that multiple-site thin-film electrodes may be a viable alternative to traditional, single-site electrodes.

29.2.3   General Method for Chronic Implantation of Microelectrode Arrays

Rats are chronically implanted with microelectrode arrays into the target region of the brain [9, 16, 28]. The animal is initially anesthetized and maintained throughout the surgery by intraperitoneal injections of sodium pentobarbital. The animal is then placed on a sterile pad where the scalp and facial hair adjacent to the whisker pad are shaven. The area is cleaned with 70% isopropyl alcohol and 10% povidone–iodine solution is applied. The animal is placed on a heating pad and secured in a stereotaxic apparatus by ear bars and a tooth plate. A nose clamp is used to further hold the head firmly in place. A head-mounting tool on the stereotax is adjusted to ensure the rat's skull is planar both dorsoventrally and mediolaterally.

An incision is made in the skin overlying the scalp. The skull is kept exposed by clamping the skin with four Kelly forceps, hemostats with box-lock joints. The periosteal surface is rinsed with saline solution to clear it of blood and debris. A hand-held dental drill is then used to make several small craniotomies in the skull. Up to four microsurgical screws are inlayed into these skull cavities. These screws serve not only as an anchor for the acrylic electrodecap but also as a ground for the electrodes. Other small craniotomies are created over specific brain regions targeted for the electrode. To perform the craniotomies, the bone circumnavigating the specific coordinates is drilled through to the underlying dura. The severed core of bone can then be removed with microsurgical forceps. Using this method the dura is often left undamaged. The exposed area of brain is continuously kept moist with saline solution. Any hemorrhaging can be controlled locally with small saline-doused cotton balls. The animal is now ready to be implanted.

An electrode is connected to a headset and mounted vertically in a stereotaxic arm. The ground wire of the electrode array is connected to a screw on the skull and the grounding pin on the headset. The electrode is aligned to bregma and digital measurements (lateral and posterior) are taken from bregma to ensure accurate implantation. The precision stereotaxic apparatus provides digital measurements in 10-µm increments. An electrode is held vertical in a stereotaxic arm, positioned directly above the craniotomy, and lowered to the surface of the brain. The digital measurement for depth is set to zero. The electrode is gradually lowered through the dura and into the brain (Fig. 29.2). Depending on the electrode design, the dura can be removed if the tip of the electrode is not sharp enough to penetrate the dura. Neural signals and depth measurements are monitored while the electrode is lowered into the brain to reach the desired location.

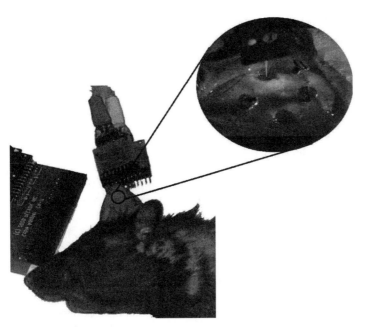

Figure 29.2   Chronically implanted rat. This photograph illustrates the acrylic cap used to hold the electrodes chronically in place, the head stages used to connect the implanted electrodes with the MNAP system (Fig. 29.1), and a CBMSE being lowered into the brain through a trephination in the skull (insert).

If characteristic neuronal responses are observed (Fig. 29.3), the electrode is affixed in place. After a recording site is identified for the electrode, the skull is dried and methyl methacrylate (dental cement) is slowly applied to the skull area surrounding the electrode crest. The acrylic cement is applied up and around any exposed surface of the electrode shank and around the skull grounding screw. Once dry, this skull cap chronically

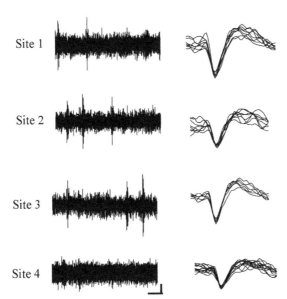

Figure 29.3   Single-neuron action potentials recorded from CBMSE. The electrode had four recording sites. The left column shows continuous waveform simultaneously recorded. The right column shows enlargement of action potential waveform.

secures the electrode in place and any secondary implant can be performed. After the final electrode is implanted, a uniform coating of cement is applied. To prevent infection, bacitracin ointment is applied to all opened skin flaps and the wound is sutured. All procedures are always performed in accordance with local animal welfare committee regulations and National Institutes of Health (NIH) guidelines.

29.2.4   Sensory Maps

After the animals are implanted, recording sessions can be performed under various experimental conditions. Many techniques exist for single-neuron discrimination [17], and, hence, these will not be discussed further. As a common procedure, a template-matching algorithm provides the timing of the action potentials for all discriminated neurons that are recorded during the experiments.

Sensory maps are employed to study the receptive fields and the coding strategies used by populations of sensory neurons. The animals are anesthetized and several locations ideally covering the entire receptive field of the neurons are stimulated. A common paradigm largely used in neurophysiology is to deliver "ramp-and-hold" (i.e., step) stimuli. For each stimulus, transistor–transistor–logic (TTL) pulses are saved together with the neural data to provide a precise temporal reference for the analyses. Step stimuli are given at a sufficiently low frequency (typically 0.5 Hz) and long duration (typically 100 ms) to allow the system to reach steady state. During these passive cutaneous stimulations, the activity of the single neurons discriminated is recorded and stored for further analyses.

29.3   PSTH-BASED CLASSIFICATION OF SENSORY STIMULI

29.3.1   Decoding Neural Code

One of the biggest challenges of neuroscience is to understand how neurons code sensory information. Scientists have been trying to answer this question for many years, but despite recent progresses, the neural code is still a largely unsolved problem. From an experimental point of view, the neural code can be studied by recording activity of populations of neurons in vivo [4, 5, 29] in response to known events or stimuli; the performance of a specific algorithm in *decoding* information from the neural response allows inferences to be made about the coding strategies actually employed by the recorded neurons [1, 30–39]. In the somatosensory system, this decoding problem has been recently defined as a classification problem [10, 40, 41]: Given the response of a population of neurons to a set of stimuli, which stimulus generated the response on a single-trial basis?

In the literature, the above classification problem for studying the neural code has been approached with multivariate techniques [42] such as linear discriminant analysis (LDA) [10, 40, 41, 43–45] and artificial neural networks (ANNs), specifically backpropagation (BP) [41, 46] and linear vector quantization (LVQ) [10, 41]. Also different types of preprocessing stages, such as principal-component analysis (PCA) [47, 48] and independent-component analysis (ICA) [49, 50], have been tried in order to improve the classification performance by reducing the dimensionality of the data [41].

Somewhat surprisingly, the classification performance seems to be almost independent of the method or the preprocessing step employed [41]. The ANN approach has often

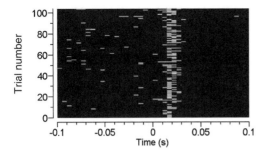

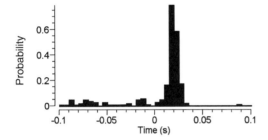

Figure 29.4   Examples of PSTH (bottom, 5 ms bin size) and corresponding perievent rasters (top, 100 trials) of cortical somatosensory neuron responding to ramp-and-hold skin contact (time 0).

been preferred over LDA because the former does not *require* dimension reduction and does not impose particular assumptions (e.g., normality) on the data [51]. However, to our knowledge, significantly greater performance of ANNs over LDA on neural data has not been demonstrated. Because of this performance invariance, it is reasonable to develop new methods to maximize computational efficiency rather than classification performance.

In has long been known that the peristimulus time histogram (PSTH; Fig. 29.4) provides an efficient way to describe the dynamics of a neuron's response to a stimulus [52]. It has been recently shown with a rigorous theoretic approach that the average neural responses (i.e., the PSTHs) describe more than 80% of the information about stimulus location carried by somatosensory cortical neurons [53]. We therefore developed a PSTH-based method to solve the above classification problem [54].

29.3.2   Method Definition

Like other classification methods, the PSTH-based method consists of two stages: training and testing. Similarly to LDA analysis [42], in the training stage a weight is assigned to every variable in correspondence to every stimulus (i.e., possible output of the classification). In LDA, this weighting process aims to minimize intrastimulus variance and maximize interstimulus variance, hence maximizing discriminability. This minimization–maximization procedure is optimal if the variables in the data set have Gaussian distributions, but it becomes unsatisfactory and computationally unstable when the number of trials is not large enough (e.g., smaller than the number of variables) to allow reliable estimation of the covariance matrix, as is often the case when using neural data. We therefore propose to avoid this minimization–maximization procedure by employing not a more complex method (e.g., artificial neural networks) but a simpler one by using directly the neurons' PSTHs as weights. For every stimulus, a weight vector is defined by calculating the average of the $B \times N$ variables over the training set trials corresponding to that stimulus, where B corresponds to the number of bins per neuron and N is the number of neurons in the population. Each weight vector is essentially a population PSTH.

As in LDA, these weights can be seen as S templates (S = number of stimuli) with $B \times N$ dimensions. The testing stage of the classification then consists of assigning every single trial in the testing set to one of the templates obtained in the training stage. Here we use the sum of the square differences, that is, the Euclidean distance, as a measure of how "close" a single trial is to the templates. A single-trial response of the neural population is then classified as being generated by a given stimulus if the distance between the single trial and the template corresponding to that stimulus is minimal compared to all the other distances.

29.3.3 Computational Efficiency

Due to its simplicity, the PSTH-based classification method is computationally more efficient than either LDA or methods based on ANNs, and it is particularly effective in a complete cross-validation design. In fact, as the templates are constructed with average responses (the PSTHs), the training stage can be run just once on the entire data set, and in the testing stage a single trial can be excluded from the training set by simply subtracting it from the template average, with negligible increase of computational time. In other words, in complete cross validation the complexity of PSTH-based method is proportional to the total number of trials (ST).

On the other hand, using an ANN classifier such as LVQ—or even the simple LDA—in complete cross validation the network would have to be retrained at every trial using all the other trials. More precisely, in the classification of S stimuli with T trials per stimulus (ST total trials), the ANN learning algorithm (i.e., a sequence of scalar products, additions and subtractions) has to be run ST (ST-1) times (assuming that the network would converge with only one iteration). In other words, the complexity of ANN classification in complete cross validation is proportional to the square of the total number of trials. For example, in the classification of 20 whiskers, with 100 trials per whisker in complete cross validation, the PSTH-based method is (at least) 2000 times faster than an ANN classifier.

To avoid the "squared" computational cost, when using the neural network approach complete cross validation is usually not used. Instead, the number of available trials is increased to 300–400 [10]. This procedure ensures two nonoverlapping groups of trials, one for the training set and the other for the testing set, but it considerably increases the time to perform the experiments. The PSTH-based method minimizes both the experimental and the processing times.

29.3.4 PSTH-Based Classification on Raw Neural Data

Since the rat trigeminal somatosensory system has received the greatest amount of attention, the PSTH-based method was validated by applying it to populations of neurons from the rat trigeminal somatosensory barrel field cortex (whiskers). However, this method is suitable for many types of sensorimotor stimuli. The objective was to classify which whisker was contacted on a single-trial basis, knowing the exact contact time ($<$1 ms precision) and the single-trial response of the neural population. The rat whisker system was chosen, because its discrete topological organization represents an elegant animal model which has previously been used to study coding strategies of tactile stimulus location [7, 55–57].

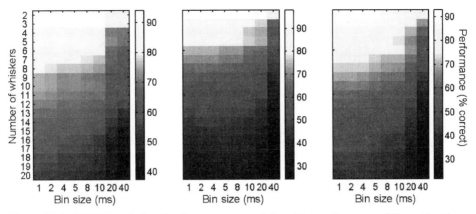

Figure 29.5   PSTH-based classification on raw neural data. The performance of the method is presented as a function of the bin size used in the classification (*x*-axes) and the number of whiskers to be discriminated (*y*-axes). Every element of the three matrices is the average classification performance (% correct) respectively of three populations of 24 (left), 22 (center), and 17 (right) cortical neurons recorded from the barrel field somatosensory cortex of three different rats. Colors are light-graded from minimal performance (black) to maximal performance (white). In general classification performance decreases by increasing either the bin size or the number of whiskers to be discriminated [54].

Even though the method was designed to optimize computational efficiency, the PSTH-based classification performance is surprisingly high. For example, with a population of 24 cortical somatosensory neurons it was possible to correctly classify 20 stimulus locations in the rat whisker pad in 52.1% of the trials (Fig. 29.5). These results are comparable to the LVQ approach [10]. In addition, the performance degradation observed with the PSTH-based method when increasing the bin size is in agreement with the LVQ results [10] and with recent results obtained with information theory [53], supporting the role of action potential timing for stimulus representation [58, 59] in the primary somatosensory cortex.

Interestingly, another way to interpret the PSTH-based classification method is to see the PSTHs of the neurons as the weights of a trained LVQ-like fully connected neural network with only one layer of S artificial neurons (S = number of stimuli to be discriminated) whose transfer function is the Euclidean distance between the input and the weights. A single trial is then classified in a competitive way to the neuron (i.e., the stimulus) at the minimum distance to the input.

29.3.5   Comparison Between PSTH-Based Classification and LDA

The PSTH-based classification method was directly compared with LDA [54]. Here we present an example from one population of cortical neurons (Fig. 29.6). Linear discriminant analysis has been chosen as a reference method because it is the classification most commonly employed in the neuroscience literature [10, 40, 41, 43–45]. A three-way repeated-measure analysis of variance (ANOVA) was used, obtaining three major results.

The first result is that, although overall the two methods are not significantly different, the PSTH-based classification method performs significantly better than LDA when the bin size is reduced down to 1 ms. At all the other bin sizes tested the performances of the two methods were not significantly different (Fig. 29.6*b*,*c*). This result is due to

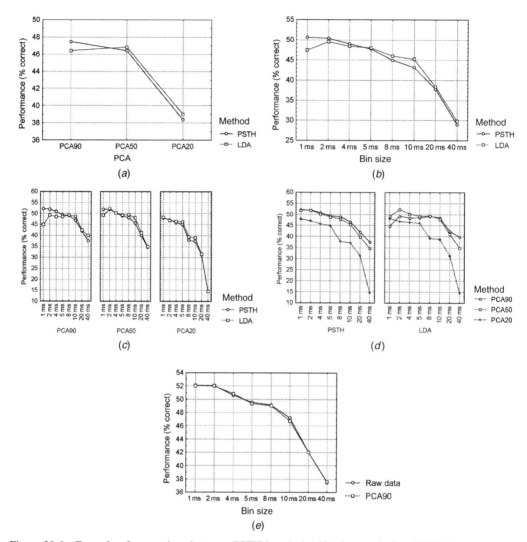

Figure 29.6   Example of comparison between PSTH-based classification method and LDA in one population of 24 neurons recorded from the rat barrel field somatosensory cortex. (*a*) Two-way interaction between classification method employed (PSTH, blue line with circles; LDA, red line with squares) and level of PCA dimension reduction, namely retaining 90% of variance (PCA90, left column), 50% of variance (PCA50, middle column), or 20% of variance (PCA20, right column). (*b*) Two-way interaction between classification method (PSTH, blue line; LDA red line) and time resolution (i.e., the bin size used to divide the poststimulus time window, ranging from 1 to 40 ms on the *x* axis). (*c*) Three-way interaction between the method (PSTH, blue line with circles; LDA, red line with squares), the dimension reduction (PCA90 left, PCA50 middle, PCA20 right), and the bin size (from 1 to 40 ms on the *x* axis). (*d*) Different visualization of three-way interaction: This time the colors represent different levels of dimension reduction (PCA90 in blue with circles, PCA50 in red with squares and PCA20 in green with diagonal squares), the methods are separated in the two plots (PSTH left, LDA right), and the bin size is always on the *x* axis (1–40). (*e*) Comparison between PSTH-based classification method applied on raw data (blue line with circles) and after PCA dimension reduction, retaining 90% of the variance (red line with squares) for all the binsizes (from 1 to 40 on the *x* axis).

the fact that when the bin size is decreased, the number of variables increases and the estimation of the covariance matrices in LDA deteriorates.

The second major result is that the PSTH-based classification performance increases as the percentage of variance retained after PCA dimension reduction increases (i.e., with the number of principal components; Fig. 29.6d). On the other hand, LDA performance as a function of the retained variance is apparently less predictable (Fig. 29.6d): For example, 90% of the variance provides the bests results at the maximum bin size (40 ms) and the poorest results at the minimum bin size (1 ms). These observations can be interpreted in terms of a trade-off between increasing the amount of information employable for the classification and deteriorating the covariance estimate, increasing the number of variables. This trade-off renders LDA less suitable for studying fine differences of temporal resolution (i.e., different bin sizes) in the neural code, which is instead feasible using the PSTH-based classification method.

Finally, the third major result is that there is no difference in the performance of the PSTH-based method applied after 90% of the variance is retained or on raw neural data (Fig. 29.6e). Importantly, dimension reduction was a necessary step for LDA. This means that the PSTH-based method, in contrast with LDA, can be optimally applied on the raw neural data.

29.4   CONCLUSIONS AND FUTURE DIRECTIONS

We described methods to simultaneously record ensembles of single neurons for longitudinal studies of population activity. Experimentally induced changes in parameters can be assessed by recording through electrode arrays chronically implanted in the neural tissue. Arrays with 32–64 electrodes yield 50–100 relatively stable single units that can be recorded over long time periods in both anesthetized and awake conditions, including behaving tasks. Such data can now be routinely obtained from the same animals before and after injury, pharmacological manipulations, and other interventions. Deficits and recovery can be assessed through behavior testing and monitoring changes in the firing patterns of these cells over time.

These longitudinal studies necessarily generate large data sets for which nonlinear analysis techniques such as nonlinear regression and neural networks are too computationally intensive to be practically applied. The PSTH-based method for classifying neuronal responses presented here is one of the simplest—and most computationally efficient—classifiers that can be designed; nevertheless the performance obtained by applying it on highly dimensional and sparse neural data is surprisingly high. We therefore suggest that the PSTH-based method is an efficient alternative to more sophisticated methods such as LDA and ANNs to study the neural code as a classification problem, especially when large data sets are used and when the time resolution of the neural code is one of the factors of interest.

A long-term goal of this work is to apply the methods presented here to the plastic changes after spinal injury. A spinal cord injury is also a brain injury. The neurons in the motor cortex send their axons down the spinal cord. These axons will be severed or injured after a spinal injury, thereby creating changes in the neural circuits used to code for sensory and motor events. While it has been well established that an adult brain undergoes reorganization in response to changes in inputs from the periphery, surprisingly little information exists on the kind of changes that take place in the brain after spinal injury. For example, previous studies have shown that after lesions to the dorsal columns there is either no change, a reduction, elimination, or even increase in the response properties of neurons in different cortical areas and to different stimulus paradigms [60–63]. The classical justification for these conflicting observations was based on the difference in the lesions, which is indeed a great source of variability. However, technological limitations (i.e., the restriction to record very few neurons at a time in multiple acute penetrations) confined most of the cited works to the study of anatomy or passive response properties after the lesion. To assess changes in brain circuitry, neurons must be recorded before and after the spinal injury, hence chronic recording of single neurons is essential. A further advantage to chronic neural recordings is that we can not only study reorganization after the injury but also measure how these ensembles of neurons respond functionally to therapeutic intervention. We suggest that the methods outlined here are ideally suited to study how ensembles of neurons code for sensori motor events, the role of plasticity in these circuits, and most importantly how these circuits in the brain are modified in pathological conditions, in particular after spinal cord injury.

ACKNOWLEDGMENTS

We would like to thank Larry Andrews of NB Labs for the microwire electrodes used in this work and Spire Corporation for assistance in manufacturing the ceramic-based microelectrodes arrays. We would also like to thank Harvey Wiggins of Plexon, Inc. for the multineuron acquisition system. This work was supported by NIH grant 2P50NS24707.

REFERENCES

1. Georgopoulos, A. P., Schwartz, A. B., and Kettner, R. E. Neuronal population coding of movement direction. *Science*, 1986; 233:1416–1419.

2. Gerstein, G. L. and Perkel, D. H. Simultaneously recorded trains of action potentials: Analysis and functional interpretation. *Science*, 1969; 164:828–830.

3. Gerstein, G. L., Perkel, D. H., and Subramanian, K. N. Identification of functionally related neural assemblies. *Brain Res.*, 1978; 140:43–62.

4. Kralik, J. D., Dimitrov, D. F., Krupa, D. J., Katz, D. B., Cohen, D., and Nicolelis, M. A. Techniques for long-term multisite neuronal ensemble recordings in behaving animals. *Methods*, 2001; 25:121–150.

5. Nicolelis, M. A., Dimitrov, D., Carmena, J. M., Crist, R., Lehew, G., Kralik, J. D., and Wise, S. P., Chronic, multisite, multielectrode recordings in macaque monkeys. *Proc. Natl. Acad. Sci. USA*, 2003; 100:11041–11046.

6. Freeman, W. J. The physiological basis of mental images. *Biol. Psych.*, 1983; 18:1107–1125.

7. Nicolelis, M. A., Baccala, L. A., Lin, R. C., and Chapin, J. K. Sensorimotor encoding by synchronous neural ensemble activity at multiple levels of the somatosensory system. *Science*, 1995; 268:1353–1358.

8. Eggermont, J. J. Functional aspects of synchrony and correlation in the auditory nervous system. *Concepts Neurosci.*, 1993; 4:105.

9. Moxon, K. A., Gerhardt, G. A., Bickford, P. C., Rose, G. M., Woodward, D. J., and Adler, L. E. Multiple single units and populations responses during inhibitory gating of hippocampal auditory response in freely-moving rats. *Brain Res.*, 1999; 825:75–85.

10. Ghazanfar, A. A., Stambaugh, C. R., and Nicolelis, M. A. Encoding of tactile stimulus location by somatosensory thalamocortical ensembles. *J. Neurosci.*, 2000; 20:3761–3775.

11. Evarts, E. V. Precentral and postcentral cortical activity in association with visually triggered movement. *J. Neurophysiol.*, 1974; 37(2):373.

12. Mountcastle, V. B., Lynch, J. C., Georgopoulus, A., Sakata, H., and Acuna, C. Posterior parietal association cortex of the monkey: Command functions for operations within extrapersonal space. *J. Neurophysiol.*, 1975; 38(4):871.

13. Wilson, M. A. and McNaughton, B. L., Dynamics of the hippocampal ensemble code for space. *Science*, 1996; 261:1055.

14. Wilson, M. A. and McNaughton, B. L. Reactivation of hippocampal ensemble memories during sleep. *Science*, 1996; 265:6761.

15. Carmena, J. M., Lebedev, M. A., Crist, R. E., O'Doherty, J. E., Santucci, D. M., Dimitrov, D. F., Patil, P. G., Henriquez, C. S., and Nicolelis, M. A. L. Learning to control a brain-machine interface for reaching and grasping by primates. *PLoS Biol.*, 2003; 1(2):1–16.

16. Moxon, K. A. Multichannel electrode design: Considerations for different applications. In *Methods for Simultaneous Neuronal Ensemble Recordings*,

M. A. L. NICOLELIS, Ed., CRC Press, Boca Raton, FL, 1999, pp. 25–45.

17. WHEELER, B. C. Automatic discrimination of singe units. In *Methods for Neural Ensemble Recordings*, M. A. L. NICOLELIS, Ed., Boca Raton, FL, CRC Press, 1999, p.61–77.

18. WISE, K. D. AND WEISSMAN, R. H. Thin films of glass and their application to biomedical sensors. *Med. Biol. Eng.*, 1971, 9:339–350.

19. POCHAY, P., WISE, K. M., ALLARD, L. F., AND RUTLEDGE, L. T., A multichannel depth array fabricated using electron-beam lithography. *IEEE Trans. Biomed. Eng.*, 1979; 26(4):199–206.

20. BLUM, N. A., CARKHUFF, B. G., CHARLES, H. K., EDWARDS, R. L., AND MEYER, R. A. Multisite microprobes for neural recordings. *IEEE Trans. Biomed. Eng.*, 1991; 38(1):68–74.

21. PROHASKA, O. J., OLCAYTUG, F., PFUNDNER, P., AND DRAGUAN, H. Thin-film multiple electrode probes: Possibilities and limitations. *IEEE Trans. Biomed. Eng.*, 1986; 33(2):223–229.

22. EICHMAN, H. AND KUPERSTEIN, M. Extracellular neural recording with multichannel microelectrodes. *J. Electrophysiol. Tech.*, 1986; 13:189.

23. DRAKE, D. L., WISE, K. D., FARRAYE, J., ANDERSON, D. J., AND BEMENT, S. L. Performance of planar multisite microarrays in recording extracellular single-unit intracortical activity. *IEEE Trans. Biomed. Eng.*, 1988; 35:719–732.

24. HOOGERWERF, A. C. AND WISE, K. D. A three-dimensional microelectrode array for chronic neural recording. *IEEE Trans. Biomed. Eng.*, 1994; 41(12): 1136–1146.

25. CARTER R. R. AND HOUK, J. C. Multiple single-unit recordings from the CNS using thin-film electrode arrays. *IEEE Trans. Rehab. Eng.*, 1993; 1:173–184.

26. WILLIAMS, J. C., RENNAKER, R. L., AND KIPKE, D. R., Long-term recording characteristics of wire microelectrode arrays implanted in cerebral cortex. *Brain Res. Protocols*, 1999; 4:303–313.

27. BRAGIN, J. H., WILSON, C. L., ANDERSON, D. J., ENGLE, JR., J., AND BUZSAKI, G. Multiple site silicon-based probes for chronic recordings in freely moving rats: Implantation, recording and histological verification. *J. Neurosci. Methods*, 2000, 98:77–82.

28. MOXON, K. A., LEISER, S. C., GERHARDT, G. A., BARBEE, K., AND CHAPIN, J. K. Ceramic based multisite electrode arrays for electrode recording. *IEEE Trans. Biomed. Eng.*, 2004.

29. NICOLELIS, M. A., GHAZANFAR, A. A., FAGGIN, B. M., VOTAW, S., AND OLIVEIRA, L. M. Reconstructing the engram: Simultaneous, multisite, many single neuron recordings. *Neuron*, 1997; 18:529–537.

30. LEE, C., ROHRER, W. H., AND SPARKS, D. L. Population coding of saccadic eye movements by neurons in the superior colliculus. *Nature*, 1988; 332:357–360.

31. BIALEK, W., RIEKE, F., DE Ruyter VAN STEVENINCK, R. R., AND WARLAND, D. Reading a neural code. *Science*, 1991; 252:1854–1857.

32. YOUNG, M. P. AND YAMANE, S. Sparse population coding of faces in the inferotemporal cortex. *Science*, 1992; 256:1327–1331.

33. DECHARMS, R. C. AND MERZENICH, M. M. Primary cortical representation of sounds by the coordination of action-potential timing. *Nature*, 1996; 381:610–613.

34. DE RUYTER VAN STEVENINCK, R. R., LEWEN, G. D., STRONG, S. P., KOBERLE, R., AND BIALEK, W. Reproducibility and variability in neural spike trains. *Science*, 1997; 275:1805–1808.

35. LEWIS, J. E. AND KRISTAN, W. B., JR. Representation of touch location by a population of leech sensory neurons. *J. Neurophysiol.*, 1998; 80:2584–2592.

36. MCALPINE, D., JIANG, D., AND PALMER, A. R. A neural code for low-frequency sound localization in mammals. *Nat. Neurosci.*, 2001; 4:396–401.

37. PETERSEN, R. S., PANZERI, S., AND DIAMOND, M. E. Population coding of stimulus location in rat somatosensory cortex. *Neuron*, 2001; 32:503–514.

38. PASUPATHY, A. AND CONNOR, C. E. Population coding of shape in area V4. *Nat. Neurosci.*, 2002; 5:1332–1338.

39. NIRENBERG, S. AND LATHAM, P. E. Decoding neuronal spike trains: How important are correlations? *Proc. Natl. Acad. Sci. USA*, 2003; 100:7348–7353.

40. NICOLELIS, M. A., LIN, R. C., AND CHAPIN, J. K. Neonatal whisker removal reduces the discrimination of tactile stimuli by thalamic ensembles in adult rats. *J. Neurophysiol.*, 1997; 78:1691–1706.

41. NICOLELIS, M. A., GHAZANFAR, A. A., STAMBAUGH, C. R., OLIVEIRA, L. M., LAUBACH, M., CHAPIN, J. K., NELSON, R. J., AND KAAS, J. H. Simultaneous encoding of tactile information by three primate cortical areas. *Nat. Neurosci.*, 1998; 1:621–630.

42. KRZANOWSKI, W. J. *Principles of Multivariate Analysis*. Oxford University Press, Oxford, 1988.

43. GOCHIN, P. M., COLOMBO, M., DORFMAN, G. A., GERSTEIN, G. L., AND GROSS, C. G. Neural ensemble coding in inferior temporal cortex. *J. Neurophysiol.*, 1994; 71:2325–2337.

44. SCHOENBAUM, G. AND EICHENBAUM, H. Information coding in the rodent prefrontal cortex. II. Ensemble activity in orbitofrontal cortex. *J. Neurophysiol.*, 1995; 74:751–762.

45. DEADWYLER, S. A., BUNN, T., AND HAMPSON, R. E. Hippocampal ensemble activity during spatial delayed-nonmatch-to-sample performance in rats. *J. Neurosci.*, 1996; 16:354–372.

46. MIDDLEBROOKS, J. C., CLOCK, A. E., XU, L., AND GREEN, D. M. A panoramic code for sound location by cortical neurons. *Science*, 1994; 264:842–844.

47. JACKSON, J. E. *A User's Guide to Principal Components.* Wiley, New York, 1991, pp. 1–25.

48. CHAPIN, J. K. AND NICOLELIS, M. A. L., Principal component analysis of neuronal ensemble activity reveals multidimensional somatosensory representations. *J Neurosci. Methods*, 1999; 94:121–140.

49. BELL, A. J. AND SEJNOWSKI, T. J. An information-maximization approach to blind separation and blind deconvolution. *Neural Comput.*, 1995; 7:1129–1159.

50. LAUBACH, M., SHULER, M., AND NICOLELIS, M. A. Independent component analyses for quantifying

neuronal ensemble interactions. *J. Neurosci. Methods*, 1999; 94:141–154.

51. KOHONEN, T. *Self-Organization and Associative Memory*, 2nd ed. Springer-Verlag, Berlin, 1987.

52. GERSTEIN, G. L. AND KIANG, N. Y. S. An approach to the quantitative analysis of electrophysiological data from single neurons. *Biophys. J.*, 1960; 1:15–28.

53. PANZERI, S., PETERSEN, R. S., SCHULTZ, S. R., LEBEDEV, M., AND DIAMOND, M. E. The role of spike timing in the coding of stimulus location in rat somatosensory cortex. *Neuron*, 2001; 29:769–777.

54. FOFFANI, G. AND MOXON, K. A. PSTH-based classification of sensory stimuli using ensembles of single neurons. *J. Neurosci. Methods*, 2004; 135:107–120.

55. SIMONS, D. J. AND LAND, P. W. Early experience of tactile stimulation influences organization of somatic sensory cortex. *Nature*, 1987; 326:694–697.

56. AHISSAR, E., SOSNIK, R., AND HAIDARLIU, S. Transformation from temporal to rate coding in a somatosensory thalamocortical pathway. *Nature*, 2000; 406:302–306.

57. PETERSEN, R. S., PANZERI, S., AND DIAMOND, M. E. Population coding in somatosensory cortex. *Curr. Opin. Neurobiol.*, 2002; 12:441–447.

58. HOPFIELD, J. J. AND HERZ, A. V. Rapid local synchronization of action potentials: Toward computation with coupled integrate-and-fire neurons. *Proc. Natl. Acad. Sci. USA*, 1995; 92:6655–6662.

59. HOPFIELD, J. J. Pattern recognition computation using action potential timing for stimulus representation. *Nature*, 1995; 376:33–36.

60. SCHWARTZ, A. S., EIDELBERG, E., MARCHOK, P., AND AZULAY, A. Tactile discrimination in the monkey after section of the dorsal funiculus and lateral lemniscus. *Exp. Neurol.*, 1972; 37(3):582–596.

61. TOMMERDAHL, M., WHITSEL, B. L., VIERCK, C. J. JR., FAVOROV, O., JULIANO, S., COOPER, B., METZ, C., AND NAKHLE, B. Effects of spinal dorsal column transection on the response of monkey anterior parietal cortex to repetitive skin stimulation. *Cereb. Cortex.*, 1996; 6(2):131–55.

62. JAIN, N., CATANIA, K. C., AND KAAS, J. H. Deactivation and reactivation of somatosensory cortex after dorsal spinal cord injury. *Nature*, 1997; 386(6624):495–498.

63. VIERCK, C. J. Impaired detection of repetitive stimulation following interruption of the dorsal spinal column in primates. *Somatosens. Mot. Res.*, 1998; 15(2):157–63.

CHAPTER 30

MODEL OF MAMMALIAN VISUAL SYSTEM WITH OPTICAL LOGIC CELLS

J. A. Martín-Pereda and A. González Marcos

30.1 INTRODUCTION

The retina of the vertebrates has been, from the pioneering work by Cajal [1] at the end of nineteenth century, a main topic for biologists, physiologists, neurologists, physicists, and engineers. The reason for this interest was due to several factors. The first one is related to the importance of the vision processes in the relationship between the living beings and their environment. The second one is the large number of functions carried out by the retina. Besides the above facts, the retina possesses other characteristics that justify its study. These reasons are those derived from the previously mentioned great amount of information that is processed and the way it is done. The form of processing images in parallel is superior to the way in which it is done by most of the present artificial vision systems. Lessons obtained from the way the retina works can be a very good source of new ideas. These lessons could be adopted and implemented in other systems. This is the aim of this chapter.

30.2 PREVIOUS WORKS

Many works have been carried out this way in the last years following the above indicated lines. Detailed journal articles describing a variety of neurovision system architectures for preattentive vision, visual perception, object recognition, color vision, stereovision, and image restoration are included in [2]. One of the more interesting results was obtained by Mead [3] in his work on a silicon retina. He built a configuration that was modeled on the distal portion of the vertebrate retina. The resulting chip generates, in real time, outputs that correspond directly to signals observed in the corresponding levels of biological retinas. Moreover, it shows a tolerance for device imperfections that is characteristic of a collective system. Many other works are related with to very large scale integrated (VLSI) technologies. In this way, a compact VLSI neuroprocessor which includes neurons and synapse cells, able to estimate the motion of each pixel of a system with 25 different velocities, has been reported [4]. Image detection and smoothing have been obtained with an active resistive mesh containing both positive and negative resistors to implement a Gaussian convolution in two dimensions [5]. Image processors that are

Handbook of Neural Engineering. Edited by Metin Akay

based on charge-coupled devices (CCDs) have been reported in several works [6]. In general, three broad areas of neural network implementation are present: computer-based software, electronic hardware, and optical/optoelectronic hardware. Information about them is given in [2].

Most of the operations handled by the biological retina are performed with very sophisticated hardware, with no software present. The architectures are based on a high number of neural circuits with same basic, the neuron. Each neuron presents its own behavior according to its working conditions. Two similar neurons can give rise to different outputs depending on the way their synapses are arranged. These outputs will determine the processes in the following stages.

The present work will present an emulation of the vertebrate retina. The main block for this emulation is an optical logic cell, employed by the authors to perform some parts of an optical computer [7–10]. As will be shown, this simple configuration allows emulation of many possible outputs of ganglion cells and to process information about light motion in front of the retina. The information obtained from this model is based on different characteristics of the output pulses. These characteristics determine the properties of the impinging light.

An important point needs to be considered. Our model of the retina is a photonic model. It is true that many of the functions to be implemented can be carried out by electronic methods. This kind of implementation should be easier, in some aspects, than the optoelectronic one. As a matter of fact, the previously reported retina model by Mead [3] is able to perform most of the needed functions with an electronic chip. Microelectronics is currently much more developed than photonics and gives a higher versatility than optical systems. But some particular aspects are very difficult to develop with electronic circuits. The visual system is, as was indicated, a highly parallel system. The quantity of information carried out from retina to brain is huge. Living beings do it by a large number of channels running along the visual track. This number is much higher than the larger interconnection channels in supercomputers. To try to emulate this architecture is almost impossible with present-day electronic technology. But optical methods, as for example wavelength domain multiplexing (WDM), allow carrying a large number of channels through the same physical link without any possible crosstalk. A certain number of parallel physical links are equivalent to the same number of logical links, each with different wavelength, and go through the same physical space. This is one of the possible advantages of working with light instead of electrons. Some others will appear in this work.

30.3 NEUROPHYSIOLOGICAL BASIS OF MODEL

The retina is quite different from any of other mammalian sense organ in that a good deal of the neural processing of the afferent information has already occurred before it reaches the fibers of the optic nerve. The fibers of the optic nerve are in fact two synapses removed from the retinal receptors, and particularly as far as the rods and cones in the periphery are concerned, there is considerable convergence of information from large groups of receptors. What happens is that receptors synapse with bipolar cells, and these in turn synapse with the million or so ganglion cells whose axons form the optic nerve. These two types of neurons form consecutive layers on top of the receptor layer and are mingled with two other types of interneurons that make predominantly sideways connections. These are the horizontal cells at the bipolar/receptor level and the amacrine cells at the ganglion cell/bipolar level.

The way neurons work also needs to be pointed out. First, three of them, namely photoreceptor, horizontal, and bipolar cells, respond to light by means of hyperpolarization. These neurons do not produce action potentials. The cells in the second group, amacrine and ganglion cells, show a large variety of responses. They are action potentials in every case as well as depolarizing. Different types are reported in the literature [11, 12]. The amacrine cells show transient depolarizing responses, including what are apparently all-or-nothing action potentials, at the onset or cessation of light. Finally, ganglion cells can be divided into three types according to their response to illumination. "On" units respond to the onset of illumination, "off" units to the cessation of illumination, and on–off units to both onset and cessation. Another classification is made on the basis of the action potentials produced in response to stimuli. Most cells produce "transient" responses, with just a few action potentials immediately after a change in illumination.

Several retina models appear in the literature following these lines [13, 14]. The studied configuration in this chapter appears in Figure 30.1 and it was partially reported

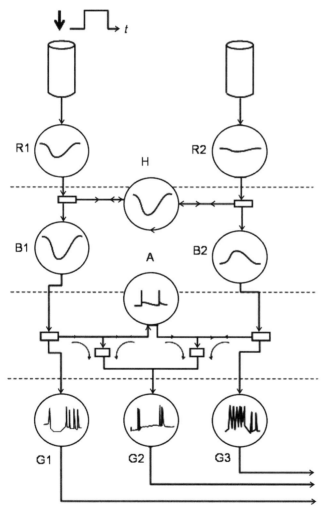

Figure 30.1   Intracellular responses of various retinal cells and their connections with each other. The receptor of the left is illuminated with a brief flash of light. R, receptors; H, horizontal cell; B, bipolar cells; A, amacrine cell; G, ganglion cells.

by us [15–19]. Just two photoreceptors have been taken. This configuration is similar the one proposed by Dowling [13] to summarize the activity of the various retinal cells. As can be seen, the receptor on the left is illuminated with a brief flash of light imposed on a dim background that illuminates both receptors, R1 and R2. A large response is observed in the stimulated receptor whereas the adjacent receptor that is not illuminated (right receptor) shows only a small response that probably reflects mainly the electrical coupling between the photoreceptor cells. Bipolar and horizontal cells are both activated by the receptors. The scheme of Figure 30.1 shows that bipolar cell B1 is polarized strongly in a graded and sustained fashion by direct contacts with receptor R1. Moreover, this bipolar cell potential is antagonized by horizontal–bipolar cell B2 interaction. Bipolar cell B2 responds to indirect (surround) illuminations by depolarizing. As can be seen, the switch from hyperpolarizing to depolarizing potentials along the surrounding illumination pathway occurs at the horizontal–bipolar junction.

Amacrine cell, A, responds to light mainly with transient depolarizing potentials at the onset and cessation of spot illumination. The responses of the two basic types of ganglion cells found in the vertebrate retinas appear to be closely related to the responses of the input neurones to the ganglion cells. The G1 ganglion cell has a receptive field organization very similar to that of bipolar cells. Central illumination hyperpolarizes both the bipolar and ganglion cells, B1 and G1, in a sustained fashion, and surrounding illumination depolarizes the bipolar B3 and ganglion G3 cell in a sustained fashion. This type of ganglion cell appears to receive most of its synaptic input directly from the bipolar cell terminals through excitatory synapses. The ganglion cells illustrated in Figure 30.1 are off-center cells but there are some other types present in the vertebrate retinas. Ganglion cell G2 responds transiently to retinal illumination, much as the transient amacrine cells do. This type of response is the one adopted in our model.

Although this model is a very simple one, it is very useful to implement most of the functions performed at the mammalian retina. More complicated models can be obtained directly from this one.

30.4   STRUCTURE OF FUNDAMENTAL BUILDING BLOCK

As has been pointed out before, a simple cell has been the basis for the architecture of the proposed system. This cell has been employed by us as the main block for some structures in optical computing [7–10]. Its characteristics have been presented in several places and they may be synthesized as follows. Figure 30.2 shows the real internal configuration of the optically programmable logic cell (OPLC) with its two basic nonlinear optical devices. The main characteristics of these nonlinear devices are also shown. Device Q corresponds to a thresholding or switching device. Device P response is similar to that achieved by a self-electro-optic-effect device (SEED), a very common device in optical computing.

Additional details, ampler than those presented here as well as its physical implementation, can be found in [7]. Their characteristics are shown in the insets of Figure 30.2. It has two signal inputs, I_1 and I_2, as well as two signal outputs, O_1 and O_2, and two control signals, h and g. Because the cell works with optical signals, it allows an internal multilevel processing. In this way, input and output data signals are binary but the control is performed by multilevel signals. Table 30.1 lists the different programming combinations available from our OPLC. At least in one of the outputs, the eight

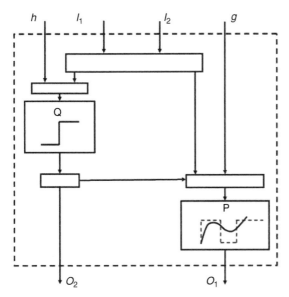

Figure 30.2   Basic configuration of the OPLC: I, input signals; O, output signals; h, g, control signals. Q, on-off device; P, SEED.

Boolean functions—AND, OR, XOR, ON and the negative NAND, NOR, XNOR and OFF—can be obtained. When some feedback is added, for instance connecting a part of the output from O_1 as control signal to the P device, as well as a multilevel periodic input, some nonlinear behavior is obtained. Under certain conditions and with a certain value of the feedback time delay, it is possible to obtain even a chaotic signal. Moreover, with no input, the frequency of the periodic signal is a function of the feedback delay time.

A detail of each block in Figure 30.2 appears in Figure 30.3. The study reported here was performed by computer simulation of the OPLC with the Simulink application from MATLAB. It has been considered the ideal response for the nonlinear devices. This fact relates to the output level but not to the possible hysteresis. The hysteresis appears in any real behavior of a nonlinear device and should be the origin of noncontrolling responses when the signal level is beyond the tolerance range. An example of the logic behavior of this cell appears in Table 30.1.

TABLE 30.1

Control signal to Q ———————> Control signal to P ↓	Output from Q–O_2: AND 0–0.4 Output from P–O_1	Output from Q–O_2: OR 0.5–0.9 Output from P–O_1	Output from Q–O_2: ON 1.0–2.0 Output from P–O_1
0–0.4	XOR	XOR	NAND
0.5	NAND	NOR	NOR
0.6–0.9	ON	XNOR	XNOR
1.0	XNOR	XNOR	AND
1.1–1.4	XNOR	ON	OR
1.5	AND	OR	OR
1.6–2.0	OR	OR	ON
2.0–2.5	ON	ON	ON

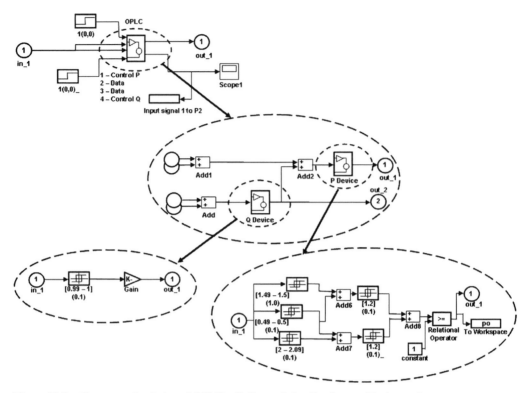

Figure 30.3   Computer simulation of OPLC cell. Internal details of every block are shown.

Some comments are needed to justify the employed cell. Several approaches have been adopted with respect to the behavior of the neural units. The most common has been assigning them a single logic function. Moreover, some nonlinear activation operation is added in order to provide a bounded scalar neural output. These operators range from linear to Gaussian. The most common are unipolar and bipolar sigmoidal functions. These two operations perform the two key elements for signal processing in the biological neuron, synapse, and soma. They are the basis for most of the present neural networks systems. The McCulloch-Pitts and Fukushima models follow these lines. The adaptive linear neuron, Adaline, has $n + 1$ inputs and two outputs: an analog one and a binary one. Logic gates such as NOT, OR, AND, and MAJ can be realized with a single Adaline. But the simplest nonlinearly separable binary logic functions, the exclusive-or (XOR) logic function, needs two Adaline "neurons." An artificial neural network composed of many Adalines is more flexible than a standard logic circuit employing logic gates since such a network can be trained to realize different logic functions. But this use of two cells should be improved to another with just a single unit.

As we have indicated, our present approach adopts a different point of view. The basic unit employed is a logic unit with two signal inputs, two outputs, and two more inputs for control signals. This configuration allows, as we have indicated previously, any logic Boolean function from the input signals. The unit cell is more complex than previous units for simple logic gates, but its main advantage is that only a single cell needs to be employed for any logic function. Moreover, the presence of two control inputs gave the possibility of implementing different functions and working operations at both outputs.

Electronic and optoelectronic implementations are possible because, as will be shown, they are based on photonic devices. This fact gives a new possible working philosophy: to carry different signals through the same physical channel if these signals have different optical wavelengths.

30.5 RETINA MODEL SIMULATION

The adopted architecture has the configuration shown in Figure 30.4. Each blackbox has the internal structure of Figure 30.2. In most cases, only one or two inputs receive signals from previous layers. Other inputs get no signals. With respect to outputs, just one of the two that are possible is connected to the following cell. This situation gives the opportunity to employ the nonoperative inputs and outputs in other possible processes. We report here the simplest structure and maintain other possibilities for further work.

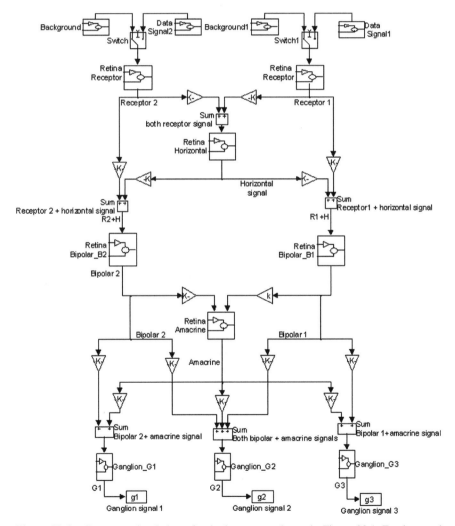

Figure 30.4   Computer simulation of retinal structure shown in Figure 30.1. Fundamental blocks are OPLCs (Figs. 30.2 and 30.3).

We have adopted in most of the cases, according to the way photoreceptors work, the level of a logic 1 as the resting potential when there is no light impinging onto the photoreceptors and a logic 0 when an optical signal is present. Because of the binary character of our model, there is no possibility to register different light levels. This point will be considered in future work.

The above situation means that the receptor cells must operate with a NAND function. And so the OPLC was programmed for this task. In our model, when there is some light acting on receptor R1, it switches from 1 to 0. No light in receptor R2 implies keeping the 1 level.

The photoreceptor response goes then to the horizontal and bipolar cells. A NAND function is performed by the horizontal cell. In bipolar cells B1 and B2, the situation is somehow different in both. Dowling's model gives a hyperpolarization response for B1 and a depolarization for B2. This response is obtained with AND and NAND functions on cells B1 and B2, respectively. This behavior can be achieved with two different configurations of the OPLC. These configurations correspond not to changes in the internal structure but to the way the connections are taken. This means adopting a particular arrangement for how data and control signals reach the inputs of the OPLC.

There are several possible outputs obtained from amacrine and ganglion cells. The case reported here corresponds to a nonsymmetrical way of working. Namely, light on receptor R1 and no light on R2 give outputs at ganglion cells G1, G2, and G3 different from the case when there is no light at R1 and light at R2.

30.6 RESULTS

The results obtained for the above-indicated configuration correspond to the behavior of the ganglion cells in the Dowling model. They are shown in Figure 30.5. Amacrine cells must give a different output signal than the ones appearing at receptor, horizontal, and bipolar cells. Its resting state is, in our case, a logic 1 and switches to a logic 0 at the beginning and the end of light pulses. This switch is in the form of a brief impulse. After this peak, it returns to the resting state. Hence, its behavior is in the form of action potentials with a burst of just one pulse.

An important point to be said here is that, with this arrangement, the amacrine is in charge of detecting the illumination time on receptors. This aspect is not considered in the literature. The more important function customarily assigned to amacrine cells corresponds to spatial aspects of visual signals. We believe that this proposed temporal aspect needs to be considered in future work.

The composite signals from bipolar and amacrine cells reach the ganglion cells. There are several possible responses from these ganglion cells. Three of them appear in Figure 30.5. They have been obtained with our present computer simulation. The first result corresponds to ganglion G1; its resting state is a logic 1. It shows pulses, going to 0, at the beginning and at the end of the input flash. The resting state of ganglion G3 is again a logic 1, but this resting state changes to 0 during the whole light flash interval; moreover, a new pulse appears now in the transition from light to darkness. Finally, ganglion G2 has a resting state of logic a 0 when no light is present and changes to a 1 when light appears, with pulses at the beginning and end of the light interval.

The above-indicated behavior corresponds, in a certain way, to hyperpolarizing bipolar cells and off-center ganglion cells (G1 and G3) that respond to direct central illumination (left side in Figure 30.4) by hyperpolarizing and to indirect (surround) illumination (right side) by depolarizing. The switch from hyperpolarizing to depolarizing

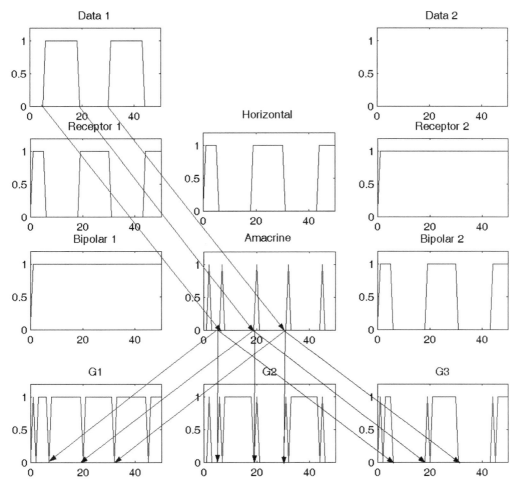

Figure 30.5   Results obtained from computer simulation of retinal structure (Fig. 36.4).

potentials along the surrounding illumination pathway occurs at the horizontal–bipolar junction. The on–off ganglion cell (G2) receives strong inhibitory input from amacrine cells; these cells receive their excitatory input from both amacrine and bipolar cells. In the reported simulation, just single pulses appear in the final signals. No pulses are present as in Dowling's model. But this result may be easily obtained with the possibility offered by the OPLC to respond with a periodic signal when adequate feedback is added. This possibility has been reported by us and it may be easily implemented. It has not been applied in this case in order to obtain a clearer result.

This structure has the characteristic of being nonsymmetric. This means that the behavior of the total retina layer is not the same if the light is coming from the right path or from the left path.

30.7   EXTENSION TO HIGHER PROCESSING LEVELS

Although some invertebrates process most of the information they get in their retinal layers, most higher order living beings process this information in different layers of their visual cortex. Hence, the visual data have to be transferred to a different region to be analyzed. This is the case in mammal and the one presented here.

Once details from a certain image have been processed at the retina, the next step is to transfer the obtained information to the primary visual cortex, V1. This transferrence is done, in mammals, via the lateral geniculate nucleus (LGN) in an organized, retinopic fashion. The visual cortex analyzes information in several ways, and one possible model is provided by Hubel and Wiesel [21]. The most interesting part of this model conceives pattern analysis as being a process that breaks down the visual scene into line segments, with different populations of neurons responsible for reacting to the different orientations of these external contours. This process is the result of the many information channels coming from the ganglion cells through the optic nerve to the visual cortex. A possible model to implement this transferrence of information should have as many physical channels as conducting nerves going from the retina to the V1 zone. Each channel would carry the response of each ganglion cell. This situation, from a theoretical point of view, is straightforward to establish but, from an experimental configuration, is very difficult to implement. The main problem arises from the huge number of channels going between these two regions.

A possible solution to this problem has been developed by us as a possible way to transfer simple images between two distant areas [20]. A technique based on WDM can be implemented. A simple scheme appears in Figure 30.6. This scheme shows a simple way to carry information from ganglion cells to the visual cortex through a limited number of physical channels. This model is not the way living beings perform this task, but it offers a possible way to implement an equivalent behavior in an artificial system. It is based on the transmission through the same physical channel of the signals to be processed with different optical wavelengths. This allows the use of just one physical channel, namely an

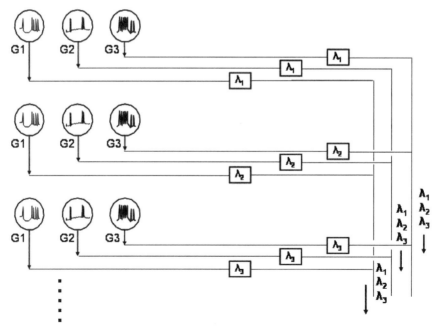

Figure 30.6   Signals transmission from ganglion cells to first layers of visual cortex. Signals generated at the same type of ganglions are multiplexed into a single channel with different wavelengths.

optical fiber, to transmit a large number of different signals. Figure 30.6 shows an example corresponding to the case of sending the responses of different equivalent blocks from the retina to the cortex. Each signal from the same type of ganglion cells is carried out by a different optical frequency. In this way, the responses overlapped without crosstalk. A simple wavelength multiplexer can be employed in this step. A demultiplexer will perform the inverse operation in the region corresponding to the cortex.

To apply this method, we have implemented a way to analyze a characteristic not usually analyzed in common artificial pattern recognition systems. This system analyzes possible symmetries in linear configurations. Moreover, in order to simplify the result, data are light pulse inputs similar to the ones employed in the previous retina model. This model may be easily enlarged to obtain more complex results.

A linear configuration has been adopted for the first layer of the model. It corresponds to the possible signals coming out from ganglion cells and transmitted to a region of the cortex through a transmission channel. These signals are, as we have shown, the final result of the input data impinging on the receptors. They give information about the time length of the input light pulse and, in our linear case, number of excited receptors. This number has a close correspondence with the spatial length of the signal. These signals are compared at a second layer, in the way shown in Figure 30.7. Boxes correspond to the previously mentioned OPLCs. Output data are compared at the third and following layers in the same way as before. The process keeps going until it reaches layer 6, where there is just a single output.

The possibilities from this architecture are as many as possible logic functions are able to perform the OPLCs. This situation gives the possibility of implementing any type of synapses between neurons, both excitatory and inhibitory, as well as oscillatory behaviors. The reported architecture is just a two-dimensional configuration without lateral branches corresponding to other possible functions. Because the whole information corresponding to the input image is transferred to a train of pulses in each type of ganglion cells, a different type of architecture will have to be implemented for each of the input data to analyze (for example, colors, shapes, motion, and edges).

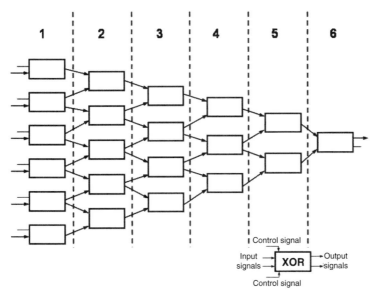

Figure 30.7   Simplified model for first layers of visual cortex.

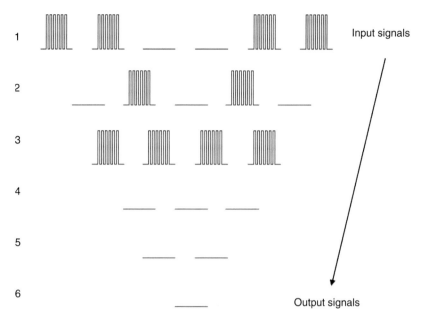

Figure 30.8   Signals generated at different layers of proposed model in Figure 30.7 when there is symmetry in initial linear image.

We have considered just one of the two possible outputs from the OPLCs. Moreover, just one type of logic function was adopted, namely, XOR. With these boundary conditions, our structure is able to recognize symmetries or asymmetries in the input signals. If some type of symmetry is present on the signal going onto the first layer of cells, a 0 will be obtained at the output. On the contrary, asymmetries will give rise to

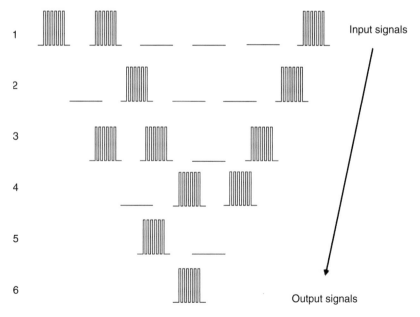

Figure 30.9   Signals generated at different layers of proposed model in Figure 30.7 when there is an asymmetry in initial linear image.

a 1 at the final cell. Some other features could be implemented with similar configurations. Some examples appear in Figures 30.8 and 30.9. Figure 30.8 corresponds to symmetry and Figure 30.9 to asymmetry.

30.8 FUTURE TRENDS

A model of the mammalian retina and some zones of the visual cortex has been reported in this work. It is based on optically programmable logic cell as basic units for the five different types of retinal neurons, namely photoreceptor, horizontal, bipolar, amacrine, and ganglion cells, and for the neurons at the V1 region. Every one has been modeled with the same elemental unit. Just some minor changes in the internal or external connections, as well as different weight functions, have allowed the simulation of different behaviors. This possibility provides a new method to model neural networks with just a single type of logic cell. Moreover, it allows easy implementation, with both electronic and optoelectronic methods. The reported structure is able to emulate almost any possible output signal from ganglion cells or any processing at higher levels of the visual cortex.

Although the presented retina model is composed of just two receptors, one horizontal, two bipolars, one amacrine, and three ganglion cells, it could be extended to larger configurations. The main advantage of this structure is the possibility of building almost any retinal architecture with the use of a single basic structure as well as the possibility of obtaining different ganglion outputs from a very simple structure. A further remark needs to be pointed out. According to the reported model, amacrine cells are in charge of detecting the illumination time on receptors. This fact has not been taken into account in any previous emulation and deserves further study.

With respect to the processing signals in the visual cortex, the reported example is just a very simple one. This model needs to be extended to more complex concepts than the one analyzed here. A possibility is using methods similar to the one employed by us in optical communications [20] to study some image characteristics as lines or shapes.

ACKNOWLEDGMENTS

This work was partly supported by CICYT, grant TEC2006-13874-C02-02/MIC, and CAM "Comunidad Autónoma de Madrid," grant 07T/0037/2000.

REFERENCES

1. S. RAMÓN Y CAJAL, "La rétine des vertébrés," *La céllule* 9, 119–257 (1892).
2. *Neuro-Vision Systems. Principles and Applications*, Eds. M. GUPTA AND G. K. KNOPF, IEEE Press, New York (1993).
3. C. MEAD, *Analog VLSI and neural Systems*, Addison-Wesley, Reading, MA (1989).
4. J.-C. LEE, B. J. SHEU, W.-C. FANG, AND R. CHELLAPPA, "VLSI neuroprocessors for video motion detection," *IEEE Trans. Neural Networks* 4, 178–191 (1993).
5. H. KOBAYASHI, J. L. WHITE, AND A. A. ABIDI, "An active resistor network for Gaussian filtering of images," *IEEE J. Solid-State Circuits* 25, 738–748 (1991).
6. A. M. CHIANG AND M. L. CHUANG, "A CCD programmable image processor and its neural network applications," *IEEE J. Solid-State Circuits* 26, 1894–1901 (1991).
7. A. GONZALEZ-MARCOS AND J. A. MARTIN-PEREDA, "Digital chaotic output from an optically processing element," *Opt. Eng.* 35, 525–535 (1996).
8. A. GONZALEZ-MARCOS AND J. A. MARTIN-PEREDA, "Analysis of irregular behaviour on an optical computing logic cell," *Opt. Laser Technol.* 32, 457–466 (2000).

9. J. A. MARTÍN-PEREDA and A. GONZALEZ-MARCOS, "Some connections between neurophysiology and optical computing based on the theory of complexity," in *Fluctuation Phenomena: Disorder and Nonlinearity*, Eds. J. BISHOP AND L. VÁZQUEZ, World Scientific Press, Singapore, pp. 107–113 (1995).

10. A. GONZALEZ-MARCOS AND J. A. MARTÍN-PEREDA, "Method to analyze the influence of hysteresis in optical arithmetic units," *Opt. Eng.* 40, 2371–2385 (2001).

11. J. G. NICHOLLS, A. ROBERT, AND B. G. WALLACE, *From Neuron to Brain*, Sinauer Associates, Inc., Sunderland, MA (1992).

12. G. M. SHEPHERD, *Neurobiology*, Oxford University Press, New York (1994).

13. R. A. BARKER, *Neuroscience*, Ellis Horwood, New York (1991).

14. J. E. DOWLING, *The Retina: An Approachable Part of the Brain,* Belknap Press of Harvard University Press, Cambridge, MA (1987).

15. J. A. MARTÍN-PEREDA AND A. GONZÁLEZ-MARCOS, "An approach to visual cortex operation: optical neuron model," *Presented at the Conference on Lasers and Electro-Optics*, CLEO/Europe'94, Technical Digest 355–356, Amsterdam (1994).

16. J. A. MARTÍN-PEREDA, A. GONZÁLEZ-MARCOS, AND C. SÁNCHEZ-GUILLÉN, "A model of the mammalian retina and the visual cortex. disorders of vision," in *Proc. IEEE Conference on Engineering in Medicine and Biology*, Technical Digest CD-ROM 138–140, Amsterdam (1996).

17. J. A. MARTÍN-PEREDA AND A. GONZÁLEZ-MARCOS, "Image processing based on a model of the mammalian retina," in *Applications of Photonics Technology, vol. 3: Closing the Gap between Theory, Development, and Application*, Eds. G. A. LAMPROPOULOS AND R. A. LESSARD. *Proc. SPIE*, 3491, 1185–1190 (1998).

18. J. A. MARTIN-PEREDA AND A. P. GONZALEZ-MARCOS, "Image characterisation based on the mammalian visual cortex," in *Applications of Digital Image Processing XXII*, Ed. A. G. FESCHER, *Proc. SPIE*, 3808, 614–623 (1999).

19. A. GONZALEZ-MARCOS AND J. A. MARTIN-PEREDA, "Image characterisation based on the human retina and the mammalian visual cortex," presented at the 2nd European Medical and Biological Engineering Conference EMBEC'02, Wien (2002).

20. J. A. MARTIN-PEREDA AND A. P. GONZALEZ-MARCOS, "Photonic processing subsystem based on visual cortex architecture," in *Active and Passive Optical Components for WDM Communications III*, Eds A. K. DUTTA, A. S. AWWAL, N. K. DUTTA, AND K. FUJIURA. *Proc. SPIE*, 5246, 676–685 (2003).

21. D. H. HUBEL AND T. N. WIESEL, "Functional architecture of macaque monkey visual cortex (Ferrier Lecture)," *Proc. R. Soc. London B.*, 198, 1–59 (1977).

CHAPTER 31

CNS REORGANIZATION DURING SENSORY-SUPPORTED TREADMILL TRAINING

I. Cikajlo, Z. Matjačić, and T. Bajd

31.1 INTRODUCTION

Within the last 10 years novel methods of neurological rehabilitation were developed which take into account the plasticity of the injured central nervous system (CNS). Numerous studies [1] have demonstrated that when the impaired CNS is subject to repeated practice in repeatable conditions a major functional improvement of the practiced task is achieved. A general methodological framework was developed which emphasizes that the level of sensory and motor augmentation should be set at the level that enables execution of the practiced task, thus facilitating the reorganization of the injured CNS, leading to functional improvement. When the functional task is walking, treadmill training with partial body weight support (BWS) is the method that matured over the years and is today used in most of the rehabilitation environments. The method meets the above-mentioned demands regarding the necessary conditions facilitating neurological recovery by supporting part of the body weight according to current sensory motor abilities of the training subject and is at the same time simple enough to be used in everyday clinical practice. Several large-scale controlled clinical studies [2, 3] have proven the efficacy of the approach in acute as well as chronic conditions. Unfortunately, the major deficiency of the proposed method is that it requires a therapist to assist the walking subject. The therapist sits alongside the treadmill assisting the stepping movements of the paretic leg, which is strenuous work in ergonomically unfavorable conditions. Additionally, the assistance provided by the therapist is rather variable and is even more pronounced with the elapsed time of training and increased fatigue of both the therapist and the walking subject. Functional electrical stimulation (FES) has a long tradition as an orthotic and therapeutic aid in the rehabilitation of walking after paraparesis [4].[1] Direct stimulation of motoneurons (artificial activation of spinal neural circuits) has been successfully employed to artificially augment the movement of the affected lower extremity, most frequently during the swing phase. Years of clinical practice have shown that a single-channel peroneal stimulation is an adequate aid to assist the flexion of the hip, knee, and ankle in the population of incomplete spinal cord injury (SCI) persons. Consequently

---

[1]In the 1970s and 1980s the FES emerged as a new hope for patients with spinal cord injury. The nerve stimulation restored the movement of the extremities and a significant contribution to the standing and walking after spinal cord injury has been presented [5, 6].

Handbook of Neural Engineering. Edited by Metin Akay
Copyright © 2007 The Institute of Electrical and Electronics Engineers, Inc.

most research combined the method of treadmill BWS training with single-channel FES. Therefore the physiotherapist is relieved of manually assisting the movement. The weight bearing of paretic extremity during the stance phase is assisted by BWS. Hesse et al. [3] and Field-Fote [2] have successfully demonstrated that the combined approach is much more effective than classical treadmill plus BWS walking. The effectiveness takes place when the physiotherapist manually controls the timing and intensity of the stimulation, visually estimates the quality of the performed swings, and provides verbal feedback to the walking subjects. According to these findings we have been motivated to develop a system [7] that consists of several sensors for walking parameter assessment and provides cognitive feedback. The goal was to make the training process repeatable. The system comprises two two-axial accelerometers and a gyroscope. As supplemental sensors knee and ankle goniometers were used. A computer algorithm [8] integrates all the assessed signals and estimates the performed swing quality that is provided as three state audio feedback. The patient was taught to maximize his efforts according to the feedback. The timing and intensity of the FES were crucial to assure repeatability that is needed for neurorehabilitation. Therefore we presented a fully automated sensory-driven FES augmentation system [9] that would also determine the proper timing of triggering and intensity of the electrical stimulation. The neurorehabilitation training with an automated FES augmentation where patients' effort plays an important role could improve the efficacy of the treadmill training.

31.2 METHODOLOGY

An incomplete SCI patient's walking disturbances or mobility deficiencies can be overcome by FES assistance. In our case we have used single-channel surface peroneal nerve stimulation that provokes flexion response and enables the patient to walk. The walking cycle has been divided into stance phase and swing phase. The swing phase of the right lower extremity during walking on a treadmill is presented in Figure 31.2. His/her lower extremity movement has been monitored by a multisensor device attached to the front of the shank and goniometers in ankle and knee joint. Using the efficient algorithm [8] we were able to estimate the swing phase. It is based on correlation with the reference or desired swing phase time course and represents a swing-phase quality, which is provided as an audio cognitive feedback to the patient at the end of the performed swing phase. The patient, aware of the quality of the performed swing phase, can improve the following swing-phase performance by voluntary action. Depending on the success of the voluntary action, the FES assistance intensity is adjusted. If the swing phase was performed successfully, the FES assistance was decreased; in the opposite event it was increased. To avoid continuous changing of stimulation intensity after every swing phase, the counter was implemented to assure adjustment after several performed steps [9]. A treadmill allows the patient a repeatable training of walking that leads to the reorganization of the CNS [10].

31.2.1   Gait Reeducation System Design

Gait reeducation system consists of a sensor, a process unit, a stimulator with electrodes, and audio cognitive feedback Figure 31.1. All these units form two control loops which also involve a patient walking on a treadmill.

One of the control loops handles the FES assistance. Data obtained from the sensor were used to estimate the swing-phase quality and to determine the shank angle.

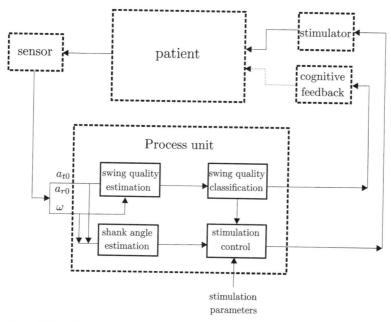

Figure 31.1   Block diagram. The process unit provides cognitive feedback to the patient and controls the needed electrical stimulation according to the assessed data. The patients' voluntary activity is gained by the cognitive feedback.

Considering that the preswing phase (second photo in Fig. 31.2) can be determined by the shank angle, this information was used to trigger the FES. The duration of FES was with reference to the duration of the swing phase. From another processing block the swing-phase quality coefficient was provided and classified as "good," "sufficient," and "poor" to be provided as cognitive feedback. The swing quality counter has been implemented to supervise the FES assistance adjustment. It counted the number of consecutive good and consecutive poor swing phases. When a certain number of consecutive swing phases were performed, the intensity of the FES was adjusted. In a case of good swings the intensity was decreased and vice versa.

The other control loop involved the patient voluntary action. Data assessed from lower extremity during treadmill walking were used for the swing-phase quality estimation (Section 31.2.1.1, [8]). On basis of the estimated quality the cognitive feedback (Section 31.2.1.2) was formed. The patient was focused on feedback only and was trying to improve the swing-phase quality by voluntary action.

Figure 31.2   The picture presents the swing phase performance with FES support. FES was decreased due to the marked improvement in swing phase performance.

31.2.1.1 Swing-Phase Estimation   The swing-phase estimation algorithm was a key for both control loops described in the previous section. Data needed for estimation were assessed by the multisensor device (Fig. 31.5) placed at the front of the shank. The angular velocity ω of the shank was used in combination with the neural network to identify the swing phase during walking. When the swing-phase presence was detected, the time course of the absolute value of the shank acceleration $|a_m|$ was being stored into buffer. The reference acceleration $|a_r|$ time course was assessed during treadmill walking of the healthy person or from the less affected extremity of the incomplete SCI patient that will participate in the training or measurement process. After the swing phase, also detected by the same neural network algorithm, was completed, the correlation between the actual buffered acceleration time course and the stored reference took place:

$$\varphi_{m,r}(t,\tau) = \frac{1}{n}\sum_{k=1}^{n} a_m(t) \cdot a_r(t+\tau) = E[a_m(t) \cdot a_r(t+\tau)] \tag{31.1}$$

where τ is a time delay and n a number of samples included in the computation. The correlation coefficient $\rho_{m,r}$ is derived as

$$\rho_{m,r} = \frac{E[(a_m(t) - m_m(t)) \cdot (a_r(t+\tau) - m_r(t+\tau))]}{\sqrt{E[(a_m(t) - m_m(t))^2 \cdot (a_r(t+\tau) - m_r(t+\tau))^2]}} \tag{31.2}$$

and presents the quantitative estimation of swing-phase quality (0 for poor, 1 for perfect). Here m represents the mean of the signal.

The estimated swing-phase quality was available for the analysis of treadmill walking and was classified for the purpose of cognitive feedback.

31.2.1.2 Cognitive Feedback   The cognitive feedback was based on information from the swing-phase estimation algorithm. While the swing phase was evaluated quantitatively, it was not suitable for representation in real time because of its complexity. The comprehension of the cognitive feedback is usually a cause for diverting the patient's attention from walking [11]. Regarding our research [8] the patient was able to distinguish between three or two different levels of cognitive feedback depending on treadmill speed during walking. Those levels were determined from the swing-phase quality using a threshold classification algorithm:

```
cognitive_feedback algorithm
{
    if (swing phase quality > 0 and < 0.1)  ⟹ swing phase 'poor'
    if (swing phase quality > 0.1 and < 0.6)  ⟹ swing phase
    'sufficient'
    if (swing phase quality > 0.6)  ⟹ swing phase 'good'
}
```

The threshold levels were variable, depending on the patient's deficit and the physiotherapist's decision, allowing settings to be preset before the training process takes place.

As presented in Figure 31.3 the algorithm classifies the swing-phase quality into three levels: good, sufficient, and poor. Also the thresholds are presented. Each classified cognitive feedback level was provided to the patient during treadmill walking as a different sound.

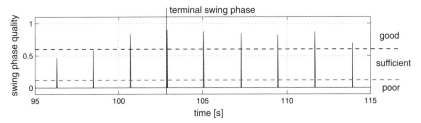

Figure 31.3   The cognitive feedback provides the quality of the swing phase in discrete levels: poor, sufficient, and good. The criterion for the swing quality classification is adjustable.

31.2.1.3 *Computer-Controlled FES*   After the cognitive feedback has been provided to the patient and enabled him or her to take voluntary action, the counter made a note of it. If the swing phase was good, the counter counted upward until the following swing phases were good. If a poor or sufficient swing phase took place, the counter was reset. On the other hand the counter was counting poor swing phases until a good or sufficient swing phase appeared and reset the counter. When the counter had reached the preset number of phases, the FES intensity was adjusted and the counter reset. The consecutive good swing phases decreased the stimulation intensity for 10% and poor swing phases increased the stimulation intensity for 10%.

The FES assistance was provided to the patient during the swing-phase performance. The stimulation had to be present before the swing phase; otherwise the swing phase might not have taken place due to the patient deficit. In our previous research [8] the knee angle was used to determine the moment of FES triggering. Due to problems with reliability and attachment, the goniometer was replaced by an algorithm that provided the shank angle according to the vertical plane. This parameter turned out to eliminate these problems and goniometer was not needed anymore. The obtained parameter determined the preswing phase (second photo in Fig. 31.2), and considering the user preset time delay and stimulation time, the computer was able to determine the moment of triggering.

31.2.2 Instrumentation

The developed system (Fig. 31.4) is intended to be used with incomplete SCI patients. Therefore it was applied and tested at the rehabilitation center. The data assessment was done with the sensory device (Fig. 31.5). It is equipped with a microcontroller that deals with assessment, processing, and serial communication with a PC (Pentium III, 500 MHz, 256 Mb RAM). The device comprises two two-axial accelerometers (Analog Devices ADXL, Horwood, MA, USA) and a gyroscope (Murata ENC, Japan). Additionally knee and ankle goniometes (Biometrics, Ltd.) were used. The user interface, as for all other process and control software, is written in MATLAB/Simulink (MathWorks Inc., Hatick, MA, USA) and C++. The user interface enables the parameter change, setting the number of consecutive swings required for stimulation intensity control and many other stimulation parameters. Clinically well-accepted single-channel surface peroneal nerve stimulation with the following parameters was used: stimulation frequency 20 Hz, pulse width 200 μs, and current 35 mA.

31.2.3 Measurement Procedure

In the case study an incomplete SCI patient (Table 31.1) was involved. His major deficit was a lack of foot dorsiflexion and lower extremity swing. Therefore he was a regular FES

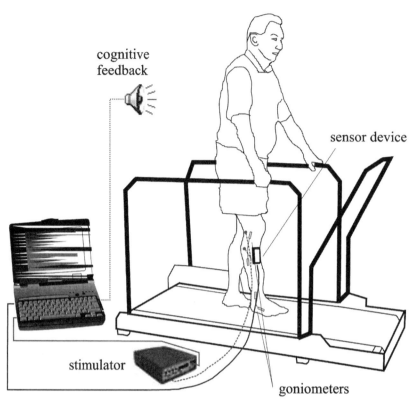

Figure 31.4   The FES swing reeducation set-up. Sensor assesses data from the affected extremity and sends it to the process unit (PC). The process unit controls the electrical stimulator and provides the cognitive feedback.

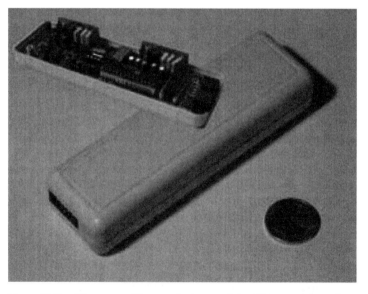

Figure 31.5   The new multisensor device consists of 2 two-axial accelerometers and a single-axial gyroscope. Assessed information (sampling frequency is 100 Hz) is processed by microcontroller that provides data to PC via RS232.

TABLE 31.1   Patient Medical Data[a]

Parameter	Patient BK male
Age	30 y
Weight	84 kg
Height	175 cm
Injury type	C4-5[b]

[a]Institute for Rehabilitation, R. Slovenia.
[b]American Spinal Injury Association (ASIA) classification C.

user for 1 month using MicroFES™ (The Jozef Stefan Institute, Slovenia) peroneal nerve stimulator to provoke flexion response. The MicroFES is a single-channel electrical stimulator triggered by the shoe insole with a pair of electrodes applied over the common peroneal nerve. Development, testing, and validation of our approach took place at the rehabilitation center.

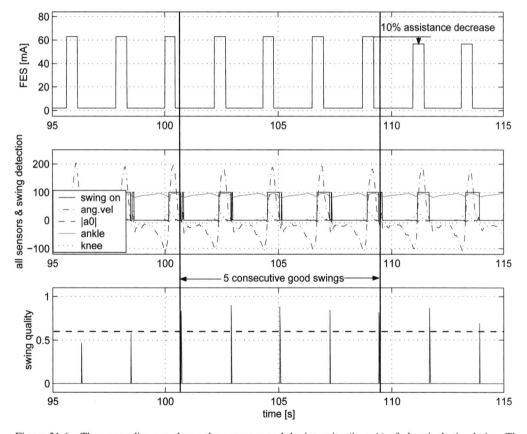

Figure 31.6   The upper diagram shows the presence and the intensity (in mA) of electrical stimulation. The middle diagram shows all the assessed signals. "Swing on" presents the swing phase detection which is based on gyroscope signal ("ang.vel") that assesses the angular velocity of the shank. The diagram shows both goniograms, the "knee" and the "ankle", and the acceleration ("$|a_0|$") signal. In the lower diagram the swing quality is presented. After pre-set number-5-of consecutive good swings the stimulation intensity was decreased.

Goniometers were placed at knee and ankle joints to measure knee flexion and ankle dorsiflexion, respectively. The multisensor device was attached with Velcro to the shank in the orientation, where tangential and radial acceleration of the movement in the sagittal plane could be assessed. Also the shank angular velocity was recorded in the plane of progression. The patient was walking on a treadmill with FES support on both lower extremities. Using the algorithm [8], the swing phase was detected, and during the phase the time course of the absolute value of the ankle joint acceleration was recorded. The time course presented the reference swing phase, which we wanted to be performed by the patient during the training process.

As expected, the repeatable training process during which the patient has an important active role by voluntary activity of lower extremities might influence on reorganization of the CNS. The MicroFES was replaced by a computer-controlled electrical stimulator assisting the movement of the right lower extremity. The assistance provided by this stimulator was minimized according to the patient's effort and success in swing-phase performance. We asked the patient to focus only on cognitive feedback that notified us about the quality of the performed swing phase. The performed swing phase was considered good when the time course of the assessed ankle joint acceleration was adequate to the reference swing phase. If the patient was able to perform five consecutive good swings, the assistance provided by FES was decreased for 10% (Fig. 31.6). In the case of poor consecutive swings the FES support was increased, while the alternation of good, poor, or sufficient swings had no effect on FES intensity.

The training process lasted 4 min and was performed twice in one day.

31.3 RESULTS

The results of the repeatable training process are expected to show a higher level the patient's voluntary activity. According to the described method, the patient's voluntary activity consequently provokes a decrease of the needed FES assistance. Figure 31.6 presents the FES intensity retrenchment after the patient had performed five consecutive good swing phases. The patient was aware of the quality of every swing-phase he had performed via cognitive feedback. The cognitive feedback was the classified output of the swing-phase quality estimation that is presented on the lower diagram. The user-defined states are presented: The swing phase was considered poor when the swing-phase quality was bellow 0.1 and over 0.6 when good. Between the defined states it was deemed sufficient. The diagram also presents the swing-phase detection from the gyroscope signal.

In the beginning the patient was not able to walk or even perform the swing on the treadmill without electrical stimulation; therefore he was an FES user. After the computer-controlled electrical stimulator had been used to provoke flexion response, the patient was asked to focus on cognitive feedback and his swing-phase performance during treadmill walking. The swing-phase quality was better the less FES assistance was provided or, on the other hand, with all emphasis on the patient's voluntary activity. Figure 31.7 presents the patient's treadmill walking during a 4 min-session. The physiotherapist preset the stimulation parameters (stimulation frequency 20 Hz, pulse width 200 μs, and output current 35 mA and time delays) in a graphical user interface on a PC to the level that provides enough FES assistance. During walking the patient was able to perform five consecutive good swing phases (lower diagram) and therefore the assistance has been decreased for 10% (upper diagram). There were also several good swings between 100 and 150 s and no FES intensity alteration, since the algorithm demands five consecutively performed swing phases, either good or poor to change the stimulation

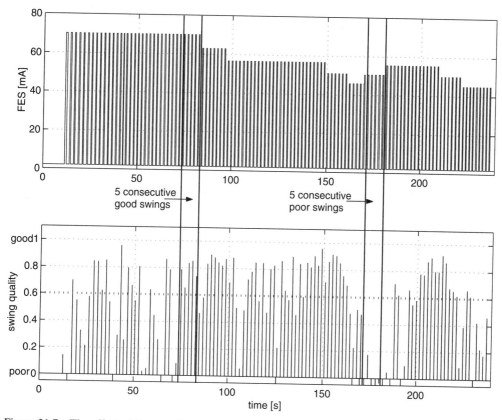

Figure 31.7 The effect of the good/poor swings on the stimulation intensity. When the patients' swing phase became good/poor the need of FES assistance was decreased/increased.

intensity. Between 170 and 180 s the patient performed five consecutive poor swing phases and the FES assistance was consequently increased for 10%. After 4 min we were able to notice that the patient's voluntary activity had increased as compared to the beginning of the session as the FES assistance was decreased for 35%.

31.4 FUTURE TRENDS

In our earlier development of a gait reeducation system for use in incomplete paraplegia the basic idea was to involve the patient in the rehabilitation process [4] by using cognitive feedback supported by an artificial multisensor system. Triggering of the single-channel FES system augmenting the swing phase was left to either the patient or the therapist. Our earlier work [7] and the experiences of Hesse et al. [3] and Field-Fote [2] exposed the need for automated triggering and intensity adjustment of the FES system augmenting the swing for reasons of improving the repeatability of the training conditions where the level of FES support is adjusted at a suitable level in accordance with the walking subject's performance. Therefore the main focus of the work presented in this chapter was to develop the sensory-driven automated FES augmentation system and explore the functioning of such a system in a clinical setting involving a neurologically impaired subject.

The results of our case study clearly show that the proposed approach of sensory-driven automated FES augmentation of the swinging extremity can be successfully incorporated into gait treadmill training. The recordings presented in Figure 31.7 clearly demonstrate the interaction of the walking subject and the state of the sensory-driven FES augmentation system. From the gait performance it appears that the walking subject was indeed an active participant of the gait reeducation process as in the first phase of the training session the level of FES augmentation was reduced and thereafter varied around the level that appeared to be needed. The main achievement of such training is that the walking subject receives not only the cognitive feedback information, which is needed to improve performance on a conscious level, but also relevant sensory feedback that is input to the spinal neural circuits at the spinal level as the level of FES support is related to the gait performance. All together rather comprehensive and repeatable gait-training conditions are achieved, which should have impact on the increased quality and shortened duration of needed gait reeducation treatment. By developing and testing the described gait reeducation system we are in a position to evaluate the presumed effectiveness of the approach in the controlled clinical trials, which is the objective of our further work.

ACKNOWLEDGMENTS

The authors wish to acknowledge the financial support of the Republic of Slovenia Ministry of Education, Science and Sport. Special thanks to the patient and Pavla Obreza, PT. Authors also wish to thank their colleague, Janez Šega, Bsc EE.

REFERENCES

1. FIELD-FOTE, E. C. (2000). Spinal Cord Control of Movement: Implication for Locomotor Rehabilitation Following Spinal Cord Injury. *Phys. Therapy*, 80:477–484.
2. FIELD-FOTE, E. C. (2001). Combined Use of Body Weight Support, Functional Electric Stimulation, and Treadmill Training to Improve Walking Ability in Individuals with Chronic Incomplete Spinal Cord Injury. *Arch. Phys. Med. Rehabil.*, 82:818–824.
3. HESSE, S., MALEŽIČ, M., SCHAFFRIN, A., AND MAURITZ, K. H. (1995). Restoration of Gait by Combined Treadmill Training and Multichannel Electrical Stimulation in Non-Ambulatory Hemiparetic Patients. *Scand. J. Rehab. Med.*, 27:199–204.
4. BAJD, T., KRALJ, A., TURK, R., BENKO, H., AND ŠEGA, J. (1989). Use of Functional Electrical Stimulation in the Rehabilitation of Patients with Incomplete Spinal Cord Injuries. *J. Biomed. Eng.*, 11:96–102.
5. KRALJ, A. AND GROBELNIK, S. (1973). Functional Electrical Stimulation—A New Hope for Paraplegic Patients? *Bull. Prosthesis Res.*, 10–20:75–102.
6. KRALJ, A. AND BAJD, T. (1989). *Functional Electrical Stimulation: Standing and Walking after Spinal Cord Injury*. Boca Raton, FL: CRC Press.
7. BAJD, T., CIKAJLO, I., ŠAVRIN, R., ERZIN, R., AND GIDER, F. (2000). FES Rehabilitative Systems for Re-Education of Walking in Incomplete Spinal Cord Injured Persons. *Neuromodulation*, 3:167–174.
8. CIKAJLO, I. AND BAJD, T. (2002). Swing Phase Estimation in Paralyzed Persons Walking. *Technology & Health Care*, 10:425–433.
9. CIKAJLO, I., MATJAČIĆ, Z., AND BAJD, T. (2003). Development of Gait Re-education System in Incomplete Spinal Cord Injury. *J. Rehabil. Med.*, 35:213–216.
10. MEDVED, V. (2001). *Measurement of Human Locomotion*. Boca Raton, FL: CRC Press.
11. MULDER, T. AND HOCHSTENBACH, J. (2002). *Handbook of Neurological Rehabilitation, Motor Control and Learning: Implications for Neurological Rehabilitation*. Hillsdale, NJ: Erlbaum.

CHAPTER 32

INDEPENDENT COMPONENT ANALYSIS OF SURFACE EMG FOR DETECTION OF SINGLE MOTONEURONS FIRING IN VOLUNTARY ISOMETRIC CONTRACTION

Gonzalo A. García, Ryuhei Okuno, and Kenzo Akazawa

The study of the electrical signal generated by contracting muscle fibers (electromyogram, EMG) is relevant to the diagnosis of motoneuron diseases and to neurophysiological research. In general, clinical diagnosis methods for measuring EMG make use of invasive, painful needle electrodes. In this chapter, we describe a noninvasive alternative to these methods, a newly developed algorithm consisting of three major steps: (i) signal preprocessing; (ii) independent component analysis (ICA); and (iii) template-matching, that decompose surface EMG (s-EMG) recorded by an electrode array.

32.1 ORIGIN, MEASUREMENT, AND ANALYSIS OF EMG SIGNALS

The central nervous system (CNS) conveys commands to the muscles by trains of electrical impulses through alpha-motoneurons (α-MN), whose bodies (*soma*) are located in the spinal cord. The α-MN axons terminate in the motor endplate, where they make contact with a group of muscle fibers. A motor unit (MU) consists of an α-MN and the muscle fibers it innervates (see Fig. 32.1a).

When an α-MN fires and its electrical impulse reaches the end plate, it propagates along the muscle fibers eliciting their mechanical contraction (twitch). The electrical activity of a firing MU (motor unit action potential, MUAP) can be detected by intramuscular or surface electrodes, and the obtained signal is called an electromyogram [1].

The CNS regulates the force exerted by the muscle using two mechanisms: recruitment of inactive MUs [2] and modulation of their firing rate [1]. As the firing rate increases, more twitch contractions are added and thus greater contraction is achieved.

The study of the firing pattern of single α-MNs and some features of their MUAP waveform such as shape or amplitude is relevant to both neurophysiological basic

Handbook of Neural Engineering. Edited by Metin Akay

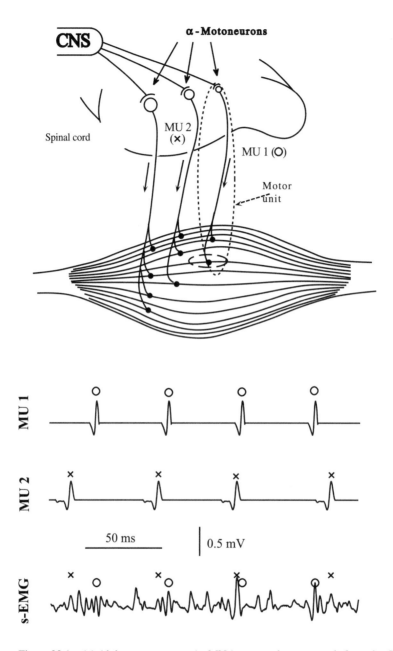

Figure 32.1 (*a*) Alpha-montneurons (α-MNs) convey the commands from the CNS to the muscles. A motor unit (MU) consists of an α-MN and the muscle fibers it innervates. (*b*) An example of signal obtained with surface electrodes showing the superposed activity of several MUs.

studies (e.g., CNS motor control strategies) and diagnosis of motoneuron diseases [3]. In general, these investigations have been carried out with EMG obtained using needle electrodes placed inside the muscle (intramuscular recordings). The MUAP waveform of needle EMG recordings is generally sharp enough to identify single MUs. However, this technique is painful for patients, time consuming for physicians, and a possible cause of infections and tissue damaging.

A noninvasive, painless alternative is to study surface EMG picked up by electrodes placed on the skin over the target muscle. The acquisition and processing of s-EMG signals are well-founded techniques [4–6]; however, as shown in Figure 32.1b, the obtained signal is a time and spatial summation of several MUAPs that overlap (the so-called interference pattern). For this reason, classical techniques used for intramuscular recordings [7] cannot guarantee an effective decomposition of the s-EMG. In addition, s-EMG has a very low signal-to-noise ratio (SNR) due to the filtering effect of the tissues crossed by the signals before arriving at the measurement system [8–10] and the high-level noise picked up by the electrodes [11].

There are several reports on automatic decomposition of s-EMG. Xu et al. [12] developed an artificial neural network that worked up to 5% of subjects' maximal voluntary contraction (MVC). Sun et al. [13] made use of a decomposition algorithm for MU size estimation, but reliable firing rate calculation seems to be impossible due to numerous misses in identifying MUAPs. Bonato et al. [14] proposed to decompose s-EMG in the time–frequency domain. Maekawa et al. [15] applied blind deconvolution to decompose s-EMG at very weak levels of contraction. Hogrel [16] developed a semi-automatic algorithm for MU recruitment study, focusing only on the first 12 firings of each newly recruited MU.

The above-mentioned methods are applicable to s-EMGs recorded at low force levels, but they are not suitable when s-EMGs are recorded at high force levels. There are other alternative approaches that estimate the MU firing rate at higher contraction levels [17–20], but their goal is to give a MU general activity approximation, not a full s-EMG decomposition.

Reucher et al. developed a high-spatial-resolution s-EMG (HSR-EMG) electrode array based on a Laplacian spatial filter that allows the detection of single MU activity in a noninvasive way [21].

The purpose of our studies was to reliably decompose s-EMG into their constitutive MUAP trains (MUAPTs) at higher levels of contraction. To this end, we developed an algorithm that uses signal conditioning filters, ICA [22–24], and a template-matching technique [25, 26]. As has been shown previously [27–29], ICA can be used successfully for solving spike overlaps. In each iteration of the algorithm, the action potentials of one MU could generally be separated from the others. After that, using the template-matching technique, we were able to identify the action potential trains of the target MU.

The algorithm decomposed satisfactory s-EMG into their constitutive MUAPTs up to 30%, 50%, and even 60% MVC for six of the nine experimental subjects. The obtained results were in agreement with the generally accepted behavior of MU firing rate.

For a detailed treatment of ICA, see [30] and [31]; for general blind source separation (BSS) techniques, see [32]. A complete study of pattern recognition and classification can be found in [33, 34].

32.2 METHOD

32.2.1 Data Acquisition

Surface EMG signals were acquired from nine healthy, right-handed subjects (eight males, one female) aged between 22 and 35 who gave their informed consent.

Figure 32.2 shows the experimental setting. Each subject sat comfortably on a chair and was strapped to it with nylon belts. His or her forearm was put in a cast fixed to the table with a shoulder angle of approximately 30°. A strain gauge was placed between forearm cast and the table fixation to measure the torque exerted by the subject.

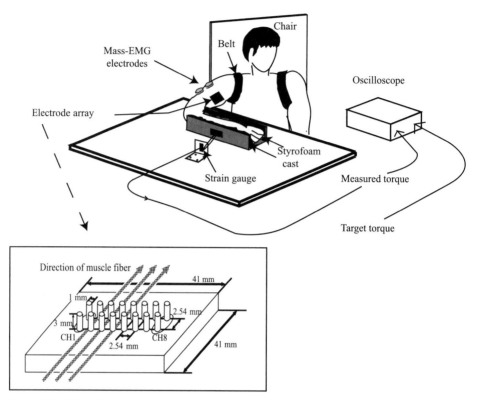

Figure 32.2   Experimental setting. Eleven biomechanical channels were recorded during constant-force isometric contractions; eight-channel s-EMG, one channel of mass-EMG from the antagonist muscle, the target elbow torque and its measured value. The inset shows the s-EMG electrode array consisting of 16 stainless steel electrodes. Eight-channel EMG bipolar signals (CH 1 to CH 8) were measured in simple differential mode.

An electrode array was placed on the biceps short head muscle belly, out of the innervation zone [35, 36], transversal to the muscle fiber direction after cleaning the skin with alcohol and SkinPure (Nihon Kohden Corp., Tokyo, Japan). The electrode array [37] (inset in Fig. 32.2) consisted of 16 stainless steel electrodes of 1 mm diameter, 3 mm height, 2.54 mm spaced in both plane directions attached to an acrylic rigid base. The eight-channel, bipolar, simple differential EMG signal was measured [input impedance above 10^{12} Ω, common-mode rejection ratio (CMRR) 110 dB], filtered with a first-order, band-pass Butterworth filter (cutoff frequencies 70 Hz and 1 Hz), and then amplified (gain 80 dB). To reduce the impedance electrode-skin, Gelaid electrode paste (Nihon Kohden Corp.) was applied at the tip of the electrodes.

In addition, a pair of bipolar, surface electrodes (Ag–AgCl, 1 cm in diameter) were placed with a center-to-center distance of about 2 cm between one another along the muscle fiber direction of the antagonist muscle (triceps lateral head) to check whether there was involuntary co-contraction during the recordings. This signal was amplified (gain 58.8 dB, CMRR 110 dB) to the range ± 10 V, full-wave rectified, and smoothed with a second-order low-pass filter (cutoff frequency 2.7 Hz).

The target torque, set by a personal computer (PC), was displayed in an oscilloscope as a thick line and the measured torque was displayed as a thin line. The subject was asked to match both lines by watching the oscilloscope, trying to avoid co-contraction. If co-contraction occurs, the force measure will not be accurate because it is defined with respect to the MVC of only one muscle.

Eight channels from the s-EMG, one from the antagonist muscle's EMG, and two related to the elbow torque (target torque and measured torque) were sampled at a frequency of 10 Hz, 12 bits per sample by a PCI-MIO-16E-1 analog-to-digital (A/D) converter card (National Instruments, Austin, TX) installed in the PC. These sampled signals were sent to a recorder (8826 Memory HiRecorder, Hioki, Ueda, Japan) for real-time displaying and also to a PC equipped with LabView Version 5 (National Instruments), which controlled the experiments.

At the beginning of each experiment, the subjects were asked to exert for 3 s their maximal contraction, trying to avoid co-contraction. Subject's MVC was calculated as the average around the peak of the maximum detected value. Recordings at 0% MVC were carried out as well to estimate the noise level of the measurement system. Both initial recordings were performed three times by each subject.

We recorded the EMG from the surface electrode array and the elbow torque during 5 s of isometric, constant contractions [38] at different levels between 5 and 60% of each subject's MVC. These measurements were repeated five times at each contraction level allowing a pause between recordings to avoid fatigue.

32.2.2 Signal Processing

Data were transferred from the PC to a more powerful workstation where we applied an algorithm consisting of several independent modules implemented using MATLAB (The MathWorks Inc., Natwick, MA). Figure 32.3. shows the flowchart of this algorithm.

1. The raw EMG signal was preprocessed using a modified dead-zone filter (MDZF) developed by us. This filter removed both the noise and the low-amplitude EMG signals that did not reach a threshold set with the aid of a graphical user interface. An example of the MDZF performance is shown in Figure 32.4. In this example, the upper threshold was set at 20% of the channel maximal value, and the lower threshold was set at 15% of the channel minimum value. The waveform between the consecutive zero crossings A and B had a peak P_1 that lied under the lower threshold; therefore, it passed the MDZF, preserving its original features. The same was applicable to the waveform between B and C because its peak P_2 was above the upper threshold. However, the waveforms contained between C and D disappeared from the signal because their respective peaks did not reach the thresholds. Only when the SNR of the raw EMG was too low, we applied a digital weighted low-pass differential (WLPD) filter [39] defined by

$$y[n] = \sum_{i=1}^{N} w[i](x[n+i] - x[n-i]) \tag{32.1}$$

   where $\mathbf{x}$ corresponds to the digitalized input signal, $\mathbf{y}$ to the output of the filter, and $\mathbf{w}$ to the window function whose weights are given by $w[i] = \sin(i\pi/N)$, $0 < i \leq N$, N being the width (expressed in number of samples) of the window function. Since the estimated average length of the MUAP central waveform was 4 ms, we used $N = 10$ so that the central frequency of the filter was 253 Hz, and its cutoff frequencies (at -3 dB) were 164 and 349 Hz. The WLPD filter was designed by Xu and Xiao specifically for enhancing the MUAP waveform of s-EMG [39]. This filter is the weighted version of the simple LPD widely used for needle electrodes [40].

2. The blind-identification algorithm by joint approximate diagonalization of eigenmatrices (JADE) developed by Cardoso and Souloumiac [41] was applied to solve the overlaps and cancellations between action potentials of different MUs.

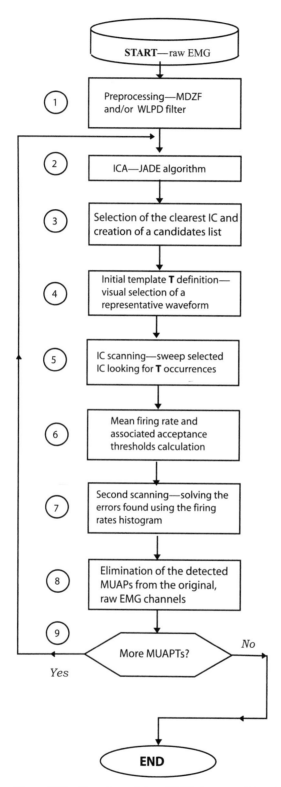

Figure 32.3 Flowchart of the s-EMG decomposition algorithm. The main steps are: signal preprocessing (1), decomposition (2), template matching (3 to 5), and post-processing (6 to 8).

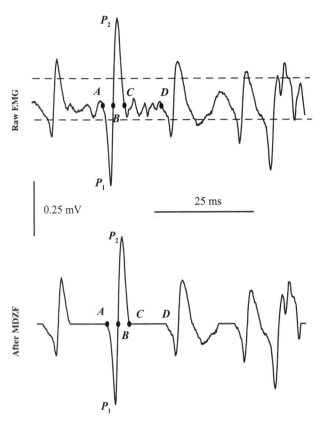

Figure 32.4   Modified Dead-Zone Filter (MDZF) performance over one s-EMG channel recorded at 30% MVC. Waveforms between points *A* and *B*, and between *B* and *C* passed the filter because their respective peaks P_1 and P_2 were above the acceptance thresholds; whereas waveforms between *C* and *D* did not pass the filter.

This algorithm takes as input a vector **v** containing the signals picked up by *m* detectors $(v_1, v_2, \ldots, v_m)^T$, which are the linear combination of *n* unknown independent components (ICs) $s_1, s_2, \ldots, s_n$. Denoting by **s** the $n \times 1$ vector containing the ICs and by **A** the $m \times n$ mixing matrix, the linear relationship between original sources and recorded signals is given by $\mathbf{v} = \mathbf{As}$. Thus, the problem of recovering the original signals **s** is limited to finding the matrix **A** so that $\mathbf{s} = \mathbf{A}^{-1}\mathbf{v}$, $\mathbf{A}^{-1}$ being the inverse of **A**.

The output of JADE is a matrix $\tilde{\mathbf{s}}$ containing the estimated original signals. Like for other ICA algorithms, the JADE requirements that enable us to estimate **s** are the following: (a) the original signals has to be stationary and statistically independent, (b) their number has to be lower or equal to the number of detectors, and (c) the mixing process has to be stationary and without any time delay.

In our case, the original signals were the MUAPTs generated by each MU during a sustained contraction (see Section 32.4 for more details about the applicability of the ICA model to s-EMG). After examining the performance of three ICA algorithms (including FastICA [42]) as a preprocessing tool for s-EMG decomposition, we decided to use JADE.

3. We selected the IC where one of the MUAPTs appeared clearest. Then we applied to this IC the MDZF to obtain a list of MUAP candidates.

4. The most representative and clearest MUAP waveform of the MUAPT was selected as initial template **T** by visual inspection.

5. A scanning of the selected IC was carried out shifting **T** over each candidate **C** of the list. They were compared using the following matching-error metric:

$$\varepsilon = \frac{\sqrt{\sum_{i=1}^{N} g_i \cdot (T_i - C_i)^2}}{\sqrt{\sum_{i=1}^{N} g_i \cdot T_i^2}} \tag{32.2}$$

which gave a normalized weighted measurement of the area left between the waveforms of the template **T** and the candidate **C**. Here, T_i and C_i are the ith elements of vectors **T** and **C** respectively. The length of the template, N, is given in number of samples, its value in our case was between 60 and 80, that is, a duration between 6 and 8 ms. We found that the most representative feature of a MUAP waveform is its curve between the negative peak point i_n and the positive peak point i_p; we therefore set the weight value $g_i = 2$ for the samples between i_n and i_p, being $g_i = 1$ otherwise.

The upper threshold for the matching-error ε was set generally to approximately 0.3 (minimum similarity of 70%). If ε was below the toleration threshold, the candidate was accepted and annotated in a MUAP list; then, template **T** was updated by the following weighted average:

$$T_i^{\text{new}} = \frac{\omega \cdot T_i^{\text{old}} + (1 - \varepsilon) \cdot C_i}{\omega + (1 - \varepsilon)} \qquad i = 1, \ldots, N \tag{32.3}$$

The parameter ω was set to 2 so that the current template **T** (obtained by averaging previous MUAPs) had more weight than each newly accepted MUAP. Since ε is lower bounded by 0, we can use $(1 - \varepsilon)$ as an upper-bounded coefficient of similarity between **C** and **T**.

6. With the MUAP list obtained in the previous step, the instantaneous firing rate (FR) was calculated as the inverse of the inter-peak interval (IPI) between every consecutive pair of identified MUAPs. The maximal coefficient of FR variation during isometric contractions is generally reported to be up to 0.3 [43, 44]. From this information, we calculated the upper and lower FR acceptance thresholds. Firing rates outside those thresholds were considered unusual.

7. The candidates accepted as MUAPs were detected in the original EMG signal and the MUAPT fading ratio was calculated as the interchannel quotient of the median MUAP amplitude. If a candidate already accepted showed an FR above the upper threshold, a low similarity measure, and a different fading ratio from the rest of the MUAPs, we eliminated it from the list. In order to solve the FRs below the lower threshold we performed a second IC scanning only around the points of a supposedly missed MUAP.

8. The candidates finally accepted as MUAPs were eliminated from the original EMG signal.

9. To estimate the power of a signal we used its root-mean-squared (RMS) value [45], defined as

$$\text{RMS} = \sqrt{\frac{1}{N} \sum_{n=1}^{N} x^2[n]} \tag{32.4}$$

where **x** is the signal and N is its length measured in number of samples. We defined the SNR of an s-EMG recording as the quotient of this signal RMS and the RMS of the signal recorded at 0% MVC (noise),

$$\text{SNR} = \frac{\text{RMS}_{\text{signal}}}{\text{RMS}_{\text{noise}}} \qquad (32.5)$$

If the SNR of the cleaned EMG signal was above 2 and if by visual inspection we decided that there were further MUAPTs to be identified, the process was iterated, using in each loop the signal cleaned by the former iteration (peel-off approach).

32.3   RESULTS

We examined the quality of the s-EMG recordings obtained from nine subjects performing different contraction levels, ranking from 5 or 10% MVC to 30, 40, 50, or 60% MVC. The recordings of six subjects were further investigated. The recordings of the remaining three subjects were discarded: in two cases (Subjects 5 and 9) because the recordings had a SNR below 2, in the other (Subject 2) because there were too many active MUs even at low levels of contraction, producing too many MUAP overlaps to be solved by JADE (in other words, there were more source signals than electrode pairs).

A total of 29 decompositions of s-EMGs into their constitutive MUAPTs were carried out successfully using our algorithm.

An example of the performance of the ICA algorithm used in this study (JADE) is shown in Figure 32.5. Figure 32.5*a* shows a WLPD filtered s-EMG signal recorded with the electrode array at 30% MVC contraction level. Six active MUs were found. We focused on the four major MUs (MU 1 to MU 4), whose action potentials appeared clear over the noise. The MUAPs of MU 3 (□ in the figure) and MU 4 (△) were identifiable in channels 1–4, those corresponding to MU 1(×) and MU 2 (○) in channels 5–8. It was straightforward to visually distinguish them and measure their IPI to estimate their average FR. However, at approximately $t = 30$ ms and $t = 75$ ms, overlaps of actions potentials (MU 2 with MU 3 and MU 1 with MU 3, respectively) occurred. At $t = 140$ ms, there was a destructive summation of MUAPs that seemed to correspond to MU 1, MU 2, and MU 4. In these conditions, neither visual inspection nor a template-matching program would be able to identify correctly the MUAPs.

After applying the JADE algorithm to these signals, we obtained the five independent components (IC 1 to IC 5) shown in Figure 32.5*b*. MU 1 activity appeared as the main component in IC 1, with medium and little power in IC 2 and IC 3, respectively. MU 2 power decreased from IC 3 to IC 1. The action potentials of MU 3 appeared independently in IC 4. In this way, the overlaps at $t = 30$ ms and $t = 75$ ms could be solved. Since MU 4 was detected separately in IC 5, the cancellation observed in the raw data at $t = 140$ ms was also clarified.

Figure 32.6 shows a decomposition example of a 30% MVC recording; Figure 32.6*a* shows the final templates and Figure 32.6*b* the original raw s-EMG (only odd channels are shown). This signal was composed of two major MUs (MU 1 and MU 2). Circles indicate where a MU 1 firing was detected; crosses indicate the firings of MU 2. MU 1 appeared clearest in channel 3 and then faded through the other channels. MU 2 showed stronger action potentials in channel 7. This fading is characteristic of all our electrode array s-EMG recordings; for this reason we applied JADE only to the group of channels where the target MUAPT appeared stronger.

Figure 32.7 shows the histogram of the instantaneous FR calculated in step 6 of the algorithm for a 30% MVC recording. Since the median FR was 21.4 firings/s, the

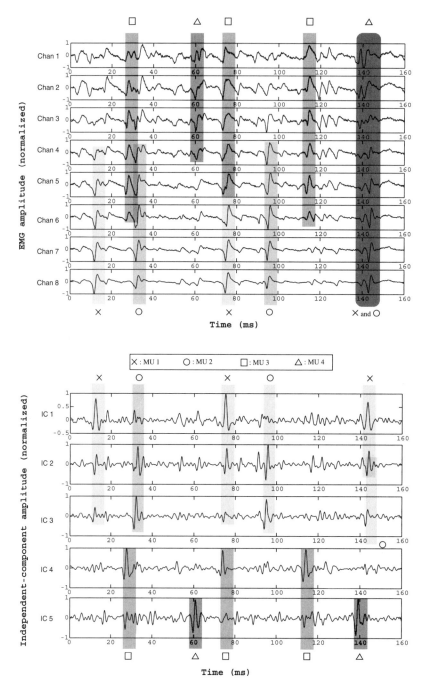

Figure 32.5 JADE performance. (*a*) Eight-channel (Chan 1 to Chan 8) raw EMG signal obtained at 30% of MVC isometric contraction level. The activity of four major MUs (MU 1 to MU 4) is highlighted. Approximately at $t = 30$ ms and $t = 75$ ms an overlapping of two action potentials occurred (MU 2 with MU 3, and MU 1 with MU 3 respectively). At $t = 140$ ms, there was a destructive summation of MUAPs (possibly, MU 1, MU 2, and MU 4). (*b*) Output of JADE algorithm for the above signal after WLPD filtering. The action potentials of MU 1, MU 2, MU 3, and MU 4 appeared nearly alone in different independent components (ICs), solving the overlaps and cancellation observed in the raw data.

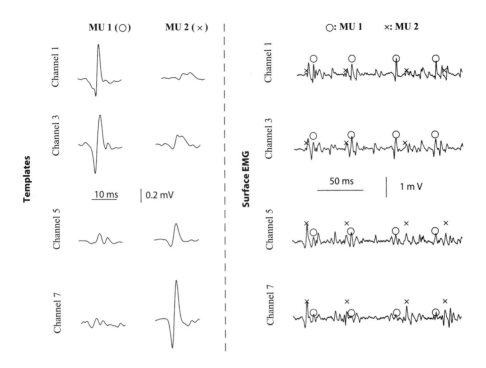

Figure 32.6   An example of s-EMG decomposition; (*a*) Final templates for each s-EMG channel. (*b*) Identified MUAPTs marked over the original s-EMG.

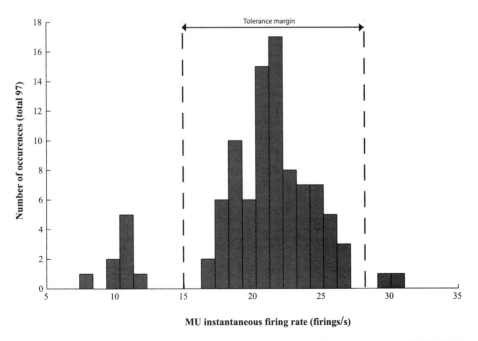

Figure 32.7   Instantaneous MU firing rate histogram for a recording performed at 30% MVC. Median frequency was 21.4 firings/s. Vertical dashed lines mark the upper and lower acceptance thresholds.

acceptance FR upper threshold was set to 27.8 firings/s and the lower to 15.0 firings/s (marked with vertical dashed lines). From this graph, we calculated that the average number of outliers for all of the 29 decompositions in the first scanning was 12.8, with standard deviation (SD) 5.3, which means a 12.5% (SD 3.6) of the total MUAPs.

After the second scanning of the IC, the median of the instantaneous FR was calculated for each contraction level for each of the six subjects. In Figure 32.8 the final results are plotted. Vertical thin lines indicate the SD associated to each point (mean of the standard deviations: 2.5, SD 1.0). In subjects 4 and 8, it was possible to decompose two MUAPTs. As reported previously [17, 46–48], as the isometric

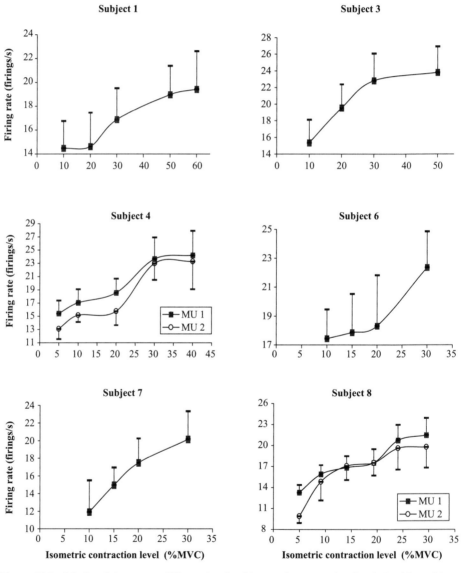

Figure 32.8 Median firing rate at different levels of isometric contraction for six healthy subjects. Vertical thin lines indicate the standard deviation (s.d.) associated to each point (mean SD: 2.45). In the cases of Subject 4 and Subject 8, two MUAPTs were decomposed at each contraction level.

contraction level increased, the FR of individual MU also increased. The recruitment threshold of all the identified MUs was between 5 and 10% MVC. Therefore, we can assume that all of them were of small size [2]. The FR evolution as the force level increases could be different for bigger MUs.

32.4   DISCUSSION

In order to solve MUAP overlaps, we have used the ICA algorithm JADE. Ideally, in each independent component given as output by this algorithm we should have obtained the MUAPT corresponding to one, and only one, MU. However, this was not achieved in all cases. The possible causes are the following: (a) the non statistical independence between the original source signals (MUAPTs), (b) the number of source signals is greater than the number of electrodes, and (c) nonlinearities and time delays of the mixing process.

Generally, we applied JADE to a group of between two and four s-EMG channels. When JADE was applied to a higher number of s-EMG channels, it was necessary to correct the delay existing between consecutive channels using an alignment module based on least mean-squared error.

In order to confirm whether an identified MU in a new contraction level was the same as the one decomposed in the previous level, we compared their templates using the similarity metric given in Eq. (32.2). We also compared their respective fading ratios calculated in step 7 of the algorithm. The final decision was taken after visual inspection of the original raw s-EMG.

The algorithm we developed is not fully automatic; it was necessary to set the thresholds of the MDZF and of the scanning module. In addition, visual inspection was necessary to decide to which channels to apply the ICA algorithm as well as to interpret its output. The initial template decision was, in fact, manual. Optionally, the scanning process can be graphically supervised in real time, so that the operator is allowed to accept candidates that are below the similarity threshold. We now aim at supplementing the algorithm with a specific module that will set such parameters.

32.5   FUTURE TRENDS

As s-EMG acquisition and processing techniques evolve, more reliable s-EMG decomposition algorithms will be developed, making possible noninvasive neurological basic investigations as well as diagnosis and follow-up of motoneuron diseases. However, it will still be necessary to use needle electrodes for deep-muscle evaluation and for definitive assessments.

ACKNOWLEDGMENTS

The authors would like to thank R. Nishitani for technical assistance. Thanks also to Dr. S. Rainieri for helpful discussion on the original manuscript. This work was partially supported by the Ministry of Education, Culture, Sports, Science, and Technology of Japan (Grant-in-Aid for Scientific Research). G. A. G. is supported by a grant from the same ministry.

REFERENCES

1. J. V. Basmajian and C. J. De Luca, *Muscle Alive. Their Functions Revealed by Electromyography*, 5th ed., Williams & Wilkins, Baltimore, 1985.

2. E. Henneman, G. Somjen, and D. O. Carpenter, "Functional Significance of Cell Size in Spinal Motoneurons," *J. Neurophysiol.*, vol. 28, pp. 560–580, 1965.

3. A. M. Halliday, S. R. Butler, and R. Paul, Ed., *A Textbook of Clinical Neurophysiology*, Wiley, New York, 1987.

4. B. Freriks, H. J. Hermens, Roessingh Research and Development b.v., "European Recommendations for Surface ElectroMyoGraphy," Results of the SENIAM Project, 1999, (CD-ROM).

5. G. Rau and C. Disselhorst-Klug, "Principles of High-Spatial-Resolution Surface EMG (HSR-EMG): Single Motor Unit Detection and Application in the Diagnosis of Neuromuscular Disorders," *J. Electromyogr. Kinesiol*, vol. 7(4), pp. 233–239, 1997.

6. R. Merletti, D. Farina, and A. Granata, "Non-Invasive Assessment of Motor Unit Properties with Linear Electrode Arrays," *Electroencephalography Clinical Neurophysiology* 50(Supplement), pp. 293–300, 1999.

7. D. Stashuk, "EMG Signal Decomposition: How Can It Be Accomplished and Used?," *J. Electromyogr. Kines.*, vol. 11, pp. 151–173, 2001.

8. E. J. De La Barrera and T. E. Milner, "The Effect of Skinfold Thickness on the Selectivity of Surface EMG," *Electroencephalog. Clin. Neurophysiol.*, vol. 93, pp. 91–99, Dec. 1993.

9. D. Farina, C. Cescon, and R. Merletti, "Influence of Anatomical, Physical, and Detection-System Parameters on Surface EMG," *Biol. Cybern.*, vol. 86, pp. 445–456, 2002.

10. N. Nordander, J. Willner, G.-Å. Hansson, B. Larsson, J. Unge, L. Granquist, and S. Skerfving, "Influence of the Subcutaneous Fat Layer, as Measured by Ultrasound, Skinfold Calipers and BMI, on the EMG Amplitude," *Eur. J. Appl. Physiol.*, vol. 89, pp. 514–519, 2003.

11. E. Huigen, A. Peper, and C. A. Grimbergen, "Investigation into the Origin of the Noise of Surface Electrodes," *Med. Biol. Eng. Comput.*, vol. 40, pp. 332–338, 2002.

12. Z. Xu, S. Xiao, and Z. Chi, "ART2 Neural Network for Surface EMG Decomposition," *Neural Comput. Appl.*, vol. 10, pp. 29–38, 2001.

13. T.-Y. Sun, T.-S. Lin, and J.-J. Chen, "Multielectrode Surface EMG for Noninvasive Estimation of Motor Unit Size," *Muscle & Nerve*, vol. 22, pp. 1063–1070, 1999.

14. P. Bonato, Z. Erim, and J. A. Gonzalez-Cueto, "Decomposition of Superimposed Waveforms Using the Cross Time Frequency Transform," in *Proceedings of the 23rd Annual International Conference of the IEEE Engineering in Medicine and Biology Society*, Istanbul, October 25–28, 2001, vol. 2, pp. 1066–1069.

15. S. Maekawa, T. Arimoto, M. Kotani, and Y. Fujiwara, "Motor Unit Decomposition of Surface EMG Using Multichannel Blind Deconvolution," in *Proc. of XIVth Congress of the ISEK*, Vienna, Austria, June 2002, pp. 38–39.

16. J.-Y. Hogrel, "Use of Surface EMG for Studying Motor Unit Recruitment during Isometric Linear Force Ramp," *J. Electromyog. Kinesiol.*, vol. 13(5), pp. 417–23, Oct. 2003.

17. D. Stashuk and Y. Qu, "Robust Method for Estimating Motor Unit Firing-Pattern Statistics," *Med. Biol. Eng. Comput.*, vol. 34, pp. 50–57, Jan. 1996.

18. S. Karlsson, J. Yu, and M. Akay, "Time-Frequency Analysis of Myoelectric Signals during Dynamic Contractions: A Comparative Study," *IEEE Trans. Biomed. Eng.*, vol. 47 (2), pp. 228–238, Feb. 2000.

19. P. Zhou and W. Z. Rymer, "Estimation of the Number of Motor Unit Action Potentials in the Surface Electromyogram," in *Proceedings of the First IEEE EMBS Conference on Neural Engineering*, Capri Island, Italy, March 20–22, 2003, pp. 372–375.

20. P. Zhou and W. Z. Rymer, "Motor Unit Action Potential Number Estimation in the Surface Electromyogram: Wavelet Matching Method and Its Performance Boundary," in *Proc. of the 1st International IEEE EMBS Conference on Neural Engineering*, Capri Island, Italy, March 20–22, 2003, pp. 336–339.

21. H. Reucher, J. Silny, and G. Rau, "Spatial Filtering of Noninvasive Multi-Electrode EMG: Part I—Introduction to Measuring Technique and Applications," *IEEE Trans. Biomed. Eng.*, vol. 34, pp. 98–105, 1987.

22. P. Comon, "Independent Component Analysis, A New Concept?" *Signal Process.*, vol. 36, pp. 287–314, 1994.

23. J.-F. Cardoso, "Blind Signal Separation: Statistical Principles," *Proc. IEEE*, vol. 86(10), pp. 2009–2025, Oct. 1998.

24. A. Hyvärinen and E. Oja, "Independent Component Analysis: Algorithms and Applications," *Neural Networks*, vol. 13, pp. 411–430, 2000.

25. B. Mambrito and C. J. De Luca, "A Technique for the Detection, Decomposition and Analysis of the EMG Signal," *Electroencephalogr. Clin. Neurophysiol.*, vol. 58, pp. 175–188, 1984.

26. R. Okuno and K. Akazawa, "Recruitment and Decruitment of Motor Units Activities of M. Biceps Brachii during Isovelocity Movements," *Paper Presented at IEEE-EMB Annual Conference*, Istanbul, Turkey, Sept. 2001.

27. H. Nakamura, T. Moritani, M. Kotani, M. Yoshida, and K. Akazawa, "The Independent Component Approach to the Surface Multi-Channel EMGs Decomposition Method," in *Proc. of the XVIII Int. Congress of the Society of Biomechanics*, Zurich, Switzerland, July 8–13, 2001.

28. G. A. García, R. Nishitani, O. Ryuhei, and K. Akazawa, "Independent Component Analysis as Preprocessing Tool for Decomposition of Surface Electrode-Array Electromyogram," in *Proc. of the 4th Int. Symp. on ICA and BSS (ICA2003)*, Nara, Japan, April 1–4, 2003, pp. 191–196.

29. S. Takahashi, Y. Anzai, and Y. Sakurai, "New Approach to Spike Sorting for Multi-Neuronal Activities Recorded with a Tetrode—How ICA Can Be Practical," *Neurosci. Res.*, vol. 46(3), pp. 265–272, July 2003.

30. T.-W. Lee, *Independent Component Analysis: Theory and Applications*, Kluwer, Boston, 1998.

31. A. Hyvärinen, J. Karhunen, and E. Oja, *Independent Component Analysis*, Wiley Interscience, New York, 2001.

32. A. Cichocki and S. Amari, *Adaptative Blind Signal and Image Processing*, Wiley, New York, 2002.

33. K. Fukunaga, *Introduction to Statistical Pattern Recognition*, 2nd ed., Academic Press, Tokyo, 1990.

34. R. O. Duda, P. E. Hart, and D. G. Stork, *Pattern Classification*, 2nd ed., Wiley, New York, 2001.

35. R. Okuno and K. Akazawa, "Bipolar Surface Electrode Location for Measuring Upper-Arm Electromyograms in Elbow Movements," in *Proceedings of the 22nd Annual International Conference of the IEEE EMBS*, Chicago, July 23–28, 2000.

36. L. Nannucci, A. Merlo, R. Merletti, A. Rainoldi, R. Bergamo, G. Melchiorri, D. Lucchetti, I. Caruso, D. Falla, and G. Jull, "Atlas of the Innervation Zones of Upper and Lower Extremity Muscles," in Proc. of *XIVth Congress of the ISEK*, Vienna, Austria, June 2002, pp. 353–354.

37. T. Masuda, H. Miyano, and T. Sadoyama, "A Surface Electrode Array for Detecting Action Potential Trains of Single Motors Units," *Electroencephalog. Clin. Neurophysiol.*, vol. 60, pp. 435–443, 1985.

38. W. F. Brown, *The Physiological and Technical Basis of Electromyography*, Butterworth, Boston, 1984, pp. 241–248.

39. Z. Xu and S. Xiao, "Digital Filter Design for Peak Detection of Surface EMG," *J. Electromyogr. Kinesiol.*, vol. 10, pp. 275–281, 2000.

40. S. Usui and I. Amidror, "Digital Low-Pass Differentiation for Biological Signal Processing," *IEEE Trans. Biomed. Eng.*, vol. 29(10), pp. 686–693, 1982.

41. J. F. Cardoso and A. Souloumiac, "Blind Beamforming for Non-Gaussian Signals," *IEE Proc.*, vol. 140(6), pp. 362–370, Dec. 1993.

42. A. Hyvärinen and E. Oja, "A Fast Fixed-Point Algorithm for Independent Component Analysis," *Neural Computat.*, vol. 9, pp. 1483–1492, 1997.

43. H. P. Clamann, "Activity of Single Motor Units during Isometric Tension," *Neurology*, vol. 9, pp. 254–260, 1970.

44. D. F. Stegeman, J. H. Blok, H. J. Hermens, and K. Roeleveld, "Surface EMG Models: Properties and Applications," *J. Electromyogr. Kinesiol.*, vol. 10, pp. 313–326, 2000.

45. R. M. Rangayyan, *Biomedical Signal Analysis. A Case-Study Approach*, IEEE Press & Wiley Series in Biomedical Engineering, M. Akay Series Ed., Wiley, New York, 2002; pp. 256–263.

46. H. S. Milner-Brown and R. B. Stein, "The Orderly Recruitment of Human Motor Units during Voluntary Isometric Contractions," *J. Physiol. (Lond.)*, vol. 230, pp. 359–370, 1973.

47. K. Kanosue, M. Yoshida, K. Akazawa, and K. Fujii, "The Number of Active Motor Units and Their Firing Rates in Voluntary Contraction of Human Brachialis Muscle," *Jpn. J. Physiol.*, vol. 29, pp. 427–443, 1979.

48. C. J. De Luca, "Physiology and Mathematics of Myoelectric Signals," *IEEE Trans. Biomed. Eng.*, vol. 26(6), pp. 313–325, 1979.

CHAPTER 33

RECENT ADVANCES IN COMPOSITE AEP/EEG INDICES FOR ESTIMATING HYPNOTIC DEPTH DURING GENERAL ANESTHESIA

Erik Weber Jensen, Pablo Martinez, Hector Litvan, Hugo Vereecke,
Bernardo Rodriguez, and Michel M. R. F. Struys

33.1 INTRODUCTION

The objective of this chapter is to provide an overview of the most recent advances in the development of auditory-evoked potentials (AEPs) and electroencephalographic (EEG) based indices for estimating the hypnotic depth during general anesthesia. In order to accomplish this objective a comprehensive description of both AEP and EEG is included as well as a section explaining different methods for rapid extraction of the AEP.

33.2 DEPTH OF ANESTHESIA

In order to ensure patient safety and well-being during surgery, general anesthesia aims to produce three effects: hypnosis (sleep), analgesia (decreased responses to pain), and muscular relaxation (reduced muscular tone, lack of movement). Accordingly, anesthetic drugs are administered to the patient in order to achieve each effect. For example, hypnotic drugs such as thiopental and propofol may produce sleep without suppressing movement whereas opioids produce analgesia with only small hypnotic effect. Neuromuscular blocking agents (NMBAs) are administered to achieve muscular relaxation.

Nevertheless, the variety of stimuli exerted during surgery as well as NMBA use may obscure the signs of inadequate anesthesia. Of particular importance is the need to prevent episodes of awareness, where the patient is awake during part of the operation without it being noticed by the anesthesiologist due to the effect of muscular relaxants. The patient may look totally comfortable; however, in the worst case situation he or she is able to follow the action of the surgery.

A number of case reports describe patients experiencing psychological problems, such as difficulties to sleep or nightmares, derived from explicit or implicit recall caused by awareness during anesthesia [1, 2].

Handbook of Neural Engineering. Edited by Metin Akay

Depth of anesthesia can be defined as the measure of the balance between the anesthetic, which causes depression of the central nervous system, and the surgical stimulation that can cause increased level of arousal. The effects of these two factors, one decreasing and the other increasing the arousal level, exhibit large interindividual variation, and hence the actual depth of anesthesia is difficult to predict from administered dose and the surgical stimulation.

Classical indicators such as blood pressure or movement do not necessarily indicate the need for analgesics or hypnotics nor are they good indicators of depth of anesthesia, and hence the need for a system to measure the effects of anesthetic drugs on-line without the influence of muscular relaxants. This will contribute not only to predict the onset of awareness but also to allow the medical practitioner to titrate anesthetic drugs and prevent oversedation of patients, allowing a faster and better recovery.

Jones and Koniezko [3] categorized four levels of anesthetic depth and introduced the term *awareness*:

Level 1: conscious awareness without amnesia

Level 2: conscious awareness with amnesia

Level 3: subconscious awareness with amnesia

Level 4: no awareness

The first level is also termed *spontaneous recall* or *explicit memory*, and it is the level which is most easy to detect as it corresponds to "fully awake." Levels 2 and 3 represent levels of consciousness where there is no spontaneous recall but where the patient will respond to verbal instructions given during the anesthesia either immediately (level 2) or at a later time (level 3). A number of psychological methods are described to detect levels 2 and 3; however, the results have been ambiguous and hence their clinical value remains uncertain [4]. Therefore only levels 1 and 4 are well defined in Jones and Konieczko's scale.

33.3   AUDITORY-EVOKED POTENTIALS

The AEP is defined as the passage of electrical activity from the cochlea to the cortex, which produces a waveform consisting of 11 waves in the EEG. These can be divided into three parts: brainstem auditory-evoked potential (BAEP), middle-latency auditory-evoked potential (MLAEP), and long-latency auditory-evoked potential (LLAEP). These parts indicate the sites in the brain from which the various waves are thought to originate (Fig. 33.1). The BAEP is represented by the Roman numerals I–VI and extends from 0 to 10 ms after the stimulus. These waves represent the process of stimulus transduction in the brainstem: acoustic nerve (I), cochlea nucleus (II), superior olivary complex (III), ventral nucleus of the lateral lemniscus and preolivary region (IV), inferior colliculus (V), and medial geniculate body (VI) [5–7]. The early cortical or middle-latency auditory-evoked potentials, marked by the waves N_0, P_0, N_a, P_a, and N_b, are thought to originate from the medial geniculate body and the primary auditory cortex [8]. These waves occur from 10 to 100 ms after the stimulus. The third part, more than 100 ms after the stimulus, is a LLAEP and reflects the neural activity of the frontal cortex and association areas.

Factors known to affect the BAEP are the age and sex of the subject; the rate, amplitude, and polarity of the stimulus; the filter settings used; and the dose of certain administered anesthetic agents and other drugs. Studies have shown that the BAEP peaks change during anesthesia with volatile agents [9–13] but not during anesthesia maintained with

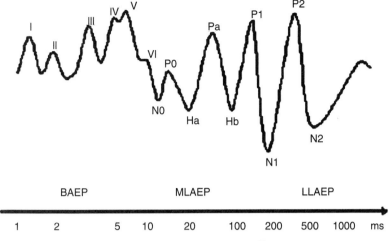

Figure 33.1   Schematic of BAEP, MLAEP, and LLAEP.

intravenous agents [14–18]. The LLAEPs show great intra- and interindividual variations because these signals, which arise in the association cortex, are heavily influenced by the drugs of choice and the subject's individual emotional state and alertness [19, 20].

The MLAEPs, on the other hand, show graded changes in their latencies and amplitudes for a variety of anesthetic agents and less intra- and interindividual variations than LLAEPs [16–18]. Therefore, MLAEPs have been widely examined as a measure of depth of anesthesia.

The AEP is clinically useful in detecting and localizing lesions in the auditory pathway and in the investigation of hearing loss in infants and in other noncommunicative subjects. The AEP is present in the first days of life. In the infant, the latencies of the peaks are prolonged and the amplitudes diminished. With advancing age, the peaks will change toward latencies and amplitudes found in a normal adult. There is good evidence that the AEP is useful in the evaluation of the state of maturation in newborn infants [21].

The first evoked potential measurement systems were based on the superposition of ink-written or photographic traces, and electronic and later digital averagers were applied to AEP-averaging processes.

The signal-to-noise ratio (SNR) of AEPs is often very poor, and hence they are more difficult to extract than, for example, visual-evoked potentials. This is a problem, especially in the operating theatre, due to the use of other electric equipment producing vast amounts of noise. Modern instrumentation amplifiers with a high common-mode rejection ratio (CMRR) can overcome the problems of a moderate level of electrical noise; however, that arising from electrical surgery (diathermy), which produces a strong electrical field, renders AEP monitoring impossible. Research applying AEP for monitoring depth of anesthesia was initiated in the beginning of the 1980s [9, 10]. Since then a vast number of papers have been published in this field of interest [10–18].

Figure 33.2 shows the changes of amplitudes and latencies in the AEP with increased doses of propofol.

The values shown above are examples of changes of AEP when administering different doses of volatile agents; however, they are not general values as there is a big interindividual variability of the AEP response to anesthetics. In this study it was concluded that thiopentone, propofol, etomidate, enflurane, and isoflurane all showed a dose–response suppression of the AEP, whereas receptor-specific agents such as midazolam, diazepam,

AEP waveform   Propofol (μg/mL)

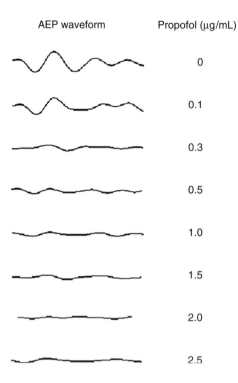

Figure 33.2   MLAEP during increased propofol concentrations.

and flunitrazepam have very little effect on the AEP. Finally it was concluded that isoflurane and isoflurane $+$ 50 N_2O (nitrous oxide) had identical effect on the AEP.

Schwender et al. [22] also correlated end-tidal concentrations of desflurane with changes in various peaks of the AEP. A tendency toward increased latency of Pa, P1, and Nb waves and decreased Na/Pa amplitude was noticed when the end-tidal concentration of desflurane was increased; however, it is not statistically significant. The study also showed that the latency of peak V of the BAEP changed very little through the entire period of anesthesia. It was concluded that an end-expiratory concentration greater than 4.5% significantly suppresses the different MLAEP components. Therefore the risk of awareness phenomena (e.g., auditory perception and intraoperative awareness) was diminished as well.

A variant of AEP monitoring is the 40-Hz auditory steady-state response (ASSR) described by Plourde [23]. The ASSR is elicited with a stimulus frequency of approximately 40 Hz. This high stimulus frequency causes overlapping of the responses to the successive stimuli. Both inhalational agents and intravenously administered drugs cause a profound attenuation of the amplitude of the 40-Hz ASSR. The author concludes that the 40-Hz ASSR depends on the level of consciousness; however, further validation is needed.

33.4   EXTRACTION OF AEP

33.4.1   Introduction

Throughout the last two decades a number of methods have been explored in order to facilitate a single-sweep or a few-sweeps extraction of the AEP. The main focus of this section is to review a series of methods with particular emphasis on the autoregressive model with exogenous input (ARX model) to extract the AEP. The performance of this method is examined extensively whereas other methods for single-sweep analysis are briefly described.

33.4.2 Wavelets

Wavelets have found use in a number of signal processing applications in biomedical engineering. The wavelet approach is often termed a "mathematical microscope" because it is able to detect transient changes in a signal with various degrees of detail. The wavelet transformation (WT) is a signal decomposition into a set of basis functions, called wavelets. These are obtained from a single standard wavelet by dilation and contraction (scaling) and time shifts. Consequently, a WT scale can be viewed as an alternative to time–frequency representation, leading to a time-scale representation.

Bartnik et al. [24] applied the WT to the AEP and concluded that an AEP from a single sweep could be extracted even though the AEP and the EEG are in the same frequency range, provided that their temporal positions differ. However, the SNR of the AEP would be improved if some preaveraging had been carried out. The WTs have been applied to visual-evoked potentials (VEPs) by Geva et al. [25]. The wavelet approach is a promising tool for single-sweep analysis of evoked potentials and it should be explored further. Recently Kochs et al. [26] have applied wavelet analysis to extraction of AEPs in clinical studies.

33.4.3 Wiener Filtering

The Wiener filter is the optimal solution in a least-mean-square (LMS) sense of the following problem:

$$s(t) = d(t) + n(t)$$

where $s(t)$ is the measured data, $d(t)$ the desired signal, and $n(t)$ the noise sequence. The exercise is to find a filter that produces the optimal estimate of d, described in the frequency domain as

$$\tilde{D}(f) = \Phi(f) \cdot S(f)$$

It can be shown that the optimal filter is obtained by

$$\Phi(f) = \frac{|D(f)|^2}{|D(f)|^2 + |N(f)|^2}$$

The signal and noise have to be additive, statistically independent, and in a wide sense stationary process. This requirement can be difficult to accomplish if d is the AEP and n the background EEG activity. A poor SNR also makes it difficult to calculate a proper estimate of $D(f)$ and $N(f)$. In its classical form a Wiener filter is only able to optimally weigh the relative power spectral density of the signal and the noise at each frequency; hence the phase of the desired signal is not properly recovered.

33.4.4 Optimal Vector Quantization

Haig et al. [27] suggest the use of global optimal vector quantization to classify single-trial event-related responses (ERPs). Vector quantization represents the signal vectors by a small number of pattern or code book vectors. Each signal vector is classified on the basis of which codebook vector it is nearest, for example, by calculating the Euclidean distance. It was concluded that only a small subset of the ERPs correlated well to the averaged ERP waveform, and so this method facilitates a more detailed analysis of the ERPs as compared with moving time average (MTA).

33.4.5 Nonlinear System Identification by *m*-Pulse Sequences

In 1982 Eysholdt and Schreiner reported on the use of *m*-pulse sequences to extract the BAEP from the ongoing EEG activity [28]. The BAEP was considered as the impulse response of the auditory pathway. The BAEP was obtained by the cross-correlation function of the overlapped response with a stimulus-related sequence. The method allows a faster stimulus rate compared with that allowed by moving time averaging and therefore decreases the processing time necessary to extract the AEP. The method was further refined by Shi and Hecox [29]. The trade-off is that the morphology of the AEP might be changed by nonlinearities in the response due to the increased stimulus rates. An examination of the effects of increased stimulus rate is found in [30]. However, as pointed out by Shi and Hecox [29], an application of the *m*-pulse sequence is that one data record simultaneously contains information about the BAEP (or MLAEP if the window is increased) as a response to a variety of stimulus rates.

33.4.6 Autoregressive with Exogenous Input (ARX) Model

33.4.6.1 Introduction The ARX model, also termed *single-input, single-output* (SISO) model [31], is applied in several distinct fields. The application of ARX models to VEPs was originally described by Cerutti [32, 33] and later by Liberati and Cerutti [34] and applied by Magni [35]. The SNR of the VEP is considerably larger than that of the AEP; hence single-sweep analysis was possible. For AEP extraction, a preaverage of 15–50 sweeps improves the SNR before applying the ARX model [36]. The concept of the ARX model is that the preaveraged sweeps contain a stochastic part (the preaveraged EEG) and a deterministic part (the AEP). Both are then modeled by calculating the coefficients of the ARX model.

33.4.6.2 Definition Several publications suggest that an EEG segment can be modeled by an autoregressive (AR) process [37, 38]. The ARX model is then obtained by adding an exogenous input to the AR model; hence the ARX model is defined by the equation

$$y(t) + a_1 \cdot y(t-1) + \cdots + a_n \cdot y(t-n) = b_1 \cdot u(t) + \cdots + b_m \cdot u(t-m+1) + e$$

where n is the order of the backward coefficients $(a_1, \ldots, a_n)$ and m the order of the forward coefficients $(b_1, \ldots, b_m)$. The output is y, u the exogenous input, and e the error.

The exogenous input u is an AEP produced by averaging the latest 256 sweeps. The output y is an average of the 18 latest collected sweeps consisting of averaged EEG background activity and the AEP. When the coefficients of the model are determined, the ARX–AEP is obtained by infinite impulse response (IIR) filtering of the exogenous input u. Figure 33.3 shows a diagram of the ARX model.

33.4.6.3 Model Order The model order is determined by considering the error function defined by

$$L = \frac{1}{N} \sum_{i=1}^{N} e(i)^2$$

where $e(i) = y(i) - \hat{y}(i)$ and N is the number of samples in one sweep.

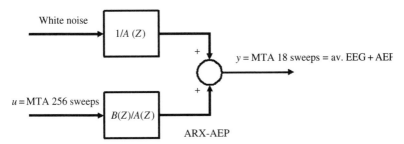

Figure 33.3   Generalized ARX model. White noise is driving the autoregressive part, defined by the averaged EEG activity. Preaveraging was done using 18 sweeps.

The variables $y(i)$ and $\hat{y}(i)$ are the real and predicted preaverage of 18 sweeps, respectively. The identification is tested by the Andersons test on whiteness of the prediction error e. If the prediction error is white at a confidence level of 95%, the identification is accepted. The optimal values of n and m are selected by minimizing the final prediction error (FPE) function defined by Akaike [39]:

$$\text{FPE} = \frac{N + n + 1}{N - n - 1} L$$

where n is the total number of coefficients of the ARX model. The FPE represents the need of minimizing the error function and the need of limiting the number of parameters of the ARX model.

The order of the ARX model should ideally be calculated for each sweep. This is a very time consuming process; hence to comply with the need of fast processing time, a fixed ARX model order for both backward and forward coefficients is chosen. The Jury criterion is applied to decide whether the poles are inside the unit circle. In the opposite case the AEP was rejected.

33.4.6.4   *Estimation of Signal-to-Noise Ratio*

In situations where the SNR is extremely low, more than 18 sweeps are necessary in order to produce a reliable output of the ARX model. Hence an estimation of the SNR is needed. The SNR can be assessed by calculating the ratio between the amplitude of a synchronous and an asynchronous average of N sweeps. If a response to the stimulus is present in the sweeps, then the synchronized average will enhance the peak amplitudes of the response. On the other hand, the asynchronous average will not enhance peak amplitudes of the response; therefore the maximal amplitude is an estimate of the noise in the averaged signal. Hence the SNR estimator is defined as

$$\text{SNR} \approx \frac{\text{max\_amp(synchronized average)}}{\text{max\_amp(asynchronous average)}}$$

The asynchronous average process is an estimation of the background noise. The SNR can be based on a single asynchronous average process or a set of several asynchronous averages using its values to have a better estimation of the noise and obtain useful statistical inferences about the estimated SNR.

A closed-loop control system, using a proportional–integral–differential (PID) controller, is applied to control the SNR of the averaging processes. A desired SNR is defined and reflects the adequate AEP extraction. The controller acts on the averaging processes in the following manner: If the estimated SNR of the extracted AEP is low (less than the desired AEP), the number of averages is increased in order to improve the SNR. This has the disadvantage of reducing the system's ability to track fast changes on the AEP. In contrast, if the estimated SNR is higher than desired, the number of averages of the averaging processes is reduced to improve the detection of fast changes of the AEP. This control system balances the quality of the extraction with the speed of the averages and is applied to the averaging processes of the ARX model.

33.4.6.5 *Control of Click Volume*

Most studies done with the AEP have used a fixed click intensity in the range 60–80 dB hearing limit (HL). However, the hearing threshold has large interindividual variations; therefore a fixed volume level has two disadvantages. If the click intensity is too high, the patient can produce a startle response, which is a muscular peak time locked to the acoustic stimulus that superimposes itself on the peak of neurological origin, the AEP. If the click intensity is too low, the patient will be incapable of eliciting a reliable AEP. A click volume controller will then be ideal in order to provide each patient with the proper stimulation. In the case of overstimulation, the muscular peak produced by the startle effect is of much larger amplitude than the AEP; therefore the SNR increases to approximately 4–10. The controller should then decrease the volume until the SNR has reached the target considered as an adequate stimulation. The minimum volume is 45 dB. On the other hand, a very low SNR could be caused by a hearing deficiency. In this case the click intensity should be raised, but with the constraint that the intensity should not exceed the maximum level (e.g., 70 dB HL).

A proposed volume controller that follows these guidelines is shown in Figure 33.4. This control system is similar to the control system described in the previous section, based on an estimated and a desired SNR value (one which is considered to correspond to an adequate stimulation volume). A fuzzy subsystem is incorporated into the control system in order to modify the desired SNR (SNRd). The fuzzy inference system combines the information about how the different parts of the AEP are modified by click intensity (BAEP, MLAEP) and also may describe the situations where the AEP is affected in a different way by the volume depending on patient state. For example, when the suppression of the EEG is high (long periods with amplitude less than around 5 µV, burst suppression) related sometimes with a very deep anesthesia, the AEP physiologically is almost completely suppressed, so the desired SNR must be reduced to avoid an unnecessary increase of the volume. At this point the fuzzy model also modifies the global SNR to

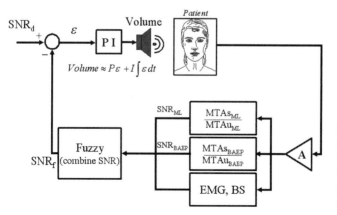

Figure 33.4 Volume of acoustic stimulus regulated by SNR. An SNR target value has been applied and shown to be adequate for extraction of the AEP.

incorporate this feature into the control system. The EMG is also an important parameter included into the fuzzy system (the startle effect is stronger when the EMG is high), so the system must give the adequate SNR estimation to carry out the volume control under this situation.

Both control systems based on the SNR, to control the number of averages and to control the stimulation volume, can be used alone or together in order to give the adequate stimulation and the optimum number of averages to extract an AEP that better describes the patient's hypnotic state.

33.5   EEG-BASED INDICES

33.5.1   Clustering Analysis

Clustering analysis of EEG segments was described by Thomsen et al. [40, 41]. Initially, 11 features were extracted, 10 normalized autocorrelation coefficients and the root-mean-square (RMS) amplitude of segments with 2 s duration. The features were transformed to power spectra by AR modeling using the Durbin algorithm and stored in a feature vector. Unsupervised learning based on hierarchical clustering of representative sets of segments was performed, with the aim of establishing a taxonomy where a group of EEG segments could correlate to a certain depth of anesthesia. It was concluded that the clustering of EEG segments correlated better with the clinical signs of depth of anesthesia than did single parameters such as spectral edge frequency (SEF), median frequency (MF), and power ratio.

A classification by neural networks was attempted by Eckert et al. [42]. A number of EEG epochs (channel C3 P3, duration 30 s segments) yielded patterns which were used as the training input of a self-organizing neural network. A total number of 25549 EEG epochs recorded in 196 patients in the period beginning 5 min before induction of anesthesia until extubation were used for the training. In comparison with the SEF and MF of the EEG, neural discriminant analysis achieved better discrimination between awake and deeply anesthetized stages (reclassification of 96% vs. 70%). According to the authors both hierarchical clustering and neural discriminant analysis produce better results than do single parametric methods. It should be remembered, though, that none of these methods have been reproduced by any other research groups, partly because the complexity of these algorithms is higher than in single parametric methods.

33.5.2   Bispectral Analysis of EEG (BIS)

Bispectral analysis is a double-frequency analysis. It is defined as the double fast Fourier transform (FFT) of the third-order cumulant of a signal in the time domain. Conventional power spectral analysis assumes that all sine waves present in the signal are independent, whereas bispectral analysis quantifies the degree of phase coupling between every pair of frequencies. The degree of phase coupling at each frequency pair is represented with the bicoherence index (normalized bispectrum), which is a continuous variable. The triple correlation (third-order cumulant) of a zero-mean time signal is defined as the integral

$$R(a, b) := \int_{-\infty}^{\infty} f(t)f(t + a)f(t + b)\, dt$$

where $f(t)$ is a time signal.

The bispectrum can now be calculated as the two-dimensional Fourier transform of the triple correlation

$$B(u, v) := \int_{-\infty}^{\infty} \int_{-\infty}^{\infty} R(a, b)e^{-2\pi j(ua+vb)} \, da \, db$$

The amplitude of the bispectrum is dependent not only on the phase coupling but also on the amplitude of the analyzed signal. The amplitude influence is omitted by normalizing the bispectrum and the result is termed the bicoherence.

Bispectral analysis is implemented in a commercially available monitor, the A-2000 BIS monitor, (Aspect Medical Systems, Inc., Newton, MA). The A-2000 calculates an index based on the bicoherence, called the BIS. The BIS index is an interval from 0 to 100 where increasing value is assumed to correspond to a higher probability of consciousness. The exact algorithm is not published and this makes it impossible to describe physiological relationship of the BIS to depth of anesthesia. The clinical relevance of BIS is a database, including a very high number of patients undergoing general anesthesia. The BIS index version 4.0 is a weighted sum of four subparameters: beta ratio, bispectral analysis, quasi-burst suppression, and burst suppression [43].

Flaishon et al. [44] used BIS to predict the recovery time from a thiopentone or a propofol bolus. They concluded that a patient with a BIS value of less than 55 would have an extremely small probability of consciousness. Doi et al. [45] reported recovery of consciousness at bispectral values from 50 to 85 *with an increasing probability at a higher value of the index.* The BIS values before and after eye opening were not significantly different, and in this particular study it was concluded that BIS could not detect the transition from unconsciousness to consciousness, where consciousness was defined as a response to a well-defined command.

33.5.3 Patient State Index (PSI)

The patient state analyzer (PSA 4000) is a four-channel processed EEG monitor (Physiometrix, Hospira, Inc., Lake Forest, IL). The PSI algorithm was constructed by using gradual discriminant analysis based upon multivariate combinations of quantified EEG (qEEG) variables found to be sensitive to changes in the level of anesthesia but insensitive to the specific substances producing such changes. It includes changes in power in various EEG frequency bands, changes in symmetry, and synchronization between critical brain regions and the inhibition of regions of the frontal cortex. The PSI is computed from continuously monitored changes in the qEEG during surgery using statistical analysis to estimate the likelihood that the patient is anesthetized. The computed PSI is displayed in numeric form on a 0–100 scale. The PSA performance during propofol, alfentanil, and nitrous oxide anesthesia was described by Drover et al. [46]. The number of publications and the prevalence of this device are low compared to that of the BIS monitor.

33.5.4 Entropy

A number of techniques have been applied to analyze the "irregularity" or "disorder" of the EEG signal in order to determine depth of anesthesia. The rationale behind this concept lies in the notion that the EEG changes from irregular to more regular patterns as the anesthetic level increases.

To this effect, a variety of parameters have been studied. These include the spectral and approximate entropies, fractal spectrum, and hierarchic Lempel–Ziv complexity [47–49].

Possibly the most developed of these methods is the one applied by Viertiö-Oja et al. [47]. The system, based on spectral entropy calculations, uses a set of time windows of different lengths which are selected according to the corresponding frequencies in the EEG in order to shorten the response time, which could be excessive for real-time applications. Two spectral entropy values are then calculated in the frequency bands 0–32 Hz (E_{EEG}) and 0–47 Hz ($E_{EEG+EMG}$). The E_{EEG} entropy evaluates the frequencies where the EEG has most of its energy, while the $E_{EEG+EMG}$ analyzes also the frequencies where EMG is present in order to detect muscular activity that could interfere with the EEG measurements. This technology has been implemented in the M-module from Datex-Ohmeda, Finland.

33.5.5   Cerebral State Index

A new index, the cerebral state index (CSI, Danmeter A/S, Odense, Denmark) is using four subparameters from the EEG: beta ratio, alpha ratio, difference between those two, and burst suppression, defining an index from 0 to 100 as a fuzzy weighted sum of those. The beta and alpha ratios were defined as

$$\beta\text{-ratio} = \log\frac{E_{30-42.5\,\text{Hz}}}{E_{11-21\,\text{Hz}}} \qquad \alpha\text{-ratio} = \log\frac{E_{30-42.5\,\text{Hz}}}{E_{6-12\,\text{Hz}}}$$

where E indicates the energy in the specific frequency band.

During burst suppression (BS) the alpha and beta ratios are no longer monotonously decreasing as a function of the anesthetic depth and therefore cannot be used in the calculation of the final index. The novelty of the CSI is that a fuzzy inference system was used to define the index. The particular method used was the adaptive neuro fuzzy inference system (ANFIS) [50]. Figure 33.5 shows the structure of ANFIS, with N inputs and one output, the CSI. The therapeutic window for adequate surgical anesthesia was defined as 40–60, lower than 40 indicating deep anesthesia and above 90 being awake.

The CSI was validated in a number of studies; one of these will be highlighted here. The data used for this study was originally published in *Anesthesiology* [51], where data was obtained from 20 ASA I (American Society of Anesthesiologists State I) female patients aged 18–60 years and scheduled for ambulatory gynecologic surgery and monitored with EEG and AEP. The clinical signs were assessed every 4 min using the Observers Assessment of Alertness and Sedation (OAAS) scale, which ranges from 5 to 0, where 5 is awake and 0

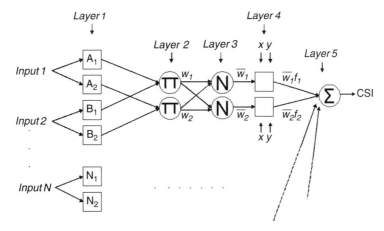

Figure 33.5   Structure of ANFIS system for calculation of CSI.

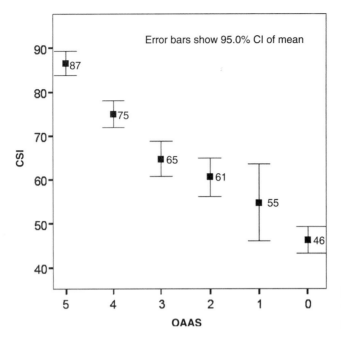

Figure 33.6 The OAAS versus CSI during propofol anesthesia.

corresponds to anesthesia with no response to painful stimuli. Figure 33.6 shows that a high correlation between the CSI and the OAAS existed. This is also reflected by the prediction probability (P_K value) which was 0.92 [standard error (SE) = 0.01]. The CSI has also been tested during target-controlled infusion of propofol [52].

33.6 AEP-BASED INDICES

33.6.1 Introduction

The AEP contains several peaks; hence it is necessary to conclude on which of these values to base clinical inference. In 1988 McGee et al. published a paper illustrating the use of different strategies (e.g., area of a peak and trough complex) for mapping the AEP waveform into a single number [53]. One of the conclusions of the study is that detection of amplitude and latency of a single peak is subject to a high degree of variability.

Single peak latency (Nb) as a measure of loss of consciousness has been described by Schwender et al. [54]. Mantzaridis suggested an index based on the sum of square roots of differences between samples in a sweep [55].

Jensen et al. [36] have proposed an index, the A-line ARX index (AAI) that includes only the middle-latency part of the AEP to achieve increased robustness. A didactic study and statistical analysis describing how the measured depth of anesthesia and the actual depth of anesthesia should correlate was published by Smith et al. [56, 57].

33.6.2 Requirements

The requirements of an AEP index for detection of depth of anesthesia are outlined below:

1. Least possible delay in order to show rapid transitions from asleep to awake.

2. Independent of patient conditions such as skin surface resistance, sex, and age.

3. Drug independence. The index should remain unchanged when administering an equipotent dose of all common anesthetic drugs used in clinical practice.

4. Sufficient SNR to distinguish index changes from noise.

The first requirement is accomplished if the delay in the extraction of the AEP is minimal, for example, by using the ARX model.

The second requirement contains several parts. The change in skin resistance can alter the energy (amplitude) of the collected sweeps. The skin resistance should be below 5 kΩ, and ideally this parameter should be measured frequently with an impedance meter. Increasing skin impedance will in general increase the noise in each sweep; therefore the energy will increase as well. The influence of age was described by Hegerl et al. [58] to increase the peak latency of the P300 with 0.92 ms/year (mean peak latency was 280 ms in subjects aged 20 years) whereas the P200 (160 ms in subjects aged 20 years) was not influenced by age. Amenedo and Diaz [59] found that age correlated positively with Na, Na–Pa, and Nb–Pb amplitude; also they did not find any gender differences. A study by Philips et al. [60] concluded that Pa and Pb amplitudes and latencies did not differ between healthy male and female subjects.

The third requirement is related to the use of different anesthetics. Volatile and intravenous anesthetic agents are known to produce different effects on the BAEP, whereas both types of anesthetics produce a dose–response influence on the MLAEP.

The fourth requirement mostly depends on the quality of the extracted AEP. A high SNR (within an adequate range, a very high SNR could be a symptom of overstimulation, not related to the state of the patient) makes inference on the AEP more reliable. It is desirable that the ratio between index values while awake and while asleep be as large as possible. This can also be expressed by requiring the minimum index value while awake to be larger than the maximum index value while asleep. This is the ideal performance characteristic, meaning that the sensitivity and specificity are both 100%. However, this has not been obtained by any depth of anesthesia parameter so far.

33.6.3 AEPex Index

Mapping of the two-dimensional AEP into a scalar was described by Mantzaridis and Kenny [61]. The index was defined in such way that a decrease in amplitude or an increase in the latency of the AEP, as it is observed when the patient is anesthetized, results in a decrease in the index. The AEP used for the calculation of the index had a latency of 144 ms, containing 256 samples. It was extracted by averaging 256 sweeps, resulting in an update delay of approximately 30 s.

The formula of the index is shown below:

$$\text{AEP}_{\text{idx}} = k \cdot \sum_{i=1}^{N-1} \sqrt{|x_i - x_{i+1}|}$$

33.6.4 AEP-Derived A-Line ARX Index: AEP Index with Correction for BAEP, LLAEP, and Muscular Response

The AEPex described in the previous section weighs all parts of the AEP, with latency 0–144 ms, equally. Like the AEPex, the AAI is also based on the assumption that the peak amplitude decreases and the peak latency increases when anesthetizing the subject. In order to fulfill requirement 3 it was proposed to exclude the BAEP part of the AEP to obtain an index behavior with less variation in the response to an equipotent

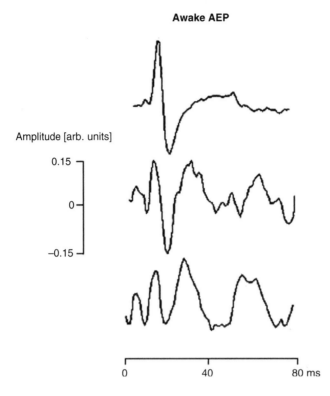

Figure 33.7 Examples of AEP from awake subjects with different amplitudes of muscular response.

dose of different anesthetic drugs. The acoustic stimulus may produce a muscular response, which can be detected in the latency interval 10–20 ms [5].

Figure 33.7 shows the AEP from three different awake subjects. Figure 33.7a has a very dominating muscular artifact with amplitude much larger than the rest of the AEP, whereas the subjects in Figure 33.7b,c present an AEP with less muscular artifact. This phenomenon was described by Picton [5], who showed that tension in the scalp musculature profoundly altered the early part of the MLAEP recording. Furthermore, the muscular response is suppressed by administration of NMBA; hence NMBA interferes with the changes occurring due to the hypnotic effect of the anesthetics.

The LLAEP is known to have larger interindividual variability than the MLAEP. During physiological sleep (stages 3 and 4) it has been reported that the amplitudes of the P2 peak (with latency 168 to 248 ms in that particular study) and the P3 peak (252–408 ms) were larger than while the subjects were awake [62]. Similar results were obtained by Van Hoof et al. during anesthesia [63]. Furthermore, that study also showed amplitude increases for the P1–N1 complex during anesthesia. The P1–N1 complex has latency down to 100 ms. Therefore the AAI has been designed to incorporate only the MLAEP in a window from 20 to 80 ms in order to exclude BAEP, LLAEP, and muscular artifacts. The BAEP and LLAEP are effectively excluded but the startle response (muscular artifact) can in some occasions surpass 20 ms if the volume of the acoustic stimulation is not attenuated. The formula is

$$AEP_{index} = K \sum_{i=k}^{N-1} |x_i - x_{i+1}|$$

where x_i is the ith sample (the value of x_i is the amplitude of the AEP at the time equal to i multiplied by the sampling period) of the AEP and N the number of samples in one sweep.

The parameter K is used for normalization such that the AEP index equals 100 when the patient is awake. With the applied sampling frequency $f_s = 900$ Hz, the values of k and N were 18 and 72, respectively, in order to achieve the 20–80-ms window for the AAI. The AAI index is currently in use and implemented on the A-line monitor (Danmeter A/S, Odense, Denmark).

33.7 COMPOSITE AEP AND EEG INDEX

The objective of this new approach is to define a hybrid index (HI) derived from AEPs and EEG parameter(s). The index should reliably differentiate awake and asleep states in a graduated manner.

In order to test the method, 120 patients from our database scheduled for elective surgery were included in the analysis. Seventy patients underwent cardiac surgery and were anesthetized with propofol using a target-controlled infusion (TCI) Diprifusor pump (target 5 µg/mL blood propofol concentration during 5 min, steps 1 µg/mL every 4 min). Propofol was administered until loss of response to noxious stimuli, defined as loss of a meaningful response to a verbal command. The remaining 45 patient data set was recorded at Gent University Hospital for two previous studies [51, 64]. These were scheduled for elective gynecological surgery. Three groups were then defined: group A only receiving propofol, group B receiving first 2-ng/mL effect site concentration of remifentanil, and group C receiving 4 ng/mL effect site concentration of remifentanil before the start of propofol infusion. The AEP and EEG were collected using the AEP monitor (Danmeter A/S, Odense, Denmark). During the anesthesia clinical signs were recorded and the state of the patient was defined based on an expert's evaluation using the OAAS.

33.7.1 EEG Parameter

The EEG is recorded from the same electrodes as the AEP. During anesthesia the energy in the EEG typically shifts from higher to lower frequencies; therefore indices based on the ratio between the energy in two frequency bands should carry information about the level of consciousness during anesthesia. The particular component of the EEG used here was the beta ratio, that is, the logarithm to the ratio of energies in the 30–47-Hz band and that of the 10–20 Hz.

33.7.2 Hybrid Index

The HI is defined as a linear weighting of the AAI and the EEG index (EI). The SNR of the AEP is calculated. If the SNR value is less than 1.5, then the weight on the EI was increased and vice versa. The AAI version 4.1 implemented in the AEP monitor/2 is a hybrid index combining the AEP, EEG, and BS.

The obtained indices were compared to the OAAS scale values. The ability of the different indicators to describe the level of consciousness was evaluated using prediction probability (P_K), which compares the performance of indicators having different units of measurements, as developed by Smith et al. Then, a P_K of 1 for the BIS or AAI indicator would mean that the BIS or AAI always increases (decreases) as the patient gets lighter (deeper) according to the "gold standard" depth measure. Such an indicator can perfectly predict anesthetic depth. Alternatively, a P_K value of 0.5 would mean that the indicator is not better than tossing a fair coin for predicting anesthetic depth.

TABLE 33.1   The Prediction Probability (P_K) Values for the Three Indices

	AAI (ver 4.0)	EI (EEG)	HI (AAI ver 4.1)
P_K (SE)	0.89 (0.01)	0.88 (0.01)	0.92 (0.01)

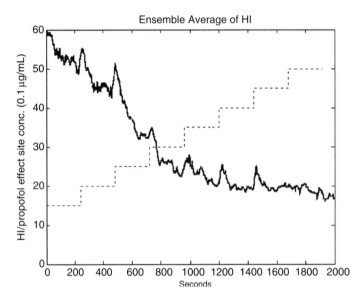

Figure 33.8   The HI values (solid curve) during increased target effect site concentration of propofol (dashed curve). The OAAS 0, the deepest level of anesthesia, corresponded to a HI of 18 at 1900 s after start of infusion.

The results of the P_K analysis are shown in Table 33.1. A t-test showed that the P_K of the HI was significantly larger than the AAI and EI ($p < 0.05$). Figure 33.8 shows an example of the behavior of the HI during induction of propofol.

The results show that there is significant difference between HI while awake and while anesthetized. The AEP is in general abolished during very deep anesthesia while there still are changes in the complexity of the EEG. But the AEP has in general a better performance than the EEG in the lighter levels of anesthesia. In conclusion, an index combining information from both the AEP and EEG seems to have a better prediction of the level of consciousness at all ranges of the scale.

REFERENCES

1. MEYER, B. C. AND BLACHER, R. S. A traumatic neurotic reaction induced by succinylcholine chloride. *NY State J. M.* 1961;61:1255–1261.
2. BERSTRØM, H. AND BERGSTEIN, K. Psychic reactions after analgesia with nitrous oxide for caesarean section. *Lancet* 1968;2:541–542.
3. JONES, J. G. AND KONIECZKO, K. Hearing and memory in anaesthetised patients. *Br. Med. J.* 1986;292:1291–2931.
4. SEBEL, P. S., BONKE, B., AND WINOGRAD, E. *Memory and Awareness in Anesthesia*. Englewood Cliffs, NJ: Prentice-Hall, 1993.

5. PICTON, T. W., HILLYARD, S. A., KRAUSZ, H. I., AND GALAMBOS, R. Human auditory evoked potentials. I: Evaluation of components. *Electroenc. Clin. Neurophysiol.* 1974;36:179–190.

6. JEWETT, D. L. Volume-conducted potentials in response to auditory stimuli as detected by averaging in the cat. *Proc. San Diego Biomed. Symp.* 1970;28:609–618.

7. LEV, A. AND SOHMER, H. Sources of averaged neural responses recorded in animal and human subjects during cochlear audiometry (electrocochleogram). *Arch. Klin. Exp. Ohren. Nasen. Kehlkopfheil.* 1972;201:79–90.

8. WOODS, D. L., CLAYWORTH, C. C., KNIGHT, R. T., SIMPSON, G. V., AND NAESER, M. A. Generators of middle- and long-latency auditory evoked potentials: Implications from studies of patients with bitemporal lesions. *Electroencephalogr. Clin. Neurophysiol.* 1987;68:132–148.

9. THORNTON, C., HENEGHAN, C. P. H., JAMES, M. F. M., AND JONES, J. G. The effects of halothane and enflurane with controlled ventilation on auditory evoked potentials. *Br. J. Anesth.* 1984;56:315–323.

10. DUBOIS, M. Y., SATO, S., CHASSY, J., AND MACNAMARA, T. E. Effects of enflurane on brainstem auditory evoked responses in humans. *Anesth. Analg.* 1982;61:898–902.

11. MANNINEN, P. H., LAM, A. M., AND NICHOLAS, J. F. The effects of isoflurane and isoflurane-nitrous oxide anesthesia on brainstem auditory evoked responses in humans. *Anesth. Analg.* 1985;64:43–47.

12. SCHMIDT, J. F. AND CHRAEMMER-JORGENSEN, B. Auditory evoked potentials during isoflurane anesthesia. *Acta Anesthesiol. Scand.* 1986;30:378–380.

13. HENEGHAN, C. P. H., THORNTON, C., NAVARATNARAJAH, M., AND JONES, J. G. Effects of isoflurane on the auditory evoked responses in man. *Br. J. Anaesth.* 1987;59:277–282.

14. BERTOLDI, G., MANNO, E., BRUERA, G., GILLI, M., AND VIGHETTI, S. The influence of etomidate, flunitrazepam and ketamine on the BAEP of surgical patients with no audiological or neurological alterations. *Minerva Anesthesiol.* 1983;49:349–356.

15. THORNTON, C., HENEGHAN, C. P. H., NAVARATNARAJAH, M., JONES, J. G., AND BATEMAN, P. E. The effect of etomidate on the auditory evoked response in man. *Br. J. Anaesth.* 1985;57:554–561.

16. THORNTON, C., HENEGHAN, C. P. H., NAVARATNARAJAH, M., AND JONES, J. G. Selective effect of Althesin on the auditory evoked response in man. *Br. J. Anaesth.* 1986;58:422–427.

17. THORNTON, C., KONIECZKO, K. M., KNIGHT, A. B. ET AL. The effect of propofol on the auditory evoked response and on esophageal contractility. *Br. J. Anaesth.* 1989;63:411–417.

18. SAVOIA, G., ESPOSITO, C., BELFIORE, F., AMANTEA, B., AND CUOCOLO, R. Propofol infusion and auditory evoked potentials. *Anesthesia* 1988;43:46–49.

19. VAN HOOFF, J. C., DE BEER, N. A. M., BRUNIA, C. H. M., CLUITMANS, P. J. M., AND KORSTEN, H. H. M. Event-related potential measures of information processing during general anaesthesia. *Electroencephalog. Clin. Neurophysiol.* 1997;103:268–81.

20. DE BEER, N. A., VAN HOOFF, J. C., BRUNIA, C. H., CLUITMANS, P. J., KORSTEN, H. H., AND BENEKEN, J. E. Mid-latency auditory evoked potentials as indicators of perceptual processing during general anaesthesia. *Br. J. Anaesth.*, 1996;77:617–624.

21. MAURER, K., LEITNER, H., AND SCHAFER, E. Detection and localization of braistem lesions with auditory braintem potentials. In BARBER, C., Ed., *Evoked Potentials*, Lancaster, England: MTP Press, 1980; pp. 391–398.

22. SCHWENDER, D., KLASING, S., CONZEN, P., FINSTERER, U., PÖPPEL, E., AND PETER, K. Midlatency auditory evoked potentials during anaesthesia with increasing endexpiratory concentrations of desflurane. *Acta Anaesthesiol. Scand.* 1996;40:171–176.

23. PLOURDE, G. Clinical use of the 40-Hz auditory steady state response. *Int. Anesthesiol. Clin.* 1993;31:107–120.

24. BARTNIK, E. A., BLINOWSKA, K. J., AND DURKA, P. J. Single evoked potential reconstruction by means of wavelet transform. *Biol. Cybernet.* 1992;62:175–181.

25. GEVA, A. B., PRATT, H., AND ZEEVI, Y. Y. Spatio temporal source estimation of evoked potentials by wavelet type decomposition. In GATH, I. AND INBAR, G. Eds., *Advances in Processing and Pattern Analysis of Biologocal Signals*, Plenum, New York, 1996.

26. KOCHS, E., STOCKMANNS, G., THORNTON, C., NAHM, W., AND KALKMAN, C. J. Wavelet analysis of middle latency auditory evoked responses: Calculation of an index for detection of awareness during propofol administration. *Anesthesiology* 2001;95(5):1141–1150.

27. HAIG, A. R., GORDON, E., ROGERS, G., AND ANDERSON, J. Classification of single-trial ERP sub-types: Application of globally optimal vector quantization using simulated annealing. *Electroencephalogr. Clin. Neurophysiol.* 1995;94:288–297.

28. EYSHOLDT, U. AND SCHREINER, C. Maximum length sequences—A fast method for measuring brain-stem evoked responses. *Audiology* 1982;21:242–250.

29. SHI, Y. AND HECOX, K. E. Nonlinear system identification by m-pulse sequences: Application to BAEP. *IEEE Trans. Biomed Eng.* 1991;38:834–845.

30. DON, M., ALLEN, A., AND STARR, A. Effect of click rate on the latency of ABR in humans. *Ann. Otol. Rhinol. Lar.* 1977;86:186–195.

31. ZHU, Y. AND BACKX, T. *Identification of Multivariable Industrial Processes*. London: Springer Verlag, 1993.

32. CERUTTI, S., BASELLI, G., AND LIBERATI, D. Single sweep analysis of visual evoked potentials through a model of parametric identification. *Biol. Cybern.* 1987;56:111–120.

33. CERUTTI, S., CHIARENZA, G., LIBERATI, D, MASCELLANI, P., AND PAVESI, G. A parametric method of identification of single-trial event-related potentials in the brain. *IEEE Trans. Biomed. Eng.*, 1988;35:701–711.

34. LIBERATTI, D. AND CERUTTI, S. The implementation of an autoregressive model with exogenous input in a single sweep visual evoked potential analysis. *J. Biomed. Eng.* 1989;11:285–292.

35. MAGNI, R., GIUNTI, S., BIANCHI, B., RENI, G., BANDELLO, F., DURANTE, A., CERUTTI, S., AND BRANCATO, R. Single sweep analysis using an autoregressive model with exogenous input (ARX) model. *Documenta Opthalmol.* 1994;86:95–104.

36. JENSEN, E. W., LINDHOLM, P., AND HENNEBERG, S. Auto regressive modeling with exogenous input of auditory evoked potentials to produce an on-line depth of anaesthesia index. *Methods Inform. Med.* 1996;35:256–260.

37. COHEN, A. *Biomedical Signal Processing*, Vols I–II. Boca Raton, FL: CRC Press, 1986.

38. JANSEN, B. H. Analysis of biomedical signals by means of linear modeling. *CRC Crit. Rev. Biomed. Eng.* 1985;12:4.

39. AKAIKE, H. Statistical predictor identification. *Ann. Inst. Statist. Math.* 1970;22:203–217.

40. THOMSEN, C. E. AND PRIOR P. F. Quantitative EEG in assessment of anaesthetic depth: Comparative study of methodology. *Br. J. Anaesth.* 1996;77(2): 172–178.

41. THOMSEN, C. E., ROSENFALCK, A., AND NØRREGAARD CHRISTENSEN, K. Assessment of anaesthetic depth by clustering analysis and autoregressive modelling of electroencephalograms. *Comput. Methods Prog. Biomed.* 1991;34:125–138.

42. ECKERT, O., WERRY, C., NEULINGER, A., AND PICHLMAYR, I. Intraoperative EEG monitoring using a neural network. *Biomed. Tech. (Berl.)* 1997;42(4):78–84.

43. SIGL, J. C. AND CHAMOUN, N. G. An introduction to bispectral analysis for the electroencephalogram. *J. Clin. Monit.* 1994;10:392–404.

44. FLAISHON, R., WINDSOR, A., SIGL, J., AND SEBEL, P. S. Recovery of consciousness after thiopenthal or propofol. *Anesthesiology* 1997;86:613–619.

45. DOI, M., GAJRAJ, R. J., MANTZARIDIS, H., AND KENNY, G. N. C. Relationship between calculated blood concentration of propofol and electrophysiological variables during emergence from anaesthesia: Comparison of bispectral index, spectral edge frequency, median frequency and auditory evoked potential index. *Br. J. Anaest.* 1997;78:180–184.

46. DROVER, D., LEMMENS, H. J., PIERCE, E. T., PLOURDE, G., LOYD, G., ORNSTEIN, E., PRICHEP, L. S., CHABOT, R. J., AND GUGINO, L. Patient state index titration of delivery and recovery from propofol, alfentanil, AND nitrous oxide anesthesia. *Anesthesiology* 2002;97:1013–1014.

47. VIERTIÖ-OJA, H., MERILÄINEN, P., SÄRKELÄ, M., TALJA, P., TOLVANEN-LAAKSO, H., AND YLI-HANKALA, A., Spectral entropy, approximate entropy, complexity, fractal spectrum and bispectrum of EEG during anesthesia. Paper presented at 5th International Conference on Memory, Awareness and Consciousness, New York, June 2001.

48. BRUHN, J., RÖPCKE, H., AND HOEFT, A., Approximate entropy as an electroencephalographic measure of anesthetic drug effect during desflurane anesthesia. *Anesthesiology* 2000;92:715–726.

49. BRUHN, J., LEHMANN, L. E., RÖPCKE, H., BOUILLON, T. W., AND HOEFT, A. Shannon entropy applied to the measurement of the electroencephalographic effects of desflurane. *Anesthesiology* 2001;95:30–35.

50. JANG, J. S. R. ANFIS: Adaptive-network-based fuzzy inference system. *IEEE Trans. Sys. Man Cybernet.* 1993;23:665–685.

51. STRUYS, M. R. F, JENSEN, E. W., SMITH, W., SMITH, T., RAMPIL, I., DUMORTIER, F., AND MORTIER, E. P. Performance of the ARX-derived auditory evoked potential index as an indicator of anesthetic depth. A comparison with BIS and hemodynamic measures during propofol administration. *Anesthesiology* 2002;96:803–816.

52. ZHONG, T., GUO, Q. L., PANG, Y. D., PENG, L. F., AND LI, C. L. Comparative evaluation of the cerebral state index and the bispectral index during target-controlled infusion of propofol. *Br. J. Anaesth.* 2005;95:798–802.

53. McGEE, T., KRAUS, N., AND MANFREDI, C. Towards a strategy for analyzing the auditory middle latency response waveform. *Audiology* 1988;27:119–130.

54. SCHWENDER, D., DAUNDERER, M., MULZER, S., KLASING, S., FINSTERER, U., AND PETER, K. Midlatency auditory evoked potentials predict movements during anaesthesia with isoflurane or propofol. *Anesth. Analg.* 1997;85:164–173.

55. MANTZARIDIS, H. Closed-loop control of anaesthesia. PhD thesis, Strathclyde University, Strathclyde, 1996.

56. SMITH, W. D., DUTTON, R. C., AND SMITH, N. T. Measuring performance of anaesthetic depth indicators. *Anesthesiology* 1996;84:38–51.

57. SMITH, W. D., DUTTON, R. C., AND SMITH, N. T. A measure of association for assessing prediction accuracy that is a generalization of non-parametric ROC area. *Statist. Med.* 1996;15:1199–1215.

58. HEGERL, U., KLOTZ, S., AND ULRICH, G. Late acoustically evoked potentials—effect of age, sex and different study conditions. *EEG EMG Z Elektroenzephalogr. Verwandte Geb.* 1985;16(3):171–178.

59. AMENEDO, E. AND DÍAZ, F. Effects of aging on middle-latency auditory evoked potentials: A cross-sectional study. *Biol. Psychiatry*, 1998;43:210–219.

60. PHILLIPS, N. A., CONNOLLY, J. F., MATE KOLE, C. C., AND GRAY, J. Individual differences in auditory middle latency responses in elderly adults and patients with Alzheimer's disease. *Int. J. Psychophysiol.* 1997;27:125–136.

61. MANTZARIDIS, H., AND KENNY, G. N. C. Auditory evoked potential index: A quantitative measure of changes in auditory evoked potentials during general anaesthesia. *Anaesthesia* 1997;52:1030–1036.

62. BUCHSBAUM, M., CHRISTIAN, G., AND PFEFERBAUM, A. Effect of sleep stage and stimulus intensity on auditory average evoked responses. *Psychophysiology* 1975;12:707–712.

63. Van HOOFF, J. C., DE BEER, N. A. M, BRUNIA, C. H. M, CLUITMANS, P. J. M, AND KORSTEN, H. H. M. Event-related potential measures of information processing during general anaesthesia. *Electroencephalogr. Clin. Neurophysiol.* 1997;103:268–281.

64. STRUYS, M. M. R. F., VEREECKE, H., MOERMAN, A., JENSEN, E. W., VERHAEGHEN, D., DE NEVE, N., DUMORTIER, F., AND MORTIER, E. P. Ability of the bispectral index, autoregressive modeling with exogenous input-derived auditory evoked potentials, AND predicted propofol concentrations to measure patient responsiveness during anesthesia with propofol and remifentanil. *Anesthesiology* 2003;99(4):802–812.

CHAPTER 34

ENG RECORDING AMPLIFIER CONFIGURATIONS FOR TRIPOLAR CUFF ELECTRODES

I. F. Triantis, A. Demosthenous, M. S. Rahal, and N. Donaldson

34.1 INTRODUCTION

Cuff electrodes are appropriate for recording electroneurogram (ENG) signals chronically after implantation [1]. Because the amplitude of the neural signal inside the cuff is usually only a few microvolts, it may easily be overwhelmed by interference or noise. A major source of interference is the myoelectric activity of nearby muscles (the electromyogram, or EMG) [2]. The cuff is an insulating tube that serves to increase the signal amplitude and, if perfectly symmetrical, prevents pick-up of signals from unwanted potentials outside the cuff. Tripolar cuff amplifier configurations include the *quasi-tripole* (QT), the *true tripole* (TT) [3], and the recently introduced *adaptive tripole* (AT). The latter was developed to compensate for actual asymmetry in the cuff which introduces interference [4, 5]. We expect that neural signals will be used as feedback and command sources in future neuroprosthetic systems, such as muscle stimulators for correcting gait, providing hand-grasp or controlling the urinary bladder after spinal cord injury [6–8]. This chapter offers an insight into recording cuff electrodes and amplifier configurations used with tripolar cuffs.

34.2 CUFF ELECTRODE DESCRIPTION AND SIGNAL CHARACTERISTICS

The ENG recording techniques for peripheral nerves using cuff electrodes were introduced by Hoffer et al. [9] and Stein et al. [2]. The increase in signal amplitude due to the cuff in recording the ENG extraneurally was shown in [10] to be the result of the local restriction of the extracellular space which, in effect, increases the resistance of the extracellular return path of the fiber's action currents. The optimum cuff length was shown experimentally in [2, 11] to be between 20 and 30 mm with the signal amplitude decreasing when the length is smaller than 15 mm. The amplitude of the signal was also shown in [12] to decrease with the square of the cuff diameter for diameters up to 4 mm, indicating that the effectiveness of cuff electrodes improves for tighter fitting cuffs, as shown experimentally in [1, 9, 11] where single fiber action potentials ranging from 3 to 20 μV were recorded using cuffs with inner diameters of 2.6 and 0.3 mm, respectively.

Handbook of Neural Engineering. Edited by Metin Akay

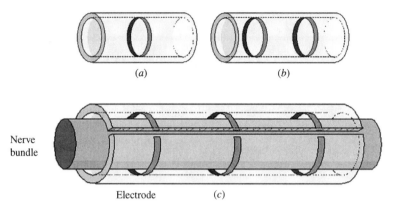

(a)   *(b)*

Nerve
bundle

Electrode   *(c)*

Figure 34.1   (*a,b*) Ideal monopolar and bipolar cuffs. (*c*) Tripolar cuff placement around nerve. The remaining slit is usually glued or covered by a flap to ensure that the tube is closed along its length.

Typical cuffs consist of silicone tubes with ring electrodes made of platinum–iridium or stainless steel attached to the inside wall, covering most of the circumference but leaving a gap for a longitudinal opening so that the nerve may be inserted into the cuff (Fig. 34.1*c*). Originally, closure of the slit along the cuff was achieved by tying threads around [13]. Later methods include injecting a liquid polymer into the cuff to fill the space between the nerve and the cuff to make the fitting tighter [14]. Naples et al. introduced the "spiral cuff" in [15], which curls up to fit the nerve tightly but is still able to expand if the nerve swells, perhaps due to inflammation. This design has been used for stimulation and recording and is made to curl by gluing an unstretched sheet onto a prestretched sheet of rubber. A cuff made by dip coating a mandrel and pre formed platinum electrodes was presented in [16], where the cuff closure mechanism was implemented using interdigitating cubes. Typically, the slit along the tube is glued or covered by a silicone flap attached to the cuff [17].

The placement of the cuff around the nerve causes some changes in the nerve after a period of implantation, as connective tissue covers and invaginates the cuff. The nerve may also change shape to adapt to the internal space [18]. To avoid compression neuropathy, the cuff diameter has to be at least 20% larger than that of the nerve bundle [11], and its length must be at least 10 times larger than its inner diameter [19]. Increasing the length significantly reduces interference; however, there are limitations depending on the location of implantation.

The dimensions of the cuff affect the spectrum of the recorded ENG and therefore the cuff acts as a spatial filter [20]. The characteristics of this linear filter depend on the number and position of the electrodes inside the cuff and are affected by the tissue impedances and the cuff length but not so much by its diameter. The spatial filter has a broad-bandpass characteristic with a peak in the spectrum for each conduction velocity and, therefore, fiber diameter [20]. As pointed out by Upshaw [21], this means that the signal-to-noise ratio should be improved by using narrow-bandpass filters, centered on the spectral peak corresponding to the propagation velocity for the fibers from which one wishes to record.

In general neural frequencies (ENG), seen at the tripolar electrodes, lie in the frequency range of 500 Hz–10 kHz, usually with maximum power between 1 and 3 kHz. Frequencies at the low end are often difficult to record due to electrical interference from the surrounding muscles. Their myoelectric (EMG) potentials are of millivolt

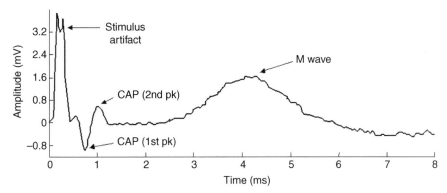

Figure 34.2 Typical stimulus artifact, CAP, and M-wave signals recorded from tripolar cuff [25].

amplitudes with a frequency range of 1 Hz–3 kHz, peaking at approximately 250 Hz [22]. There is also interference from thermal noise, generated from the electrodes and the amplifiers, which, with a good amplifier, can be about 0.7 μV root-mean-square (RMS) [1, 23]. With signals of such low level, interference is also likely to appear at power line frequencies and possibly from radio sources if care is not taken to prevent rectification before the high-gain stages of the amplifiers.

If the nerve is stimulated, the neural signal recorded has greater amplitude (few tenths of a millivolt), as the neurons conduct a synchronous volley of action potentials or *compound action potential* (CAP). The reaction of the muscles activated by the stimulated nerves is an *M-wave*, which may be two or three times larger than the naturally occurring EMG and peaks at higher frequencies. There is then also an additional source of interference, the *stimulus artifact*, due to the field from the pulse current. This appears at the electrodes as a spike that may reach several millivolts [22, 24]. Typical stimulation-induced signals are shown in Figure 34.2.

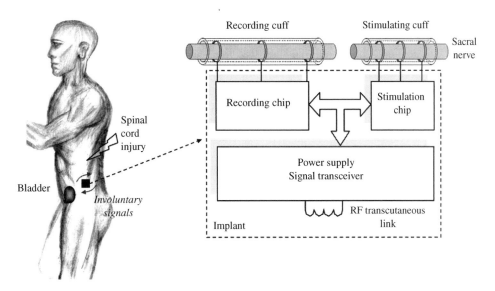

Figure 34.3 Example of bladder stimulation implant [39].

There are three main application areas for recording cuffs [13]. They can be used in chronic studies of physiology and pathology of the neuromuscular system by measuring naturally occurring signals in animals, as in [26–28]. They can also be used for studying the regeneration of axotomized nerve fibers and the status of the nerves after damage [29, 30]. Finally, sensory signals recorded using cuff electrodes can be used for feedback in neuroprostheses, as shown by animal studies in [22, 31–34] and in humans, where recorded signals have been used for control of foot-drop stimulation in hemiplegics [6, 35], and in hand-grasp applications in tetraplegics [36–38]. An example for the latter is shown in Figure 34.3, which shows the block diagram of an implant for ENG recording and for stimulation of the sacral nerve roots [39].

34.3 INTERFERENCE REDUCTION AND CUFF ELECTRODE PARAMETERS

In monopolar cuffs (Fig. 34.1a), the recording is carried out between a ring electrode inside the cuff and a reference electrode outside: The inner electrode has to be close to the center of the cuff to maximize the amplitude of the measured ENG. In bipolar cuffs (Fig. 34.1b), where the measurement is taken differentially between the two electrodes, the amplitude of the measured signal is proportional to the separation between the electrodes; however, as the electrodes approach the edges of the cuff, the amplitude drops, as the low-resistivity area of the fluid outside the cuff comes into effect. In tripolar cuffs measurements are taken between the middle and the two end electrodes. In order to maximize the amplitude of the signal, the end electrodes have to be placed close to the ends of the cuff. However, the main factor affecting measurements in tripolar cuffs is the spacing between the electrodes and not their actual position or the length of the cuff. The monopolar and tripolar cuffs have similar frequency responses and their lengths affect their properties as spatial filters. Overall the tripolar cuff gives higher signal amplitudes than the monopolar and bipolar counterparts, and measurements are less sensitive to the cuff length, as the most significant parameter is the electrode pitch. A *screened tripolar cuff* has two extra electrodes nearer the ends than the outer electrodes of the tripole and shorted together. This arrangement further decreases the amplitude of EMG interference, but for best effect, the additional electrodes should not be placed right at the ends of the cuff, and so this tends to increase the overall length. However, considering interference from external fields, using a screened tripole has the same effect as increasing the length of the cuff as it effectively decreases the gradient of the interfering field inside it [4].

According to [40] the tissue inside the cuff approximates an ideal distributed resistance, and therefore, there are no significant phase variations between interfering potentials caused by a single external source. This is of great significance: The *linearizing property* of the cuff [41] makes it appear like a potential divider, and the constant phase means that the equivalent model has no capacitors (though the electrode models do have capacitance). However, end effects reduce the linearity of the interference field inside the cuff [4]. Longer cuffs result in greater interference reduction, as the resistance of the tissue inside the cuff becomes higher and therefore less current flows through the cuff due to external fields. In practice, the cuff length is likely to be determined by anatomical and surgical considerations such as the unbranched length of the nerve that can be exposed, the amount of movement that is expected in the tissue, the blood supply, and so on. For a given length of cuff, the interference from an unscreened tripole will be minimized when the outer electrodes are near but not right at the end of the cuff. The interference will then depend on the impedance balance for the two halves of the tripole.

34.4 TRIPOLE CUFF AMPLIFIERS

34.4.1 General Requirements

The designer of ENG amplifiers to be used with cuff electrodes should consider several requirements that are shared with other biomedical amplifiers. The differential gain must be high, usually to raise the signal level from microvolts to the appropriate range for analog-to-digital conversion, but lower gains may be necessary if the same system is to be used for stimulation-induced signals. The front-end noise of the amplifiers must be very low, and care must be taken to choose the optimal technology [42]. Although transformer-coupled ENG recordings have been reported in the literature (e.g., in [22]) for scaling up the signal before introducing amplifier noise from subsequent stages, the use of transformers in implants is completely impractical [34] and is not necessary. Minimal noise will be achieved if the so-called *noise resistance* of the amplifier equals (or, in practice, approximates) the source resistance [43]. Low-power *integrated circuit* design is desirable for a recording system that is to be integrated into an implant [22]. If the reference electrode for the differential amplifier is not close to the cuff, there will be common-mode voltages (perhaps from the ECG). These will produce differential voltages at the inputs, which will then be amplified with the signal, due to two effects:

1. If capacitors are used in series with the electrodes, to prevent direct current flow, differences between the impedances of the capacitors (due to tolerances) will cause interference. This effect is greater at low frequencies.

2. The source resistances of the electrodes in the cuff are not equal (see below), and if there is a cable from the cuff to the amplifier, the capacitances from signal wires to the body are likely to be unequal. These resistances and capacitances will generally make unequal time constants that produce interference at high frequencies [23].

Both effects are reduced by increasing the common-mode impedance of the amplifier. When testing ENG amplifiers, the effects cause a reduction in *common-mode rejection ratio* to values that may be much lower than will be possible for the amplifier when operating with low source impedances [23].

The differential input impedance need not be very high. The common-mode impedance should be $>10^6 \, \Omega$ but the differential need only be high compared to the source impedances (10^2–$10^3 \, \Omega$) for the attenuation to not be important (i.e., $\sim 10^4 \, \Omega$).

34.4.2 Amplifier Configurations

Figure 34.4 shows the interface impedances [4, 22] of the cuff and the internal potential fields of the ENG and interference sources due to its linearizing property [41]. This schematic diagram allows an analysis of the voltages appearing between the electrodes in terms of the currents of the nerve and muscle sources and the tissue and electrode impedances. The tissue impedance outside the cuff is Z_o and inside it is separated into Z_{t1} and Z_{t2} between the middle and each of the end electrodes, resulting in $Z_{t1} + Z_{t2} = Z_t$. The electrode impedances are Z_{e1}, Z_{e2}, and Z_{e3} and it is assumed here

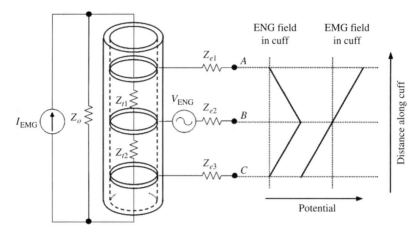

Figure 34.4   Basic cuff interface impedances and linearized ENG and interference fields (not to scale). The ENG is shown as an equivalent voltage source. Typical values: $Z_o = 200\ \Omega$, $Z_t = 2.5\ k\Omega$, $Z_e = 1\ k\Omega$, $I_{EMG} = 1\ \mu A$, $V_{ENG} = 3\ \mu V$ [4, 13, 22].

that $Z_{e1} = Z_{e2} = Z_{e3} = Z_e$. The current source I_{EMG} produces the interference signal from its field outside and inside the cuff and the ENG is shown as an equivalent voltage source in series with the middle electrode. The interfering voltages across points AB and BC appear as antiphase while the respective ENG signals appear in phase.

The QT (Fig. 34.5a) and the TT (Fig. 34.5b) are the two main amplifier configurations used with tripolar cuffs for ENG recordings. In the QT the short-circuited

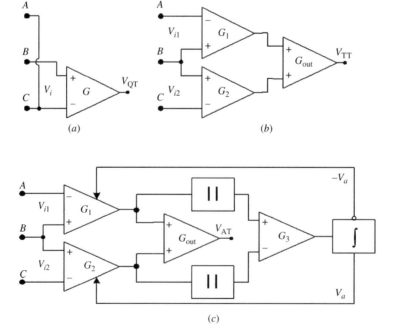

Figure 34.5   (a) The QT and its cuff connectivity (based on Fig. 34.4). (b) The TT [3]. (c) The AT. Gains G_1 and G_2 are controlled by feedback signals V_a and $-V_a$. The input stage is preceded by low-noise preamplifiers, not shown.

end electrodes give their average voltage provided their impedances are equal. If the cuff is perfectly sealed along the slit and the central electrode is exactly in the middle of the outer electrodes and the tissue impedances of its two halves are equal, the potential recorded by the middle electrode [41] is the same as the average potential recorded by the shorted end electrodes. Therefore, when a differential amplifier is connected across the middle electrode and this virtual terminal, as in the QT, the interference potentials at the two inputs are equal and cancel out at the output. In the TT the two outer cuff electrodes are connected to two separate differential amplifiers and the middle electrode to the other input of both amplifiers. The outputs from the two amplifiers are then summed in a third amplifier where ideally the two EMG potentials are of opposite polarity and will cancel out [3]. One benefit of the TT is that the ENG amplitude recorded is about twice that in the QT.

Analysis of the thermal noise for the QT and the TT (unpublished results) must take account of the voltage and current noise from the amplifiers but also the Johnson noise from the electrode impedances Z_{e1}–Z_{e3} and from the tissue in the cuff Z_{t1} and Z_{t2}. Assume that the external resistance (Z_o) is negligible and the input impedance of the amplifiers is infinite. The relative amplitudes for the four sources are shown in Table 34.1. Given that the signal amplitude ratio is 2 : 1, this shows that the signal-to-noise ratio of the TT will be better than the QT if amplifier voltage noise or cuff noise dominates, similar if amplifier current noise dominates, and inferior if electrode noise dominates.

Despite the use of tripole electrodes, significant artifacts are still present in the recorded signal for three main reasons [44]:

1. Inhomogeneous tissue growth inside the cuff causes the impedances Z_{t1} and Z_{t2} to change unevenly with time [21] varying the imbalance inside the cuff.
2. Tripolar electrode structures require a high degree of symmetry. Manufacturing tolerances of perhaps ± 0.5 mm in the electrode positions will destroy the symmetry required and result in large artifacts.
3. The linearization of the EMG field inside the cuff is not ideal due to end effects.

The performance of the QT and the TT can be compared in terms of interference reduction and cuff imbalance sensitivity. Using the parameters of Figure 34.4, the residual EMG at the output of the QT and the TT can be expressed in terms of tissue and electrode impedances, ignoring any cuff end effects [4]. If the gain of the QT amplifier is unity (i.e., $G = 1$), the residual EMG at its output is [44]

$$V_{QT} = I_{EMG} \frac{Z_o(Z_{t2} - Z_{t1})}{2Z_o + 2Z_t + Z_t Z_o / Z_e} \tag{34.1}$$

For interference elimination V_{QT} should be zero; therefore Z_{t1} and Z_{t2} should be equal. If there is an imbalance so that $Z_{t1} \neq Z_{t2}$, the QT does not provide any means of

TABLE 34.1 Main Noise Sources in QT and TT Amplifier

Noise source	Relative amplitude, T T : QT
Amplifier voltage noise	$\sqrt{2}$: 1
Amplifier current noise	2 : 1
Thermal noise from electrodes Z_e	5 : 1
Thermal noise from cuff Z_t	1 : 1

correcting this imbalance externally. However, it achieves interference reduction due to the screening provided by short circuiting the outer electrodes [41].

The TT doubles the ENG amplitude (V_{i1} is added to V_{i2}) but reduces EMG interference using *only* the linearizing property of the cuff [4]. Ideally, EMG signals would still be eliminated because their two inputs V_{i1} and V_{i2} are equal and opposite, while the respective ENG signals, appearing at the same points, are in phase. Hence, the ideal TT amplifies the ENG while completely eliminating the interference. Generally, assuming unity gain for G_{out}, the residual EMG at its output is [44]

$$V_{TT} = I_{EMG} \frac{Z_o(G_1 Z_{t1} - G_2 Z_{t2})}{Z_o + Z_t} \tag{34.2}$$

Therefore, if $G_1 = G_2 = 1$, $V_{TT} \geq 2V_{QT}$. So the doubling of the ENG output of the TT is less than the increase in interference for a given imbalance (proportional to value of $(Z_{t1} - Z_{t2})$, defined in [25] and later re-defined in [39] and [51]). This shows that the TT is more sensitive to imbalance than the QT.

However, a modification of the TT may provide externally controlled imbalance correction using adjustable-gain amplifiers for G_1 and G_2. The AT shown in Figure 34.5c was developed to automatically compensate for the cuff imbalance and minimize EMG interference [4, 5, 45]. The input amplifiers have variable gain controlled by the differential feedback signals V_a and $-V_a$. The control stage performs a comparison of the amplitudes from the two channels by first rectifying the signals and then amplifying the difference of their absolute values using amplifier G_3. The output of G_3 is applied to a long-time-constant integrator, which produces the feedback signals V_a and $-V_a$. The variable-gain amplifiers counterbalance the cuff imbalance by equalizing the amplitudes of the composite signal at the outputs of the two channels that are applied to summing amplifier G_{out}, similar to the TT. As a result the equal and antiphase interfering signals from the two channels are eliminated and in-phase ENG signals are added and amplified.

34.5 COMPARISON OF CONFIGURATIONS

34.5.1 In Vitro

In order to test a tripolar amplifier configuration in the laboratory a saline bath can be used to provide realistic conditions [46, 47]. In the configuration illustrated in Figure 34.6, the EMG field is generated by two electrodes in various positions inside the bath and the ENG signal is injected by a transformer in series with the middle electrode. Cuff asymmetry is present, indeed it is difficult to avoid, and an obstacle is moved inside the cuff [48] to introduce further imbalance, similar to that introduced by tissue growth. This allows a comparison of the performance of the amplifier configurations for various positions of the obstacle inside the cuff. Other saline bath experiments can be performed with an insulated wire through the cuff, with a very thin strip of insulation removed [46, 49], representing one node of Ranvier. Such experiments can be performed for evaluating both cuff and amplifier designs.

Figure 34.7 shows the results from an experiment performed in [48] using the saline bath setup of Figure 34.6. Comparative measurements were taken from a discrete-component amplifier system allowing the use of both the TT and the AT. The EMG was represented by a random signal band limited from 50 to 300 Hz, with a maximum amplitude of approximately 1.5 mV peak superimposed to a 2-μV, 1-kHz sinusoidal signal representing the ENG. The waveform at the top (Fig. 34.7a) illustrates a recording

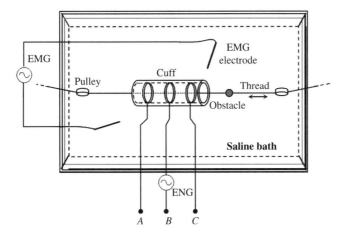

Figure 34.6   Saline bath experimental setup.

of the system output for 20 s. The TT was employed for the first 10 s of the recording and then the AT was switched on, reducing the interference substantially. The imbalance applied by the obstacle was approximately 12% (imbalance definition in [25]) and the outputs of the two configurations are magnified over 40 ms at the bottom of Figure 34.7, with the TT output (Fig. 34.7b), having a signal-to-interference (S/I) ratio

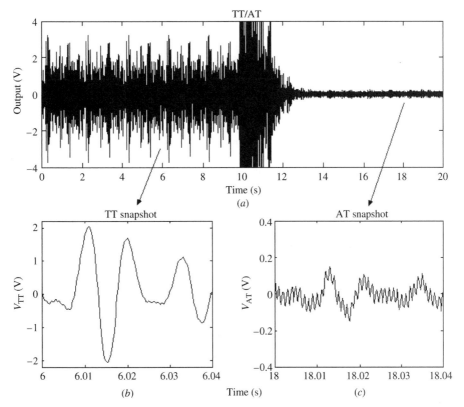

Figure 34.7   (a) System output for 12% imbalance. Obstacle position: 15 mm inside cuff. System operates for 10 s as TT, then is switched to AT. (b,c) The 40-ms snapshots of TT and AT, respectively.

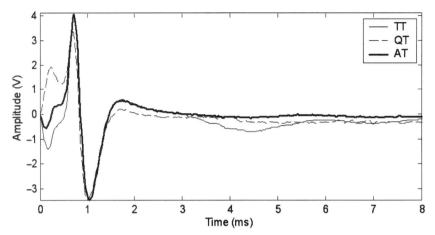

Figure 34.8   The QT, TT, and AT outputs scaled to the same CAP amplitude.

of approximately 1 : 44 and the AT in Figure 34.7c with S/I ratio of 1 : 3, where the higher frequency sinusoidal ENG is visible.

34.5.2   In Vivo

A comparison of all the three tripolar cuff amplifier configurations was performed through in vivo experiments on rabbits described in [45]. The recording system used was configured to enable manual selection between the QT, TT, and AT configurations. The signals were produced by bipolar stimulation applied to the sciatic nerve and the recording cuff was fitted around the tibial nerve. Figure 34.8 illustrates the responses of the three tripolar configurations for one of the sessions of the experiment, scaled to have the same CAP amplitude to allow a visual comparison of the output waveforms. In this case the QT has the largest and the AT the smallest stimulus artifact peak, while the M wave is smallest for the AT and largest for the TT, measured between 3 and 5 ms. The main parts of a typical input waveform to the system during this experiment are pointed out in Figure 34.2b. The performance of the recording systems in this experiment was measured in terms of S/I ratios considering only the M wave as interference in the calculations. The presence of the stimulus artifact degraded somewhat the performance of the AT, which was nonetheless found to be statistically better than both the QT and the TT.

34.6   FUTURE TRENDS

It is not at all easy to get satisfactory neural signals from electrodes in nerve cuffs due to their small amplitude relative to thermal noise and to interference from outside the cuff. However, it has been shown to be possible to obtain satisfactory signals chronically from a tripolar cuff in a patient using the QT configuration with a cable connecting the cuff to a commercial instrumentation amplifier [23, 50]. Compared to the QT, the TT should be slightly better in signal-to-noise ratio (SNR) and, if made adaptive (i.e., AT), should be better at rejecting interference despite the unpredictability of the impedances in the cuff. We expect that these are significant advantages. From a technical point of view, the implementation of an AT requires that low-noise amplifiers and the feedback system shown in Figure 34.5c must be integrated to make them small enough to be included in an implant and yet draw little current from the power supply. Based on these requirements, an integrated AT has been fabricated and tested and was shown in [51] to achieve up to about to 300 higher SNR than the TT and the QT for high imbalance values. At present, the challenge is to incorporate it into the implantable system shown in Figure 34.3, which can monitor ENG from recording cuffs and stimulate at the same or other cuffs. Such implanted devices, with inputs from nerve and outputs to nerves, should be able to perform neuroprosthetic functions that have not been possible in the past.

However, until such devices are available for use with patients, discussing the viability of using neural signals for feedback or command inputs is speculative. We expect to see trials of various neuroprosthetic systems that have ENG inputs and anticipate seeing improvements to the cuff design (e.g., mounting the amplifiers on the cuff [52]), improved electronic design, and new ways to enhance the signal (e.g., [53]), which will make it possible to extract more information without resorting to electrodes that penetrate the nerve.

REFERENCES

1. STEIN, R. B., NICHOLS, T. R., JHAMANDA, J., DAVIS, L., AND CHARLES, D. "Stable long-term recordings from cat peripheral nerves." *Brain Res.*, 28: 21–38, 1977.

2. STEIN, R. B., CHARLES, D., DAVIS, L., JHAMANDA, J., MANNARD, A., AND NICHOLS, T. R. "Principles underlying new methods for chronic neural recording." *Le J. Can. Sci. Neur.*, 2: 235–244, 1975.

3. PFLAUM, C., RISO, R. R., AND WIESSPEINER, G. "Performance of alternative amplifier configurations for tripolar nerve cuff recorded ENG." Paper presented at 18th Int. Conf. IEEE EMBS, Amsterdam, 1996, IEEE.

4. RAHAL, M. S. "Optimisation of nerve cuff electrode recordings for functional electrical stimulation." PhD thesis, Electronic and Electrical Engineering, University College London, 2000.

5. TRIANTIS, I. F., Rieger R., TAYLOR, J., AND DONALDSON, N. "Adaptive interference reduction in nerve cuff electrode recordings." *IEEE International Conference on Electronics Circuits, and Systems (ICECS)*, 2: 669–672, 2001.

6. HAUGLAND, K. AND SINKJAER, T. "Cutaneous whole nerve recordings used for correction of footdrop in hemiplegic man." *IEEE Trans. Rehab. Eng.*, 3: 307–317, 1995.

7. HAUGLAND, M., LICKEL, A., HAASE, J., AND SINKJAER, T. "Control of FES thump force using slip information obtained from the cutaneous electroneurogram in quadriplegic man." *IEEE Trans. Rehab. Eng.*, 7: 215–227, 1999.

8. JEZERNIK, S. AND SINKJAER, T. "Detecting sudden bladder pressure increases from the pelvic nerve afferent activity." *Proc. 20th Int. Conf. IEEE EMBS*, 20: 2532–2535, 1998.

9. HOFFER, J. A., MARKS, W. B., AND RYMER, W. Z. "Nerve fiber activity during normal movements." *Soc. Neurosci. Abstr.*, 4: 300, 1974.

10. STEIN, R. B. AND PEARSON, K. G. "Predicted amplitude and form of action potentials recorded from unmyelinated nerve fibres." *J. Theor. Biol.*, 32: 539–558, 1971.

11. HOFFER, J. A. "Techniques to study spinal-cord, peripheral nerve, and muscle activity in freely moving animals." *Neurometh.*, 15: 65–145, 1990.

12. STRUIJK, J. J. "The extracellular potential of a myelinated nerve fiber in an unbounded medium and in nerve cuff models." *Biophys. J.*, 72: 2457–2469, 1997.

13. STRUIJK, J. J., THOMSEN, M., LARSEN, J. O., AND SINKJAER, T. "Cuff electrodes for long-term recording of natural sensory information." *IEEE EMBS.*, 18: 91–98, 1999.

14. JULIEN, C. AND ROSSIGNOL, S. "Electroneurographic recordings with polymer cuff electrodes in paralyzed cats." *J. Neurosci. Meth.*, 5: 267–272, 1982.

15. NAPLES, G. G., MORTIMER, J. T., SCHEINER, A., AND SWEENEY, J. D. "A spiral nerve cuff electrode for peripheral nerve stimulation." *IEEE Trans. Biomed. Eng.*, 35: 905–916, 1988.

16. KALLESOE, K., HOFFER, J. A., AND STRANGE, K. "Implantable cuff having improved closure." U.S. patent no. 5,487,756, 1996.

17. ANDREASEN, L. AND STRUIJK, J. "On the importance of configuration and closure of nerve cuff electrodes for recording." *Proc. 20th Int. Conf. IEEE EMBS*, 20: 3004–3007, 1998.

18. LARSEN, J. O., THOMSEN, M., HAUGLAND, M., AND SINKJAER, T. "Degeneration and regeneration in rabbit peripheral nerve with long-term nerve cuff electrode implant: A stereological study of myelinated and

unmyelinated axons." *Acta Neuropathol.*, 96: 365–378, 1998.

19. HOFFER, J. "Techniques to study spinal-cord, peripheral nerve, and muscle activity in freely moving animals." In *Neuromethods, Neurophysiological Techniques: Applications to Neural Systems*, A. A. BOULTON, G. B. BAKER, AND C. H. VANDERWOLF, Eds, Humana, Clifton, NJ, 1990, pp. 65–145.

20. STRUIJK, J. J. "On the Spectrum of Nerve Cuff Recordings." *Proc. 19th Int. Conf. IEEE/EMBS*, Chicago, Oct. 30–Nov. 2, 6:3001–3003, 1997.

21. UPSHAW, B. M., "Real-time digital processing algorithms and systems for the application of human sensory nerve signals in neuroprostheses." PhD thesis, Center for Sensory-Motor Interaction, Aalborg University, 1999.

22. NICOLIC, Z. M., POPOVIC, B. D., STEIN, B. R., AND KENWELL, Z. "Instrumentation for ENG and EMG Recordings in FES Systems." *IEEE Trans. Biomed. Eng.*, 41: 703–706, 1994.

23. DONALDSON, N. D. N., ZHOU, L., PERKINS, T. A., MUNIH, M., HAUGLAND, M., AND SINKJAER, T. "Implantable telemeter for long term electroneurographic recordings in animals and man." *Med. Biol. Eng. Comput.*, 41: 654–664, 2003.

24. ANDREASEN, L., "The nerve cuff electrode and cutaneous sensory feedback." PhD thesis, Center for Sensory-Motor Interaction, Aalborg University, 2002.

25. TRIANTIS, I. F., DEMOSTHENOUS, A., DONALDSON, N., AND STRUIJK, J. J. "Experimental assesment of imbalance conditions in a tripolar cuff for ENG recordings." Paper presented at 1st Int IEEE EMBS Conf. Neural Engineering, Capri, Italy, 2003, pp. 380–383.

26. HOFFER, J. A., LOEB, G. E., AND PRATT, C. A. "Single unit conduction velocities from averaged nerve cuff electrode records in freely moving cats." *J. Neurosci. Meth.*, 4: 211–225, 1981.

27. STEIN, R. B., GORDON, T., OGUZTORELI, AND LEE, R. G. "Classifying sensory patterns and their effects on locomotion and tremor." *Can. J. Physiol. Pharmacol.*, 59: 645–655, 1981.

28. SAHIN, M., HAXHIU, M. A., DURAND, M. D., AND DRESHAJ, I. A. "Spiral nerve cuff electrode for recording of respiratory output." *J. Appl. Physiol.*, 83: 317, 1997.

29. DAVIS, L. A., GORDON, T., HOFFER, J. A., JAHMANDAS, J., AND STEIN, R. B. "Compound action potentials recorded from mammalian peripheral nerves following ligation or resuturing." *J. Physiol.*, 285: 543–559, 1978.

30. GORDON, T., HOFFER, J. A., JHAMANDAS, J., AND STEIN, R. B. "Long-term effects of axotomy on neural activity during cat locomotion." *J. Physiol.*, 303: 243–263, 1980.

31. HAUGLAND, M. K., HOFFER, J. A., AND SINKJAER, T. "Skin contact force information in sensory nerve signals recorded by implanted cuff electrodes." *IEEE Trans. Rehab. Eng.*, 2: 18–28, 1994.

32. HAUGLAND, M. K., HOFFER, J. A., AND SINKJAER, T. "Slip information provided by nerve cuff signals: Application in closed-loop control of functional electrical stimulation." *IEEE Trans. Rehab. Eng.*, 2: 29–36, 1994.

33. HOFFER, J. A. AND SINKJAER, T. "A natural force sensor suitable for closed-loop control of functional neuromuscular stimulation." In *Proc. 2nd Vienna Int. Workshop on Functional Electrostimulation*, University of Vienna Bioengineering Laboratory, Ed., Vienna, Austria, 1986 pp. 47–50.

34. POPOVIC, D. B., STEIN, R. B., JONANOVIC, K. L., RONGCHING, D., KOSTOV, A., AND ARMSTRONG, W. W. "Sensory nerve recording for closed-loop control to restore motor functions." *IEEE Trans. Biomed. Eng.*, 40: 1024–1031, 1993.

35. UPSHAW, B. AND SINKJAER, T. "Digital signal processing algorithms for the detection of afferent nerve activity recorded from cuff electrodes." *IEEE Trans. Rehab. Eng.*, 6:172–181, 1998.

36. HAUGLAND, M. K., LICKEL, A., HAASE, J., AND SINKJAER, T. "Hand neuroprosthesis controlled by natural sensors." *IEEE Trans. Rehab. Eng.*

37. SINKJAER, T., HAUGLAND, M. K., AND HAASE, J. "Natural neural sensing and artificial muscle control in man." *Exp. Brain Res.*, 98: 542–545, 1994.

38. SINKJAER, T., HAUGLAND, M. K., AND HAASE, J. "Neural cuff electrode recordings as a replacement of lost sensory feedback in paraplegic patients." *Neurobionics*, 1: 267–277, 1993.

39. TRIANTIS, I. F. "An adaptive amplifier for cuff imbalance correction and interference reduction in nerve signal recording", Ph.D. dissertation, University College London, London, 2005.

40. STEIN, R. B., CHARLES, D., GORDON, T., HOFFER, J. A., AND JHAMANDAS, J. "Impedance properties of metal electrodes for chronic recording from mammalian nerves." *IEEE Trans. Biomed. Eng.*, 25: 532–536, 1978.

41. STRUIJK, J. J. AND THOMSEN, M. "Tripolar nerve cuff recording: Stimulus artifact, EMG, and the recorded nerve signal." Paper presented at IEEE—EMBS 17th Ann. Int. Conf., Montreal, 1995.

42. RIEGER, R., TAYLOR, J., DEMOSTHENOUS, A., DONALDSON, N., AND LANGLOIS, P. "Design of a low noise preamplifier for nerve cuff electrode recording." *IEEE J. Solid State Circ.*, 38: 1373–1379, 2003.

43. HOROWITZ, P. AND HILL, W. *The Art of Electronics*. 2nd ed. Cambridge University Press, 1989.

44. RAHAL, M. S., WINTER, J., TAYLOR, J., AND DONALDSON, N. "An improved configuration for the reduction of EMG in electrode cuff recordings: A theoretical approach." *IEEE Trans. Biomed. Eng.*, 47:1281–1284, 2000.

45. TRIANTIS, I. F., DEMOSTHENOUS, A., AND DONALDSON, N. "On cuff imbalance and tripolar ENG amplifier configurations." *IEEE Trans. on Biomed. Eng.* 52: 314–320, 2005.

46. ANDREASEN, L., STRUIJK, J., AND HAUGLAND, M. "An artificial nerve fiber for evaluation of nerve cuff electrodes." Paper presented at 19th Int. Conf. IEEE/EMBS, Chicago, IL, 1997.

47. TRIANTIS, I. F., DONALDSON, N., AND DEMOSTHENOUS, A. "Saline-bath testing of a system for removing

artifact from ENG signals." *FESnet '02*, University of Strathclyde, Glasgow, UK, 2002. *1st An. Conf. Network for Rehabilitation using Functional Electrical Stimulation for Therapy and Function Restoration.* http://fesnet.eng.gla.ac.uk/conference/prog.html.

48. TRIANTIS, I. F., DEMOSTHENOUS, A., AND DONALDSON, N. "An ENG amplifier for EMG cancellation and cuff imbalance removal." *EPSRC/IEEE Postgraduate Research Conf. in Electronics Photonics Communications and Software (PREP'03)*, pp. 105–106, ed. EPSR/IEEE, Exeter, UK, 2003.

49. ANDREASEN, L. N. S., STRUIJK, J. J., AND LAWRENCE, S. "Measurement of the nerve cuff electrodes for recording." *Med. & Biol. Eng. & Comput.*, 38:447–453, 2000.

50. Hansen M., HAUGLAND, M., SINKJAER, T., AND DONALDSON, N. "Real time foot drop correction using machine learning and natural sensors." *Neuromod.*, 5: 41–53, 2002.

51. DEMOSTHENOUS, A. AND TRIANTIS, I. F. "An adaptive ENG amplifier for tripolar cuff electrodes." *IEEE J. of. Solid-State Circuits*, 40: 412–421, 2005.

52. www.md.ucl.ac.be/sens/.

53. SAHIN, M. AND DURAND, M. D. "Improved nerve cuff electrode recordings with subthreshold anodic currents." *IEEE Trans. BME*, 45: 1044–1050, 1998.

CABLE EQUATION MODEL FOR MYELINATED NERVE FIBER

P. D. Einziger, L. M. Livshitz, and J. Mizrahi

35.1   INTRODUCTION

The cable equation is a simple and elegant one-dimensional (1-D) tool capable of handling analytically linear and nonlinear nonmyelinated axon models [1–4]. It relies on two fundamental physical concepts, namely Kirchhoff's current and voltage laws, leading to a transmission line model. This model has been shown, for nonmyelinated axons surrounded by fluid of infinite conductivity, to be the dominant mode of the exact solution of a 3-D potential problem [5, 6]. Recently, it has been shown that this solution is still dominant even in the case where the outer conductivity is finite [7].

Unfortunately, the cable equation has not yet been extended analytically for myelinated axon model, which is crucially important for applications involving vertebrates. In attempting to extend the nonmyelinated axon model, we start here again with Kirchhoff's laws while imposing periodicity on the membrane conductivity. The resultant modified transmission line problem leads to a linear second-order ordinary differential equation with periodic coefficients, known as Hill's equation [8–10]. Interesting applications of the theory of Hill's equation can be found in many branches of mathematical physics. Hence, the features and characteristics of our novel excitation model are well studied and tabulated.

The general internal source response, expressed via repeated convolutions, uniformly converges provided that the periodic membrane is passive. The solution can be interpreted as an extended source response in an equivalent nonmyelinated axon (i.e., the response is governed by the classical cable equation). The extended source consists of the original source and a novel activation function, replacing the periodic membrane in the myelinated axon model. Furthermore, the conductivity of the equivalent axon is the precise average of the periodic myelinated axon conductivity. Hill's formulation is further reduced into Mathieu's equation [8–10] for the specific choice of sinusoidal conductivity, thereby resulting in an explicit closed-form expression for the transmembrane potential (TMP). Floquet's modes, recognized as the nerve fiber activation modes, can also be incorporated in our excitation model, provided that the periodic membrane pointwise passivity constraint is properly modified. Indeed, the modified condition, enforcing the periodic membrane passivity constraint on the average conductivity only, leads to the inclusion of the nerve fiber activation modes in our linear model [11–13].

The validity of the generalized cable equation model, presented herein for myelinated nerve fibers, is verified through rigorous Green's function formulation for TMP induced in 3-D myelinated cylindrical cells. As previously [5–7], it is shown that

Handbook of Neural Engineering. Edited by Metin Akay

the dominant pole contribution of the exact modal expansion is the TMP solution of our myelinated model.

35.2 CABLE EQUATION MODEL

35.2.1 Myelinated Nerve Fiber

The myelinated axon model, depicted in Figure 35.1, consists of two cylindrical regions, the axoplasmic core and the surrounding fluid, separated by a thin myelinated membrane of radius a. Assuming that the core and the outer fluid are homogeneous, isotropic, ohmic conductors, their electrical parameters are denoted by σ_i and $\sigma_e \to \infty$, respectively. The myelinated membrane is characterized by periodic conductivity $G_m(z)$.

Let the distance between two adjacent nodes of Ranvier be denoted by L_p; then

$$G_m(z) = G_m(z + L_p) \tag{35.1}$$

The membrane conductivity $G_m(z)$ is most effectively represented as

$$G_m(z) = G[1 + \alpha \mathcal{G}(z)] > 0 \tag{35.2}$$

where G denotes the average conductance

$$G = \frac{1}{L_p} \int_0^{L_p} G_m(z)\, dz \tag{35.3}$$

and α is a dimensionless parameter. Obviously $\mathcal{G}(z)$ is a dimensionless and periodic function

$$\mathcal{G}(z) = \mathcal{G}(z + L_p) \tag{35.4}$$

with zero mean

$$\int_0^{L_p} \mathcal{G}(z)\, dz = 0 \tag{35.5}$$

Without loss of generality, assuming that $\mathcal{G}(z)$ is also symmetric,

$$\mathcal{G}(z) = \mathcal{G}(-z) \tag{35.6}$$

then it can be represented by a Fourier series expansion

$$\mathcal{G}(z) = \sum_{n=1}^{\infty} A_n \cos(nk_p z) \qquad k_p = \frac{2\pi}{L_p} \tag{35.7}$$

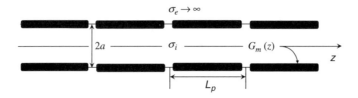

Figure 35.1 Myelinated axon model.

or alternatively via Fourier integral transform

$$\mathcal{G}(z) = \frac{1}{2\pi} \int_{-\infty}^{\infty} \gamma(k) \exp(jkz) \, dk \tag{35.8}$$

where

$$\gamma(k) = \pi \sum_{-\infty}^{\infty} A_n \delta(k - nk_p) \qquad A_n = A_{-n} \qquad A_0 = 0 \tag{35.9}$$

Similarly, the membrane conductivity $G_m(z)$ in (35.2) has to be symmetric, that is,

$$G_m(z) = G_m(-z) \tag{35.10}$$

and thus can also be represented by the Fourier series expansion

$$G_m(z) = G\left[1 + \alpha \sum_{n=1}^{\infty} A_n \cos(nk_p z)\right] \tag{35.11}$$

or alternatively via the Fourier integral transform

$$G_m(z) = \frac{1}{2\pi} \int_{-\infty}^{\infty} \Gamma(k) \exp(jkz) \, dk \tag{35.12}$$

where

$$\Gamma(k) = G[2\pi\delta(k) + \alpha\gamma(k)] \tag{35.13}$$

As $\alpha \to 0$ our model reduces into the conventional nonmyelinated model. However, in the reverse limit as $\alpha \to \infty$, the passivity condition in (35.2) may be violated, that is, $G_m(z) < 0$ for some z, and then it corresponds to an activated fiber (negative conductance corresponds to inward Na$^+$ current).

35.2.2 Hill's Equation

Since $\sigma_e \to \infty$, the outer potential is identically zero and the transmembrane voltage coincides with the internal voltage. The steady-state internal voltage and current differential relations along the axon are given via Kirchhoff's voltage and current laws [1, 3, 4],

$$\frac{dV(z)}{dz} = -R_s I(z) \tag{35.14}$$

and

$$\frac{dI(z)}{dz} = -G_p(z)V(z) + i_s(z) \tag{35.15}$$

respectively, where $i_s(z)$ represents an internal current source distribution. A resistive transmission line model for the myelinated axon in (35.14) and (35.15) is depicted in Figure 35.2.

The distributed elements R_s and $G_p(z)$ denoting the series (axoplasmic core) resistance and the parallel (myelinated membrane) conductance distributions are given by

$$R_s = \frac{1}{\pi a^2 \sigma_i} \tag{35.16}$$

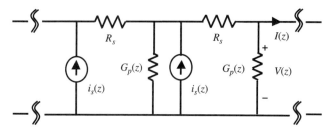

Figure 35.2   Resistive transmission line model for myelinated axon.

and

$$G_p(z) = 2\pi a G_m(z) = 2\pi a G[1 + \alpha \mathcal{G}(z)] \tag{35.17}$$

respectively. The first-order system of differential equations (35.14) and (35.15) can be represented alternatively by an equivalent single second-order differential equation, also known as the cable equation,

$$\frac{d^2 V(z)}{dz^2} - R_s G_p(z) V(z) = -v_0(z) \tag{35.18}$$

The voltage source distribution $v_0(z)$ and the periodic function $R_s G_p(z)$ are given by

$$v_0(z) = R_s i_s(z) \tag{35.19}$$

and

$$R_s G_p(z) = \frac{1 + \alpha \mathcal{G}(z)}{L_c^2} \tag{35.20}$$

respectively, where L_c denotes the characteristic length constant

$$L_c = \sqrt{\frac{\sigma_i a}{2G}} \tag{35.21}$$

Equation (35.18) is a nonhomogeneous Hill equation which is well-known in mathematical physics [9, 10].

In attempting to solve Hill's equation for all α's in (35.20), we assume solutions of the form

$$V(z) = \sum_{i=0}^{\infty} \alpha^i V_i(z) \tag{35.22}$$

$$I(z) = \sum_{i=0}^{\infty} \alpha^i I_i(z) \tag{35.23}$$

Substituting (35.22) into (35.18) results in recursive differential equations for $V_i(z)$,

$$\frac{d^2 V_i(z)}{dz^2} - \frac{V_i(z)}{L_c^2} = -v_i(z) \tag{35.24}$$

where

$$v_i(z) = -\frac{\mathcal{G}(z) V_{i-1}(z)}{L_c^2} \qquad i \geq 1 \tag{35.25}$$

The current $I_i(z)$ is readily obtained via (35.14),

$$I_i(z) = -\frac{1}{R_s}\frac{dV_i(z)}{dz} \tag{35.26}$$

Utilizing (35.22) and (35.24), Hill's equation (35.18) can be replaced with a simpler but equivalent equation, namely the conventional cable equation

$$\frac{d^2 V(z)}{dz^2} - \frac{V(z)}{L_c^2} = -v(z) \tag{35.27}$$

but with modified source

$$v(z) = \sum_{i=0}^{\infty} \alpha^i v_i(z) = v_0(z) - \frac{\alpha \mathcal{G}(z) V(z)}{L_c^2} \tag{35.28}$$

This source is composed of the original source function $v_0(z)$ and an additional distributed source corresponding to the summation over $i \geq 1$ and accounting for the replacement of the periodic membrane conductance $G_m(z)$ with its average conductance G [Eq. (35.3)]. Equation (35.27) could be obtained from a first-order system of differential equations, equivalent to Eqs. (35.14) and (35.15), namely (Fig. 35.3)

$$\frac{dV(z)}{dz} = -R_s I(z) \tag{35.29}$$

$$\frac{dI(z)}{dz} = -G_e V(z) + i_s(z) + i_e(z) \tag{35.30}$$

where

$$G_e = 2\pi a G \tag{35.31}$$

and

$$i_e(z) = -\alpha G_e \mathcal{G}(z) V(z) \tag{35.32}$$

Note that the voltage sources v_0 and $-\alpha \mathcal{G}(z) V(z)/L_c^2$ on the right-hand side (RHS) of (35.28) could be obtained directly via the RHS of (35.30) as $R_s i_s(z)$ and $R_s i_e(z)$, respectively.

The solution of (35.24) is facilitated with Green's function,

$$g(z) = \frac{L_c \exp\left(-|z|/L_c\right)}{2} \tag{35.33}$$

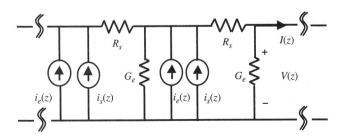

Figure 35.3 Equivalent conventional transmission line model, but with modified source $i_e(z)$.

satisfying

$$\frac{d^2 g(z)}{dz^2} - \frac{g(z)}{L_c^2} = -\delta(z) \tag{35.34}$$

leading to

$$V_i(z) = v_i(z) \otimes g(z) \tag{35.35}$$

where $\otimes$ denotes the convolution operation. Substituting (35.19) and (35.25) into (35.35) leads to a recursive closed-form expression for $V_i(z)$, containing $i+1$ convolutions,

$$V_i(z) = \frac{L_c R_s}{2} \left(\frac{-1}{2L_c}\right)^i \left\{ \left[\left(\left[i_s(z) \otimes \exp\left(-\frac{|z|}{L_c}\right) \right] \mathcal{G}(z) \right) \right. \right.$$
$$\left. \left. \otimes \exp\left(-\frac{|z|}{L_c}\right) \right] \mathcal{G}(z) \cdots \mathcal{G}(z) \right\} \otimes \exp\left(-\frac{|z|}{L_c}\right) \tag{35.36}$$

The recursive structure of (35.36) aids in obtaining an explicit bound on $|V_i(z)|$,

$$|V_i(z)| \leq L_c^2 R_s I M^i \tag{35.37}$$

where I and M are bounds on $i_s(z)$ and $\mathcal{G}(z)$, respectively,

$$|i_s(z)| \leq I \qquad \text{and} \qquad |\mathcal{G}(z)| \leq M \tag{35.38}$$

It should be noted that (35.37) guarantees a uniform convergence of (35.22) and (35.23) if

$$|\alpha M| < 1 \tag{35.39}$$

Note that condition (35.39) is identical with the passivity constraint $G_m(z) > 0$ [i.e., $|\alpha \mathcal{G}(z)| < 1$] in (35.2).

35.2.3   Point Source Response

The impulse response of Hill's equation (35.18) is now obtained by setting

$$i_s(z) = I_s \delta(z) \tag{35.40}$$

Substituting (35.40) into (35.36) and utilizing (35.26) result in

$$V_0(z) = \frac{L_c I_s R_s}{2} \exp\left(-\frac{|z|}{L_c}\right) \tag{35.41}$$

and

$$I_0(z) = \frac{I_s}{2} \exp\left(-\frac{|z|}{L_c}\right) \tag{35.42}$$

respectively. Expressions (35.41) and (35.42) are the impulse responses of conventional nonmyelinated fiber. The number of convolutions contained in (35.36) is reduced here to i, that is,

$$
V_i(z) = \frac{L_c R_s I_s}{2} \left(\frac{-1}{2L_c}\right)^i \left\{ \left[\left(\exp\left(-\frac{|z|}{L_c}\right) \mathcal{G}(z) \right) \right. \right.
$$

$$
\left. \left. \otimes \exp\left(-\frac{|z|}{L_c}\right) \right] \mathcal{G}(z) \cdots \mathcal{G}(z) \right\} \otimes \exp\left(-\frac{|z|}{L_c}\right) \quad (35.43)
$$

35.2.4 Mathieu's Equation

The periodic function $\mathcal{G}(z)$ can generally be expanded into a Fourier series expansion, as outline in (35.7). Thus, a single sinusoidal function is the most elementary choice for $\mathcal{G}(z)$. Selecting $A_1 = 1$ and $A_n = 0$, $n \geq 2$, in (35.7), results in

$$
\mathcal{G}(z) = \cos(k_p z) \quad (35.44)
$$

Hill's equation in (35.18) is reduced to Mathieu's equation [9, 10],

$$
\frac{d^2 V(z)}{dz^2} - \frac{1 + \alpha \cos(k_p z)}{L_c^2} V(z) = -v_0(z) \quad (35.45)
$$

Explicit evaluation of $V_1(z)$ can be carried out by substituting (35.44) into (35.43), leading to

$$
V_1(z) = -V_0(z) \frac{1 + \cos(2\pi z/L_p) + [L_p/(\pi L_c)] \sin(2\pi |z|/L_p)}{4[1 + (\pi L_c/L_p)^2]} \quad (35.46)
$$

The function $V_1(z)$ is a myelinated axon correction of order α for the conventional nonmyelinated point source response $V_0(z)$ in (35.41). Additional terms $V_i(z)$ $(i > 1)$ can usually be neglected for realistic axon parameters. Thus, $v_1(z)$ in (35.25),

$$
v_1(z) = -\frac{\mathcal{G}(z) V_0(z)}{L_c^2} = -\frac{R_s I_s}{2L_c} \cos\left(\frac{2\pi z}{L_p}\right) \exp\left(-\frac{|z|}{L_c}\right) \quad (35.47)
$$

may be regarded as an explicit activation function [3], accounting for myelin contribution. Normalized transmembrane voltages, namely, $V_0(z)/V_s$ and $V_1(z)/V_s$ $(V_s = L_c R_s I_s/2)$ corresponding to the conventional cable equation distribution and to the myelinated axon correction, respectively, are depicted in Figure 35.4. Since the correction term $V_1(z)/V_s$ is negligibly small for $L_p/L_c \to 0$, the myelinated axon performs as a conventional nonmyelinated axon with average membrane conductivity G_e. However, for $L_p/L_c \to \infty$ the correction term significantly increases, since the average membrane conductivity is now approaching either $G_e(1 - |\alpha|)$ or $G_e(1 + |\alpha|)$ depending on whether the point source location is under a myelinated sector or a Ranvier node, respectively. In this case the inclusion of higher order terms $(i > 1)$ in (35.22) may be required, depending on α values.

35.2.5 Floquet's Modes

As indicated at the end of Section 35.2.2, the uniform convergence of the series expansions in (35.22) and (35.23) is guaranteed by (35.39) or equivalently by the passivity constraint in (35.2). This constraint will be shown subsequently, in Section 35.3.2, to be also a sufficient condition for the existence of a unique 3-D Green's function.

Herein, we explore solutions of a source-free version of the cable equation in (35.18), under less restricted constraint, by allowing the passivity condition to be violated,

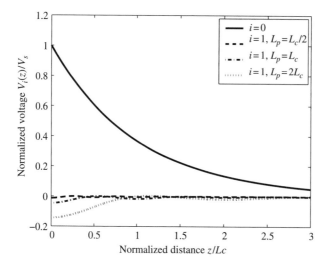

Figure 35.4 Normalized transmembrane voltages: conventional cable equation distribution $V_0(z)/V_s$ and myelinated axon correction $V_1(z)/V_s$ ($V_s = L_cR_sI_s/2$), for three different L_p/L_c values ($L_p/L_c = 1/2, 1, 2$).

that is, $G_m(z) < 0$, for some z, corresponding to an activated fiber (negative conductance corresponds to inward Na$^+$ current). Solutions of the homogeneous and symmetric Hill's equation

$$\frac{d^2V(z)}{dz^2} - \frac{1 + \alpha\mathcal{G}(z)}{L_c^2}V(z) = 0 \qquad (35.48)$$

obtained by setting $v_0(z) = 0$ in (35.18), are governed by the Floquet theorem [9, 10] Floquet's theorem states: (i) There exists a solution of the form

$$V_\beta(z) = \exp(j\beta z)P_\beta(z) \qquad (35.49)$$

where β, known as the characteristic exponent or the propagation constant, is not zero or integral multiple of $\pi/L_p = k_p/2$ and depends on L_p/L_c and α. The function $P_\beta(z)$ is a periodic function with the same period as $\mathcal{G}(z)$ in (35.48), namely L_p. Similarly

$$V_\beta(-z) = \exp(-j\beta z)P_\beta(-z) \qquad (35.50)$$

satisfies (35.48) whenever (35.49) does. (ii) In the special case where β is zero or an integral muliple of $\pi/L_p = k_p/2$, $P_\beta(z)$ is proportional to $P_\beta(-z)$ and has L_p or $2L_p$ periodicity depending on whether the integral multiplicity of $k_p/2$ is even or odd, respectively.

Equations (35.49) and (35.50), termed the Floquet solutions or alternatively Floquet modes, are capable of activating the undamped and quasi-periodic TMP along a myelinated axon provided that

$$\Im[\beta] = 0 \qquad (35.51)$$

Note that the myelinated axon supports transient TMP propagation in addition to the steady-state direct-current response considered herein if condition (35.51) holds for a broader frequency range. The modes are damped away from their sources for

$$\Im[\beta] \neq 0 \qquad (35.52)$$

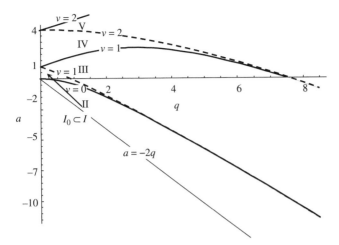

Figure 35.5   Regions in the a–q plane for Floquet modes associated with Mathieu's equation. Activation regions II and IV are associated with undamped and quasi periodic TMP (Eq. 35.51). Regions I, III, and V are associated with Floquet modes that damped away from their sources (Eq. 35.52). Subregion I_0 and the lower half plane $a < 0$ are associated with point-wise passive and average passive membranes Eq. 35.2 and Eqs. 35.21, 35.53, respectively.

The above discussion is well illustrated and demonstrated via the well-tabulated Mathieu functions [11–13]. Indeed, setting

$$a = -\left(\frac{L_p}{\pi L_c}\right)^2 \qquad q = -\frac{a\alpha}{2} \qquad v = \frac{2\beta}{k_p} \qquad v = \frac{\pi z}{L_p} \qquad (35.53)$$

one obtains an activation region map [11–13] as depicted in Figure 35.5 for $v = 0, 1, 2$. Note that the straight line $a = -2q$ determines the subregion I_0 in which the passivity constraint in (35.2) is satisfied. This line establishes an asymptote for all the characteristic lines enclosing the activation regions in the lower $a < 0$ plane. In this plane, the average membrane conductivity is always passive [i.e., $G > 0$ in (35.21) and (35.53)].

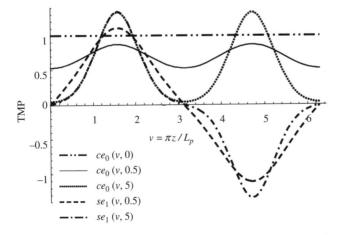

Figure 35.6   Even and odd periodic Mathieu functions of order 0 and 1, respectively.

Finally, the periodic functions $P_\beta(z)$ associated with the second part (ii) of the Floquet theorem, namely

$$P_0(z) = ce_0(v, q) \qquad P_{k_p/2}(z) = se_1(v, q) \tag{35.54}$$

corresponding to $v = 0$ and the lower $v = 1$ boundaries (Fig. 35.5), respectively, are depicted in Figure 35.6. As expected, the even and odd periodic Mathieu functions $ce_0(v, q)$ and $se_1(v, q)$ have periods π and 2π, respectively [11–13]. It should be noted that the recently proposed piecewise constant conductivity profile [14], opposing to the smooth sin profile in (35.44), leads to an explicit integration of Hill's equation in terms of trigonometric functions.

35.3 TMP INDUCED ALONG 3-D MYELINATED NERVE FIBER

35.3.1 Formulation

The 3-D myelinated nerve fiber configuration, depicted in Figure 35.7, consists of an internal source point S, an observation point P (located on the membrane surface), and two cylindrical regions, the axoplasmic core and the surrounding fluid, separated by a thin myelinated membrane of radius a. Assuming that the core and the outer fluid are homogeneous, isotropic, ohmic conductors, their electrical parameters are denoted by σ_i and $\sigma_e \to \infty$, respectively. The myelinated membrane conductivity $G_m(z)$ is defined and characterized in Section 35.2.1.

35.3.2 Integral Representation for TMP

The evaluation of the electrode current distributions and potentials is carried out within the quasi-static (low-frequency) regimen. The TMP, that is, the difference between the internal and external potentials on the membrane surface, is given for a point source excitation as

$$V(\phi, z, \mathbf{r}') = \frac{I_s}{\sigma_i} G(\mathbf{r}_a, \mathbf{r}') \tag{35.55}$$

where $G(\mathbf{r}_a, \mathbf{r}')$ denotes the point source response (Green's function). The coordinates $\mathbf{r}' = (\rho', 0, 0)$ and $\mathbf{r}_a = (a, \phi, z)$ correspond to locations of the source point S and the observation point P, respectively. The subscript a represents the quantity evaluated at

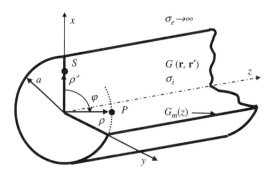

Figure 35.7 Physical configuration.

TABLE 35.1   Poisson Equations and Boundary Conditions for $G(\mathbf{r}, \mathbf{r}')$ and $g_n(\rho, \rho', k)$

	$G(\mathbf{r}, \mathbf{r}')$	$g_n(\rho, \rho')$
Differential equation	$\nabla^2 G(\mathbf{r}, \mathbf{r}') = -\delta(\mathbf{r} - \mathbf{r}')$	$\left(\dfrac{1}{\rho}\dfrac{d}{d\rho}\rho\dfrac{d}{d\rho} - k^2 - \dfrac{n^2}{\rho^2}\right)g_n(\rho, \rho', k)$ $= -\dfrac{1}{\rho}\delta(\rho - \rho')$
Source conditions	$\int_{V \to 0}\nabla^2 G(\mathbf{r}, \mathbf{r}')\,dV = \oint_{A \to 0}\nabla G(\mathbf{r}, \mathbf{r}')\,d\mathbf{A} = -1$	$\dfrac{dg_n(\rho'^+, \rho', k)}{d\rho} - \dfrac{dg_n(\rho'^-, \rho', k)}{d\rho} = -\dfrac{1}{\rho'}$
Decay at infinity	$zG(\mathbf{r}, \mathbf{r}')\Big|_{\rho<a, z\to\infty} < \infty, \quad G(\mathbf{r}, \mathbf{r}')\Big|_{\rho>a} = 0$	$g_n(\rho, \rho', k)\Big|_{\rho>a} = 0$
Membrane condition	$G_m(z)G(\mathbf{r}_a, \mathbf{r}') = -\sigma_i\partial G\frac{(\mathbf{r}_a, \mathbf{r}')}{\partial\rho}, \quad G_m(z) > 0$	$\dfrac{1}{2\pi}\Gamma(k) \otimes g_n(a, \rho', k) = -\sigma_i\dfrac{dg_n(a, \rho', k)}{d\rho}$

the inner surface of the membrane, and I_s denotes the point source current in (35.40). The Green's function in (35.55) can be expressed in terms of cylindrical harmonics [15], leading to

$$G(\mathbf{r}, \mathbf{r}') = \frac{1}{2\pi^2}\sum_{n=0}^{\infty}\epsilon_n\cos(n\phi)\int_0^{\infty} g_n(\rho, \rho', k)\cos(kz)\,dk \qquad (35.56)$$

where $\epsilon_0 = 1$ and $\epsilon_n = 2$, $n > 0$. Both $G(\mathbf{r}, \mathbf{r}')$ and $g_n(\rho, \rho', k)$ in (35.56) are symmetric with respect to z and k, respectively, due to (35.10), and satisfy 3-D and 1-D Poisson equations and appropriate constraints as summarized in Table 35.1.

The membrane condition, in the last row of Table 35.1, establishes a generalization of the conventional membrane condition (e.g., [5–7]); that is, multiplication with periodic function $G_m(z)$ in (35.2) results in convolution with its Fourier transform $\Gamma(k)$ in (35.13). Furthermore, it guarantees the uniqueness of $G(\mathbf{r}, \mathbf{r}')$ if $G_m(z) > 0$, consistently with the passivity requirement in (35.2). Substitution of $G(\mathbf{r}, \mathbf{r}')$ in (35.56) into (35.55) results in an integral representation for the TMP,

$$V(\phi, z, \mathbf{r}') = \frac{I_s}{2\pi\sigma_i}\sum_{n=0}^{\infty}\epsilon_n V_n(z, \rho')\cos(n\phi) \qquad (35.57)$$

where $V_n(z, \rho')$, the TMP (Fourier's) coefficient, is expressed as

$$V_n(z, \rho') = \frac{1}{\pi}\int_0^{\infty} v_n(\rho', k)\cos(kz)\,dk \qquad (35.58)$$

and $v_n(\rho', k)$, the TMP spectral coefficient, is given via

$$v_n(\rho', k) = g_n(a, \rho', k) \qquad (35.59)$$

The characteristic Green's function $g_n(\rho, \rho', k)$ is given via [15]

$$g_n(\rho, \rho', k) = I_n(k\rho_<)[A_n(k)I_n(k\rho_>) + K_n(k\rho_>)] \qquad \rho < a \qquad (35.60)$$

where $\rho_> \equiv \max(\rho, \rho')$ and $\rho_< \equiv \min(\rho, \rho')$. The coefficient $A_n(k)$ can be found by applying the membrane condition, in the last row of Table 35.1, leading to

$$\frac{1}{2\pi}\Gamma(k) \otimes [A_n(k)I_n(ka) + K_n(ka)] = -\sigma_i k[A_n(k)I'_n(ka) + K'_n(ka)] \tag{35.61}$$

Here I_n and K_n are the modified Bessel functions of order n and the prime means differentiation with respect to the argument. The coefficient $A_n(k)$, solution of (35.61), can be obtained by successively shifting k in (35.61) through an infinite number of shifts $k \to k - nk_p$, $n = 0, \pm 1, \pm 2, \ldots$ (e.g., [9, 10]). Alternatively, it can be solved iteratively as outlined in (35.22). Indeed, assume representations of the form

$$G(\mathbf{r}, \mathbf{r}') = \sum_{i=0}^{\infty} \alpha^i G_i(\mathbf{r}, \mathbf{r}') \tag{35.62}$$

$$g_n(\rho, \rho', k) = \sum_{i=0}^{\infty} \alpha^i g_{n,i}(\rho, \rho', k) \tag{35.63}$$

and

$$A_n(k) = \sum_{i=0}^{\infty} \alpha^i A_{n,i}(k) \tag{35.64}$$

where

$$G_i(\mathbf{r}, \mathbf{r}') = \frac{1}{2\pi^2} \sum_{n=0}^{\infty} \epsilon_n \cos(n\phi) \int_0^{\infty} g_{n,i}(\rho, \rho', k) \cos(kz)\, dk \tag{35.65}$$

and

$$g_{n,0}(\rho, \rho', k) = I_n(k\rho_<)[A_{n,0}(k)I_n(k\rho_>) + K_n(k\rho_>)]$$
$$g_{n,i}(\rho, \rho', k) = A_{n,i}(k)I_n(k\rho')I_n(k\rho) \qquad i > 0 \tag{35.66}$$

Substituting $G(\mathbf{r}, \mathbf{r}')$ [Eq. (35.62)] and $g_n(\rho, \rho', k)$ [Eq. (35.63)] in the first row of Table 35.1 results in

$$\nabla^2 G_0(\mathbf{r}, \mathbf{r}') = -\delta(\mathbf{r} - \mathbf{r}') \qquad \nabla^2 G_i(\mathbf{r}, \mathbf{r}') = 0 \qquad i > 0 \tag{35.67}$$

and

$$\left(\frac{1}{\rho}\frac{d}{d\rho}\rho\frac{d}{d\rho} - k^2 - \frac{n^2}{\rho^2}\right)g_{n,0}(\rho, \rho', k) = -\frac{1}{\rho}\delta(\rho - \rho')$$
$$\left(\frac{1}{\rho}\frac{d}{d\rho}\rho\frac{d}{d\rho} - k^2 - \frac{n^2}{\rho^2}\right)g_{n,i}(\rho, \rho', k) = 0 \qquad i > 0 \tag{35.68}$$

respectively. Similarly, the membrane condition in the last row of Table 35.1, can be expressed as

$$Gg_{n,0}(a, \rho', k) = -\sigma_i \frac{dg_{n,0}(a, \rho', k)}{d\rho}$$

$$Gg_{n,i}(a, \rho', k) + \frac{1}{2\pi} G\gamma(k) \otimes g_{n,i-1}(a, \rho', k) = -\sigma_i \frac{dg_{n,i}(a, \rho', k)}{d\rho} \qquad i > 0 \qquad (35.69)$$

Next, substituting (35.66) into (35.69) results in

$$G[A_{n,0}(k)I_n(ka) + K_n(ka)] = -\sigma_i k[A_{n,0}(k)I'_n(ka) + K'_n(ka)]$$

$$GA_{n,i}(k)I_n(ka) + \frac{1}{2\pi} G\gamma(k) \otimes g_{n,i-1}(a, \rho', k) = -\sigma_i k A_{n,i}(k)I'_n(ka) \qquad i > 0 \qquad (35.70)$$

The coefficients $A_{n,i}(k)$ can now be expressed explicitly as

$$A_{n,0}(k) = -\left[\frac{Ga}{\sigma_i}\frac{K_n(ka)}{I_n(ka)} + ka\frac{K'_n(ka)}{I_n(ka)}\right]\Big/ q_n(k)$$

$$A_{n,i}(k) = -\frac{1}{2\pi}\frac{Ga}{\sigma_i}\frac{\gamma(k) \otimes g_{n,i-1}(a, \rho', k)}{I_n(ka)q_n(k)} \qquad i > 0 \qquad (35.71)$$

where

$$q_n(k) = \frac{Ga}{\sigma_i} + ka\frac{I'_n(ka)}{I_n(ka)} \qquad (35.72)$$

Finally, substituting (35.71) into (35.66) leads to a recursive closed-form expression $g_{n,i}(a, \rho', k)$ containing i convolutions,

$$g_{n,0}(a, \rho', k) = \frac{I_n(k\rho')}{I_n(ka)q_n(k)}$$

$$g_{n,i}(a, \rho', k) = \left(\frac{-Ga}{2\pi\sigma_i}\right)^i \left\{\left[\left(\frac{I_n(k\rho')}{I_n(ka)q_n(k)} \otimes \gamma(k)\right)\frac{I_n(k\rho')}{q_n(k)}\right]\right.$$

$$\left. \otimes \gamma(k) \cdots \otimes \gamma(k)\right\}\frac{I_n(k\rho')}{q_n(k)} \qquad (35.73)$$

The TMP (Fourier's) coefficients in (35.58) can now be expressed as

$$V_n(z, \rho') = \sum_{i=0}^{\infty} \alpha^i V_{n,i}(z, \rho') \qquad (35.74)$$

where, utilizing (35.58) and (35.59),

$$V_{n,i}(z, \rho') = \frac{1}{\pi}\int_0^{\infty} v_{n,i}(\rho', k)\cos(kz)\, dk \qquad (35.75)$$

and

$$v_{n,i}(\rho', k) = g_{n,i}(a, \rho', k) \qquad (35.76)$$

respectively. It should be noted that $q_n(k)$ in (35.72) depends on the intrinsic characteristics of the axon only, namely, its geometry and the associated conductivities (radius a and σ_i, G, respectively). Therefore, the zeros of (35.72), which are essential for the alternative representation of TMP to be carried out next, are independent of the source strength and location.

35.3.3 Alternative Representation: TMP Evaluation

The integral in (35.58) representing a continuous summation over each spectral component of the TMP spectrum [the slow convergence is of $O(1/ka)$, since $v_{n,0}(\rho',k) = O(1/ka)$, in (35.76), as $ka \to \infty$ and $\rho' \to a$] can be expressed alternatively via an equivalent but discrete spectrum which converges much faster (of exponential order for $z > 0$). The alternative representation utilizes $\phi-\rho$ eigenfunction expansion guided along the z direction rather than $\phi-z$ expansion guided along the ρ direction. This is definitely more appropriate for axon propagation problems [5–7].

When the alternative representation is performed via contour deformation in the complex k plane, as depicted in Figure 35.8, it is preferable to represent $V_{n,0}(z,\rho')$ in (35.75) as

$$V_{n,0}(z,\rho') = \frac{1}{\pi} \Re \left[\int_0^\infty v_{n,0}(\rho',k)e^{jk|z|} \, dk \right] \tag{35.77}$$

The continuous spectrum representation (in k space) can be converted into an alternative representation containing discrete eigenvalues contributions,

$$V_{n,0}(z,\rho') = j \sum_{m=1}^\infty \frac{J_n(\lambda_{n,0,m}\rho')}{J_n(\lambda_{n,0,m}a)} \frac{e^{-\lambda_{n,0,m}|z|}}{q'_n(j\lambda_{n,0,m})} \tag{35.78}$$

where J_n is a Bessel function of the first kind and order n. The terms in (35.78) corresponding to the pole (residue) contribution at $q_n(k = j\lambda_{n,0,m}) = 0$ in (35.72), explicitly

$$q_n(j\lambda_{n,0,m}) = \frac{Ga}{\sigma_i} + \lambda_{n,0,m}a \frac{J'_n(\lambda_{n,0,m}a)}{J_n(\lambda_{n,0,m}a)} = 0 \tag{35.79}$$

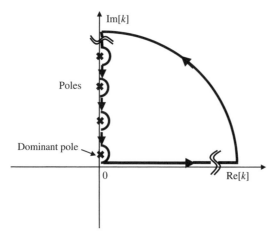

Figure 35.8   Contour deformation in the complex k-plane.

are given via the residue theorem, where $\lambda_{n,0,m}$, the complex roots of (35.78), lie along the real λ axes (Fig. 35.8), leading to

$$q'_n(j\lambda_{n,0,m}) = \frac{(G_m a/\sigma_i)^2 + (\lambda_{n,0,m}a)^2 - n^2}{\lambda_{n,0,m}} \qquad (35.80)$$

The eigenvalue expansion can be well approximated by its dominant contribution, namely the lowermost eigenvalue, which corresponds to the dominant pole, lying in close vicinity to the origin $k \to 0$, as $(Ga/\sigma_i) \to 0$, that is,

$$k_{0,0,1}a = j\lambda_{0,0,1}a = j\sqrt{\frac{2Ga}{\sigma_i}} = \frac{ja}{L_c} \qquad (35.81)$$

Hence, the residue expansion in (35.78) is well approximated, as $\lambda_{0,0,1}a \to 0$, by

$$V_{0,0}(z,\rho') \sim \frac{e^{-\lambda_{0,0,1}|z|}}{\lambda_{0,0,1}a^2} = \frac{L_c e^{-|z|/L_c}}{a^2} = \frac{2g(z)}{a^2} \qquad (35.82)$$

where $g(z)$ is given in (35.33).

Note that when substituting $g_{n,i}(a,\rho',k)$ in (35.73) into the cosine transform (35.75), via (35.76), the i multiplications and i convolutions in k space are interchanged into i convolutions and i multiplications in z space, respectively. Utilizing (35.57), (35.58), (35.74), and (35.75) in conjunction with the complex contour deformation depicted in Figure 35.8, and the setting $n = 0$ and $m = 1$, results in the point source response in (35.43), thereby verifying the myelinated cable equation contribution (35.43) as the dominant pole (residue) contribution of the exact modal expansion (35.57).

35.4 SUMMARY

The novel cable equation model for myelinated nerve fiber presented herein assumes periodic membrane conductivity. The resultant cable equation is represented by Hill's and Mathieu's equations. The general internal source response, expressed via repeated convolutions, uniformly converges provided that the periodic membrane is passive. The solution can be interpreted as an extended source response in an equivalent nonmyelinated axon. The extended source consists of the original source and a novel activation function [3], replacing the periodic membrane in the myelinated axon model.

Rigorous Green's function formulation for TMP induced in 3-D myelinated nerve fiber verifies that the dominant pole contribution of the exact modal expansion is the TMP of the myelinated nerve fiber.

The complete modal expansion of Hill's and Mathieu's equations, governed by Floquet's theorem, is well known and extensively discussed in mathematical physics [9, 10]. It can be utilized for linear and nonlinear myelinated axon models.

Floquet's modes, recognized as the nerve fiber activation modes, can also be incorporated in our excitation model, provided that the periodic membrane pointwise passivity constraint is properly modified [14]. Thus, the modified condition, enforcing the periodic membrane passivity constraint on the average conductivity only, leads to the inclusion of the nerve fiber activation modes in our linear model.

REFERENCES

1. J. J. B. JACK, D. NOBLE, AND R. W. TSIEN, *Electric Current Flow in Excitable Cells*, Clarendon Press, Oxford, 1988.

2. R. PLONSEY, *Bioelectric Phenomena*, New York, McGraw-Hill, 1969.

3. F. RATTAY, *Electrical Nerve Stimulation*, Springer-Verlag, Wien, 1990.

4. J. P. REILLY, *Applied Bioelectricity*, Springer, New York, 1998.

5. R. H. ADRIAN, "Electrical Properties of Striated Muscle," in *Handbook of Physiology, Section X, Skeletal Muscle*, Ed. L. PEACHY. American Physiological Society, Bethesda, MD, 1983, pp. 275–300.

6. A. PESKOFF, "Green's Function for Laplace Equation in an Infinite Cylindrical Cell," *J. Math. Phys.* 15, 2112–2120, 1974.

7. L. M. LIVSHITZ, P. D. EINZIGER, AND J. MIZRAHI, "Rigorous Green's Function Formulation for Transmembrane Potential Induced along a 3-D Infinite Cylindrical Cell," *IEEE Trans. Biomed. Eng.* 49, 1491–1503, 2002.

8. W. MAGNUS AND S. WINKLER, *Hill's Equation*, Dover Publications, New York, 1979.

9. H. HOCHSTADT, *The Functions of Mathematical Physics*, Wiley, New York, 1971.

10. E. T. WHITTAKER AND G. N. WATSON, *A Course of Modern Analysis*, Cambridge University Press, London, 1927.

11. E. W. WEISSTEIN, "Mathieu Function," From MathWorld—A Wolfram Web Resource, http://mathworld.wolfram.com/MathieuFunction.html.

12. M. ABRAMOWITZ AND I. A. STEGUN, *Handbook of Mathematical Functions*, Dover Publication, New York, 1972.

13. D. FRENKEL AND R. PORTUGAL, "Algebraic Methods to Compute Mathieu Functions," *J. Phys. A: Math. Gen.* 34, 3541–3551, 2001.

14. P. D. EINZIGER, L. M. LIVSHITZ, AND J. MIZRAHI, "Generalized Cable Equation Model for Myelinated Nerve Fiber," *IEEE Trans. Biomed. Eng.* 52, 1632–1642, 2005.

15. J. D. JACKSON, *Classical Electrodynamics*, 3rd ed. Wiley, New York, 1999.

CHAPTER *36*

BAYESIAN NETWORKS FOR MODELING CORTICAL INTEGRATION

Paul Sajda, Kyungim Baek, and Leif Finkel

36.1   INTRODUCTION

Much of our perceptual experience is constructed by the visual cortex. Architecturally, the visual cortex appears to be designed for integrating multiple sources of information that come through top-down, bottom-up, and horizontal connections, yet the computational mechanisms by which the integration and resulting inferences occur remain mostly unexplained. A challenge in neural engineering is to reverse engineer the visual cortex to reveal the essence of the computation within this distributed neural network.

Researchers from multiple fields of study have been developing theoretical models to describe the computational mechanisms underlying visual perception. Such efforts date back to the 1850s when Helmholtz described perception as "unconscious inference," a process by which the perceiver progresses from experiencing sensations evoked by an object to recognizing its properties. The information present in the images formed on the retina is incomplete and ambiguous and therefore the visual cortex must integrate previous or independently available knowledge about the environment to perform inferential reasoning and construct a scene. This process is made automatically and unconsciously, and eventually one does not even notice the sensations on which it is based [15, 19, 26].

Recently, much focus has been on developing probabilistic frameworks for understanding the neural mechanisms and computational principles underlying inference within the brain. Information present in raw sensory data is inherently complex and noisy and the challenge is to find models capable of dealing with "complex uncertainty." One particular framework which appears promising uses graphical models for representing, learning, and computing multivariate probability distributions. These "Bayesian" graphical models, also termed Bayesian networks, represent a marriage between probability theory and graph theory and provide a network architecture for dealing with uncertainty and complexity. In Bayesian networks, evidence is accumulated by discrete nodes and "passed" between nodes within the network, resulting in a representation of the joint distribution of the constituent variables. These network models provide the computational means for developing probabilistic representations and associated algorithms for inference and learning [22, 35]. Together with their network architecture, Bayesian networks provide a natural theoretical framework for developing distributed computational models of perceptual inference.

Handbook of Neural Engineering. Edited by Metin Akay

Though attractive as a mathematical framework, an open question is whether neurons in the brain are organized to form Bayesian networks. To address this question one must consider the specifics of the anatomy and physiology in visual cortex. A strong vertical organization is one of architectural features of the cortex, an organization principle uncovered by Mountcastle's discovery of the cortical columns [31]. This columnar structure exists for neurons tuned to different stimulus dimensions with the result being ocular dominance, orientation, and position columns. These findings suggest that columns might form a fundamental functional unit for a variety of visual processes. The sets of columns referred to as cortical "hypercolumns" by Hubel and Wiesel [21] appear to include circuitry necessary to process information from a particular location of visual space across a variety of stimulus dimensions. Hypercolumns are connected via long-range horizontal connections [8, 16, 17, 40] which typically span several hypercolumn lengths, enabling communication between nearby hypercolumns. The span of these connections suggests a locally connected network. In addition, there are extensive feedback projections from extrastriate areas, terminating in multiple layers of V1 [6, 44]. Integration of the bottom-up, horizontal, and top-down inputs into a V1 hypercolumn is therefore likely mediated by geniculate, cortico-cortical, and extrastriate feedback connections which are rather sparse and local in nature.

We argue for the view that the cortical hypercolumn serves as a fundamental functional module for integration and inference in the visual system. Furthermore, we propose that the activity of populations of neurons in cortical hypercolumns explicitly represents probability distributions of attributes presented in the visual stimuli and that the anatomical connections between hypercolumns communicate these probability distributions. This leads to the hypothesis that inference in the cortical network is performed based on the Bayesian network framework. Advantages of the Bayesian network framework in the context of cortical integration include:

1. An explicit representation of uncertainty in the stimulus
2. A common language of probability for integration across cue and modality
3. A sparse, locally connected network architecture

In this chapter we describe a Bayesian network for integrating spatial cues and streams (form and motion) for inferring intermediate-level representations for constructing the visual scene. We show, through simulations, that the Bayesian network leads to results consistent with human perception. In its current form, our model does not attempt to be biophysically realistic, instead focusing on the network-level computation and isomorphisms with cortical networks. For example, nodes in our Bayesian network are not composed of spiking neurons; instead they represent the presumed collective response of a population of neurons, likely in specific cortical layers.

In the following section we describe the integration problem in more detail. We then briefly describe the cortical circuitry in which the problem arises in order to more clearly demonstrate isomorphisms between Bayesian and cortical networks. We then describe a Bayesian network model consisting of Bayesian hypercolumns, collections of nodes processing a local aperture of the visual input space, computing probability distributions for local scene features and integrating these local features via message passing "beliefs" to neighboring hypercolumns. Following the model description, we present simulation results for several simple example problems related to the construction of intermediate-level visual representations.

36.2 PROBLEM OF CORTICAL INTEGRATION

The classical view of information processing in visual cortex is that of a bottom-up process in a feedforward hierarchy. However, the bottom-up information that encodes physical properties of sensory input is often insufficient, uncertain, and even ambiguous for visual processing (e.g., the picture of a Dalmatian dog [48] and Rubin's vase [42]). Psychophysical, anatomical, and physiological evidence suggests that bottom-up factors interact with top-down effects which play a crucial role in the processing of input stimuli. The top-down factors can modulate the expected output of the visual system.

In addition to the bottom-up and top-down inputs, information flows horizontally between sets of cortical neurons—that is, the hypercolumns. This is due to the fundamental aperture problem that the visual system must deal with; Each hypercolumn "sees" a small patch of the visual field decomposed over the input space (orientation, color, motion etc.). The hypercolumns capture information locally in the input space, which is too small to cover visual objects at a global scale. To form coherent representations of objects, nonlocal dependencies in the image space must be captured and exploited. This implies the need for hypercolumns to communicate information about the local visual field to other hypercolumns distributed throughout the cortex. Communication of such nonlocal dependencies can be thought of as a form of *contextual integration*.

One model for visual processing assumes three stages—early, intermediate-, and high-level vision—involving different types of visual representations. When we see the world, the visual system decomposes the scene into surfaces using various cues. This implies that there exist neural representations of surfaces, and it has been thought to be developed at intermediate-level processing [14]. Previous studies suggest that such intermediate-level representations play an important role in integration [32, 47, 52]. For example, Sajda and Finkel focus on the construction of the neural representations of surfaces in the intermediate-level cortical process and present a network-based model [43]. They propose forming surface representations through ownership of contours and argue that such intermediate-level representations serve as a medium for integration.

The different types of information that flow through these bottom-up, top-down, and horizontal connections must be easily and effectively integrated. Anatomical studies indicate that the architecture of the visual cortex is designed for the integration task; however, the computational mechanisms continue to remain largely unknown. In our current model, we assume that the intermediate-level visual representations such as ownership (or direction of figure [43]) might be a substrate for the integration process and focus on developing a computational model for integration.

36.3 CORTICAL CIRCUITRY

The organization of columnar circuitry in the visual system has been extensively studied. For example, Callaway has proposed a generic model of vertical connectivity connecting layers within columns in primary visual cortex of cats and monkey [8]. Although the majority of connections are within columns, physiological and simulation results indicate that there exist horizontal, long-range connections between sets of columns and that these connections give rise to complex, modulatory neuronal responses.

Since the term *hypercolumn* was coined by Hubel and Wiesel [21], it has been used to describe the neural machinery necessary to process a discrete region of the visual field. Typically, a hypercolumn occupies a cortical area of ~ 1 mm^2 and contains between 10^5 and 10^6 neurons. Current experimental and physiological studies have revealed substantial complexity in neuronal response to multiple, simultaneous inputs, including contextual influence, as early as V1 [17, 24, 25]. Such contextual influences, considered to be mediated by modulatory influences of the extraclassical receptive field, are presumed to arise from the horizontal inputs from neighboring hypercolumns and/or from feedback from extrastriate areas [7, 10, 36, 37, 46].

It appears that in the visual system information computed locally across the input space by individual hypercolumns is propagated within the network and integrated, perhaps in a modulatory way, to influence neural responses which ultimately directly correlate with perception. In the next section we describe how such integration could be done within a probabilistic framework, leading to what we term *Bayesian hypercolumns.*

36.4 BAYESIAN INFERENCE IN NETWORKS

36.4.1 Bayesian Hypercolumns

The Bayesian hypercolumn framework provides a structure for integrating top-down, bottom-up, and horizontally transmitted information and results in a neural analog of Bayesian inference. In the brain, information is most likely encoded by populations of neurons rather than by a single neuron. Since the initial studies on the statistical properties of population codes that first used Bayesian techniques to analyze them [34], researchers have suggested that it might be probability distributions instead of single values that the population codes actually represent [1, 39, 54]. A computational model proposed by Deneve et al. [11] uses a biologically plausible recurrent network to decode population codes by maximum-likelihood estimation. Therefore the network essentially implements optimal inference and the simulation results suggest that cortical areas may functions as ideal observers. If it is true that information is represented as probability distributions in cortical areas, it means that the brain may perform Bayesian inference that effectively deals with uncertainty commonly arising in visual tasks.

We argue that the pattern of activity distributed across each hypercolumn instantiates a conditional probability density function (PDF) of some random variables (feature, object, etc.). In striate visual cortex, for example, the PDF might specify an estimate of local orientation as well as the uncertainty in this estimate. Horizontal and top-down inputs modulate this activity, thus expressing the conditional probability of the output given the current links to the population. Anatomical connections communicate the PDFs between hypercolumns. The connection weights are learned through training but capture statistical regularities in the image domain. In visible-domain spatial imagery, these statistical regularities would correspond to the Gestalt laws of perceptual organization.

As Zemel et al. [54] have observed, "a population of neurons is a terrible thing to waste representing a scalar." Accordingly, the PDF representation endows the hypercolumn with computational power far surpassing an equivalent number of "grandmother cells." As environmental conditions change, due to noise or novelty, each hypercolumn recalculates the likelihood of its own distribution based on the state of other hypercolumns. This represents a form of inference, analogous to that carried out in formal Bayesian networks. Recognition and classification are carried out by projecting the current stimulus onto the prior probabilities stored in the cortical architecture.

36.4.2   Bayesian Networks as Model for Cortical Architecture

Bayesian networks are directed graphs in which the nodes represent variables of interest (e.g., stimulus features, target classes) and the links represent conditional probability distributions between possible values of these variables. Bayesian networks are "deliberative" rather than passive reactive agents; thus, unlike traditional artificial neural networks, they respond to novel or changing stimulus conditions instantaneously without relearning [35]. If the environment changes or an input channel is lost, the joint PDF of features and targets is simply recomputed based on the new inputs and stored a priori knowledge. Two features of Bayesian networks account for this advantage over traditional neural networks: The messages passed are probability distributions and message passing is multithreaded so that the influence from different streams is not corrupted. Figure 36.1 illustrates the isomorphism between cortical hypercolumns and the Bayesian network architecture which we propose.

Bayesian networks enable a direct way to integrate different data types (top down, bottom up), since all variables are all mapped to the same probability space. Use of a probabilistic format provides a measure of the degree of confidence together with the most likely classification. Most importantly, Bayesian networks are inference engines, and they can be used to detect relationships among variables as well as a description of these relationships upon discovery.

Over the last several years, substantial progress has been made in Bayesian network theory, probabilistic learning rules, and their application to image analysis [38, 41, 45, 49]. Recent work by Hinton and Brown [20] has been directed at trying to understand the relationship between Bayesian and neural processing. Application of this neuro-Bayesian

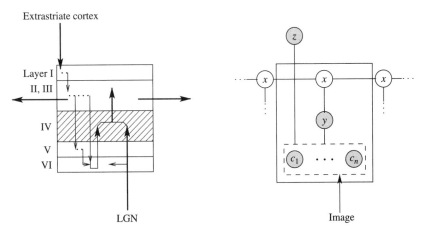

Figure 36.1   Simplified diagram of cortical connections focusing on laminar structure of a hypercolumn (left) and isomorphic architecture of proposed Bayesian network model (right). The lateral geniculate nucleus (LGN) provides bottom-up input via strong connections to layer IV and indirect connections passing through layer VI. The top-down feedback signals from extrastriate cortex pass indirectly into layer VI, then are projected onto layer IV. The feedforward connections from layer IV subsequently provide activation to layer II/III, which forms intrinsic horizontal, long-range connections between hypercolumns. In our network model images correspond to the bottom-up LGN input, from which a set of features are extracted (nodes c_i). Those features are combined and also with the top-down, prior knowledge represented by node z to form PDFs for observations (node y). Then the observations are passed into hidden nodes x, corresponding to layer II/III in the hypercolumn structure. The contextual integration is performed by passing PDFs between neighboring hidden nodes.

approach to image processing has been significantly advanced by Weiss [50], who has demonstrated that when applied to difficult image processing problems, Bayesian networks converge orders of magnitude more rapidly than current relaxation-based algorithms. Processing time is only limited by the time (iterations) required for information to propagate between all units focused on the target. This is consistent with Marr's dictum [29] that visual processing should only take as long as required for all relevant information about the image to be transmitted across cortex, and no further iterations should be necessary once the information has arrived at the appropriate hypercolumn. In contrast, most current algorithms, which are based on relaxation and regularization methods, do not converge at multiple times this number of iterations. Thus Bayesian methods offer the hope of matching the time constraints posed by human visual recognition.

Weiss has also demonstrated the ability of Bayesian networks to infer intermediate-level representations, for example, to integrate local and more global shape information to determine the direction of figure (DOF) [50]. Weiss points out that the updating rules used in Bayesian networks require three building blocks: a distributed representation, weighted linear summation of probabilities, and normalization [51]. These processes have clear ties to cortical networks [18], and such processes have been implemented in retinomorphic and cortex-based chips [4] and in multichip systems with field-programmable connections [5].

36.4.3 Belief Propagation

As mentioned previously, in graphical models (or Bayesian networks, in particular) a node represents a random variable and links specify the dependency relationships between these variables [22]. The states of the random variables can be hidden in the sense that they are not directly observable, but it is assumed that they may have observations related to the state values. The states of hidden variables are inferred from the available observations, and the belief propagation (BP) algorithm solves inference problems based on local message passing. In this section, a BP algorithm is described on undirected graphical models with pairwise potentials. It has been shown that most graphical models can be converted into this general form [53].

Let x be a set of hidden variables and y a set of observed variables. The joint probability distribution of x given y is given by

$$P(x_1, \ldots, x_n \mid y) = c \prod_{i,j} T_{ij}(x_i, x_j) \prod_i E_i(x_i, y_i)$$

where c is a normalizing constant, x_i represents the state of node i, $T_{ij}(x_i, x_j)$ captures the compatibility between neighboring nodes x_i and x_j, and $E_i(x_i, y_i)$ is the local interaction between the hidden and observed variables at location i. An approximate marginal probability of this joint probability at node x_i over all x_j other than x_i is called the local *belief* $b(x_i)$.

The BP algorithm iterates a local message computation and belief updates [55]. The message $M_{ij}(x_j)$ passed from a hidden node x_i to its neighboring hidden node x_j represents the probability distribution over the state of x_j. In each iteration, messages and beliefs are updated as follows:

$$M_{ij}(x_j) = c \int_{x_i} dx_i \, T_{ij}(x_i, x_j) E_i(x_i, y_i) \prod_{x_k \in N_i/x_j} M_{ki}(x_i)$$

$$b(x_i) = cE_i(x_i, y_i) \prod_{x_k \in N_i} M_{ki}(x_i)$$

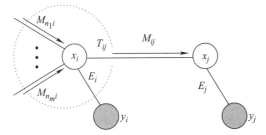

Figure 36.2   Illustration of local message passing from node x_i to node x_j. Open circles are hidden variables, while shaded circles represent observed variables. The local belief at node x_j is computed by combining the incoming messages from all its neighbors and the local interaction E_j.

where N_i/x_j denotes a set of neighboring nodes of x_i except x_j. Here, M_{ij} is computed by combining all messages received by x_i from all neighbors except x_j in the previous iteration and marginalizing over all possible states of x_i (Fig. 36.2). The current local belief is estimated by combining all incoming messages and the local observations.

It has been shown that, for singly connected graphs, BP converges to exact marginal probabilities [53]. Although how it works for general graphs is not well understood, experimental results on some vision problems, such as motion analysis, also show that BP works well for graphs with loops [13].

36.5 SIMULATION RESULTS

We demonstrate the application of a Bayesian network model, having locally connected nodes each computing a distribution over multiple stimulus dimensions (i.e., cues), for two example problems involving intermediate-level surface representations: computing DOF and integrating form and motion cues. For the DOF example, multiple spatial cues are combined to infer the relative figure–ground relationships. The nodes corresponding to hidden variables represent the beliefs (or probabilities) for the DOF. These beliefs are constructed by integrating probabilities based on the observed cues and beliefs at neighboring nodes which are integrated via the network of horizontal connections.

The model for computing DOF (form stream) is then extended by adding a second stream for motion processing. Each node in the motion stream computes probability distributions for local direction of motion in the same way as the DOF is estimated, except that it takes into account the influence of the local DOF estimated in the form stream. Simulations show the two network streams can interact, through Bayesian mechanisms, to qualitatively account for motion coherence phenomena uncovered via psychophysical experiments. Additional details can be found in [58].

36.5.1 Inferring Direction of Figure

"Direction of figure" represents the contour ownership of an object's occluding boundary to a region which represents the object's surface [43]. Therefore DOF specifies a figure–ground relationship directly for occluding contours. In studies on visual perception, the contour ownership is hypothesized to be assigned in early stages of visual processing, which received support from recent neurophysiological findings reported by Zhou et al. [55], showing that more than half of the neurons in extrastriate cortex (areas V2 and V4) are selective to contour ownership. One can consider the assignment of DOF as essentially a problem in probabilistic inference, with DOF being a hidden variable that is not directly observed but can potentially be inferred from local observations and some form of message passing via horizontal dependencies.

As shown in many perceptual studies, various cues are used in the figure–ground discrimination [23, 33, 43]. In our model [3] local figure convexity and similarity/proximity cues are combined to form initial observations. The local interaction $E_{i,\text{cvx}}$ between hidden variable x_i^{dof} and observed variable $y_{i,\text{cvx}}$ specified by the convexity at point i is determined by the local angle of the contour at the location. At the same time, the local interaction $E_{i,\text{sim}}$ between x_i^{dof} and $y_{i,\text{sim}}$ is computed from the similarity/proximity cue by looking for points having similar local tangent angle (i.e., orientation) that lie in a direction orthogonal to the contour at point i. The interaction $E_{i,\text{cvx}}$ prefers smaller angles and $E_{i,\text{sim}}$ favors shorter distances with similar orientations.

The two local interactions are combined based on a *weak fusion model* [9]. The weak fusion model is a simple scheme which suggests that a property of the environment is estimated separately by each independent cue and then combined in a weighted linear fashion to compute the overall effect. Following the weak fusion scheme, the total local interaction $E_i(x_i^{\text{dof}}, y_i^{\text{dof}})$ is computed by the weighted average of the interactions made by two separate observations $y_{i,\text{cvx}}$ and $y_{i,\text{sim}}$:

$$E_i(x_i^{\text{dof}}, y_i^{\text{dof}}) = w_{\text{cvx}} E_{i,\text{cvx}} + w_{\text{sim}} E_{i,\text{sim}}$$

There are several classic examples in which the discriminating figure from the background is not immediately clear. The first two square spiral figures of Figure 36.3a are

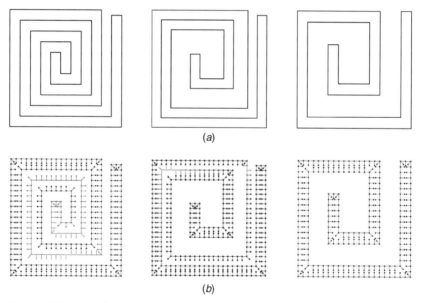

(a)

(b)

Figure 36.3   (a) *Left*: spiral figure in which figure–ground discrimination is ambiguous unless serially tracing the contour. *Middle*: increasing spiral width as it winds around toward center generates incorrect perception of figure surface. *Right*: figure in which correct decision of figure surface can be made immediately. All figures are reproduced from [43]. (b) DOF estimate by the model for the three spiral images shown on top. Resulting DOF is represented by small bars with varying intensity, showing the degree of certainty for the corresponding direction. Note the low certainty (light shading) for the central part of the first spiral, indicating ambiguity of DOF, consistent with human perception. The high certainty (dark shading) for the central part of the second spiral shows an incorrect percept, which is also consistent with human interpretation.

such examples [43]. Discrimination of the figure's surface in the first spiral is difficult, with difficulty increasing for regions close to the center of the spiral.[1] On the other hand, discriminating the figure surface in the second spiral seems to be straightforward. Immediately, we tend to perceive the thin strip in the center as the figure, but this is in fact incorrect. In this case, the width of the spiral increases as it winds around toward the center, generating an incorrect perception of the figure–ground. Unlike the first two figures, figure–ground can be correctly inferred almost instantly for the third spiral.

Figure 36.3b shows that the network model predicts DOF very close to human perception. The DOF figures illustrate increasing ambiguity and/or incorrect interpretation for the center region of the first two spirals and perfect figure segmentation for the last spiral. Although they are not shown here, results indicate that it takes longer for the network model to converge for the first two spirals, with more oscillations in the DOF assignment, as compared to the third example. The results are slightly different, especially near the periphery of the spirals, from those obtained by the neural computational models described in [43]. This is likely due to the fact that the current model does not exploit observations from closure cues.

Certain figures are perceptually ambiguous in that figure and ground can shift or even oscillate. One of the most famous figures that demonstrates this perceptual ambiguity is Rubin's vase, shown in Figure 36.4a. In Figure 36.4, one can perceive either faces or a vase (never both simultaneously), and whenever the perceived object is shifted, the contour ownership is also changed accordingly.

One can bias the interpretation of the ambiguous figures by providing prior information. For example, prior cues might emerge from recognition of distinct visual features (e.g., nose, eyes, chin). In our probabilistic network model, prior information can be considered as another cue for DOF and therefore combined in the same way as described above. In this case, the network model integrates bottom-up (convexity), horizontal (similarity/proximity), and top-down (prior information) cues using a simple combination mechanism.

Figure 36.4b shows the DOF estimation for the Rubin's vase image without using prior information. The network assigns the contour to faces mainly because of the influence from the convexity cue; however, the confidence level is low throughout the entire contour. Prior information is added by explicitly specifying figure direction at several locations on the contour (Fig. 36.4c). The results of combining prior cues are shown on the right. Even if a small weight is assigned for the prior cue,[2] it strongly biases the DOF.

36.5.2 Integrating Form and Motion Streams

The motion of a homogeneous contour is perceptually ambiguous because a neuron responding to the moving contour sees only a limited extent of the visual scene. Due to this "aperture problem," a single local measurement along the contour cannot determine the object's motion, and therefore motion integration across space is required. There are also several features that may provide unambiguous local motion and dominate the motion integration. Examples of such features are line terminators and junctions.

The degree of certainty in DOF at junctions inferred in the form stream can be used to distinguish between *intrinsic terminators*, which are due to the natural end of a line of an

---

[1]Note that all DOF relationships can be determined if one traces the contours of the spiral and "propagates" high-confidence DOF assignments to low-confidence regions. However, in this discussion we ignore such "tracing," though it could be integrated as another form of prior information/cue.

[2]The ratio of w_{prior} over $(w_{\mathrm{cvx}} + w_{\mathrm{sim}})$ is $1 : 8$.

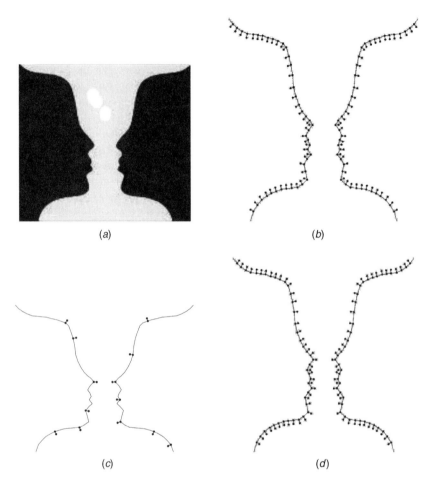

Figure 36.4 (*a*) Rubin's vase (from [2]). (*b*) DOF prediction without using prior information. (*c*) Prior cues for vase features. (*d*) DOF estimate of network model integrating prior information. Figure shows the bias of DOF toward the vase, reflecting strong influence of the prior cue.

object, and *extrinsic terminators*, which are not created by the end of the line itself but rather are caused by occlusion by another surface. Intrinsic terminators provide an unambiguous signal for the true velocity of the line, while extrinsic terminators provide a locally ambiguous signal which must be suppressed for accurate motion computation [32]. The DOF is an explicit surface representation, and therefore the degree of certainty (i.e., belief) in DOF at junctions can be used to represent the strength of the evidence for surface occlusion, which determines the terminator type.

The hidden node in the motion stream represents a velocity of corresponding location along the contour. We assume that both the pairwise compatibility $T_{ij}(x_i^{\text{mot}}, x_j^{\text{mot}})$ and the local interaction $E_i(x_i^{\text{mot}}, y_i^{\text{mot}})$ that models the velocity likelihood at apertures are Gaussian. Also, T_{ij} is set manually and E_i is defined by a mean of the normal velocity at point i and a local covariance matrix Cov_i. Before the BP algorithm starts, the variance at junction points is modulated by a function of DOF belief $b(x_i^{\text{dof}})$ as follows:

$$\text{Cov}_i = e^{\alpha\{b(x_i^{\text{dof}})-0.5\}} \cdot \text{Cov}_i$$

Initially, the covariance matrices of hidden variables are set to represent infinite uncertainty, and mean vectors are set to zero. When the BP algorithm converges, the motion integration is performed by computing the mixture of Gaussians:

$$p(v) = \sum_i p(v \mid i)p(i)$$

where $p(v \mid i)$ is the probability of velocity v from the resulting Gaussian of hidden variable x_i^{mot} and $p(i)$'s are the mixing coefficients.

Figure 36.5a shows the two stimuli used for the experiments. They are $90°$ rotated versions of the diamond stimuli described in McDermott et al. [30], but the perceptual

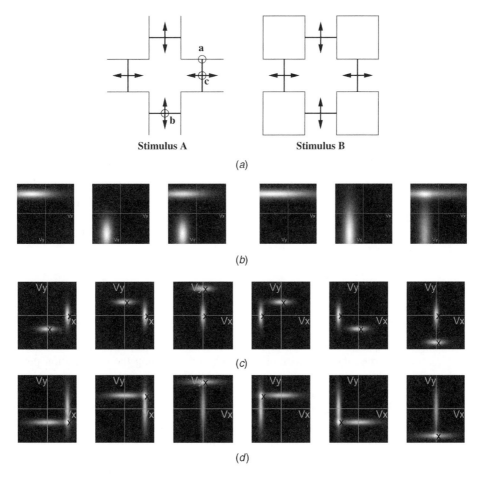

Figure 36.5   (a) Stimuli generated from four line segments that move sinusoidally, $90°$ out of phase, in vertical and horizontal directions. The line segments are presented with L-shaped occluders (stimulus A) and closed square occluders (stimulus B). The presence of the occluding surface alters the motion perception. (b) Velocity estimation results for stimulus A (first three figures) and stimulus B (last three figures). *First and fourth*: Estimation at location **b**. *Second and fifth*: Estimation at location **c**. *Third and last*: Velocity computed by combining the two estimations using mixture of Gaussian. (c,d) A sequence of resulting velocity estimation for six successive frames sampled from a period of sinusoidal motion with a regular interval. (c) Two separate motions oscillating in the direction normal to the line segment for stimulus A. The sequence in (d), on the other hand, shows a coherent motion forming a circular trajectory.

effect is basically the same. The motion of moving line segments is identical between the two stimuli. A pair of vertical line segments move together sinusoidally in a horizontal direction while the two horizontal lines move simultaneously in a vertical direction. The vertical and horizontal motions are 90° out of phase. The difference between the two stimuli is the shape of the occluders, which alters the perceived motion. A single coherent rotation is perceived for stimulus B, while we are more likely to see two separate motions of the line segments for stimulus A [30].

Figure 36.5b shows the resulting velocity estimated by the network model for stimulus A (first three figures) and stimulus B (last three figures). Since the estimated motions along the line segments that move simultaneously are almost identical after convergence, the first two figures for each stimulus show only the motions at locations **b** and **c** (shown in Fig. 36.5a in the velocity space). In the third and last figures in Figure 36.5b, which display the integrated motion, we clearly see the bimodal distribution for stimulus A, while a single peak is formed at the intersection of two distributions for stimulus B. This implies that we perceive two separate motions for stimulus A and a single coherent motion for stimulus B. Figures 36.5c,d illustrate the resulting velocity estimation for six successive frames sampled from a period of the sinusoidal motion. The maximum a posterior in each frame follows a circular trajectory for stimulus B, which is consistent with perceiving rotation.

A prior cue can easily be added for inferring DOF in stimulus A. Psychophysically, this might simulate priming subjects or integrating disparity information to add stereoscopic depth so that the subjects are biased to see the occlusion of L-shaped figures more strongly or weakly. Adding priors for stronger occluders strengthens the belief for DOF in the indicated direction, and consequently more coherent motion would be expected. Although not shown here, as the weights on the prior cue are increased, the solution becomes more unimodal and with a strong prior it produces a single peak at the intersection similar to stimulus B.

36.6 CONCLUSION

In this chapter we have attempted to demonstrate how Bayesian networks might represent an appropriate theoretical framework for understanding the computation performed by cortical hypercolumns for inferring intermediate-level visual representations and integrating cues across form and stream. The hypercolumn architecture is prevalent at all levels of cortical visual processing, not just in primary visual cortex. Thus one would expect the framework for Bayesian integration might also occur at other levels of processing, such as object recognition, occurring in the hypercolumn networks of inferotemporal cortex.

We have not addressed the issue of learning and adaptation, which is critical in cortical networks. The parameters for our model are fixed. We are now exploring methods for learning the emission and transition probabilities in the network given an ensemble of natural spatiotemporal imagery. If the Bayesian network framework is consistent with cortical network processing, we would expect this to provide insight into the organization of receptive fields and contextual interactions seen in the biological networks.

We have only demonstrated the use of Bayesian networks in the inference mode. However, since they represent a generative model, such networks have a much broader utility. For example, one can use the model to synthesize novel "scenes." In the context of cortical networks, this is consistent with the ideas that early visual cortical areas are involved in generating visual mental imagery [27, 28]. The ability to use the framework to, for example, infer, discriminate, synthesize, and compress makes it attractive as a model of general-purpose visual processing.

The implementations we have presented are not meant to be complete models, instead providing the skeleton of a framework suggesting how spatial cues from multiple streams might be integrated using a common framework to construct a scene. Others have attempted to develop more biologically realistic models of form and motion integration, with mechanisms that are perhaps more neural. However such models are difficult to analyze, particularly in terms of understanding the general role/effect of uncertainty in the observations and reversal or ambiguity in perception. Although we do not claim that our current implementation is a biologically realistic model, it is worth noting that it is constructed as a network model, with no requirements for global connectivity or consistency. Further work by our group is aimed at developing biophysically realistic models of cortical hypercolumns and the cortico-cortical connectivity to investigate what specific neural machinery could be used to locally construct distributions of observations and integrate these via local message passing. Ultimately our goal is a "reverse neuroengineering" of these hypercolumn networks.

ACKNOWLEDGMENTS

This work was supported by the DoD Multidisciplinary University Research Initiative (MURI) program administered by the Office of Naval Research under Grant N00014-01-1-0625 and a grant from the National Imagery and Mapping Agency, NMA201-02-C0012.

REFERENCES

1. C. H. ANDERSON AND D. C. VAN ESSEN, "Neurobiological computational systems," in J. M. ZURADA, R. J. MARKS II, AND C. J. ROBINSON (Eds.), *Computational Intelligence Imitating Life*, pp. 213–222, IEEE Press, New York, 1994.

2. T. J. ANDREWS, D. SCHLUPPECK, D. HOMFRAY, P. MATTHEWS, AND C. BLAKEMORE, "Activity in the fusiform gyrus predicts conscious perception of Rubin's vase-face illusion," *NeuroImage*, vol. 17, pp. 890–901, 2002.

3. K. BAEK AND P. SAJDA, "A probabilistic network model for integrating visual cues and inferring intermediate-level representations," Paper presented at Third International Workshop on Statistical and Computational Theories of Vision, Nice, France, October 2003.

4. K. A. BOAHEN, "Retinomorphic chips that see quadruple images," Paper presented at MicroNeuro'99: 7th International Conference on Neural, Fuzzy, and Bio-Inspired Systems, Granada, Spain, IEEE Computer Society Press, 1999.

5. K. A. BOAHEN, "Point-to-point connectivity between neuromorphic chips using address events," *IEEE Transactions on Circuits and Systems-II*, vol. 47, pp. 416–434, 2000.

6. J. M. L. BUDD, "Extrastriate feedback to primary visual cortex in primates: A quantitative analysis of connectivity," *Proceedings of the Royal Society of London B*, vol. 265, pp. 1037–1044, 1998.

7. J. BULLIER, J. M. HUPÉ, A. JAMES, AND P. GIRARD, "Functional interactions between areas V1 and V2 in the monkey," *Journal of Physiology (Paris)*, vol. 90, pp. 217–220, 1996.

8. E. M. CALLAWAY, "Local circuits in primary visual cortex of the macaque monkey," *Annual Review of Neuroscience*, vol. 21, pp. 47–74, 1998.

9. J. J. CLARK AND A. L. YUILLE, *Data Fusion for Sensory Information Processing Systems*, Kluwer Academic, Boston, 1990.

10. S. MARTINEZ-CONDE, J. CUDEIRO, K. L. GRIEVE, R. RODRIGUEZ, C. RIVADULLA, AND C. ACUÑA, "Effect of feedback projections from area 18 layers 2/3 to area 17 layers 2/3 in the cat visual cortex," *Journal of Neurophysiology*, vol. 82, pp. 2667–2675, 1999.

11. S. DENEVE, P. E. LATHAM, AND A. POUGET, "Reading population codes: A neural implementation of ideal observers," *Nature Neuroscience*, vol. 2, no. 8, pp. 740–745, 1999.

12. M. DIESMANN, M. O. GEWALTIG, AND A. AERTSEN, "Stable propagation of synchronous spiking in cortical neural networks," *Nature*, vol. 402, pp. 529–533, 1999.

13. W. T. FREEMAN, E. C. PASZTOR, AND O. T. CARMICHAEL, "Learning low-level vision," *International Journal of Computer Vision*, vol. 40, no. 1, pp. 25–47, 2000.

14. L. H. FINKEL AND P. SAJDA, "Constructing visual perception," *American Scientist*, vol. 82, pp. 224–237, 1994.

15. W. S. GEISLER AND D. KERSTEN, "Illusions, perception and Bayes," *Nature Neuroscience*, vol. 5, no. 6, pp. 508–510, 2002.

16. C. D. GILBERT AND T. N. WIESEL, "Clustered intrinsic connections in cat visual cortex," *Journal of Neuroscience*, vol. 3, pp. 1116–1133, 1983.

17. C. D. GILBERT, "Horizontal integration and cortical dynamics," *Neuron*, vol. 9, pp. 1–13, 1992.

18. D. J. HEEGER, E. P. SIMONCELLI, AND J. A. MOVSHON, "Computational models of cortical visual processing," *Proceedings of the National Academy of Sciences*, vol. 93, pp. 623–627, January 1996.

19. H. HELMHOLTZ, *Physiological Optics*, Vol. III: *The Perceptions of Vision* (J. P. Southall, Trans.), Optical Society of America, Rochester, NY, 1925.

20. G. E. HINTON AND A. D. BROWN, "Spiking Boltzmann machines," *Advances in Neural Information Processing Systems*, MIT Press, Cambridge, MA, pp. 122–128, vol. 12, 2000.

21. D. H. HUBEL AND T. N. WIESEL, "Functional architecture of macaque monkey visual cortex," *Proceedings of the Royal Society of London B*, vol. 198, pp. 1–59, 1977.

22. M. JORDAN, *Graphical Models: Foundations of Neural Computation*, MIT Press, Cambridge, MA, 2001.

23. G. KANIZSA, *Organization in Vision*, Praeger, New York, 1979.

24. M. K. KAPADIA, M. ITO, C. D. GILBERT, AND G. WESTHEIMER, "Improvement in visual sensitivity by changes in local context: Parallel studies in human observers and in V1 of alert monkeys," *Neuron*, vol. 15, pp. 843–856, 1995.

25. M. K. KAPADIA, G. WESTHEIMER, AND C. D. GILBERT, "Spatial distribution of contextual interactions in primary visual cortex and in visual perception," *Journal of Neurophysiology*, vol. 84, pp. 2048–2062, 2000.

26. D. C. KNILL, D. KERSTEN, AND A. YUILLE, "Introduction: a Bayesian formulation of visual perception," in D. C. KNILL AND W. R. RICHARDS (Eds.), *Perception as Bayesian Inference*, pp. 1–21, Cambridge University Press, Cambridge, 1996.

27. S. M. KOSSLYN, "Visual mental images and re-presentations of the world: A cognitive neuroscience approach," in J. S. GERO AND B. TVERSKY (Eds.), *Visual and Spatial Reasoning in Design*, MIT Press, Cambridge, MA, 1999.

28. S. M. KOSSLYN, A. PASCUAL-LEONE, O. FELICIAN, J. P. KEENAN, W. L. THOMPSON, G. GANIS, K. E. SUKEL, AND N. M. ALPERT, "The role of area 17 in visual imagery: Convergent evidence from PET and rTMS," *Science*, vol. 284, pp. 167–170, 1999.

29. D. MARR, *Vision: a Computational Investigation into the Human Representation and Processing of Visual Information*, W. H. Freeman, San Francisco, 1982.

30. J. MCDERMOTT, Y. WEISS, AND E. H. ADELSON, "Beyond junctions: Nonlocal form constraints on motion interpretation," *Perception*, vol. 30, pp. 905–923, 2001.

31. V. MOUNTCASTLE, "Modality and topographic properties of single neurons of cat's somatic sensory cortex," *Journal of Neurophysiology*, vol. 20, pp. 408–434, 1957.

32. K. NAKAYAMA, S. SHIMOJO, AND G. H. SILVERMAN, "Stereoscopic depth: Its relation to image segmentation, grouping, and the recognition of occluded objects," *Perception*, vol. 18, pp. 55–68, 1989.

33. K. NAKAYAMA, "Binocular visual surface perception," *Proceedings of the National Academy of Sciences USA*, vol. 93, pp. 634–639, 1996.

34. M. PARADISO, "A theory for the use of visual orientation information which exploits the columnar structure of striate cortex," *Biological Cybernetics*, vol. 58, pp. 35–49, 1988.

35. J. PEARL, *Probabilistic Reasoning in Intelligent Systems: Networks of Plausible Inference*, Morgan Kaufmann, San Francisco, CA, 1988.

36. U. POLAT AND A. M. NORCIA, "Neurophysiological evidence for contrast dependent long-range facilitation and suppression in the human visual cortex," *Vision Research*, vol. 36, no. 14, pp. 2099–2109, 1996.

37. U. POLAT, K. MIZOBE, M. W. PETTET, T. KASAMATSU, AND A. M. NORCIA, "Collinear stimuli regulate visual responses depending on cell's contrast threshold," *Nature*, vol. 391, no. 5, pp. 580–584, 1998.

38. J. PORTILLA, V. STRELA, M. J. WAINWRIGHT, AND E. P. SIMONCELLI, "Image denoising using Gaussian scale mixtures in the wavelet domain," Technical Report TR2002–831, Computer Science Department, Courant Institute of Mathematical Sciences, New York University. September 2002.

39. A. POUGET, P. DAYAN, AND R. ZEMEL, "Information processing with population codes," *Nature Reviews Neuroscience*, vol. 1, pp. 125–132, 2000.

40. K. S. ROCKLAND AND J. S. LUND, "Intrinsic laminar lattice connections in primate visual cortex," *Journal of Comparative Neurology*, vol. 216, pp. 303–318, 1983.

41. J. K. ROMBERG, H. CHOI, AND R. G. BARANIUK, "Bayesian tree-structured image modeling using wavelet-domain hidden Markov models," *IEEE Transactions on Image Processing*, vol. 10, no. 7, pp. 1056–1068, 2001.

42. E. RUBIN, *Visuell Wahrgenommene Figuren*, Gyldenalske Boghandel, Copenhagen, 1915.

43. P. SAJDA AND L. H. FINKEL, "Intermediate-level visual representations and the construction of surface perception," *Journal of Cognitive Neuroscience*, vol. 7, no. 2, pp. 267–291, 1995.

44. P. SALIN AND J. BULLIER, "Corticocortical connections in the visual system: Structure and function," *Physiological Reviews*, vol. 75, no. 1, pp. 107–154, 1995.

45. C. SPENCE, L. PARRA, AND P. SAJDA, "Hierarchical image probability (HIP) models," in *IEEE International Conference on Image Processing*, vol. 3, pp. 320–323, 2000.

46. D. D. STETTLER, A. DAS, J. BENNETT, AND C. D. GILBERT, "Lateral connectivity and contextual

interactions in macaque primary visual cortex," *Neuron*, vol. 36, pp. 739–750, 2002.

47. G. R. STONER AND T. D. ALBRIGHT, "Image segmentation cues in motion processing: Implications for modularity in vision," *Journal of Cognitive Neuroscience*, vol. 5, no. 2, pp. 129–149, 1993.

48. J. B. THURSTON AND R. G. CARRAHER, *Optical Illusions and the Visual Arts*, Reinhold, New York, 1966.

49. M. J. WAINWRIGHT AND E. P. SIMONCELLI, "Scale mixtures of Gaussians and the statistics of natural images," *Advances in Neural Information Processing Systems*, vol. 11, pp. 855–861, 1999.

50. Y. WEISS, "Bayesian belief propagation for image understanding," Paper presented at Workshop on Statistical and Computational Theories of Vision, 1999.

51. Y. WEISS, "Correctness of local probability propagation in graphical models with loops," *Neural Computation*, vol. 12, pp. 1–42, 2000.

52. M. WHITE, "A new effect of pattern on perceived lightness," *Perception*, vol. 3, no. 4, pp. 413–416, 1979.

53. J. S. YEDIDIA, W. T. FREEMAN, AND Y. WEISS, "Understanding belief propagation and its generalizations," in G. LAKEMEYER AND B. NEBEL (Eds.), *Exploring Artificial Intelligence in the New Millennium*, Morgan Kaufmann, San Francisco, CA, pp. 239–269, 2003.

54. R. S. ZEMEL, P. DAYAN, AND A. POUGET, "Probabilistic interpretation of population codes," *Neural Computation*, vol. 10, pp. 403–430, 1998.

55. H. ZHOU, H. S. FRIEDMAN, AND R. VON DER HEYDT, "Coding of border ownership in monkey visual cortex," *Journal of Neuroscience*, vol. 20, no. 17, pp. 6594–6611, 2000.

56. P. SAJDA AND K. BACK, "Integration of force and motion within a generative model of visual cortex," *Neural Networks*, vol. 17, no. 5–6, pp. 809–821, 2004.

CHAPTER 37

NORMAL AND ABNORMAL AUDITORY INFORMATION PROCESSING REVEALED BY NONSTATIONARY SIGNAL ANALYSIS OF EEG

Ben H. Jansen, Anant Hegde, Jacob Ruben, and Nashaat N. Boutros

37.1 INTRODUCTION

The spontaneous electrical activity generated by the brain can be recorded from the scalp using macroelectrodes of about 1.0 cm in diameter. The resulting signal is referred to as the electroencephalogram (EEG). The EEG amplitude is typically between $10 \, \mu V$ and several hundred microvolts and has a bandwidth of 0–80 Hz, with most of the activity below 30 Hz. The EEG is thought to reflect the postsynaptic potentials, summed over hundreds of millions of neurons [1]. Due to volume conduction effects, the spatial resolution of the EEG is rather poor, but it provides perfect time resolution. Multichannel EEGs are routinely recorded in clinical practice using 16–32 electrode placed on the scalp according to the 10–20 system of electrode placement [2], which ensures replicability.

The EEG reflects activity related to internal processes and to events occurring in the external world. The latter activity is referred to as an event-related potential (ERP). For example, visual or auditory stimulation would result in EEG changes. Some EEG changes may occur without external stimulation, but they are related to some event. For example, EEG changes may be seen when a stimulus is missing from a long series of regularly presented stimuli. The event-related changes in the spontaneous EEG are typically very small, and signal enhancement procedures need to be used to retrieve the ERPs from the spontaneous EEG. The method of choice is ensemble averaging. This involves repeatedly applying the stimulus and recording the EEG over the interval of interest, starting from some point before the stimulus application through the actual response. The responses to all stimuli are then summed and divided by the number of stimuli (N). Assuming that each stimulus produces an identical, time-locked response which is additive to and independent of the spontaneous EEG, a signal-to-noise improvement by a factor of $\sqrt{N}$ may be obtained. The resulting ensemble-averaged evoked potentials (EPs) are generally characterized by a sequence of positive and negative waves, which are identified by their polarity (P for positive, N for negative) and their latency (e.g., P300 is a positive wave occurring 300 ms poststimulus) or the order in which they were

Handbook of Neural Engineering. Edited by Metin Akay
Copyright © 2007 The Institute of Electrical and Electronics Engineers, Inc.

generated following stimulus delivery (e.g., N2 is the second negative wave). The amplitudes and latencies of the various components are the primary quantitative measures derived from EPs. Components can be classified as exogenous, that is, due to the primary processing of the stimulus-related activity in the brainstem, thalamus, and primary sensory cortex, or endogenous, that is, resulting from higher level or cognitive processing. The latter have a latency of 250 ms or more, while the former occur within 50 ms of stimulus presentation. The characteristics of the exogenous components correlate with stimulus characteristics such as loudness and pitch, while the latency and amplitude of the exogenous components covary with the degree of cognitive processing required. Components between 50 and 250 ms have both exogenous and endogenous characteristics.

The ensemble-averaging method for EP enhancement is plagued by several problems. First, there is mounting evidence that the EEG and EPs are not two independent processes. Specifically, a relationship between prestimulus EEG activity and visual-evoked response morphology has been demonstrated in [3, 4]. Başar and colleagues [5–8] were able to show that the prestimulus EEG within a given bandwidth is directly related to the poststimulus activity filtered in the same band. Second, the stimulus-related afferent activity may not be additive to the ongoing activity. Rather, the event-related activity may cause a phase change in all or some of the frequency components of the spontaneous EEG. This hypothesis was first raised by Sayers et al. [9]. Recently, evidence of stimulus-induced phase synchronization has been reported by several groups, using a variety of EP analysis methods [10–14]. Using the piecewise Prony method [10], we have recently presented evidence that phase reordering plays an important role in the generation of the N100 and P200 components of the auditory EP [15]. In this context, the term "phase synchronization" denotes that the phase of the spontaneous EEG activity is changed to a specific value upon stimulus presentation. Consequently, poststimulus EEG is aligned ("in phase") across trials. The instances at which maximal intertrial phase synchronization occurs correspond to the latencies of the components seen in the averaged EP.

Advanced signal analysis tools designed to track the stimulus-induced EEG changes on a single-trial basis can shed light on the mechanism underlying EP generation. In the present study, we show how one such method (the piecewise Prony method [10]) can be used to uncover phase reordering abnormalities in a population of schizophrenia patients, as compared to a normal control group. Schizophrenia is a disorder of the brain characterized by a variety of symptoms (delusions, hallucinations, disorganized thinking, catatonia, social withdrawal, absence of emotion). We also show, that these phase abnormalities explain the decreased amplitude of the P50, N100, and P200 components observed in schizophrenia [16, 17] and the decreased ability to attenuate the amplitude of the responses with stimulus repetition (sensory gating) [18, 19]. Utilizing a paired-click paradigm, where two identical stimuli (S_1 and S_2) separated by a brief interval are repeatedly presented, sensory gating can be operationally defined by an amplitude ratio. Stronger attenuation of irrelevant input and thus better gating capability are reflected by a lower ratio of the amplitude of averaged EPs to S_2 stimuli to the amplitude of the S_1 responses.

37.2 METHODS

Twenty patients with chronic schizophrenia who were diagnosed based on the Structured Clinical Interview for DSM-IIIR (*SCID) were recruited for the study.

All patients were on stable doses of atypical antipsychotic agents (except clozaril). Patients with a history of substance abuse had at least one month of verified abstinence and a negative urine test for drugs of abuse within 24 hours from the time of the study. None of the subjects had a history of head injury leading to any length of loss of consciousness. Twenty healthy control subjects were recruited from hospital personnel or from newspaper ads. Healthy control subjects were age and gender matched to the schizophrenia group in general and not on one-to-one bases. Specifically, the normal subjects had an average age of 43 years (range 25–55 years) and comprised 14 males and 6 females. The schizophrenia subjects' average age was 42.5 years (range 32–56 years), with 15 males and 5 females. Subjects with a history of hearing problems were excluded.

Auditory-evoked potentials (AEPs) were recorded from the scalp of the normal subjects and the schizophrenia patients using a double-stimulus paradigm, where two identical stimuli are presented. Tone bursts of 1000 Hz, with a duration of 4 ms plus a 1-ms rise and fall time and 85 dB sound pressure level, as measured at the ear using a measure-and-hold digital sound meter (Tandy Corp, Ft. Worth, TX), were produced by a Grass auditory stimulator. Sounds were delivered using ear inserts (Etymotic Research, Elkgrove, IL) in pairs, with 400–600 ms between the first and second tone bursts. The interstimulus intervals were randomized by the computer. Sixty to 100 pairs of tone bursts, separated by at least 8 s (subject to a 100-ms computer-imposed jitter), were presented. The subjects were instructed to reduce movement, relax, stay awake, and avoid eye movement. No cognitive task was given. Monopolar recordings were made from gold-disk electrodes affixed to the Fz, Cz, Pz, F1, F2, T3, and T4 locations, referenced to linked ears, but only the Cz lead was analyzed. In addition, eye movements were recorded with a bipolar lead connecting one electrode placed 1 cm above and lateral to the right outer canthus to a second electrode placed 1 cm below and lateral to the left outer canthus. Electrical signals were amplified 20,000 times by Grass amplifiers with band pass filters set at 0.05 and 300 Hz.

All data were inspected for artifacts, and trials were rejected when activity in any channel exceeded 75 μV. A large number of the trials obtained from the schizophrenia subjects did not meet this criterion. However, 19 of the 20 patients produced at least 25 artifact-free trials, which we considered the minimal number of trials for meaningful data analysis, and hence we ended up with 20 normal and 19 schizophrenia subjects. A sampling rate of 1000 Hz was used for digitization, but all data were downsampled to 250 Hz before computer analysis. This downsampling was required for the proper functioning of the piecewise Prony method (PPM) algorithm (see below).

Averages for the S_1 and S_2 responses were computed for each subject as well as grand averages for each population and the amplitudes of the P50, N100, and P200 components were measured for each subject.

The PPM [10] was applied to single AEPs. This method decomposes single trials into exponentially increasing and/or decreasing sinusoidal components ("Prony components"). The decomposition is done iteratively, modeling lower frequencies first, which are then removed from the signal. The effect of the stimulus on the ongoing EEG can be observed directly by graphing the components found in a number of trials in a specific frequency band.

The PPM was applied from 500 ms prestimulus to 500 ms poststimulus for S_1 and S_2 separately. The single trials were decomposed to obtain components in the 0–2, 2–4, 4–8, 8–12, and 12–16-Hz bands.

The degree of synchronization was assessed by means of histograms of the instantaneous phase of the components across an ensemble of trials. The phase histograms of the components found in a particular frequency band in 25 trials were generated once every 20 ms. In the absence of phase alignment, the histograms resemble uniform density functions, becoming less so if a phase preference develops. The Kuiper statistic κ [20] was used to quantify the degree with which the phase histograms resembled a uniform density function. Kuiper's statistic is defined as

$$\kappa = \ln\left(2\sum_{j=1}^{\infty}(4j^2V^2 - 1)e^{-2j^2V^2}\right) \tag{37.1}$$

with

$$V = \left(\sqrt{N} + 0.155 + \frac{0.24}{\sqrt{N}}\right) \cdot \max_{i=1,\dots,L}[S_N(x_i) - F_0(x_i)]$$
$$+ \max_{i=1,\dots,L}[F_0(x_i) - S_N(x_i)] \tag{37.2}$$

where N is the sample size, L is the number of bins in the histogram, x_i is the upper bound of bin i, and $F_0(x_i)$ and $S_N(x_i)$ are the theoretical and actually observed cumulative distribution functions, respectively. Because of the natural log function, κ will be negative only, with values close to zero indicating that $S_N(x_i)$ matches $F_0(x_i)$.

The procedure is explained graphically in Figure 37.1. Shown are the decompositions in the 4–8-Hz band of 25 trials (S_1 responses) from one normal subject. The phase histograms computed at 20-ms intervals starting 60 ms prestimulus and

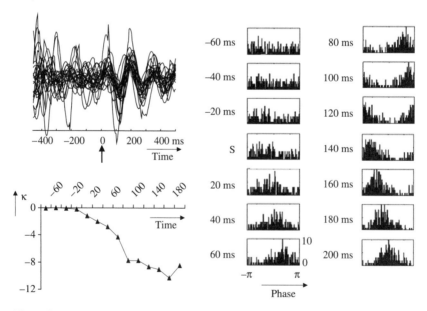

Figure 37.1 *Top left*: PPM components in 4–8-Hz band derived from 25 trials of one healthy subject. Arrow indicates moment of stimulation. Vertical axis is in arbitrary units. *Right*: Phase histograms for these components. Histograms are computed at 20-ms intervals. Horizontal axis ranges from $-\pi$ through π. Time proceeds from top of column to bottom of column. S denotes stimulus presentation. *Bottom left*: Kuiper statistic κ at 20-ms intervals from 60 ms prestimulus through 200 ms poststimulus.

continuing through 200 ms poststimulus are shown as well. As one may observe, the relatively random phase organization seen prestimulus is reflected in the almost perfectly uniform phase histograms seen from 60 ms prestimulus through the moment of stimulus presentation. Poststimulus, a preferred phase becomes evident, as seen in the unimodal phase distributions, and also is readily seen from the single-trial decompositions. The κ values undergo a big change, starting from essentially zero before stimulus presentation to large negative values poststimulus, indicating that the phase histograms no longer resemble uniform distributions.

Amplitude and degree of phase synchronization differences between populations and/or stimuli were assessed for statistical significance using t-tests for the sample mean. Significant differences were indicated if $p < 0.05$. When evaluating phase synchronization differences, 14 t-tests (one for each 20-ms interval) were conducted. Because the κ values for adjacent time intervals will be correlated and following the advice given by Perneger [21], no Bonferroni adjustments of the critical value for p were made. However, where feasible, we provide both the t and p values, together with the actual data input to the statistical procedure.

The relationship between EP amplitude and degree of phase synchronization was assessed by means of multiple linear regression analysis (least-squares method). Specifically, κ_{40}, that is, the degree of phase synchronization at 40 ms poststimulus for the 0–2, 2–4, 4–8, and 8–12-Hz frequency bands was used as an independent measure to predict the P50 amplitude. The corresponding κ_{100} and κ_{200} values were used to predict the N100 and P200 amplitudes, respectively. These κ values were selected because they correspond to the latency of the three components being studied. Multiple linear regression analysis returns the residual variance R^2, which is the sum of squared differences between actual and predicted amplitudes. The F-tests on R^2 were conducted to assess the significance of the correlation between degree of phase synchronization and EP amplitude. In case a significant correlation was found ($p < 0.05$), the regression coefficients were checked to determine if they differed significantly from zero, and the regression analysis was repeated using these coefficients only.

37.3   RESULTS

The grand averages of the S_1 and S_2 responses for the normal and schizophrenia population were computed. In both populations, the P50, N100, and P200 components had significantly larger amplitudes for the S_1 than for the S_2 responses ($p < 0.008$). The amplitude differences between the two populations were not significant with the exception that the normal population had a larger S_1 response than the schizophrenia group ($p < 0.03$) for N100.

The gating index, defined as the ratio of the S_2 amplitude over the S_1 amplitude, was computed for the P50, N100, and P200 components. Significant differences between the two subject populations were observed for the N100 ($p < 0.02$) and the P200 components ($p < 0.01$), with the normal controls showing more gating (i.e., smaller ratios) than the patient population. A relatively large number of subjects have a P50 gating ratio greater than 1, suggesting that the S_2 response is larger than the S_1 response. This may in part be due to the difficulty identifying P50, as all but one of the normal subjects had a ratio smaller than 1 for N100, and no normal control exceeded a ratio of 1 for P200. The latter two components are readily identified in the ensemble-averaged ERP. Interestingly, five schizophrenia subjects had ratios greater than 1 for both N100 and P200, with

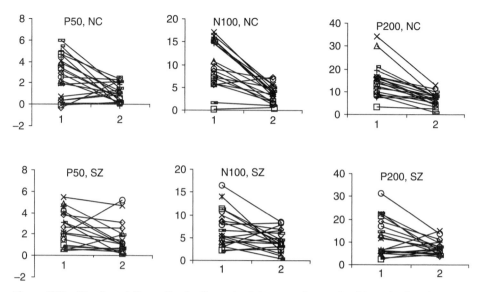

Figure 37.2 The S_1 and S_2 amplitudes for each of the normal control subjects (top) and schizophrenia patients (bottom) for P50 (left), N100 (middle), and P200 (right).

two patients having a ratio greater than 1 for N100 or P200. A detailed view of the S_1 and S_2 amplitudes is provided in Figure 37.2. It appears that the normal subjects are more "consistent" gaters than the schizophrenia group, but it is also evident that substantial gating takes place among the schizophrenia population.

For the normal population, it was generally observed that phase synchronization occurs in the delta and theta bands. Specifically only four subjects did not show any phase resetting at all in these bands. This phase synchronization was much more pronounced for the S_1 stimulus than for S_2. While some of the schizophrenia patients behaved much like normal subjects, in general, schizophrenia patients displayed much less phase synchronization than normal subjects; only eight subjects phase synchronized in response to S_1 in the delta/theta bands. The schizophrenia patients showed much more variation in their responses and six showed more phase resetting for the S_2 stimulus than for S_1, something which was not seen in any of the normal subjects.

We present graphs of the components found in the various frequency bands for representative subjects. The first case shows phase synchronization in the 0–2, 2–4, and 4–8-Hz bands (see Fig. 37.3) in a normal subject. The 0–2-Hz synchronization starts around 150 ms poststimulus and continues beyond 500 ms. The 2–4-Hz band synchronization starts quite early comparatively, but its effect dies out much sooner. Stimulus S_2 has very little or no effect on the phase. A certain amount of variability is also observed among the different trials in the same band. For example, in the theta band (4–8 Hz), there are some trials which seem unaffected by the S_1 stimulus.

A second normal subject displayed strong S_1 phase synchronization in all but the 0–2- and 12–16-Hz bands (see Fig. 37.4). Also S_2 had a phase synchronizing effect, but to a lesser degree and primarily confined to the 4–12-Hz range.

One subset of the schizophrenia subjects did not show phase synchronization in response to either stimulus. See, for example, the overlay plots for schizophrenia subject B shown in Figure 37.5.

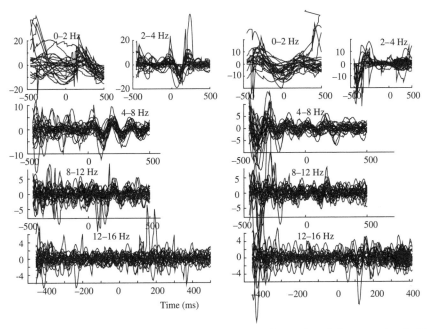

Figure 37.3   Components from 25 trials of normal subject A for all frequency bands. Upper left panel is 0–2-Hz band, upper right is 2–4 Hz. Second from top is 4–8 Hz, next to bottom is 8–12 Hz, and 12–16 Hz is at the bottom. *Left*: stimulus S_1; *Right*: stimulus S_2. Stimulus is presented at 0 ms, and 500 ms pre- and poststimulus activity is presented in each frequency band.

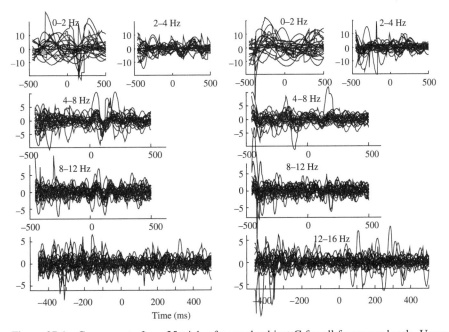

Figure 37.4   Components from 25 trials of normal subject C for all frequency bands. Upper left panel is 0–2-Hz band, upper right is 2–4 Hz. Second from top is 4–8 Hz, next to bottom is 8–12 Hz, and 12–16 Hz is at the bottom. *Left*: stimulus S_1; *Right*: stimulus S_2. Stimulus is presented at 0 ms, and 500 ms pre- and poststimulus activity is presented in each frequency band.

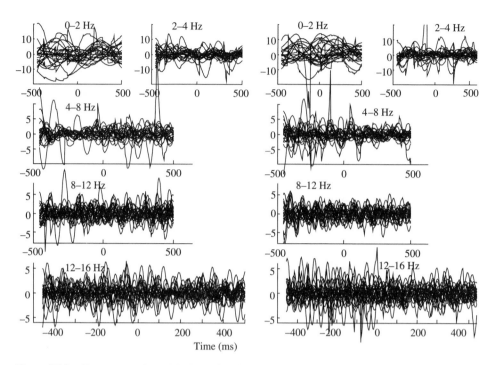

Figure 37.5   Components from 25 trials of schizophrenia subject B for all frequency bands. Upper left panel is 0–2-Hz band, upper right is 2–4 Hz. Second from top is 4–8 Hz, next to bottom is 8–12 Hz, and 12–16 Hz is at the bottom. *Left*: stimulus S_1; *Right*: stimulus S_2. Stimulus is presented at 0 ms, and 500 ms pre- and poststimulus activity is presented in each frequency band.

On the other hand, some of the schizophrenia patients behaved much like normal subjects. For example, schizophrenia subject D showed much more phase synchronization for S_1 than for S_2, as one can see from Figure 37.6.

The temporal evolution of the population averages of the Kuiper statistic κ for the S_1 and S_2 stimuli in the 2–4- and 4–8-Hz frequency bands is shown in Figure 37.7. As one can see, pre-S_1 values are close to zero in all frequency bands, indicating uniform (random) phase. The κ measure decreases in these frequency bands following presentation of S_1 stimuli, which is a sign of phase synchronization. Repeating stimuli produced κ values that were consistently larger than for S_1 stimuli. Also, the post-S_2 κ values remained close to their prestimulus values. This suggests that S_2 stimuli induced less phase synchronization. The differences between the S_1 and S_2 κ values were evaluated using t-tests. It was found that, in the normal population, the differences were significant ($p < 0.05$) from 0 through 200 ms in the 2–4-Hz band and from 40 through 200 ms in the 4–8-Hz band. The t-test results for the schizophrenia population were similar, in the sense that most differences were seen in the 2–4-Hz band (from 40 through 180 ms) and in the 4–8-Hz band (140–200 ms). It appears that the schizophrenia subjects reach maximum synchronization in response to S_1 later than the normal subjects and at a lower level.

Figure 37.8 shows that large differences in the S_1 response exist between the normal and schizophrenia population. Significant differences, as determined by a t-test, in the 2–4-Hz band were obtained between 0 and 180 ms ($p < 0.05$) and between 20 and 160 ms in the 4–8-Hz band. The healthy individuals and schizophrenia patients had very similar κ values for the S_2 responses; significant differences were seen only between 60 and 100 ms in the 4–8-Hz band.

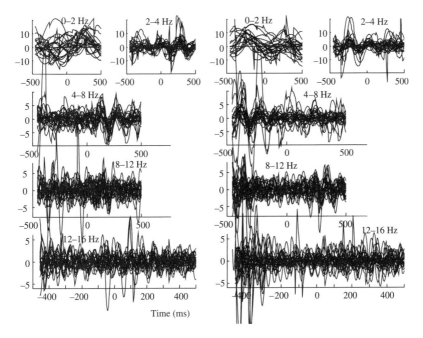

Figure 37.6   Components from 25 trials of schizophrenia subject D for all frequency bands. Upper left panel is 0–2-Hz band, upper right is 2–4 Hz. Second from top is 4–8 Hz, next to bottom is 8–12 Hz, and 12–16 Hz is at the bottom. *Left*: stimulus S_1; *Right*: stimulus S_2. Stimulus is presented at 0 ms, and 500 ms pre- and poststimulus activity is presented in each frequency band.

The multiple linear regression analysis conducted to investigate the relationship between EP amplitude and degree of phase synchronization indicated that P50 amplitude and κ_{40} were not correlated. However, a significant correlation between κ_{100} and N100 was found for the normal population ($R^2 = 0.58$, $p < 0.008$). The correlation between

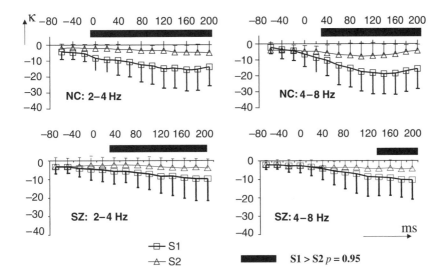

Figure 37.7   Temporal evolution of average Kuiper statistic for normal control group (NC, top) and schizophrenia patients (SZ, bottom) for S_1 and S_2 responses in 2–4-Hz (*left*) and 4–8-Hz (*right*) bands.

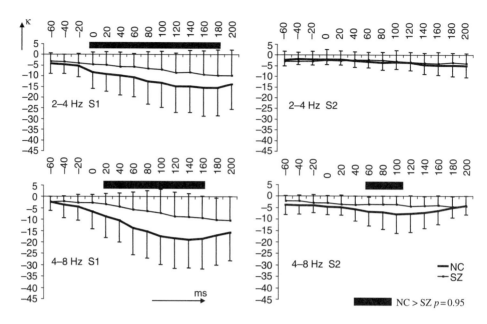

Figure 37.8 Same data as in Figure 37.7 but ordered to show population differences.

κ_{200} and P200 was significant for the normal group ($R^2 = 0.55$, $p < 0.01$) and the patient population ($R^2 = 0.52$, $p < 0.03$).

37.4 FUTURE TRENDS

Using a signal analysis tool specifically designed for single trial evoked potential analysis, we have been able to show that phase synchronization is a primary mechanism by which EP components with a latency between 50 and 250 ms are generated. We have also demonstrated that novel (S_1) stimuli produce less phase synchronization of the ongoing EEG in schizophrenia subjects than in normal subjects. No, or much smaller, population differences in the ability to phase synchronize the ongoing EEG in response to repeating (S_2) stimuli were found. These results are in concordance with earlier reports, using different methods, that the capability to phase reorganize is deficient in certain groups of psychiatric subjects as compared to normal subjects [14, 22]. Our observation that most of the phase synchronization differences are seen for S_1 responses matches prior work, suggesting that a significant part of the sensory gating abnormality can be attributed the aberrations seen in the response to S_1 stimuli [23–25]. However, while our data show that the major abnormality in schizophrenia patients is in the response to S_1 stimuli, additional deviations can also be detected in response to S_2 stimuli in the 4–12-Hz frequency range.

Recently, Makeig et al. [13] reported evidence for visual stimulus-induced phase resetting in all scalp channels and EEG frequencies below 20 Hz. Combined with our findings for auditory responses, one may conclude that the EP generation mechanism is similar across the various sensory modalities.

The mostly exogenous components accessed using the auditory double-click paradigm utilized here differ from the endogenous components such as the P300. The latter can be observed using an "oddball" paradigm which entails the presentation of a sequence of frequently and less frequently occurring stimuli ("targets"). If the subject is asked to attend to the target stimuli by, for example, pushing a button, a positive component with a 300–400-ms latency may be observed in the ensemble average of the responses to the targets. This component is not present in the nontarget stimuli. We have recently presented preliminary results, obtained using the PPM, indicating that the P300 component arises from an increase in power in the 0–4-Hz band [26]. This finding is in partial agreement with the results for visual ERPs reported in [27], where it is shown that endogenous components are generated by a different mechanism than exogenous EP components.

Future work will have to focus on identifying the neural mechanisms by which stimulus-induced phase synchronization of the spontaneous EEG is effected and why less phase resetting occurs for the repeating stimuli. Several hypotheses have been proposed. For example, an elementary neuronal mechanism within the hippocampus was proposed in [28]. This model comprises an inhibitory neuron which forms a synaptic loop between a sensory neuron and a major output neuron. The initial sensory stimulus excites both the major output neuron (pyramidal cell) and the inhibitory interneuron. Subsequent sensory stimulation would produce a diminished response if the inhibitory neuron is still active from the initial stimulation. This model was further elaborated in [29], where it was shown that recurrent activity within the CA3 region of the hippocampus could be one mechanism to cause suppression of the second response. We have recently proposed that the interaction between thalamic and cortical excitatory feedback loops and the inhibitory effect of the thalamic reticular formation is, at least in part, responsible for the smaller response to the repeating stimulus [30]. Under this scenario, specific afferent activity, associated with the novel stimulus, excites the primary sensory thalamic relay nucleus, which in turn provides excitation of the primary sensory cortex. This activity is fed back to the thalamus. The mutual excitation of thalamus and cortex results in oscillatory activity, the amplitude of which is gradually decreased by the inhibitory action of the thalamic reticular formation, which receives excitatory collaterals from the thalamocortical and corticothalamic cells. Consequently, the thalamic relay nuclei will be in a hyperpolarized state following the novel stimulus, and therefore, the gain of the thalamic relay nuclei will be reduced when the repeating stimulus arrives shortly after the first, effectively blocking the transfer of specific sensory afferent activity to the cortex.

Identifying the exact neural mechanism of stimulus-induced phase synchronization will require matching the predictions made by the various models with actual observations. Nonstationary signal analysis tools such as the one used here will be indispensable for this task.

37.5 SUMMARY

The results of a detailed investigation of the changes in the brain's electrical activity following auditory stimulation have been presented. The detailed study of single-trial responses was made possible by a newly designed signal analysis tool, the PPM. The PPM decomposes single-trial AEPs into sinusoidal, exponentially decaying/increasing components employing a curve-fitting method. The AEPs from 20 normal subjects and 19 schizophrenia patients were obtained using a double-click paradigm, in which two identical stimuli were delivered with an interstimulus interval of 500 ms and an interpair interval of 8 s. Pre- and poststimulus phase histograms were compared to determine the degree of phase synchronization produced by auditory stimulation in the two populations.

The results show that auditory stimulation results in a phase synchronization of the spontaneous EEG. Much less phase synchronization occurs to repeating stimuli than to novel ones. It was found that schizophrenia patients have a phase synchronization deficiency, as compared to a normal control group, especially in the 4–12-Hz frequency range. Several hypotheses for the neural mechanism that would give rise to stimulus-induced phase synchronization are presented as well.

ACKNOWLEDGMENTS

The research was supported in part by grants R01 MH58784 from the National Institute of Mental Health, and THECB-003652-0001-1999 from the Texas Higher Education Coordinating Board.

REFERENCES

1. CREUTZFELDT, O. AND HOUCHIN, J. Neuronal basis of EEG waves. In: RÉMOND, A., Ed., *Handbook of Electroencephalography and Clinical Neurophysiology*, Vol. 2, Part C. Amsterdam, The Netherlands: Elsevier, 1974, pp. 5–55.

2. JASPER, H. H. The ten-twenty electrode system of the International Federation. *Electroenceph. Clin. Neurophysiol.* 1958;10:371–375.

3. BRANDT, M. E., JANSEN, B. H., AND CARBONARI, J. P. Pre-stimulus spectral EEG patterns and the visual evoked response. *Electroencephal. Clin. Neurophysiol.* 1991;80:16–20.

4. JANSEN, B. H. AND BRANDT, M. E. The effect of the phase of pre-stimulus alpha activity on the averaged visual evoked response. *Electroencephal. Clin. Neurophysiol.* 1991;80:241–250.

5. BAŞAR, E. *EEG-Brain Dynamics: Relation between EEG and Brain Evoked Potentials.* Amsterdam, The Netherlands: Elsevier Biomedical Press, 1980.

6. BAŞAR, E., FLOHR, H., HAKEN, H., AND MANDELL, A. J. (eds.). *Synergetics of the Brain.* Berlin, Germany: Springer-Verlag, 1983.

7. BAŞAR, E. Are brain's sensory-evoked and event-related potentials controlled by general EEG dynamics? *Int. J. Neurosci.* 1986;29:160–161.

8. KARAKAŞ, S., ERZENGIN, Ö. U., AND BAŞAR, E. A new strategy involving multiple cognitive paradigms demonstrates that ERP components are determined by the superposition of oscillatory responses. *Clin. Neurophysiol.* 2000;111:1719–1732.

9. SAYERS, B. McA., BEAGLEY, H. A., AND HESHALL, W. R. Objective evaluation of auditory evoked EEG responses. *Nature* 1974;251:608–609.

10. GAROOSI, V. AND JANSEN, B. H. Development and evaluation of the piecewise Prony method for evoked potential analysis. *IEEE Trans. Biomed. Eng.* 2000;47:1549–1554.

11. KOLEV, V. AND YORDANOVA, J. Analysis of phase-locking is informative for studying event-related EEG activity. *Biol. Cybern.* 1997;76:229–235.

12. LASKARIS, N. A. AND IOANNIDES, A. A. Exploratory data analysis of evoked response single trials based on minimal spanning tree. *Clin. Neurophysiol.* 2000;112:698–712.

13. MAKEIG, S., WESTERFIELD, M., JUNG, T.-P., ENGHOFF, S., TOWNSEND, J., COURCHESNE, E., AND SEJNOWSKI, T. J. Dynamic brain sources of visual evoked responses. *Science* 2002;295:690–694.

14. WINTERER, G., ZILLER, M., DORN, H., FRICK, K., MULERT, C., WUEBBEN, Y., HERRMANN, W. M., AND COPPOLA, R. Schizophrenia: Reduced signal-to-noise ratio and impaired phase-locking during information processing. *Clin. Neurophysiol.* 2000;111:837–849.

15. JANSEN, B. H., AGARWAL, G., HEGDE, A., AND BOUTROS, N. N. Phase synchronization of the ongoing EEG and auditory EP generation. *Clin. Neurophysiol.* 2003;114:79–85.

16. BOUTROS, N. N., RUSTIN, T., ZOURIDAKIS, G., PEABODY, C. A., AND WARNER, M. D. The P50 auditory evoked responses and subtypes of schizophrenia. *Psychiatry Res.* 1993;7:243–254.

17. BUCHSBAUM, M. S. The middle evoked response components and schizophrenia. *Schizophr. Bull.* 1977;3:93–104.

18. ADLER, L. E., PACHTMAN, E., FRANKS, R. D., PECEVICH, M., WALDO, M. C., AND FREEDMAN, R. Neurophysiological evidence for a defect in neuronal mechanisms involved in sensory gating in schizophrenia. *Biol. Psychiatry* 1982;17:639–654.

19. SIEGEL, C., WALDO, M., MIZNER, G., ADLER, L. E., AND FREEDMAN, R. Deficits in sensory gating in schizophrenic patients and their relatives. *Arch. Gen. Psychiatry* 1984;41:607–612.

20. FISHER, N. I. *Statistical Analysis of Circular Data.* Cambridge, Great Britain: Cambridge University Press, 1993.

21. PERNEGER, T. V. What's wrong with Bonferroni adjustments. *Br. Med. J.* 1998;316:1236–1238.

22. ZOURIDAKIS, C., JANSEN, B. H., AND BOUTROS, N. N. A fuzzy clustering approach to study the auditory P50 component in schizophrenia. *Psychiatry Res.* 1997;69:169–181.

23. BLUMENFELD, L. AND CLEMENTZ, B. A. Response to the first stimulus determines reduced auditory evoked response suppression in schizophrenia: Single trials analysis using MEG. *Clin. Neurophysiol.* 2001;112: 1650–1659.

24. JIN, Y., POTKIN, S. G., PATTERSON, J. V., SANDMAN, C. A., HETRICK, W. P., AND BUNNEY, W. E. Effects of P50 temporal variability on sensory gating in schizophrenia. *Psychiatry Res.* 1997;70:71–81.

25. PATTERSON, J. V., JIN, Y., GIERCZAK, M., HETRICH, W. P., POTKIN, S., BUNNEY, JR., W. E., AND SANDMAN, C. A. Effects of temporal variability on P50 and the gating ratio in schizophrenia. *Arch. Gen. Psychiatry* 2000;57:57–64.

26. JANSEN, B. H., HEGDE, A., RUBEN, J., DOSHI, H., AND BOUTROS, N. N. Non-stationary signal analysis tools to study evoked potential generation. *Clin. Electroencephal.* 2002;33:148.

27. DEMIRALP, T., ADEMOGLU, A., COMERCHERO, M., AND POLICH, J. Wavelet analysis of P3a and P3b. *Brain Topogr.* 2001;13:251–267.

28. FREEDMAN, R., WALDO, M., Bickford-WINNER, P., AND NAGAMOTO, H. Elementary neuronal dysfunction in schizophrenia. *Schizophr. Res.* 1991;4:233–243.

29. FLACH, K. A., ADLER, L. E., GERHARDT, G. A., MILLER, C., BICKFORD, P., AND MACGREGOR, R. J. Sensory gating in a computer model of the CA3 neural network of the hippocampus. *Biol. Psychiatry* 1996;40:1230–1245.

30. LIU, C., JANSEN, B. H., AND BOUTROS, N. N. Modeling of auditory evoked potentials. Paper presented at Second Joint EMBS-BMES Conference, Houston, Oct. 23–26, 2002.

CHAPTER 38

PROBING OSCILLATORY VISUAL DYNAMICS AT THE PERCEPTUAL LEVEL

H. Fotowat, H. Öğmen, H. E. Bedell, and B. G. Breitmeyer

38.1 INTRODUCTION

The goals of neuroengineering are twofold. First, it aims to develop and apply engineering know-how to problems in neuroscience, that is, a "forward-engineering" approach tailored to neuroscience. Second, it aims to "reverse engineer" nervous systems in order to develop novel engineering concepts and techniques. These concepts and techniques can, in turn, be used in forward neuroengineering as well as in the design of biomimetic systems. In this chapter we present research aimed at reverse engineering the human visual system. Light impinging on the photoreceptors is transformed into neural activity which propagates to various parts of the nervous system. The nervous activity in turn gives rise to our subjective experience, that is, our perception of the environment. Studies of brain and mind are conducted at different levels. At the neurophysiological level, the focus is on the relationship between stimuli and neural activity. At the outer psychophysical level, the focus is on the relationship between stimuli and percepts. To complete the loop, a relationship between neural activity and percepts needs to be established; this is the goal of inner psychophysics (see Fig. 38.1). While there have been extensive studies of the primate visual system at the neurophysiological and at the outer psychophysical levels, our understanding of the relationship between neural activities and percepts has been very limited. For example, high-frequency oscillatory neural activities have been observed in the visual system of several species for about a hundred years [e.g., 1, 2], but their functional role in vision, if any, remains one of the fundamental questions in cognitive neuroscience.

It has been theorized that high-frequency oscillatory activity can form the neural basis of visual awareness, focused arousal, attention, and feature binding [e.g., 3–10]. The validation of these proposals requires evidence from the perceptual level so as to establish a causal relationship between neurophysiological oscillations and perceptual states. However, psychophysical tests of these proposals produced discrepant results [11–17]. These studies examined the effect of adding high-frequency components to the stimulus itself. It was reasoned that neural oscillations entrained by these high-frequency input components would enhance the functional role played by the oscillatory activity at the corresponding frequency. A tacit assumption behind this rationale is a functional equivalence between externally driven, that is, entrained, oscillations and intrinsic oscillations. However, as early as the beginning of the twentieth century, a distinction

Handbook of Neural Engineering. Edited by Metin Akay
Copyright © 2007 The Institute of Electrical and Electronics Engineers, Inc.

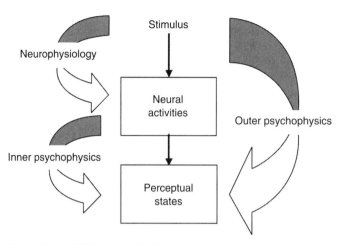

Figure 38.1   Different levels of investigation in probing workings of perceptual systems.

between rhythmic responses to nonrhythmic stimuli (intrinsic oscillations) and rhythmic responses to rhythmic stimuli (entrained oscillations) was made (Fröhlich, cited in [18]). More recent research showed that the visual system possesses parallel pathways that differ in their sensitivity to stimulus characteristics. For example, the magnocellular (or M) pathway responds more strongly to stimuli of lower spatial frequencies and higher temporal frequencies and the parvocellular (or P) pathway exhibits a complementary behavior [19]. Accordingly, the intrinsic oscillations generated by cortical cells will be accompanied by a different pattern of activity in the afferent pathways depending on whether one uses a rhythmic or nonrhythmic stimulus. It is therefore highly desirable to measure perceptual correlates of intrinsic oscillations generated by nonrhythmic stimuli. The human visual system integrates signals over space and time [20] and thus acts as a "low-pass filter," making it highly challenging to observe high-frequency activities at the psychophysical level. Recently, based on an analysis of a computational model of retino-cortical dynamics (RECOD) [21], we designed psychophysical experiments that provided evidence for perceptual correlates of oscillatory neural activity [22, 23]. In this chapter, we will introduce our approach and use it to examine how the perceptually measured oscillatory dynamics depend on stimulus and background luminance. Overall, this research provides an outer psychophysical technique which, when combined with neurophysiological recordings, can serve as a basis for inner psychophysical studies.

38.2   THEORETICAL BASIS OF STUDY

The general structure of the RECOD model is shown in Figure 38.2. The lower two ellipses represent retinal ganglion cells with fast-phasic (transient) responses and those with slower tonic (sustained) responses [24–27]. These populations of retinal ganglion cells project to distinct layers of the lateral geniculate nucleus (LGN) forming two parallel afferent pathways, called the magnocellular and parvocellular pathways.

The M and P pathways form the primary inputs of different cortical visual areas subserving various functions such as the computation of motion, form, and brightness [19, 28]. The model uses a lumped representation for the cortical targets of M and P pathways. The main cortical targets of the M pathway represent the areas that play a major role in the computation of motion and temporal change. The main cortical targets of the P pathway

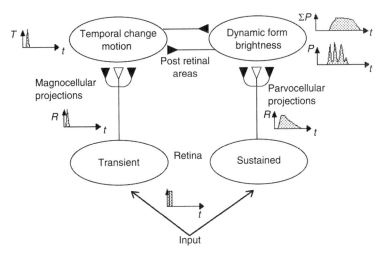

Figure 38.2   General architecture of RECOD model. Filled and open synaptic symbols represent inhibitory and excitatory connections, respectively.

represent the areas that play a major role in the computation of dynamic form and brightness (upper ellipses in Fig. 38.2). In terms of cortical interactions between these pathways, the model postulates reciprocal inhibition, as shown by the connections between the upper ellipses in the figure. The lumped representation for the areas involved in the computation of dynamic form and brightness contains recurrent connections (not shown in Fig. 38.2) to represent the extensive feedback observed in postretinal areas. This recurrent circuit produces oscillatory responses (denoted by P in the figure) that represent the neurophysiologically observed oscillations [e.g., 29–35]. To link neural responses to visual perception, the model postulates that the oscillatory activity is temporally integrated (ΣP in Fig. 38.2) according to the psychophysically determined temporal integration characteristics of the human visual system and that this temporally integrated activity underlies the perception of brightness and form. The mathematical equations of the model can be found in [22].

In the metacontrast masking paradigm [36–38], the visibility of a target stimulus is measured in the presence of a spatially nonoverlapping mask stimulus. The mask stimulus is turned on after the target stimulus with a delay between the onset of mask and target, called the stimulus onset asynchrony (SOA). An analysis of the RECOD model in the context of metacontrast masking produced a novel prediction, namely that spatiotemporally localized suprathreshold stimuli (briefly flashed dots) should produce a new type of metacontrast masking function that exhibits oscillations [22].

The prediction of the model is illustrated in Figure 38.3. The plots at the top of Figure 38.3 depict the postretinal activities that the target (T) and mask (M) would generate if they were presented in isolation. The filled triangle depicts the fast transient response in postretinal areas driven by the M pathway and the sawtooth shaped activity depicts the oscillatory response in postretinal areas driven by the P pathway. The arrows between the transient and oscillatory activities depict the mutual inhibition between transient and sustained postretinal neural populations. When the SOA is such that the mask-generated transient activity overlaps in time with the bursting phase of the target-generated oscillatory activity (SOA_1 on the top left), this burst of activity will be suppressed due to the inhibition exerted by the transient activity. This in turn will cause a decrease in the integrated activity generated by the target. As a result, the visibility

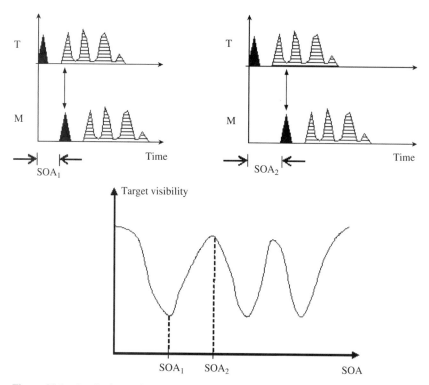

Figure 38.3   Predictions of RECOD model for spatiotemporally localized stimuli in metacontrast masking. The top two panels depict the postretinal activities that the target (T) and mask (M) stimuli would generate if they were presented in isolation. The arrows between transient (filled triangle) and sustained oscillatory (sawtooth-shaped activity) activities illustrate mutual inhibition between transient and sustained oscillatory activities. The bottom panel depicts the metacontrast masking function.

of the target will decrease for SOA_1, as shown in the metacontrast function at the bottom of Figure 38.3. When the SOA is such that the mask-generated transient activity occurs between two bursting phases of the target-generated oscillatory activity (SOA_2 on the top right), inhibition will not significantly suppress the oscillatory activity in the sustained channel. This will result in a relatively high level of perceived target brightness, as shown in the metacontrast function at the bottom of Figure 38.3. When a range of SOA values is considered, one can see that the shape of the metacontrast function will reflect the shape of the oscillatory activity. Psychophysical experiments and model simulations provided support for this prediction [22, 23]. Earlier observations [39, 40] suggested that oscillatory activities should be prominent when the stimulus detectability is high. In this chapter, we extend our previous study to examine the dependence of oscillatory metacontrast functions on stimulus and background luminance.

38.3   METHODS

38.3.1   Materials and Procedures

Stimuli consisted of three horizontally aligned dots on a uniform background field. The dot stimuli were generated by ultrabright light-emitting diodes (LEDs) and were viewed through an optical apparatus designed to superimpose a uniform background on the light emitted by the LEDs and to optically reduce the retinal size of the LEDs to 1 arcmin diameter. The horizontal separation between adjacent LEDs was 9 arcmin. The luminance of each LED was calibrated by a Minolta LS-110 photometer. The central LED served as the fixation stimulus. It was briefly turned off (200 ms) before the presentation of the target–mask pair in each experimental trial to prevent its interactions with the target–mask stimuli. The other two LEDs served as the target and mask stimuli. They were turned on sequentially for a duration of 10 ms each and separated by the SOA value. To prevent predictive eye movements, for each trial the target LED was randomly selected to be either the left or the right LED. After this target–mask sequence, the central LED was turned on as a match stimulus. The task of the observer was to indicate which of the two, the target or the match stimulus, appeared brighter. The luminance of the match stimulus was modified during the experiment using a double-staircase procedure to converge to the point of subjective equality (PSE). All experimental methods were identical to those in [22] with the following exceptions: The sampling period of the SOA was 1 ms. Seven overlapping SOA intervals ($<$30,40$>$; $<$35,50$>$; $<$45,60$>$; $<$55,70$>$; $<$65,80$>$; $<$75,90$>$; $<$85,100$>$) were run in different sessions. There were four experimental conditions corresponding to target and background luminance pairs of: C1: 2.5 log units (LU) above detection threshold, 13.5 cd/m^2; C2: 2.5 LU, 0.5 cd/m^2; C3: 0.5 LU, 0.5 cd/m^2; C4: 0.5 LU, 13.5 cd/m^2. One of the authors and two naïve observers participated in the experiments.

38.3.2 Data Processing

The metacontrast function spanning the SOA values from 30 to 100 ms was constructed by combining the data from the seven intervals. The overlapping portions of the intervals were used to estimate the intrinsic variations in neural response timing. To compensate for these intrinsic variations, the different SOA segments were shifted in time until the overlapping portions of the data produced a minimal normalized-squared-error, GoSF, across all test intervals. The error function was computed as

$$\text{GoSF}_k = \frac{1}{N} \sum_{i=1}^{N} (dp_{1i} - dp_{2i})^2$$

where N is the number of overlapping peaks and dips through the full range of the metacontrast masking function, dp_{1i} and dp_{2i} are the values of the ith dip or peak within overlapping intervals, and the index k corresponds to the kth shift combination. In order to test that this compensation process did not create artifactual structure in the metacontrast function, we calculated the normalized squared difference of the overlapping metacontrast masking data points,

$$\text{GoFF} = \frac{1}{N_t} \sum_{i=1}^{N_t} (m_{1i} - m_{2i})^2$$

where GoFF stands for goodness-of-fit factor, N_t is the total number of overlapping metacontrast data points, and m_{1i} and m_{2i} are the values of the metacontrast function at each point i in the overlapping intervals. The comparison of this value to the distribution of goodness-of-fit factors obtained from the randomly scrambled versions of the data showed that the result produced by our compensation procedure was statistically significant. Finally, the resulting metacontrast function was filtered by a zero-phase low-pass filter with a cutoff frequency of 200 Hz.

38.4 RESULTS

As in our previous study [22], for a given experimental condition, the magnitude of the masking function was variable across observers but consistent within each observer. The observed frequency of oscillations was higher in this study than our previous study [22], which is attributable to higher temporal sampling. Examples of a relatively strong and a relatively weak masking function are shown in Figures 38.4 and 38.5.

In Figure 38.4, the magnitude of masking is about 1 LU. In Figure 38.5, the magnitude of masking is about 0.1 LU. Because different observers and experimental conditions

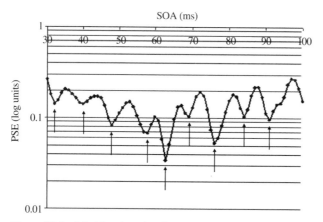

Figure 38.4 Masking function for observer HF, C1 (2.5-LU target on 13.5-cd/m^2 background). The dips are indicated by arrows.

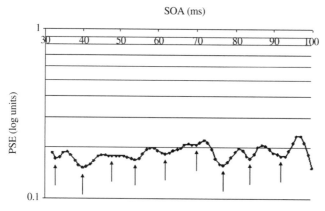

Figure 38.5 Masking function for observer JR, C3 (0.5-LU target on 0.5-cd/m^2 background). The dips are indicated by arrows.

produced different magnitudes of the masking functions, in order to compare the pattern of oscillations across observers and conditions, we plotted the timing of the dips in the masking function as a function of the dip number.

Figure 38.6 shows the dip timings for C1. The plot shows a very good agreement across the observers in the locations of early dips (1–3). As the dip number increases, the locations of the dips become more variable across the observers. An inspection of Figure 38.7 reveals a similar finding for C2. However, the variability in this case is slightly higher than that for C1.

The location and distribution of the dips in C3 (Fig. 38.8) are similar to C2, suggesting that under mesopic viewing conditions target visibility does not considerably affect the patterning of the dips. However, comparing C1 and C4 (Figs. 38.6 and 38.9) shows that under photopic viewing conditions target visibility plays a major role in the patterning of dips. As can be seen in Figure 38.9, the timing of the dips is highly scattered among the three observers in C4.

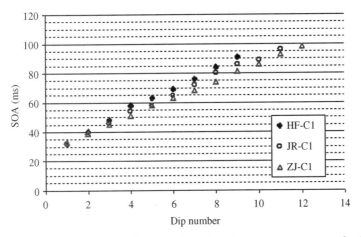

Figure 38.6   Timing of dips in masking function for three observers for C1 (2.5-LU target on 13.5-cd/m^2 background).

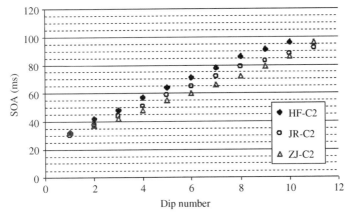

Figure 38.7   Timing of dips in masking function for three observers for C2 (2.5-LU target on 0.5-cd/m^2 background).

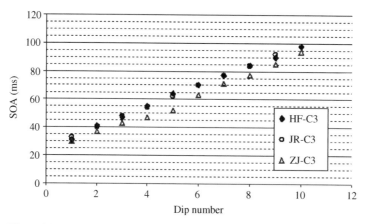

Figure 38.8   Timing of dips in masking function for three observers for C3 (0.5-LU target on 0.5-cd/m^2 background).

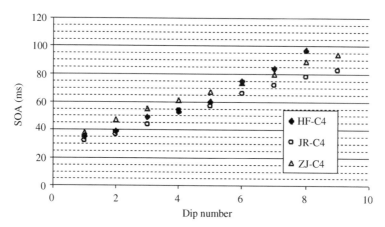

Figure 38.9   Timing of dips in masking function for three observers for C4 (0.5-LU target on 13.5-cd/m^2 background).

38.5   DISCUSSION

Without the occurrence of a contrast polarity reversal in the perception of the target, the maximum possible masking magnitude is equal to the visibility of the target, namely, 2.5 LU in C1 and C2 and 0.5 LU in C3 and C4. As a result, the dynamic range to observe oscillatory masking functions becomes more limited as the target luminance decreases. Indeed, previous observations suggest that oscillatory perceptual phenomena are more salient at high luminance levels [39, 40]. Our results show that while the patterning of oscillations is more stable across observers when target visibility is high, one can observe a similar patterning for targets of relatively lower visibility provided that the background is dim.

We also found that the locations of early dips are more consistent across observers than the locations of later dips. A possible interpretation of this finding is that the early cortical response components arise mainly from feedforward signaling and relate more directly to exogenous variables. On the other hand, late-response components are likely to carry information conveyed by feedback signaling and thus are more likely to represent endogenous variables, such as attention.

38.6 FUTURE TRENDS

Understanding the internal codes used by the nervous system is one of the most fundamental aspects of neuroscience and neuroengineering. As illustrated in Figure 38.1, this requires investigations at different levels so as to establish causal relationships between stimuli, neural responses, and perceptual and cognitive states. An ubiquitous aspect of neural dynamics is the oscillatory character of neuronal activity. The oscillations are particularly prominent in recordings at a large population level, such as those obtained by electroencephalograms (EEGs) [1, 2]. Understanding whether and how the oscillatory dynamics relate to perceptual/cognitive states has been a challenging puzzle in neuroscience. Are synchronized oscillatory firings used as a code to bind different aspects, such as color, motion, and shape, of the same object [e.g., 3, 4, 6, 7]? Or, are they simply a byproduct of the combined positive- and negative-feedback loops found in the cortical neurocircuitry? From the engineering perspective, positive feedback is generally avoided in analog systems; however, the complexity of information processing carried out by the nervous systems seems to necessitate extensive positive-feedback signaling. Given the observation that signal timing varies significantly among different cortical areas [41–43], a combination of positive feedback and delays is likely to generate oscillatory activities.

Accordingly, one hypothesis is that the oscillatory responses are a byproduct of the cortical circuitry and are "filtered" through a temporal integration process at later stages, as illustrated in Figure 38.2. Given the outer psychophysical results presented in this chapter and previous work [22, 23], insights at the inner psychophysical level can be gained by combining the psychophysical method outlined here with neurophysiological recordings. For example, changes in perceived brightness of the target in metacontrast have been shown to have correlated changes in EEG waves [44, 45]. Thus, oscillatory metacontrast masking functions may be probed by a combination of perceptual and EEG techniques. Previous studies also reported a correlation between neuronal firing and psychophysically determined metacontrast functions in monkeys [46]. Accordingly, a complementary approach would be to combine perceptual and intracerebral recordings in the monkey visual system. Finally, EEG and masking data provide converging evidence that schizophrenic patients show deficits in generating or maintaining oscillatory neural responses [47]. Thus, the technique discussed in this chapter may also be useful in clinical applications.

ACKNOWLEDGMENTS

We thank Saumil Patel for help on the experimental setup and Ben Jansen for comments. This research is supported by National Institutes of Health Grants R01-MH49892, R01-EY05068, and P30-EY07551.

REFERENCES

1. BAŞAR, E. AND BULLOCK, T. H. (1992). *Induced Rhythms in the Brain*. Boston: Birkhäuser.

2. BAŞAR, E. (1998). *Brain Function and Oscillations*. Berlin: Springer.

3. MILNER, P. M. (1974). A model for visual shape recognition. *Psychol. Rev.*, 81, 521–535.

4. VON DER MALSBURG, C. (1981). The correlation theory of brain function. Internal report 81-2, Max Planck Institute for Biophysical Chemistry, Göttingen.

5. SHEER, D. E. (1984). Focused arousal, 40-Hz EEG, and dysfunction. In: ELBERT, T., ROCKSTROH, B., LUTZENBERGER, W., AND BIRBAUMER, N. (Eds.),

Self-Regulation of the Brain and Behavior. Berlin: Springer, pp. 64–84.

6. ECKHORN, R., BAUER, R., JORDAN, W., BROSCH, M., KRUSE, W., MUNK, M., AND REITBOECK, H. J. (1988). Coherent oscillations: A mechanism of feature linking in the visual cortex? *Biol. Cybern.*, 60, 121–130.

7. GRAY, C. M., AND SINGER, W. (1989). Stimulus-specific neuronal oscillations in orientation columns of cat visual cortex. *Proc. Natl. Acad. Sci. USA*, 86, 1698–1702.

8. CRICK, F. AND KOCH, C. (1990). Towards a neurobiological theory of consciousness. *Semin. Neurosci.*, 2, 263–275.

9. LLINAS, R. R. (1992). Oscillations in CNS neurons: a possible vote for cortical interneurons in the generation of 40 Hz oscillations. In: E. BAŞAR AND T. H. BULLOCK (Eds.) *Induced Rhythms in the Brain.* Boston: Quine-Woodbine, vol. 4, pp. 269–286.

10. ENGEL, A. K., FRIES, P., KÖNIG, P. BRECHT, M., AND SINGER, W. (1999). Temporal binding, binocular rivalry and consciousness. *Consciousness and Cognition*, 8, 128–151.

11. FAHLE, M. AND KOCH, C. (1995). Spatial displacement, but not temporal asynchrony, destroys figural binding. *Vision Res.*, 35, 491–494.

12. KIPER, D. C., GEGENFURTNER, K. R., AND MOVSHON, J. A. (1996). Cortical oscillatory responses do not affect visual segmentation. *Vision Res.*, 36, 539–544.

13. LEONARDS, U., SINGER, W., AND FAHLE, M. (1996). The influence of temporal phase differences on texture segmentation. *Vision Res.*, 17, 2689–2697.

14. BLAKE, R. AND YANG, Y. (1997). Spatial and temporal coherence in perceptual binding. *Proc. Natl. Acad. Sci. USA*, 94, 7115–7119.

15. ELLIOTT, M. A. AND MÜLLER, H. J. (1998). Synchronous information presented in 40-Hz flicker enhances visual feature binding. *Psychol. Sci.*, 9, 277–283.

16. ALAIS, D., BLAKE, R., AND LEE, S.-H. (1998). Visual features that vary together over time group together over space. *Nature Neurosci.*, 1, 160–164.

17. USHER, M. AND DONNELLY, N. (1998). Visual synchrony affects binding and segmentation in perception. *Nature*, 394, 179–182.

18. ANSBACHER, H. L. (1944). Distortion in the perception of real movement. *J. Exp. Psychol.*, 34, 1–23.

19. Van ESSEN, D. C., ANDERSON, C. H., AND FELLEMAN, D. J. (1992). Information processing in the primate visual system: An integrated systems perspective. *Science*, 255, 419–423.

20. DAVSON, H. (1990). *Physiology of the Eye*, 5th ed. New York: Pergamon.

21. ÖĞMEN, H. (1993). A neural theory of retino-cortical dynamics. *Neural Networks*, 6, 245–273.

22. PURUSHOTHAMAN, G., ÖĞMEN, H., AND BEDELL, H. E. (2000). Gamma-range oscillations in backward-masking functions and their putative neural correlates. *Psychol. Rev.*, 107, 556–577.

23. PURUSHOTHAMAN, G., ÖĞMEN, H., AND BEDELL, H. E. (2003). Supra-threshold intrinsic dynamics of the human visual system. *Neural Comput.*, 15, 2883–2908.

24. GOURAS, P. (1968). Identification of cone mechanisms in monkey ganglion cells. *J. Physiol.*, 199, 533–547.

25. DE MONASTERIO, F. M. AND GOURAS, P. (1975). Functional properties of ganglion cells of the rhesus monkey retina. *J. Physiol.*, 251, 167–195.

26. CRONER, L. J. AND KAPLAN, E. (1995). Receptive fields of P and M ganglion cells across the primate retina. *Vision Res.*, 35, 7–24.

27. KAPLAN, E. AND BENARDETE, E. (2001). The dynamics of primate retinal ganglion cells. *Prog. Brain Res.*, 134, 1–19.

28. LIVINGSTONE, M. AND HUBEL, D. (1988). Segregation of form, color, movement, and dept: Anatomy, physiology, and perception. *Science*, 240, 740–749.

29. DOTY, R. W. AND KIMURA, D. S. (1963). Oscillatory potentials in the visual system of cats and monkeys. *J. Physiol.*, 168, 205–218.

30. MAUNSELL, J. H. R. AND GIBSON, J. R. (1992). Visual response latencies in striate cortex of the macaque monkey. *J. Neurophysiol.*, 68, 1332–1344.

31. LIVINGSTONE, M. S. (1996). Oscillatory firing and interneuronal correlations in squirrel monkey striate cortex. *J. Neurophysiol.*, 75, 2467–2485.

32. Friedman-HILL, S., MALDONADO, P. E., AND GRAY, C. M. (2000). Dynamics of striate cortical activity in the alert macaque: I. Incidence and stimulus-dependence of gamma-band neuronal oscillations. *Cerebral Cortex*, 10, 1105–1116.

33. MALDONADO, P. E., FRIEDMAN-HILL, S., AND GRAY, C. M. (2000). Dynamics of striate cortical activity in the alert macaque: II. Fast time scale synchronization. *Cereb. Cortex*, 10, 1117–1131.

34. FRIES, P., NEUENSCHWANDER, S., ENGEL, A. K., GOEBEL, R., AND SINGER, W. (2001). Rapid feature selective neuronal synchronization through correlated latency shifting. *Nature Neuroscience*, 4, 194–200.

35. ROLS, G., TALLON-BAUDRY, C., GIRARD, P., BERTRAND, O., AND BULLIER, J. (2001). Cortical mapping of gamma oscillations in areas V1 and V4 of the macaque monkey. *Vis. Neurosci.*, 18, 527–540.

36. BREITMEYER, B. G. (1984). *Visual Masking: An Integrative Approach.* New York: Oxford University Press.

37. BACHMANN, T. (1994). *Psychophysiology of Visual Masking: The Fine Structure of Conscious Experience.* New York: Nova Science Publishers.

38. BREITMEYER, B. G. AND ÖĞMEN, H. (2000). Recent models and findings in visual backward masking: A comparison, review, and update. *Percept. Psychophys.*, 62, 1572–1595.

39. BIDWELL, S. (1899). *Curiosities of Light and Sight.* London: Swan Sonnenschein and Co.

40. McDOUGALL, W. (1904). The sensations excited by a single momentary stimulation of the eye. *Br. J. Psychol.*, 1, 78–113.

41. NOWAK, L. G. AND BULLIER, J. (1998). The timing of information transfer in the visual system. *Cereb. Cortex*, 12, 205–241.

42. SCHMOLESKY, M. T., WANG, Y., HANES, D. P., THOMPSON, K. G., LEUTGEB, S., SCHALL, J. D., AND LEVENTHAL, A. G. (1998). Signal timing across the macaque visual system. *J. Neurophysiol.*, 79, 3272–3278.

43. MAUNSELL, J. H. R., GHOSE, G. M., ASSAD, J. A., MCADAMS, C. J., BOUDREAU, C. E., AND NOERAGER, B. D. (1999). Visual response latencies of magnocellular and parvocellular LGN neurons in macaque monkeys. *Vis. Neurosci.*, 16, 1–14.

44. VAUGHAN, H. G. AND SILVERSTEIN, L. (1968). Metacontrast and evoked potentials: A reappraisal. *Science*, 160, 207–208.

45. BRIDGEMAN, B. (1988). Visual evoked potentials: Concomitants of metacontrast in late components. *Percept. Psychophys.*, 43, 401–403.

46. BRIDGEMAN, B. (1980). Temporal response characteristics of cells in monkey striate cortex measured with metacontrast masking and brightness discrimination. *Brain Res.*, 196, 347–364.

47. GREEN, M. F., MINTZ, J., SALVESON, D., NUECHTERLEIN, K. H., BREITMEYER, B., LIGHT, G. A., AND BRAFF, D. L. (2003). Visual masking as a probe for abnormal gamma range activity in schizophrenia. *Biol. Psychiatry*, 53, 1113–1119.

CHAPTER 39

NONLINEAR APPROACHES TO LEARNING AND MEMORY

Klaus Lehnertz

39.1 INTRODUCTION

The electroencephalogram (EEG) reflects brain electrical activity owing to both intrinsic dynamics and responses to external or internal stimuli. To examine pathways and time courses of information processing under specific conditions, several experiments have been developed controlling sensory inputs and/or higher cognitive functions. Depending on the applied paradigm, EEG time series recorded during these experiments can be classified as *triggered* and *nontriggered* (cf. Fig. 39.1).

In order to detect and extract relevant information from these EEG time series, a variety of analysis techniques that operate in the time or frequency domain are usually applied [1]. With the advent of the theory of nonlinear deterministic dynamics [2] new concepts and powerful algorithms were developed to analyze complex behavior, a distinctive feature of the EEG. When applied to EEG signals these *nonlinear time-series analysis techniques* [3] were shown to allow reliable characterization of normal and pathological brain function [4–7]. Moreover, findings presented here indicate that nonlinear approaches may also improve understanding of neuronal processes underlying learning and memory in the human brain.

39.2 NONLINEAR ANALYSES OF TRIGGERED EEG

Repeated presentation of well-defined stimuli is assumed to induce synchronized neural activity in specific regions of the brain, occurring as voltage changes in the EEG. These *evoked potentials* (EPs) often exhibit multiphasic peak amplitudes within the first hundred milliseconds after stimulus onset. They are specific for different stages of information processing, thus giving access to both temporal and spatial aspects of neural processes. Other classes of experimental setups are used to investigate higher cognitive functions. For example, subjects are required to remember words or they are asked to respond to specific target stimuli, for example, by pressing a button upon their occurrence. The neural activity induced by this kind of stimulation also leads to potential changes in the EEG. These *event-related potentials* (ERPs) can extend over a few seconds, exhibiting peak amplitudes mostly later than EPs (for an overview on event-related synchronization and desynchronization see [8]).

When compared to the ongoing EEG, EP and ERP possess very low peak amplitudes which, in most cases, are not recognizable by visual inspection. Thus, to improve their low

Handbook of Neural Engineering. Edited by Metin Akay
Copyright © 2007 The Institute of Electrical and Electronics Engineers, Inc.

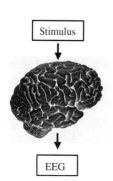

Stimulus

Triggered EEG activity:
- Time locked to each stimulus
- Time scale: msec–sec
- Evoked potentials (EP)
- Event-related potentials (ERPs)
- Event-related synchronization (ERS)
- Event-related desynchronization (ERD)

Nontriggered EEG activity:
- Related to specific task
- Time scale: task dependent (sec–min)
- Ongoing activity
- Characterization via dynamical measures

EEG

Figure 39.1   External or internal stimuli induce characteristic changes in the EEG that can be divided into the classes triggered (timelocked to each stimulus) and nontriggered (related to a given task).

signal-to-noise ratio, EPs/ERPs are commonly averaged, assuming synchronous, time-locked responses not correlated with the ongoing EEG. Let $s_i(t)$, $i = 1, \ldots, N$ (where N denotes the number of presented stimuli), be the measured stationary EEG time series consisting of an additive superposition of the a priori unknown EP/ERP [denoted as $x_i(t)$] and some background activity $\eta_i(t)$. Averaging will reduce the amplitude of $\eta_i(t)$ by N while the amplitude of $x_i(t)$ will be reduced by $\sqrt{(N)}$ only. Apart from the fact that a large number of trials N might be necessary, these assumptions may be inaccurate and, as a result of averaging, variations of EP/ERP latencies and amplitudes are not accessed. In particular, short-lasting alterations which may provide relevant information about cognitive functions are probably smoothed or even masked by the averaging process. Therefore, investigators are interested in a *single-trial analysis* that allows extraction of reliable signal characteristics out of single EP/ERP sequences.

Various signal processing techniques have been developed to perform single-trial analyses (see [9] for an overview). Among these, wavelet-based methods are particularly suitable because the wavelet transform provides several useful properties for the analysis of transient and time-variant signals [10, 11]:

- Wavelets are local, which makes most coefficient-based algorithms naturally adapted to inhomogeneities.
- They can represent smooth signals as well as singularities.
- They can represent a variety of functions.

For these reasons, wavelets have become a popular tool for the analysis of brain electrical activity [12], especially for denoising and classification of single-trial ERPs [13–18].

We have recently exploited geometrical properties of short and transient signals like ERP in state space for nonlinear de-noising by wavelets [19, 20]. This method combines the wavelet transform with noise reduction techniques originally developed for a characterization of nonlinear deterministic systems [21]. With the latter techniques it is assumed that the dynamics of the clean signal is confined to some low-dimensional manifold in state space (cf. Fig. 39.2a) which is blurred due to the noise. A reconstruction of this manifold from scalar time series $\{s_t\}$, where $t = 1, \ldots, T$, can be achieved by means of delay coordinates $\vec{s}_i = (s_i, s_{i-\tau}, \ldots, s_{i-(m-1)\tau})$ with $i = 1, \ldots, M = T - m\tau$, where $\{\vec{s}_i\}$ defines the reconstructed state space trajectory (cf. [3] for details). Here τ is

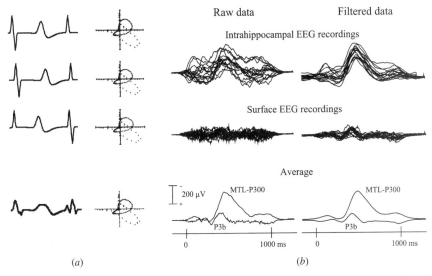

Figure 39.2 (*a*) Similarity in state space: three exemplary artificial signals using time-domain representation (left) and corresponding two-dimensional state space plots (right) reconstructed by time delay embedding. The last row depicts the respective average. (*b*) Nonlinear denoising in state space: raw and denoised single trials of P300 potentials recorded simultaneously from within the hippocampal formation and from the scalp from an epilepsy patient undergoing presurgical evaluation. The last row depicts the respective averages. (Modified from [20].)

a time delay and m is the embedding dimension which, according to Whitney's theorem, must be chosen as $m \geq 2d + 1$ (where d is the dimension of the manifold formed by the genuine trajectory in state space) if any exact determinism present in the original system is to be preserved. Denoising can be achieved by a local projection onto this manifold using singular-value decomposition (SVD). However, estimating relevant signal components using the inflexion of ordered eigenvalues is not always applicable to EEG because eigenvalues may decay almost linearly. In this case, an a priori restriction to a fixed embedding dimension m is in need, running the risk either to discard important signal components or to remain noise of considerable amplitude if only little is known about the signal. In addition, SVD stresses the direction of highest variances, so that transient signal components may be smoothed by projection. Finally, the number of signal-related directions in state space may alter locally, which is also not concerned by SVD.

To circumvent these problems we calculate wavelet transforms of delay vectors $\{\vec{s}_i\}$ instead and determine signal-related components by estimating variances separately for each state space direction. Scaling properties of wavelet bases allow very fast calculation as well as focusing on specific frequency bands. We applied our method to ERP-like test signals contaminated with white and isospectral noise [19] as well as to ERPs that were recorded intra- and extracranially in epilepsy patients during a number of different paradigms investigating higher cognitive functions [19, 20, 22] and achieved convincing results (cf. Fig. 39.2*b*).

39.3 NONLINEAR ANALYSES OF NONTRIGGERED EEG

Another approach consists of analyzing the *ongoing, nontriggered* EEG recorded during specific neuropsychological tasks that control sensory inputs and/or higher cognitive functions [23]. In this case EEG time series comprise both potentials that are and others that are not time locked to the events of a task. No further differentiation between these potentials is intended since a task is usually considered as an entity and any time resolution within a task is waived. Characterizing measures are extracted for the whole task and eventually compared between different tasks/conditions and/or between different locations in case of multichannel EEG recordings.

A number of studies have found power spectral changes predominantly in the alpha (8–13-Hz), theta (4–7-Hz) and more recently gamma (around 40-Hz) bands as well as changing topographical EEG structure to be related to learning and memory processes [e.g., 23–29]. Since the early 1990s these studies are complemented by new approaches invoking nonlinear time-series analysis techniques. Among others particularly estimates of an effective correlation dimension [30] have been used to characterize changes in brain electrical activity recorded from healthy subjects performing neuropsychological tasks [31–38]. These studies have shown an increased dimensional complexity of the

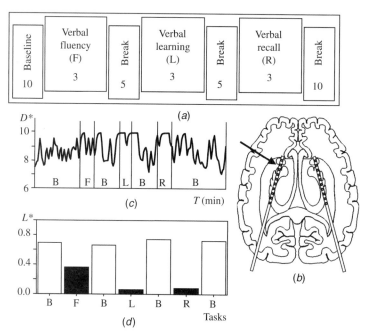

Figure 39.3 (*a*) Example of a neuropsychological test schedule: During the fluency task (F) patients are required to produce as many words as possible starting with the predefined letter; during the learning task (L) patients are required to learn 15 highly frequent words that are orally presented five times in randomized order; during the retrieval task (R) patients are required to freely recall words memorized during L. The number of correctly retrieved words during R and the absolute number of words minus perseverations produced during F serve as measures of verbal memory and fluency, respectively. Numbers denote duration (in minutes) of each task. (*b*) Schematic view of intrahippocampally placed depth electrodes. (*c*) Exemplary time course of an effective correlation dimension D^* estimated from depth EEG signals recorded from the left entorhinal cortex [marked by an arrow in (*b*)] of an epilepsy patient performing the test schedule with a high performance rate (fluency task 25 words; recall task 12 words). (*d*) Alterations of the neuronal complexity loss L^* (cf. [44] for details) during each phase of the test schedule. Low values of L^* indicate higher degrees of freedom in the competitive interactions between neuronal assemblies.

ongoing surface EEG during memory tasks, reflecting higher degrees of freedom in the competitive interactions among neuronal assemblies. However, due to a missing or only weak correlation with task performance nonlinear EEG measures are currently considered to be relatively nonspecific. This shortcoming can in part be attributed to the poor spatial resolution of surface EEG recordings to resemble electrical signals from deeper layers. This holds true particularly for mesial temporal structures that are significantly involved in declarative (episodic/semantic) memory processing [39, 40].

Intracranial EEG recordings in patients with temporal lobe epilepsy, in whom invasive recordings are indicated for medical reasons, offer a unique opportunity to circumvent this problem. Depth electrodes have to be placed in structures relevant for memory processes in both hemispheres (cf. Fig. 39.3b) and the high signal-to-noise ratio of intracranial recordings makes it possible to correlate nonlinear EEG measures during memory-related paradigms with the graded memory deficits of temporal lobe epilepsy patients [41].

We analyzed stereo-electroencephalographic data recorded intrahippocampally in patients with temporal lobe epilepsy who participated in a neuropsychological test schedule (cf. Fig. 39.3a). The schedule consisted of tasks mainly involving verbal-memory-related processes [42, 43]. Specific changes in time of the dimensional complexity (neuronal complexity loss L^* [44], cf. Figs 39.3c,d) proved a valuable parameter reflecting spatially distributed neuronal activity during verbal learning and memory processes. In line with current neurobiological models of declarative memory [39, 40], L^* allowed to reliably predict recall performance of both intentionally [45] and incidentally [46] learned verbal material. In addition, when analyzing interdependencies [47] between different brain areas, we observed specific regions of mesial temporal structures to participate in specific cognitive tasks. Particularly interhippocampal interactions turned out to be of high relevance for verbal learning and memory processes [48].

39.4 CONCLUSION

Although nonlinear time-series analysis of brain electrical activity is still at its beginning, these approaches generate new ways of viewing brain function. Results obtained so far emphasize the high value of nonlinear EEG analysis techniques to characterize neuronal activity during learning and memory processes. Recent developments including techniques taking into consideration the highly nonstationary character of brain electrical activity [49] as well as techniques disentangling subtle spatially distributed neuronal synchronization phenomena [29, 50–53] may further improve understanding of brain dynamics. Thus, nonlinear EEG analyses are expected to offer supplementary information about the integrated, coordinated activity of multiple neuronal systems underlying higher cognitive processes.

ACKNOWLEDGMENT

I gratefully acknowledge valuable discussions with and contributions of Ralph G. Andrzejak, Jochen Arnhold, Wieland Burr, Peter David, Arndt Effern, Christian E. Elger, Jürgen Fell, Guillen Fernandez, Peter Grassberger, Thomas Grunwald, Christoph Helmstaedter, Peter Klaver, Thomas Kreuz, Fernando Lopes da Silva, Florian Mormann, Christoph Rieke, Thomas Schreiber, Bruno Weber, Guido Widman, and Heinz-Gregor Wieser. This work was supported by the Deutsche Forschungsgemeinschaft.

REFERENCES

1. F. H. LOPES DA SILVA, W. STORM VAN LEEUWEN, AND A. RÉMOND, *Clinical Applications of Computer Analysis of EEG and Other Neurophysiological Signals.* Amsterdam: Elsevier, 1986.

2. E. OTT, *Chaos in Dynamical Systems.* Cambridge: Cambridge University Press, 1993.

3. H. KANTZ AND T. SCHREIBER, *Nonlinear Time Series Analysis.* Cambridge: Cambridge University Press, 1997.

4. E. BAŞAR, *Chaos in Brain Function.* Berlin: Springer, 1990.

5. D. DUKE AND W. PRITCHARD, *Measuring Chaos in the Human Brain.* Singapore: World Scientific, 1991.

6. B. H. JANSEN AND M. E. BRANDT, *Nonlinear Dynamical Analysis of the EEG.* Singapore: World Scientific, 1993.

7. K. LEHNERTZ, J. ARNHOLD, P. GRASSBERGER, AND C. E. ELGER, *Chaos in Brain?* Singapore: World Scientific, 2000.

8. G. PFURTSCHELLER AND F. H. LOPES DA SILVA, Event-related EEG/MEG synchronization and desynchronization: Basic principles. *Clin. Neurophysiol.,* vol. 110, pp. 1842–1857, 1999.

9. A. EFFERN, K. LEHNERTZ, T. GRUNWALD, G. FERNANDEZ, P. DAVID, AND C. E. ELGER, "Time adaptive denoising of single trial event related potentials in wavelet domain." *Psychophysiology,* vol. 37, pp. 859–865, 2000.

10. C. S. BURRUS, R. A. COPINATH, AND H. GUO, *Wavelets and Wavelet Transforms.* Englewood cliffs, NJ: Prentice-Hall, 1998.

11. C. K. CHUI, *Introduction to Wavelets.* San Diego: Academic Press, 1992.

12. V. J. SAMAR, A. BOPARDIKAR, R. RAO, AND K. SWARTZ, "Wavelet analysis of neuroelectric waveforms: A conceptual tutorial." *Brain Lang.,* vol. 66, pp. 7–60, 1999.

13. E. A. BARTNIK AND K. J. BLINOWSKA, "Wavelets: New method of evoked potential analysis." *Med. Biol. Eng. Comput.,* vol. 30, pp. 125–126, 1992.

14. O. BERTRAND, J. BOHORQUEZ, AND J. PERNIER, "Time frequency digital filtering based on an invertible wavelet transform: An application to evoked potentials." *IEEE Trans. Biomed. Eng.,* vol. 41, pp. 77–88, 1994.

15. D. L. DONOHO, I. M. JOHNSTONE, AND B. W. SILVERMAN, "De-noising by soft-thresholding." *IEEE Trans. Inform. Theor.,* vol. 41, pp. 613–627, 1995.

16. T. DEMIRALP, A. ADEMOGLU, M. COMERCHERO, AND J. POLICH, "Wavelet analysis of P3a and P3b." *Brain Topogr.,* vol. 13, pp. 251–267, 2001.

17. J. FELL, P. KLAVER, K. LEHNERTZ, T. GRUNWALD, C. SCHALLER, C. E. ELGER, AND G. FERNANDEZ, "Rhinal-hippocampal coupling is essential for human memory formation." *Nature Neurosci.,* vol. 4, pp. 1259–1264, 2001.

18. D. BOTTGER, C. S. HERRMANN, AND D. Y. VON CRAMON, "Amplitude differences of evoked alpha and gamma oscillations in two different age groups." *Int. J. Psychophysiol.,* vol. 45, pp. 245–251, 2002.

19. A. EFFERN, K. LEHNERTZ, T. SCHREIBER, P. DAVID, AND C. E. ELGER, "Nonlinear denoising of transient signals with application to event related potentials." *Phys. D,* vol. 140, pp. 257–266, 2000.

20. A. EFFERN, K. LEHNERTZ, G. FERNANDEZ, T. Grunwald, P. DAVID, AND C. E. ELGER, "Single trial analysis of event related potentials: Nonlinear denoising with wavelets." *Clin. Neurophysiol,* vol. 111, pp. 2255–2263, 2000.

21. P. GRASSBERGER, R. HEGGER, H. KANTZ, C. SCHAFFRATH, AND T. SCHREIBER, "On noise reduction methods for chaotic data." *Chaos,* vol. 41, pp. 127–140, 1993.

22. G. FERNANDEZ, A. EFFERN, T. GRUNWALD, N. PEZER, K. LEHNERTZ, M. DÜMPELMANN, D. VAN ROOST, AND C. E. ELGER, "Real-time tracking of memory formation in the human rhinal cortex and hippocampus." *Science,* vol. 285, pp. 1582–1585, 1999.

23. R. M. CHAPMAN, J. C. ARMINGTON, AND H. R. BRAGDEN, "A quantitative survey of kappa and alpha EEG activity." *Electroenceph. Clin. Neurophysiol.,* vol. 14, pp. 858–868, 1962.

24. A. GUNDEL AND G. F. WILSON, "Topographical changes in the ongoing EEG related to the difficulty of mental tasks." *Brain Topogr.,* vol. 5, pp. 17–25, 1992.

25. W. KLIMESCH, "EEG alpha and theta oscillations reflect cognitive and memory performance: A review and analysis." *Brain Res. Rev.,* vol. 29, pp. 169–195, 1999.

26. F. ROSLER, M. HEIL, AND E. HENNIGHAUSEN, "Exploring memory functions by means of brain electrical topography: A review." *Brain Topogr.,* vol. 7, pp. 301–313, 1995.

27. C. J. STAM, "Brain dynamics in theta and alpha frequency bands and working memory performance in humans." *Neurosci. Lett.,* vol. 286, pp. 115–118, 2000.

28. K. SAUVÉ, "Gamma-band synchronous oscillations: Recent evidence regarding their functional significance." *Conscious. Cogn.,* vol. 8, pp. 213–224, 1999.

29. F. J. VARELA, J. P. LACHAUX, E. RODRIGUEZ, AND J. MARTINERIE, "The brainweb: Phase synchronization and large-scale integration." *Nat. Rev. Neurosci.,* vol. 2, pp. 229–239, 2000.

30. P. GRASSBERGER, T. SCHREIBER, AND C. SCHAFFRATH, "Non-linear time sequence analysis." *Int. J. Bifurcation Chaos,* vol. 1, pp. 521–547, 1991

31. R. A. GREGSON, L. A. BRITTON, E. A. CAMPBELL, AND G. R. GATES, "Comparisons of the nonlinear dynamics of electroencephalograms under various task loading conditions: A preliminary report." *Biol. Psychol.,* vol. 31, pp. 173–191, 1990.

32. W. LUTZENBERGER, N. BIRBAUMER, H. FLOR, B. ROCKSTROH, AND T. ELBERT, "Dimensional analysis

of the human EEG and intelligence." *Neurosci. Lett.*, vol. 143, pp. 10–14, 1992

33. R. A. GREGSON, E. A. CAMPBELL, AND G. R. GATES, "Cognitive load as a determinant of the dimensionality of the electroencephalogram: A replication study." *Biol. Psychol.*, vol. 35, pp. 165–178, 1993.

34. W. LUTZENBERGER, T. ELBERT, N. BIRBAUMER, W. J. RAY, AND H. SCHUPP, "The scalp distribution of the fractal dimension of the EEG and its variation with mental tasks." *Brain Topogr.*, vol. 5, pp. 27–34, 1992.

35. C. J. STAM, T. C. A. M. VAN WOERKOM, AND W. S. PRITCHARD, "Use of non-linear EEG measures to characterize EEG changes during mental activity." *Electroenceph. Clin. Neurophysiol.*, vol. 99, pp. 214–224, 1996.

36. G. SAMMER, "Working memory load and dimensional complexity of the EEG." *Int. J. Psychophysiol.*, vol. 24, pp. 173–182, 1996.

37. M. MÖLLE, L. MARSHALL, W. LUTZENBERGER, R. PIETROWKSY, H. L. FEHM, AND J. BORN, "Enhanced dynamic complexity in the human EEG during creative thinking." *Neurosci. Lett.*, vol. 208, pp. 61–64, 1996.

38. A. MEYER-LINDENBERG, U. BAUER, S. KRIEGER, S. LIS, K. VEHMEYER, G. SCHULER, AND B. GALLHOFER, "The topography of non-linear cortical dynamics at rest, in mental calculation and moving shape perception." *Brain Topogr.*, vol. 10, pp. 291–299, 1998.

39. L. R. SQUIRE AND S. ZOLA-MORGAN, "The medial temporal lobe memory system." *Science*, vol. 253, pp. 1380–1386, 1991.

40. R. JAFFARD AND M. MEUNIER, "Role of the hippocampal formation in learning and memory." *Hippocampus*, vol. 3 (special issue), pp. 203–218, 1993.

41. K. LEHNERTZ, "Nonlinear time series analysis of intracranial EEG recordings in patients with epilepsy—An overview." *Int. J. Psychophysiol.*, vol. 34, pp. 45–52, 1999.

42. K. LEHNERTZ, B. WEBER, C. HELMSTAEDTER, H. G. WIESER, AND C. E. ELGER, "Neuronal complexity in mesial temporal lobes during cognitive tasks and relations to task performance." *Brain Topogr.*, vol. 10, pp. 73–74, 1997.

43. K. LEHNERTZ, B. WEBER, C. HELMSTAEDTER, H. G. WIESER, AND C. E. ELGER, "Alterations in neuronal complexity during verbal memory tasks index recruitment potency in temporo-mesial structures." *Epilepsia*, vol. 38 (Suppl. 3), p. 238, 1997.

44. K. LEHNERTZ AND C. E. ELGER, "Spatio-temporal dynamics of the primary epileptogenic area in temporal lobe epilepsy characterized by neuronal complexity loss." *Electroenceph. Clin. Neurophysiol.*, vol. 95, pp. 108–117, 1995.

45. K. LEHNERTZ, "Nonlinear time series analysis of intrahippocampal EEG recordings in epilepsy patients during verbal learning and memory processes." *Int. J. Psychophysiol.*, vol. 30, p. 42, 1998.

46. C. HELMSTAEDTER, K. LEHNERTZ, G. WIDMAN, B. WEBER, AND C. E. ELGER, "Neuronal complexity loss in temporo-mesially recorded EEG predicts recall performance of incidentally learned material." *Int. J. Psychophysiol.*, vol. 30, p. 30, 1998.

47. J. ARNHOLD, P. GRASSBERGER, K. LEHNERTZ, AND C. E. ELGER, "A robust method for detecting interdependences: Application to intracranially recorded EEG." *Phys. D*, vol. 134, pp. 419–430, 1999.

48. J. ARNHOLD, P. GRASSBERGER, K. LEHNERTZ, AND C. E. ELGER, "Nonlinear interdependence in intracranial EEG recordings during verbal learning and memory tasks." *Epilepsia*, vol. 40 (Suppl. 7), p. 170, 1999.

49. C. RIEKE, K. STERNICKEL, R. G. ANDRZEJAK, C. E. ELGER, P. DAVID, AND K. LEHNERTZ, "Measuring nonstationarity by analyzing the loss of recurrence in dynamical systems." *Phys. Rev. Lett.*, vol. 88, doi 244102, 2002.

50. J.-P. LACHAUX, E. RODRIGUEZ, J. MARTINERIE, AND F. J. VARELA, "Measuring phase synchrony in brain signals." *Hum. Brain Mapp.*, vol. 8, pp. 194–208, 1999.

51. F. MORMANN, K. LEHNERTZ, P. DAVID, AND C. E. ELGER, "Mean phase coherence as a measure for phase synchronization and its application to the EEG of epilepsy patients." *Physica D*, vol. 144, pp. 358–369, 2000.

52. S. MICHELOYANNIS, M. VOURKAS, M. BIZAS, P. SIMOS, AND C. J. STAM, "Changes in linear and nonlinear EEG measures as a function of task complexity: Evidence for local and distant signal synchronization." *Brain Topogr.*, vol. 15, pp. 239–247, 2003.

53. C. J. STAM, M. BREAKSPEAR, A. M. VAN CAPPELLEN VAN WALSUM, AND B. W. VAN DIJK, "Nonlinear synchronization in EEG and whole-head MEG recordings of healthy subjects." *Hum. Brain Mapp.*, vol. 19, pp. 63–78, 2003.

CHAPTER 40

SINGLE-TRIAL ANALYSIS OF EEG FOR ENABLING COGNITIVE USER INTERFACES

Adam D. Gerson, Lucas C. Parra, and Paul Sajda

40.1 INTRODUCTION

Establishment of robust direct brain–machine communication channels has enhanced the function and performance of both impaired and normal nervous systems. The creation of direct links between brain and machine has been most strongly motivated by the need to provide a mode of communication for those silenced by conditions such as amyotrophic lateral sclerosis (ALS). Such brain–computer interfaces (BCIs) [1] translate cortical activity involved with motor planning and motor response to select letters and words on a computer display or control a robotic prosthesis. Perhaps the most striking results have been demonstrated through invasive systems based on multiunit micro-electrode implants. Microelectrode arrays implanted in motor cortex have enabled monkeys to reach and grasp in three dimensions using a robotic arm and gripper without using their hands [2]. In fact, invasive systems have already allowed humans with ALS or high-level spinal injury to communicate [3].

Invasive brain–machine interfaces need not be limited to monitoring cortical activity. Representing an astonishing paradigm shift, neuroscientists in Brooklyn have assumed the role of animal trainer, wirelessly guiding rats using cortical microstimulation [4]. Rats were implanted with microelectrodes in left and right somatosensory (barrel) cortex and the medial forebrain bundle (MFB). Rats following navigational cues provided by stimulating somatosensory cortex were subsequently rewarded through stimulation of MFB. Both navigational and reward stimuli were controlled by the rat trainers; however, one can envisage a system in which motor cortex is monitored so reward centers are automatically stimulated if navigational cues are correctly executed. Clinically motivated cortical stimulation systems have successfully treated those suffering with Parkinson's disease [5] and epilepsy [6] by stimulating the subthalamic nucleus and vagus nerve, respectively. Cochlear implants have restored hearing by stimulating the auditory nerve [7], and visual cortical implants promise to restore vision [8].

Invasive brain–machine interfaces are clearly well motivated, improving quality of life by restoring communication, motor control, hearing, and possibly vision to a sizable segment of the population. Brain–machine communication channels also enable cognitive user interfaces (CUIs) designed to enhance healthy nervous system function, augmenting user performance during cognitively demanding tasks by boosting

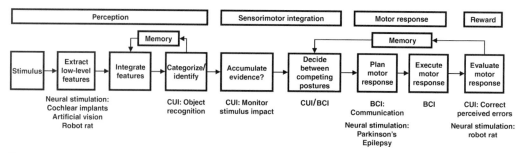

Figure 40.1 Stimulus response processing stages for CUI design.

information processing capacity and reducing stress-induced deficiencies of executive control functions.[1] Widespread adoption of invasive cognitive interfaces is highly unlikely; as such, any CUI must rely on noninvasive neuroimaging modalities. Functional magnetic resonance (fMRI) imaging has revolutionized neuroscience by delineating the anatomical origin of scores of cognitive processes. While fMRI studies are critical to the design of any CUI, the size, cost, and more importantly temporal resolution of scanners prohibit their use in any practical application. Millisecond temporal resolution and low cost make electroencephalography (EEG) the most viable modality for cognitive interface design.

Robust classification algorithms are critical in order to realize practical benefits of EEG. Sensitivity to environmental noise is of obvious concern; however, recent advances in sensor technology[2] suggest classifier development should focus on identifying neural sources. Traditionally, electrophysiological research relies on averaging across hundreds of stimulus or response-locked trials to uncover the dynamics of neural activity from EEG. The underlying assumption has been that by averaging across trials the contribution of background EEG activity is minimized relative to neural activity correlated with a stimulus. Such event-related potential (ERP) analysis, however, masks variability between trials that may be of critical importance when striving to understand underlying cortical interactions. For example, one study describes an alpha-phase distribution across trials that comprises averaged visual stimulus-evoked ERPs [9]. The authors of this study suggest that visual stimuli in fact reset the phase of dynamic neural activity resulting in variability of alpha phase across stimulus-locked trials. Any practical EEG-based interface must account for intertrial variability.

Immediate applications for cognitive user interfaces can be derived by teasing apart processing stages between sensory stimuli and motor response as outlined in Figure 40.1. Contemporary models indicate that sensory information is decomposed into discrete fundamental features that are subsequently integrated under constraints imposed by adjacent features and higher order areas associated with memory and reward. The process of sensorimotor integration is still subject to debate; however, one hypothesis proposes that perceptual evidence accumulates lending weight to competing neuronal populations poised to execute motor plans for distinct postures. Motor responses are subsequently monitored and the outcome influences reward centers which feed back to influence subsequent decisions. Monitoring neural populations associated with perception, sensorimotor integration, and reward immediately suggests CUI interfaces for high-throughput object recognition and error correction. After reviewing linear classification methods,

[1] Visit DARPA's vision for Cognitive User Interfaces: http://www.darpa.mil/ipto/programs/augcog/.

[2] For example, noncontact bioelectrodes: http://www.quasarusa.com.

we present two examples of CUIs designed expressly to augment cognition by enhancing throughput during an object recognition task and correcting response errors during stressful motor-intensive tasks.

40.2 MACHINE LEARNING: LINEAR METHODS

The merits of linear and nonlinear methods were the subject of a debate held during the Second International Meeting on BCIs [10]. This discussion reviewed important factors to consider when fitting a learning algorithm to the problem of classifying EEG such as the degree of prior information, nature of data distributions, amount of training data, and computational costs [11] for online real-time cognitive interfaces. At the present time we prefer linear methods for EEG classification since a principled approach defining the origin of nonlinearities in EEG has yet to be clearly defined. Compared with nonlinear methods, linear methods are consistent with the linearity of volume conduction, are less likely to overfit noise, and have a significantly lower computational cost.

Linear methods for analyzing multichannel EEG can be categorized as supervised or unsupervised. Traditionally implemented as an unsupervised method, independent-component analysis (ICA) [12] decomposes signals into several independent components with time series that are viewed and analyzed separately. Unsupervised methods do not leverage truth labels associated with experimental events such as stimulus type. Supervised methods, on the other hand, exploit training labels given knowledge of the task and/or subject responses. The method of common spatial pattern (CSP) [13] is an example of supervised source recovery widely used in EEG. This method weights electrodes according to the power captured for two classes of data. It finds orientations in the sensor space in which the power is simultaneously maximized for one class and minimized for the other. An alternative to maximizing power is to maximize the discrimination between two classes. Parra et al. [14] proposed a linear discrimination method that spatially integrates sensor values in well-defined temporal windows to recover sources that maximally discriminate two classes given labeled EEG.

All three methods linearly transform the original signals as $Y = WX$, where X are the observations (the original EEG signal matrix), W is the transform matrix (or vector) that is calculated using the different linear approaches, and Y is the resulting source matrix (or vector) representing the recovered sources. Note the recovery of sources Y given an underlying linear mixture of observations in X.

40.2.1 Independent Component Analysis

Independent-component analysis is a method of finding a linear transformation of input vectors X that maximizes the statistical independence of output source vectors Y such that $Y = WX$. Principal-component analysis (PCA) finds a transformation that decorrelates input vectors by finding a linear transformation resulting in orthogonal output vectors such that the inner product of the vectors is zero. Traditionally, the PCA transformation matrix consists of the normalized eigenvectors of the input vectors' covariance matrix. In contrast, ICA finds a transformation such that the mutual information between output vectors tends to zero. The ICA transformations account for higher order statistical properties of sources as opposed to PCA, which is based on second-order statistics.

Electroencephalography is well suited for ICA analysis [12]. The application of ICA to any problem requires compliance with several assumptions. Most importantly, the underlying sources responsible for generating a set of observations must be statistically independent. Applying ICA to EEG generates a set of statistically independent hypothetical neural or artifactual sources from sensor observations. The ICA places no restrictions on the spatial structure of such sources, which might be widely distributed or spatially isolated. For example, a widely distributed source could correspond to 60-Hz environmental noise affecting all sensors, while an ideal isolated source might correspond to activity in the somatosensory homunculus associated with tactile stimulation of the right index finger. However, ICA will not generate sources corresponding to different neural processes that share the same underlying statistics. This limitation can most likely be overcome only by incorporating prior information concerning the functional significance of neural anatomy.

The information maximization algorithm [15] commonly used to apply ICA to EEG imposes two additional assumptions, namely that the probability density distributions of sources closely resemble the gradient of a generalized logistic sigmoid and that the number of independent sources is equal to the number of sensors. In [15], the gradient of a generalized logistic regression function is shown to be a probability density function with high kurtosis. The statistics of EEG, as is the case of most natural signals, do in fact resemble a highly kurtotic Gaussian distribution so this assumption is approximately satisfied. The dimension of neural sources is unknown and so one of the key challenges in ICA of EEG lies in determination of the optimal number sensors and source channels.

40.2.2 Common Spatial Patterns

The CSP approach [13] finds an optimal set of spatial filters that produce features ideal for binary classification. Essentially, optimal spatial filters are determined through joint diagonalization of two covariance matrices derived from each task-related class. The normalized covariance matrix of each single-trial $N \times T$ matrix $\mathbf{X}$, where N is the number of channels and T is the number of samples, is determined as $\mathbf{C} = \mathbf{X}\mathbf{X}'/\text{trace}(\mathbf{X}\mathbf{X}')$. The average of covariance matrices from class 1 ($\mathbf{C}_1$) and class 2 ($\mathbf{C}_2$) trials are then summed to produce a composite covariance matrix $\mathbf{C}_c = \mathbf{C}_1 + \mathbf{C}_2$. The eigenvectors and eigenvalues of this spatial covariance matrix yield a whitening transformation $\mathbf{P} = (\lambda_c)^{-1/2}U_c'$, where $\mathbf{C}_c = U_c\lambda_c U_c'$. Transforming the average covariance matrices corresponding to the two classes $\mathbf{S}_1 = \mathbf{P}\mathbf{C}_1\mathbf{P}'$ and $\mathbf{S}_2 = \mathbf{P}\mathbf{C}_2\mathbf{P}'$, assures that $\mathbf{S}_1$ and $\mathbf{S}_2$ share common eigenvectors such that $\mathbf{S}_1 = \mathbf{B}\lambda_1\mathbf{B}'$ and $\mathbf{S}_2 = \mathbf{B}\lambda_2\mathbf{B}'$, where $\lambda_1 + \lambda_2 = I$. The first and last eigenvectors of $\mathbf{B}$ then represent optimal projections associated with class 1 and class 2, respectively. In other words projecting whitened EEG data along the vectors defined by the first and last eigenvectors of $\mathbf{B}$ will yield feature vectors ideal for discrimination between EEG data associated with the two classes. The projection matrix is then defined as $\mathbf{W} = (\mathbf{B}'\mathbf{P}')$ and an EEG trial is transformed as $\mathbf{Y} = \mathbf{W}\mathbf{X}$. The columns of $\mathbf{W}^{-1}$ are the common spatial patterns and can be interpreted as time-invariant EEG source distribution vectors.

40.2.3 Linear Discrimination

Linear discrimination is also a supervised method and can be used to compute the optimal spatial integration of a large array of sensors for discrimination between two classes [14]. As with ICA and CSP, linear discrimination finds a transformation $\mathbf{Y} = \mathbf{W}\mathbf{X}$, where $\mathbf{X}$ is a two-dimensional (2-D) $N \times (M \times T)$ matrix representing M trials ($M = I + J$, I trials for class 1 and J trials for class 2) of EEG data at T time points and N recording electrodes.

Determined with logistic regression, W is a spatial weighting coefficient vector defining a hyperplane maximally separating two classes. Timing information is exploited by discriminating and averaging within a short time window relative to a known external event such as stimulus presentation or motor response. Discrimination accuracy can be assessed via receiver operating characteristic (ROC) analysis.

40.3 COGNITIVE USER INTERFACE FOR ERROR CORRECTION

Error-related negativity (ERN) in EEG has been linked to perceived response errors and conflicts in decision making. Single-trial detection of the ERN has been proposed as a means of correcting communication errors in a BCI system [16]. We have developed single-trial ERN detection to predict task-related errors. The system can be used as an automated real-time decision checker for time-sensitive control tasks. This open-loop error correction paradigm represents the first application of real-time cognitive event detection and demonstrates the utility of real-time EEG brain monitoring.

There has been a recent spike in studies of cortical regions associated with conflict monitoring during motor response. Functional MRI and EEG studies of

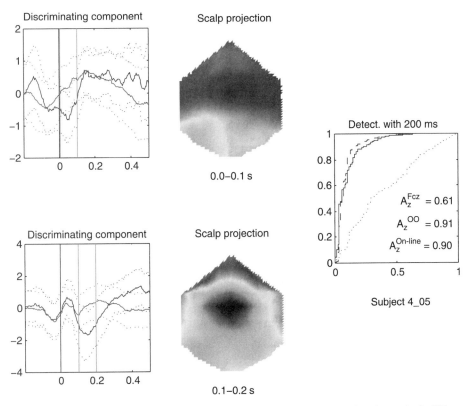

Figure 40.2 ERN detection using linear discrimination combining the time intervals 0–100 ms and 100–200 ms after response. (left) Discriminating component and (center) scalp projection graphs results were obtained with off-line linear classification. Similar results are obtained with on-line adaptation (right). Single-trial ROC results compare A_z for off-line (LOO), on-line, and using only F_{cz} electrode. Note that on-line and off-line are comparable.

TABLE 40.1   Summary of On-Line Error Correction for Each of Seven Subjects

Subject ID	Original error rate (%)	Error reduction (%)[a]
1	6	23
2	10	−6
3	15	−1
4	15	49
5	13	27
6	14	47
7	18	12

[a]Negative values indicate degradation in performance.

interference tasks report significant increases in anterior cingulate cortex (ACC) activity preceding [17] and following [18] response selection that is believed to be associated with conflict monitoring [19] and emotional evaluation of errors [20]. Referred to as error-related negativity or medial frontal negativity (MFN) in EEG studies, such electrical activity is observed during flanker [19], Stroop [21], rapid serial visual presentation (RSVP) [17, 22], and gambling [20] tasks. During an Eriksen flanker task [23] subjects are instructed to indicate by button-press the class of a central character flanked by distractor characters (e.g., " $<<<<<<$ " or " $<<><<$ "). Errors in motor response are readily generated during this task and are accompanied by ERN [24] associated with ACC.

Using a linear classifier, we have reported up to 79% correct detection of the ERN within 100 ms of the erroneous response [14]. More interestingly, we have described a set of adaptive, linear algorithms for artifact removal and ERN detection optimized for high-throughput real-time single-trial correction of human errors during an Eriksen flanker task [25]. The results obtained for a typical subject with 90% correct detection are shown in Figure 40.2. The previously described frontocentral ERN is observed within 100 ms following the response. In addition, a more prolonged bilateral posterior positivity is observed for correct trials, which further improves discrimination. This system is capable of significant improvement in human–machine performance as summarized in Table 40.1.

40.4   COGNITIVE USER INTERFACE FOR IMAGE SEARCH

The appearance of a target image during a RSVP task elicits an EEG response associated with target recognition. We have demonstrated that the detection of these EEG signals on a single-trial basis can be used to replace the slow manual response of a human operator, thereby significantly increasing the throughput of image search tasks [26]. This paradigm has the potential to improve the performance of image analysts and radiologists who need to routinely survey large volumes of aerial imagery or medical images (e.g., mammograms) within short periods of time. In addition, the approach looks to measure the "bottleneck" between constant-delay perceptual processing and more variable delay cognitive processing. Thus the detected signatures can be used to "gauge" if cognitive systems are capable/incapable of assimilating perceptual input for fast decision making.

40.4.1 Rapid Object Recognition

Current models of visual object recognition propose information flows through a series of feedforward processing stages in which low-level features are extracted from a visual scene, then integrated under constraints imposed by adjacent and top-down connections [27]. The true nature of cortical circuits responsible for perception and recognition remains a mystery. In fact there is much debate as to whether recognition relies on information flow through corticocortical feedback loops or rather one feedforward sweep through the visual system [28].

While direct functional imaging of cortical circuits is not yet feasible, indirect evidence from single-unit recordings, ERPs, and psychophysical studies describes macroscopic cortical regions comprising the visual system in terms of both anatomical spatial constraints and functional temporal constraints. The challenge in any such study is designing experiments that tease apart cortical processing stages involved with object recognition and delineate the spatial extent and temporal order, latency, duration, and influence of each stage in response to specific classes [29] of visual stimuli.

One experimental task that simulates natural saccadic scene acquisition is RSVP [29]. During an RSVP task a continuous sequence of images is presented in a static location. Electrophysiological studies of macaque monkey cortical cell response to RSVP stimuli indicate processing required for object recognition is completed within 150 ms of stimulus onset [30]. Face-selective neurons in superior temporal sulcus (STSa) were monitored while an image sequence of seven differently oriented faces was presented (14–222 ms/image). Neurons consistently responded selectively to target face images of a specific orientation approximately 108 ms following target onset, regardless of presentation rate. Response duration was proportional to stimulus duration.

This study does not necessarily reflect visual processing time required for all classes of natural images. Face-selective neuronal activity reflects responses of highly specialized cortical pathways that may not participate in processing particular subsets of natural scenes. Neuronal response times may also be related to early stages of the visual processing pathway dedicated to low-level feature extraction. In addition, since cortical response latencies are shorter in macaque than humans [31], these findings do not directly translate to visual object recognition processing time of humans. A seminal EEG-based RSVP study [32] established a significant difference between trial-averaged frontal electrode ERPs approximately 150 ms following presentation of target versus distractor images. Lateral motor-response-related activity was observed approximately 375 ms after stimulus onset. Target images contained an animal in a random location within a natural scene while distractor images were natural scenes. More recently, a similar experiment reported EEG activity correlated with image categorization begins within 80 ms of image presentation [33]. These results demonstrate that EEG signatures of rapid object recognition/categorization can be seen, with a very short latency following stimulus presentation, by averaging across multiple trials.

While the early onset of differential ERP activity noted in these studies suggests recognition is achieved following a single feedforward sweep through the visual system [34], another study reports this activity is due to low-level feature recognition rather than object categorization [28]. A cued-target paradigm was designed to test for differences in ERPs resulting from target and nontarget visual stimuli with contextual rather than featural differences. Target categories (e.g., animal, furniture, dog) were presented about half a second before each image. Target and nontarget image sets were identical ensuring no differences in low-level features. Event-related potentials for the cued-target task are markedly different than the original single-category go/no-go task. A presentation-locked ERP difference similar to that previously reported is only present when there are low-level featural differences between targets and nontargets. There is a later differential component about 150–300 ms following image presentation arising from contextual differences, the latency of which is correlated with reaction time. If this component is in fact associated with object recognition, its latency does permit inclusion, albeit brief, flow of information through corticocortical feedback loops. Of course this task requires interaction between visual stimulus response and verbal memory, which may add an additional processing stage. One hypothesis of interest is that the variable latency results from integration of ambiguous signals to reach a decision in posterior parietal cortex [35].

We have reported single-trial detection of spatial signatures in EEG related to visual target recognition within 200 ms of image onset during an RSVP task [36]. This RSVP task differed from Thorpe's original experiment in that subjects were asked to detect target images within sequences (barrages) of 100 images [37] that had a 50% chance of containing a single target image. Target images consisted of a person/people comprising no more than 25% of a natural scene while distractor images were natural scenes. Subjects were instructed to press a button at the beginning of a sequence and release it if a target appeared. Electroencephalography from target and distractor trials was compared on a single-trial basis using linear discriminant analysis [14, 38] and a forward linear model was used to determine sensor projections of the discriminating source activity [14]. As shown in Figure 40.3, this forward model indicated that discriminating activity began approximately 200 ms following image presentation moving anteriorly over sensory motor areas 300–400 ms following image presentation. Since these signatures are learned/detected single trials, it is possible to analyze variability between trials as well as determine classification performance on new trials.

This experiment required that subjects make a motor response immediately after detecting the target. In order to decouple motor response from EEG activity related to target recognition, we revised our experimental protocol and instructed subjects to indicate target appearance by pressing a button after all images in the sequence were presented. Using the same linear discrimination method and forward model, optimal linear spatial EEG signatures were computed at multiple time windows of 50 ms duration reflecting cortical activity related to target recognition in the absence of motor activity. Figure 40.4 shows an example of applying optimal spatial linear discriminators learned at the peaks of a temporal classification performance curve. In Figure 40.4b we see the result of applying a discriminator learned at a 190-ms time window to all 60 EEG channels 1 s before and after the stimulus presentation. As would be expected, we see a peak in the trial-averaged value near 190 ms, indicating the detection of the target. In addition we see a strong negative peak near 330 ms. A strong negative peak represents high negative correlation between the discriminator and sensors and indicates identical spatial activity but opposite sign. Since the mean value for distractor trials is approximately zero, a high negative correlation would correspond to a peak on the plots of Figure 40.4a; that is, the fluctuation periods we see in Figures 40.4b,c,d should be roughly twice that seen in Figure 40.4a. Figure 40.4d shows the results when using a discriminator learned at 490 ms. As expected, a positive peak occurs near 490 ms and, in addition, a small negative peak is present near 330 ms. However, there is no positive peak at 190 ms, which might be expected if the signature were simple fluctuations of the same spatially localized areas. Likewise there is no strong positive peak in Figure 40.4b at 490 ms. Thus these signatures

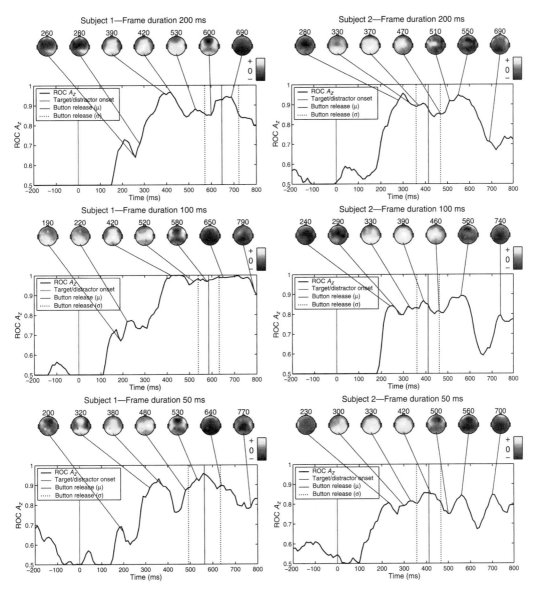

Figure 40.3   Area under ROC curve (A_z) for leave-one-out test of optimal linear discriminating source as a function of the center of a 100-ms discrimination window, together with scalp plots showing sensor projections of discriminating sources, for both subjects at each image presentation rate. By sliding the window used to train a linear discriminator, we are able to study the temporal sequence of neuronal responses evoked by visual stimuli. Due to the high temporal resolution afforded by EEG this method provides an intuitive description of communication between visual and sensorimotor cortex. Results show multiple loci for discriminating activity (e.g., motor and visual). Left: subject 1 for images presented at 200, 100, and 50 ms per image. Right: subject 2.

appear to be significantly different, though they share some negative correlation with the activity near 330 ms. Figure 40.4c is a discriminator learned at 330 ms. It has a small negative peak at 220 ms but no significant peak at 490 ms. Figure 40.4 also shows the scalp projections of the learned signatures. Here we see a sign reversal between the signatures at 190 and 330 ms.

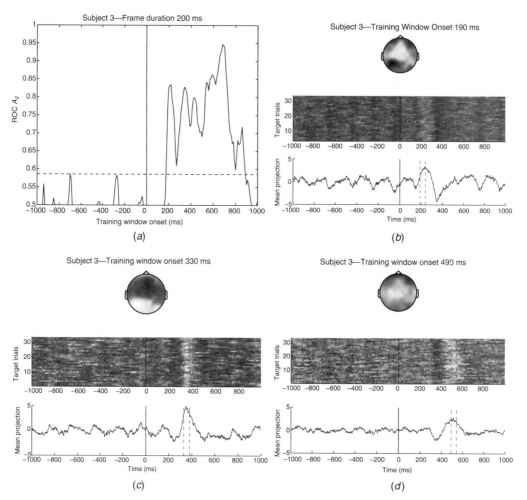

Figure 40.4 Optimal discriminating activity related to object recognition in the absence of motor activity. (*a*) ROC performance results (A_z). Shown are leave-one-out discrimination results for the optimal linear discriminator learned within a 50-ms window and applied within the same window. Results are computed 1 s before and after stimulus-locked presentation. Note that this RSVP task includes a barrage of images and therefore multiple images are presented during this 2-s interval. The dashed horizontal line indicates maximum value before stimulus-locked image presentation and therefore all values less than this value are considered noise. (*b*–*d*) Results of applying the optimal spatial linear discriminator learned at three different times— (*b*) 190 ms, (*c*) 330 ms, and (*d*) 490 ms—to the EEG data. Shown are scalp projections for the discriminator, single-trial results of applying the discriminator across the 2-s interval, and the trial-averaged response of the discriminator. Vertical dotted lines indicate time interval in which each discriminator was trained. In the scalp plots, dark indicates strong positive coupling to the sensors and light strong negative coupling. In the single-trial plots, dark indicates large positive values when the signals are projected onto the discriminator (i.e., high probability of target) and light large negative values (i.e., low probability of target).

These studies indicate significant differential activity associated with object categorization arising as early as 190 ms after image presentation within a barrage of images. Single-trial analysis of spatial activity in high-density EEG leads to fluctuations in discriminating performance with a period that is independent of stimulus presentation rate. Further analysis decoupling motor response seems to suggest that optimal discriminating components learned at different times are often negatively correlated with activity at other times, indicating a strong spatially overlapped fluctuation or oscillation of EEG activity, while other discriminating components have signatures with little correlation with such activity. This appears to indicate that different signatures are present at different times for discriminating target from distractor.

Such reports provide evidence that object recognition is achieved through a series of activations in distinct cortical regions and places temporal limits on processing time in the absence of motor response. As yet studies do not indicate the degree of interaction between these processing stages, the spatial extent of each region, or the nature of the underlying cortical circuits. Our current exploration with simultaneous EEG/fMRI should provide additional clues; however, it is clear that more sophisticated experimental designs and imaging modalities are necessary to clarify the nature of neural activity responsible for human object recognition.

40.4.2 Image Reprioritization Interface

These findings affect the design of an interface to increase search speeds of image analysts. The EEG signatures of object recognition detected via the linear discriminator can be used to reprioritize an image sequence, placing detected targets in the front of an image stack, as shown in Figure 40.5. Image sequences are reordered based on single-trial classification of EEG. Images with classifier output exceeding a threshold are classified as targets and moved to the beginning of the sequence.

The onset of the temporal window selected to train the linear classifier has a significant impact on resequencing performance. Figure 40.6 shows reprioritization performance across three presentation rates resulting from 100-ms discrimination training windows centered 250 ms, 300 ms, and 400 ms following stimulus onset. For comparison, sequences were reordered according to button releases. For this case, target images were classified as targets if they preceded button releases by the mean latency between stimulus onset and button release. Mean button release latencies were determined across trials for each frame duration. Also plotted is the optimal resequence, which

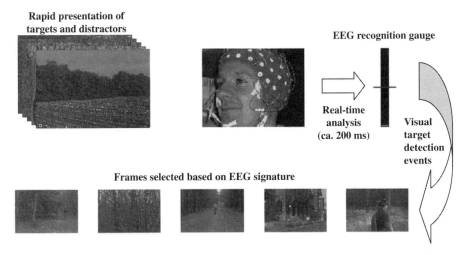

Figure 40.5 Image reprioritization interface. During sequence reprioritization, images are moved to the beginning of the image deck if classifier output associated with a robust EEG signature is greater than an optimal threshold.

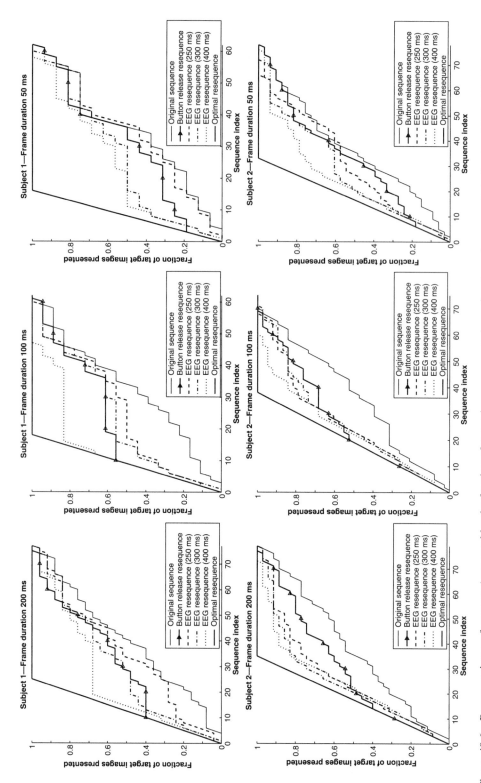

Figure 40.6 Resequencing performance (measured by the fraction of target images presented as a function of sequence index) using detected EEG signatures and button release. Top: subject 1 for images presented at 200, 100, and 50 ms per image. Bottom: subject 2.

would place all targets before distractors. For subject 1, the best performance was for using a discrimination window centered at 400 ms (200 ms before the mean button press). Windows at earlier times, for example at 250 ms, were often worse than the button release results. The late response, together with the scalp plots of Figure 40.3, suggest the most robust signature for subject 1 to be generated via motor planning activity. Subject 2 results were more consistent across the three time windows, with efficient reprioritization for 250, 300, and 400 ms. Together with Figure 40.3, this suggests that both a visual and motor component provide robust signatures for detection of targets and reprioritization of the image sequence. In most cases, the detected EEG signatures result in a reprioritization for more efficient image search compared to the overt response. Additional details and results can be found in [26].

40.5 CONCLUSION

Cognitive user interfaces are redefining the limits of human prowess by enriching perception, cognition, and action. From the perspective of neuroscience, intertrial variability in EEG derived from CUI design may explain the source of psychophysical observations such as response time variability and provide clues describing interaction of processing stages between stimulus and motor response within the context of ongoing EEG activity. Understanding factors that contribute to differences between trials suggests robust classification schemes that enable development of a myriad of next-generation interfaces.

Immediate applications stem from our understanding of motor and visual systems. While suffering from low bit rate, BCIs for communication and control provide an outlet for those battling motor neuron diseases such as ALS. In fact, communication interfaces based on attentional modulation of a well-studied ERP known as the P300 have demonstrated significant improvements in classification accuracy resulting in higher throughput[3] [39, 40]. Modulation of P300 latency and amplitude provides a means to assess perceptual load [41] and processing capacity for a subset of tasks [42]. The ERN-based interfaces designed for error correction may help prevent disasters during crisis situations that require quick decisions during motor-intensive tasks in stressful environments such as military cockpits. The ability to monitor human object recognition leads to the development of interfaces designed to improve search speeds of imagery analysts or assess the impact of visual stimuli [43]. In fact, the timing of visual stimulus presentation with respect to the phase of ongoing alpha observed in EEG can alter perception and affect stimulus salience [44–46].

Realization of clinically motivated EEG-based interfaces will very soon become accessible. Electroencephalography has proven effective in detection and characterization of neurological disorders such as epilepsy, schizophrenia [47], depression, and Parkinson's and Alzheimer's diseases [48]. Discovery of changes in EEG up to several hours prior to epileptic seizure onset [49] has motivated development of seizure prediction algorithms [50] that could permit the implementation of devices capable of preventing seizures from occurring [51]. Detection of motor planning in premotor cortex will permit development of thought-enabled robotic prostheses and muscle stimulators for amputees or patients with spinal cord injury [52]. The interfaces illustrated here only hint at the possibilities.

---

[3]For a description of the P300 speller visit: http://ida.first.fraunhofer.de/projects/bci/competition_ii/

ACKNOWLEDGMENTS

This work was supported by the Defence Advanced Research Projects Agency (DARPA) under contract N00014-01-C-0482 and the DoD Multidisciplinary University Research Initiative (MURI) program administered by the Office of Naval Research under Grant N00014-01-1-0625.

REFERENCES

1. J. R. WOLPAW, N. BIRBAUMER, W. J. HEETDERKS, D. J. MCFARLAND, P. H. PECKHAM, AND G. SCHALK "Brain-computer interface technology: A review of the first international meeting," *IEEE Transactions on Rehabilitation Engineering*, vol. 8, pp. 164–173, June 2000.

2. J. M. CARMENA, M. A. LEBEDEV, R. E. CRIST, J. E. O'DOHERTY, D. M. SANTUCCI, D. F. DIMITROV, P. G. PATIL, C. S. HENRIQUEZ, AND M. A. L. NICOLELIS, "Learning to control a brain-machine interface for reaching and grasping by primates," *PLoS Biology*, vol. 1, no. 2, pp. 193–208, Oct. 13, 2003.

3. P. R. KENNEDY AND B. KING, "Dynamic interplay of neural signals during the emergence of cursor related cortex in a human implanted with the neurotrophic electrode," in J. K. CHAPIN AND K. A. MOXON (Eds.), *Neural Prostheses for Restoration of Sensory and Motor Function*, CRC Press, Boca Raton, FL, 2000, pp. 221–233.

4. S. K. TALWAR ET AL., "Rat navigation guided by remote control," *Nature*, vol. 417, pp. 37–38, May 2, 2002.

5. P. KRACK, A. BATIR, N. VAN BLERCOM, S. CHABARDES, V. FRAIX, C. ARDOUIN, A. KOUDSIE, P. D. LIMOUSIN, A. BENAZZOUZ, J. F. LEBAS, A.-L. BENABID, AND P. POLLAK, "Five-year follow-up of bilateral stimulation of the subthalamic nucleus in advanced Parkinson's disease," *New England Journal of Medicine* vol. 349, pp. 1925–1934, Nov. 13, 2003.

6. D. K. NGUYEN AND S. S. SPENCER, "Recent advances in the treatment of epilepsy," *Archives of Neurology*, vol. 60, no. 7, pp. 929–935, July 2003.

7. B. E. PFINGST, "Auditory prostheses," in J. K. CHAPIN AND K. A. MOXON (Eds.), *Neural Prostheses for Restoration of Sensory and Motor Function*, CRC Press, Boca Raton, FL, 2000, pp. 3–43.

8. R. A. NORMANN, E. M. MAYNARD, P. J. ROUSCHE, AND D. J. WARREN, "A neural interface for a cortical vision prosthesis," *Vision Research*, vol. 39, no. 15, pp. 2577–2587, 1999.

9. S. MAKEIG, M. WESTERFIELD, T.-P. JUNG, S. ENGHOFF, J. TOWNSEND, E. COURCHESNE, AND T. J. SEJNOWSKI, "Dynamic brain sources of visual evoked responses," *Science*, vol. 295, pp. 690–694, Jan. 2002.

10. K.-R. MÜLLER, C. W. ANDERSON, AND G. E. BIRCH, "Linear and nonlinear methods for brain-computer Interfaces" *IEEE Transactions on Neural Systems and Rehabilitation Engineering*, vol. 11, no. 2, pp. 165–169, 2003.

11. T. HASTIE, R. TIBSHIRANI, AND J. FRIEDMAN, *The Elements of Statistical Learning; Data Mining, Inference, and Prediction*, Springer, New York, 2003.

12. S. MAKEIG, A. BELL, T. JUNG, AND T. SEJNOWSKI, "Independent component analysis of electroencephalographic data," *Advances in Neural Information Processing Systems*, vol. 8, pp. 145–151, 1996.

13. H. RAMOSER, J. MUELLER-GERKING, AND G. PFURTSCHELLER, "Optimal spatial filtering of single trial EEG during imagined hand movement," *IEEE Transactions Rehabilitation Engineering*, vol. 8, no. 4, pp. 441–446, 2000.

14. L. PARRA, C. ALVINO, A. TANG, B. PEARLMUTTER, N. YEUNG, A. OSMAN, AND P. SAJDA, "Linear spatial integration for single trial detection in encephalography," *NeuroImage*, vol. 17, pp. 223–230, 2002.

15. A. J. BELL AND T. J. SEJNOWSKI, "An information-maximization approach to blind separation and blind deconvolution," *Neural Computation*, vol. 7, pp. 1129–1159, 1995.

16. G. SCHALK, J. WOLPAW, D. MCFARLAND, AND G. PFURTSCHELLER, "EEG-based communication: Presence of an error potential," *Clinical Neurophysiology*, vol. 111, pp. 2138–2144, 2000.

17. S. NIEUWENHUIS, N. YEUNG, W. VAN DEN WILDENBERG, AND K. R. RIDDERINKHOF, "Electrophysiological correlates of anterior cingulate function in a go/no-go task: Effects of reponse conflict and trial type frequency," *Cognitive, Affective, and Behavioral Neuroscience*, vol. 3, no. 1, pp. 17–26, 2003.

18. V. VAN VEEN, J. D. COHEN, M. M. BOTVINICK, V. A. STENGER, AND C. S. CARTER, "Anterior cingulate cortex, conflict monitoring, and levels of processing," *NeuroImage*, vol. 14, pp. 1302–1308, 2001.

19. M. BOTVINICK, L. E. NYSTROM, K. FISSELL, C. S. CARTER, AND J. D. COHEN, "Conflict montiroing versus selection-for-action in anterior cingulate cortex," *Nature*, vol. 402, pp. 179–181, Nov. 11, 1999.

20. W. J. GEHRING AND A. R. WILLOUGHBY, "The medial frontal cortex and the rapid processing of monetary gains and losses," *Science*, vol. 295, pp. 2279–2282, Mar. 22, 2002.

21. W. J. GEHRING AND D. E. FENCSIK, "Functions of the medial frontal cortex in the processing of conflict and errors," *Journal of Neuroscience*, vol. 21, no. 23, pp. 9430–9437, Dec. 1, 2001.

22. H. BOKURA, S. YAMAGUCHI, AND S. KOBAYASHI, "Electrophysiological correlates for response inhibition in a Go/NoGo task," *Clinical Neurophysiology*, vol. 112, pp. 2224–2232, 2001.

23. B. A. ERIKSEN AND C. W. ERIKSEN, "Effects of noisy letters upon the identification of a target letter in nonsearch task," *Perception and Psychophysics*, vol. 16, pp. 143–149, 1974.

24. M. FALKENSTEIN, J. HOORMAN, S. CHRIST, AND J. HOHNSBEIN, "ERP components on reaction erros and their functional significance: A tutorial," *Biological Psychology*, vol. 52, pp. 87–107, 2000.

25. L. PARRA, C. SPENCE, A. GERSON, AND P. SAJDA, "Response error correction—A demonstration of improved human-machine performance using real-time EEG monitoring," *IEEE Transactions on Neural Systems and Rehabilitation Engineering*, vol. 11, no. 2, pp. 173–177, 2003.

26. A. D. GERSON, L. C. PARRA, AND P. SAJDA, "Cortically-coupled computer vision for rapid image search," *IEEE Transactions on Neural Systems and Rehabilitation Engineering*, vol. 4, no. 2, pp. 174–179, 2006.

27. K. BAEK AND P. SAJDA, "A probabilistic network model for integrating visual cues and inferring intermediate-level representations," paper presented at Third International Workshop on Statistical and Computational Theories of Vision, Nice, France, Oct. 2003.

28. J. JOHNSON AND B. OLSHAUSEN, "Timecourse of neural signatures of object recognition," *Journal of Vision*, vol. 3, pp. 499–512, 2003.

29. E. ROSCH, C. B. MERVIS, W. D. GRAY, D. M. JOHNSON, AND P. BOYES-BRAEM, "Basic objects in natural categories," *Cognitive Psychology*, vol. 8, pp. 382–439, 1976.

30. C. KEYSERS, K.-K. XIAO, P. FLDIK, AND D. I. PERRETT, "The speed of sight," *Journal of Cognitive Neuroscience*, vol. 13, no. 1, pp. 90–101, 2001.

31. J. LIU, A. HARRIS, AND N. KANWISHER, "Stages of processing in face perception: An MEG study," *Nature Neuroscience*, vol. 5, no. 9, pp. 910–916, Sept. 2002.

32. S. THORPE, D. FIZE, AND C. MARLOT, "Speed of processing in the human visual system," *Nature*, vol. 381, No. 6582, pp. 520–522, June 6, 1996.

33. R. VANRULLEN AND S. THORPE, "The time course of visual processing: From early perception to decision-making," *Journal of Cognitive Neuroscience*, vol. 13, no. 4, pp. 454–461, 2001.

34. R. VANRULLEN AND S. J. THORPE, "Surfing a spike wave down the ventral stream," *Vision Research*, vol. 42, pp. 2593–2615, 2002.

35. J. I. GOLD AND M. N. SHADLEN, "Banburismus and the brain: Decoding the relationship between sensory stimuli, decisions, and reward," *Neuron*, vol. 36, pp. 299–308, 2002.

36. P. SAJDA, A. GERSON, AND L. PARRA, "High-throughput image search via single-trial event detection in a rapid serial visual presentation task," paper presented at 1st International IEEE EMBS Conference on Neural Engineering, Capri, Italy, March 2003.

37. J. VAN HATEREN AND A. VAN DER SCHAAF, "Independent component filters of natural images compared with simple cells in primary visual cortex," *Proceedings of the Royal Society of London B*, vol. 265, pp. 359–366, 1998.

38. M. I. JORDAN AND R. A. JACOBS, "Hierarchical mixtures of experts and the EM algorithm," *Neural Computation*, vol. 6, pp. 181–214, 1994.

39. L. A. FARWELL AND E. DONCHIN, "Talking off the top of your head: Toward a mental prosthesis utilizing event-related brain potentials," *Electroencephalography and Clinical Neurophysiology*, vol. 70, no. 6, pp. 510–523, 1988.

40. E. DONCHIN, K. M. SPENCER, AND R. WIJENSINGHE, "The mental prosthesis: Assessing the speed of a P300-based brain-computer interface." *IEEE Transactions on Rehabilitation Engineering*, vol. 8, pp. 174–179, 2000.

41. J. B. ISREAL, C. D. WICKENS, G. L. CHESNEY, AND E. DONCHIN, "The event-related brain potential as an index of display-monitoring workload," *Human Factors*, vol. 22, no. 2, pp. 211–224, Apr. 22, 1980.

42. A. KOK, "On the utility of P3 amplitude as a measure of processing capacity," *Psychophysiology*, vol. 38, pp. 557–577, 2001.

43. M. WELLS, "In search of the buy button," *Forbes*, Sept. 1, 2003.

44. R. VANRULLEN AND C. KOCH, "Is perception discrete or continuous?" *Trends in Cognitive Sciences*, vol. 7, no. 5, pp. 207–213, 2003.

45. F. J. VARELA, A. TORO, E. R. JOHN, AND E. L. SCHWARTZ, "Perceptual framing and cortical alpha rhythm," *Neuropsychologia*, vol. 19, no. 5, pp. 675–686, 1981.

46. M. GHO AND F. J. VARELA, "A quantitative assessment of the dependency of the visual temporal frame upon the cortical rhythm," *Journal of Physiology*, vol. 83, no. 2, pp. 95–101, 1988.

47. J. M. FORD, "Schizophrenia: The broken P300 and beyond," *Psychophysiology*, vol. 36, pp. 667–682, 1999.

48. S. BAILLET, J. C. MOSHER, AND R. M. LEAHY, "Electromagnetic brain mapping," *IEEE Signal Processing Magazine*, pp. 14–30, Nov. 2001.

49. B. LITT, R. ESTELLER, J. ECHAUZ, M. DALESSANDRO, R. SHOR, T. HENRY, P. PENNELL, C. EPSTEIN, R. BAKAY, M. DICHTER, AND VACHTSEVANOS, "Epileptic seizures may begin hours in advance of clinical onset: A report of five patients," *Neuron*, vol. 30, pp. 51–64, 2001.

50. S. D. CRANSTOUN, H. C. OMBAO, R. VON SACHS, W. GUO, AND B. LITT, "Time-frequency spectral estimation of multichannel EEG using auto-SLEX method," *IEEE Transactions on Biomedical Engineering*, vol. 49, no. 9, pp. 988–996, Sept. 2002.

51. B. LITT, "Engineering devices to treat epilepsy: A clinical perspective," in *2001 Proceedings of the*

23rd Annual EMBS International Conference, October 25–28, Istanbul, Turkey, IEEE Press, pp. 4124–4127, October 2001.

52. W. CRAELIUS, "The bionic man: Restoring mobility," *Science*, vol. 295, pp. 1018–1021, Feb. 2002.

53. A. KADNER, E. VIIRRE, D. C. WESTER, S. F. WALSH, J. HESTENES, A. VANKOV, AND J. A. PINEDA, "Lateral inhibition in the auditory cortex: An EEG index of tinnitus?" *Neuro Report*, vol. 13, no. 4, pp. 443–446, Mar. 25, 2002.

INDEX

Handbook of Neural Engineering. Edited by Metin Akay
Copyright © 2007 The Institute of Electrical and Electronics Engineers, Inc.

ABOUT THE EDITOR

Metin Akay is a professor of bioengineering and interim department chairman of the Harrington Department of Bioengineering at the Fulton School of Engineering, Arizona State University at Tempe. He received his Bachelor of Science and Master of Science in Electrical Engineering from the Bogazici University, Istanbul, Turkey in 1981 and 1984, respectively, and a Ph.D. from Rutgers University in 1990.

He is the author/coauthor/editor of 14 books and has given more than 50 keynote, plenary, and invited talks at international meetings including the first, second and third Latin American Conference on Biomedical Engineering in 1998, 2001, and 2004.

Dr. Akay is the founding chair of the Annual International Summer School on Biocomplexity from System to Gene sponsored by the NSF and Dartmouth College and technically cosponsored by the IEEE EMBS of the Satellite Conference on Emerging Technologies in Biomedical Engineering. In 2003, he was also the founding chair of the International IEEE Conference on Neural Engineering and the first chair of the steering committee of the *IEEE Transaction on Computational Biology and Bioinformatics* sponsored by the IEEE (CS, EMBS, NN, etc.) and non-IEEE societies. He was the invited guest editor for the special issues of *Bioinformatics: Proteomics* and *Genomics Engineering* of the *Proceedings of IEEE*, one of the most highly cited IEEE journal.

Prof. Akay is a recipient of the IEEE EMBS Service Award, an IEEE Third Millennium Medal, and the IEEE Engineering in Medicine and Biology Society Early Career Achievement Award 1997. He also received the Young Investigator Award of the Sigma Xi Society, Northeast Region in 1998 and 2000.

Dr. Akay is a fellow of Institute of Physics, senior member of the IEEE, a member of BMES, Eta Kappa, Sigma Xi, Tau Beta Pi. He also serves on the editorial or advisory board of several international journals including the IEEE T-BME, IEEE T-NSRE, IEEE T-ITIB, *Proceedings of IEEE*, *Journal of Neural Engineering*, NIH Neural Engineering and Bioengineering partnership study sections and several NSF review panels.